Molecular Luminescence Spectroscopy

Methods and Applications: Part 1

CHEMICAL ANALYSIS

A SERIES OF MONOGRAPHS ON
ANALYTICAL CHEMISTRY AND ITS APPLICATIONS

VOLUME 77

A WILEY-INTERSCIENCE PUBLICATION

JOHN WILEY & SONS

New York / Chichester / Brisbane / Toronto / Singapore

Molecular Luminescence Spectroscopy

Methods and Applications: Part 1

STEPHEN G. SCHULMAN

College of Pharmacy
University of Florida
Gainesville, Florida

A WILEY-INTERSCIENCE PUBLICATION

JOHN WILEY & SONS

New York / Chichester / Brisbane / Toronto / Singapore

Chem
Sep/oe

Library of Congress Cataloging in Publication Data:

Main entry under title:

Molecular luminescence spectroscopy.
  (Chemical analysis, ISSN 0069-2883 ; v. 77)
  "A Wiley-Interscience publication."
  Includes index.
  Contents: Pt. 1. Methods and applications.
  1. Luminescence spectroscopy. I. Schulman,
Stephen G. (Stephen Gregory), 1940–   . II. Series.

QD96.L85M65 1985   543′.085   84-21880
ISBN 0-471-86848-5 (v. 1)

Printed in the United States of America

10 9 8 7 6 5 4 3 2 1

To Gail, Barbara and Anneke

CONTRIBUTORS

WILLY R. G. BAEYENS
Department of Pharmaceutical
  Chemistry and Drug Analysis
Faculty of Pharmaceutical Sciences
State University of Ghent,
Ghent, Belgium

HARRY G. BRITTAIN
Department of Chemistry
Seton Hall University
South Orange, NJ

ALBERTO FERNANDEZ-GUTIERREZ
Department of General Chemistry
Faculty of Sciences
University of Granada,
Granada, Spain

A. HULSHOFF
Pharmaceutical Laboratory
Department of Analytical Pharmacy
State University of Utrecht
Utrecht, The Netherlands

H. THOMAS KARNES
College of Pharmacy
University of Florida
Gainesville, FL

H. LINGEMAN
Pharmaceutical Laboratory
Department of Analytical
  Pharmacy
State University of Utrecht
Utrecht, The Netherlands

ARSENIO MUÑOZ DE LA PEÑA
Department of Analytical
  Chemistry
Faculty of Sciences
University of Extremadura
Badajoz, Spain

JEFFREY S. O'NEAL
College of Pharmacy
University of Florida
Gainesville, FL

STEPHEN G. SCHULMAN
College of Pharmacy
University of Florida
Gainesville, FL

O. S. WOLFBEIS
Institute for Organic Chemistry
Division of Analytical Chemistry
Karl Franzens University
Graz, Austria

PREFACE

Luminescence spectroscopy, in slightly over three decades, has transcended its origins as a curiosity in the physical laboratory to become a firmly established and widely employed branch of analytical chemistry. Fluorimetry and phosphorimetry are now routinely used in the quantitation, qualitative identification, and structural characterization of inorganic and organic compounds and even of cellular structures.

Slightly over a decade ago it appeared that fluorescence and phosphorescence spectroscopy had reached something of an impasse as an area of active research in analytical chemistry. It was recognized that although luminescence spectroscopy was a very sensitive means of quantitation, it was subject to serious interferences by contaminants in solution and would not likely be useful unless analytes were first separated from their matrices. The features of luminescence spectra were, for the most part, far too general for qualitative identification to be based on simple fluorescence and phosphorescence spectra alone. Moreover, overlapping spectra of analytes with identical fluorophores made analysis of multicomponent samples nearly impossible. Phosphorescence spectroscopy required the use of very low temperatures and the equipment used to measure phosphorescence under this circumstance created great difficulty in reproducible sampling. Fluorescent probes, used by bioanalytical chemists to examine the structure and functions of biological macromolecules and cellular organelles, gave only the vaguest information about the "polarity" of binding sites and even this information was largely restricted to a single fluorescent probe and a single kind of protein.

In the past decade much of this has changed. Owing to elegant new instrumentation and especially to new techniques, some of which are entirely new and some of which are borrowed from other disciplines, fluorescence, chemiluminescence, and phosphorescence spectroscopies can be routinely applied to real analytical problems. In liquid chromatography fluorescence detection represents an interface between the selectivity of an elegant separation method and an ultrasensitive detection method. In immunoassay, fluorescence and chemiluminescence spectroscopies have lent their sensitivities to the selectivity of the immune response to allow sensitive quantitation, often without chemical or physical separation of the analyte from an extremely "dirty" matrix. Currently, analytes having nearly identical fluorescence spectra can often be determined simultaneously by taking advantage of differences in their fluorescent decay times rather than of their fluorescence spectral band intensities. Novel fluorescent probes such as the luminescent lanthanide ions, which have the ability to replace endogenous metal ions in biological structures without altering structure or function, have provided a fairly specific insight into some of the chemical details of proteins. The list of recent advances in analytical applications of luminescence spectroscopy goes on and on.

In this and its companion volume several applications of luminescence spectroscopy will be considered. Some of these applications would not have been possible a decade ago and some are recent developments in long existing methods of application. The list of subjects is not exhaustive but provides a range that should include something of interest for all analytical chemists. The various chapters of this work have been written by authors who are active investigators in their areas of contribution and the material is as up to date as the temporal problems of assembling a multiauthored reference work will allow.

The editor would like to express his gratitude to Ms. Susie Hansen and Ms. Gail Schulman for proofreading the manuscripts and to Ms. Vada Taylor and Ms. Lily Chan for technical assistance with preparation of the manuscript.

STEPHEN G. SCHULMAN

Gainesville, Florida
January 1985

CONTENTS

Molecular Luminescence Spectroscopy

Methods and Applications: Part 1

CHAPTER

1

LUMINESCENCE SPECTROSCOPY: AN OVERVIEW

STEPHEN G. SCHULMAN

College of Pharmacy
University of Florida
Gainesville, Florida 32610

1.1. INTRODUCTION

The emission of light by electronically excited molecules is the basis of molecular luminescence spectroscopy. Because conventional photodetectors respond with high sensitivity to visible and ultraviolet light, luminescence spectroscopy is widely employed in the quantitative analysis of a great variety of organic and inorganic substances. With the recent advent of using luminescence spectrophotometers as detectors in liquid chromatography, analytical interest in luminescence spectroscopy has intensified. Quantitative luminescence techniques rely on the relationships between analyte concentration and luminescence intensity and are, therefore, similar in concept to most other physicochemical methods of analysis. However, such features of luminescence spectra as position in the spectrum, band-shape, luminescence decay time, and excitation spectrum can also be used for qualitative as well as quantitative analysis. These features are often extremely sensitive to solvation, complexation, and so on, and luminescent molecules are often used as probes to report information to the scientist about their microenvironments. This book and its companion volume will deal with the application of molecular luminescence spectroscopy to problems of interest to a variety of analytical chemists and perhaps other scientists. The purpose of this chapter is to provide a brief introduction to luminescence spectroscopy for those readers who lack a strong background in the area. This chapter, then, will briefly consider the origin, nature, and measurement of molecular luminescence spectra and the dependence of luminescence upon molecular structure, reactivity, and interactions with the environment. Let us begin by dividing molecular luminescence spectroscopy into two broad areas: that in which the potentially luminescent molecule was initially excited by way of the absorption of visible or ultraviolet light and that in which the potentially luminescent molecule was excited by virtue of being produced in a thermal chemical reaction in which the products or intermediates are so energetic that they are electronically excited. The former process is termed *photoluminescence* and the latter, *chemiluminescence*.

1.2. PHOTOLUMINESCENCE SPECTROSCOPY

1.2.1. Absorption and Emission of Light by Molecules

Photoluminescence is the emission of light subsequent to the absorption of light by molecules. The absorption or excitation process entails the interaction of the electric field associated with the exciting light with the loosely bound π or nonbonded electrons of the absorbing molecule. This interaction distorts the electronic distribution of the absorbing molecule and causes energy to be absorbed from the electric field of the light wave.

The intensity (I_a) of light absorbed (in units of, say, photons s^{-1} cm^{-2}) is given by the Beer–Lambert law:

$$I_a = I_0 - I_t = I_0 (1 - 10^{-\epsilon Cl}) \qquad (1.1)$$

where I_0 and I_t are, respectively, the incident and transmitted light intensities, ϵ is the molar absorptivity (a property of the absorbing species), C is the concentration of absorber, and l is the optical depth of the sample.

The energy of the light absorbed is given by the Planck frequency relation:

$$E = Nh\nu = \frac{Nhc}{\lambda} \qquad (1.2)$$

where E is the energy associated with light of frequency ν or wavelength λ, c is the velocity of light, and N is Avogadro's number. Because of the quantization of molecular energy levels, electronic excitation of a particular molecular species occurs only if E corresponds to the difference in energy between the ground electronic state and an electronically excited state of the absorber.

In molecules, each electronic state has several associated vibrational states. In the ground electronic state, almost all molecules occupy the lowest vibrational level at room temperature. By exciting a particular absorption band with UV or visible light it is possible to promote the molecule of interest to one of several vibrationally as well as electronically excited levels (Fig. 1.1). The loss of excess vibrational and electronic energy then ensues with the excess energy taken up by solvent molecules in inelastic collisions with the excited solute. This rapid (10^{-13}–10^{-12} s) thermal deactivation of the highly excited molecule is called *vibrational relaxation* when the molecule loses vibrational energy within a given electronic state and *internal conversion* when the molecule undergoes a radiationless transition from an upper electronic state to a lower one. The processes of vibrational relaxation and internal conversion carry the molecule to the lowest vibrational level of the lowest excited singlet state (or excited state of the same spin multiplicity as the ground electronic state). Certain molecules may return all the way to the ground electronic state in the same time frame by thermal deactivation. However, in some molecules, for a variety of reasons the return from the lowest vibrational level of the lowest excited singlet state to the lowest vibrational level of the ground state by internal conversion and vibrational relaxation is forbidden (i.e., of low probability or long duration). In these molecules return to the ground electronic state occurs by one of two alternative pathways,

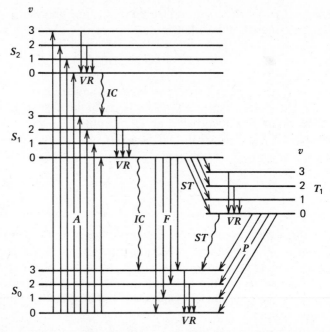

Fig. 1.1 Photophysical processes of conjugated molecules. Electronic absorption (A) from the lowest vibrational level ($v = 0$) of the ground state (S_0) to the various vibrational levels ($v = 0, 1, 2, 3$) of the excited singlet states (S_1 and S_2) is followed by rapid, radiationless internal conversion (IC) and vibrational relaxation (VR) to the lowest vibrational level ($v = 0$) of S_1. Competing for deactivation of the lowest excited singlet state S_1 are the radiationless internal conversion and singlet–triplet intersystem crossing (ST) as well as fluorescence (F). Fluorescence is followed by vibrational relaxation (VR) in the ground state. Intersystem crossing (ST) is followed by vibrational relaxation (VR) in the triplet state (T_1). Phosphorescence (P) and nonradiative triplet–singlet intersystem crossing (ST) return the molecule from the triplet state (T_1) to the ground state (S_0). Vibrational relaxation in S_0 then thermalizes the "hot" ground-state molecule. (Reprinted courtesy of Marcel Dekker, Inc.)

the simpler being the direct emission of ultraviolet or visible radiation whose frequency or wavelength is governed by the gap between the lowest excited singlet state and the ground electronic state. The radiative transition between excited and ground states of the same spin multiplicity occurs in a time frame of 10^{-11}–10^{-7} s after excitation, is called fluorescence, and is one of the two major aspects of photoluminescence. Owing to the fact that the ground electronic state of a molecule has several vibrational levels associated with it, fluorescence emission does not occur at a single wavelength, as suggested by Eq. (1.2), but rather occurs over a range of wavelengths corresponding to several vibrational transitions as components of a single electronic transition.

Several processes may compete with fluorescence for deactivation of the lowest excited singlet state. Consequently, the intensity of fluorescence I_f is obtained by multiplying I_a [Eq. (1.1)] by the fraction of excited molecules ϕ_f that actually fluoresce.

$$I_f = \phi_f I_a \tag{1.3}$$

ϕ_f is called the quantum yield of fluorescence or the fluorescence efficiency. It is usually a fraction but may be unity in some exceptional cases and is related to the rates of fluorescence (k_f) and competitive deactivating processes (k_d) by

$$\phi_f = \frac{k_f}{k_f + k_d} \tag{1.4}$$

Thus, the greater the numbers or rates of processes competing with fluorescence for deactivation of the lowest excited singlet state, the lower the value of ϕ_f and hence from Eqs. (1.1) and (1.3) it follows that

$$I_f = \phi_f I_0 \, (1 - 10^{-\epsilon Cl}) \tag{1.5}$$

Equation (1.5) may be expanded in series to

$$\phi_f I_0 \, (1 - 10^{-\epsilon Cl}) = (2.3\epsilon Cl - \frac{(2.3\epsilon Cl)^2}{2l} + \frac{(2.3\epsilon Cl)^3}{2l} - \cdots)\phi_f I_0 \tag{1.6}$$

For values of $\epsilon Cl < 0.02$, the higher terms in (1.6) may be neglected so that

$$I_f = 2.3\phi_f I_0 \epsilon Cl \tag{1.7}$$

Thus, at very low absorbance I_f is linear with analyte concentration.

Another important property of fluorescing molecules is the lifetime of the lowest excited singlet state (τ_f). If the mean rate of fluorescence is the number of fluorescence events per unit time, the mean lifetime of the excited state is the reciprocal rate, or the mean time per fluorescence event. The quantum yield of fluorescence and the lifetime of the excited state are related by

$$\tau_f = \phi_f \tau_N \tag{1.8}$$

where τ_N is called the natural lifetime of the excited state and represents the lifetime the fluorescing molecule would have if fluorescence was the sole pathway for deactivating the lowest excited singlet state. While τ_f and ϕ_f are determined largely by the kinds and rates of processes competing with fluorescence, τ_N is a function only of molecular electronic structure. It is one of the very few parameters that can be used to deduce information about the structure of the thermally equilibrated fluorescing species and how it might differ from the thermally equilibrated absorbing molecule.

The second pathway for deactivation of the lowest excited singlet state competes with fluorescence and is called *intersystem crossing* from the lowest excited singlet to the lowest excited triplet state. Intersystem crossing entails a change in spin angular momentum, which, classically, violates the law of conservation of angular momentum. Although it is about a million times less probable (slower) than a typical singlet–singlet vibrational process (e.g., internal conversion), it is of rate comparable to spin-allowed luminescence (fluorescence). Hence, it is competitive with fluorescence for deactivation of the lowest excited singlet state. Subsequent to intersystem crossing, the molecules populating the lowest triplet state undergo vibrational relaxation to the lowest vibrational level of the lowest triplet state. Molecules in the lowest triplet state can return to the ground state radiationlessly by triplet–singlet intersystem crossing or by the emission of light. The emission of light accompanying the transition from the triplet state to the ground singlet state is also a forbidden transition and is characterized by a very long duration (10^{-4}–10 s).

The radiative transition from an upper electronic state to a lower electronic state of different spin is called *phosphorescence*. Phosphorescence spectroscopy is the second of the two major branches of photoluminescence spectroscopy. Because triplet states are so long-lived, chemical and physical processes in solution compete effectively with phosphorescence for deactivation of the lowest excited triplet state. Except for the shortest-lived phosphorescences, collisional deactivation by solvent molecules, quenching by paramagnetic species (e.g., oxygen), photochemical reactions, and certain other processes preclude the observation of phosphorescence in fluid media unless the potentially phosphorescent molecules are protected from their environment by inclusion in micelles. Rather, phosphorescence is normally studied in glasses at liquid nitrogen temperature or, more recently, at room temperature with the phosphorescent species adsorbed on filter paper or other solid supports. Although some molecules will fluoresce or phosphoresce exclusively, some will demonstrate both fluorescence and phosphorescence. In this case the band occurring at longer wavelength will be the phosphorescence band because the lowest triplet state always lies lower in energy than the lowest excited singlet state. This is a consequence of the lower repulsive energy in the triplet state than in the singlet state of the same orbital configuration that arises from the greater average separation of the electrons in the triplet state.

The relationship between phosphorescence intensity I_0 and analyte concentration is similar to that for fluorescence.

$$I_p = \phi_{ST}\phi_P I_0 (1 - 10^{-\epsilon Cl}) \tag{1.9}$$

In Eq. (1.9), I_0, ϵ, C, and l have the same meanings as in the case of absorption or fluorescence, but it is now necessary to define ϕ_{ST}, the quantum yield of intersystem crossing and ϕ_p, the quantum yield of phosphorescence. The former is the fraction of excited molecules that undergo intersystem crossing from the lowest excited singlet state to the lowest triplet state, and the latter is the fraction of excited molecules having undergone intersystem crossing that are actually deactivated by phosphorescence. As in the case of fluorescence, at low absorbance Eq. (1.9) reduces to the linear form

$$I_P = 2.3\phi_{ST}\phi_P I_0 \epsilon Cl \tag{1.10}$$

which is convenient in analytical phosphorimetry.

1.2.2. The Influence of Molecular Structure on Photoluminescence

Although many molecules are capable of absorbing ultraviolet and visible light, only a few return to the ground state radiatively. Whether or not a molecule employs fluorescence or phosphorescence to rid itself of electronic excitation energy is largely a function of the molecular structure.

Both fluorescence and phosphorescence are most often observed in highly conjugated organic molecules with rigid molecular skeletons. The less vibrational and rotational freedom that the molecule or its parts have, the greater is the probability that the energy gap between the lowest excited singlet or triplet state and the ground electronic state will be large and require deactivation by luminescence. Aromatic hydrocarbons fall into this category and are characterized by high quantum yields of fluorescence and moderate quantum yields of phosphorescence. (The absence of luminescence is indicated by a vanishingly small quantum yield of luminescence.)

Aromatic molecules containing freely rotating substituents or lengthy aliphatic side chains usually tend to luminesce less intensely than those without those substituents. This results from the introduction of a large number of rotational and vibrational degrees of freedom by the exocyclic substituents. At very low temperatures, the quantum yields of both fluorescence and phosphorescence tend to be much greater than at ambient temperatures as a result of restricted vibrational and rotational freedom and, consequently, a lower efficiency of internal conversion.

There is a tendency for most substituents to increase the quantum yields of photoluminescence. This tendency arises from the increases of rate constants for radiative decay accompanying extension of an aromatic system and breakdown of its symmetry that result from substitution of a highly symmetrical arylhydrocarbon with a strongly interacting group such as —NH$_2$ or —OH. The reasons for the increases in the radiative rate constants are rooted in quantum mechanics and will not be discussed here. However, with certain substituents, the fluorescence quantum yields of aromatic molecules are diminished. This is especially so in the case of heavy atoms, such as Br and I, and certain groups, such as —NO$_2$, —CHO, and nitrogen in six-membered heterocyclic rings (e.g., quinoline) and other groups having sp^2 hydridized nonbonding electrons. Each of these substituents has the ability to cause mixing of the spin and orbital electronic motions of the aromatic system (spin-orbital coupling). Spin-orbital coupling destroys the concept of molecular spin as a well-defined property of the molecule and thereby enhances the probability or rate of singlet → triplet intersystem crossing. This process favors population of the lowest triplet state at the expense of the lowest excited singlet state and thus decreases the fluorescence quantum yield. However, the efficient population of the lowest triplet state favors a high yield of phosphorescence. Consequently, nitro compounds, bromo and iodo derivatives, aldehydes, ketones, and N-heterocyclics tend to fluoresce very weakly or not at all, although many of them phosphoresce quite intensely.

Molecular structure can have a profound effect on the positions in the spectrum where fluorescence and phosphorescence occur, as well as on their intensities. According to Eq. (1.2), the smaller the energy gap between the ground and lowest excited states, the longer will be the wavelengths of photoluminescence. It can be shown by quantum mechanics that the more extended a conjugated system is, the smaller will be the separations between the ground state and the lowest excited singlet and triplet states. It is not surprising, therefore, that benzene, naphthalene, and anthracene fluoresce maximally at 262 nm, 320 nm, and 379 nm and phosphoresce maximally at 339 nm, 470 nm, and 670 nm, respectively.

Similarly, the affixment of conjugating substituents onto a conjugated system extends the conjugation of the latter and causes the photoluminescence maxima of the substituted derivatives to lie at wavelengths longer than those of the parent compound. Hence, the fluorescence and phosphorescence of aniline lie at 340 nm and 373 nm, respectively, while those of the parent hydrocarbon benzene lie at 262 nm and 339 nm, respectively.

1.2.3. The Influence of the Environment on Photoluminescence

Solvent Effects

Although the capacity for photoluminescence is primarily a function of molecular structure, the environment, particularly the solvent, of a potentially fluorescing or phosphorescing molecule can have a dramatic effect on the appearance of fluorescence or phosphorescence.

Solvent interactions with solute molecules are largely electrostatic, and it is usually the differences between the electrostatic stabilization energies of ground and excited states that contribute to the relative intensities and spectral positions of fluorescence and phosphorescence in different solvents.

The changes in π electron distribution that take place on transition from the lowest excited singlet or triplet state to the ground state, during the photoluminescence, cause changes in the dipolar and hydrogen-bonding properties of the solute. If the solute is more polar in the excited state than in the ground state, fluorescence or phosphorescence occurs at longer wavelengths in a polar solvent than in a nonpolar solvent, because the more polar solvent stabilizes the excited state relative to the ground state. Moreover, the fact that photoluminescence originates from an excited state that is in equilibrium with its solvent cage and terminates in a ground state that is not causes the photoluminescences to lie at longer wavelengths the more strongly interacting the solvent (i.e., the more polar or more strongly hydrogen-bonding). This effect is generally prominent in fluorescence spectra taken in fluid solutions but is not important in phosphorescence spectra or fluorescence spectra taken in rigid media because in rigid media the solvent is not free to resolvate the ground state after emission so that the differences in energy between the equilibrium and nonequilibrium solvent cages are lost.

In molecules having atoms with sp^2 hybridized nonbonded electron pairs (e.g., carbonyl compounds and certain N-heterocyclics) the lowest excited singlet state is formed by promoting a lone-pair electron to a vacant π orbital. (This is called an n,π^* state.) These molecules tend to phosphoresce but show very little fluorescence in aprotic solvents such as aliphatic hydrocarbons because the n,π^* excited singlet state is efficiently deactivated by intersystem crossing. However, in protic solvents, such as ethanol or water, these molecules become fluorescent. This results from the destabilization of the lowest singlet n,π^* state by hydrogen bonding. If this interaction is sufficiently strong, the fluorescent π,π^* state drops below the n,π^* state, allowing intense fluorescence. Isoquinoline, for example, fluoresces in water but not in cyclohexane, where it phosphoresces intensely.

Solvents containing atoms of high atomic number (e.g., alkyl iodides) also have a substantial effect on the intensity of fluorescence of solute molecules. Atoms of high atomic number in the solvent cage of the solute molecule enhance spin-orbital coupling in the lowest excited singlet state of the solute. This favors the radiationless population of the lowest triplet state at the expense of the lowest excited singlet state. Thus in "heavy atom solvents," all other things being equal, fluorescence is always less intense than that in solvents of low molecular weight. On the other hand, phosphorescence is often enhanced by heavy atom solvents.

The Effects of Acids and Bases

The influences of solution acidity and basicity on photoluminescence spectra are derived from the dissociation of acidic functional groups or protonation of basic functional groups associated with the aromatic portions of fluorescing and phosphorescing molecules. Protonation or dissociation can alter the natures and rates of nonradiative processes competing with photoluminescence and thereby affect the quantum yields of emission. For example, salicylaldehyde, is nonfluorescent and intensely phosphorescent as a result of an n,π^* lowest excited singlet state that is rapidly deactivated by intersystem crossing. However, in alkaline solutions, where the phenolic group is dissociated, and in concentrated mineral acid solutions, where the carbonyl group is protonated, salicylaldehyde is intensely fluorescent and nonphosphorescent. Evidently, in the cation and anion forms, the fluorescent π,π^* state lies below the nonfluorescent n,π^* state in salicylaldehyde.

Protonation and dissociation alter the relative separations of the ground and excited states of the reacting molecules and thereby cause shifting of the luminescence spectra. The shifts tend to be greater in fluorescence spectra than in phosphorescence spectra and are due to the electrostatic stabilizations or destabilizations of the excited states, relative to the ground state, produced by protonation and dissociation. The protonation of electron-withdrawing groups, such as carboxyl, carbonyl, and pyridinic nitrogen, results in shifts of the luminescence spectra to longer wavelengths, while the protonation of electron-donating groups, such as the amino group, produces spectral shifts to shorter wavelengths. The protolytic dissociation of electron-donating groups such as hydroxyl, sulfhydryl, or pyrrolic nitrogen produces spectral shifts to longer wavelengths, while the dissociation of electron-withdrawing groups, such as carboxyl, produces shifting of the photoluminescence spectra to shorter wavelengths.

One of the more interesting aspects of acid–base reactions of luminescent or potentially luminescent molecules in fluid solutions is derived from the occurrence of protonation and dissociation during the lifetime of the excited state. Since phosphorescence rarely is observed in fluid solutions, this is confined to fluorescence and the lowest excited singlet state. This phenomenon affects the dependence of fluorescence upon pH in ways that are not anticipated on the basis of ground-state acid–base chemistry and must be considered in the development of a fluorometric analysis in aqueous solutions.

The lifetimes of molecules in the lowest excited singlet state are typically of the order of 10^{-11}–10^{-7} s. Typical rates of prototropic reactions are 10^{11} M^{-1} s^{-1} or lower. Consequently, excited-state proton transfer reactions may be competitive with radiative deactivation of the excited molecules. If excited-state proton transfer is too slow to compete with fluorescence for deactivation of the lowest excited singlet state, the fluorescence intensity of an emitting acid or base will depend on the absorbance at the wavelength of excitation, which in turn will depend on the pK_a of the acid or base in the ground state.

In the event that excited-state proton transfer and radiative deactivation are temporally competitive, the quantum yield of fluorescence will demonstrate a different kind of dependence on the pH of the solution. Because the electronic distribution of a molecule is usually much different in an excited state than in the ground electronic state, the variation in fluorescence intensity caused by the pH dependence of the quantum yield of fluorescence will occur in a pH region different from the pH region in which the fluorescence intensity depends on the absorbance of the analyte. This means that in, say, a phenolic molecule, which becomes more acidic in the lowest excited singlet state, one might observe fluorescence from the conjugate base at a pH as low as 3, even though the pK_a might be as high as 10. This subject will be discussed at length in Part 2.

Quenching of Photoluminescence

The fluorescence or phosphorescence of luminescent species may be decreased or even eliminated by interactions with other chemical species. This phenomenon is called quenching of luminescence.

Two kinds of quenching are distinguished. In static quenching, interaction between the potentially luminescent molecule and the quencher takes place in the ground state, forming a nonluminescent complex. The efficiency of quenching is governed by the formation constant of the complex as well as by the concentration of the quencher. The quenching of the fluorescence of o-phenanthroline by Fe(II) is an example.

In dynamic quenching the quenching species and the potentially luminescent molecule react during the lifetime of the excited state of the latter. Dynamic quenching is also called diffusional quenching and its efficiency depends on the viscosity of the solution, the lifetime of the excited state (τ_0) of the luminescent species, and the concentration of the quencher $[Q]$. This is summarized in the Stern–Volmer equation:

$$\frac{\phi}{\phi_0} = \frac{1}{1 + k_Q\tau_0[Q]} \tag{1.11}$$

where k_Q is the rate constant for encounters between quencher and potentially luminescing species and ϕ_0 and ϕ are the quantum yields of luminescence in the absence and presence of concentration $[Q]$ of the quencher, respectively.

k_Q is typical of diffusion-controlled reactions ($\sim 10^{10}\,M^{-1}\,s^{-1}$). In phosphorescent molecules, $10^{-4}\,s < \tau_0 < 10\,s$ so that $10^6\,M^{-1} < k_Q\tau_0 < 10^{11}\,M^{-1}$. Hence if the concentration of molecular oxygen is $10^{-3}\,M$ in water at atmospheric pressure, ϕ/ϕ_0 is vanishingly small for all phosphorescent molecules. This is in agreement with the observed efficient quenching of phosphorescence by dissolved oxygen.

In the case of fluorescence, $10^{-11}\,s < \tau_0 < 10^{-7}\,s$ so that $10^{-1}\,M^{-1} < k_Q\tau_0 < 10^3\,M^{-1}$. At $10^{-3}\,M$, dissolved molecular oxygen is capable of quenching only the longer-lived fluorescences.

The Influence of Metal Ions

The coordination of conjugated ligands by metal ions can affect photoluminescence in several ways. Nontransition metals will shift the wavelengths of fluorescences and phosphorescence of luminescing ligands to which they are coordinated. This results from the positive polarization, caused by the metal ion, at the sites of coordination on the ligand. In this regard the metal ion acts somewhat as a "superproton." The luminescence of a ligand may be somewhat enhanced or quenched by coordination, depending on the influence the metal ion has on the nonradiative processes competing with luminescence. In the case of the heaviest nontransition metal ions, for example, Hg(II) and Bi(III), static quenching of fluorescence, and sometimes of phosphorescence, often results from a heavy atom effect.

The fluorescences and phosphorescences of luminescing ligands are usually quenched by complexation of the ligand with main group transition metal ions. This is believed to occur by way of a process known as paramagnetic quenching, in which the unpaired electrons of the metal ion interact intimately with the π electrons of the ligand, thereby producing a pathway for intersystem crossing from the directly excited singlet states of the ligand to states of higher multiplicity introduced by interaction with the metal ion. At very low temperatures, certain transition metal complexes, notably those of Cr(III), phosphoresce, supporting the hypothesis of intersystem crossing as a fluorescence quenching mechanism. Certain transition metal complexes do fluoresce, most notably the bipyridyl complexes of Ru(II). These will be discussed at length in a subsequent chapter.

In certain transition metal complexes—namely, those of the lanthanides and actinides—the unpaired electrons are, for the most part, isolated from the ligand. Excitation of the ligand results in transfer of the excitation energy to the metal ion, where it is taken up in exciting an unpaired f electron to a higher f orbital. The resulting emissions lie in the red, for example, for Eu(III) and Tb(III) complexes. These emissions have been used to analyze coordinating analytes and to probe binding sites normally occupied by Ca(II) in biologically significant molecules.

Photoluminescence of Molecular Aggregates

At higher solute concentrations, there is a tendency for molecules to form aggregates in both ground and electronically excited states. Aggregation can substantially affect the photoluminescences of small molecules.

When two or more conjugated molecules approach closely, their π-electron systems interact, displacing the relative separations of the ground and lowest excited states with respect to the monomers. In addition, depending on the geometry of the aggregation, the electronic dipole moments of the monomers, which are affected by the excitation or emission processes, may couple (add) constructively or destructively. If two or more molecules aggregate in a plane (head to tail), the dipole moments of their lowest excited singlet states will usually reinforce each other in such a way that intense fluorescence will usually occur at longer wavelengths than in the isolated molecule. If aggregation takes place by two or more molecules overlapping their aromatic rings (stacking or sandwiching), the excited state dipole moments will tend to cancel, making excitation to the lowest excited singlet state somewhat forbidden. In this case (as in the case of the n,π^* singlet state lying lowest), intersystem crossing to the triplet state will be probable and phosphorescence will be favored over fluorescence in rigid media, while quenching of fluorescence alone will prevail in fluid media. These phenomena occur in molecular crystals as well as in small aggregates in solution, and the quantum mechanical formalism used to treat them is called *exciton theory*.

Aggregation in solution sometimes occurs subsequent to the excitation of monomeric species. In this case, the excited monomer encounters another (unexcited) monomer during the lifetime of its excited state and fluoresces as a dimer. This phenomenon was first described for concentrated solutions of pyrene and was observed as a shift of the ultraviolet fluorescence of pyrene to the blue region of the spectrum with increasing concentration. In the same concentration region the absorption spectrum of pyrene did not change, indicating that aggregation took place in the excited state but not in the ground state.

An emitting aggregate that is composed of two or more identical molecules is called an *excimer*. One composed of two or more dissimilar molecules is called a *heteroexcimer* or *exciplex*.

Energy Transfer

Energy transfer entails the excitation of a molecule that during the lifetime of the excited state passes its excitation energy to a nearby molecule. The loss of excitation energy from the species excited results in quenching of the luminescence of the energy donor and may result in luminescence from the energy acceptor, which becomes excited in the process.

Energy transfer occurs by one of two acceptor concentration-dependent processes. In the resonance excitation transfer mechanism or dipole mechanism, the donor and acceptor molecules are not in contact with one another and may be separated by as much as 10^{-6} cm (although transfer distances closer to 10^{-7} cm are more common). In the classical sense, the excited energy donor molecule may be thought of as an electrical dipole that creates an electrical field in its vicinity. Potential acceptor molecules within the range of this electrical field may absorb energy from the field, resulting in electronic excitation of the latter. The probability of and thus the rate of resonance energy transfer decreases as the sixth power of the distance between the donor and acceptor dipoles according to

$$k_{ET} = \frac{1}{\tau_D} \left(\frac{R_0}{R} \right)^6 \qquad (1.12)$$

where k_{ET} is the rate constant for resonance energy transfer, τ_D is the lifetime of the excited state of the donor molecule, R is the mean distance between the centers of the donor and acceptor dipoles, and R_0 is a constant for a given donor–acceptor pair (the critical separation) corresponding to the mean distance between the centers of the donor and acceptor dipoles for which energy transfer from donor to acceptor and luminescence from the donor are equally probable. Experimentally,

$$R_0 = \left(\frac{3000}{4\pi N[A]_{1/2}}\right)^{1/3} \qquad (1.13)$$

where N is Avogadro's number and $[A]_{1/2}$ is the concentration of acceptor molecules that reduces the quantum yield of fluorescence of the donor to half of its value measured in the absence of the acceptor species. R_0 and $[A]_{1/2}$ are therefore the two important constants of any donor–acceptor pair between which energy is transferred by the resonance mechanism. The higher the concentration of the acceptor, the smaller is R and the more probable resonance energy transfer becomes. The rate of resonance energy transfer is also determined by the lifetimes and therefore the spin multiplicities of the excited electronic states of the donors as well as by the probabilities (i.e., molar absorptivities) of the electronic transitions involved in the acceptors. Donor molecules in the lowest excited singlet state have short lifetimes, whereas donors in the triplet state have long lifetimes. On the other hand, transition to the lowest excited singlet state usually has a high probability, whereas transition from the ground state to the lowest triplet state has very low probability. The high probability of singlet $\rightarrow$ singlet transition in most acceptor molecules usually compensates for the short lifetime of the excited donor, so that singlet $\rightarrow$ singlet excitation of acceptors by excited singlet donors is a probable phenomenon. The long lifetime of the triplet state and the high probability of singlet $\rightarrow$ singlet transition in an acceptor also favor triplet $\rightarrow$ singlet resonance energy transfer. Because of the very low probability of singlet $\rightarrow$ triplet transitions in acceptor molecules, both singlet $\rightarrow$ triplet and triplet $\rightarrow$ triplet resonance energy transfer have extremely low probabilities. As a result, singlet- or triplet-sensitized phosphorescences originating from the resonance energy transfer mechanism are not generally observed. Another general requirement for the occurrence of resonance energy transfer is the overlap of the fluorescence (or phosphorescence) spectrum of the donor and the absorption spectrum of the acceptor. Any degree of overlap of these spectra will satisfy the quantization requirements for the energy of the thermally equilibrated donor molecule to promote the acceptor to a vibrational level of its excited singlet state. The greater the degree of overlap of the luminescence spectrum of the donor and the absorption spectrum of the acceptor, the greater is the probability that energy transfer will take place.

The exchange mechanism of excitation energy transfer is important only when the electron clouds of donor and acceptor are in direct contact. In this circumstance, the highest-energy electrons of the donor and acceptor may change places. Thus the optical electron of, say, a triplet donor molecule may become part of the electronic structure of an acceptor molecule originally in the ground singlet state, while the donor is returned to its ground singlet state by acquiring an electron from the acceptor. Exchange energy transfer is also most efficient when the fluorescence spectrum of the donor overlaps the absorption spectrum of the acceptor. While the resonance mechanisms preclude energy transfer resulting in a spin-forbidden transition of the acceptor molecule triplet $\rightarrow$ triplet energy transfer commonly, and singlet $\rightarrow$ triplet energy transfer occasionally occur by the exchange mechanism, provided that the donor triplet in the former case and the donor singlet in the latter case are higher in energy than the acceptor triplet state. Singlet $\rightarrow$ singlet and triplet $\rightarrow$ singlet transfer are also permitted by the exchange mechanism. Exchange energy transfer is a diffusion-controlled process (i.e., every collision between donor and acceptor leads to energy transfer) and as such its rate depends on the viscosity of the medium. Resonance energy transfer, on the other hand, is not diffusion controlled, does not depend on solvent viscosity, and may be observed at lower concentrations of acceptor species. In fluid solutions of relatively low viscosity, diffusion-controlled rate constants are of the order of 10^{10} mol^{-1} s^{-1}. The lifetime of the lowest excited singlet state of a donor molecule is $< 10^{-7}$ s. Thus if $[A]_{1/2}$ for a given acceptor species is less than 10^{-4} M, singlet $\rightarrow$ singlet and singlet $\rightarrow$ triplet exchange energy transfer are too slow to compete appreciably with fluorescence for deactivation of the excited donor. In rigid solutions, where phosphorescence is usually observed, exchange energy transfer is normally precluded unless the donor and acceptor species form a contact complex. However, in some solid systems, especially in organic crystals containing trace impurities (acceptors), where there is a high degree of order in the matrix, energy transfer may be observed as a result of a process known as exciton migration. In this process, the energy absorbed by the donor is rapidly delocalized throughout the matrix and is eventually localized by the acceptor species, which then emits fluorescence or phosphorescence.

Biological systems are highly suited for the process of resonance energy transfer because of the presence of high local concentrations of absorbing groups in these systems. Dilution of the system, however, may increase the average intermolecular distances beyond the range of dipole–dipole coupling, thereby reducing or eliminating the amount of transfer. The probability of energy transfer is greatly increased when two absorbing species are combined into a single molecule. When the two parts of the molecule are combined within the range of dipole–dipole coupling (which is responsible for the process), energy transfer will always be observed, even in dilute solution.

1.3. CHEMILUMINESCENCE

Chemical reactions are usually attended by energy changes of a few kilocalories per mole of product formed. Immediately after formation, the excess energy of the products is lost by collisions with the solvent, a process that takes $\sim 10^{-12}$ s. However, in a few exceptional cases, notably the oxidation of highly reduced species by strong oxidizing agents, the products of reaction may be 40–70 kcal mol^{-1} away from equilibrium. In this case, the reaction products may, if their electronic structures permit, become electronically excited, lose a small amount of their excess energy by vibrational relaxation, and subsequently fluoresce. Fluorescence that is initiated by chemical reaction is called chemiluminescence.

Chemiluminescence is a common phenomenon in flames where free radicals are energetically oxidized by molecular oxygen. In liquid solutions, however, it is a relatively rare occurrence, perhaps because few of the highly reactive kinds of species, such as free radicals, which are necessary for the generation of chemiluminescence, have appreciable lifetimes in liquid solutions. Because of the efficiency of quenching of the long-lived triplet state in the liquid phase, chemiluminescence as a phosphorescent phenomenon is unknown in solution.

Although the lifetimes of the excited singlet state that are responsible for chemiluminescence have durations typically of nanoseconds, the emission of light in chemiluminescent reactions often persists for several minutes after the reactions are initiated. The explanation for this observation is chemical rather than physical. The emission of chemiluminescence from a solution persists for as long as the chemiluminescent reaction takes place, and therefore is indicative of the kinetics of the reaction rather than of the photophysics of the electronically excited products of the reaction.

Probably the best-known chemiluminescent reaction is the oxidation of 3-aminophthalhydrazide (luminol). Its anion is oxidized in 0.03 N sodium hydroxide solution, by hydrogen peroxide, in the presence of potassium ferricyanide as a catalyst. The emission is the blue fluorescence of the ultimate product, the 3-aminophthalate dianion.

    Chemiluminescence occurs in a variety of living organisms, wherein it is called *bioluminescence*. A variety of fungi, algae, marine dinoflagellates, and other coelenterates are capable of the chemical generation of light. Perhaps the best-known example of bioluminescence is that which occurs with the firefly *Photinus pyralis*. Firefly bioluminescence involves the stepwise oxidation of a derivative of 5-hydroxybenzothiazole called luciferin. Luciferin, or simple derivatives thereof, is probably responsible for most bioluminescence observed in the plant and animal kingdoms. Luciferin is oxidized by molecular oxygen in a controlled series of steps regulated by the enzyme luciferase and adenosine triphosphate (ATP). The ATP forms a complex with luciferase and apparently supplies the energy to reduce oxidized luciferin to the form whose oxidation is ultimately responsible for the production of light.

    The development of chemiluminescence as an analytical technique has been hindered by the paucity of molecules that demonstrate the phenomenon in solution and the economic unimportance of these molecules themselves. However, analytical use has been made of the dependence of the rate of chemiluminescence of luciferin on ATP concentration to develop an extremely sensitive assay for ATP. The fact that chemiluminescent reactions are usually catalyzed by small concentrations of ions has resulted in the development of chemiluminescence assays for these ions. Cu^{2+} and Co^{2+}, for example, have been determined from their catalytic effect on the oxidation of luminol.

1.4.   INSTRUMENTATION

1.4.1.   Fluorescence Spectrophotometers

A spectrophotometer designed for the measurement of fluorescence spectra consists of (a) an excitation source, which is usually a lamp; (b) a monochromator, which disperses the polychromatic exciting light into its component wavelengths; (c) a sample compartment; (d) a monochromator, which disperses the radiation emitted from the sample into its component wavelengths; (e) a photodetector, which converts the emitted light into an electrical signal; and (f) a readout system such as a recorder, a galvanometer, or an oscilloscope (Fig. 1.2).

Excitation Sources

The most often used excitation sources are gas discharge lamps. The radiation emitted by these lamps arises from electrical ionization of the

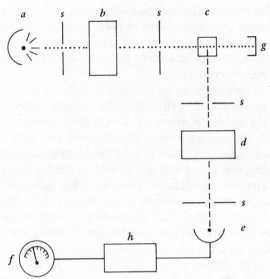

Fig. 1.2. Schematic diagram of a fluorescence spectrophotometer: (a) source of exciting light; (b) excitation monochromator; (c) sample cell; (d) emission monochromator; (e) photodetector; (f) readout system; (g) light absorbing stop for excitation beam; (h) amplifier for photodetector signal; (s) slits.(Reprinted courtesy of Marcel Dekker, Inc.)

gas by a high-voltage spark. Low-pressure discharge lamps radiate only certain characteristic wavelengths of light (characteristic of the gas used), which usually form line spectra. As the pressure and/or temperature of the gas filling the lamp is increased, these spectral lines broaden considerably into bands and the output of the lamp becomes continuous over a broad wavelength interval.

The mercury-vapor lamp is widely used because of its stability and intense line emission at 366 nm, 313 nm, and several other wavelengths. the 366-nm resonance line of the mercury-vapor lamp is the most commonly used wavelength for excitation in instruments employing these types of lamps. Low-pressure mercury-vapor lamps can be modified to provide energy at wavelengths other than the resonance lines by coating the inside of the tube envelope with phosphor crystals. The thin layer of crystalline phosphor absorbs the resonance radiation of the mercury vapor and reemits the energy at a different wavelength, which depends on the nature of the phosphor.

High-pressure xenon lamps provide a continuum of wavelengths in the near ultraviolet and visible regions and therefore allow a wide range of selectivity in the choice of the wavelength for excitation of the sample.

Most spectrofluorometers utilize the high-pressure 150-W or 250-W xenon lamp. These lamps are under 5 atm of pressure when cold and almost 20 atm when heated. Although the xenon arc lamp has a broad wavelength continuum output, it does not have the intensity of the mercury-vapor lamp at any single wavelength.

The choice of an excitation wavelength is made by setting the emission monochromator at a wavelength at which the sample exhibits fluorescence and scanning the entire spectrum of wavelengths shorter than the wavelength at which the emission monochromator is set, with the excitation monochromator. This procedure is used to obtain an excitation spectrum of the fluorescing species, which is a display of the fluorescence intensity versus the excitation wavelength. The excitation peaks will correspond roughly to the absorption spectrum of the sample. The choice of an excitation wavelength for analytical purposes is usually made on the basis of fluorescence intensity and isolation from excitation peaks of interfering substances. The xenon arc lamp allows a greater degree of freedom of the choice of an excitation wavelength but lacks the intensity of the mercury-xenon lamp. Hence the choice between selectivity and sensitivity will be governed by the excitation spectrum of the sample and the availability of the instrumentation.

Recently, lasers have been employed in place of lamps as excitation sources. The monochromaticity of laser light obviates the need for an excitation monochromator, and the increased intensity of the laser sources over conventional excitation sources has led to great improvements in detection limits for a few organic compounds.

One might think that the sensitivity of fluorescence assay could be improved by orders of magnitude by laser excitation, since the intensity of fluorescence is dependent on the intensity of the exciting radiation. However, by increasing the excitation energy, light scatter, localized heating and photodecomposition are also dramatically increased. These factors have resulted in only slight improvements in signal-to-noise ratios when laser excitation is used in place of lamp excitation. Hence, it is doubtful that lasers will replace lamps and monochromators in the excitation optics of fluorescence instrumentation, for some time.

Monochromators

Most spectrophotometers employ either diffraction gratings or quartz prisms as monochromators. The prism or grating is mounted on a turntable, which when rotated allows light of different wavelengths to be focused through a collimating slit onto the sample. The emission monochromator is also mounted on a turntable and when rotated allows light that is emitted by the sample to be focused onto the detector. Gratings are more widely used in spectroscopic instrumentation than prisms. A prism must be very large to give excellent resolution and if it is to be made of quartz to pass ultraviolet light, its cost might be prohibitive. Reflection gratings, which are widely employed, are comprised of a mirrored surface in which a series of closely spaced grooves is used to diffract the radiant energy. A typical grating will have 600–1200 lines per millimeter etched in the surface. Gratings have approximately uniform resolution and linear dispersion at all wavelengths. All wavelengths can be dispersed because none are absorbed by the reflection gratings. The principal disadvantage of a diffraction grating is that it will allow several orders of spectra to pass. This can complicate an analysis.

Between the lamp and the excitation monochromator and between the emission monochromator and the detector are two slits. These slits, which can be fixed or adjustable, focus light onto the excitation monochromator and the detector and regulate the ranges of wavelengths that excite the sample and pass onto the detector. The smaller the slit width employed, the narrower is the range of spectral bandpass and the greater is the analytical selectivity. Changing of slit widths, in the case of fixed slits, entails the replacement of the slit; in the case of variable slits, it involves the adjustment of slit width by use of a micrometer. Variable slits are more convenient to use than fixed slits. However, a major disadvantage is that with extensive use, the knife edges of the variable slits become worn and lose the ability to be calibrated. For weakly fluorescing samples, it may be necessary to increase the intensity of light exciting the sample and fluorescent light falling on the detector. This can be achieved by increasing the slit width, which generally diminishes selectivity of excitation and monitoring of fluorescence. On the other hand, for intensely fluorescing analytes it may be possible to use very narrow slits and carry out the analysis in the presence of several interfering substances.

Sample Compartments

Samples to be analyzed are contained and excited in the sample compartment. In most instruments the sample compartment is situated so that the emitted light from the sample is measured at right angles to the line of entry of the exciting light. This arrangement minimizes the interference by stray exciting light. Fluorescence is emitted from the sample in all directions. Hence only a small fraction of the total fluorescence from the sample is collected by the detector; the rest is absorbed by the walls of the sample compartment.

Sample cells most commonly employed in room temperature fluorescence spectroscopy are usually 1 cm^2 fused silica or quartz cells. The silica cells are usually better because they demonstrate less background fluorescence than the quartz cells. Cylindrical quartz tubing is used in low-temperature work because the square cells usually cannot withstand the rapid cooling of the sample.

Phosphorescence Spectrophotometers

There is no difference between a fluorescence and a phosphorescence spectrophotometer except in the sample compartment. Most spectrofluorometers can be easily adapted for phosphorimetry. Low-temperature fluorescence studies may be carried out by removing the conventional sample compartment and replacing it with one that contains a Dewar flask with a quartz window or nipple that is aligned with the excitation and emission optics. The flask is then filled with a cryogenic fluid such as liquid nitrogen. The sample is then placed in a cylindrical quartz tube and lowered into the filled flask so that the sample is positioned in the quartz window. If the sample is then illuminated, both fluorescence and phosphorescence may be detected. If the sample phosphoresces but does not fluoresce this setup may be employed to measure the phosphorescence. However, if the sample fluoresces and phosphoresces and it is desired to monitor only the phosphorescence, the fluorescent light may be prevented from reaching the detector by taking advantage of the much longer lifetime of phosphorescence relative to fluorescence. A mechanical chopper (usually a rotating can) that has lateral apertures 180° apart may be mounted in the sample compartment (Fig. 1.3). The can encloses the quartz nipple, which is aligned with the apertures in the side of the can. The can is then rotated at several thousand revolutions per minute. Its angular velocity can be controlled by a variable-speed motor. Since the apertures in the can are 180° apart and the excitation-emission optics are in a 90° configuration, when one aperture is open to the excitation source the sample is excited. Emitted light can only exit to the detector at 90° with respect to the emission optics; hence, no emitted light falls on the detector. When the can rotates through 90°, the exciting light is cut off from the sample. Since the lifetime of fluorescence is 10^{-11}–10^{-7} s, the fluorescent light from the sample will have vanished completely during the course of one-quarter turn of the can. However, the phosphorescence, which has a lifetime from 10^{-4} to several seconds, persists and is registered when the can aperture faces the detector. By use of the mechanical chopper, the phosphorescence spectrum or the phosphorescence excitation spectrum of the sample may be taken without interference from fluorescence or the

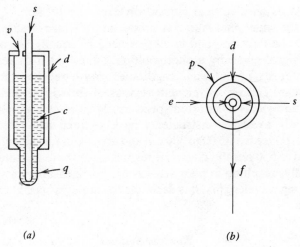

(a)   *(b)*

Fig. 1.3. Schematic diagrams of (a) Dewar flask for low-temperature fluorescence or phosphorescence studies and (b) top view of sample compartment for phosphorescence studies: (c) cryogenic liquid (liquid nitrogen); (d) Dewar flask; (e) excitation beam; (f) emission beam; (p) rotating can; (q) quartz nipple; (s) sample cell; (v) vent for vaporized nitrogen.(Reprinted courtesy of Marcel Dekker, Inc.)

fluorescence excitation spectrum. In room temperature phophorimetry the sample is dried on a solid support, which is then mounted in the sample compartment at 45° to both the excitation and emission optical paths. The optical arrangement is the same as if one wished to reflect the exciting light onto the detector. As in the case of low-temperature phosphorimetry, the rotating can chopper may be used to isolate the phosphorescence from any undesirable fluorescences or scattered light.

Chemiluminescence Instrumentation

Since chemiluminescence does not require photoexcitation, a spectrophotometer designed to measure chemiluminescence would not need a lamp or an excitation monochromator. Chemiluminescence can be studied in a conventional fluorescence spectrophotometer without employing the excitation optics. Several simple devices for quantitative chemiluminescence photometry are available that employ a sample compartment, a filter in place of a monochromator, a detector, and a galvanometer.

Detectors

Photomultiplier tubes are the most commonly employed detection devices used in modern fluorescence spectrophotometers. The photomultiplier tube produces an electrical current on exposure to light. An amplifier increases the small photocurrents produced, to a measurable level. Photomultipliers do not respond to all wavelengths equally. As a result, the electrical signal given by a photomultiplier does not truly represent the relative intensity of the wavelength spectrum it receives. Rather, it is biased toward certain wavelength regions of the visible and ultraviolet spectrum and yields a distorted spectrum. Most photomultiplier tubes in commercially available instruments have optimal and relatively even spectral response from 300 to 500 nm and are useful for most fluorescence work. Above 550 nm the electrical responses of these phototubes fall off dramatically, resulting in poor sensitivity. For measurements of fluorescence at long wavelengths, it is desirable to employ a special red-sensitive phototube.

Readout Systems

Signal readout in luminescence spectroscopy may be accomplished with recorders, galvanometers, or oscilloscopes. The galvanometers are satisfactory for luminescence measurements at a fixed wavelength but are cumbersome for determining spectra as the spectrum must be plotted manually from individual readings on the galvanometer. Recorders supply a permanent record of the fluorescence emission and excitation spectra. The strip chart recorder displays fluorescence intensity as the ordinate and wavelength as the abscissa. The wavelength is traversed at a constant speed and may be coupled to the monochromator drive motor, which is also operated at a constant speed. Several scanning speeds are usually available on strip chart recorders. These allow the spectra to be expanded or compressed to a desired scale. In the x-y recorders, the x-axis corresponds to the fluorescence intensity and the y-axis is continuously synchronized with the scanning speed of the monochromators. Because of this an undistorted spectrum can be taken even if the scanning rate, over a certain wavelength region, is changed during the entire scan. Moreover, a spectrum can be retraced by scanning in the opposite direction with the x-y recorder. This is not possible with a strip-chart recorder. The greatest limitation of recorders is the relatively long response time of the recorder pen (0.1–0.5 s). This limits the rate at which an accurate spectrum can be scanned. An oscilloscope traces the spectrum electronically and circumvents the slow response of a recorder pen. However, to obtain a permanent record of the spectrum from an oscilloscope, a photograph of the oscilloscope screen must be taken. Oscilloscopes are commonly employed for kinetic measurements of fast reactions in solution as in the study of reaction rates by stopped-flow methods using fluorimetric detection.

Instrumental Distortion of Spectra

The luminescence excitation and emission spectra taken on conventional spectrofluorometers are not "true" excitation and emission spectra (i.e., they are not representations of ϵ versus λ_{ex} or $\epsilon\phi_f$ or $\epsilon\phi_p$ versus λ_f or λ_p). These spectra are apparent excitation and emission spectra that are distorted by instrumental response such as the wavelength variable output of the lamp and responses of the monochromators and phototube. For quantitative analysis this is unimportant. Since the "true" luminescence excitation spectrum is identical to the absorption spectrum of the analyte (unless the emission is sensitized by energy transfer), the ability to take true excitation spectra is desirable because it facilitates identification of organic molecules at concentrations far lower than possible by other means. The "true" fluorescence spectrum is necessary for the accurate calculation of quantum yields. Moreover, because the instrumental distortion of the spectra varies from instrument to instrument, "true" excitation and emission spectra are highly desirable when comparing results taken on different instruments. The outputs of luminescence spectrophotometers can be corrected for instrumental response by means of quantum counters or thermopiles that calibrate the instrumental response over the entire wavelength range of the instrument. They then allow for its adjustment against a standard by means of potentiometers that control the wavelength variable output over the entire spectrum.

1.4.2. Measurement of Fluorescence and Phosphorescence Decay Times

Up to the present, all the photoluminescence techniques considered have been based on the measurement of the intensity of photoluminescence produced under "steady-state" conditions. "Steady-state" fluorimetry is derived from the excitation of the sample with a continuous temporal output of exciting radiation. The lamps and their power supplies, used in conventional fluorimeters, are sources of continuous radiation. After a short period of initial excitation of the sample, a steady state is established in which the rate of excitation of the analyte is equal to the sum of the rates of all processes deactivating the lowest excited singlet state (fluorescence, internal conversion, and intersystem crossing). When the steady state is established, the observed fluorescence intensity becomes time invariant and produces the temporally constant signal that is measured by the photodetector.

With the development of modern electro-optics it has become possible to excite a potentially fluorescent sample with a thyratron-pulsed flash lamp that emits its radiation in bursts of 2–10 ns duration with about 0.2 ms between pulses or with a pulsed laser whose emission bursts are a few picoseconds in width. A fluorescent sample excited with such a pulsed source will not fluoresce continuously. Rather, its fluorescence intensity, excited by a single pulse, will decay exponentially until the next pulse again excites the sample. The fluorescence from the sample excited by the pulsed source can be represented, after conventional detection, as a function of time on a fast sampling oscilloscope. Alternatively, after the emission pulses are adjusted in intensity so that only one photoelectron is produced at the photocathode of a fast high-gain photomultiplier, the output is plotted as a function of time using an x-y plotter in conjunction with a multichannel analyzer. The former approach is called stroboscopic fluorimetry and the latter, time-correlated single-photon counting. In either case fluorescences with decay times much longer than the lamp pulse characteristics can be treated in the same way that radioactive decay curves are analyzed. A semilogarithmic plot of fluorescence intensity against time will yield a straight line (or a series of overlapping lines if several fluorophores have comparable but not identical decay times) whose slope is proportional to the decay time and whose vertical axis intercept can be compared with that of a standard solution of the fluorophore for quantitative analysis. If, however, the lamp pulse time and the decay time of the fluorophore are comparable, the lamp characteristics must be subtracted from the observed signal to obtain the fluorophore's decay characteristics. This is usually accomplished by using a computer to solve a deconvolution integral representing the composite temporal characteristics of the lamp and the fluorophore output.

The pulsed-source (time-resolved) method, then, effects spectroscopic separation of the emissions of several fluorescing species by taking advantage of differences in their decay times rather than their fluorescence intensities. This means that several overlapping fluorescences can be quantitated simultaneously. In this regard it should be mentioned that although the stroboscopic approach is useful when a single species of interest has a decay time much longer than other species in the solution, it does not give good results when several species in solution have fairly close (within an order of magnitude) decay times. In this regard time-correlated single-photon counting is considerably superior.

A third technique, phase fluorimetry, which has gained considerable popularity of late, entails exciting the sample with sinusoidally modulated light. The fluorescence emitted from the sample will be phase-shifted by some angle (θ) with respect to the exciting light and thereby demonstrate demodulation with respect to the exciting light. The demodulation (cos θ) of the exciting light due to absorption and emission by the fluorophore is

$$\cos \theta = [1 + (2\pi\nu\tau_f)^2]^{-1/2} \tag{14}$$

where θ is the angle of phase-shift between the exciting light and the fluorescence, ν is the frequency of the exciting light and τ_f is the lifetime of the fluorescing state. Measurement of the fluorescence with a phase-sensitive detector permits determination of τ_f. Although phase fluorimetry has proved superior to stroboscopic fluorimetry, it has not yet demonstrated the strength of single-photon counting in resolving several overlapping fluorescences.

Phosphorescence decay times are much longer than fluorescence decay time and are much easier to measure. Several methods are employed to measure phosphorescence decay times. The simplest entails tripping a shutter that closes off the sample from the exciting light. The decrease in phosphorescence intensity with time, after the tripping of the shutter, is monitored directly. For lifetimes longer than a second, the shutter may be tripped mechanically (by hand) and a strip chart recorder may be used to record the decay. For lifetimes less than 1 s an electronic shutter may be used and a wideband oscilloscope may be used to display the decay as a function of time. The mean lifetime is defined as the time required for the measured luminescence signal to decay from any given value to $1/e$ (.368) of that value. This can be determined easily from the graphical representation of the logarithm of the phosphorescence intensity as a function of time. Nonlinearity of the latter plot implies the occurrence of several overlapping phosphorescences and forms the basis of time-resolved phosphorimetry.

In the ensuing chapters the principles discussed briefly in this section will be seen to have generated an analytical methodology of great sensitivity, versatility, and sophistication.

BIBLIOGRAPHY

Bowen, E. J. (ed.), *Luminescence in Chemistry*, Van Nostrand, London, 1968.

Guilbault, G. G. (ed.), *Practical Fluorescence: Theory, Methods, and Techniques*, Dekker, New York, 1973.

Hercules, D. M. (ed.), *Fluorescence and Phosphorescence Analysis*, Wiley-Interscience, New York, 1965.

Parker, C. A., *Photoluminescence of Solutions*, Elsevier, Amsterdam, 1968.

Pesce, A. J., C. G. Rosen, and T. L. Pasby, *Fluorescence Spectroscopy*, Dekker, New York, 1971.

Schulman, S. G., *Fluorescence and Phosphorescence Spectroscopy: Physicochemical Principles and Practice*, Pergamon, London, 1977.

Wehry, E. L. (ed.), *Modern Fluorescence Spectroscopy*, Plenum, New York, Vols. 1 and 2, 1976; Vols. 3 and 4, 1981.

Winefordner, J. D., S. G. Schulman, and T. C. O'Haver, *Luminescence Spectrometry in Analytical Chemistry*, Wiley-Interscience, New York, 1972.

Zander, M., *Phosphorimetry*, Academic Press, New York, 1968.

CHAPTER

2

FLUORESCENCE AND PHOSPHORESCENCE OF PHARMACEUTICALS

WILLY R. G. BAEYENS

Department of Pharmaceutical Chemistry and Drug Analysis
Faculty of Pharmaceutical Sciences
State University of Ghent
Ghent, Belgium

This chapter should cover the most important articles that have been published, mainly from 1970 on, in the field of fluorescence and phosphorescence analysis in pharmaceutical drug analysis with emphasis on drug quality control and on the methods that are applicable to the determination of drugs and metabolites in biological fluids. Because of the large number of publications in this broad area as well as the fact that several articles have been published in foreign journals that often were not available, only most representative articles have been cited. Undoubtedly there will be omissions of work on specific applications of fluorometry and phosphorimetry for various pharmacological classes. This is inevitable, because the number of publications in this field is increasing tremendously. With the growing importance of high-performance liquid chromatographic techniques in pharmaceutical analysis and the development of good fluorescence detectors, more and more scientists are using literature data on the fluorescence behavior of different drugs—intrinsic or when necessary after chemical reaction—as a potent tool for increasing selectivity and sensitivity of their chromatographic system. A separate chapter in this book is devoted to fluorescence detection in chromatography (TLC, HPTLC, and HPLC).

Apart from the biennial reviews on fluorometric and phosphorimetric analysis that have appeared in *Analytical Chemistry* (1–5), a thorough computerized literature search of *Chemical Abstracts* was carried out from 1970 onward. Enhanced selectivity or detectivity through laser excitation is gaining importance, as stated in the literature on various occasions (5–7). No separate entry was inserted for vitamins or fluorescence immunoassay methods.

2.1. INTRODUCTION: ORIGIN AND NATURE OF FLUORESCENCE AND PHOSPHORESCENCE (8–23)

Absorption of ultraviolet and visible light by molecules of an irradiated sample generates a population of molecules in excited electronic states. Absorption of energy takes place in discrete units called quanta. The quanta–energy relationship may be expressed by the formula $E = h\nu$ or $E = hc/\lambda$ where E is energy; h, Planck's constant (6.62×10^{-27} erg s); c, the velocity of light; ν, the frequency; and λ, the wavelength. Each excited electronic state has many different vibrational energy levels, and excited molecules will be distributed in the various vibrational energy levels of an excited state.

In the ground states of unsaturated organic compounds the orbitals of lowest energy are each occupied by two π electrons with antiparallel spins. The excited states arise from the promotion of an electron from the highest occupied orbital to one that is unoccupied. The spin-angular momenta of the two electrons that are now in singly occupied orbitals are no longer restricted by the Pauli principle, and their spins may therefore be either antiparallel (*singlet state*) or parallel (*triplet state*).

It should be emphasized that at the instant of excitation, only electrons are reorganized; the heavier nuclei retain their ground-state geometry. The statement of this limitation is referred to as the *Franck–Condon principle*. The triplet state, in which the unpaired electrons both have the same spin, a result of the *intersystem crossing* out of the excited state, will adopt a new minimum-energy molecular geometry. The average life-time of a molecule in the singlet excited state has been calculated to be of the order of 10^{-8} s. The triplet state lives much longer than the cor-responding singlet state with lifetimes of 10^{-4}–10 s.

Molecules at each vibrational level of the excited state theoretically could lose energy by emitting photons and as a result fall to the original condition of the ground state. The energy and therefore the wavelength of emitted light then would be exactly the same as that absorbed. Such a process is termed *resonance fluorescence* and is rarely encountered in solution chemistry. Rather, molecules initially undergo a more rapid pro-cess, a radiationless loss of vibrational energy, and so quickly fall to the lowest vibrational energy level of the excited state, known as *vibrational relaxation*. This loss of energy occurs by collisions with the solvent mol-ecules, or, if the molecules are in a solid matrix, by exchange with the vibrational modes of the crystal. This vibrational relaxation is a rapid process, being complete in about 10^{-12} s. From the lowest vibrational level of the excited state a molecule can decay to the ground state either by photoemission, a *radiative transition*, or by *radiationless processes*.

When the molecules in the upper singlet state return to the ground singlet state by emission of radiation, this process is termed *fluorescence*. The transition may be to any one of a number of vibrational levels of the ground singlet state, subject of course to the constraints of the Franck–Condon principle. The emitted radiation thus covers a wide range of fre-quencies. The transition of lowest energy in the absorption spectrum is the same as the transition of highest energy in the emission spectrum, being the transition between the ground vibrational levels of the two elec-tronic states. This transition is not always observed, since it may be for-bidden by the Franck–Condon principle. If the vibrational levels are sim-ilarly spaced in the lower and upper singlet states, the emission spectrum will be the mirror image of the absorption spectrum, the O–O transition giving the plane of symmetry.

Another possible fate of the upper singlet state is a radiationless transfer to a triplet state in a high vibrational level. Since this intersystem crossing is a forbidden transition, owing to the change in spin multiplicity, usually only a small fraction of the molecules arrive in the triplet state. These undergo rapid vibrational relaxation down to the ground vibrational level of the triplet state. Since the transition down to the ground singlet state is forbidden, because of the change in spin multiplicity, the triplet state has a potentially much greater lifetime than the excited singlet state. In solution the triplet state is generally quenched by collision with solvent molecules or dissolved oxygen, but in a rigid matrix it can survive for relatively long periods, up to several seconds, losing energy only by the emission of radiation. This process is termed *phosphorescence*. It is obvious that phosphorescence emission occurs at lower frequencies (longer wavelengths) than fluorescence, which in turn occurs at lower frequencies than the absorption of radiation by that species.

Excited states can also be *quenched* when the photoexcited state of the reactant is deactivated by transferring its energy to another molecule in solution. This substance is called a *quencher*. Quenchers can be subdivided into two classes, paramagnetic and nonparamagnetic quenchers. In the case of quenching by nonparamagnetic molecules (X), the molecules X that have lower energy levels than the excited molecule M^* can quench the latter by an energy transfer mechanism described by

$$M^* + X_0 \rightarrow M_0 + X^*$$

where "0" refers to the ground state. This quenching process may proceed by a collision of X_0 and M^*, in which case M^* may be either a singlet or a triplet. Quenching of fluorescence can also occur by a charge transfer mechanism in which an electron is transferred between M^* and X_0; it may be caused as well by intersystem crossing catalyzed by external heavy atoms as xenon, iodide ion, bromobenzene, and so on. As concerns quenching by paramagnetic molecules, molecular oxygen is the best-known quencher of singlet and triplet excited states of aromatic hydrocarbons. In the ground state $^3O_2$ is a triplet, but on quenching triplets of aromatic hydrocarbons, it appears to be excited to the singlet ($^1O_2$) state:

$$^3M^* + \, ^3O_2 \rightarrow \, ^1M_0 + \, ^1O_2$$

As other processes involving the excited state can occur to compete with luminescence emission, the extent to which such other processes occur is characterized by the parameter *quantum efficiency* (*yield*) ϕ, defined as the ratio of the number of light quanta emitted per unit time to the number of quanta absorbed in the same unit time.

In solution, most quenching processes are diffusion-controlled. The fluorescence quantum yield of an organic molecule is reduced in the presence of a quenching molecule such as oxygen or a halogen. If the magnitude of the quenching depends linearly on the concentration of the quencher, then the quenching is said to obey *Stern–Volmer* kinetics and follows the Stern–Volmer equation:

$$\frac{\Phi_0}{\Phi_q} = 1 + K_q \cdot [Q]$$

where K_q is the Stern–Volmer quenching constant, Φ_0 is the fluorescence quantum yield when no quencher is present, and Φ_q is the quantum yield in the presence of a quencher concentration of Q moles per liter. When the quenching process is a collisional process and diffusion controlled, the Stern–Volmer equation can be written in the form

$$\frac{\Phi_0}{\Phi_q} = 1 + \tau_M \cdot k_{diff} \cdot [Q]$$

where τ_M is the fluorescence lifetime in the absence of Q and k_{diff} is the rate parameter of the diffusion-controlled quenching process. In the case of concentration quenching or *self-quenching*, a diffusion-controlled process as well, the solute fluorescence is decreased with increase of solute concentration. The Stern–Volmer law can be applied here and written as

$$\frac{\Phi_0}{\Phi_q} = 1 + k \cdot [c]$$

where k is the Stern–Volmer coefficient of concentration quenching and c is the solute concentration.

Only a limited number of species exhibit fluorescence, since for most species, internal quenching is a more rapid process, that is, radiationless deactivation from the ground vibrational level of the upper singlet state to the ground state will occur. The lifetime of the upper singlet state of most fluorescent molecules is in the range of 1–100 ns. Fluorescent molecules are typically large and flat with extensive π conjugation (e.g., derivatives of polycyclic aromatic hydrocarbons). Internal rotation seems to facilitate internal quenching. Phosphorescence is favored by the presence of a paramagnetic center in the molecule or if the upper singlet is of comparatively long lifetime. Fluorescence generally occurs if the lowest excited singlet state is of π,π^* type (i.e., the excitation transition is $\pi^* \leftarrow \pi$). For species (e.g., carbonyl compounds) in which the lowest excited singlet is of n,π^* type, intersystem crossing to the triplet state is favored, and phosphorescence is observed rather than fluorescence.

2.2. APPLICATIONS OF FLUOROMETRY

2.2.1. Introduction (24–34)

The determination of therapeutic and toxic drug concentrations in biological media must often reach nanogram and sometimes picogram levels, as many potent drugs of specific action show clinical effectiveness at milligram and even lower amounts.

Advances in selective extraction and separation methods together with the development of specific and highly sensitive instrumental techniques have recently allowed residue analysis in biological media at these low concentrations to be performed in a precise way.

Fluorometric analysis is being used for the assay of a large and growing number of drugs. The extreme sensitivity of fluorescence measurements is well suited to the low levels of many drugs encountered in biological materials (e.g., identification of drugs of abuse in human urine) and in some dosage forms. For example, fluorescence assay is the method of choice in many of the unit dose analyses required by the FDA. When higher levels of drugs are encountered the simplicity, versatility, and economy of fluorescence techniques are desirable. Spectrofluorometry is often used as a detection method after extraction in the automated analysis for various drugs in urine. Luminescence analysis has offered many applications in forensic science as well, as reviewed by Lloyd (35) and Gibson (36). The intent of these reviews is to provide the analyst with a ready, up-to-date source of references to the fluorometric analysis of drugs. Vitamins are not included.

2.2.2. Analytical Drug Quality Control and Applications for Drugs and Metabolites in Biological Materials

General Discussion

Fluorescence techniques are widely used in analytical drug quality control and drug monitoring because of their potentially high sensitivity. Fluorometric methods may be classified according to whether the fluorescence measured is that of the drug in its native state or is induced by chemical transformations. Many drugs as well as compounds of biochemical interest are not native fluorophors but can be labeled in some way to yield a fluorescing product. Fluorescamine, for example, has found widespread use as a fluorescent label for primary amines, notably amino acids. Some general reactions for functional group derivatizations are described later. It is not yet entirely possible to predict fluorescence behavior from a knowledge of the structure of a molecule, although some generalizations can be made. The fluorescence emission is largely dependent on the molecular environment, that is, the presence of other molecules (e.g., solvent and quenching agents), temperature, and exciting light intensity (i.e., photodecomposition effects), and these factors have to be controlled or allowed for in analytical work. The use of internal standards is particularly important in assays of biological samples, as they tend to contain variable amounts of quenching agents. Extraneous matter (e.g., detergents, stopcock grease, ion-exchange column effluents, residues from plastic containers) may also interfere with fluorometric assays by producing spurious fluorescence or intense light scattering (37).

In general, the more extensive the conjugated system, the longer will be the wavelength of the absorption band. Light absorption will be greater in most cases when there are polar substituents in the molecule. An outstanding structural requirement for fluorescence appears to be the presence on the resonating nucleus of at least one electron donating substituent. If fluorescence is considered as the natural consequence of absorption and if lack of such fluorescence is attributable to quenching, then the absorption characteristics are of first importance and these are somewhat more predictable. Although several general considerations now serve as structural bases of fluorescence, many exceptions exist and the selection of fluorochromes often has been empirical. Many fluorochromes have been used successfully as fluorescent histological stains, but few of these are suitable for chemical attachment to proteins. The main requirement for fluorochromes in fluorescence microscopy, for instance, is high fluorescence emission in the visible spectrum and good color contrast with the autofluorescence of the background. The modification of the dye molecule that may be necessary to make a stable link with a protein molecule often results in considerable if not complete loss of fluorescence. On the other hand, as the label is partly chosen for its fluorescence properties, intrinsically nonfluorophors that have been labeled can be detected quantitatively at even lower concentrations than many intrinsic fluorophors. Criteria for judging a fluorochrome as a suitable fluorescent label are described in Nairn's book on fluorescent protein tracing (38).

Difficulties often arise in the determination of drugs because of the low therapeutic dose concentrations and the spectral similarities between drugs and their metabolites. Although the sensitivity of fluorescence techniques has long been a recognized advantage for quantitative work at low concentrations, spectral selectivity has not always been achievable, especially for initial qualitative identification of structurally similar species. Quite frequently, as can be expected, metabolites or series of drugs differ only in an attached functional group, thereby retaining the same fluorophor center, which gives rise to the same spectrum for both it and its drug parent or homolog (39).

The use of lasers as the irradiation source for analytical fluorometry has greatly extended the minimum detectability of many classes of compounds (5,40). Subpart-per-trillion detection was achieved for several compounds [e.g., rhodamine 6G (41), riboflavine (42) and polycyclic aromatic hydrocarbons (43)]; the quantitative determination by laser-induced molecular fluorescence of several vitamins, tryptophan, and fluorescamine-labeled arginine and aniline was reported as well (44). Croteau and Leblanc (45) used a laser spectrofluorometer to observe the fluorescence from colchicine, α-methoxytropone, and tropolone in aqueous solutions at room temperature. Selective laser excitation was used for the luminescence analysis of complex organic materials based on resolved line spectra (46). Strojny and de Silva (47) described laser-induced fluorometric analysis of various drugs among which were intrinsically fluorescent compounds such as quinine sulfate, salicylic acid, carprofen (a carbazole), 2-methoxy-11-oxo-11H-pyrido-[2,1-b]-quinazoline-8-carboxylic acid, and fluorescent derivatives of nonfluorescent compounds such as the quinazolinone produced by the photolysis of demoxepam, the 9-acridanone derivative of flurazepam, and the fluorescamine derivative of amphetamine. Voigtman et al. (48) compared laser-excited fluorescence and photoacoustic characteristics of a windowless flow cell intended for liquid chromatographic applications with respective characteristics of a static cuvette cell. Laser fluorometric systems for ultratrace analysis were proposed (49); pulsed laser excitation and time-filtered detection for the quantitation of fluorophors in solution was recently suggested (50), and it seems that new directions in spectroscopy (51) lead to an increasing use of the potential inherent in laser techniques, which has not yet been fully realized.

Schulman and Sturgeon (52) recently devoted a comprehensive chapter on the theory and pharmaceutical applications of fluorescence and phosphorescence spectroscopy with extensive literature references and specific examples illustrating practical problems.

A spectrophotofluorometric study of a large number of organic compounds of pharmacological interest was published by Udenfriend et al. in 1957 (53). In this pioneering work the authors investigated the fluorescence behavior of many drugs and reported for each compound the wavelengths of maximal activation and fluorescence, the optimum medium, and relative sensitivities. For a representative group of these compounds, partition characteristics at the submicrogram level have also been determined spectrofluorometrically, and quantitative double extraction procedures have been evolved that should be applicable to tissue homogenates and body fluids.

Drug determination in biological fluids was treated by Reid (54). Hajdú and Damm (55) gave a description of a new method for the fluorometric determination of drugs in biological material. The assay is based on photochemical transformation of nonfluorescent compounds into fluorophors by shortwave UV light; the pharmacokinetics of compounds with different chemical structures (e.g., clobazam, fenbendazol) can be studied in this way. The pharmacokinetics of 2-[4-(4'-chlorophenoxy)-phenoxy]-propionic acid serves as an example to demonstrate the usefulness of this method. Liquid chromatographic determination of drugs in urine by direct injection onto the reversed-phase column and fluorescence-UV detection was described by Lagerstroem (56).

A spectrofluorometric determination of pharmaceuticals with the cerium(IV)–cerium(III) system was reported (57) based on the observation of Armstrong et al. (58), who noticed the characteristic fluorescence of cerium(III) ions in dilute sulfuric acid. A wide range of pharmaceuticals such as analgesics of the morphine series, pyrazolone, salicylic acid derivatives, phenothiazines, local anesthetics, sulfonamides, purines, and various organic acids was oxidized by cerium(IV) either in solution or on TLC plates that had been immersed in a solution of cerium(IV) sulfate and the concentration of cerium(III) formed was measured spectrofluorometrically. Detection limits of some substances are 15 ng ml^{-1} for the solution method and 5 ng per spot for the TLC method.

A simple and sensitive system was developed for the detection of oxidizable compounds that are eluted from chromatographic columns (59), based on the system mentioned earlier, that is, production of fluorescent cerium(III) by the reaction of eluted compounds with cerium(IV) solution. Excellent resolution and sensitivity was obtained when the system was coupled with a modified UV-monitored anion-exchange chromatograph for the analytical determination of organic acids in body fluids.

Cline Love and Upton (60) used the fluorescence lifetimes of members of drug series as a distinguishing property. According to these authors, many drugs fluoresce and have a characteristic fluorescence lifetime in the 1–200 ns range, which is easily measurable by the time-correlated single-photon technique. Examples show that in some cases, the fluorescence lifetimes are useful in the identification of compounds. In addition, for some drugs the measured fluorescence lifetime may be the only selective fluorescence property. The fluorescence lifetimes of antidepressant drugs (amitriptyline, protriptyline), benzimidazoles (2-furanyl- and 2-thiophene-benzimidazole), barbiturates (barbital, butethal, butalbital, phenobarbital), and antimalarial drugs (atabrine derivatives) are reported for several different chemical environments. The quantum yields and natural fluorescence lifetimes are also reported for the latter two drug families. Structurally similar members of each drug group were qualitatively distinguished through temporal resolution, and the sensitivity of fluorescence was enhanced by proper choice of chromophore environment.

Spectrofluorometric quantitation on thin-layer chromatograms has been described in literature on various occasions (61–66). Guilbault (67) gave a review on solid-surface fluorescence analysis in 1977 and discussed the use of paper chromatograms, TLC plates, potassium bromide disks, and silicone rubber pads. The use of spectrofluorometry for the quantitative TLC determination of various compounds has recently become very important, as described by Segura and Navarro (68), for the estimation of lipid classes using submicroliter amounts of plasma and by Zhou et al. (69) for the quantitative determination of lecithin and sphingomyelin at nanogram levels by HPTLC.

Wagner and Lehmann (70) identified N-heterocyclic compounds on chromatograms by fluorescence. Gianelli et al. (71) evaluated a multichannel image detector for in situ analysis of fluorescent materials on thin-layer plates.

Methods for Drugs from Various Pharmacological Groups

Analgesics.   Chalmers and Wadds (72) described in 1970 a spectrofluorometric analysis of mixtures of the principal opium alkaloids, including morphine, codeine, narcotine, and papaverine. Morphine and codeine are determined in aqueous media by a differential method, and narcotine and papaverine are determined in chloroform solution after extraction. Trichloroacetic acid is used for the simultaneous quenching of the fluorescence of papaverine and intensification of that of narcotine. The use of 2-aminopyridine as a working standard facilitates rapid and repeatable results.

A sensitive and specific automated method for the fluorometric determination of morphine in urine was developed in 1972 (73). The method is based on extraction of morphine from urine and oxidation with potassium ferricyanide in alkali to the fluorescent pseudomorphine dimer. Concentrations as low as 0.2 μg free morphine per milliliter of urine are easily detected and the results show excellent agreement with determinations on the same samples by thin-layer chromatography.

Rollen et al. (74) reported formaldehyde interference with the fluorescence determination of morphine in urine. False-negative results were obtained for morphine-containing urine samples as a result of the use of small amounts of formaldehyde solution as a urine preservative.

A simple, sensitive, and specific method was proposed for the fluorometric determination of morphine and quinine extracted from biological material (urine, plasma, tissue homogenate) (75). The fluorophor of morphine was prepared by simply treating the extract with acid, alkali, and heat. Quinine in the organic phase was reextracted into 0.1 N H_2SO_4 and determined fluorometrically. Concentrations of 0.22 μg ml^{-1} of morphine and 0.10 μg ml^{-1} of quinine in urine were easily detected. The authors applied the automated turret spectrofluorometer to this method, which allowed their laboratory to monitor large numbers of urine samples daily for morphine. Their method provides for effective surveillance of heroin abuse within a narcotic control, treatment, or aftercare program.

A search of the literature on methods for opium alkaloids reveals further a spectrofluorometric method for determining morphine and codeine in opium poppy pods (76), a description of luminescence characteristics of morphine derivatives (77), and a reversed-phase HPLC detection system for morphine in biological samples through postcolumn derivatization with alkaline potassium ferricyanide into the fluorescent dimer pseudomorphine (78).

Acetylsalicylic acid fluoresces with much less intensity than salicylic acid, its hydrolysis product. On the other hand, acetylsalicylic acid phosphoresces sufficiently to allow trace determinations at liquid nitrogen temperatures; salicylic acid also phosphoresces, but to a significantly lesser extent. A fluorescence method for acetylsalicylic acid in solution and its measurement in the presence of salicylic acid was developed by Miles and Schenk in 1970 (79). Two years later, Schenk et al. (80) reported the effects of aliphatic carboxylic acids on the fluorescence emission of acetylsalicylic acid and salicylic acid. Analysis of mixtures of these compounds is best performed in 1% acetic acid–chloroform solvent.

Acetylsalicylic acid can be separated quantitatively from salicylic acid by Sephadex gel filtration (81); an activating wavelength of 305 nm and a fluorescence wavelength of 405 nm can be used for the fluorometric effluent assay. Acetylsalicylate does not fluoresce but can be measured after hydrolysis. Shane and Miele (82) described a rapid fluorometric method for the determination of salicylic acid in aspirin products in combination with noninterfering fluorescent substances as phenyltoloxamine citrate, glyceryl guaiacolate, and caffeine, interfering fluorescent materials such as hydrogenated castor oil and salicylamide, and buffering agents, such as aluminum glycinate and magnesium carbonate. Accurate and precise results were obtained. Their procedure is based on a simple dissolution of the crushed tablet in a pH 4.0 buffer solution and reading of the filtrate at λ_{ex} = 310 nm, λ_{em} = 410 nm. Shane and Stillman (83) described a rapid one-step fluorometric method for salicylic acid in buffered aspirin products, without extractions or liquid column chromatographic separations.

A method for the fluorometric determination of salicylic acid esters (phenyl, methyl, ethyl, isopropyl, and benzyl) in solution of N,N-dimethyl formamide was proposed in 1971 (84), based on their strong fluorescence in this solvent containing a small amount of base. Total salicylate in plasma can be determined directly in small plasma quantities by fluorometry, without time-consuming extractions, when zero plasma (plasma from the test person without salicylate) is available (85). Kanter and Horbaly (86) modified a fluorometric procedure for the direct measurement of aspirin in which interference because of salicylic acid and its conjugates was eliminated by reaction with ceric ammonium nitrate. Trace amounts of acetylsalicylic anhydride can be analyzed in acetylsalicylic acid (87). Kleinerman (88) reported a rapid fluorometric method for the determination of salicylate in serum while Bekersky et al. (89a) were able to determine simultaneously salicylic acid and salicyluric acid in plasma and urine by high-pressure liquid chromatography.

Street and Schenk (89b) presented fluorometric procedures for the determination of salicylamide, acetylsalicylic acid, and salicylic acid, as an impurity, in pharmaceutical preparations containing salicylamide, acetylsalicylic acid, acetaminophen, caffeine, and phenacetin as major constituents. Determination of salicylic acid in the 10^{-7} M concentration range is allowed after separation from salicylamide, acetaminophen, and caffeine.

Salicylamide was determined in blood serum and urine by a sensitive spectrophotofluorometric method involving the simultaneous determination of salicylamide and salicylic acid at pH 11 after acid hydrolysis of the salicylamide metabolites (90). The activation and emission wavelengths of salicylic acid at the optimum pH of 11 are 310 and 435 nm, respectively. Salicylamide at activation and emission wavelengths of 340 and 435 nm, respectively, also has maximum fluorescence at pH 11. It is possible to determine salicylamide and salicylic acid at pH 11 and calculate a value for total salicylamide. This may be accomplished since both compounds have maximum fluorescence (435 nm) at pH 11 of nearly equal intensity at their respective activation and similar but lesser fluorescence (435 nm) at each other's activation. Hence, in a mixture of both compounds, the sum of the fluorescence (435 nm) at the two activation wavelengths will be directly proportionate to the sum of their concentrations and independent of their ratios. The precision and accuracy of the method are comparable to the regular ferric-ion complex procedure. Specificity and sensitivity allow monitoring salicylamide bioavailability provided by different dosage forms.

Schulman and Underberg (91) studied the excitation wavelength dependence of prototropic dissociation and tautomerism of salicylamide in the lowest excited singlet state. They state that the similarity in spectral position between the fluorescence of *o*-ethoxybenzamide and the short wavelength fluorescence of salicylamide strongly suggests that the short wavelength emission of salicylamide originates from one or more neutral species (a,b, or c); while, as in methyl salicylate, the anomalous long wavelength emission originates from a zwitterion (d or f) or a nonphenolic electromer (e) of the zwitterion (d).

a   b   c

Stallings and Schulman (92) studied the solvent and acidity dependences of the electronic spectra of gentisic acid (2,5-dihydroxybenzoic acid), a minor metabolite of salicylic acid and aspirin.

d   e   f

Rutledge and Schulman (93) reported fluorescence from phenyl salicylate (salol, 2-hydroxybenzoic acid phenyl ester), which shows most intense fluorescence in chloroform. In the presence of other salicylate-type drugs, however, which mostly fluoresce at 400–450 nm, the short wavelength of fluorescence of phenyl salicylate in ethanol may be advantageous for fluorometric analysis because of selectivity.

Benhepazone (1-benzylcycloheptimidazol-2(1H)-one), a drug with potent analgesic and antiphlogistic activities, was measured fluorometrically in tissues, urine, and feces after quantitative extraction into chloroform in alkaline conditions from a mixture containing benhepazone and related compounds among which were its metabolites (94). Activation occurs at 395 nm; fluorescence is measured at 410 nm. Neither metabolic products nor native substances in biological samples interfered with the benhepazone determination.

Structure of benhepazone and related compounds.
(Benhepazone : R_1 = benzyl, R_2=R_3=H)

Changes occurring in γ-oxophenylbutazone on standing were determined by isolating the drug from decomposition products by TLC on silica gel HF 254 with diethyl ether saturated with water followed by fluorescence scanning (254 nm excitation, Wratten 155 emission filter) (95).

Dal Cortiva, De Mayo, and Weinberg (96) reported the formation of a fluorophor when meperidine is heated in a solution of formaldehyde and concentrated sulfuric acid followed by addition of water. It has been found (97) that methadone also forms a fluorophor using the same reagents under slightly different conditions. The procedure is suitable for determining microgram quantities of methadone. No prior separation step is required to measure methadone in the presence of morphine, heroin, codeine, or cocaine. Amphetamine, meperidine, and quinine fluoresce under the conditions of the assay and must be separated from methadone prior to determination.

A simple, rapid, sensitive, and specific fluorometric method for flufenamic acid, a potent anti-inflammatory agent, was reported in 1970 (98). The proposed method is based on ethyl acetate extraction of the acidic samples and a fluorescent complex is obtained on the subsequent addition of aluminium chloride to the ethanolic solution of the extract residue. Sensitivity of the method is 4 ng ml^{-1} (λ_{ex} = 358 nm; λ_{em} = 440 nm) and application to biological materials should be possible. Dell and Kamp (99) reported fluorometric analysis for flufenamic acid, derivatives, and analogous compounds. Reaction of flufenamic acid with formaldehyde (Fig. 2.1) or concentrated H_2SO_4 (Fig. 2.2) gives 1-(m-trifluoromethylphenyl)-4-oxo-1,2-dihydro-3,1,4-benzoxazine, and (4) (2)-trifluoromethylacridone, respectively, allowing submicrogram analysis. The authors used 16 different compounds for the investigation of the influence of substituents on fluorescence characteristics and presented determinations from biological material.

Native fluorescence of analgesics derived from N-phenylanthranilic acid was reported by Mehta and Schulman (100). Mefenamic acid, flu-

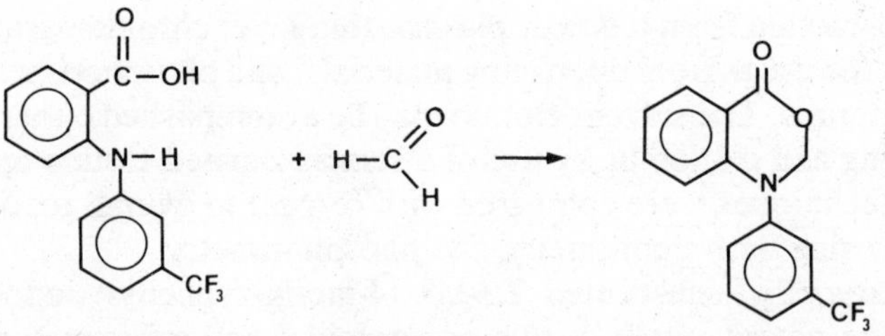

Fig. 2.1. Cyclization of flufenamic acid with formaldehyde yielding the fluorescent 1-(m-trifluoromethylphenyl)-4-oxo-1,2-dihydro-3,1,4-benzoxazine. (From ref. 99.)

fenamic acid, and meclofenamic acid show native fluorescence in organic solvents (e.g., dioxan, chloroform), which is more useful for the determination of these drugs than is the fluorescence of the derivative substituted acridones and benzoxazines obtained from these drugs by treatment with sulfuric acid or formaldehyde, respectively.

Meilinck (101) reported a fluorometric assay of carbamazepine and its metabolites in blood. Carbamazepine is separated from its metabolites by TLC after extraction from serum or plasma with dichloromethane in basic medium. Fluorescence is induced by treatment with ammonium ceric sulfate in phosphoric acid. Without the thin-layer chromatographic step the concentration of carbamazepine and that of its metabolites are determined together. The author examined different parameters in the analysis and applied the method to blood of patients taking carbamazepine.

The luminescence properties of some anti-inflammatory agents and of 6-chloro-1,2,3,4-tetrahydrocarbazole-2-carboxylic acid, 6-chloro-9-[2-(2-methyl-5-pyridyl)ethyl]-1,2,3,4-tetrahydrocarbazole-2-methanol·HCl, and (d,l)-6-chloro-α-methyl-carbazole-2-acetic acid have been examined by Strojny and de Silva (102) at both ambient (fluorescence at 25°C) and cryogenic temperature (phosphorescence at 77 K). They developed assays for indomethacin, the tetrahydrocarbazoles, and the carbazoles based on

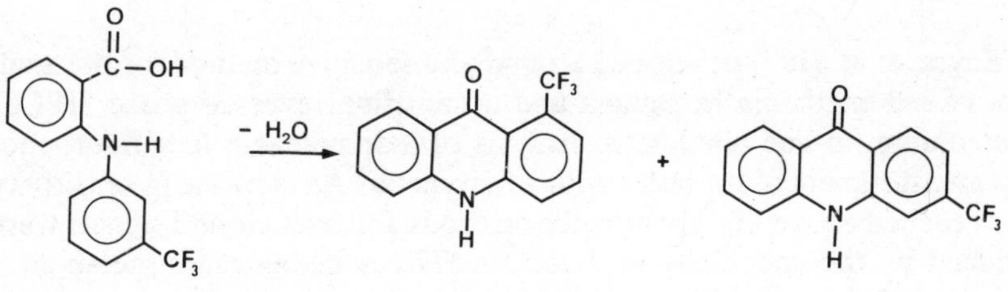

Fig. 2.2. Cyclization of flufenamic acid into trifluoromethylacridone isomer mixture: (a) 4-trifluoromethylacridon and (b) 2-trifluoromethylacridon (65:35). (From ref. 99.)

solvent extraction from blood or plasma, thin-layer chromatographic separation of the drugs from interfering materials, and phosphorimetry of the eluted materials. The solvent elution may be accomplished either by manual scraping and elution or by use of a semiautomated elution apparatus. The two techniques were compared with respect to overall recovery and precision using both fluorometry and phosphorimetry.

Trifluoromethyl-substituted 2,3-bis (4-methoxyphenyl) indoles were found to be potent, orally active nonsteroidal anti-inflammatory drugs. Hence, a method had to be developed for their determination in serum, urine and feces in order to permit studies of their absorption, metabolism and excretion. Kaiser et al. (103) noticed useful fluorescence characteristics from the 5-,6-, and 7-trifluoromethyl-substituted 2,3-bis(4-methoxyphenyl) indoles (I, II, III, respectively) and developed a simple, rapid, sensitive, and specific procedure based on an ethyl acetate extraction of alkaline specimens and subsequent fluorometric analysis of ethanolic solutions of the extract residues (λ_{ex} = 320–330 nm; λ_{em} = 415–420 nm). Their method is sensitive to 0.1 μg ml^{-1}, 0.5 μg ml^{-1}, and 0.7 μg/100 mg of these compounds in serum, urine, and feces, respectively. Previous reports from the same author described fluorometric procedures for the determination of indoxole (104) (IV) (2,3-bis(4-methoxyphenyl)indole), 4,5-bis(4-methoxyphenyl)-2-phenylpyrrole-3-acetonitrile (105) and 2,3-bis(4-methoxyphenyl)imidazo[1,2-a]pyrimidine (106) in biological materials.

(I : R=5-CF$_3$; II : R=6-CF$_3$; III : R=7-CF$_3$; IV : R=H)

Bayne et al. (107) developed a rapid and sensitive method for the analysis of indomethacin in plasma and urine using reversed-phase HPLC, postcolumn, in-line alkaline hydrolysis of indomethacin to a fluorophor and measurement of the latter with a flurometer. An increase in sensitivity was reached above UV absorbance methods and background signals were reduced by the specificity in detection. The nonconjugated metabolites of indomethacin, aspirin, and salicylate were resolved from indomethacin and the internal standard, α-methylindomethacin. A lower detection limit of 1.5 ng ml^{-1} of plasma was obtained and the authors applied their method for analyzing indomethacin in biological samples for bioavailability studies.

Glafenine (2-[(7-chloro-4-quinolinyl)amino] benzoic acid 2,3-dihydroxypropyl ester), a frequently used analgesic, shows fluorescence properties, partly caused by the anthranilic acid nucleus of the molecule, as reported by Baeyens and De Moerloose (108a). The authors provide correlations between fluorescence capacities and UV absorbance. Analytical determinations (e.g., in tablets) can be carried out in different solvents (e.g., ether, benzene, and ethanol). Fluorometric measurement in diethyl ether at λ_{ex} = 327 nm and λ_{em} = 400 nm can be performed with a limiting detectable sample concentration of 8.10^{-2} μg ml^{-1} with a linear response up to 8 μg ml^{-1}. The native fluorescence of glafenine lies by the undissociated, free base.

A bioanalysis was reported for the antiinflammatory and analgesic naproxen by high-performance reversed-phase liquid chromatography with photometric and fluorometric detection (108b).

Meptazinol [m-(3-ethyl-1-methylhexahydro-1H-azepin-3-yl)phenol], a drug used for the treatment of pain in man, was determined in plasma by high-performance liquid chromatography with fluorescence detection based on its native fluorescence (λ_{ex} = 282 nm; λ_{em} = 300 nm) (109)·m-(1-Cyclopropylmethyl-3-ethylhexahydro-1H-azepin-3-yl)phenol was used as an internal standard. A lower limit of detection of 3 ng ml^{-1} was reported and no interference from metabolites was encountered.

Pacha et al. (110) developed a rapid and sensitive fluorometric assay for the quantitative determination of fluproquazone (4-(p-fluorophenyl)-1-isopropyl-7-methyl-2(1H)-quinazolinone), a drug with analgesic and anti-inflammatory action, in plasma and urine. They extracted unchanged drug from alkalinized plasma or urine into n-heptane containing 0–1.5% isoamyl alcohol followed by a back-extraction into 5 N HCl. Oxidation is carried out with potassium persulfate and fluorescence is measured at 520 nm on excitation at 326 nm. Detection limits were reached of about 15 ng ml^{-1} in plasma and 6 ng ml^{-1} in urine, using 1 ml plasma and 2 ml urine, respectively. Fluproquazone shows strong intrinsic fluorescence (λ_{ex} = 326 nm, 420 nm; λ_{em} = 486 nm, in acid); treatment of the final measuring solution with potassium persulfate causes a slight increase of the emission wavelength to 520 nm and a lowering of the blanks in the extracts from biological materials. The automated fluorometric assay was applied to several pharmacokinetic and bioavailability studies in animals and man.

Amino Acids. Appearance of fluorescence on treatment of histidine residues with N-bromosuccinimide was reported in 1963 (111). The luminescence of tryptophan and phenylalanine derivatives was examined in 1968 (112). For tryptophan compounds, the luminescence is much more sensitive to chemical modification at 25° than at liquid nitrogen temperature. Luminescence studies of the phenylalanine peptides revealed differences between peptides that persisted at liquid nitrogen temperatures.

Electronic spectra of indoles are sensitive to solvents, particularly pronounced in the case of their fluorescence spectra. Indole, indole-5-carboxylic acid, indole-3-carboxylic acid, indole-2-carboxylic acid, 5-cyano- and 5-bromo-indoles were investigated and showed a substantial increase in dipole moment on excitation to the emitting state (113).

Roth (114) developed in 1971 a fluorescence reaction for amino acids. Starting from o-diacetylbenzene as a possible model reagent, he found that addition of a strongly reducing agent to a solution of alanine and o-diacetylbenzene produced a bright blue fluorescence, which occurred as well with some other amino acids. When replacing o-diacetylbenzene by o-phthalaldehyde, a fluorescence was obtained with all common amino acids except cysteine and the imino acids proline and hydroxyproline. This method, based on the reaction of o-phthaldialdehyde with amino acids in alkaline medium in the presence of a reducing agent such as 2-mercaptoethanol, giving rise to strongly fluorescing compounds (λ_{ex} = 340 nm; λ_{em} = 455 nm), permits fluorometric amino acid assay down to the nanomole range. No heating is required, and the fluorescence signal may be measured 5 min after mixing of the reagents. His reaction is applicable to the automatic determination of amino acids after ion exchange fractionation. High-performance liquid chromatographic determination of subpicomole amounts of amino acids by precolumn fluorescence derivatization with o-phthaldialdehyde was described by Lindroth and Mopper (115) while Cooper and Turnell (116) used this reagent for cystine detection using HPLC.

A sensitive procedure was developed for the determination of amino acids based on their interaction with pyridoxal and Zn^{2+} ion in pyridine–methanol to yield highly fluorescent chelates (117). The authors state that N-pyridoxylidene amino acid – Zn(II) chelate contains a 1:1 molar ratio of ligand to metal ion.

Bayer et al. (118) separated amino acids as dansyl derivatives on silica gel with high-performance liquid chromatography allowing detection of concentrations in the eluent below pmol ml^{-1} range. Froehlich and Yeats (119) stated that the fluorescence of a number of indoles, tyrosine, tryptophan, and some of their metabolites is greater in ethanol–water and DMSO–water systems than in pure water and suggested that analytical procedures that include detection via fluorescence should be performed in mixed aqueous solvent systems where possible. The addition of nonaqueous solvents is believed to diminish the formation of exciplexes between the excited state of the fluorophor and water, increasing the fluorescence quantum yield. The effect depends on the nature of the groups attached to the aromatic ring and on the aliphatic side chain in the case of the amino acid metabolites. Froehlich and Murphy (120) reported that the fluorescence intensity of the amino acid derivatives obtained with a number of reagents can be increased by consideration of the solvent system used for their formation. They state that the use of mixed solvent systems such as DMSO–water in place of water for the formation and detection of o-phthaldehyde derivatives of amino acids can raise the fluorescence intensity significantly; similar observations were found with dansyl amino acids and fluorescamine amino acids.

Among the numerous publications on the use of fluorescamine for the determination of amines can be cited its optimum use for in situ TLC quantitation of amino acids (121), the unique fluorescence of fluorescamine derivatives of histidine, histamine, and certain related imidazoles after heating in acid (122), TLC of histidine, histamine, and histidyl peptides at picomole level (123), the characterization of selected fluorescamine–amino acid reaction products by HPLC (124), and the fluorometric scanning of fluorescamine-labeled peptides in polyacrylamide gels (125).

Udenfriend and Stein (126) described fluorescent techniques for ultramicro peptide and protein chemistry. Their review contains 36 references illustrating the use of fluorescamine and 2-methoxy-2,4-diphenyl-3(2H)-furanone as reagents for use in fluorometric techniques for peptide chemistry. Stein (127) gave a review with 72 references on the use of fluorescamine and 2-methoxy-2,4-diphenyl-3(2H)-furanone in fluorometry of peptides and amines. A review with 35 references on fluorescent methods for isolation, characterization, and assay of peptides and proteins was published by Udenfriend (128).

The use of fluorescent spectroscopic techniques can greatly contribute to the study of drug protein interactions (129). The blue fluorescence and cross-linking of photo-oxidized proteins was reported by Fujimori (130).

Hadley et al. (131) modified an enzymatic–fluorometric assay for reduced glutathione and oxidized glutathione through the use of heat precipitation of sample protein and hydrogen peroxide oxidation of reduced glutathione to oxidized glutathione. Glutathione reductase reduces oxidized glutathione in the presence of reduced adenine dinucleotide phosphate. One mole of reduced nicotine adenine dinucleotide phosphate is required for conversion of each mole of oxidized glutathione to 2 mol of reduced glutathione. This relationship was used by Halprin and Ohkawara (132) to develop an enzymatic assay for reduced glutathione and oxidized glutathione. The amount of oxidized glutathione present in the sample was determined by following the decrease in reduced nicotine adenine dinucleotide phosphate fluorescence. The modifications by Hadley et al. (131) were made to eliminate two critical pH adjustments present in the original assay and the possibility of interference with the assay of oxidized glutathione by the perchloric acid protein precipitant and the cupric ion oxidant used in the original assay. The authors applied their modified assay for the determination of the reduced and oxidized glutathione content of Harding–Passey melanoma tissues.

Tryptophan was determined in plasma of chicks with a highly sensitive and highly resolving spectrofluorometer system (133) with a detection sensitivity of 5×10^{-3} nmol.

Kirschenbaum and Glantz (134) have reported a method for the 2,3-diphenyl-1-indanone spectrofluorometric analysis of amino acids by the synthesis of the furan derivatives of sulfoindonyl amino acids, allowing quantitative analysis at 10^{-9} M concentrations ($\lambda_{ex} = 365$ nm; $\lambda_{em} = 484$ nm).

Ahnoff et al. (135) used the derivatization reaction with 4-chloro-7-nitrobenzofurazan (NBD-Cl) for the liquid chromatographic determination of hydroxyproline in collagen hydrolysate.

Watanabe and Imai (136) reported on the use of 7-fluoro-4-nitrobenzo-2-oxa-1,3-diazole as derivatizing reagent for amino acids in high-performance liquid chromatography.

Antibiotics.   The fluorescent method of Kohn (137) for tetracyclines reported in 1961 is based on the formation of a calcium–barbital–tetracycline complex. Ibsen et al. (138) presented in 1963 a method for the fluorometric determination of oxytetracycline and other tetracyclines in biological material. Oxytetracycline is extracted from acidic aqueous solutions into amyl alcohol, and then from amyl alcohol into basic solutions; the concentration of oxytetracycline is proportional to the difference in fluorescence between a magnesium-oxytetracycline chelate, and the free compound, liberated by adding ethylenediaminetetraacetate to the solution. Spock et al. (139) developed in the same year a chemical method for the determination of chlortetracycline in premixes based on the degradation of chlortetracycline in alkaline media to form isochlortetracycline. The fluorescence of the isochlortetracycline ($\lambda_{ex} = 350$ nm; $\lambda_{em} = 420$ nm) is directly proportional to the concentration of chlortetracycline originally present prior to alkaline degradation. Kelly and Hoyt (140) developed a sensitive fluorometric procedure for the analysis of the individual components of a mixture of three tetracycline antibiotics (tetracycline, chlortetracycline, and demethylchlortetracycline) in biological fluids. In their procedure chlortetracycline is determined by conversion to the fluorescent isochlortetracycline. Tetracycline is then determined on the same aliquot by conversion to anhydrotetracycline and measurement of the fluorescence of an aluminum complex of this derivative. Demethylchlortetracycline is determined by difference after application of a method that measures total tetracyclines (141). Tetracycline was determined fluorometrically in blood serum of man and animals by incubating the serum with beryllium ions and measurement of the fluorescence at 460 nm (142). The antibiotic–Be complex contained 6 parts Be to 1 part tetracycline. The method at pH 7.0 was accurate to 0.01 µg per ml serum. A simplified tetracycline assay was reported (143) based on Kohn's method (137), allowing a rapid procedure for tetracycline concentrations above 0.1 µg ml^{-1} in 0.1-ml samples.

Katz and Fassbender (144) presented in 1973 a simple fluorometric method for oxytetracycline in premixes based on extraction with acid–methanol and dilution with ammonium hydroxide to form an oxytetracycline fluorophor (λ_{ex} = 390 nm; λ_{em} = 520 nm) that is representative of microbiological activity. Willekens (145) described a rapid and sensitive partition TLC method with fluorescence detection on pretreated kieselguhr plates for the determination of oxytetracycline and its degradation products. Tetracyclines are sensitive to both acidic and alkaline degradation media. The first stage of acidic degradation is known to yield anhydro compounds with potential toxicity. The author states that little information has been published about the quantitative evaluation of oxytetracycline impurities, among which are anhydrooxytetracycline, epioxytetracycline, α-apooxytetracycline, and β-apooxytetracycline. His method allows quantitation of traces of these impurities and is sensitive to 10^{-2} µg of these substances. Van Hoeck et al. (146) achieved a satisfactory TLC separation of tetracycline in the presence of anhydrotetracycline and their epimers and were able to perform a direct fluorometric estimation of the four substances. Ragazzi and Veronese (147) quantitated tetracyclines by direct fluorometry after TLC on cellulose plates.

Schatz (148) described a fluorometric determination of tetracycline in serum and urine by adsorption onto Al_2O_3, thus separating from interfering substances, followed by quantitative assay based on the strong fluorescence (λ_{ex} = 335 nm; λ_{em} = 414 nm) of a degradation product of tetracycline formed on warming in alkaline solution. Concentrations of 0.2–5 µg tetracycline per ml serum and 5–50 µg tetracycline per ml urine can be determined. Hall (149) used the formation of strongly fluorescent aluminum–tetracycline complexes for the simultaneous assay in plasma of a mixture containing chlortetracycline, demethylchlortetracycline, and tetracycline, in concentrations down to 0.1 mg liter^{-1}. The same author reported on the fluorometric assay of tetracycline mixtures in plasma (150) and on a rapid fluorometric assay of minocycline in plasma or serum (151). Day et al. (152) studied tetracycline complexation with calcium and organic ligands using fluorescence, circular dichroism, and solvent extraction methods in order to interpret the mechanism of the commonly used fluorometric methods for the analysis of tetracycline in biological fluids. Tetracycline forms ternary calcium complexes with barbital sodium and 1-tryptophan in alkaline solutions. The circular dichroism studies indicated that the calcium ion in these complexes is bound to the C-4 dimethylamino and the C-12a hydroxyl groups of tetracycline. These ternary complexes are strongly fluorescent (λ_{ex} = 390 nm; λ_{em} = 512 nm) and can be easily extracted into 1-butanol or ethyl acetate. Based on the characteristics of the ternary complexes and of the tetracycline degradation products, Day et al. (152) concluded that only the active form of tetracycline can be complexed and extracted for fluorescence analysis. Van den Bogert and Kroon (153) reported a fluorometric determination of tetracyclines in small blood and tissue samples. Their method consists of a modification of the method described by Poiger and Schlatter (154) that is based on buffer extraction of tetracyclines from the matrix, deproteinization and elimination of phosphates, extraction into an organic solvent as ion pairs with calcium and trichloroacetate ions and fluorescence induction by the addition of a base and magnesium ions. In the modified procedure, the tetracyclines are extracted in the presence of ethylene–diaminetetraacetic acid, and the phosphate precipitation steps are omitted. Detection of small amounts of the tetracyclines in as little as 0.2 ml of serum or 50 mg (wet weight) of tissue can be achieved. Regosz (155) used fluorometry for the determination of the active form of doxycycline and methacycline in partially decomposed solutions.

Dihydrostreptomycin was determined in a complex pharmaceutical suspension (156) after cation exchange separation by alginic acid and subsequent condensation with ninhydrin (λ_{ex} = 390 nm; λ_{em} = 495 nm). Aminoglycoside antibiotics have been determined by an extraction–fluorometric method (157) and by an HPLC method with fluorometric detection using fluorescamine as derivatizing agent (158).

The luminescence of actinomycin antibiotics in cultures of actinomycetes was reported in 1972 (159), and it was assumed that actinomycins exist in various forms that differ by their spectral luminescence characteristics. Fluorometric–colorimetric amino acid analysis of actinomycins using fluorescamine was mentioned by Felix et al. (160).

Tserng and Wagner (161) proposed a simple solvent extraction method for quantitative separation of erythromycin and erythromycin propionate from whole blood or plasma. After separation, each of the compounds is quantitatively measured by a fluorometric method based on the coupling of erythromycin with an acidic fluorescent dye, Tinopal GS (sodium salt of 2-(stilbyl-4")-(naphtho-1′,2′:4,5)-1,2,3-triazole-2″,6′-disulfonic acid). The authors investigated also the stability of erythromycin propionate in whole blood stored at temperatures from −20 to 37°C. Tsuji (162) reported a fluorometric determination for erythromycin and erythromycin ethylsuccinate in serum by a high-performance liquid chromatographic postcolumn, on-stream derivatization and extraction method, capable of detecting less than 0.01 μg ml⁻¹ of these drugs in serum. The postcolumn derivatization technique uses naphthotriazole disulfonate, which forms an ion pair with molecules protonated at low pH. Erythromycin and its esters contain one secondary amine in the desosamine moiety and can be readily protonated. As both naphthotriazole disulfonate and its ion-paired compounds have identical excitation (360 nm) and emission (>440 nm) characteristics, they must be separated before quantification, which is accomplished by an on-stream extraction with chloroform, leaving an excess of the reagent in the aqueous layer. The author used the method to analyze erythromycin and its ethylsuccinate in sera from subjects orally treated with erythromycin ethylsuccinate. In addition to a small amount of anhydroerythromycin ethylsuccinate and 8,9-anhydro-6,9-hemiketal erythromycin ethylsuccinate, at least two other metabolites were detected in sera of which one was identified as erythralosamine.

Sturgeon and Schulman (163) studied the electronic absorption spectra and protolytic equilibria of doxorubicin and concluded that the protonated amino sugar group is slightly more acidic than the phenolic group, stressing the importance of careful choice of pH and the analytical wavelength in developing any method of analysis for doxorubicin.

Csiba (164) presented a spectrofluorometric method for the quantitative determination of aminoglycoside antibiotics (e.g., tobramycin) in biological fluids, based on an ion-exchange chromatographic separation of aminoglycosides in serum and urine. The clearcut separation of aminoglycosides is achieved by eluting the column with sulfuric acid solution followed by a fluorometric determination based on the formation of a fluorescent dihydrolutidine derivative (λ_{ex} = 421 nm; λ_{em} = 488 nm) on condensation of the primary amino groups with acetylacetone and formaldehyde, under acidic conditions (pH 2.6). Sensitivity of the method is 0.05 μg ml⁻¹.

A polarization fluoroimmunoassay has been developed for the routine determination of gentamicin levels in serum (165) using fluorescein-labeled gentamicin. A homogeneous reactant-labeled fluorescent immunoassay for gentamicin determination in human serum was developed by Burd et al. (166). A derivative of umbelliferyl-β-galactoside was coupled covalently to the drug, and this conjugate was found to be nonfluorescent under assay conditions. The drug–dye conjugate was a substrate for bacterial β-galactosidase and yielded a fluorescent product. When the drug–dye conjugate was bound to antigentamicin antibody, it was inactive as an enzymatic substrate; this inactivation was relieved by the presence of gentamicin in competitive binding reactions. Hence, the rate of production of fluorescence was proportional to the gentamicin concentration. The assay requires only 1 μl of serum and offers several advantages over existing techniques: sensitivity, specificity, simplicity, and obviation of radioisotopes. Kabasakalian et al. (167) described a quantitation method for the individual components of gentamicin during the progress of a fermentation. Fluorometric measurements are carried out in situ on the 4-chloro-7-nitrobenzo-2-oxa-1,3-diazole (NBD) derivatives formed after TLC of the clarified fermentation broth. The fluorometric procedure is reported to be 800 times as sensitive as that with the ninhydrin analog.

A comparative evaluation of three methods for measuring gentamicin and tobramycin in serum (168) and a simple and rapid assay of serum gentamicin concentration by a fluorescence polarization technique (169) were reported in 1981.

Sinsheimer et al. (170) reported a specific and sensitive (nanogram range) fluorescent TLC analysis of penicillins using a nonacidic fluorescent labeling agent, 9-isothiocyanatoacridine, that reacts with the secondary amino function of the penicilloic acid after penicillin hydrolysis with base or β-lactamase.

Ampicillin forms a strongly fluorescent yellow product in acid solution during hydrolysis at elevated temperature as reported by Jusko in 1971 (171). The quantity and rate of formation of the fluorescent material are enhanced appreciably by formaldehyde, and the fluorescent product exhibits extraction properties of an organic acid. In alkaline solution the fluorescent compound has uncorrected excitation and emission maxima at 346 and 422 nm, respectively. The author suggests an assay procedure based on these observations; less than 0.05 μg ml^{-1} of ampicillin and/or α-aminobenzylpenicilloic acid can be detected. By combining fluorometric and microbiological methods, measurement can be made of ampicillin degradation products as well as the unchanged antibiotic in serum and urine. The assay procedure was tested using several penicillin and cephalosporin derivatives; hetacillin forms about as much of the fluorescent product as does ampicillin; cephalexin yielded a small amount of the fluorescent material. According to the author, the aminobenzyl group and a cleaved β-lactam ring are necessary for the formation of the fluorescent product. Addition of formaldehyde, a type of formol titration, enhances the reaction by reducing the basicity of the amino group. The formation of a 3,6-disubstituted diketopiperazine (Fig. 2.3) from either ampicillin or

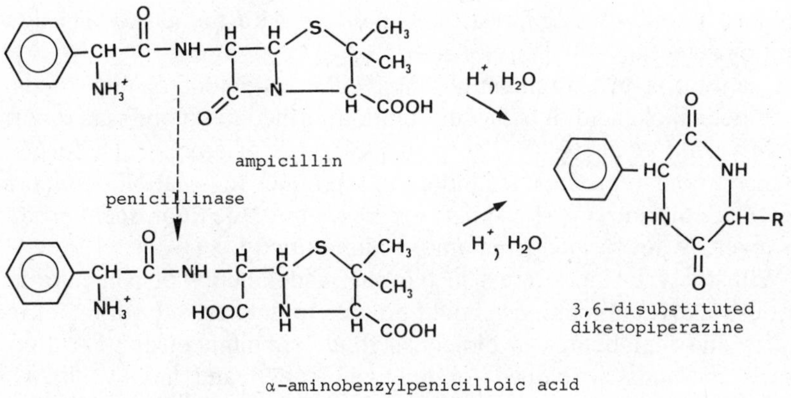

ampicillin

penicillinase

α-aminobenzylpenicilloic acid

3,6-disubstituted diketopiperazine

Fig. 2.3. Formation of a 3,6-disubstituted diketopiperazine from ampicillin and/or α-aminobenzylpenicilloic acid according to the method of Jusko (171).

α-aminobenzylpenicilloic acid (from an experiment with penicillinase) is consistent with the ability of amino acid esters to condense to form diketopiperazines. The imino protons in this structure could dissociate in solution above pH 10 to produce a more fluorescent species, which would account for the obtained pH–fluorescence profile. The remaining portion *R* of the piperazine molecule still can undergo a variety of other chemical reactions.

Miyazaki et al. (172) reported in 1974 that an intensive and reproducible fluorescent product can be obtained from α-aminobenzylpenicilloic acid in neutral solution containing mercuric chloride at 40°. Aminobenzylpenicilloic acid is readily transformed from ampicillin in alkaline solution. The fluorescent product has uncorrected excitation and emission maxima at 340 and 420 nm, respectively, and is readily extracted into ethyl acetate and chloroform from acidic or neutral media and can then be back-extracted into alkaline media. An assay based on these observations permitted detection of less than 0.1 µg ml^{-1} of ampicillin and/or aminobenzylpenicilloic acid. As the fluorescent product is not directly formed from ampicillin, separate measurement can be made of aminobenzylpenicilloic acid as well as unchanged ampicillin in aqueous solution, urine, and blood.

Barbhaiya et al. (173) developed a simple and rapid fluorometric assay of amoxycillin in plasma, based on the formation of a fluorescent derivative in basic medium that is measured in the presence of 2-methoxyethanol (λ_{ex} = 345 nm; λ_{em} = 425 nm). Drugs commonly administered together with amoxycillin (e.g., theophylline, codeine, paracetamol, phenobarbitone, diazepam) do not interfere with the assay. The lower limit of detection is 0.24 μmol L^{-1}.

Quantitative measurement of amoxycillin and its metabolite, assumed to be penicilloic acid, in aqueous solution, urine, and blood was reported by Miyazaki et al. (174). The penicilloic acid of amoxycillin forms an intensive and reproducible fluorescent product in solutions containing mercuric chloride (λ_{ex} = 345 nm; λ_{em} = 420 nm), the fluorescent product, however, is not formed from amoxycillin directly.

With the growing interest in the pharmacokinetics of penicillin antibiotics, analytical methods should be able to determine both intact molecules and metabolites in biological fluids simultaneously. The fluorometric methods developed for ampicillin (172) and amoxycillin (174) satisfy this requirement, but these methods are essentially sensitive only to α-amino-penicillins and are not applicable to many other penicillins. Tsuji et al. (175) described a new and sensitive fluorometric method for the analysis of both the intact penicillin and its biotransformation product, penicilloic acid. Their method is based on the formation of a fluorescent Schiff base resulting from the reaction between dansyl hydrazine and penilloaldehyde, which can be produced from the reaction of penicilloic acid and mercuric chloride in acidic medium; λ_{ex} = 320 nm; λ_{em} = 510 nm.

Kušnír et al. (176) prepared and purified the fluorescamine derivative of ampicillin and examined some basic chemical and biological properties of this product such as chemical stability, electrophoretic mobility, uniformity during thin-layer chromatography, ion-exchange chromatography, and antimicrobial activity.

A fluorometric method was developed for the determination of cephradine in plasma (177). A fluorescent product is formed when samples of deproteinized plasma containing cephradine are heated for 3 h at 100° and pH 1. The fluorescence (λ_{ex} = 350 nm; λ_{em} = 445 nm) is determined in sodium hydroxide solution (pH 13.5); only 0.1 ml of plasma is required, and concentrations of cephradine down to 0.1 μg ml^{-1} may be determined. The method was derived from Jusko's fluorometric assay for ampicillin (171) but originally lacked sensitivity, probably because of the greater stability of the β-lactam ring in cephalosporins which first has to be hydrolyzed in acid medium. Increasing the severity of the hydrolysis conditions yielded good results.

A rapid and sensitive fluorometric analysis for cephalosporins, which can also be applied to penicillins (e.g., ampicillin, amoxycillin, penicillin G), was presented by Yu et al. (178). The method involves reaction with O.1 N sodium hydroxide at 100°, producing stable fluorescent products.

Cephalexin and ampicillin can be detected at concentrations as low as 0.01 μg ml^{-1} using this method.

Crombez et al. (179,180) proposed a fast, specific, and sensitive HPLC procedure for the determination of cefatrizine, an orally active cephalosporin, in serum and urine. Cephradine is used as the internal standard. The separation is carried out on a reversed-phase column, filled with octadecylsilane chemically bonded microparticles. The eluent consists of a mixture of acetonitrile with 0.025 M sodium phosphate buffer pH 7. Quantitation is effected by fluorescence detection of the fluorophores formed after postcolumn derivatization with fluorescamine in a packed-bed reactor (395-nm excitation filter; 460-nm emission filter). The method has been applied to serum and urine samples, which were analyzed after deproteinization with trichloroacetic acid and injection of the clear supernatant. The authors checked accuracy and reproducibility of their procedure by determining the cefatrizine content in spiked serum and urine samples.

Erythromycin and erythromycin ethylsuccinate were determined fluorometrically in serum by an HPLC postcolumn, on-stream derivatization and extraction method (181); fluorometric and microbiological assays of erythromycin concentrations in plasma and vaginal washings were described in 1981 (182).

Preliminary evaluation of luminescent techniques for the determination of bleomycin was performed by Thompson and Williams (183).

A fluorometric method to assay chloramphenicol was reported by Clarenburg and Rao (184).

The luminescence of the antibiotic griseofulvin was reported to be the most sensitive method for detection; Neely and McDuffie (185) reported corrected ambient temperature fluorescence excitation and emission spectra, an uncorrected phosphorescence emission spectrum, and values for quantum efficiency of fluorescence, fluorescence lifetime, and phosphorescence decay times. The authors state that phosphorescence emission may be the most sensitive spectral parameter; measurable phosphorescence can be obtained from solutions as dilute as 10^{-8} M, and no photodecomposition was apparent under reasonable conditions of light exposure.

Mycophenolic acid, an antibiotic from penicillium cultures and reported to have antibacterial, antifungal, antiviral, and antitumor activity, was determined in plasma by GLC and fluorometric procedures by Bopp et al. (186). The fluorescence method consists of an extraction followed by measurement of the fluorescence of the compound in pH 10.0 buffer (λ_{ex} = 350 nm; λ_{em} = 438 nm). The method has the advantage of being faster than the GLC method (after silylation) and is therefore better suited

to handling large numbers of samples; the GLC method was less susceptible to interference from other compounds.

Hirauchi and Masuda (187) established a simple and rapid fluorodensitometric method for simultaneous determinations of siomycin A, siomycin B, and siomycin D_1 in siomycins. Their procedure is based on the
fluorescence reaction of siomycins with sulfuric acid after separation on
thin-layer chromatographic plates, and can be applied satisfactorily to
simultaneous determinations in the concentration range of 5–1000 ng per
spot.

Alkaloids.   Atropine and hyoscyamine can be determined by a flurometric method in the presence of other belladonna constituents (188).
Atropine and/or hyoscyamine is extracted with chloroform from basic
solution, and an aliquot of the extract is added to a chloroform solution
of eosine yellowish. This solution, appropriately diluted, is read on a
fluorometer at an excitation wavelength of 475 nm and an emission wavelength of 552 nm. The acid form of eosine yellowish fluoresces to a negligible extent. However, the salt form is highly fluorescent, and atropine
and hyoscyamine are sufficiently strong bases to convert eosine yellowish
to a salt in chloroform, resulting in the production of fluorescence.

Harman (1-methyl-9H-pyrido[3,4-b]-indole), a main Passiflora alkaloid, can be determined fluorometrically after TLC separation based on
its strong fluorescence in UV light ($\lambda_{ex} = 360$ nm; $\lambda_{em} = 378$ nm) (189).

A simple, sensitive, quantitative method for physostigmine in tissue
samples and in solutions utilizing its powerful native fluorescence was
reported by Taylor in 1967 (190).

Fluorescence studies on the variation of fluorescence intensity of
methyl esculetin solutions with the concentration and pH of the medium
are reported by Gomes (191). 4-Methylesculetin shows properties of a pH
indicator for pH changes from 5.5 (nonfluorescing) to 7.5 (fluorescing).

The determination of derivatives of 8-hydroxyquinoline as chelate compounds with Sn^{2+}, 4- and 8-aminoquinolines, and cinchoninic acid by
fluorescence was studied by Khabarov et al. (192).

Ajmalicine (δ-yohimbine), on treatment with nitric acid in perchloric
acid solution, gives a yellow color with absorption maximum at 355 nm
and fluorescence at 510 nm, allowing spectrophotometric (5–50 μg ml^{-1}
range) and fluorometric (0.1–3 μg ml^{-1} range) determinations (193).

Dombrowski et al. (194) described a sensitive and specific method for
17-monochloroacetylajmaline and its metabolite, ajmaline, in plasma by
TLC fluorescence detection after reaction with nitric and acetic acids.
Method specificity is accomplished by combining an ion-pair extraction

with chromatography; detection limit for the parent compounds is 0.06 μl per 2 ml of plasma.

LSD and other indole alkaloids can be identified by ultraviolet degradation products (195). Controlled degradation of the alkaloids in a chloroform solution under UV irradiation and TLC of the degraded material yield a series of colored spots under UV. The irradiated portion of each of the compounds tested produces a fluorescing spot at the origin, which is one of the major spots in most cases. Steinigen (196a) reported on the identification and quantitative determination of LSD and describes the use of its intense bluish native fluorescence in solution under long-wavelength UV light. LSD was analyzed in human body fluids by high-performance liquid chromatography, fluorescence spectroscopy, and radioimmunoassay by Twitchett et al. (196b). LSD detection in biological fluids presents much difficulty as the drug is used in low dose (1 μg kg^{-1} orally) and because of its extensive metabolism. Moreover, the fluorescence spectrum of LSD is indistinguishable from that of many other ergot alkaloids. The authors describe a complete analytical scheme for the detection and measurement of the drug based on a rapid preliminary screening by radioimmunoassay to reject negative samples, followed by a quantitative analysis using HPLC in combination with fluorometry and radioimmunoassay, allowing determinations of levels down to 0.5 ng of LSD per ml.

Härtel and Harjanne (197) compared two different methods for the determination of serum quinidine concentrations. Extraction of quinidine with benzene and second transfer to sulfuric acid gave about 40% lower values than protein precipitation with alcohol–acetone and direct fluorometric determination with and without the addition of acetic and sulfuric acid. This difference is due to a lower solubility of quinidine metabolites in the less polar solvent benzene. Quinidine metabolites are considered of remarkably lower toxicity and antiarrhythmic activity than quinidine. The authors state that because of its greater specificity for quinidine, the double extraction method of Cramer and Isaksson (198) is regarded as more suitable for the control of quinidine treatment than other methods.

Quinidine determinations in serum may yield erroneously increased results when the patient receives other fluorescent drugs at the same time. Triamterene serves as an example. Two methods were described (199) allowing the quinidine determination in serum in the presence of triamterene. With the first method the fluorescence of both substances is determined while the pH value is increased from 0 to 13; the difference is very clear and the quinidine content can be calculated from this difference. The second method is based on the fact that the fluorescence of quinidine decreases with the concentration of halogen ions, whereas that of triamterene does not. Thus the fluorescence of the serum is measured at low pH after dilution with water and, simultaneously, in the same condition after dilution with 2% NaCl solution, the quinidine content is calculated from the difference.

Broussard (200) proposed a fluorometric method for quinidine based on its extraction from alkalinized serum or plasma into benzene, followed by reextraction into a sulfuric acid solution in which quinidine fluoresces with an intensity proportional to its concentration.

Schulman et al. (201) did extensive work on the absorption and fluorescence spectra of 6-methoxyquinoline, quinine, and quinidine as a continuous function of pH. They state that 6-methoxyquinoline is a stronger base in the lowest excited singlet state, as reflected by the pH dependence of its fluorescence. The latter is used to calculate the rates and equilibria of proton exchange in the excited state. Quinine and quinidine, however, do not protonate during the lifetimes of their excited states, an observation attributed to the lower basicity of their aromatic heterocyclic nitrogen than that in 6-methoxyquinoline. The differences in basic properties between the alkaloids and 6-methoxyquinoline were employed to infer the ketonic character of the quinolinemethanol group of the alkaloids and the existence of an intramolecular hydrogen bond in quinine but not in quinidine.

Guentert and Riegelman (202) examined in 1978 the specificity of some frequently used fluorescence determination methods and a specific normal-phase HPLC assay that allows simultaneous but separate quantitation of quinidine and its major metabolites in plasma. The fluorescence values for quinidine are consistently higher than those obtained by the specific assay, which can be explained by metabolite carryover in the extraction procedure involved in the fluorescence assays.

Reserpine, an important agent in the treatment of hypertension, yields a fluorescent, greenish-yellow color by treating an alcoholic solution with dilute sulfuric acid and sodium nitrite. This reaction was adapted in 1956 to the quantitative determination of reserpine in various dosage forms, free from interference by similar alkaloids (203). A quantitative fluorometric method for the determination of reserpine in feeds at the microgram level based on the same reaction with nitrous acid, producing a yellow fluorescence, was reported later (204).

Haycock et al. (205) studied in 1966 the kinetics of the reaction of reserpine with nitrous acid. Their results show a two-step process with nitrous acid reacting with protonated reserpine to form an intermediate complex that under the influence of hydrogen ions forms a colored fluorescent product, identified as 3-dehydroreserpine. Kabadi et al. (206) described semiautomated fluorometric and colorimetric methods for the determination of reserpine in 0.25–0.50-mg tablets by measuring the fluorescence (λ_{ex} = 390 nm; λ_{em} = 510 nm) and absorbance (390 nm) of the reaction product with nitrous acid in an automatic analyzer system. A semiautomated method suitable for the determination of reserpine and related compounds in tablet formulations based on the classical nitrite procedure using vanadium pentoxide as a reagent was reported later (207). Frijns (208) performed a thin-layer chromatographic separation of reserpine and rescinnamine in tablets and injections on silica gel with chloroform–methanol (93 + 7) as eluent. After development of the chromatogram, the alkaloids were exposed to direct daylight to effect spontaneous air oxidation to the yellow green fluorescent 3-hydro compounds. The fluorescence maximum was reached in 2 h and the intensity was measured directly from the plate (λ_{ex} = 365 nm; λ_{em} = 490 nm). A semiautomated nitrous acid fluorometric method for the determination of reserpine and Rauwolfia alkaloids in unit dose (λ_{ex} = 390 nm; λ_{em} = 510 nm) was developed as well (209), without interference from reserpidine, yohimbine, and bendroflumethiazide. Gurkan (210) reported a fluorometric analysis of the alkaloids serpentine, yohimbine, and boldine. A collaborative study of a spectrofluorometric assay for *Rauwolfia serpentina* tablets and powdered root by column chromatography followed by fluorometry of the nitrous acid treated eluate was reported by Smith and Clark (211). Smith (212) described in the same way an analogous study using a vanadium pentoxide–phosphoric acid determinative step.

Caille et al. (213) were able to detect concentrations of 0.02 μg ml^{-1} of tetrahydroberberine, 0.2 μg ml^{-1} of berberine, and 10 μg ml^{-1} of hydrastine by spectrophotofluorometry.

A fluorescent thin-layer material (Kieselgel GF 254 Merck) was examined for drug analysis in 1970 (214). It was shown that the R_f values of 26 drugs (mainly alkaloids and steroids) are unaffected by the inclusion of a fluorescent indicator in the silica gel layers.

Photodensitometric determination of alkaloids after thin-layer chromatography by fluorescence reflectance readings of the unsprayed spots was favored as more sensitive and precise for the estimation of alkaloids in plant extracts, for complex drugs containing degradation products, and for production control (215).

A spectrophotofluorometric determination of some alkaloids (pilocarpine, pholcodine, cocaine, strychnine) was developed using the base-catalyzed condensation of mixed anhydrides of organic acids (malonic acid–acetic anhydride), where a tertiary amine group functions as the basic catalyst (216). The product of the condensation reaction is highly fluorescent and allows the fluorometric determination of alkaloids containing a tertiary amine group, in the nanogram per milliliter range. According to the author, the mixed anhydride system is simple to prepare and the fluorogenic reaction is completed within 15 min at 80°. The reaction is specific for but does not distinguish between different tertiary amines.

Frijns (217) determined ergometrine and ergotamine in pharmaceutical dosage forms through TLC on silica gel using chloroform–methanol (75 + 25) and chloroform–methanol (90 + 10), respectively. Fluorescence intensity after TLC development is stabilized by impregnating the chromatogram with liquid paraffin–ether (10 + 90) (λ_{ex} = 313 nm; λ_{em} = 420 nm). Bos and Frijns (218) described a method for the quantitative unit dose analysis of dihydroergotamine in tablets, injection liquids, and solutions for oral use. After extraction, TLC is performed on aluminum oxide plates using chloroform–methanol (96 + 4) followed by 1 h irradiation at 254 nm. The fluorescence intensity is increased to the tenfold value by impregnation of the plate with liquid paraffin–ether (10 + 90). The fluorescence of the spots is read in a TLC spectrophotometer with respect to two reference standards (λ_{ex} = 365 nm; λ_{em} = 470 nm).

Dihydroergotamine mesylate was determined fluorometrically by Gelbcke and Dorlet (219), and recently a method was proposed for ergot alkaloids in plasma by HPLC and fluorescence detection (220).

Antihistamines.   Jensen and Pflaum (221) studied in 1964 the reactions of 16 antihistamines with a series of oxidants. The fluorescence characteristics of the products formed on treatment with hydrogen peroxide were investigated and the corresponding studies on the cyanogen bromide systems were carried out for comparative purposes. A selective and sensitive fluorometric determination was described for tripelennamine hydrochloride in mixed anticold pharmaceutical preparations based on the bromocyanide reaction, and a mechanism for this reaction was suggested (222). Thonzylamine was determined fluorometrically with o-aminothiophenol after oxidation with $NaIO_4$; the structure of the fluorescing derivative was proposed and concentrations of 0.1–10 µg ml^{-1} could be determined (223).

Wilson et al. (224) correlated the fluorescence and phosphorescence excitation and emission spectra of four antihistamines having the 2-aminopyridine chromophore (tripelennamine citrate, pyrilamine maleate, methapyrilene hydrochloride, and thenyldiamine hydrochloride) and one having the 2-aminopyrimidine chromophore (thonzylamine hydrochloride) with the spectra of 2-aminopyridine and 2-aminopyrimidine. They studied the effect of pH on the fluorescence and phosphorescence by means of titration curves. One of their conclusions is that the significant fluorescence of this group of antihistamines appears to originate from the cation in which only the dimethylamine nitrogen is protonated (I) and that

$$\text{(I)} \qquad\qquad \text{(II)}$$

the decrease of fluorescence with increasing acidity corresponds to protonation of the ring nitrogen in the ground state (II). Fluorescence and phosphorescence calibration curves are linear over several orders of magnitude of concentration, and fluorometric and phosphorimetric analysis of single components seems to be feasible. Wirz et al. (225) reported similar fluorescence and phosphorescence investigations on antihistamines having the diphenylmethane and/or p-chlorotoluene chromophore, including diphenhydramine hydrochloride, bromodiphenhydramine hydrochloride, chlorocyclizine hydrochloride, meclizine hydrochloride, phenindamine tartrate, and pyrrobutamine phosphate. They also studied the effect of halide ions on the fluorescence and phosphorescence of these antihistamines in order to evaluate the possibility of fluorescence quenching and phosphorescence enhancement.

Triprolidine ((E)-2-[1-(4-methylphenyl)-3-(1-pyrrolidinyl)-1-propenyl] pyridine hydrochloride), a potent antihistaminic used for the treatment of various allergic conditions, was determined in human plasma by quantitative TLC with fluorescence detection by DeAngelis et al. (226). Plasma extracts were spotted on silica gel, eluted with methanol-ammonium hydroxide-chloroform (10 + 1 + 89), the plate air-dried for 15 min and then sprayed with a 2 M aqueous solution of ammonium bisulfate. After about 1 h air-drying, the fluorescent triprolidine spot was quantitated with a spectrodensitometer (λ_{ex} = 300 nm; λ_{em} > 405 nm). A sensitivity limit of 0.8 ng ml^{-1} could be reached.

Drugs Affecting the Adrenergic Nervous System. Comparison of improved fluorometric methods used to quantitate plasma catecholamines (227) revealed in 1969 that the ethylenediamine and the trihydroxyindole method are suitable and accurate for assessing sympathoadrenal activation under various stressful conditions. Excitation and fluorescence spectra of methoxy derivatives of catecholamines (228) and a fluorometric micromethod for plasma catecholamine assay (229) were reported. Griffiths et al. (230) described a fluorometric determination for the assay of human plasma catecholamine levels by selective adsorption onto alumina at pH 8.5, followed by extraction and analysis as the trihydroxyindole derivatives. Renzini et al. (231) improved the sensitivity of the trihydroxyindole method, allowing differential determination of noradrenaline and adrenaline at picogram levels. Normetanephrine, metanephrine, and 3-methoxytyramine were determined fluorometrically in hydrolyzed urine using a sulfonated resin (Dowex 50-W 200–400 mesh) (232), and a sensitive fluorometric method was developed for the estimation in urine of 4-hydroxy-3-methoxy-phenylglycol, the major metabolite of noradrenaline in brain, by separation of the glycol after preliminary enzymatic hydrolysis, solvent extraction, ion exchange chromatography, and flurometric glycol estimation after oxidation by ferric chloride (233). Direct quantitative evaluation of thin-layer chromatograms by remission- and fluorescence measurements for adrenaline, noradrenaline, and dopamine from urine via ethylenediamine conversion of their triacetyl derivatives was described in 1971 (234) as well as the quantitative analysis of catecholamine and serotonine metabolites on thin-layer plates (235). Sapira et al. (236) determined plasma catecholamines fluorometrically by a modification of the standard alumina-trihydroxyindole method in which radioactive tracer catecholamines are used to follow the chromatographic behavior of the endogenous catecholamines on alumina. Other catecholamine determinations of that period include a rapid fluorometric method of adrenaline in the adrenal glands (237), a fluorometric method for adrenaline in the presence of an aromatic primary amine (238), a fluorometric method for catecholamines in gastric juice after chromatographic isolation (239), and a fluorometric determination of catecholamines after dansylation using a three-phase of thin layers (alumina, cellulose, and silica gel) (240). In 1974 an improved autoanalyzer fluorescence method for urinary catecholamines (241) was reported, as were an extraction method for adrenaline as trihydroxyindole without interference by noradrenaline (242), a simultaneous spectrofluorometric analysis of 3,4-dihydroxyphenylalanine and the total amount of adrenaline–noradrenaline in urine (243), a simultaneous automated fluorometric estimation of noradrenaline and dopamine in brain tissue (244), and a correction for separate quenching effects in the fluorometric determination of total adrenaline plus noradrenaline (245). In 1975 fluorescence derivatization with 1-dimethylaminonaphthalene-5-sulfonyl chloride was described for adrenaline and some alkaloids (246), as phenolic groups can be derivatized with this reagent almost as easily as primary and secondary amino groups. Two other papers of this period give a description of the trihydroxyindole method for catecholamines in urine (247) and in plasma and urine (248); the trihydroxyindole reaction was applied to the methanesulphonanilides soterenol and mesuprine (249). Karasawa (250) reported a fluorometric determination of some metabolites of dopamine, norepinephrine, and serotonin. A semiautomated catecholamine determination in plasma was reported with adrenolutine and noradrenolutine fluorescence measurement on treatment with strong alkali (251); Schwedt (252) described a method for the separation and fluorometric determination of adrenaline and noradrenaline. A simple fluorometric method for determining catecholamines based on their native fluorescence in the presence of their decomposition products was reported (253), as was an automated fluorometric analysis of adrenaline in lidocaine hydrochloride injection based on an adaptation of the official USP XIX procedure (254). An improved fluorometric catecholamine assay was compared with an enzymic method (255); an estimation of plasma catecholamines in man was described in 1976 (256). De Vente et al. (257) contributed in 1981 to the fluorescence spectroscopic analysis of the hydrogen bonding properties of catecholamines, resorcinol amines, and related compounds with phosphate and other anionic species in aqueous solution. The inhibitory effect of some monoamine oxidase inhibitors on the fluorescence assay for the o-methylated metabolites of the catecholamines was described in the same year (258). Urinary free catecholamines were determined by liquid chromatography and fluorometry (259), and liquid chromatographic determinations of biogenic amines as dansyl derivatives was suggested by Schwedt and Bussemas (260–262). Kagel et al. (263) used dansyl chloride as precolumn derivatizing agent and a C8-bonded phase and isocratic elution in 80% CH_3CN–H_2O permitted simultaneous determination of a large number of amines in biological fluids and tissue extracts.

Amphetamine was determined fluorometrically in urine by an automated procedure through the reaction with formaldehyde and acetylacetone, producing the dihydrolutidine derivative ($\lambda_{ex} = 410$ nm; $\lambda_{em} = 476$ nm) (264). Thin-layer chromatographic-spectrophotofluorometric analysis of amphetamine and amphetamine analogs was described after reaction with 4-chloro-7-nitrobenzo-2,1,3-oxadiazole (NBD-Cl) (265). Klein et al. (266) reported the use of fluorescamine to detect amphetamine in urine by thin-layer chromatography, while Hopen et al. (267) detected amphetamine- and methamphetamine-type materials in pharmaceutical and biological fluids by fluorometric labeling.

An excellent report on the fluorescence and phosphorescence of several phenylethylamines and barbiturates was made by Miles and Schenk (268) to provide analytical methods based on native luminescence, rather than the chemically induced luminescence of these two classes of pharmacologically important compounds. The phenylethylamines include compounds with the general structure (I) among which amphetamine ($R_1 = R_2 = R_3 = R_5 = H; R_4 = CH_3$), epinephrine ($R_1 = R_2 = R_3 = OH$;

$$R_1 \underset{R_2}{\overset{R_3}{\underset{}{\bigcirc}}} CH - \underset{NH-R_5}{\overset{}{CH}} - R_4$$

(I)

R_4 = H; R_5 = CH_3) and ephedrine (R_1 = R_2 = H; R_3 = OH; R_4 = R_5 = CH_3). Many of the phenylethylamines exhibit similar fluorescence characteristics at 282–313 nm when excited at 260–284 nm. Fluorescence quantum efficiencies in air have been measured together with molar absorptivities to give a practical indication of the fluorescence intensities of these molecules as well as those of several barbiturates.

In 1972 a simple quantitative method for determining 3-methoxy-4-hydroxyphenylacetic acid and 3,4-dihydroxyphenylacetic acid in urine using thin-layer chromatography and fluorometric estimation (269) was described; a fluorometric method (potassium ferricyanide-ethylenediamine) combined with thin-layer chromatography for the determination of norepinephrine, epinephrine, and dopamine in human urine (270), and a semimechanized fluorometric method for 5-hydroxytryptamine (serotonine) in blood plasma and platelets (271).

Epinephrine was determined fluorometrically in low-dosage injections and in combinations with lidocaine hydrochloride by a modification of the original trihydroxyindole reaction following ion pairing with di(2-ethylhexyl)phosphoric acid (272). Schwedt (273) compared automated fluorometric determinations of the o-methylcatecholamines, metanephrine, and normetanephrine by different methods of oxidation and stabilization and concluded that oxidation with hexacyanoferrate(III) and final stabilization of the fluorophor with ascorbic acid yields the best results. A differentiated fluorometric determination for metanephrine and normetanephrine in urine based on the oxidation with potassium hexacyanoferrate(III) in the presence of zinc ions and different pH values in an autoanalyzer was described as well (274). Jackman (275) described a simple method for the assay of urinary metanephrines using HPLC with fluorescence detection. Davis et al. (276) developed a high-performance liquid-chromatographic separation and fluorescence measurement of biogenic amines in plasma, urine, and tissue. Histamine, noradrenaline, octopamine, normetanephrine, dopamine, serotonine, and tyramine are precolumn derivatized with o-phthalaldehyde, which stabilizes the molecules, facilitates extraction, and improves detection of nanogram amounts. Excitation and emission wavelengths of 340 nm and 480 nm, respectively, were used.

A semiautomated fluorometric method was described for measurement of dopa (3-(3,4-dihydroxyphenyl)-*l*-alanine) in plasma with ferricyanide as an oxidant after separation from interfering catecholamines through adsorption on alumina, elution with 0.1 N HCl, and adsorption on and elution from a cation-exchange column (277). Türler and Käser (278) determined urinary dopa in a similar way with iodine as an oxidant and discussed the significance of the method for the diagnosis of neural crest tumors.

Urinary dopamine (3,4-dihydroxyphenethylamine) can be determined semiautomatically by adsorption on alumina at pH 8.5, elution with acetic acid, eluate adjustment at pH 6.0, oxidation with sodium metaperiodate followed by alkaline sodium sulfite treatment and measurement of the fluorescent derivative (279). Oberman et al. (280) developed a quantitative method for urinary dopamine based on the reaction with dansyl chloride, TLC on silica gel with triethylamine benzene (1:8), and fluorescence reading of the eluted spot of dansyl dopamine at λ_{ex} = 335 nm, λ_{em} = 485 nm. Their method is only suitable for high dopamine excretion, in the range of 5–100 mg per day. Greenland and Michaelson (281) stressed the importance of the buffer preparation according to the original procedure of Anton and Sayre (282–284) using KH_2PO_4 and adjustment to the proper pH with NaOH; the latter authors have studied exhaustively the factors affecting the aluminum oxide–trihydroxyindole procedure for analysis of catecholamines. Recently, a comparison was made for two fluorometric methods for the determination of adrenaline, noradrenaline, dopamine, dopa, and dopac in urine (285). Fluorescence detection was recently used for the high-performance liquid chromatographic assay of dopamine β-hydroxylase in rat serum and adrenal medulla (286).

Carbidopa (*l*-(−)-α-hydrazino-3,4-dihydroxy-α-methylhydrocinnamic acid), an inhibitor of aromatic amino acid decarboxylase, was determined in plasma by a spectrofluorometric method (287). A fluorophor is formed with *p*-dimethylaminobenzaldehyde and the dimethylaminobenzalazine formed fluoresces intensely in chloroform as trichloroacetic acid ion pair (λ_{ex} = 466 nm; λ_{em} = 546 nm). The assay was applied to the analysis of carbidopa in monkey plasma. Eighty nanograms of carbidopa per 2 ml of plasma could be detected by this method.

A method for the fluorometric determination of 4-(2-hydroxy-3-isopropylaminopropoxy)-indole, a β-blocking agent, in plasma and urine was reported by Pacha (288). By condensation with *o*-phthalaldehyde in acidic solution an intensely fluorescent product is obtained, attributed to the reaction with the indole nucleus. The fluorophor (λ_{ex} = 390 nm; λ_{em} =

(II) Bufuralol   (III)

440 nm) is not very stable but can be stabilized by the addition of ascorbic acid solution, which also lowers the blank readings.

A spectrofluorometric assay for 6,7-dimethyl-α-[(isopropylamino)methyl]-2-benzofuranmethanol hydrochloride (I), a β-adrenergic active benzofuran derivative was developed and used in pharmacokinetic studies in normal volunteers (289). de Silva et al. (290) developed a spectrofluorometric assay for the determination of another member of the benzofuran series, bufuralol (7-ethyl-α[(*tert*-butylamino)methyl]-2-benzofuran-methanol hydrochloride (II), in blood and urine by a modification of a procedure published for (I) (289). The modified procedure includes a TLC separation step to resolve the parent drug (II) from its major metabolite (III) in blood and urine, thereby imparting specificity to the assay. Compounds II and III fluoresce in 0.1 N HCl with excitation at 250 nm and emission at 300–305 nm. The sensitivity limit of the assay is 2 ng of compound per ml of blood, using a 2,5-ml specimen per analysis.

A spectrophotometric and spectrofluorometric determination of 4-benzyl-6,7-dimethoxy-isoquinoline and 2-benzyl-6,7-dimethoxytetrahydroisoquinoline derivatives was reported in 1971 (291). These compounds show interesting pharmacological properties as papaverine-like antispasmodics and as adrenolytics, respectively, and possess interesting native fluorescence properties that are used for their quantitative estimations.

Garrett and Schnelle (292) developed a sensitive spectrofluorometric assay of the β-adrenergic blocker sotalol(d,l-4-(2-isopropylamino-1-hydroxyethyl)methanesulfonanilide in blood and urine. The compound fluoresces in alkali (λ_{ex} = 250 nm; λ_{em} = 350 nm) and in acid (λ_{ex} = 235 nm, λ_{em} = 309 nm), about three times more intense in alkali than in acid. The maximum extractability was found at pH 9, where the substance is primarily the zwitterion with potentiometric pK_a' values of 8.30 and 9.80, where the former was spectrophotometrically assigned to the dissociation of the conjugated sulfanilino group. A mixture of n-amylalcohol and chloroform (1:3, v/v) at a volume ratio of organic solvent and pH 9 buffer solution of 8:1 extracted 85% of the compound, which was completely reextracted into 5.2 N HCl. The fluorescence of the monocharged molecule was measured at 235/309 nm in the acid solution. The sensitivity of the assay was 0.1 μg ml^{-1} in plasma and 2.5 μg ml^{-1} in urine. The proposed assay is applicable for monitoring blood levels and urinary excretion of the drug when administered in therapeutic amounts.

Thymoxamine (5-(2-dimethylaminoethoxy)carvacrol acetate) hydrochloride, a specific competitive α-adrenoreceptor blocking drug, and its deacetylated derivative have fluorescent properties (λ_{ex} = 295 nm; λ_{em} = 335 nm) that may be exploited for the purposes of recognition and measurement in body fluids and in pharmaceutical formulations (293).

A fluorometric method has been devised for determining practolol ([1-(4-aminophenoxy)-3-isopropylaminopropan-2-ol]), a β-adrenergic receptor antagonist, in blood and urine (294). Such a reliable method is important for evaluation of blood levels, clinical response relations, and pharmacokinetic studies. The method requires acid hydrolysis of practolol and extraction of the deacetylated derivative, followed by reaction of this material with nitrosonaphthol, producing a highly fluorescent derivative in weak acid (λ_{ex} = 460 nm; λ_{em} = 550 nm).

Acebutolol (N-[3-acetyl-4-[2-hydroxy-3-[(1-methylethyl) amino]propoxy]phenyl]butanamide, a β-adrenergic blocker, can be determined in biological fluids by HPLC with fluorescence detection (295). Acebutolol and its N-acetyl metabolite are extracted in alkaline medium with a mixture of dichloromethane-n-butanol and determined on a Lichrosorb Si60, 5-μm column in the presence of DL(acetyl-2-decylamido-4-phenoxy)-1-isopropyl-amino-3-propanol-2 as internal standard. Detection occurs in an 8-μl microcuvette at λ_{ex} = 342 nm, λ_{em} = 438 nm.

A spectrofluorometric method for the determination of the beta-adrenergic receptor blocking propranolol (1-(isopropylamino)-3-(1-naphthyloxy)-2-propanol in plasma (296) and a clinical evaluation of plasma propranolol levels and β-adrenergic blockade in humans (297) were reported in 1970. A simple fluorometric method has been developed to determine blood levels of propranolol based on native fluorescence measurements after extraction (λ_{ex} = 293 nm; λ_{em} = 344 nm) (298). Short excitation wavelength fluorometric detection of the native fluorescence of propranolol and its 4-hydroxy analog was used for the high-pressure liquid chromatography of these compounds and of indole peptide and phenol compounds (299). Rosseel and Bogaert (300) reported an HPLC method for the simultaneous determination of propranolol and its metabolite 4-hydroxypropranolol in plasma. Their method involves plasma extraction at basic pH with ethyl acetate, two brief back-extractions, chromatography on a reversed-phase column and fluorescence detection (λ_{ex} = 290 nm; emission cutoff filter with 90% transmission at 340 nm). 4-Methyl propranolol was used as an internal standard and 1 ng of each compound per ml of plasma should be detectable according to the authors. Recently, Albani et al. (301) developed a simple and rapid determination method for propranolol and its active metabolite, 4-hydroxypropranolol, in human plasma by liquid chromatography with fluorescence detection.

Methanesulfonanilides, like soterenol and mesuprine, which are bioisosteric with adrenergic catecholamines, form analytical useful fluorescent species when subjected to the trihydroxyindole reaction, presumably because of adrenolutinlike species formed from aminochrome intermediates (302).

5-Hydroxytryptophan and its metabolites 5-hydroxytryptamine (serotonine) and 5-hydroxyindoleacetic acid were estimated in plasma based on solvent extraction and the use of a liquid cation exchange reagent to separate 5-hydroxytryptamine and 5-hydroxytryptophan; fluorometric determination is carried out after derivatization of the 5-hydroxyindoles with o-phthalaldehyde in strong acid (303). Reversed phase high-performance liquid chromatography and fluorometric detection was used for the quantitative determination of 5-hydroxyindole-3-acetic acid in urine and cerebrospinal fluid (304). β-Phenylethylamine, a biogenic monoamine considered to play the role of a neuromodulator or a neurotransmitter in the central nervous system, was determined fluorometrically in a sensitive and specific way in human urine by the ninhydrin reaction after n-heptane extraction (305).

Methyldopa, a widely used antihypertensive agent with sedative and aromatic L-aminodecarboxylase-inhibiting properties, can be analyzed in pharmaceutical preparations and biological fluids by various methods, including fluorometry (306–308). This technique mostly involves the production of a fluorescent indole derivative (lutin) by oxidation and rearrangement. Kim and Koda (309) described in 1977 an improved analytical method in which methyldopa is adsorbed on alumina, eluted with an organic solvent, and then oxidized to form the fluorophor dihydroxyindole. The lower limit of sensitivity was 100 ng ml^{-1}.

The antihypertensive indapamide [3-(aminosulfonyl)-4-chloro-N-(2,3-dihydro-2-methyl-1H-indol-1-yl) benzamide] can be sensitively determined in urine by a fluorometric procedure (310). Reaction of indapamide with sodium hydroxide at 100° yields a fluorescent product of which the fluorescence intensity is increased by a factor of 3 by the addition of formaldehyde. Concentrations of 0.05 μg ml^{-1} can be detected in dogs given 20 mg of the drug. A semiautomated method for this drug in biological fluids was described by the same authors via preextraction from the biological sample followed by continuous flow analysis (311).

A sensitive high-performance liquid chromatographic method for the determination of labetalol in human plasma using fluorometric detection was presented by Oosterhuis et al. (312). Labetalol (2-hydroxy-5-[1-hy-droxy-2-(1-methyl-3-phenylpropylamino)ethyl] benzamide hydro-chloride), an α- and β-adrenoceptor antagonizing antihypertensive, is sep-arated by ion-pair reversed-phase high-performance liquid chromatog-raphy and detected through its native fluorescence (detector installed at λ_{ex} = 335 nm; λ_{em} = 370 nm); chloroquine was used as an internal stan-dard.

Kamimura et al. (313) described an HPLC method for the determination of the α,β- adrenoceptor blocker 5-{1-hydroxy-2-[2-(o-methoxyphen-oxy)ethylamino]ethyl}-2-methyl-benzenesulphonamide hydrochloride in plasma. Fluorescence labeling is achieved with 5-di-n-butylaminona-phthalene-1-sulfonyl chloride as a reagent and a detection limit of 20 ng ml^{-1} was obtained, allowing plasma level determinations in man and in dogs after oral administration.

Chemotherapeutics and Related Compounds. Lapachol (2-hydroxy-3-(3-methyl-2-butenyl)-1,4-naphthoquinone) and related naphthoquinones have shown antimalarial activity and an influence on oxidative phos-phorylation, thus preventing the oxidative synthesis of adenosine tri-phosphate. Finkel and Harrison (314) described a fluorometric method based on the conversion of lapachol under partial vacuum to a fluorescent product by reduction with sodium hydrosulfite. As the fluorescent product is sensitive to oxygen, it is extracted into benzene and measured in a sample cell from which air is excluded. The method was applied for the determination of lapachol in serum.

Three procedures for fluorescence assay of pure hydroxynaphthoqui-nones (e.g., juglone, plumbagin, lawsone, lapachol, phthiocol) were re-ported (315), allowing determinations at the submicrogram level. The most general method involves biphasic reduction with sodium dithionite so-lutions and determination of the fluorophor in butyl acetate. The method is highly sensitive but not specific as all hydroxynaphthoquinones react. Another procedure is based on warm reduction with stannous chloride in an acid medium followed by determination of the fluorophor formed in chloroform, a specific and sensitive method for 5-hydroxy-1,4-naphtho-quinones (juglone series). A third procedure based on the reaction of Guilbault and Kramer (nucleophilic addition of potassium cyanide) ap-pears to be specific for 5-hydroxy- and 5,8-dihydroxy-1,4-naphthoqui-nones with both quinonoid positions free.

Sodium fluorescein was determined in ophthalmic solutions by a fluo-rometric method (316). The dye concentration was determined in a pH 9 buffer (λ_{ex} = 460 nm; λ_{em} = 515 nm) using a standard curve prepared from hydrolyzed fluorescein diacetate. Identification of sodium fluorescein, detection of possible degradation products, and differentiation of acriflavine from hydrochloride are achieved by TLC on fluorescent silica gel sheets.

Detection and fluorescent measurement of the widely used antimalarial pyrimethamine in urine was described in 1968 (317). Simmons and DeAngelis quantitated pyrimethamine and related diaminopyrimidines by enhancement of fluorescence after thin-layer chromatography (318). After development of the fluorescent silica gel G plates with chloroform–methanol (75 + 25), they were allowed to air-dry for 10–15 min before spraying with 2.0 M aqueous NH_4HSO_4. After 1 h air-drying, allowing the generated fluorophor to fluoresce from the solid state, spectrodensitometric measurement of the fluorescence at 300 nm excitation is performed. The proposed method may be used to measure very low levels of pyrimethamine in biological fluids.

Ormetoprim, used in combination with sulfadimethoxine in chicken feeds, can be assayed in feeds (319) at levels down to 0.001% by a method that involves extraction by blending with alkali and chloroform after brief enzyme digestion, cleanup of the extract by solvent transfers, a carefully timed oxidation with permanganate, destruction of excess permanganate with bisulfite, acidification, and transfer to chloroform for fluorescence reading ($\lambda_{ex} = 301$ nm; $\lambda_{em} = 348$ nm).

Nalidixic acid can be estimated in chicken muscle and liver homogenates at levels of 100 ppb by a fluorometric method (320) involving extraction, cleanup, and measurement of fluorescence of the isolated drug in 60% H_2SO_4 ($\lambda_{ex} = 325$ nm; $\lambda_{em} = 408$ nm). Prototropic and metal complexation equilibria of nalidixic acid in the physiological pH region were recently studied by Vincent et al. (321).

Trimethoprim (5-[(3,4,5-trimethoxyphenyl)methyl]-2,4-pyrimidinediamine), a frequently used antibacterial with therapeutic plasma concentrations of 1 μg ml^{-1}, can be determined in plasma (322) by extraction with chloroform in basic medium and back-extraction in sulfuric acid solution followed by oxidation with potassium permanganate into the fluorescing trimethoxybenzoic acid. Limit of detection: 50 ng ml^{-1}; reproducibility about 6%. Sigel et al. (323,324) developed a new fluorescence assay for trimethoprim and metabolites in man using quantitative thin-layer chromatography.

Isoniazid and metabolites in biological fluids were determined in 1974 by Boxenbaum and Riegelman (325) and reported as part of a Ph.D. dissertation submitted by the former at the University of California, San Francisco Medical Center. Isoniazid is considered to be one of the most effective commonly employed antituberculosis drugs. A fluorometric procedure based on formation of the isoniazid–salicylaldehyde hydrazone, of which the reduced form, extracted in isobutanol, is highly fluorescent, is applied for the determination of isoniazid in whole blood. Bisulfite is used to react with excess salicylaldehyde, thus making it nonfluorescent. Isoniazid and hydrazones, acetylisoniazid, isonicotinic acid, and isonicotinuric acid are determined in urine by colorimetry; each method is specific and accurate. Data are presented on these compounds following intravenous doses of isoniazid to both rapid and slow inactivator subjects of isoniazid.

Anaerobic photodecomposition of an acridan drug (2-chloro-9-(3-dimethylaminopropyl)acridan phosphate (1:1)) through energy transfer resulting in a quantitative conversion to its acridine derivative was described by Digenis et al. (326).

Reaction of sulfonamides and/or local anesthetics with 9-chloroacridine yielding 9-substituted aminoacridines results in a decrease in fluorescence of the reaction mixture compared to that of the blank 9-chloroacridine solution (327). This quenching fluorometry enables one to analyze various sulfonamides, local anesthetics, and mixtures of pharmaceuticals containing these drugs in the 10^{-5}–10^{-6} M range, monitoring the fluorescence at activation and emission wavelengths of 385 and 420 nm, respectively. According to the authors, the procedure should be applicable in the analysis of any primary aromatic amine. Sulfonamides such as sulfanilamide, sulfathiazole, sulfamethylthiazole, and disulfanilamide monoacetylate can be determined quantitatively in galenic forms by extraction, TLC separation on silica gel 60F254, and fluorescence quenching measurement in a densitometer (328). Hsiao et al. (329) investigated the possibility of using a fluorescent probe technique for the study of drug–povidone interactions. Povidone can interact with several drugs, which may influence their availability, efficacy, and transport. 1-Anilino-8-naphthalenesulfonate was used as the probe; sulfanilamide, sulfacetamide, and sulfabenzamide were used as the binding competitors.

A fluorometric method was developed for the quantitative and qualitative determination of quinacrine hydrochloride in drug preparations, based on its native fluorescence in 0.1 N HCl (330). Rosenberg and Schulman (331) studied the tautomerism of singly protonated chloroquine and quinacrine. Their findings indicate that the singly charged cations of the drugs exist in two tautomeric forms, which may have significance in the pharmacodynamics of both antimalarials.

Cline Love et al. (332) investigated the analytical utility of selected excited state characteristics in terms of their use to improve both sensitivity and selectivity of analysis for the atabrine-based series. The ultraviolet and visible spectra, the fluorescence spectra, fluorescence lifetime, and quantum yields were determined in different polarity solvents and for both the acid and the base forms of the molecular species. The excited state properties of atabrine and its homologs were used to illustrate what molecular properties permit differentiation between similar species.

The absolute fluorescence quantum efficiency of quinine bisulfate in aqueous medium utilizing quenching of the fluorescence by halide ions was determined by optoacoustic spectrometry by Adams et al. (333).

Cardiovascular Drugs.   Digitoxine was determined after TLC separation and reaction with p-toluenesulfonic acid in the presence of ascorbic acid and hydrogen peroxide in hydrochloric acid medium (334), producing a fluorescent derivative. This method was worked out later for the direct determination of digitaloids by fluorescence on thin-layer plates (335). Direct fluorometric estimation of the main cardenolides of extracts of *Digitalis lanata*, *Strophantus kombé*, and *S. gratus* on thin-layer chromatograms was described by Hammerstein and Kaiser in 1971 (336).

A sensitive and specific automated procedure for the unit dose analysis of digitoxin and digoxin in tablet formulations was reported, based on the fluorescence measurement of the dehydration products of the cardiotonic steroids, resulting from their reaction with hydrogen peroxide in concentrated hydrochloric acid (337). A new fluorometric procedure was suggested in 1975 for the determination of digoxin and digitoxin, based on the development of a fluorophor when the digitalis drugs are treated with concentrated sulfuric acid (338). It was suggested that the fluorescent product is a result of the reaction between sulfuric acid and the aglycone part of these heterosides. Applications to tablet and elixir forms are presented. Britten and Njau (339) reported a simple, specific, and sensitive fluorometric determination of digoxin and its application to tablet preparations allowing digoxin estimations in the presence of digitoxin, gitoxin, and ouabain. The reaction is based on the fluorophor formation when heating digoxin with ethanolic trichloroacetic acid ($\lambda_{ex} = 340$ nm; $\lambda_{em} = 420$ nm). The reaction should be complete within 40 min, and 20 ng ml^{-1} concentrations still can be determined with a linear range of the method from 0.02 to 1.0 µg ml^{-1}.

Gupta et al. (340) described a thin-layer chromatographic method for the estimation of procainamide and its major metabolite, N-acetylprocainamide, in plasma. The method involves a plasma extraction at alkaline pH with dichloromethane to isolate the drug and its metabolite. p-Amino-N-(2-dipropylaminoethyl)benzamide and its N-acetyl derivative were added to the plasma as internal standards prior to extraction. A solvent system (ethyl acetate–methanol–ammonia: 88 + 8 + 4) was selected that allowed excellent separation of the four compounds from each other, from

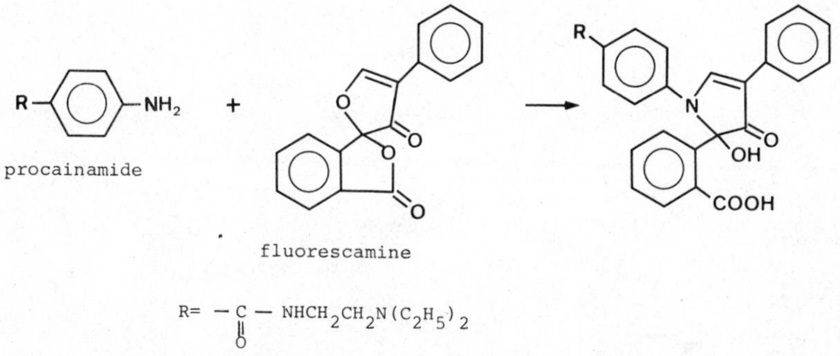

R= — $\underset{\underset{O}{\|}}{C}$ — NHCH$_2$CH$_2$N(C$_2$H$_5$)$_2$

Fig. 2.4. Reaction between procainamide and fluorescamine. (From ref. 341.)

plasma constituents, and from other drugs commonly prescribed with procainamide. The authors demonstrate that densitometric scanning in the fluorescent mode at the specific wavelength maxima is a precise and specific technique to quantitate procainamide and N-acetylprocainamide separated by TLC.

Tan and Beiser (341) developed a rapid fluorometric determination of procainamide hydrochloride dosage forms offering improvements in ease, speed, and sensitivity over the official method (diazotization titration with standard sodium nitrite at low temperatures or nonaqueous titrimetry on the evaporated and dried sample with standard perchloric acid). Their procedure is based on the reaction with fluorescamine in aqueous medium at pH 7.5 to form a fluorophor, with activation and emission wavelengths of 400 and 485 nm, respectively (Fig. 2.4).

The fluorescence is linear over the $0.04-1$ μg ml^{-1} concentration range and is stable for at least 2 h. Recovery data appear to be accurate, quantitative, and reproducible. Their method was successfully applied to commercially available dosage forms.

A thin-layer chromatographic method for the estimation of disopyramide and its metabolite, mono-N-dealkylated disopyramide, was described by Gupta et al. (342). The method involves a simple extraction of 200 μl of plasma at alkaline pH with benzene. An aliquot of the extract is quantitatively spotted on a TLC plate. The plate is developed in a solvent (ethyl acetate–methanol–ammonia: $40 + 30 + 0.5$), which allows good separation of the two compounds from one another, from plasma constituents, and from other drugs that are prescribed along with disopyramide. The plate is dipped in 20% sulfuric acid in methanol to make these compounds fluorescent. Densitometric scanning ($\lambda_{ex} = 266$ nm; $\lambda_{em} = 405$ nm) of the separated compounds provides a precise quantitative

R= -CH(CH₃)₂ : disopyramide
R= H : mono-N-dealkylated disopyramide

estimation. According to the authors, a concentration of 0.5 mg liter^{-1} of plasma for both compounds can be detected.

A fluorometric assay of the coronary active compound nifedipine (4-(2'-nitrophenyl)-2,6-dimethyl-3,5-dicarbomethoxy-1,4-dihydropyridine) in plasma was described (343) involving reduction of the nitro group to an amino group by $TiCl_3$ and the oxidation of the dihydropyridine ring to the pyridine ring by exposure at 360 nm. Coupling of this amino derivative with o-phthaldialdehyde leads to a product with increased fluorescence intensity. The differentiating determination of nifedipine and its main metabolite (4-(2'-nitrophenyl)-2-hydroxymethyl-5-methoxycarbonyl-6-methyl-pyridine-3-carboxylic acid) in plasma can be obtained by fractionated extraction of both compounds, at pH 7.5 for nifedipine and at pH 2.0 for the metabolite. The assay for the metabolite runs in the same way as that for the parent drug. This main metabolite is excreted in the urine and can be assayed there. No nifedipine is excreted in the urine.

Verapamil (5-[3,4-dimethoxyphenethyl)methylamino]-2-(3,4-dimethoxyphenyl)-2-isopropylvaleronitrile), a coronary vasodilator and antiarrhythmic agent, can be assayed fluorometrically in blood, urine, or tissue homogenates in the 0.1–10 µg ml^{-1} range after extraction into heptane and back-extraction into acid (344).

Spectrofluorometric determination of alkoxy-substituted benzylimidazolidinones in biological fluids was reported by de Silva's group in 1978

nifedipine   main metabolite

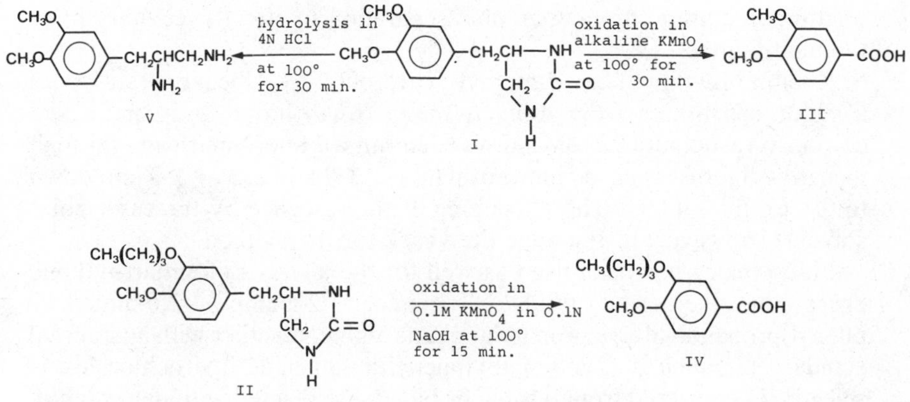

Fig. 2.5. Reaction of d,1-4-(3,4-dimethoxybenzyl)-2-imidazolidinone (I) and d,1-4-(3-butoxy-4-methoxybenzyl)-2-imidazolidinone (II) with alkaline permanganate. (From ref. 345.)

(345). Assays were developed in blood and urine for d,l-4-(3,4-dimethoxybenzyl)-2-imidazolidinone (I), a compound of clinical interest because of its cardiovascular inotropic effects as a vasodilator and for increased cardiac output and as a hypolipemic agent, and for d,l-4-(3-butoxy-4-methoxybenzyl)-2-imidazolidinone (II), a phosphodiesterase inhibitor. Although both compounds are aromatic, neither their UV absorption, intrinsic fluorescence, nor phosphorescence was sufficiently intense to enable their quantitation as the intact moiety. Investigation of possible derivatization reactions indicated that the oxidation in alkaline permanganate, involving the cleavage of the methylene bridge in these drugs, produces strongly fluorescent (and phosphorescent) benzoic acid derivatives (III) and (IV) (Fig. 2.5). The method was applied after selective extraction of each compound from buffered blood or urine. The fluorescent derivatives were dissolved in acetic acid in ethanol (1:99) and quantitated at 345 nm when exciting at 295 nm. The sensitivity limit was 0.25 μg of (I) or 0.10 μg of (II) per ml of blood or urine. The linear concentration range was 0.10–25 μg per 3 ml of final solution. Investigation of possible derivatives that could be used to measure therapeutic levels of (I) led to acid hydrolysis of (I) in 4 N HCl to 3-(3,4-dimethoxyphenyl)-2-aminopropylamine (V). This compound exhibited weak fluorescence properties similar to intact (I) and was therefore not investigated further by the authors. Alkaline hydrolysis did not yield usable derivatives. The assay for (I) or (II) in blood is specific for the respective parent compounds because of selective extraction and the absence of interfering metabolites. The analysis of urine for the free or directly extractable and the conjugated fractions required chromatographic separation to effect specificity prior to quantitation.

Khellin (4,9-dimethoxy-7-methyl-5H-furo[3,2-g]-[1]benzopyran-5-one) a major constituent from *Ammi visnaga* fruit, shows spasmolytic and coronary vasodilator action and was determined fluorometrically through its native fluorescence in chloroform (λ_{ex} = 332 nm; λ_{em} = 453 nm) down to 0.1 μg ml^{-1} (346). The quenching of fluorescence by traces of polar solvents, however, is a notable drawback for the method.

Fluorometry has been used as well for the analysis of antiarrhythmic drugs in biological fluids. Tocainide (2-amino-N-(2,6-dimethylphenyl)propanamide), a primary aliphatic amine, together with an internal standard (2-amino-N-(2,6-dimethylphenyl)butanamide hydrochloride), is selectively extracted from plasma or blood and reacted with dansyl chloride, followed by HPLC of the highly fluorescent dansyl derivatives and fluorescence detection (λ_{ex} = 360 nm; λ_{em} = 490 nm) (347). Drug concentrations of 0.1–5.0 μg ml^{-1} of plasma can be measured.

Muscle Relaxants   Dantrolene sodium (1-{[5-(p-nitrophenyl) furfurylidene]amino}hydantoin sodium hydrate), a muscle relaxant with direct and selective action on skeletal muscle, was originally determined in blood and urine by a spectrophotofluorometric procedure (348). As it was established afterward that a metabolite interfered in the original fluorometric procedure, the same authors modified their procedure and determined dantrolene in biological specimens containing drug-related metabolites (349). Method specificity for dantrolene is based on a combination of solvent extractions, chromatography, and the fluorescence characteristics of dantrolene (λ_{ex} = 395 nm; λ_{em} = 530 nm). No appreciable fluorescence is obtained when known dantrolene-related metabolites (acetylated dantrolene; metabolite A, a major nonreduced metabolite) are subjected to the method. The proposed method appears to be satisfactory for use with plasma, blood, urine, bile, tissues, and feces. The authors provide as well modifications of the method for estimating the two dantrolene-related metabolites in biological fluids. In 1977 a high-pressure liquid chromatographic method (350) was developed for the qualitative and quantitative determination of dantrolene sodium in dosage forms, detection was achieved at 375 nm and the results were compared to those of a spectrophotofluorometric method, modified from that reported by Hollifield and Conklin (349).

Stewart and Chan (351) developed a sensitive fluorometric determination for chlorzoxazone, a skeletal muscle relaxant, based on its intrinsic fluorescence in chloroform (λ_{ex} = 286 nm, λ_{em} = 310 nm). The technique is suitable for the determination of the drug in biological fluids and is also useful for the analysis of tablets. Data revealed that chlorzoxazone could be determined in plasma and urine even in the presence of 20-fold molar excesses of its major metabolite, 6-hydroxychlorzoxazone, and acetaminophen, often used in combination with this drug. In another paper (352), the authors reported the analysis of chlorzoxazone using derivatization with various fluorogenic reagents.

A rapid fluorometric method for the estimation of fazadinium bromide (1,1'-azobis-3-methyl-2-phenyl-1H-imidazo-[1,2-a]-pyridinium dibromide), a neuromuscular-relaxant drug, in dogs, was described by Pastorino (353). The drug is hydrolyzed by alkali to give 3-methyl-2-phenyl-1H-[1,2-a]-imidazopyridinium bromide, which is extracted in cyclohexane and measured for fluorescence (λ_{ex} = 325 nm; λ_{em} = 390 nm). The fluorometric assay appears to be a rapid, specific, and precise method, useful in pharmacokinetic studies.

Pancuronium bromide, a bisquaternary ammonium steroid that contains acetate ester groups at the 3- and 17-positions, is used as a muscle relaxant during anesthesia for caesarean section. Quantitation of maternal and fetal serum pancuronium bromide concentrations is therefore important for evaluating transplacental drug transfer. Its metabolism produces the 3-hydroxy, 17-hydroxy-, and 3,17-dihydroxy derivatives through ester hydrolysis. A sensitive fluorometric determination for pancuronium bromide and its metabolites in human plasma (354) and a modification to stabilize the fluorescence (355) have been reported. Wingard et al. (356) modified the procedure for pancuronium bromide ion-pair extraction into chloroform using rose bengal and subsequent fluorometric measurement by changing the extraction pH and eliminating phenol, ethanol, and acetone to give easier operation and enhanced fluorescence stability (λ_{ex} = 546 nm; λ_{em} = 570 nm). Precision, accuracy, and sensitivity of the method were evaluated.

Psychopharmaceutics.   Koechlin and D'Arconte (357) reported in 1963 the determination of chlordiazepoxide (I), a 1,4-benzodiazepine derivative, and its lactam metabolite demoxepam (II) in plasma of humans, dogs, and rats by a specific spectrofluorometric micro method. It is based on a selective extraction of the compound followed by controlled hydrolysis to the lactam derivative (II), which is also a major metabolite of chlordiazepoxide, that is converted under the influence of light to a compound exhibiting a characteristic fluorescence in alkaline solution (λ_{ex} = 380 nm; λ_{em} = 480 nm). The lactam can be measured separately by a modification of the procedure. The sensitivity limit is 0.25 μg ml^{-1} of plasma for either compound. The method was found to be adequate for measuring plasma levels of chlordiazepoxide as encountered after normal therapeutic doses

I   II   III

in man (10–30 mg). A rearrangement of the *N*-oxide function to a 4,5-epoxide under the influence of light is assumed. The method of Koechlin and D'Arconte was modified in 1966 (358) to include another metabolite, desmethylchlordiazepoxide (III). This determination is based on the quantitative conversion of (I) and (III) by controlled hydrolytic conditions, to (II), which is, as mentioned, converted with light to a compound exhibiting a characteristic fluorescence in alkaline solution. In 1976 Stewart and Williamson (359) reported a fluorometric procedure for chlordiazepoxide based on fluorophor formation with fluorescamine after acid hydrolysis of the drug, applicable to the analysis of pharmaceutical dosage forms, urine, and plasma, with an accuracy of 1–3%. Their method is free from interference from chlordiazepoxide's major metabolites as well as other drugs, such as clidinium bromide, oxazepam, and dextropropoxyphene hydrochloride. According to the authors, the procedure is subject to interference from other amine-containing drugs, such as amphetamines that might be present in analytical samples. In 1978 a sensitive fluorescence TLC-densitometric procedure was developed (360) for the specific determination of chlordiazepoxide (I) and its two metabolites, demoxepam (II) and desmethylchlordiazepoxide (III), in serum. After extraction from serum with ether, (I), (II), and (III) were separated by TLC and converted with a sulfuric acid spray to pale blue (Rf 0.63), green (Rf 0.54), and blue (Rf 0.45) fluorescent spots, respectively. Quantitation was accomplished by scanning the plate with a densitometer at 390 (I), 430 (II), and 390 (III) nm. The sensitivities were 0.05 (I), 0.01 (II), and 0.01 (III) μg ml^{-1} of serum. The author applied the method successfully to the measurement of (I)–(III) in human serum after oral administration of 20 mg of chlordiazepoxide hydrochloride. The sensitivities reached should allow the monitoring of the blood levels of each compound in bioavailability studies.

Thin-layer chromatographic separation and spectrophotofluorometric determination of 60 psychotropic pharmaceuticals of the groups of tricyclic neurothymoleptics, butyrophenones, and benzodiazepines, suitable for clinical and toxicological problems, were described by Lauffer et al. (361) in 1969.

A thin-layer chromatographic–spectrophotofluorometric determination of 7-chloro-2,3-dihydro-1-methyl-5-phenyl-1H-1,4-benzodiazepine (medazepam, Nobrium®) applicable for the analysis of gastric juice and blood plasma was described by Lauffer and Schmid (362).

Braun et al. (363) studied the fluorescence of some 1,4-benzodiazepine derivatives. In sulfuric, phosphoric, or perchloric acid, the alcoholic solutions of chlordiazepoxide, diazepam, nitrazepam, and oxazepam, develop a very intense fluorescence, peculiar to each compound, which permits a quantitative determination with a limit of sensitivity varying from 0.005 to 0.100 μg ml^{-1}. 2-Amino-5-chlorobenzophenone and 2-methylamino-5-chlorobenzophenone, acid hydrolysis products from chlordiazepoxide, diazepam, and oxazepam, do not exhibit measurable fluorescence under the analytical conditions. The N-oxilactam metabolite from chlordiazepoxide fluoresces about three times weaker than the latter, under analogous conditions.

Caille et al. (364) applied the spectrofluorometric method described by Braun et al. (363) for the quantitative determination of 1–4 benzodiazepine derivatives in several pharmaceutical dosage forms sold in Canada and compared the method with a polarographic technique. According to the authors, both methods can be compared advantageously but spectrophotofluorometry appeared to be much more specific for the determination of oxazepam. Neither method could be applied for the direct quantitative determination of the 1,4-benzodiazepine derivatives when mixed with other substances in the same dosage form.

de Silva and Strojny (365) developed a spectrophotofluorometric assay for the determination of the hypnotic flurazepam ([7-chloro-1-(2-diethylaminoethyl)-5-(2-fluorophenyl)-1,3-dihydro-2H-1,4-benzodiazepin-2-one·2HCl) (I) and its major metabolites in blood and urine. In the introduction of their excellent article a literature survey is given on the metabolism of flurazepam stating that this drug is extensively metabolized in human and in dog to yield measurable amounts of the hydroxyethyl (II) and N-desalkyl metabolite (III) in the blood, which is further metabolized to the 3-hydroxy analog (IV) present in trace amounts in urine as a glucuronide conjugate. In addition, (I) and (II) are metabolized to an acidic compound (V) in the dog by the oxidation of the alcohol side chain to a carboxylic acid (Fig. 2.6). The 1,4-benzodiazepin-2-ones all undergo acid hydrolysis to yield their respective benzophenones, which in turn are cyclized in an alkaline medium to yield the 9-acridanone derivatives. The assay described by de Silva and Strojny for the determination of flurazepam and its major metabolites in blood and urine involves the fol-

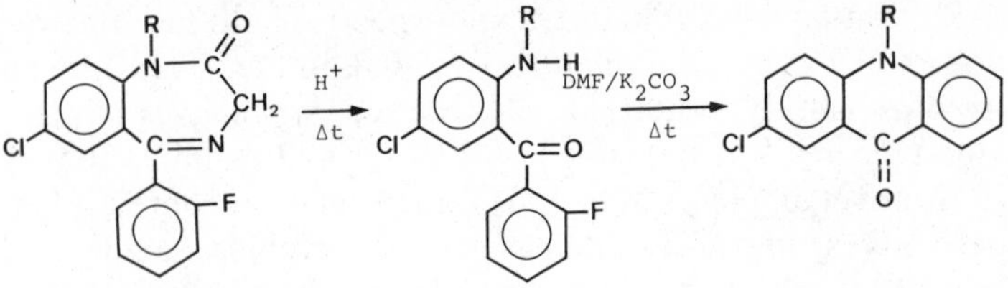

R-group   1,4-benzodiazepin-2-one   benzophenone   9-acridanone

$-(CH_2)_2-N-(C_2H_5)_2$	I	VI	X
$-CH_2-CH_2OH$	II	VII	XI
$-H$	III	VIII	XII
$-H$	IV, 3 > CHOH	VIII	XII
$-CH_2COOH$	V	IX	XIII

Fig. 2.6. Chemical reactions of N-1-alkyl-substituted 2′-fluoro-1,4-benzodiazepin-2-ones. (From ref. 365.)

lowing: selective extraction into diethyl ether from blood buffered to pH 9.0 or urine made alkaline with NaOH, and back-extraction into 4 N HC1 followed by hydrolysis at 100° for 2 h to yield the respective benzophenones. The benzophenones are extracted into diethyl ether (from the hydrolysate after making it alkaline with NaOH), the residue of which is reacted in dimethyl formamide, using K_2CO_3 as a catalyst at 100° for 2 h to effect the cyclization to the 9-acridanone derivatives. These derivatives are extracted into diethyl ether and separated by TLC, and their fluorescence is determined in 80:20 methanol–0.1 N HC1 (after elution from the silica gel) at their respective maxima of activation and emission. The fluorescence yield of the 9-acridanones is sufficiently high and linear with concentration to enable their quantitation in the range of 0.003–10.0 μg of compound per ml of blood or urine. The assay was modified to employ selective extraction procedures for the determination of flurazepam and its major metabolites in the free and bound (conjugated) forms in urine, either by spectrophotometry as their benzophenones or by spectrophotofluorometry as their respective 9-acridanones. TLC was used to separate the different reaction products, thus imparting the required specificity for the quantitation of each compound. The authors applied their method to the determination of blood levels and urinary excretion in a dog following intravenous and oral administration and in two human subjects following the oral administration of flurazepam·2HC1.

A fluorometric method for determining nitrazepam (1,3-dihydro-7-nitro-5-phenyl-2H-1,4-benzodiazepin-2-one) (I), a potent sleep inducer

I

with anticonvulsant properties, and the sum of its two main metabolites (i.e., the 7-amino and the 7-acetamido derivative) in plasma and urine was reported by Rieder (366) in 1973. After deproteinization the two metabolites are extracted from the sample, hydrolyzed with 6 N HCl, producing mainly 2,5-diaminobenzophenone, which reacts with o-phthalaldehyde, producing strong fluorescence that is measured. The unchanged nitrazepam is reduced at room temperature with zinc–HCl and the 2,5-diaminobenzophenone, produced by hydrolysis of nitrazepam, is reacted with o-phthalaldehyde. The method provides a sensitivity of 0.01 μg ml^{-1} of sample and is suited for pharmacokinetic studies with therapeutic doses of the drug used in man. In 1981 the method based on treatment with zinc–hydrochloric acid followed by reaction with o-phthalaldehyde was applied to the determination of oxazepam, lorazepam, dipotassium chloroazepate, demoxepam, and chlordiazepoxide (367).

A spectrofluorometric method for the determination of oxazepam was reported by Caille et al. in 1973 (368). In phosphoric acid solution at 90°C, oxazepam develops a very intense fluorescence that permits a quantitative determination with a sensitivity limit of 5 ng ml^{-1} (λ_{ex} = 360 nm; λ_{em} = 475 nm). The fluorescent compound formed during this treatment was isolated by precipitation of the preparative reaction mixture by the addition of water followed by recrystallization of the derivative out of ethanol–water, yielding a solid melting at about 312°C. Spectral and chemical evidence showed that the compound formed is the trimer of 6-chloro-4-phenyl-quinazoline-2-carboxaldehyde (Fig. 2.7).

Brinkman et al. (369) described in 1981 a sensitive and selective detection for low levels of demoxepam after separation by means of reversed-phase HPLC and short-wavelength UV conversion of the tranquilizer into a highly fluorescent product (370) with a detection limit of about 100 pg. They demonstrated as well that detection limits of 40-100 pg can also be obtained for the photoproducts of the phenothiazines fenergan, largactil, levopromazine, and nedaltran.

In 1974 a spectrodensitometric method was reported for the simultaneous analysis of a 1,5-benzodiazepine and its metabolite in blood and urine (371).

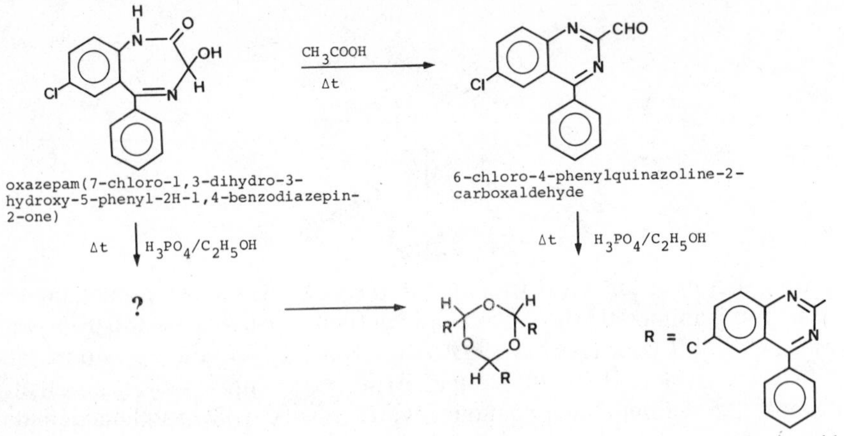

oxazepam(7-chloro-1,3-dihydro-3-
hydroxy-5-phenyl-2H-1,4-benzodiazepin-
2-one)

6-chloro-4-phenylquinazoline-2-
carboxaldehyde

Fig. 2.7. Conversion of oxazepam into a fluorescent derivative through phosphoric acid treatment in the heat. (From ref. 368.)

Clobazam   (7-chloro-1-methyl-5-phenyl-1H-1,5-benzodiazepine 2,4(3H,5H)-dione) (I), a new antianxiety agent from the 1,5-benzodiazepine class, was determined fluorometrically in human plasma by Stewart et al. (372). The method is based on fluorophor formation upon irradiation of the drug with short-wavelength UV light (254 nm) for 35 min. The obtained fluorescence is linear over a 0.1–6.4 μg ml⁻¹ range using excitation and emission wavelengths of 350 and 400 nm, respectively; the detection limit is 50 ng ml⁻¹. The procedure is specific for clobazam in samples containing its major metabolite, N-desmethylclobazam (II), and also in samples containing 1,4-benzodiazepines and other selected drugs.

Protriptyline, one of the widely used tricyclic antidepressants, was determined in plasma by a simple, sensitive, and specific fluorometric method (373), free of interference from most other commonly used psy-

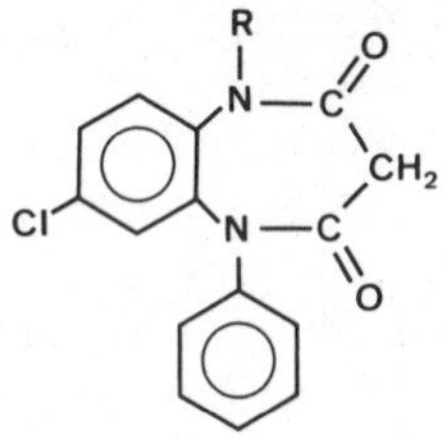

I : R=CH₃

II : R=H

methaqualone

chiatric drugs. The method is based on extraction of the drug from the matrix and measurement of the native fluorescence in dilute sulfuric acid (λ_{ex} = 240,300 nm; λ_{em} = 360 nm).

Schulman et al. (374) reported the native fluorescence of methaqualone (2-methyl-3-o-tolyl-4(3H)-quinazolinone). The authors give a summary of the existing detection and determination methods for this hypnotic drug, among which a time-consuming fluorometric procedure (375) based on reduction, under anhydrous circumstances, with lithium borohydride, to yield the fluorescent dihydromethaqualone. Although it had been reported that methaqualone does not possess native fluorescence (375), Schulman et al., in fact, did observe fluorescence in a variety of solvents, such as water, sulfuric acid, methanol, and chloroform, and gave a discussion on the nature and analytical possibilities of this phenomenon.

Meilink (376) describes a fluorometric assay of carbamazepine and its metabolites in serum or plasma using a simplified method for the conversion of carbamazepine into a fluorescent derivative. After extraction from serum or plasma, thin-layer chromatography is used to separate carbamazepine from its metabolites. Without the thin-layer chromatographic step the concentrations of carbamazepine and of metabolites are determined together. Different parameters in the analysis have been examined. Carbamazepine (Tegretol®), a 5H-dibenz-(b,f)-azepine derivative, is chemically related to imipramine, from which it differs by a double bond at the 10,11 position in the azepine ring and by the nature of the side chain attached to the nitrogen. Carbamazepine is determined by its fluorescence on ceric sulfate treatment (λ_{ex} = 400 nm; λ_{em} = 475 nm).

carbamazepine   imipramine

Linear fluorescence response is obtained in concentrations from 12.5 to 5000 ng ml^{-1}. The fluorescing conversion product was not identified.

Phenelzine (β-phenylethylhydrazine), an antidepressant, does not possess a sufficiently conjugated system to produce analytically useful fluorescence. Caddy and Stead (377) tried to improve detection limits for this drug by preparing pyridazine, phthalazinone, and some substituted phthalazinone derivatives; reagents investigated for their ability to form fluorescent products with phenelzine were p-dimethylaminobenzaldehyde and Fluram. While linear relationships between the intensity of fluorescence and concentration for the pyridazine, the phthalazinedione, and the dihydroxyphthalazinedione derivatives have been established, the limits of detection preclude the use of these compounds for the assay of the drug in urine and were therefore not investigated as procedures that might be used for the much lower levels of phenelzine expected to be present in blood. The methods described may have some application for solid dosage forms. The authors developed a fairly specific and sensitive test, using p-dimethylaminobenzaldehyde, for the identification of phenelzine in urine. A similar test based on the development of a fluorescent product with Fluram is also sensitive but lacks the specificity of the former test. Following the method of De Bernardo et al. (378), no fluorescing products could be obtained with Fluram in aqueous solutions. According to Caddy and Stead, the reason for this may lie in the fact that, whereas primary amines react with Fluram to form a stable, rigid, five-membered ring compound, hydrazines, as a result of possessing adjacent active nitrogens, can more readily form the more flexible six-membered substituted pyrazine (Fig. 2.8). If this is correct, it might be expected that the physical imposition of rigidity to this structure may produce a fluorescent species. By dropping a solution of fluram in acetone onto an air-dried sample of phenelzine in solution on a filter paper, buffered at pH 5–6, a fluorescent product can be observed when the paper is examined under long-wavelength ultraviolet light. As the compound fluoresces only on solid support and not in solution, fluorescence photodensitometry on thin-layer plates can be performed.

Butaclamol, a benzo[6,7]cyclohepta[1,2,3-de]-pyrido [2,1-a] isoquinoline neuroleptic (I), was analyzed in serum by fluorescence induction by Hasegawa et al. (379). Butaclamol does not exhibit natural fluorescence, but its oxidation products are highly fluorescent. The method involves cyclohexane extraction of serum samples followed by TLC of the concentrated extracts. The developed TLC plates (chloroform–ethyl acetate–ethanol–formic acid: 36 + 36 + 25 + 3) are sprayed with an oxidizing reagent (manganese dioxide and zinc chloride in nitric acid–hydrochloric acid mixture) and heated at 110°. Highly fluorescent spots were

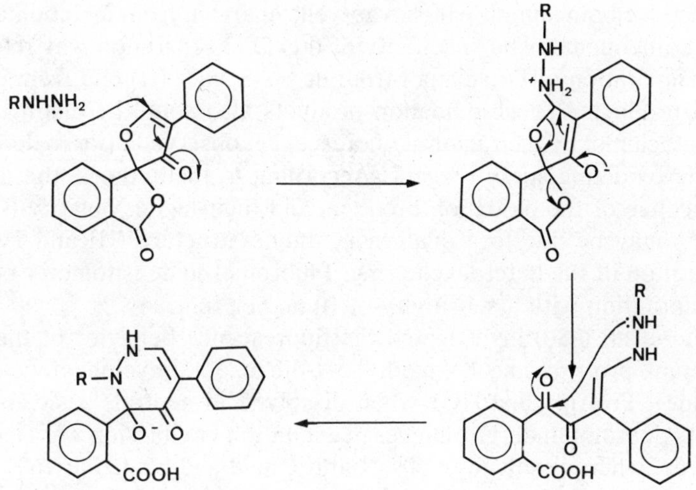

Fig. 2.8. Suggested mechanism for the reaction of hydrazines with Fluram according to Caddy and Stead (377).

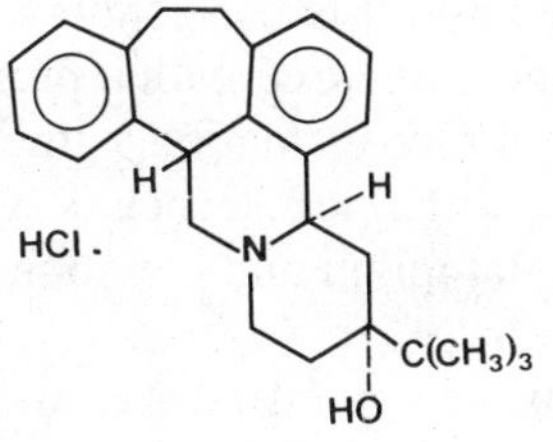

I

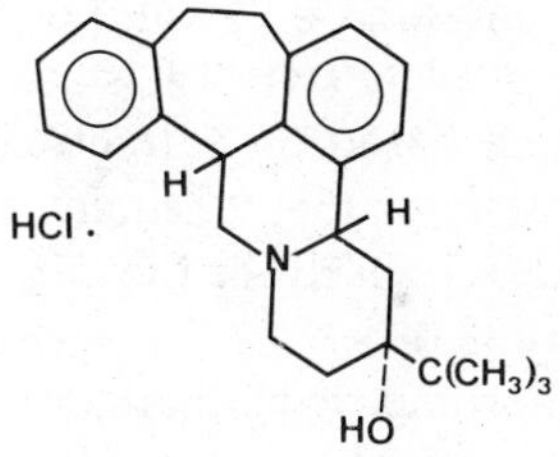

II

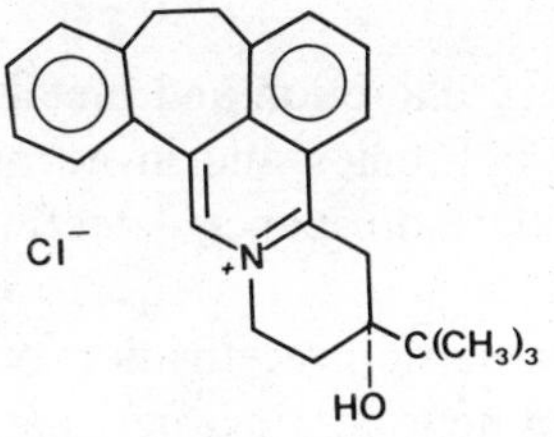

III

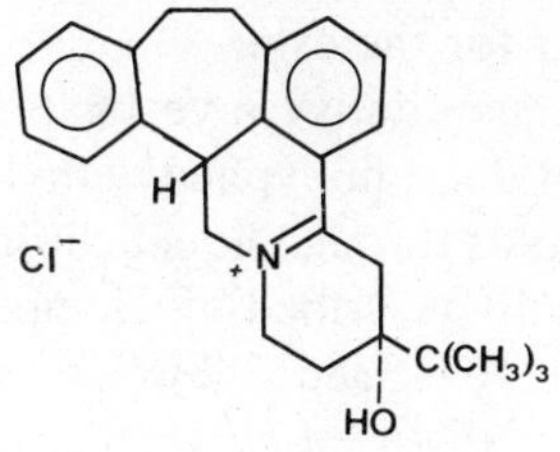

IV

produced for butaclamol, which was well separated from metabolites and serum components. The specificity of the TLC separation was reflected by the separation of butaclamol from its *cis*-isomer (II) and from its oxidation products. These oxidation products also could be distinguished from butaclamol by their natural fluorescence, observed prior to detection with the oxidizing spray reagent. According to the authors, the natural fluorescence of the oxidation products and their higher polarity (low R_f in TLC) may be due to a quaternary amine structure (III and IV) and unsaturation of the heterocyclic ring. Fluorometric densitometry permitted quantitation with a sensitivity of 10 ng per spot.

Caille et al. (380) investigated the fluorescence behavior of the antidepressant protriptyline (*N*-methyl-5H-dibenzo[a,d]cycloheptene-5-propylamine). Protriptyline·HC1 when dissolved in neutral, basic, or acid solvents does not undergo changes in excitation or emission wavelengths. However, when dissolved in 50% sulfuric acid and heated at 160° for 30 min, there is a shift toward the long wavelengths and a marked rise in the sensitivity (0.005 μg ml^{-1}, compared to 3 μg ml^{-1} for the ultraviolet spectrophotometry).

A method for the fluorescence assay of citalopram, a specific, potent serotonin reuptake inhibitor under clinical trials as an antidepressant, was developed by Overø (381). Direct scanning of the native fluorescence (λ_{ex} = 240 nm; λ_{em} = 295 nm) of drug and metabolite (Fig. 2.9) spots on the thin-layer plate (HPTLC silica gel) is used and compared with a previous technique based on extraction, TLC, and ion-pair formation with a fluorescent anion (382). Liquid chromatography with fluorescence detection was recently used for the determination of citalopram and metabolites in plasma (383).

A fluorometric method for the simultaneous and specific assay in serum of clopenthixol, its decanoate, and a clopenthixol metabolite after i.m. injection of clopenthixol decanoate was reported by Overø (384). The separation is achieved by extractions and TLC and the fluorescence is developed by H_2SO_4 treatment. The limit of detection in 3 ml serum samples is about 2 ng ml^{-1} for clopenthixol and metabolite, and somewhat higher for the ester.

A quantitative investigation on the urinary excretion and metabolism of 3,4-dimethoxyphenylethylamine in schizophrenics and in normal individuals through liquid chromatography with fluorescence detection was recently described by Hempel et al. (385).

Baeyens (386–388) made a spectrofluorometric investigation of therapeutically used butyrophenones, a series of neuroleptic drugs that show pharmacological action at low concentrations. Butyrophenones can be

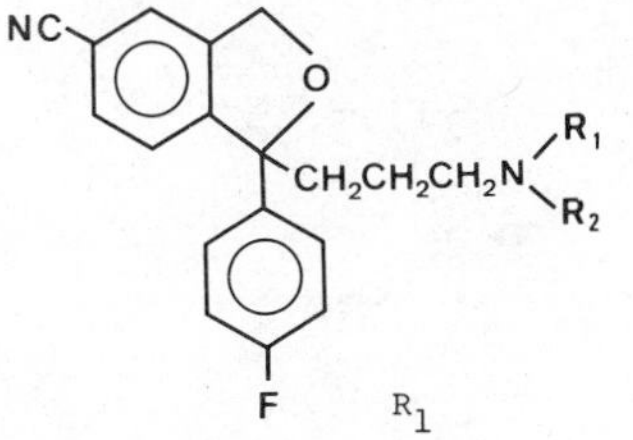

	R_1	R_2	
Citalopram	CH_3	CH_3	
Monodesmethyl metabolite	CH_3	H	
Didesmethyl metabolite	H	H	
Citalopram N-oxide	CH_3	CH_3	$-N \rightarrow O$

Fig. 2.9. Citalopam and its metabolites. (From ref. 381.)

considered as 4-aminobutyric acid derivatives with the general structure shown in Fig. 2.10 together with the most important derivatives.

The general structure of these compounds can be represented by the formula *P-H*, in which *P* is the 4'-fluorobutyrophenone moiety and *H* is a heterocyclic substituent replacing one hydrogen atom of the end methyl group. Moiety *P* shows no fluorescent properties. Moiety *H* usually fluoresces unless there is no conjugated system, as in pipamperone. Consequently, the observed fluorescence of the butyrophenones is caused by moiety *H*. Moiety *P* phosphoresces strongly (triplet state) with an emission maximum at 415 nm (τ = 0.004 s). Moiety *H*, when conjugated, phosphoresces with an emission maximum below 415 nm. All *P-H* derivatives phosphoresce at 415 nm ($\tau < 0.1$ s). The ketonic 4'-fluorobutyrophenone is strongly reactive in its π-excited triplet state. Butyrophenones are fluorescent, ranging from weakly to strongly intense, when irradiating below 350 nm. The intensity of fluorescence and the position of the maxima depend on the nature of the solvent used. The fluorescence properties of the strong butyrophenone fluorophors have been used for the determination of these compounds in pharmaceutical preparations after suitable dilution by means of buffer–solvent mixtures. The quenching interference as a result of the ultraviolet absorption by the 4-hydroxybenzoic acid ester preservatives in the liquid preparations is eliminated by extraction with diethyl ether in acidic medium. Detection limits of 0.05 μg ml^{-1} can be reached for some representatives. Trifluperidol, pipamperone, and azaperone can be converted into strong fluorophors by the action of sulfuric acid on their aqueous or alcoholic solutions.

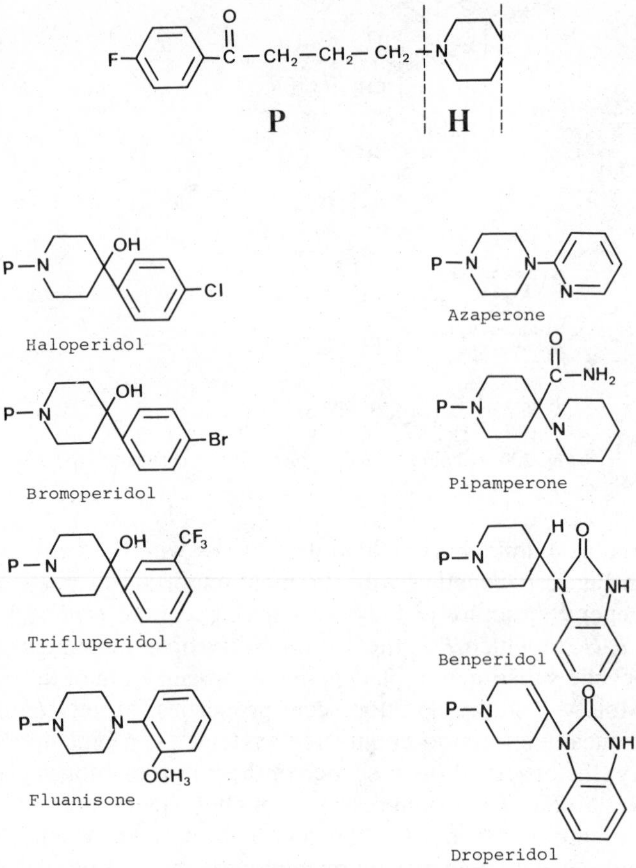

Fig. 2.10. General butyrophenone structure and structures of some important derivatives. (From ref. 386.)

Baeyens et al. (389) studied the influence of N-oxidation on the native fluorescence characteristics of azaperone (I), a butyrophenone neuroleptic frequently used in veterinary medicine. Azaperone fluoresces intensely in acid medium at pH 2 as well as in various organic solvents, which can be worked out quantitatively. By treatment of azaperone with an excess of hydrogen peroxide (Fig. 2.11), a mononitrogen oxide at the piperazine N1 atom (II) was formed with identical fluorescent properties as the parent molecule. The nucleus of electronic transitions leading to fluorescence emission is situated in the aromatic aminopyridine substituent, which was confirmed by similar experiments on pure 1-(2-pyridyl) piperazine yielding

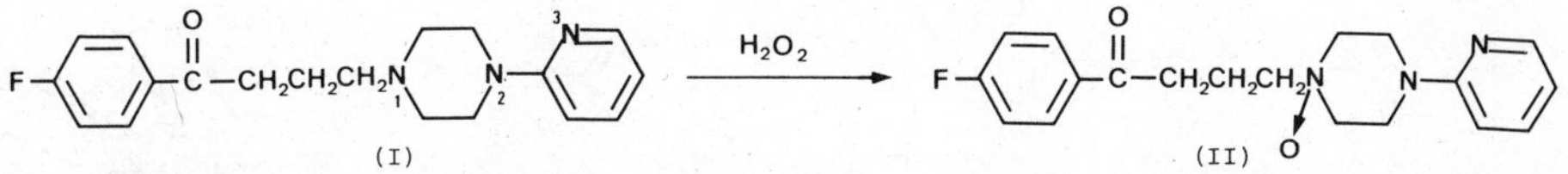

Fig. 2.11. Reaction of azaperone with excess hydrogen peroxide. (From ref. 389.)

analogous spectra and on 1-(4-fluorophenyl)-1-butanone showing no fluorescing properties at all, as expected. Modification of the aminopyridine ring system would result in an alteration of the fluorescent properties of azaperone by direct influence on the π-electrons of the aromatic system.

De Croo et al. (390) studied the influence of gases (air, nitrogen, helium) on the native fluorescence of some butyrophenones on thin-layer plates, as quenching effects due to oxygen are noticeable at lower excitation wavelengths. It was concluded that the use of nitrogen flushing onto silica gel layers leads to an increase in fluorescence, with the intensity of the effect, however, depending on the nature of the compound used. Nitrogen flushing did not affect the linearity of the measurements when scanning different concentrations; hence their suggestion may be used to enhance fluorescence of butyrophenones in quantitative work.

A fluorometric method was described by Baeyens and De Moerloose (391) for direct and sensitive determinations of azaperone, haloperidol, and bromoperidol in pharmaceutical preparations. These butyrophenones can be converted into stable and strongly fluorescing derivatives by the action of potassium permanganate on their alcoholic solutions in acid medium, the limits of detection (and wavelength maxima) being, respectively, 8×10^{-3} ($\lambda_{ex} = 309$ nm; $\lambda_{em} = 377$ nm), 5×10^{-2} ($\lambda_{ex} = 305$ nm; $\lambda_{em} = 383$ nm) and 5×10^{-2} μg ml^{-1} ($\lambda_{ex} = 310$ nm; $\lambda_{em} = 392$ nm). The authors report the liquid chromatographic isolation and structure determination of one of the main fluorophors, produced in the case of azaperone.

Baeyens (392) determined pimozide (1-{1-[4,4-bis(4-fluorophenyl)butyl]-4-piperidinyl}-1,3-dihydro-2H-benzimidazol-2-one) in oral preparations by its intense native fluorescence in diethyl ether ($\lambda_{ex} = 290$ nm; $\lambda_{em} = 319$ nm). Pimozide is used as a long-acting antipsychotic neuroleptic drug with relatively low toxicity and shows structural similarity to benperidol. Both molecules contain the benzo-1,3-diazolin-2-one fluorophor, producing intense emission at about 320 nm when excited around 285 nm. A linear response for pimozide in diethyl ether is obtained up to 10 μ ml^{-1}, with a detection limit of 10 ng ml^{-1}.

Fluorometry of phenothiazines has been reported in literature on various occasions (393–397). Manier et al. (398) developed a fluorometric method for the measurement of butaperazine in human plasma following

pimozide   benperidol

its extraction by a three-step procedure and a buffer wash to eliminate interfering metabolites; a sensitivity of 20 ng ml^{-1} plasma was achieved. An automated fluorescence method was described for the determination of perphenazine (399) in large numbers of samples based on the reaction with permanganate according to Mellinger and Keeler (393) to form a fluorophor (λ_{ex} = 274 nm; λ_{em} = 380 nm) with a sensitivity of about 0.1 μg ml^{-1}. White et al. (400) described in 1976 a rapid fluorometric determination of phenothiazines by in situ photochemical oxidation. Perphenazine, chlorpromazine, fluphenazine, trifluoperazine, and thioridazine, on exposure to an intense ultraviolet source, are photo-oxidized and the products are simultaneously excited. According to the authors this photochemical–fluorometric method is unique, as no addition of diluting reagents is needed and detection limits are lowered (10 and 1000 ppb of perphenazine, chlorpromazine, fluphenazine, and trifluoperazine). Chlorpromazine can be determined fluorometrically by the dansylation method (401,402) and 11 metabolites can be quantitated (403). Kaul et al. (404) determined nonconjugated chlorpromazine metabolites in blood of schizophrenic patients using the dansylation method and described some problems relevant to isolation of the metabolites prior to quantitation. Their optimum procedure of assay uses only a 6-ml blood sample. A new sensitive quantitative fluorometric determination was developed in 1976 for chlorpromazine and its sulfoxide (405). Reaction of the tertiary amine base with 9-bromomethylacridine forms a quaternary compound that on photolysis produces highly fluorescent products that are determined fluorometrically (λ_{ex} = 350 nm; λ_{em} = 474 nm). The method is capable of determining 15–20 ng of the drug with good accuracy and is specific for tertiary amine drugs.

Steroids.   Estrogenic steroids are chemically characterized by an aromatic ring A and a hydroxyl group at carbon 3, which is consequently phenolic. This phenolic group provides estrogenic steroids chemical and physical properties that would make their isolation from blood and urine and their quantitative estimation easier than that of other steroids; however, estrogens occur in much lower concentrations than other steroids, with the exception of the stallion and during pregnancy. Fluorometric measurements of estrogens after acid treatment have shown marked superiority over existing methods (infrared, visible, or ultraviolet absorption of pure compounds or their derivatives) with respect to the amount required for a quantitative determination. The sulfuric acid methods have been preferred over the phosphoric acid method by most workers partly because they produce compounds with much greater fluorescence output. Bauld et al. (406) studied in 1960 the sulfuric acid fluorescence and absorption spectra of eight natural estrogens (estrone, estradiol-17β, estriol, 16-epiestriol, 16-ketoestradiol-17β, 16α-hydroxyestrone, 16-ketoestrone, and 2-methoxyestrone) with respect to the influence of exciting wavelength, time of heating, acid concentration, and addition of ethanol or toluene–ethanol.

Penzes and Oertel (407) estimated estrogens after conversion into their azobenzene-4-sulfonic acid esters and densitometry of the derivatives on chromatoplates at 313 nm; an approximate sensitivity of 50 ng could be obtained. Based on the reaction of estrogens with dansyl chloride, the same authors were able to apply direct fluorometry of dansyl estrogens on chromatoplates for the quantitation of ng amounts of these steroids (408). Numazawa and Nambara (409) gave a review with 98 references on the colorimetric and fluorometric determination of estrogens, preparation and gas chromatography of their Me_3Si and related derivatives, radioimmunoassay, competitive protein binding method, and double isotope method with $p\text{-}C_6H_4\,(^{131}I)^{35}SO_2Cl$. Lee and Hähnel (410) described a rapid assay for estrogen conjugates in pregnancy urine involving extraction of the conjugates with Amberlite XAD-2, Kober reaction (hydroquinone in concd. H_2SO_4), and fluorometry of the dichloroethane extract of the reaction mixture. Howarth and Robertshaw (411) investigated several methods for urinary estrogen determination in pregnancy. Outch et al. (412) described a rapid method for the measurement of total estrogens in nonpregnancy urine by direct extraction of the estrogens into ethyl acetate following saturation with ammonium sulfate and addition of pyridinium to the urine; finally, the estrogens are measured fluorometrically using the Kober reaction and Ittrich (4-nitrophenol in chloroform) extraction. Total urinary estrogens in urine after 20 weeks of pregnancy were determined by rapid hydrolysis with enzymes of *Helix pomatia* (snail), followed by a simplified extraction procedure and estimation by Ittrich's fluorometric method using a solution of p-nitrophenol in chloroform (413). Urinary estrogens at concentrations of 10–150 μg liter^{-1} were determined fluorometrically by Ittrich's method after hydrolysis of the urine samples with β-glucuronidase, extraction with ether and with NaOH (414).

James (415) reported a fluorometric procedure for determining estradiol valerate in sesame oil or ethyl oleate and presented the results of a collaborative study on this subject (416).

Craig et al. (417) described an automated method for the determination of estrogens in pregnancy urine with the Kober reagent and compared it with a colorimetric technique.

Goebelsmann (418) determined total urinary estrogens by acid hydrolysis, solvent partition, reduction of estrone to estradiol, methylation, and chromatography with fluorescence measurement of the Kober chromogens following chloroform extraction according to Ittrich.

Ethinyl estradiol was determined in tablets by a fluorometric procedure (419), based on the Liebermann–Burchard reaction. A chloroform extract of ethinyl estradiol is reacted with acetic anhydride and sulfuric acid, and the resulting fluorophor is measured at 400 nm while exciting at 324 nm. The limit of linearity is up to 10 μg ml^{-1}; the detection limit is 0.5 μg ml^{-1}. The method is only applicable to single-component preparations; progestational steroids produce a color that quenches the fluorescence of ethynyl estradiol, necessitating an adequate separation of the two components.

Dvir and Chayen (420) described a method for the measurement of estriol in pregnancy urine. After hydrolysis of the urine, estrogens are reacted with dansyl chloride, yielding fluorescent derivatives. The estriol derivative is separated by TLC on Kieselgel G in a solvent of ethanol–chloroform (5:95), located under UV, eluted, and measured fluorometrically at 346 nm excitation and 525 nm emission wavelengths. Foster et al. (421) developed in 1971 a fluorometric procedure for estriol in pregnancy urine using Bachman's reagent (p-phenolsulfonic acid sodium salt in phosphoric acid). An improved method for estriol in urine of pregnant women via formation of its derivative with dansyl chloride followed by TLC on silica gel, elution of the dansyl–estriol spots with methanol, and fluorescence measurement (λ_{ex} = 346 nm; λ_{em} = 525 nm) was reported in 1975 (422). Fluorometric measurement of dansylated derivatives was described in 1976 for the determination of estrogenic steroids and stilbenes in the urine of calves (423). A simplified routine method for the determination of estriol in pregnancy urine (424) based on the Kober–Ittrich reaction and a simple and reliable method for estriol in nonpregnancy urine (425) by acid hydrolysis, separation through solvent partition, gel adsorption, and fluorometric measurement based on previous procedures were reported in the early 1970s. TLC fluorescence was used for the determination of estrogens in the urine of nonpregnant women (426). A rapid, specific method for diethylstilbestrol analysis using an in-line photochemical reactor with high-performance liquid chromatography and fluorescence detection was reported in 1982 (427).

James (428) assayed 17 α-ethynylestradiol-3-methyl ether (mestranol) directly in some oral contraceptive tablets utilizing native fluorescence (λ_{ex} = 284 nm; λ_{em} = 327 nm) with a sensitivity of 0.8–1.0 μg ml^{-1}. The method is applicable to single-tablet analyses; however, chlormadinone acetate interferes in the procedure. Mestranol was determined fluorometrically in contraceptive tablets using a H_2SO_4–methanol reagent (1:1) (λ_{ex} = 475 nm; λ_{em} = 515 nm) (429).

Graef (430) described a method for the fluorometric determination of oxosteroids with dansylhydrazine, allowing quantitative estimations down to 0.01 μg dioxosteroid and 0.02 μg monooxosteroid per ml chloroform solution. Dansylhydrazine was used in 1981 as a prelabeling reagent for the determination of 17-oxosteroids in serum and urine by fluorescence high-performance liquid chromatography (431).

Pal (432) described a fluorometric method for the determination of urinary conjugated testosterone and epitestosterone in human subjects after hydrolysis with β-glucuronidase and sulfatase at pH 5.2 for 72 h, fractionation on a Sephadex column, paper chromatography, and fluorometric measurement with a sulfuric acid–ethanol-induced reaction. The method is sensitive (0.01 μg of both compounds give distinct readings) and highly reproducible. Egg and Huck (433) separated testosterone and epitestosterone on thin layers of aluminum oxide followed by heating of the plates for 20 min at 180°. Testosterone and other Δ^4-3-ketosteroids are characterized in long-wave ultraviolet light (366 nm) by a light blue fluorescence (λ_{max} = 440 nm). The reaction is very sensitive (detection limit about 5 ng) and allows the determination of small amounts of urinary testosterone by means of thin-layer densitometry without preliminary purification. Eechaute et al. (434) developed a reaction for the induction of red fluorescence from testosterone, epitestosterone, androstenedione, and, to some extent, corticosterone and desoxycorticosterone. The reaction is very sensitive for testosterone, and the maximum intensity of the red fluorescence at 615 nm is obtained on activation at 598 nm. This reaction allowed the authors to develop a simple method for the fluorometric estimation of testosterone in human male plasma (435) in the following way. Five-ml heparinized plasma are added to 5 ml water and extracted with 30 ml dichloromethane. The extract is washed with 4 ml 2 N NaOH and three times with 4 ml water. The washed extract is evaporated to dryness and the residue is chromatographed on a silica gel column; elution is carried out with 1.5% ethyl alcohol in dichloromethane. The eluate fraction from 13 to 25 ml is collected and evaporated to dryness; the sulfuric acid–ethanol-induced fluorescence reaction is carried out on the residue. The fluorescence at 620 nm is measured with activation light of 600 nm; 0.1 μg testosterone per 100 ml plasma can be estimated with a sufficient degree of accuracy.

Alkalay et al. (436) determined in 1972 the relative bioavailability of methyltestosterone in man using a spectrophotofluorometric procedure based on the fluorescence induced by hydrochloric acid and stabilized by ascorbic acid. Their method is specific and capable of assaying nanograms per milliliter amounts of unchanged drug in blood plasma or serum.

Njau (437) studied the nature of acid-induced fluorescence of 17α-methyltestosterone. According to this author, successful development of a fluorometric method with total selectivity for the determination of a particular steroid ideally depends on a reaction taking place on the ring system to form a product with unique photophysical properties. Since this reaction is difficult to achieve, steroid fluorescence analysis has depended largely on chemical transformation of the steroid skeleton using concentrated mineral acids. The use of organic acids, however, may reduce the number of reaction products and possibly lead to more specific fluorophors from individual steroids. Britten and Njau (438) found trichloroacetic acid to be highly effective in locating steroidal substances as fluorescent and strongly UV-absorbing products on TLC plates. Of more than 20 steroids subjected to this reagent, the cardiac glycosides and aglycones, 17α-methylandrost-5-ene-3,17-diol (I) and 17α-methyltestosteron (II), reacted most readily. Examination of the trichloracetic acid reaction with (II) under controlled conditions (439) showed the major product formed to be 1,2,10,15,16,17-hexahydro-10,17,17-trimethylcyclopenta[a] phenanthren-3-one (III). Other phenanthrenones (IV and V) and cinnamylidene compounds (VI–VIII) with an approximately similar chromophore were synthesized (439–441) and studied to provide information on the structural requirements needed for the fluorescence of (III). Concerning the effects of pH on the absorption and emission characteristics of (IV), it is suggested (437) that in strongly acidic solvents, enolization of the chromophore occurs (Fig. 2.12) and it is the enol form that exhibits fluorescence.

Segura and Gotto (442) reported the development of stable and highly fluorescent spots from many biological materials in TLC after heating the plates in an atmosphere of ammonia liberated from ammonium hydrogenocarbonate. Levi and Reisfeld (443) extended the sensitivity of the method to the nanogram range for the determination of cholesterol, testosterone, and LSD.

Colorimetric and fluorometric analysis of steroids in general has been reviewed by Bartos and Pesez (444).

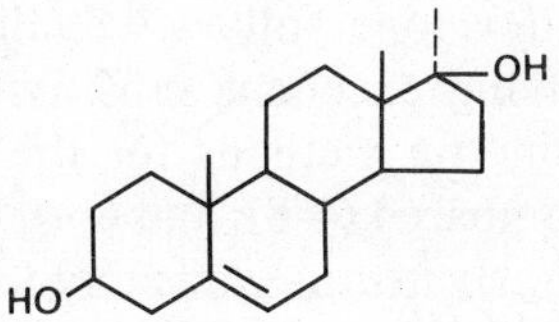

(I)

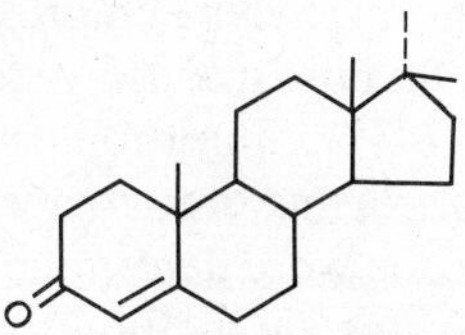

(II)

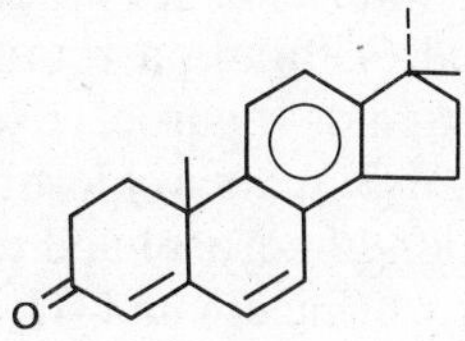

(III)

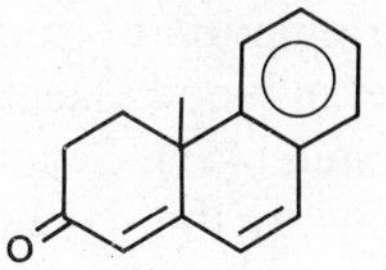

(IV)

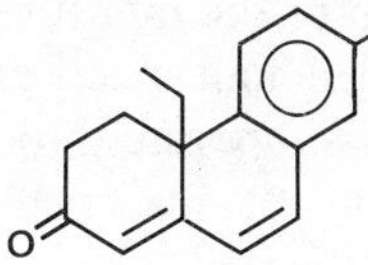

(V)

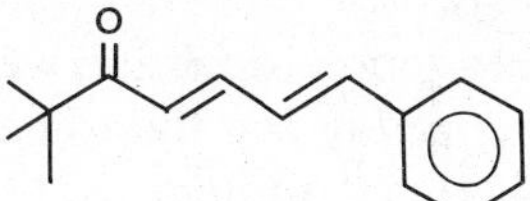

(VI)

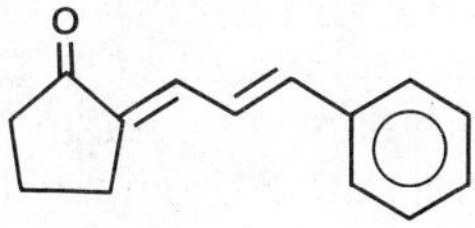

(VII)

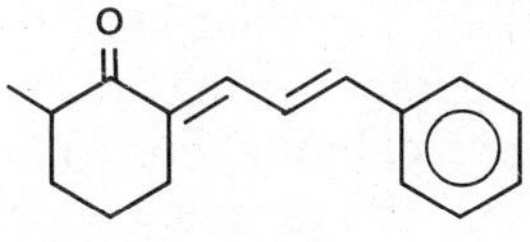

(VIII)

The specificity of fluorometric corticoid determinations was discussed in 1962 (445); sulfuric acid–ethanol fluorescence provides an efficient and reliable determination of unconjugated plasma corticoids as cortisol and corticosterone. A review on the extraction of corticosteroids from biological material, the purification of the extracts, and the determination procedure was given in 1969 (446). Hydroxycorticosteroids in plasma were separated into cortisol and corticosterone by TLC on silica gel using

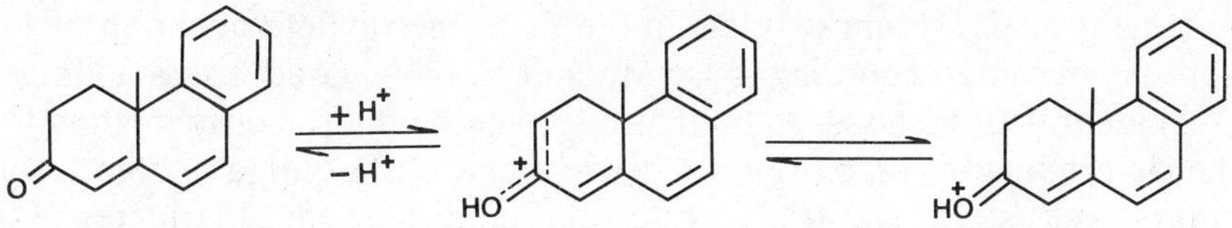

Fig. 2.12. Chromophore enolization of (IV) in strongly acidic solvents. (From ref. 437.)

chloroform–methylene chloride (90:10) as developer followed by the fluorometric ethanol–sulfuric acid method giving excellent sensitivity and specificity (447). Fluorometric and colorimetric methods for the determination of plasma corticosteroids were compared (448), and cortisol and corticosterone were determined simultaneously in urine and amniotic fluid with a fluorometric method that incorporates a correction for nonspecific fluorescence (449). Fluorometric and modified Porter–Silber method for 17-OHCS for urinary cortisol were compared (450) and a routine fluorometric determination of nonconjugated corticoids (hydrocortisone and corticosterone) in blood plasma with ethanol–H_2SO_4 after chloroform extraction was cited (451). Separate determination of cortisol and corticosterone in plasma by thin-layer chromatography using ethanol–H_2SO_4 fluorescence induction of the eluted spots was mentioned (452), while Pirke and Stamm (453) compared the specificity of simple methods (nine fluorometric methods and the Porter–Silber method) for the measurement of plasma cortisol; three fluorometric methods showed high specificity, gave satisfactory precision, and were relatively simple to operate. Nonspecific interferences in the fluorometric determination of plasma cortisol were cited (454) and an improved fluorometric estimation of Δ^4-3-ketosteroids (testosterone, cortisol, and progesterone) at the nanogram level by means of lithium hydroxide reaction was suggested (455). Kawasaki et al. (456) determined cortisol in human plasma and urine by high-performance liquid chromatography using fluorophotometric detection. After extraction with methylene chloride, cortisol is labelled with dansyl hydrazine, separated by HPLC, and detected at 350 nm excitation and 505 nm emission wavelengths. Funk et al. (457) described a high-performance thin-layer chromatographic determination of fluorescence-labeled cortisol; the fluorescence intensity of cortisol dansyl hydrazone derivatives can be increased by a factor of up to 10 by dipping the HPTLC plates into a liquid paraffin-n-hexane solution after development offering a detection limit of about 2.5 pg per spot.

11-Hydroxy corticosteroids were determined fluorometrically using the ethanol–H_2SO_4 method (458–461); the fluorometric determination of urinary unconjugated 11-hydroxycorticosteroids was applied as the screening test for the adrenocortical function (462) while Usui et al. (463) corrected for nonspecific fluorescence produced by the sulfuric acid treatment of the serum extracts in the fluorometric determination of 11-hydroxysteroids, according to De Moor et al. (464). Graef and Staudinger (465) identified the nonspecific fluorogens in the fluorometric method for the determination of plasma 11-hydroxycorticosteroids as cholesterol, cholesterol esters, and triglycerides and confirmed that the increase of nonspecific fluorescence after ACTH injection is caused by the cholesterol increase. The authors state that it is possible to remove the non-specific fluorogens by partition between water and petroleum ether. A discussion was reported on some problems associated with fluorometric determination of 11-hydroxycorticosteroids in blood, such as separation of cortisol and corticosterone, contamination, and quantitation of corti-costeroid-binding globulin (466). Mejer and Blanchard (467) mentioned a rapid procedure for clinical screening of plasma 11-hydroxycorticoster-oids using a modified sulfuric acid–ethanol method and investigated the specificity of the proposed fluorometric method (468).

A sulfuric acid fluorescence reaction was suggested as the basis for a simple method for the determination of urinary pregnanediol (469).

Miscellaneous.

☐ Determination of microamounts of acetylcholine by a spectrofluo-rometric method was reported (470), based on the intensification of the fluorescence of a mixture of solutions of quercetin and tetraphenyl hy-poborate in absolute methanol on the addition of acetylcholine in ether (λ_{ex} = 380 nm; λ_{em} = 540 nm). Linearity is observed in the 0.6×10^{-5}–4.8×10^{-5} M range.

A fluorometric determination was suggested by Gelbcke (471) for the anticholinergic trihexyphenidyl hydrochloride and some derivatives with analogous structure.

☐ Panalaks and Scott (472) detected aflatoxins B_1, B_2, G_1, and G_2 quantitatively at subnanogram levels by HPLC on a 5-μm Lichrosorb column, using a Lichrosorb-packed flow cell in the fluorometric detector. Their method is based on the observation that a number of organic com-pounds tend to fluoresce more intensely, although with modified fluo-rescence emission, when adsorbed on silica gel. Thorpe et al. (473) de-termined aflatoxins by reverse-phase HPLC and postcolumn derivatization using a fluorescence detector.

☐ Miles and Schenk (268) reported on the fluorescence and phos-phorescence of several phenylethylamines and barbiturates. Barbiturates are derived from barbituric acid differing from each other in the nature of the R group in position five:

A complete study of the fluorescence of the barbiturates was thought desirable by these authors, as this would reveal the effect of substitution on a very small conjugated π system. Investigated compounds include amobarbital, aprobarbital, barbital, butabarbital, cyclopentallyl barbituric acid, diallylbarbituric acid, pentobarbital, phenobarbital, secobarbital, and thiopental, of which corrected fluorescence data and relative quantum efficiencies in air are reported. These substances fluoresce only in alkaline solution when excited in the 250-nm region. This fluorescence must arise

(I)   (II)

keto tautomer   enol tautomer

from excitation of the barbiturate "enol anion", which must exist for significant absorption to occur in this region. The authors used fluorometric analysis to assay pharmaceutical preparations for single components and for amphetamine plus barbiturate.

King et al. (473) described the temperature dependence of fluorescence as an aid to the identification of oxybarbiturates. Phenobarbitone, seconal, and barbitone, for example, can be distinguished from one another, but amylobarbitone, butabarbitone, and pentabarbitone show a similar temperature dependence to that of barbitone. They state that in favorable cases the measurement of fluorescence intensity at only two different temperatures might be sufficient to obtain a provisional identification of a barbiturate. Gifford et al. (475) provided extensive investigations on the structure–luminescence correlation in the oxybarbiturate series. They examined and correlated the ultraviolet absorption spectra, fluorescence, and phosphorescence excitation and emission spectra of 16 barbiturates and certain model compounds and proposed the structure of the fluorescent dianion in the 5,5-disubstituted barbiturates. King and Gifford (476) reported structure–luminescence correlations in the thiobarbiturate series and stated that luminescence techniques can be used as a provisional means of identifying an unknown thiobarbiturate. In 5,5-disubstituted thiobarbiturates, two fluorescent species are present in solution, one of which is a monoanion and the other of which is a zwitterion. 5-Phenylsubstituted compounds exhibited anomalous features but did not show a luminescence characteristic of an isolated phenyl group.

Fluorometry was used for the determination of phenobarbital in tablets (477).

Dill et al. (478) reported a simplified, rapid, and low-cost benzophenone procedure for diphenylhydantoin in plasma by eliminating the need for preliminary extraction of plasma with an organic solvent. The authors perform their assays directly on 0.2 ml of plasma by treating it with alkaline permanganate to form benzophenone, extracting the benzophenone with heptane and then shaking the heptane layer with sulfuric acid and measuring the fluorescence of the acid layer ($\lambda_{ex} = 355$ nm; $\lambda_{em} = 485$

nm). The assay seems to be highly reproducible and adequately sensitive to detect 1 μg of the drug per milliliter of plasma. Fluorometric and gas–liquid chromatographic assays of 154 plasmas gave results that were not significantly different, even in the presence of phenobarbital, other barbiturates, and anticonvulsant drugs, or of various tranquilizers and other commonly used drugs.

☐ The detection, characterization, and determination of the cannabinols and their metabolic and photochemical products present an urgent problem in the forensic field. Bowd et al. (479) described the application of luminescence spectrophotometry to this problem. They determined the characteristics of the absorption, fluorescence, and phosphorescence spectra of the cannabinols and provided a basis for the detection of these compounds at concentrations down to 10 ng ml^{-1}.

Cannabinoids and their analogs can be detected in biological materials (480) (blood, plasma, urine) based on the condensation with a polycarboxylic acid and acid catalyst at 80–90° for >20 min to give a highly fluorescent derivative with maximum intensity in the pH range of 9–11.

Cannabinoid fluorescence may be induced by thermal treatment and this property may be used for the identification of cannabis constituents in human urine (481).

Bourdon (482) reported a fluorometric determination method for Δ^9-tetrahydrocannabinol and its metabolites in urine. Vinson et al. (483) described a thin-layer chromatographic method for the detection of Δ^9-tetrahydrocannabinol (THC) in blood and serum. The procedure utilizes a reagent, 2-p-chlorosulfophenyl-3-phenylindone, which reacts with the phenolic group to form a derivative. Following extraction and cleanup, the derivative is prepared and separated from the reagent and naturally occurring compounds by thin-layer chromatography. The derivative is then detected by spraying the plate with an alkoxide spray (saturated solution of 8 g of sodium metal in 100 ml of methanol and 8 ml of dimethyl sulfoxide), which produces a fluorescent spot visible under long-wave UV light. The method can detect 0.2 ng ml^{-1} of tetrahydrocannabinol in 5 ml of serum and is suitable for routine screening. The derivatizing reagent mentioned, 2-p-chlorosulfophenyl-3-phenylindone, DIS-Cl (484), has been used for derivatizing amino acids (485), amino sugars (486), and vitamin B$_6$ (487). The reaction of DIS-Cl with the phenolic group of tetrahydrocannabinol is shown in Figure 2.13; reaction of the derivative with methoxide on a TLC plate produces a fluorescent spot.

☐ A fluorometric method was described for the determination of phenprocoumon (3-(1-phenylpropyl)-4-hydroxycoumarin) in blood plasma (488). After treatment with tartaric acid–benzene, the organic

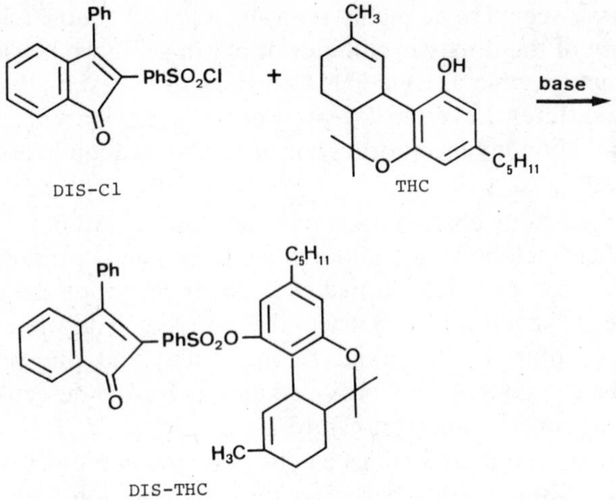

Fig. 2.13.   Reaction of DIS-Cl with tetrahydrocannabinol. (From ref. 483.)

layer is extracted with 3 N NaOH solution followed by measurement of the aqueous phase at λ_{ex} = 313 nm, λ_{em} = 400 nm; the detection limit is about 0.2 μg ml^{-1}. The method should be suitable for the determination of phenprocoumon levels during long-term treatment.

Flavonoids, coumarins, and cardiotonic heterosides were determined spectrofluorometrically by the formation of fluorescing complexes with AlCl$_3$, ZrOCl$_2$, and isoniazid; the excitation range is 320–440 nm and the emission range is 390–580 nm (489). The luminescence of polyhydroxyflavones was studied and analytical applications were suggested (490).

Schulman and Rosenberg (491) evaluated the biprotonic phototautomerism of 7-hydroxy-4-methylcoumarin in the lowest excited singlet state by fluorescence analysis, using steady-state kinetics. Haing Lee et al. (492) developed an HPLC method for the fluorometric determination of warfarin and its metabolites (diastereomeric warfarin alcohols and 4'-,6-, 7-, and 8-hydroxywarfarin). Their detection scheme utilizes a specific and sensitive postcolumn acid–base fluorescence enhancement technique that is applicable to human plasma and urine samples with nanogram range detection limits. Maes et al. (493a) performed a fluorometric analysis of the binding of warfarin to human serum albumin.

Free fatty acids in serum, together with representatives of a new class of oxirane carboxylic acids with blood-glucose-lowering activity, were determined simultaneously by Voelter et al. (493b) by fluorescence labeling with 4-bromomethyl-7-methoxycoumarin followed by HPLC separation.

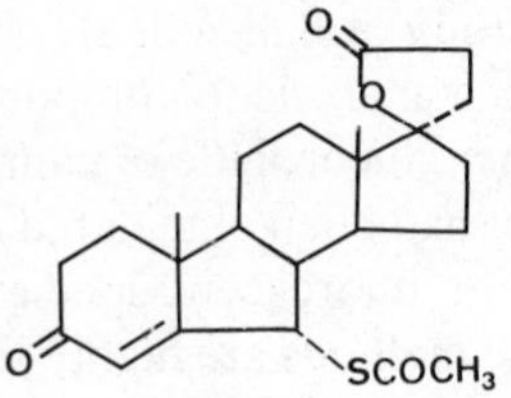

(I)

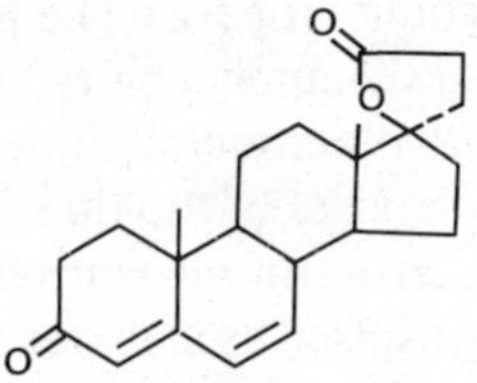

(II)

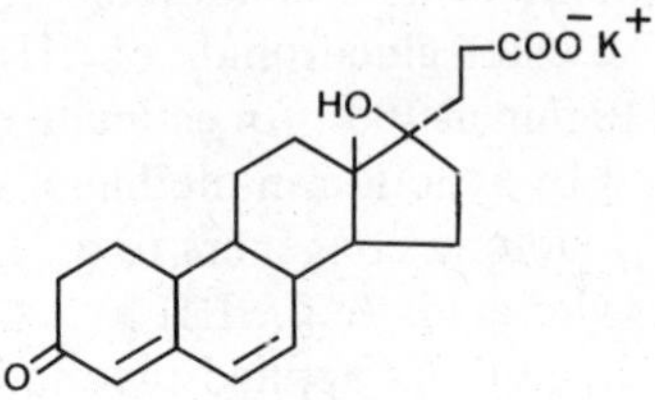

(III)

☐ Comparative studies on the fluorescence behavior of some mono- and diaminopyridines were presented by Baeyens and De Moerloose (494), as the aminopyridine nucleus often occurs in drugs or in their metabolic transformation products. The fluorescence characteristics of 2-,3-, and 4-aminopyridine; 2,3-, 3,4-, and 2,6-diaminopyridine; and 2-amino-6-picoline were investigated in various solvents and at different pH values. 4-Aminopyridine exhibits negligible fluorescence, probably because of (n,π^*) transitions. All other compounds are strong fluorophors, mainly because of the (π,π^*) nature of the lowest excited state. The authors provide practical data for the fluorometric determination of small amounts of these molecules, that is, excitation and emission maxima, relative fluorescence intensities, optimum pH, limits of linearity, detection limits, UV absorption maxima, O–O bands, and variation of wavelengths and fluorescence intensities with pH.

☐ Furosemide (4-chloro-N-furfuryl-5-sulfamoylanthranilic acid), a commonly used diuretic, can be quantitated in plasma (495,496) by acid extraction with ether, buffering at pH 7.5 followed by fluorescence measurement at 422 nm on excitation at 356 nm.

Spironolactone (I), canrenone (II), and potassium canrenoate (III) can be analyzed in the low nanogram range utilizing a fluorescence reaction in sulfuric acid (62% v/v) (497). In prior investigations canrenone was identified as a major metabolite of the antimineralocorticoid spironolactone; potassium canrenoate was also metabolized to (II). Gochman and Gantt (498) had already developed a fluorometric micromethod, which only accounts for (II). The purpose of the study of Sadée et al. (497) was to extend the applicability of the fluorescence analysis to compounds (I–III) and to other metabolites, such as the ester glucuronide of canrenoate. The mechanism of the fluorescence reaction had been elucidated (499) to exclude errors in measuring metabolites. The fluorescent species, a trienone, was used as an external standard. The methods are highly sensitive ($\sim$10 ng ml^{-1} of (I–III) in plasma) and specific for 17-alkyl-17-hydroxy-4,6-dien-3-one steroids. The 7α-thioacetyl-4-en-3-one (I) was dethioacetylated under mild alkaline or acid conditions to the 4,6-dienone moiety prior to extraction. The ester glucuronide of (III) was quantitatively hydrolyzed at pH 13 for 10 min at 100°. An estimate of acid-labile conjugates of (I–III) was achieved by reaction in methanolic 3,6% sulfuric acid for 30 min at 100°. Spiro-γ-lactone congeners (e.g., I and II) were separated from γ-hydroxycarboxylic acids (e.g., III) by extraction into methylene chloride at pH 7–8. The authors applied the methods to biological samples, such as plasma, bile, urine, and gastric fluids.

The diuretic hydroflumethiazide (3,4-dihydro-6-(trifluoromethyl)-2H-1,2,4-benzothiadiazine-7-sulfonamide 1,1-dioxide) was determined in plasma and urine by a rapid, accurate, sensitive, and reproducible assay that was developed after studies of its UV and fluorescence spectral properties and partitioning behavior (500). The assay is based on initial extraction from acidified plasma or urine into ether, back-extraction into basic solution followed by acidification to about pH 1, and measurement of the fluorescence derived from the unionized molecule (λ_{ex} = 333 nm; λ_{em} = 393 nm). According to the authors the method is convenient for routine clinical use and has sufficient sensitivity to quantify hydroflumethiazide levels after administration of therapeutic doses.

☐   Glycodiazine   (N-[5-(2-methoxyethoxy)-2-pyrimidinyl]benzenesulfonamide), an oral hypoglycemic agent, and two metabolites (demethylated and carboxylated glycodiazine) fluoresce at an excitation wavelength of 320 nm in alkaline medium with an intensely bluish tint (maximum of emission 420 nm), which allows their fluorometric determination. After extraction with chloroform the concentration of glycodiazine in plasma can be determined with good precision with a determination limit of about 1 μg ml^{-1} of plasma (501). On account of their different physicochemical properties the two metabolites (i.e., demethylated and carboxylated glycodiazine) can be determined in urine separately at pH 1.0 or 7.0, respectively.

A vibrational analysis of the electronic spectra of some hypoglycemic agents was reported in 1971 (502). In a comparison of the absorption, excitation, phosphorescence, and fluorescence spectra of $PhSO_2NH_2$ and those of PhSO₂NHCONHBu and its derivatives p-RC₆H₄SO₂NHCONHR₁ ((I),R,R₁ = H, Bu; Me, Bu; Me, H; Me, cyclopentyl; Me, cyclohexyl; Cl, Pr; MeO, Bu), no energy transfer from the benzene nucleus to the lower excited state of the -NHCONH- moiety was observed despite the differences in biological activity of the individual members of the hypoglycemic series (I).

A fluorescence densitometric method for the determination of gluconic and lactobionic acids in pharmaceutical preparations was described by Guebitz et al. (503).

☐ de Silva et al. (504) developed a sensitive and specific spectrofluorometric assay for the determination of 2-hydroxynicotinic acid and its major metabolite, the N-1-riboside in blood and urine. 2-Hydroxynicotinic acid is one of a series of analogs of nicotinic acid with greater pharmacological activity in vitro than nicotinic acid and that was also found to be more potent than the latter in rats as a hypolipidemic agent in the inhibition of cholesterol and fatty acid biosynthesis. Both 2-hydroxynicotinic acid and its N-1-riboside metabolite exhibit strong fluorescence in acidic media. Drug and metabolite are separated by thin-layer chromatography and eluted from the silica gel into methanol–0.1 M hydrochloric acid (80 + 20). The sensitivity is 0.04–0.05 μg for 2-hydroxynicotinic acid and 0.10–0.12 μg for the metabolite per ml of blood or urine.

de Silva and co-workers (505) developed a spectrofluorometric method for the determination of 6-chloro-9-[2-(2-methyl-5-pyridyl)ethyl]-1,2,3,4-tetrahydrocarbazole-2-methanol (I) hydrochloride and its 2-carboxylic acid analog (II) in blood and urine. Both compounds are of clinical interest because they have shown marked hypolipidemic activity in animals. Their method involves extraction of both compounds at neutral pH, either from blood into ethyl acetate (the residue of which is dissolved in ether) or from urine directly into ether. Both the alcohol and the acid are separated from each other by selective extraction into acid or base, respectively, and then reextracted into ether from the respective aqueous medium by appropriate pH adjustment. The residues of the ether extracts containing the compounds are dissolved separately in 0.25 N NH₄OH. Methylene blue is added to all samples, which are then exposed to UV energy for 15 min to produce the fluorophors. The fluorescence of the solutions is read at 370 nm, with excitation at 340 nm. The linear range of quantitation of both compounds is 0.02–10 μg each per ml of final solution. The authors applied their method to the determination of blood levels and urinary excretion of the alcohol and its acid metabolite in a dog. Characterization of the reaction products of (I) and (II) was attempted following their sep-

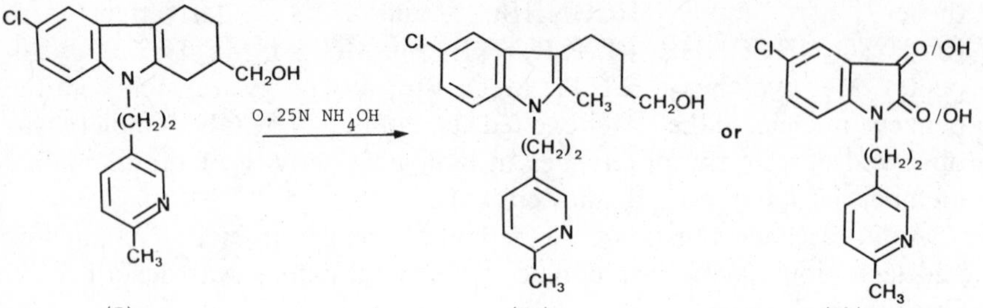

Fig. 2.14.   Fluorometric determination of substituted tetrahydrocarbazoles by a methylene blue sensitized photolytic reaction according to de Silva et al. (From ref. 505.)

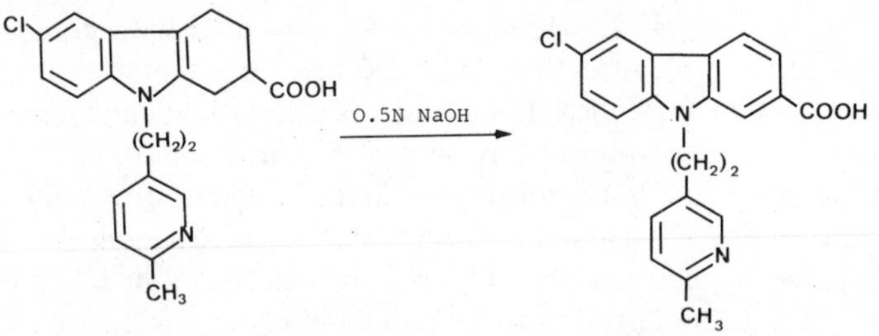

aration by TLC. The mechanism shown in Figure 2.14 was suggested and comparisons with indomethacin and indole-3-acetic acid were made.

☐   Abdel Fattah (506) has performed extensive experiments on the qualitative and quantitative aspects of the luminescence behavior of various imidazoles of pharmaceutical interest. These compounds include mebendazole and flubendazole—both broad-spectrum anthelmintics; econazole nitrate, miconazole nitrate, and imazalil—all potent fungicides; carnidazole—a trichomonacide; metomidate hydrochloride and etomidate—both hypnotics; and the anthelmintics levamisole hydrochloride (with growing importance in the treatment of cancer and viral diseases), dexamisole hydrochloride (antidepressant, psychoenergizer) and tetramisole hydrochloride. Her work is to be seen not only as an extension of the methods of analysis of these imidazole derivatives but as a contribution to the general knowledge of fluorescence and phosphorescence behavior of this series of drugs. The results of her investigation on native fluorescence properties can be summarized as follows. Only those compounds possessing the benzimidazole nucleus fluoresce, with the intensity depending on molecular substitution. 2-Aminobenzimidazole, a compound used as reference fluorescence analog in these experiments, exhibits intense fluorescence emission in ethanol, greatly depending on solvent polarity and pH. Mebendazole and flubendazole emit relatively weak fluorescence in various media, which can be attributed to the influence of their carbonyl group (lowest excited singlet of n,π^* type), although the conjugated benzimidazole system (enhanced π-π^* transition as lowest excited singlet) is present. On the other hand, critical solvent variation leads to the conclusion that dimethyl sulfoxide, in addition to its good solubility properties, enhances the intensity of fluorescence for both substances relative to other solvents. Analytical graphs of mebendazole and flubendazole in DMSO were established and detection limits from 0.5 to 1 μg ml^{-1} were calculated, meaning poor sensivity for this technique. Most other imidazoles did not exhibit fluorescence properties clearly; the necessity of conjugation of π bonds of an aromatic ring with the imidazole nucleus is demonstrated. In the final chapter the author discusses correlation between molecular structure, electron transitions and luminescence emission of mebendazole and some analogous structures: fluorescence in solution, fluorescence on filter paper, 77 K phosphorescence and room temperature phosphorescence.

☐ Dibucaine (2-butoxy-N-(2-diethylaminoethyl)cinchoninamide), a potent local anesthetic, was determined fluorometrically in cerebrospinal fluid (507) by mixing with 0.5 M acetate buffer pH 5.5 and measuring the fluorescence intensity at 410 nm with excitation at 330 nm. In case of blood contamination a modification of the method was proposed. Martucci and Schulman (508) studied the pH dependence of the electronic spectra of dibucaine as it contains several basic functional groups, in order to determine the spectroscopic properties of its various prototropic forms, thus establishing the optimum pH conditions for fluorometric analysis.

☐ A simplified fluorometric method was described for the determination of low concentrations of plasma methotrexate (509), a drug extensively used for treatment of proliferative diseases such as leukemia and carcinoma. In view of the extremely toxic nature of the drug it is important to have a reliable and reproducible method for its determination in plasma. The proposed method consists of a single fluorescence measurement on plasma protein-free filtrates (trichloroacetic acid) after oxidation of the drug with $KMnO_4$ at pH 5.0. By using permanganate the resulting fluorescence was found to be proportional to the drug concentration. Plasma methotrexate levels of 50 μg ml^{-1} (5 μg ml^{-1} filtrate) could be determined if the trichloroacetic acid-treated diluted plasma was heated and stirred at 100° prior to filtration of the precipitated proteins, causing a release of absorbed methotrexate into the supernatant solution.

$$CO-NHCH_2CH_2N\diagup^{C_2H_5}_{\diagdown C_2H_5}$$

OCH₃ ... Cl ... NH₂

metoclopramide

$$CO-NHCH_2CH_2N\diagup^{C_2H_5}_{\diagdown C_2H_5}$$

NH₂

procainamide

☐   The fluorescence properties of metoclopramide (4-amino-5-chloro-N-(2-diethylaminoethyl)-2-methoxybenzamide), a commonly used antiemetic, and its determination in pharmaceutical dosage forms were reported by Baeyens and De Moerloose (510).

Metoclopramide is to be regarded as a derivative of procainamide of which fluorescence was observed as a result of a systematic investigation of the fluorescence properties of various drugs possessing a primary aromatic amino function. Metoclopramide can be determined in 2-methylpropan-1-ol, propan-2-ol and in water (average λ_{ex} = 313 nm; λ_{em} = 361 nm). The effects of pH on the fluorescence of metoclopramide, procainamide, and synthesized metoclopramide monoacetyl derivative at the primary aromatic amine function were studied. The absence of the chlorine atom and of the methoxy group in the procainamide molecule seems to result in a much more pronounced fluorophoric structure of the free base. The metoclopramide monoacetyl derivative shows analogous fluorescence properties as its parent compound, quite unexpectedly, however, as literature generally reports a total loss of fluorescence when acetylating aromatic amino groups. Fluorometric determinations of metoclopramide can be carried out in pH 2.0 buffer solution (λ_{ex} = 310 nm; λ_{em} = 360 nm) with a linear response up to 6 μg ml^{-1} and a detection limit of 3.10^{-2} μg ml^{-1}.

Domperidone (5-chloro-1-{1-[3-(2,3-dihydro-2-oxo-1H-benzimidazol-1-yl)propyl]-4-piperidinyl}-1,3-dihydro-2H-benzimidazol-2-one), a gastrokinetic antinauseant for the treatment of dyspepsia and vomiting, was determined in pharmaceutical preparations by Baeyens and De Moerloose (511) using its native fluorescence characteristics originating from the benzo-1,3-diazolin-2-one fluorophor group. Domperidone can be determined in tablets, oral solution and drops, injection ampoules and suppositories by measurement of the fluorescence in ethanol (λ_{ex} = 283 nm; λ_{em} = 324 nm) and in 0.01 M HCl (λ_{ex} = 284 nm; λ_{em} = 329 nm) with a detection limit of 0.01 μg ml^{-1}.

☐   Fluorescence spectrometric analysis was suggested as one of the methods for hygienic monitoring of carcinogen contamination (512).

Hatano et al. (513) used a high-speed liquid chromatograph with a new spectrofluorometric detector equipped with double-beam optics and a 3-μl flow cell for the qualitative and quantitative determination of separated components such as polycyclic aromatic hydrocarbons.

Hurtubise et al. (514) reported luminescence methods for detecting polycyclic aromatic hydrocarbons on aluminum oxide after separation by dry-column chromatography. Samples of coronene, benzo[a]pyrene, triphenylene, and phenanthrene in n-hexane are added to the top of the aluminum oxide column, followed by development with n-hexane–ether (19 + 1). Detection is achieved by submerging the column in liquid nitrogen and viewing the fluorescence and phosphorescence under longwave and shortwave ultraviolet radiation.

Laser-excited Shpol'skii spectroscopy for the selective excitation and determination of polynuclear aromatic hydrocarbons was described by Yang et al. (515). Direct qualitative and quantitative characterization of complex mixtures of these compounds, including multialkylated isomers, seems to be possible by the technique. The Shpol'skii effect provides selectivity based on the inherent sharp absorption bandwidths exhibited by polynuclear aromatic hydrocarbons in appropriate frozen n-alkane hosts.

Maple and Wehry (516) explored the analytical utility of site selection techniques for the characterization of multicomponent, matrix-isolated mixtures of polar polycyclic aromatic hydrocarbon derivatives by employing hydroxyl derivatives of naphthalene as prototypes. According to the authors, the use of laser excitation for the selective excitation of polar molecules residing in different "sites" or environments results in much greater spectral resolution than can be achieved with conventional lamp sources. The same authors explored the analytical utility of fluorescence photoselection for distinguishing overlapping spectral bands in mixtures of matrix-isolated polycyclic aromatic hydrocarbons (517).

☐ The fluorometric assay of polyanions in complex foods such as carrageenan stabilizers in dairy products and heparin in hog mucosa extracts was mentioned by Murray and Cundall (518).

☐ Uchiyama et al. (519–522) developed a new fluorometric analysis of the sweetener dulcin [(4-ethoxyphenyl)urea] using sodium nitrite. They isolated the fluorescent compound from the reaction mixture, investigated its chemical structure, and studied the reaction mechanism of fluorophor formation. Dulcin produces 4-ethoxybenzenediazonium ion via p-phenetidine by the action of nitrous acid; the final fluorescent compound produced was identified as 1,3-bis(4-ethoxyphenyl)-5-tetrazolone.

☐ Fluorometry has gained considerable importance since the development of HPLC methods for the quantitative determination of various pharmaceuticals and their metabolites in biological environment required more and more sensitive and selective detection systems. Snyder and Kirkland (523) summarized derivatizing agents for enhancing fluorometric detection via precolumn reaction of the sample and gave a survey of the reactions that have been used in high-performance liquid chromatographic reaction detectors. Frei et al. (524) discussed problems and the potential of postcolumn fluorescence derivatization of peptides in high-performance liquid chromatography. Frei and Lawrence (525) edited an excellent book on chemical derivatization in chromatography, covering precolumn derivatization with emphasis on pharmaceutical analysis and reaction detectors in liquid chromatography.

The use of fluorescence spectrometry for the detection on HPLC effluents has been described in literature on various occasions (526–534b). These detectors generally consist of a filter fluorometer equipped with a small-volume flow-through cell.

Chemical derivatization in liquid chromatography for enhancing detectability of various functional groups has been treated by many authors (525, 529, 535–537). Schwedt and Reh (538) described the use of fluorometric reaction detectors for high-performance liquid chromatography purposes, using selected steroids as examples. The authors combined reversed-phase, adsorption chromatography, systems with chemically bonded phases as well as one-step reaction with isonicotinoylhydrazide in organic solvents and a multistep reaction of ketosteroids to form dihydrolutidine derivatives in aqueous solution. They also examined reactor materials and effects of variable dynamic dimensions on band spreading (539). Scholten et al. (540) described the use of polytetrafluoroethylene coils in postcolumn photochemical reactors for liquid chromatography with application to pharmaceuticals.

Hershberger et al. (541) performed real-time fluorescence monitoring of effluents at multiple wavelengths of excitation and emission by interfacing a high-performance liquid chromatograph to the video fluorometer using a laminar flow cell to minimize dead volume and scattered light. In this way they displayed a total fluorescence chromatogram, as well as two selected excitation–emission wavelength chromatograms in real time, allowing selected fluorescence monitoring of compounds whose chromatographic retention profiles overlap significantly. Appellof and Davidson (542) developed strategies for analyzing data from video fluorometric monitoring of liquid chromatographic effluents in the qualitative analysis of a multicomponent fluorescent mixture. Their method analyzes a three-dimensional matrix obtained by passing the effluent of an HPLC instrument through the video fluorometer and provides an estimate of the number of components in the mixture as well as the emission and excitation spectra and HPLC profile for each component.

Voigtman et al. (543) compared laser-excited fluorescence and photoacoustic limits of detection for static and flow cells intended for liquid chromatographic applications. Ogan and Katz (544) compared fluorescence and UV detection for the liquid chromatographic separation of alkylphenols.

A fluorescence HPLC detector with continuously variable wavelengths was presented at the 1981 Pittsburgh Conference (545), consisting of aberration-corrected concave holographic gratings for both the excitation and emission monochromators, covering UV and visible range from 240 to 650 nm, equipped with a quartz flow-through cell of 12 μl. At the same meeting an evaluation was described of a microcomputer-controlled HPLC system that has the capability of sequentially measuring UV–VIS, fluorescence, and CARS (coherent anti-Stokes Raman spectroscopy) spectra of eluents (546); an HPLC–fluorescence detection using a flow cell packed with bonded phase material was proposed as well (547). A novel reactor for photochemical postcolumn derivatization in HPLC was recently described (548).

□ Several fluorometric techniques for drug determinations in various samples are based on the formation of ion pairs with different counterions followed by extraction with an organic solvent and subsequent measurement. Some of these ion-pairing methods are described briefly in the following paragraphs.

Pentazocine, a potent analgesic, was determined quantitatively in biological samples (549–551). Borg and Mikaelsson (552) extracted the drug from the samples in unchanged form, isolated it by partition chromatography as ion pair and measured by fluorometry. They determined partition coefficients and acid dissociation constants of pentazocine and its extraction constants as chloride ion pair. The conditions of their analytical method were calculated from these constants and the selectivity of the method was estimated. The absolute recovery of samples containing 100 ng was 90% ± 5%. The minimum amount that could be detected with acceptable precision was 50 ng.

Extraction and fluorometric determination of anthracene-2-sulfonate ion pairs of ammonium compounds in the 10^{-8} M range was described by Westerlund and Borg (553). Extraction constants were determined with organic ammonium ions and sodium as cations using methylene chloride and chloroform + 1-pentanol as organic phases. Quantum yield and sensitivity of fluorescence of the anthracene-2-sulfonate ion were determined and corrected fluorescence spectra were given. The authors studied the quantitative extraction and fluorometric determination of amitriptyline as anthracene-2-sulfonate ion pair; a relative standard deviation of 5% could be obtained in the range 10–100 ng ml^{-1}. The use of their method for quantitative determination of nanogram amounts of amines used as drugs is discussed, as well as the influence of coextraction of anthracene-2-sulfonic acid as sodium salt and free acid.

Persson (554) studied the ion-pair extraction of a fluorescent secondary amine, protriptyline, in the nanogram range (10^{-7}–10^{-8} M) by a fluorometric technique. The extraction and dissociation constants were determined for ion pairs with chloride, perchlorate, and dihydrogen phosphate. Chloroform and 1,2-dichloroethane with additions of 1-pentanol were used as organic phases.

Westerlund et al. (555) used the highly fluorescent anion of anthracene-2-sulfonic acid in ion-pair extraction studies of monovalent and divalent amines used as drugs. Their extraction studies were made in the low concentration range (10^{-7}–10^{-8} M) with methylene chloride as organic phase using fluorometric measurements. The amines studied include amitriptyline, nortriptyline, protriptyline, imipramine, desipramine, dibenzepin, terodiline, prenylamine, and emetine (divalent). The authors reported extraction and dissociation constants of the ion pairs and compared them with published constants of ion pairs of the same amines with other anion components; they discussed the fluorometric measurements especially with regard to ways of normalizing the fluorescence of different ion pairs in methylene chloride.

Lagerström et al. (556) studied the quantitative ion-pair extraction of ammonium compounds with anthracene-2-sulfonate as counter-ion. Conditions for quantitative determinations in the low concentration range using fluorometric measurements were investigated. An amine and a quaternary ammonium ion were extracted with the fluorescent anion of anthracene-2-sulfonic acid. The determination of the ion pairs, after extraction to an organic phase and after back-extraction to an aqueous phase, was studied and recoveries and standard deviations for the quantitative extractions were given. The authors studied as well the standardization of the fluorometric measurements, by standard curves and by reference solutions. Their results show that the ion-pair extraction technique can be used for quantitative fluorometric determination of nonfluorescent cationic compounds in the 10^{-7}–10^{-8} M range.

Sensitive fluorometry with the fluorescent dye salt tetrabromosulfonfluorescein was used to determine long-chain secondary or tertiary amines, such as amitriptyline, chlorpromazine, and so on (557). The dye solution (200 µg ml^{-1}) was added to the test solution at less than pH 2.5 with 0.1 N HCl, the amine complex extracted by chloroform and mixed

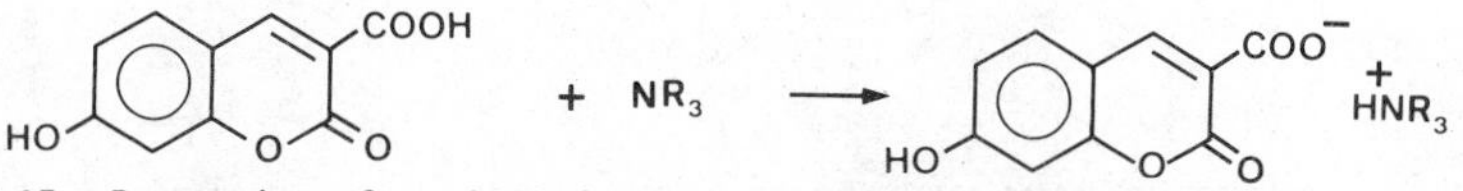

Fig. 2.15. Interaction of amphetamines with 3-carboxy-7-hydroxycoumarin resulting in highly fluorescent coumarin–amine salts. (From ref. 561.)

with alkaline methanol followed by fluorescence measurement at 550 nm on excitation at 535 nm.

Baker et al. (558a) described a reversed-phase ion-pairing liquid chromatographic separation and fluorometric detection of guanidino compounds as guanidine, methylguanidine, arginine, creatinine. On-line postcolumn derivatization of the guanidines occurs with phenanthrenequinone in an alkaline stream. Postcolumn derivatization of guanidino compounds by high-performance cation-exchange chromatography using ninhydrin was mentioned by Hiraga and Kinoshita (558b).

A reversed-phase ion-pairing HPLC separation and fluorometric detection of polyamines was reported by Simpson et al. (559). The polyamines putrescine, spermidine, and spermine were separated utilizing reversed-phase ion-pairing HPLC coupled with on-line postcolumn derivatization with Roth's reagent (o-phthalaldehyde and 2-mercaptoethanol in a 0.50 M potassium borate buffer at pH 9.0) and fluorometric detection (λ_{ex} = 340 nm; λ_{em} ⩾ 440 nm). Minimum detection limits ranged from 25 ng (spermine) to 1 ng (putrescine).

Sodium coumarin 6-sulfonate, a fluorometric ion-pairing reagent for tertiary amines, was applied to the determination of chlorpheniramine maleate by Dent et al. (560).

Stewart and Lotti (561) developed a fluorometric method of analysis for amphetamines based on the interaction between aliphatic and/or cyclic amines and 3-carboxy-7-hydroxycoumarin to yield highly fluorescent coumarin–amine salts (Fig. 2.15). This interaction results in an increase in fluorescence and is accompanied by a change in the activation wavelength of the reaction mixture. The use of 420 and 450 nm as the excitation and emission wavelengths, respectively, for the procedure can be correlated with known amphetamines run as analytical solutions and corresponds to pure synthesized coumarin–amine salts formed from various aliphatic and cyclic amines in the original method from the same authors (562). The method permits detection of various amphetamines as amphetamine sulfate, methamphetamine hydrochloride, benzophetamine hydrochloride, chlorphentermine hydrochloride, methylphenidate hydrochloride, and phendimetrazine tartrate in the presence of various medicinals found in combination with them in commercial dosage forms, even in the pres-

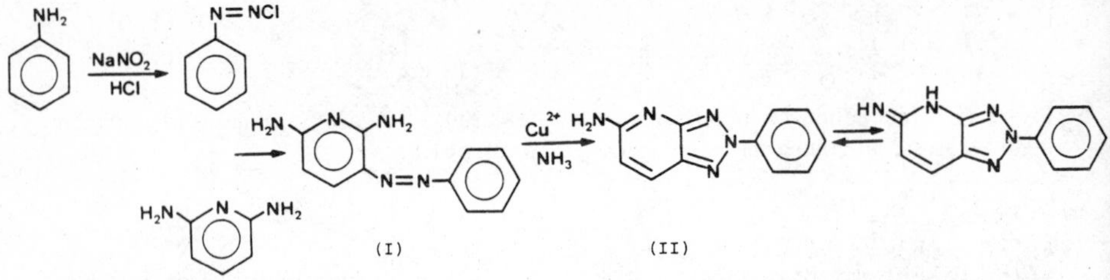

Fig. 2.16. Fluorophor formation from aniline through the 2,6-diaminopyridine method. (From ref. 564.)

ence of a 2- to 32-fold excess of fluorescent coumarin reagent. Primary, secondary, and/or tertiary aromatic amines, aromatic heterocycles, aromatic and aliphatic amides, and carbonyl-containing compounds do not interfere; the presence of other extractable strongly basic organic compounds, however, can interfere significantly with the fluorescence readings of the various amphetamines.

Dombrowski and Pratt (563) developed in 1971 a highly sensitive fluorometric method for the determination of primary aromatic amines in the nanogram range. Their procedure requires diazotization of the aniline amino group followed by coupling with 2,6-diaminopyridine and reaction of the resulting azo dye with ammoniacal cupric sulfate to produce an intensely fluorescent derivative. Various amines yield fluorogens with essentially common spectral characteristics (λ_{ex} = 360 nm; λ_{em} = 420 nm). Baeyens et al. (564) isolated the principal fluorophor in the case of aniline. Liquid chromatography on silica gel allowed the isolation of the main blue fluorescing principle extracted from a preparative reaction mixture. Combined spectroscopic methods indicated the formation of a triazole structure (II), resulting from the cyclization between the 2-amino function in the pyridine moiety and one of the nitrogens in the azo linkage (Fig. 2.16). High-resolution mass spectrometry was used to confirm the proposed structure. It was shown that this fluorophor can be obtained by oxidation of phenazopyridine (I), a drug used as an analgesic.

Functional Group Reactions and Special Techniques

Amines and Related Compounds.   Primary aliphatic amines can be analyzed fluorometrically by reaction with 9-isothiocyanatoacridine producing unstable thioureas that yield intensely fluorescent compounds, as described by Sinsheimer et al. (565). The structure elucidation and synthesis of a typical compound after the initial reaction of 9-isothiocyanatoacridine with butylamine forming N-butyl-N'-(9-acridinyl)thiourea was reported by De Leenheer et al. (566).

Monforte et al. (567) used 7-chloro-4-nitrobenzo-2-oxa-1,3-diazole (NBD chloride) for the fluorometric determination of primary and secondary amines in blood and urine after TLC and applied the reaction to the quantitative determination of amphetamine in biological extracts. Frei and Lawrence (568) described in 1973 reagents such as 4-chloro-7-nitrobenzo-2,1,3-oxadiazole, dansyl chloride and dansyl hydrazine for the fluorogenic labeling of carbamates, ureas, organophosphorous compounds, aliphatic amines, aldehydes, ketones, biphenyls, and some compounds of pharmaceutical interest, together with the column liquid chromatographic properties of these derivatives.

Watanabe and Imai (569, 570) compared the reactivities of NBD-F (fluoro derivative of NBD-Cl) with secondary amino acids in the hope of reducing reaction time in comparison with the other halogeno-NBDs (NBD-Cl and NBD-Br). They state that NBD-F seems to be superior to both NBD-Cl and NBD-Br because of its high reactivity toward secondary amino acids and because of the higher fluorescence intensities that are obtained.

Recently, ammonium 4-chloro-7-sulfobenzofurazan was proposed as a new fluorogenic thiol-specific reagent (571).

Benson and Hare (572) demonstrated that a modified formulation of Roth's reagent (114) can be used to detect α-amino acids, peptides, and proteins in the picomole range. They overcame most of the shortcomings described for Roth's method by increasing the concentration of 2-mercaptoethanol 10-fold and adding Brij to the reagent mixture. They stated that o-phthalaldehyde exhibits greater sensitivity in comparison with ninhydrin and fluorescamine (573, 574). Larsen and West (575) presented a fluorometric procedure for quantitative amino acid analysis of a peptide hydrolysate by precolumn o-phthalaldehyde derivatization and high-performance liquid chromatography with picomole detection limits.

Fluorescamine has been applied in forensic toxicological analysis (576) and for the detection and determination of polyamines in biological material via ion-pair high-pressure liquid chromatography (577).

de Silva and Strojny (578) presented a systematic examination of a series of drugs having either a primary aromatic or aliphatic amine group that were reacted with fluorescamine in solution and on thin-layer chromatographic plates. They established the analytical parameters for optimal reaction and sensitivity that can be utilized for their detection in biological fluids. The compounds investigated include amphetamine sulfate, phenylpropanol amine, p-aminobenzoic acid, procaine hydrochloride, p-aminosalicylic acid, several sulfonamides, 7-aminoclonazepam, 7-amino-3-hydroxyclonazepam, l-dopa, dopamine, and 1-adamantanamine.

A quenching fluorometric method for primary amines using sodium-β-naphthoquinone-4-sulfonate was reported by De Moerloose and Baeyens (579). This reagent possesses native fluorescence properties, depending on pH of the solution. Several compounds, mainly primary amines, react with this reagent, causing a decrease in fluorescence intensity that can be quantitated at pH 9. The method was worked out for glycine, d-phenylalanine, sulfamerazine sodium, sodium 4-aminosalicylic acid, norepinephrine bitartrate, piperazine hydrate, and acetylacetone, at $\lambda_{ex} = 345$ nm, $\lambda_{em} = 445$ nm.

The determination of primary amines by means of fluorescent Schiff base derivatives was studied by Hwang et al. (580). The reactions between amines and carbonyl compounds that lead to the formation of Schiff bases (imines) involve two reversible steps, the first involving the formation of an unstable tetrahedral carbinolamine and the second involving the dehydration of this intermediate to form the imine

$$R_1NH_2 + R_2R_3CO \rightleftharpoons \begin{bmatrix} R_2 & NHR_1 \\ & \diagdown C \diagup \\ R_3 & \diagup \quad \diagdown OH \end{bmatrix} \rightleftharpoons R_2R_3C{=}NR_1 + H_2O$$

The Schiff bases of aliphatic aldehydes are unstable and polymerize readily, whereas those derived from aromatic aldehydes are stabilized by their effective conjugation systems. For any given aldehyde the rate of Schiff base formation increases with the basicity of the amine; primary aliphatic amines thus form Schiff bases more rapidly than anilines and other weak bases. As secondary amines do not yield Schiff bases (alkylidenediamines are formed instead), the fluorogenic reactions described by these authors seem to be effectively specific for primary aliphatic amines. 1-Pyrenealdehyde and 2-fluorenealdehyde were found to be suitable fluorogenic reagents for primary aliphatic amines, forming Schiff bases that are very stable and intensely fluorescent in acidic ethanol. The derivatives of 1-pyrenealdehyde can be detected at concentrations less than 1 ng ml^{-1}. Derivatives of 1-pyrenealdehyde can be readily produced by reactions at the surface of a TLC plate. Combination of this approach with a simple deproteinizing procedure permits analysis for nanogram quantities of primary amines (amphetamine, histamine) in blood serum.

Examples of the use of fluorescent heterocycles in chemistry and biology were reported in 1976 (581), and a study of the fluorescence analysis of primary amines in nonaqueous solutions appeared one year later (582). Frei et al. (583) developed an HPLC fluorescence detection system for amines.

Khalaf and Rimpler (584) described the synthesis of 5-isothiocyanato-1,8-naphthalenedicarbox-4-methylphenylimide, a new fluorescence reagent for qualitative and quantitative analyses of compounds containing amino groups.

Guebitz et al. (585) described fluorescence derivatization of tertiary amines with 2-naphthyl chloroformate. Many drugs contain amino groups but, as is well known, apart from primary and secondary amines, practically no satisfactory reactions are known for the derivatization of tertiary amines into fluorescent active substances. The reaction of tertiary amines with chloroformates, which undergo dealkylation under suitable conditions forming carbamates with the resulting secondary amine, was taken from gas chromatography and applied for the HPLC analysis of these amines with fluorescence detection. In addition to primary and secondary amines, tertiary amines also form carbamates under dealkylation when heated with chloroformates (Fig. 2.17).

Highly fluorescent carbamates are formed, with more than 24 h fluorescence stability (λ_{ex} = 275 nm; λ_{em} = 335 nm). The authors applied their method to antihistamines, analgesics, local anesthetics, and psychotropic drugs and extended their derivatization procedure with good results to TLC chromatography with fluorodensitometric evaluation.

Fourche et al. (586) described the fluorescence emitted by 21 aminophosphonic acids (I) on reaction with o-diacetylbenzene, o-phthaldialdehyde, and fluorescamine and compared this with the fluorescence observed with the corresponding

$$R\!-\!CH\!-\!PO_3H_2$$
$$\underset{\displaystyle NH_2}{|}$$

(I)

carboxylic analogs. A good similarity has been shown in both series for excitation and emission wavelengths; the fluorescence intensities are generally lower in the phosphonic series.

Fluorometry of some imidoximes and their possible analytical applications were described in 1976 (587).

De Leenheer et al. (588) described thin-layer fluorescent scanning studies for the analysis of 9-alkylaminoacridines and 9-alkylacridine derivatives that have importance as potential fluorescent labeling agents.

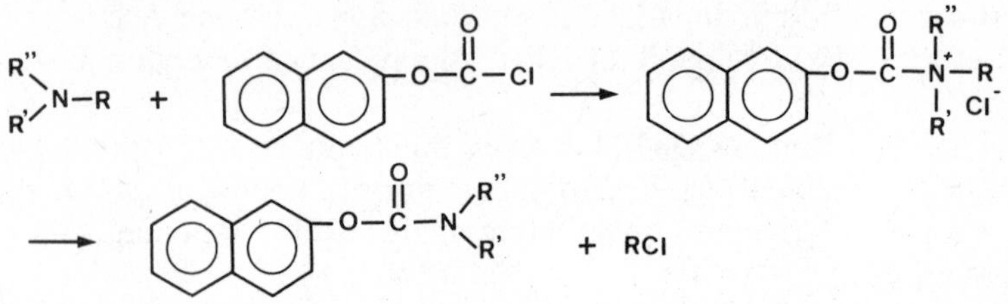

Fig. 2.17. Reaction scheme of 2-naphthyl chloroformate with tertiary amines. (From ref. 585.)

Fluorometric determination of pyridine and some substituted pyridines after hydroxylation with the Hamilton hydroxylation system has been reported by Wong and Connors (589). The system consists of hydrogen peroxide, catechol, and ferric ion. The resulting hydroxylated pyridines are measured fluorometrically (λ_{ex} = 290–340 nm; λ_{em} = 350–410 nm). The method is applicable to 10^{-6}–10^{-4} M aqueous solutions of several pyridine compounds, and linear fluorescence concentration plots are obtained.

Many methods for the determination of ketones by formation of hydrazones depend on absorption spectroscopy and do not have sufficient sensitivity for biochemical and biological analysis. In order to establish a more sensitive analytical method, some fluorescent derivatives of hydrazine were investigated (590). A reagent specifically selected for determination of α-oxo acids was 4'-hydrazino-2-stilbazole (I) (Fig. 2.18), which forms hydrazones (II) with generally strong fluorescing properties (e.g., with pyruvic acid). By the application of this method, a simple and microdetermination procedure for total α-oxo acids in blood was established (591). No interference by biological materials was noticed and the sensitivity was about 100–200-fold to that of the existing colorimetry.

Carbonyl Compounds.   Brandt and Cheronis (592) described the detection and estimation of acetone at submicrogram ranges by fluorometry through reaction with 2-diphenylacetyl-1,3-indandione-1-hydrazone.

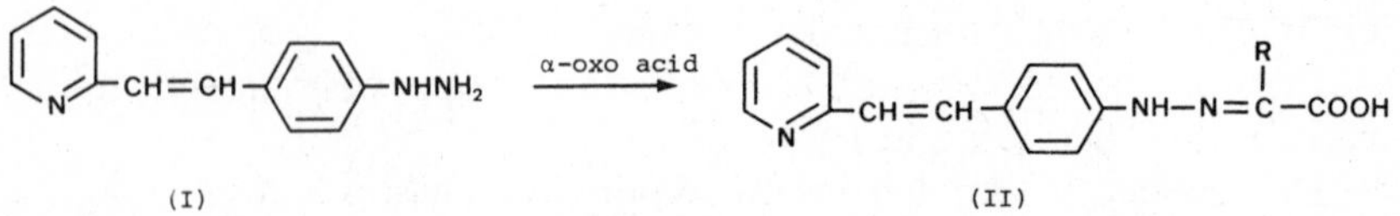

(I)   (II)

Fig. 2.18.   Fluorescence reaction for α-oxo acids with 4'-hydrazino-2-stilbazole. (From refs. 590, 591.)

DTAN (2,2'-dithiobis(1-aminonaphthalene)

Spikner and Towne (593) described the quantitative conversion of sub-microgram quantities of α-keto acids (glyoxylic, α-ketobutyric, α-keto-caproic, α-ketoglutaric, α-ketoisocaproic, α-ketoisovaleric, oxalacetic, phenylpyruvic, and pyruvic acid) to corresponding substituted quinoxa-lines by reaction with o-phenylenediamine. The fluorescing derivatives are stable and allow sensitive (0.05–0.50 μg ml^{-1}) determinations on small reaction volumes.

In 1978 Ohkura et al. (594) developed a sensitive fluorometric method for the determination of aromatic aldehydes based on their reaction in dilute sulfuric acid with a new reagent, 2,2'-dithiobis(1-aminonaphthal-ene) (DTAN), in the presence of tri-n-butylphosphine, sodium sulfite, and sodium phosphite at ambient temperature. According to the authors, the fluorescences produced are fairly characteristic of individual aldehydes and the method appears to be extremely selective for aromatic aldehydes and very sensitive, especially for p-hydroxybenzaldehyde, o-methoxy-benzaldehyde, p-methoxybenzaldehyde, and p-tolualdehyde, which can be determined at nanogram per milliliter concentrations. Ohtsubo et al. (595) isolated and characterized in 1979 the structure of the fluorescent product formed from benzaldehyde (I) or p-hydroxybenzaldehyde (II) with 2,2'-dithiobis(1-aminonaphthalene) in the presence of tri-n-butyl-phosphine, phosphite, and sulfite as the corresponding 2-arylnaphtho[1,2-d] thiazole.

Fluorescence Induction and Enhancement in Thin-Layer Chromatography. Shanfield et al. (596) found that a simple gaseous electrical discharge can induce visible fluorescence in a variety of organic compounds separated

DTAN derivative of - benzaldehyde : R=H (I)

- p-hydroxybenzaldehyde : R=OH (II)

on silica gel thin-layer chromatoplates. Exposure times ranging from 5 s to 5 min sufficed to detect submicrogram quantities with the unaided eye. As the method is simple and convenient, it appears to have broad applicability to a wide variety of organic compounds. The mechanism involved in producing fluorophors seems to be complex and many factors remain to be investigated.

Uchiyama and Uchiyama (597) investigated the enhancement of fluorescence in thin-layer chromatography by spraying viscous organic solvents and studied the elucidation of the physicochemical factors that affect this phenomenon. They state that the effect is not related to the transmissibility of light through the thin layer but to the properties of the spraying reagent (polarity, viscosity, acidity). As an example, the appropriate reagents for 5-dimethylamino-1-naphthalenesulfonamides (DANS-amines) are nonpolar, viscous, and nonacidic solvents, such as a mixture of liquid paraffin and *n*-hexane, which enhance the fluorescence intensity 10-fold. The use of a solvent such as this liquid paraffin mixture enhanced the fluorescence intensity of benzo[*a*]pyrene 35-fold. In 1980 both authors (598) reported on the physicochemical properties of the most suitable spray reagent for fluorescence enhancement in thin-layer chromatography. They investigated the relationship between fluorescence enhancement and the properties of spray reagents using 5-dimethylamino-1-naphthalenesulfondimethylamide (DNS–DMA) as a fluorescent compound. A mixture of Triton X-100 and chloroform (2:8) showed good results for DNS–DMA and DNS–amine fluorescences on a plate were enhanced about 100-fold. The Triton X-100 mixture proved its use as well for other DNS derivatives, such as DNS–amino acids and DNS–peptides.

In 1981 Segura and Navarro (599) described the use of metal salts as fluorescence-inducing reagents in thin-layer chromatography. Their investigations were the result of an effort to overcome the limitations occurring by the use of ammonium hydrogen carbonate as introduced by Segura and Gotto (600) in 1974. In their procedure (599) the formation of fluorescent derivatives is induced by thermal treatment of the chromatographic plates in the presence of a metal salt or another inorganic compound. They state that silicon tetrachloride and tin (IV) chloride can be applied in gaseous form, whereas zirconyl chloride and zirconyl sulfate have to be applied in aqueous solution. Zirconium derivatives seem to have several advantages over other fluorogenic reagents, as convenience of application (the plates can be sprayed and dipped into the reagent, or the reagent can be incorporated into the layer prior to the development), higher sensitivity of detection (nanogram–picogram range), speed cf reaction (seconds or minutes), versatility and reasonable stability of the fluorophors (weeks). They applied their method to the analysis of various compounds, such as drugs, hormones, triglycerides, phospholipids, prostaglandins, carbohydrates, and amino acids.

Zhou et al. (601) compared thermal, $ZrCl_4$, and gas discharge methods to induce fluorescence for use in high-performance thin-layer chromatography. They found that the simple procedure of exposing the chromatoplates to HCl, HBr, or HNO_3 room temperature vapors, followed by heating, is sufficient to produce stable, strong fluorescence in a wide variety of compounds, which brings to HPTLC the sensitivity advantages inherent in fluorescence as a visualization technique.

Miscellaneous. Nelis and Sinsheimer (602) described a highly sensitive fluorometric procedure for the determination of aliphatic epoxides under physiological conditions. These compounds show high chemical reactivity and have become suspect, together with other epoxides, with regard to their mutagenicity. Their procedure is based on the alkylation of nicotinamide and subsequent reaction of the resulting N-alkylnicotinamides with a ketone as acetophenone in basic medium followed by acidification. Strongly fluorescent compounds with reasonable stability are produced (λ_{ex} = 370 nm; λ_{em} = 430 nm), allowing assays in the picomole range with good reproducibility. In a later study Nelis et al. (603) compared the alkylation of nicotinamide and 4-(p-nitrobenzyl)pyridine for the determination of aliphatic epoxides and concluded that the nicotinamide procedure has the advantage that the initial alkylation can be run under more physiological conditions and that there is an increase in the stability of the final chromophore.

Avigad (604) used dansyl hydrazine as a fluorometric reagent for thin-layer chromatographic analysis of reducing sugars. Reducing sugars, particularly aldoses, react readily with dansyl hydrazine, and the fluorescent hydrazones produced can be separated by TLC and quantitatively determined by spectrofluorometry after elution from the chromatograms (λ_{ex} = 360 nm; λ_{em} = 510 nm).

Kormos et al. (605) described the determination of isocyanates in air by liquid chromatography with fluorescence detection. Aliphatic and aromatic isocyanates react readily with N-methyl-1-naphthalenemethylamine to form fluorescent urea derivatives (Fig. 2.19). Excess reagent was separated from isocyanate urea derivatives by liquid chromatography. For fluorescence detection an excitation wavelength of 226 nm and a 300-nm cutoff emission filter were used. Detection limits lower than ppb scale were obtained for several isocyanates.

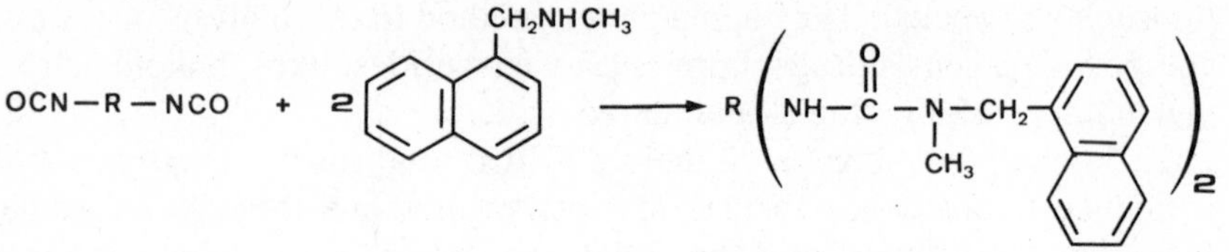

Fig. 2.19. Fluorescence reaction of isocyanates with N-methyl-1-naphthalenemethylamine. (From ref. 605.)

Other Uses of Fluorescence Spectrometry

☐ Ullman et al. (606) devised a general immunochemical method for the assay of haptens and proteins and applied it to morphine, a morphine–albumin conjugate, and human immunoglobulin G, employing a fluorescein-labeled antigen and a quencher-labeled antibody. By use of fluorescein and rhodamine as the fluorescer and quencher, respectively, dipole–dipole-coupled excitation energy transfer can occur within the antigen–antibody complex. The resulting quenching of fluorescence can be inhibited by competitive binding with unlabeled antigen. Alternatively, separate antibody samples can be labeled with fluorescein and rhodamine, respectively. Unlabeled antigen causes aggregation of the separately labeled components with resultant quenching.

☐ A review on fluorescence lifetimes of biomolecules was reported by Forster in 1976 (607).

☐ Handschin and Ritschard (608) determined fluorophor, protein, and fluorophor–protein ratios in fluorescamine and 2-methoxy-2,4-diphenyl-3(2H)-furanone fluorescent antibody conjugates.

☐ Burgett et al. (609) developed a noncompetitive method for the determination of the C3 component of human complement in serum, based on the use of a specific antibody covalently attached to derivatized polyacrylamide beads and a fluorescently labeled specific antibody.

☐ Thompson (610) determined levels of low-molecular-weight substances in human serum by homogeneous substrate-labeled fluorescent immunoassays.

☐ Kaplan et al. (611) evaluated and compared radio-, fluorescence, and enzyme-linked immunoassays for serum thyroxine.

☐ An excellent literature survey on fluorescence immunoassay was provided by Visor and Schulman (612) in 1981. They state that the major limitations of radioimmunoassay result from the biological hazards of radioactivity, the difficulties associated with various licensing requirements, the safe disposal of wastes, and the expense and inconvenience incurred by the short radioactive lifetimes of the labels. An alternative to the use of radiolabels is offered by the use of fluorescent probes, and their spectral characteristics are monitored as a function of the immunoreactions between the ligand and the antibody. The authors evaluate the methods currently in use and discuss future developments.

In the same year, Quattrone et al. (613) gave a review of approaches toward the fluorescence immunoassay of drugs, while Jolley (614) overviewed fluorescence polarization immunoassays for the determination of therapeutic drug levels in human plasma.

Hinsberg et al. (615) described the application of laser fluorometry to enzyme-linked immunoassay.

Li et al. (616) determined serum theophylline by fluorescence polarization immunoassay utilizing an umbelliferone derivative as a fluorescent label.

Collins et al. (617a) developed a solid-phase Cl_q-binding fluorescence immunoassay for detection of circulating immune complexes.

Rho and Stuart (617b) described the design and construction of an automated three-dimensional plotter for fluorescence measurements, permitting unique characterization of even very closely related structures.

Miller et al. (617c) reported on the use of derivative fluorescence spectroscopy in biochemical analysis.

☐   The number of clinical applications of fluorescence spectroscopy and fluorescence detection is growing rapidly; as a result, it is nearly impossible to cite even the most important of these approaches.

A review of clinical applications using liquid chromatography for the determination of therapeutic agents or endogenous compounds of clinical or biochemical importance was given by Vandemark and DiCesare (618).

Moules et al. (619) determined binding of amphiphatic drugs and probes to biological membranes.

A new fluorometric method was used for the study on the determination of antithrombin (III) (620). Spectrofluorodensitometry of conjugated bile acids on thin-layer chromatograms by sulfuric acid–ethanol treatment was mentioned by Touchstone et al. (621). Immunoinhibition was used for the spectrophotometric and fluorometric enzymic determination of serum creatine kinase MB isoenzyme (622). Dickson et al. (623) developed an automated fluorometric method for screening for erythrocyte glucose-6-phosphate dehydrogenase deficiency. A fluorometric assay method was suggested for examining the excretion of N-acetyl-β-glucosaminidase and β-galactosidase following surgery to the kidney (624).

Fluorometric methods have been proposed for the ultramicro determination of serum γ-glutamyltranspeptidase activity (625), for heparin in plasma (626), and for histamine in human blood through a modification of the Kremzner and Wilson method on 1 ml blood (627).

Watanabe and Hayashi (628) devised in 1972 a fluorometric method for 3-hydroxyanthranilic acid in human urine. The compound is extracted with ether at pH 3 and separated by development with benzene–ethanol–water (50 + 10 + 1) on a cellulose powder plate. After extraction with dioxane from the plate, the fluorescence was estimated ($\lambda_{ex} = 348$ nm; $\lambda_{em} = 410$ nm). The method allowed the authors to estimate urinary excretion of 3-hydroxyanthranilic acid of normal persons and patients with bladder cancer. Toseland et al. (629) developed in the same year a simple and sensitive method for the determination of 3-hydroxyanthranilic acid in human urine by thin-layer electrophoresis and fluorometry. 5-Hydroxyindoleacetic acid was determined in human urine by a simple fluorometric method (630) and by an HPLC method with fluorescence detection (631). Watanabe et al. (632) developed a fluorometric assay for 3-hydroxykynurenine based on the reaction with p-toluenesulfonyl chloride in basic condition to yield a fluorescent product. The assay has been used in combination with the TLC technique to estimate the urinary excretion of 3-hydroxykynurenine in control subjects and patients with bladder cancer.

Hinsberg et al. (633) described an enzyme immunoassay for the determination of insulin in human serum, based on fluorometric detection of enzyme activity. The authors employed a sandwich method using horseradish peroxidase as the label; the enzyme activity was measured by fluorescence using the substrate, p-hydroxyphenylacetic acid. The fluorescence measurements can be carried out by liquid chromatography–laser fluorometry or by conventional cuvette spectrofluorometry. Their procedure allows rapid and sensitive quantitation of insulin with a 45-min total incubation period and 7.9 pmol or 46 pg ml^{-1} detection limit.

Cashion et al. (634) detected nucleosides and nucleotides on paper with fluorescent thin-layer sheets.

Some other clinical applications of fluorescence spectrometry that may illustrate the importance of the technique are the estimation of phenylpyruvic acid in urine and serum by HPLC with fluorescence detection (635); the TLC separation and fluorescence densitometric determination of porphyrins (636); the fluorometric assay of sialic acid in the picomole range through a modification of the thiobarbituric acid assay (637); a study on the effects of thyroxine binding on the stability, conformation, and fluorescence properties of thyroxine-binding globulin (638); the fluorescent detection of tryptamine in the nanogram range (639), and a liquid chromatographic–fluorometric method for the analysis of picogram amounts of tryptophan metabolites in cerebrospinal fluid (640).

2.3. APPLICATIONS OF PHOSPHORIMETRY

2.3.1. Low-Temperature Phosphorimetric Methods

Historical Development

The possibility of qualitatively identifying organic molecules by their phosphorescence spectra was first suggested by Lewis and Kasha in 1944 (641). In 1957 the use of phosphorimetry as a new method of chemical analysis was reported by Keirs, Britt, and Wentworth (642). In 1958, Freed and Salmre (643) described the construction of a phosphorimeter used for analytical determinations of several drugs. The possibilities of chemical analysis by phosphorimetry were reviewed in 1962 by Parker and Hatchard (644). Freed and Vise (645) constructed a spectrophosphorimeter to determine the phosphorescence excitation and emission spectra of N-acetyl-L-tyrosine ethyl ester, whose spectrum was similar to that of tryptophan. In 1963 Winefordner and Latz (646) described the construction of a spectrophosphorimeter consisting of an unfiltered 150-W mercury arc for excitation, a rotating-can phosphoroscope, and a grating monochromator for the measurement of phosphorescence emission spectra. Latz (647) has reported a thorough study of the procedural factors that influence phosphorimetry as a means of chemical analysis. O'Haver and Winefordner (648) gave a mathematical approach on the influence of phosphoroscope design on the measured phosphorescence intensity. In 1968 Hollifield and Winefordner (649) described a rotating sample-cell method for increasing the precision of low-temperature phosphorescence measurements, with the idea of a spinning sample cell coming from NMR techniques.

Numerous papers describing the instrumentation and the qualitative and quantitative uses of phosphorimetry in nonaqueous solvents were published at the end of the 1960s, and their contents have been summarized in several reviews (650–655).

An excellent review on the various phosphorimetric techniques together with their analytical applications was published by Aaron and Winefordner in 1975 (656). During the same year O'Donnell and Winefordner (657) gave an extensive review on advances in instrumentation and methodology in phosphorimetry that should facilitate clinical analyses of those molecular species difficult or impossible to measure by conventional methods (colorimetry, fluorometry, etc.). Several of the most recent advances in phosphorimetry and several possible new techniques that offer promise for real biological system analyses, such as those found in the clinical laboratory, are considered.

Baeyens (658,659) gave a review on phosphorescence techniques considered from the analytical–pharmaceutical point of view.

Most of the work in the field of phosphorimetry has been carried out by Winefordners' research group (660–665), Gainesville, Florida, and one of the most recent reviews on uses of phosphorimetry for organic analysis comes from Ward et al. (666) from the same laboratory. These authors state that, in spite of the sensitivity of low-temperature phosphorescence (LTP) for many biologically and environmentally important molecules and the selectivity arising from time resolution in addition to spectral resolution, LTP has failed to catch on as a popular method for routine analysis because of two major inconveniences. First, in order to achieve a low temperature, some cryogenic equipment is necessary. Second, there is great inconvenience involved in introducing the sample into the system. On the other hand, conventional low-temperature phosphorimetry offers low detection limits, wide-range analytical calibration curves, and great selectivity of measurement. Modification of the environment by means of "heavy-atom solvents" containing heavy ions, such as I^- (NaI) or Ag^+ ($AgNO_3$), often increases phosphorescence intensities and decreases phosphorescence lifetime. According to the same authors, the precision of phosphorimetric measurements is typically 10% relative standard deviation; with care in cellpositioning and cleanliness, 1–2% RSD is possible, but difficult to achieve.

Wild (667) described the use of highly resolved spectra in Shpolskii matrices (microcrystalline n-alkanes) for luminescence spectroscopy at 77K. Dekkers et al. (668) reported phosphorescence polarization measurements of aromatic compounds oriented in stretched polyethylene films. The use of a synchromotor phosphoroscope was reported in 1976 (669) and the analytical possibilities of phase-resolved phosphorimetry were discussed in 1977 by Lue Yen and Winefordner (670); Williams (671) described an approach to phosphorescence measurement using a pulsed xenon source.

More extensive descriptions of the fundamental aspects of phosphorimetry and instrumentation can be found in the books by Schulman (672), Zander (650), Guilbault (673), Wehry (674), Winefordner–Schulman—O'Haver (675), Parker (652), Hercules (676) and, recently, Lumb (677).

Analytical Applications

A comprehensive survey of the native luminescence properties of antibacterial sulfonamides at room temperature and at 77K was presented by Bridges, Gifford et al. (678). They state that the fluorescence characteristics of sulfanilamide are similar to those of aniline. N^1-substituted sulfonamides containing a π-electron-deficient heterocycle are weakly or nonfluorescent, while those containing π-electron excessive heterocyclic rings, aliphatic chains, or acyl groups are strongly fluorescent. The N^4-substituted sulfonamides are found to be nonfluorescent except for N^4-acetylsulfapyridine. The influence of solvents and of pH on the fluorescence of sulfanilamide and its isomers is reported and extensive discussions of these phenomena are given. A procedure for the direct fluorometric estimation of sulfonamides in serum is presented by the authors. Serum, 0.2 ml, was diluted with 1.0 ml of water and the proteins were precipitated by addition of 1.0 ml 20% trichloroacetic acid. The resultant mixture was then shaken and centrifuged. One milliliter of the supernatant was diluted with 50 ml of buffer and the pH was adjusted to the required value by the addition of 1 M NaOH. For sulfanilamide, sulfacetamide, and sulfaguanidine, 0.6 ml of 1 M NaOH was added to 0.025 M borate buffer (pH 8.0) and for sulfamethoxazole, 1.0 ml of 1 M NaOH was added to Walpole acetate buffer (pH 4.0). The solutions were then made up to 100 ml with buffer. Instantaneous fluorescence readings are required because of the rapid photolysis of dilute sulfonamide solutions.

Gifford (679) described a device capable of scanning thin-layer chromatographic strips at 77K for directly measuring phosphorescence emission. Gifford et al. (680) published in 1975 the construction and evaluation of this scanning thin-layer single-disk multislot phosphorimeter and gave an expression relating the dimensions of the single-disk phosphoroscope to the measured phosphoroscence intensity. They demonstrated the potential application of the device by separating sulfamerazine, sulfamethazine, and sulfadiazine by thin-layer chromatography followed by qualitative and quantitative estimation. Sample strips are cooled by conduction, which is an advantage over techniques using a Dewar flask, in that the number of layers of silica through which radiation must pass is reduced; the absence of liquid nitrogen in the light path reduces scatter. The authors conclude that the combined technique of TLC and phosphorimetry is considered to have great potential for use in pharmaceutical and biological analysis. Miller et al. (681) reported on solvent enhancement effects in thin-layer phosphorimetry and described the construction and use of an improved thin-layer phosphorimeter. Their device permits flexible chromatography media to be scanned at 77K and allows a complete characterization of the luminescence properties of the chromatographically separated solutes. Various compounds were examined, among which were 4-aminobenzoic acid, phenobarbitone, N-methylphenobarbitone, chlorpromazine·HC1, procaine·HC1, and several sulfonamides. They showed that the phosphorescence intensities of a variety of adsorbed materials are greatly increased by spraying the chromatography medium with a suitable solvent (e.g., ethanol) immediately before examination. The magnitude of the effect seems to depend on the stationary phase, the structure of the adsorbed material, and the solvent sprayed. Their instrument is also suitable for examining the luminescence of adsorbed molecules at ambient temperatures.

McDuffie and Neely (682) determined griseofulvin by time-resolved phosphorimetry in mixtures with dechlorogriseofulvin based on the tenfold difference in phosphorescence lifetimes of 0.11 s and 1.16 s, respectively.

Morrison and O'Donnell (683) determined diphenylhydantoin in plasma by phosphorescence spectrometry after permanganate oxidation, based on a procedure described by Dill et al. (684). The method is based on extraction of the drug from plasma into 1,2-dichloroethane at pH 6–7, a subsequent extraction into alkaline medium followed by oxidation with basic $KMnO_4$. The oxidation product, benzophenone, is extracted into methylcyclohexane, a solvent with the ability to form a clear glass at liquid nitrogen temperatures, followed by phosphorescence measurement at 446 nm on excitation with 260 nm. A detection limit of 0.05 μg ml^{-1} with an error of less than 10% was found. The presence of phenobarbital or other commonly used anticonvulsants did not interfere with the assay.

Gifford et al. (685) described the determination of oxybarbiturates by means of their low-temperature luminescence and showed that, in certain cases, the luminescence spectra and phosphorescence lifetimes could provide information on the identity of an unknown barbiturate. In a later communication, King et al. (474) described a further procedure that facilitates the characterization of certain barbiturates by making use of the temperature dependence of their fluorescence. Gifford et al. (475) examined the phosphorescence excitation and emission spectra of 16 barbiturates and certain model compounds. Phosphorescence was observed at a lower wavelength than that of the room temperature fluorescence for those barbiturates containing a 5-phenyl substituent. They discussed the effects of steric interactions and pH on the phosphorescence of the phenyl substituted barbiturates and determined detection limits at 77 K for the barbiturates in common use, situated at 0.3–0.5 μg ml^{-1}. King and Gifford (476) correlated structure luminescence in the thiobarbiturates series.

Miles and Schenk (686) reported in 1973 on the phosphorescence of seven phenylethylamines (amphetamines, desoxyephedrine, ephedrine, phenylpropanolamine, methoxyphenamine, phenylephrine, epinephrine) and 10 barbiturates (amobarbital, aprobarbital, barbital, butabarbital, cyclopentallyl barbituric acid, diallylbarbituric acid, pentobarbital, phenobarbital, secobarbital, thiopental). Phenobarbital behaves unusually in

(I) Phenothiazines studied by Gifford et al. (687).

that its phosphorescence emission maximum at 370 nm is at higher energies than its fluorescence emission maximum at 415 nm. The authors showed that each emission arises from a different chromophore in phenobarbital.

Gifford et al. (687) provided low-temperature luminescence data for a series of commonly used phenothiazines (I). Wavelength maxima, phosphorescence lifetimes, and detection limits are given in their work. Most of the compounds studied showed typical double excitation and single phosphorescence maxima at 250–265, 305–315, and 485–505 nm. The phosphorescence lifetimes were almost all in the range 60–80 msec. The nature of the substituent at position 10 has little effect on the spectra, probably since substituents in all the phenothiazines studied have three carbon atoms separating the ring system from the nitrogen atom. The groups in position 2, which had the largest effects, were electron-withdrawing groups, such as $—CF_3$, $—SO_2N(CH_3)_2$, and $—CN$, which produced a bathochromic shift of phosphorescence and also caused detectable fluorescence.

Low-temperature luminescence properties of anti-inflammatory agents in blood or plasma following thin-layer chromatographic separation have been reported by Strojny and de Silva (688) in 1975. de Silva et al. (689) performed luminescence studies (fluorescence and phosphorescence) of pharmaceuticals from the tetrahydrocarbazole, carbazole, and 1,4-benzodiazepine class on thin-layer chromatographic plates at 77K. Strojny et al. (690) recently developed an analytical procedure for the phosphorimetric determination of nanogram concentrations of diazepam, nordiazepam, 3-hydroxydiazepam (temazepam), and 3-hydroxynordiazepam (oxazepam) in blood, plasma, or urine. Their assay is specific by virtue of selective extraction, TLC separation, and phosphorimetric spectral characteristics. 2-Amino-5-nitrobenzophenone contamination in nitrazepam was determined by low-temperature spectrophosphorimetry (691). The compound shows very intense phosphorescence in ethanol glasses at liquid nitrogen temperature; in the same conditions nitrazepam is not phosphorescent.

Sanders et al. (664) studied the phosphorescence characteristics of 37 antimetabolites of importance in plant and animal growth and indicated the possible application of phosphorimetric techniques to the determination of 17 of these compounds in biological materials.

Boutilier et al. (692) studied the use of sodium iodide and silver nitrate as inorganic probes in the phosphorimetric analysis of the nucleosides naturally occurring in DNA using methanol–water (10–90 v/v) at 77K. Both iodide and silver increased the phosphorescence quantum yield and thus improved the limits of detection, with silver nitrate being more effective for all but deoxyguanosine of the four nucleosides studied (deoxyadenosine, deoxyguanosine, thymidine, cytidine). This enhancement of phosphorescence was attributed to silver ion acting as an internally bound heavy atom and to iodide ion acting as an external heavy atom. Nanogram range minimum detectable quantities were reached for all nucleosides.

Mayer et al. (693) discussed the use of phosphorescence as a detection method for purine-containing compounds on thin-layer chromatograms. Aqueous solutions were spotted on cellulose chromatogram sheets, developed with isobutyric acid–water–ammonia (75 + 25 + 1), air-dried, and phosphorescence was induced by dipping the sheets into a Dewar filled with liquid nitrogen and simultaneously irradiating the plate with shortwave UV light. From all the adenine compounds tested (adenine, adenosine, AMP, ADP, ATP, cyclic 3',5'-AMP), excellent phosphorescence responses were obtained on cellulose chromatograms, with nanomole range detection limits.

Boutilier and Winefordner (694) studied the influence of type and concentration of external heavy atoms on phosphorescence lifetimes of phenanthrene, carbazole, quinine, 7,8-benzoflavone and thiopropazate at 77 K in ethanol–water (10–90 v/v). Iodide (KI), silver ($AgNO_3$), and thallous ($TlNO_3$) ions were tested. They observed that in all cases silver ion had a larger effect on the phosphorescence lifetime than iodide, which indicates that a "charge transfer complex" heavy atom effect is a more likely mechanism than an "exchange" heavy atom effect. Thallous ion was found to have relatively little effect on the low-temperature lifetimes and spectra, which is opposite to its effect in room temperature phosphorimetry. Both authors described the use of a pulsed laser (N_2 laser or flashlamp pumped dye laser) time-resolved phosphorimeter for phosphorescence lifetime and detection limit measurements (695). Detection limits and lifetimes are given for several polyaromatic hydrocarbons and drugs (e.g., morphine, codeine, procaine, butazolidine, vinblastine) at 77 K in ethanol–water (10–90 v/v) with no heavy atom, with 0.75 M KI, and with 0.1 M $AgNO_3$. They state that detection limits with I^- present are generally inferior to those without heavy atom present because of background noise from impurities in the KI and that detection limits with Ag^+ present are generally comparable to or better than those without Ag^+.

Acuña et al. (696) reported on the quantitative analysis of benzoic and toluic acids by phosphorimetry, as their fluorescence quantum yields are practically zero. Benzoic acid, o-, m-, and p-toluic acids undergo dimerization in nonpolar solvents. The phosphorescence excitation spectra at different acid concentrations show that two species (monomer and dimer) with different phosphorescent quantum yield coexist in solution. According to the authors, these difficulties are overcome if a polar solvent (isopentane:ether) is used because only one species, probably the solvated monomer, is then present. A rigid, transparent glass with no tendency to break up is obtained from this solvent at 77 K. Linear working curves over the concentration range $10^{-7}-10^{-3}M$ are obtained, with a lower detection limit of about 0.02 μg ml^{-1}.

Mau and Puza (697) contributed to a better understanding of the phosphorescence behavior of chlorophylls as disagreements on this subject were reported in literature.

Miscellaneous drugs have been investigated by low-temperature phosphorescence during the last decade, with varying importance from a pharmaceutical or clinical point of view. The solvent effect was investigated on the phosphorescence of cocaine and its model compound, methyl benzoate (698). 6-Mercaptopurine and related compounds were studied in acidic, neutral, and alkaline ethanolic glasses and a technique combining TLC and phosphorimetric scanning was suggested for trace analysis of these compounds in human blood plasma (699). β-Glucuronidase can be sensitively assayed in biological materials based on the phosphorimetric determination of p-nitrophenol, formed enzymatically from p-nitrophenyl β-D-glucuronide (700). 3-Methoxy-4-hydroxyphenylethyleneglycol, a major metabolite of norepinephrine, can be determined phosphorimetrically in urine at nanogram scale after separation on Amberlite CG-50, oxidation with periodate to vanillin and measurement of the latter compound in a mixture of ether and ethanolic potassium hydroxide (701). Pattern recognition techniques for low-temperature luminescence examined for different instrumental parameters was recently suggested for the identification of hazardous chemicals spilled in the environment (702). Phosphorimetric investigation of several tryptophan metabolites and the determination of kynurenic acid in urine was reported by Hood and Winefordner (703), and recently a study was reported on the phosphorescence and optically detected magnetic resonance of the tryptophan residue in human serum albumin (704).

2.3.2.   Room Temperature Phosphorimetric Methods

Historical Development

Solid-surface luminescence techniques—especially room-temperature phosphorescence (RTP)—have received considerable attention in the last few years. Room-temperature phosphorescence has recently attained much interest, as this relatively new analytical technique is based on the phosphorescence emitted at room temperature by organic compounds adsorbed on various solid supports, such as filter paper, silica gel, and sodium acetate.

Schulman and Walling (705) observed room-temperature phosphorescence from molecules adsorbed on paper and on the surfaces of thin-layer chromatography matrices as silica and alumina. Comparable observations, though, had already been made by Roth (706), who reported a study of the RTP of various compounds on several supporting media.

Lloyd and Miller (707) did a literature search on the first observations of phosphorescence of adsorbed molecules at room temperature and reported that this phenomenon had been "discovered" no less than four times, going back to 1944. For example, Brown (708) observed the yellow phosphorescence of a naphthiminazole on a paper chromatogram in ultraviolet light for several seconds after withdrawal of the light source, a report that might have been overlooked in literature as his observations were incidental to the main topic of his paper and were not mentioned in the summary.

Paynter et al. (709) developed and applied the RTP phenomena toward analytical usefulness and described the method as fast, economical, and convenient for analyzing a variety of molecules, many of biological interest.

Aaron and Winefordner (710) gave in 1975 a review of the various phosphorimetric techniques, including RTP, together with their analytical applications. They state that with necessary improvements, this method should provide a new and sensitive analytical tool, particularly useful for molecules of biological interest.

Seybold and White (711) described the use of the external heavy-atom effect in RTP analysis. Vo-Dinh et al. (712) studied the use of sodium iodide as external heavy atom perturber for increasing phosphorescence quantum yields, while Jakovljevic (713) demonstrated the analytical potential of lead and thallium salts for quantitative RTP.

Lloyd (714) mentioned the use of packed flow-through microcells for the measurement of the room-temperature phosphorescence spectra of adsorbed compounds.

Schulman and Parker (715) studied the effects of moisture, oxygen, and the nature of the support–phosphor interaction on the RTP of some polar organic compounds adsorbed on various supports.

Vo-Dinh et al. (716) designed a simple phosphorimetric instrument for RTP with a continuous filter paper device.

Lue Yen Bower et al. (717) described in 1980 the present status and future of room-temperature phosphorimetry. They state that the major limitation to RTP is the ever-present phosphorescence background (~400–600 nm) in virtually all substrates in which the analyte produces significant phosphorescence signals. They and others have evaluated substrates for RTP on a rather random, haphazard basis with relatively little success in overcoming the major limitation. They estimate the future potential of RTP to be immense in clinical, environmental, forensic, and industrial applications because of its potential simplicity, capability of automation, and the possibility of prior chromatographic separations on the same substrate, particularly if the phosphorescence background can be minimized.

Parker et al. (718) reviewed the development and physical aspects of RTP as a new technique for chemical determinations. In a second part (719), basic factors concerning the practical analytical technique are discussed, such as methods of sample preparation, special instrumentation, and quantitative capability, together with a listing of the variety of organic compounds reported to display room temperature phosphorescence.

Miller (720) gives an extensive review on the phenomenon of room-temperature phosphorescence and describes it as a promising trace analysis method with great sensitivity and selectivity. Equipment, matrices and solvents, spectra, lifetimes, and detection limits for RTP studies are discussed.

A thorough review on recent uses of phosphorimetry for organic analysis is presented by Ward et al. (666). According to these authors, an ideal substrate for analytical RTP work has not yet been found. They state that of all the substrates tried, filter paper seems to be the most promising because it has been observed to induce phosphorescence from a great number of compounds; the broadband background of filter paper, however, often interferes with quantitative measurements. References on the use of various substrates besides filter paper, the RTP observations for nonionic compounds, polyaromatic hydrocarbons and ionic compounds, the influence of heavy-atom solvents, and the comparison of LTP–RTP detection limits are mentioned by the same authors.

Hurtubise (721) recently published an excellent book on solid surface luminescence analysis in which he devotes a chapter to some interactions and proposed mechanisms that result in RTP of organic compounds. He describes RTP as a sensitive and selective analytical tool of which many theoretical and practical aspects remain to be developed. In most cases, it should be possible today to adsorb an organic compound onto a solid surface and obtain analytically useful RTP data. A chapter on RTP applications illustrates the analytical potential of RTP. According to this author, the speed, simplicity, sensitivity, and selectivity of the technique should make RTP applications appear in a variety of areas. As the RTP data can be obtained rapidly and inexpensively, RTP should be readily adaptable to routine determinations in quality control laboratories.

The most common explanation for the RTP phenomenon is that the matrix holds the adsorbed molecule rigid and thereby restricts vibrational motions necessary for nonradiative decay from the triplet state. An alternative explanation is that RTP arises because compounds become "trapped" within the matrix and are thereby protected from quenching by molecular oxygen (722).

Techniques for fluorometric or phosphorimetric detection in thin-layer chromatography continue to be developed and refined (723).

McAleese et al. (724) described in 1980 a simple, rapid, reproducible, and economical technique for RTP sample preparation that alleviates moisture and oxygen quenching. They postulated that the impregnating agent's ability to decrease moisture permeability explains the inhibition of moisture quenching.

Phosphorescence can also be observed at room temperature in aqueous solutions under certain conditions (e.g., by protecting the solute using detergent micelles, as will be described further).

Ward et al. (725) investigated the use of rinsing and heating of filter paper in an attempt to reduce phosphorescence background at room temperature. They summarized that washing or heating the paper is generally shown not to give a significant reduction in the background phosphorescence of filter paper used as a support for room temperature phosphorimetry.

Room-Temperature Phosphorescence on Solid Surfaces

The application of RTP techniques on solid surfaces in the field of drug analysis is still developing. As stated earlier, the technique is attractive because of its simplicity, speed, economical character, nondestructive properties, and the possibility that is offered in several cases for selectively increasing phosphorescence emission of specific components in mixtures using the heavy-atom effect.

In 1977 von Wandruszka and Hurtubise (726) described analytical findings for the RTP of folic acid, p-hydroxybenzoic acid, and benzocaine.

Folic acid, when adsorbed on sodium acetate, has phosphorescence excitation and emission maxima at 320 nm and 465 nm, respectively, with a detection limit of 4 ng per sample spot; the maxima for p-hydroxybenzoic acid were 285 nm and 420 nm, respectively, with a 20-ng limit of detection. Benzocaine did not show RTP when adsorbed on sodium acetate but could be readily hydrolyzed by refluxing in dilute acid, followed by neutralization with aqueous alkali producing the characteristic p-aminobenzoic acid phosphorescence when applied to sodium acetate and evaporated to dryness. The same authors published during the same year an extensive study on the room temperature phosphorescence of compounds adsorbed on sodium acetate (727). Comparisons of molecular structures and consideration of reflectance, fluorescence, and infrared spectra allowed the postulation of certain molecular criteria for room temperature phosphorescence. They considered the interactions of p-aminobenzoic acid with a sodium acetate surface in detail. The adsorption was found to involve the formation of the sodium salt of p-aminobenzoic acid on the sodium acetate surface, as well as hydrogen bonding. The RTP technique of compounds adsorbed on sodium acetate as presented by these authors has the analytical advantages of selectivity and insensitivity to moisture; analytical data for several compounds are reported.

The room-temperature phosphorescence of polyaromatic hydrocarbons was investigated by several authors. Vo-Dinh et al. (728) studied the RTP of anthracene, coronene, phenanthrene, pyrene, benzo[a]pyrene, and carbazole on paper support and stated that heavy-atom-containing solvents such as silver nitrate and sodium iodide were found to have specific effects in inducing phosphorescence emission from these compounds. Their phosphorescence data provide complementary information to the fluorometric data of these polyaromatic hydrocarbons and permit the use of time-resolved and phase-resolved methods. Lue-Yen Bower and Winefordner (729) studied RTP of numerous molecules with emphasis on several polynuclear aromatic hydrocarbons. They state that the heavy-atom effect leads to a significant enhancement of the RTP signals of the latter compounds with the trend being $Tl^+ > Ag^+ > Pb^{2+} > Hg^{2+}$; the Tl^+ also resulted in enhanced spectral features of emission bands. The authors could induce RTP from the polynuclear aromatic hydrocarbons on a sodium acetate support as well as on filter paper. Vo-Dinh and Hooyman (730) applied the technique of selective external heavy-atom perturbation (SEHAP) to improve RTP analysis of complex samples. They demonstrated that heavy atoms can selectively increase the phosphorescence emission of specific components in mixtures. They investigated the usefulness of many heavy-atom perturbers for a wide variety of polynuclear aromatic compounds of environmental interest in coal-derived products.

According to the authors, the SEHAP technique extends the analytical possibilities of RTP analysis, since targeted components in complex samples can be selectively monitored. Their experimental procedures consist of spotting 3 µl of solution on a filter paper circle (0.6 cm in diameter), followed by a 3-min drying period using an infrared heating lamp. The sample is then transferred to the sample compartment for spectroscopic measurements. Warm and dry air flows through the compartment during the measurement to ensure a dry sample and the removal of moisture present in the sample chamber. According to the authors, all the steps mentioned (from spotting to measuring) take approximately 10 min, and up to 20 samples can be prepared at the same time. Heavy atoms were inserted using three procedures. The first method consisted of delivering 3 µl of heavy-atom solution followed immediately by spotting 3 µl of sample solution on the same substrate area. The second method consisted of reversing the order in delivering sample and heavy-atom solution. The third method employed paper already pretreated with the heavy-atom solution by dipping the paper strip into a heavy-atom solution and subsequent drying at room temperature. The authors did not observe specific trend with these three different methods and chose therefore the first method for most RTP measurements. Vo-Dinh et al. (731) published in 1980 a highly sensitive RTP method for the identification and quantification of polynuclear aromatic compounds in a synthetic fuel. The selectivity of their analysis was greatly improved by using selective heavy-atom purturbation and synchronous excitation scanning. Synchronous luminescence and RTP were used by Vo-Dinh et al. (732) for the analysis of a workplace air particulate sample.

Ford and Hurtubise (733) noticed the room temperature phosphorescence of the phthalic acid isomers, p-aminobenzoic acid, and terephthalamide when adsorbed on dried silica gel. They discussed the analytical potential of determining terephthalic acid by RTP and suggest that the mode of interaction between the adsorbed molecule and the silica gel surface is mainly hydrogen bonding. Since no RTP signals generally were obtained by drying the chromatoplate first and then putting the sample on the dried plate, it was assumed that water and ethanol are adsorbed very rapidly by the dried plates and that these molecules occupy sites on silica gel that would otherwise be available to other molecules.

Meyers and Seybold (734) examined the luminescence properties of tryptophan, tyrosine, their methyl esters, a tryptophan–tyrosine dipeptide, and indole in aqueous solutions and adsorbed on filter paper at room temperature. They found that the RTP of tryptophan on filter paper is enhanced 455-fold; that of its methyl ester, 340-fold; and that of indole, 370-fold by addition of 1.0 M sodium iodide, the enhancement factors apparently being the largest ever reported for any compounds. The dipeptide exhibits RTP from its tryptophan residue, which is also strongly enhanced by added sodium iodide. According to the authors, the RTPs of adsorbed tyrosine and its methyl ester were not detectable without heavy-atom enhancement; the methyl esters of tryptophan and tyrosine were considerably less fluorescent than the parent amino acids.

Parker et al. (735) investigated in 1979 the phosphorescence properties of a number of pteridines, among which was folic acid at room temperature and 77 K. All the pteridines that exhibited phosphorescence at 77 K showed RTP when the compounds were adsorbed on sodium-acetate-impregnated paper. Many of these pteridines showed both RTP and an E-type delayed fluorescence emission. The study provides additional insight into the RTP phenomenon and indicates that the selectivity and sensitivity of the RTP technique should offer an excellent potential to avoid complex separations in biological analyses.

Ford and Hurtubise (736) described a phosphoroscope and reflection assembly for a spectrodensitometer in order to obtain RTP data from several nitrogen heterocycles, mainly quinoline and phenanthroline derivatives, adsorbed on dried silica gel and filter paper. They discussed the effects of acid and different gases on the RTP of these compounds adsorbed on silica gel and filter paper and considered the analytical potential of this new approach. Silica gel was stressed over filter paper in the RTP work because RTP can be combined with the speed and versatility of TLC on silica gel for analytical determinations.

In 1980, Ford and Hurtubise (737) investigated the RTP of benzol[f]quinoline adsorbed on several silica gel samples by luminescence, reflectance, and infrared spectroscopy. Their results showed that silica gel chromatoplates containing a polymeric binder with carboxyl groups were the best samples for inducing strong RTP from benzo[f]quinoline, the polymer itself being essential for inducing strong RTP. They postulated that benzo[f]quinoline was adsorbed flatly on the surface, with the carboxyl groups anchoring benzo[f]quinoline via hydrogen bonding with π electrons. Hurtubise (738) observed considerable enhancement of the RTP of benzo[f]quinoline and phenanthridine by spotting HBr-ethanol solutions of the compounds onto silica gel chromatoplates. Acridine showed no RTP but gave very strong fluorescence on silica gel chromatoplates. He states that if a compound does not yield RTP, it should be tested for fluorescence and vice versa; if a compound yields both RTP and fluorescence, the additional information can be useful in characterization and identification work.

Abdel Fattah et al. (739) reported room temperature phosphorescence of some imidazoles of pharmaceutical interest under various experimental conditions. Spectral data, limits of detection, decay times, and analytical possibilities for mebendazole, flubendazole, econazole nitrate, miconazole nitrate, imazalil, carnidazole, metomidate hydrochloride, etomidate, levamisole hydrochloride, dexamisole hydrochloride, and tetramisole hydrochloride were determined on filter paper with a self-constructed sample holder using NaOH, NaI, NaI-NaOH, $AgNO_3$, $Pb(CH_3COO)_4$, CH_3COOTl, and $Hg(NO_3)_2$ and subsequent drying processes. Mebendazole and flubendazole exhibited good phosphorescence emission at room temperature; the other imidazoles only produced very weak emission. In the case of mebendazole, alkaline samples ($\lambda_{ex} = 360$ nm; $\lambda_{em} = 498$ nm, $\tau = 0.61$ s) treated with lead(IV) and thallium(I) gave highest emission intensities; all heavy metals caused a hypsochromic shift in the excitation spectra. Only silver(I) and lead(IV) produced a significant blue shift of the phosphorescence wavelength. The limit of linearity for mebendazole is 100 ng/2 μl spot; the detection limit is 6 ng/2 μl spot.

Room-Temperature Phosphorescence in Micellar Solutions

Cline Love and co-workers (740) were the first to demonstrate analytical applications of micelle-stabilized room temperature phosphorescence in aqueous solution, which is based on the remarkable enhancement of room temperature phosphorescence in solution when the species is incorporated into a micellar assembly. Micelles are described as aggregates of amphiphilic molecules formed in solution resulting from hydrophobic repulsive interactions of the nonpolar portions of the molecules. Analytical data are given for naphthalene, pyrene, and biphenyl in Tl/Na and Ag/Na mixed counterion lauryl sulfate micelles. The average precision of their measurements is about 6%, with sensitivities being competitive with other phosphorescence techniques. Sample temperature had a moderate effect on RTP intensities. The ratio of heavy metal to sodium in the micelle was found to have a dramatic effect on relative luminescence intensities up to 10–20% heavy atom. They observed that the fluorescence was greatly diminished and the RTP enhanced; no RTP was noticed in the absence of heavy atoms. The authors give a perspective on the general applicability of their method in analytical chemistry. Significant advantages offered by micelle-stabilized room temperature phosphorescence are the ability to work in fluid solution and at room temperature, the possibilities for drug–receptor pathway studies and for studying high molecular nonpolar species in aqueous solution, the increased shielding of the analyte from quenchers by the rigid, structured environment of the micelle. In addition, fluorescence intensity interferences are diminished in the presence of heavy atoms, which permits the use of conventional fluorometers in selected cases. The authors mention as a disadvantage of the method the time required for degassing the samples to prevent oxygen quenching. Finally, they conclude that the vast choice of cationic, anionic, and neutral micelles, both normal and inverted, holds great promise in finding a particular micellar system to provide accurate and precise analytical measurements for practically all luminescent species.

Skrilec and Cline Love (741) reported micelle stabilized room temperature phosphorescence of substituted arenes in aqueous Tl–Na lauryl sulfate micellar solution. Ketones, aldehydes, alcohols, carboxylic acids, phenols, amines, and a large molecule of pharmaceutical interest were investigated, and limits of detection generally compared favorably with published 77 K results. They observed some selectivity in that electron-donating substituents produced larger intensities compared to electron-withdrawing groups substituted on the same molecule.

Cline Love et al. (742) evaluated the potential analytical utility of micelle-stabilized room-temperature phosphorescence lifetimes by extending a dynamic model of micelle–analyte interactions to several limiting cases of analytical significance. According to the authors, the exceptional and dynamic nature of these fluid systems provides additional parameters that can allow differentiation of similar molecules based on their triplet-state lifetimes where similar differentiation on solid substrates or at 77 K is impossible. The authors designed and constructed an instrument capable of measuring lifetimes down to 10 μs to evaluate the triplet-state decay rates of selected compounds.

Cline Love and Skrilec (743,745) described a background correction method for interfering radiation present in micelle-stabilized room-temperature phosphorimetry. By use of reversible chemically induced quenching of the excited triplet states produced in fluid solution, both scattered light and fluorescence occurring in the phosphorescence spectral region of interest can be determined and appropriate corrections can be made. They applied their method to the analysis of coal-derived hydrogen donor solvents for polycyclic aromatics.

Cline Love and Skrilec (744) communicated at the 1981 Pittsburgh Conference spectral characteristics (limit of detection, excitation and emission wavelengths, triplet state lifetime) of the micelle-stabilized RTP for a representative variety of functionally substituted aromatic compounds dissolved in aqueous solutions of sodium lauryl sulfate detergent containing 30% thallium(I) counterions. The classes of molecules studied include unsubstituted carbocyclics, derivatives of carbocyclics containing either neutral, electron-donating, or electron-withdrawing groups, and heterocyclics. The authors generally obtained excellent sensitivity and wide applicability for molecules in the first three classes mentioned; weak to nonmeasurable phosphorescence was observed for many members of the latter two classes. Generalizations are provided regarding the structural and electronic configurations of a molecule most conductive to the production of micelle-stabilized room-temperature phosphorescence. They state that the degree of preferential solubility of the analyte in the micellar aggregate determines in part the probability of triple state emission.

Room-Temperature Phosphorescence in Liquid Solutions

Gooijer et al. (746) and Donkerbroek et al. (747,748) recently examined the possibilities of room temperature phosphorescence in solution. As phosphorescence emission is generally obtained from rigid systems (frozen solutions, solid supports) where internal molecular motions are restricted and where diffusion of the known triplet quenchers as oxygen and impurities are suppressed, the authors wanted to apply the unusual phenomenon of RTP in solution, without the use of micelles, to the detection of organic species in dynamic systems such as autoanalyzers and liquid chromatographs. They presented the phosphorescence data of some model systems like halogenated naphthalenes and biphenyls and reported some results on phenothiazines. They showed that the substituted naphthalenes especially are interesting since these are appropriate acceptor molecules in sensitized phosphorescence experiments. Sensitized phosphorescence offers an alternative analytical technique, as was shown with the donor–acceptor pair benzophenone-1,4-dibromonaphthalene. The authors report, however, that this acceptor is not suitable as a universally applicable acceptor, as it shows high absorption in regions where most potential donors absorb. They conclude that further research is going on in this field to find other potential acceptor systems, with the goal of eventually applying them in analytical technology.

REFERENCES

1. "Fluorometric Analysis", C. E. White and A. Weissler, *Anal. Chem. 44*, 5, 182R (1972).

2. "Fluorometric Analysis", A. Weissler, *Anal. Chem. 46*, 5, 500R (1974).

3. "Fluorometric and Phosphorometric Analysis", C. M. O'Donnell and T. N. Solie, *Anal. Chem. 48*, 175R (1976).

4. "Fluorometric and Phosphorometric Analysis", C. M. O'Donnell and T. N. Solie, *Anal. Chem. 50*, 189R (1978).

5. "Molecular Fluorescence, Phosphorescence, and Chemiluminescence Spectrometry", E. L. Wehry, *Anal. Chem. 52*, 75R (1980).

6. J. C. Wright and M. J. Wirth, *Anal. Chem. 52,* 988A (1980).

7. J. C. Wright and M. J. Wirth, *Anal. Chem. 52,* 1087A (1980).

8. S. Udenfriend, *Fluorescence Assay in Biology and Medicine,* Academic Press, New York, 1964.

9. S. Udenfriend, *Fluorescence Assay in Biology and Medicine,* Vol. II, Academic Press, New York, 1969.

10. S. G. Schulman, *Fluorescence and Phosphorescence Spectroscopy: Physicochemical Principles and Practice,* Pergamon Press, New York, 1977.

11. M. D. Lumb, *Luminescence Spectroscopy,* Academic Press, New York, 1978.

12. J. D. Winefordner, S. G. Schulman, T. C. O'Haver, *Luminescence Spectrometry in Analytical Chemistry,* Wiley-Interscience, New York, 1972.

13. G. G. Guilbault, Dekker, *Practical Fluorescence: Theory, Methods, and Techniques,* New York, 1973.

14. D. E. Guttman, in: *Pharmaceutical Chemistry,* Vol. II, L. G. Chatten (ed.), Dekker, New York, 1969, Chap. 4.

15. C. E. White and R. J. Argauer, Fluorescence Analysis: A Practical Approach, Dekker, New York, 1970.

16. M. Zander, *Phosphorimetry,* Academic Press, New York, 1968.

17. E. L. Wehry, *Modern Fluorescence Spectroscopy,* Vol. II, Plenum, New York, 1976.

18. D. M. Hercules, *Fluorescence and Phosphorescence Analysis: Principles and Applications,* Interscience, New York, 1966.

19. W. Baeyens, *Luminescentie Spektrometrische Analysetechnieken,* State University of Ghent, Belgium, Faculty of Pharmaceutical Sciences, 1978 (Dutch).

20. W. Baeyens, *Farm. Tijdschr. Belg. 51,* 156 (1974) (Dutch).

21. J. E. Crooks, *The Spectrum in Chemistry,* Academic Press, New York, 1978.

22. S. G. Schulman, in: *Drug Fate and Metabolism,* Vol. II, E. R. Garret and J. Hirtzels (eds.), Dekker, New York, 1978, Chap. 4.

23. W.-D. Thomitzek, *Die Pharmazie, 26,* 82 (1971).

24. J. A. F. de Silva and L. D'Arconte, *J. Forensic Sci. 14,* 184 (1969).

25. D. W. Cornish, D. M. Grossman, A. L. Jacobs, A. F. Michaelis, and B. Salsitz, *Anal. Chem. 45,* 221R (1973).

26. S. Fleury, *Chim. Anal. 48,* 266 (1966) (French).

27. S. J. Mulé, *J. Chromatog. 55,* 255 (1971).

28. D. J. Blackmore, A. S. Curry, T. S. Hayes, and E. R. Rutter, *Clin. Chem. 17,* 896 (1971).

29. P. Froehlich, *Appl. Spectrosc. Rev. 12,* 83 (1976).

30. J. W. Bridges, *Methodol. Dev. Biochem. 5,* 25 (1976).

31. D. A. Terhaar and T. J. Porro, "Fluorescence Spectrophotometry (for drug analysis)," in: J. J. Thoma, P. B. Bondo, and I. Sunshine (eds.), *Guidel. Anal. Toxicol. Programs, 2,* CRC, 1977, p. 153; in *Chem. Abstr., 89,* 122786u (1978).

32. S. G. Schulman and D. V. Naik, *Drug Fate Metab. 2,* 195 (1978); in *Chem. Abstr. 88,* 163523d (1978).

33. E. L. Wehry, *Anal. Chem. 52,* 75R (1980).

34. S. G. Schulman and R. J. Sturgeon, *Drugs Pharm. Sci. 11,* 229 (1981).

35. J. B. F. Lloyd, *Chem. Br. 11,* 442 (1975).

36. E. P. Gibson, *J. Forensic Sci. 22,* 680 (1977).

37. J. W. Bridges, *Proc. Soc. Anal. Chem.* 168 (1970).

38. R. C. Nairn, *Fluorescent Protein Tracing,* E. & S. Livingstone Ltd., London, 1969.

39. L. J. Cline Love and L. M. Upton, *Anal. Chim. Acta 118,* 325 (1980).

40. J. H. Richardson and S. M. George, *Anal. Chem. 50,* 616 (1978).

41. A. B. Bradley and R. N. Zare, *J. Am. Chem. Soc. 98,* 620 (1976).

42. J. H. Richardson, B. W. Wallin, D. C. Johnson, and L. W. Hrubesh, *Anal. Chim. Acta 86,* 263 (1976).

43. J. H. Richardson and M. E. Ando, *Anal. Chem. 49,* 955 (1977).

44. J. H. Richardson, *Anal. Biochem. 83,* 754 (1977).

45. R. Croteau and R. Leblanc, *J. Lumin. 15,* 353 (1977).

46. L. A. Bykovskaya, R. I. Personov, and Y. V. Romanovskii, *Anal. Chim. Acta 125,* 1 (1981).

47. N. Strojny and J. A. F. de Silva, *Anal. Chem 52,* 1554 (1980).

48. E. Voigtman, A. Jurgensen, and J. D. Winefordner, *Anal. Chem. 53,* 1921 (1981).

49. K. Miyaishi, M. Kunitake, T. Imasaka, T. Ogawa, and N. Ishibashi, *Anal. Chim. Acta 125,* 161 (1981).

50. G. R. Haugen and F. E. Lytle, *Anal. Chem. 53,* 11, 1554 (1981).

51. M. B. Denton, *Abstracts of papers Presented at the 1981 Pittsburgh Conference on Analytical Chemistry and Applied Spectroscopy,* no. 3, Convention Hall, Atlantic City, NJ, March 1981.

52. S. G. Schulman and R. J. Sturgeon, "Fluorescence and Phosphorescence Spectroscopy", in: J. W. Munson (ed.), *Pharmaceutical Analysis,* Dekker, New York, 1982.

53. S. Udenfriend, D. E. Duggan, B. M. Vasta, and B. B. Brodie, *J. Pharmacol. Exp. Ther. 120,* 26 (1957).

54. E. Reid (ed.), *Assay of Drugs and Other Trace Compounds in Biological Fluids,* North-Holland, Amsterdam, 1976.

55. P. Hajdú and D. Damm, *Drug Res. 26,* 2141 (1976).

56. P. O. Lagerstroem, *J. Chromatogr. 225,* 476 (1981).

57. G. Amann, G. Gübitz, R. W. Frei, and W. Santi, *Anal. Chim. Acta 116,* 119 (1980).

58. W. A. Armstrong, D. W. Grant, and W. G. Humphers, *Anal. Chem. 35,* 1300 (1963).

59. S. Katz, W. W. Pitt, Jr., and G. Jones, Jr., *Clin. Chem. 19,* 817 (1973).

60. L. J. Cline Love and L. M. Upton, *Anal. Chim. Acta 118,* 325 (1980).

61. E. Sawicki, T. W. Stanley, and H. Johnson, *Microchem. J. 8,* 257 (1964).

62. B. L. Hamman and M. M. Martin, *Anal. Biochem. 15,* 305 (1966).

63. B. L. Hamman and M. M. Martin, *J. Lab. Clin. Med.* (St. Louis) *73,* 1042 (1969).

64. E. Eich, H. Geiszler, E. Mutschler, and W. Schunack, *Drug. Res. 19,* 1895 (1969) (Germ.).

65. W. Brühl, Jr., and E. Schmid, *Drug. Res. 20,* 485 (1970) (Germ.).

66. M. L. Gianelli, J. B. Callis, N. H. Andersen, and G. D. Christian, *Anal. Chem. 53,* 1357 (1981).

67. G. G. Guilbault, *Photochem. Photobiol. 25,* 403 (1977).

68. R. Segura and X. Navarro, *Proceedings of the International Symposium Advances in Chromatography,* Las Vegas, Nevada, A. Zlatkis (ed.), University of Houston Press, Houston, Texas, April 1982.

69. L. Zhou, H. Shanfield, and A. Zlatkis, *Proceedings of the International Symposium Advances in Chromatography,* Las Vegas, Nevada, A. Zlatkis (ed.), University of Houston Press, Houston, TX, April 1982.

70. H. Wagner and H. Lehmann, *Fresenius' Z. Anal. Chem. 283,* 115 (1977) (Germ.).

71. M. L. Gianelli, J. B. Callis, N. H. Andersen, and G. D. Christian, *Anal. Chem. 53,* 1357 (1981).

72. R. A. Chalmers and G. A. Wadds, *Analyst 95,* 234 (1970).

73. M. Sansur, A. Buccafuri, and S. Morgenstern, *J. AOAC, 55,* 880 (1972).

74. Z. Rollen, A. Yert, and L. Needham, *Clin. Chem. 24,* 840 (1978).

75. S. J. Mulé and P. L. Hushin, *Anal. Chem. 43,* 708 (1971).

76. A. I. Dumitrashko and F. V. Babilev, *Izv. Akad. Nauk Mold. SSR. Ser. Biol. Khim. Nauk. 1,* 83 (1976) (Russ.), *Anal. Chem. 50,* 204R (14M) (1978).

77. A. Bowd and J. H. Turnbull, *J. Chem. Soc.* (Perkin Trans. II) *1,* 121 (1977), in *Anal. Chem. 50,* 204R (8M) (1978).

78. P. E. Nelson, S. L. Nolan, and K. R. Bedord, *J. Chromatogr. 234,* 407 (1982).

79. C. I. Miles and G. H. Schenk, *Anal. Chem. 42,* 656 (1970).

80. G. H. Schenk, F. H. Boyer, C. I. Miles, and D. R. Wirz, *Anal. Chem. 44,* 1593 (1972).

81. K. H. Lee, L. Thompkins, and M. R. Spencer, *J. Pharm. Sci. 57,* 1240 (1968).

82. N. Shane and D. Miele, *J. Pharm. Sci. 59*, 397 (1970).

83. N. Shane and R. Stillman, *J. Pharm. Sci. 60*, 114 (1971).

84. T. Katio, K. Kasuya, and T. Inoue, *Bunseki Kagaku 20*, 801 (1971) (Japan.); in *Chem. Abstr. 76*, 20986y (1972).

85. S. Øie and K. Frislid, *Pharm. Act. Helv. 46*, 632 (1971).

86. S. L. Kanter and W. R. Horbaly, *J. Pharm. Sci. 60*, 1898 (1971).

87. H. D. Spitz, *J. Chromatogr. 140*, 131 (1977).

88. M. Kleinerman, *Anal. Lett. 10*, 205 (1977).

89a. I. Bekersky, H. G. Boxenbaum, M. H. Whitson, C. V. Puglisi, R. Pocelinko, and S. A. Kaplan, *Anal. Lett. 10*, 539 (1977).

89b. K. W. Street, Jr., and G. H. Schenk, *J. Pharm. Sci. 70*, 641 (1981).

90. S. A. Veresh, F. S. Hom, and J. J. Miskel, *J. Pharm. Sci. 60*, 1092 (1971).

91. S. G. Schulman and W. J. M. Underberg, *Photochem. Photobiol. 29*, 937 (1979).

92. M. S. Stallings and S. G. Schulman, *Anal. Chim. Acta 78*, 483 (1975).

93. J. M. Rutledge and S. G. Schulman, *Anal. Chim. Acta 75*, 449 (1975).

94. H. Murata and T. Wada, *Chem. Pharm. Bull. 15*, 1906 (1967).

95. L. Cere and P. Barolo, *Atti Congr. Qual.*, 6th, 1967 451–473 (1968) (Ital.); in *Chem. Abstr. 72*, 74527k (1970).

96. L. A. Dal Cortivo, M. M. De Mayo, and S. B. Weinberg, *Anal. Chem. 42*, 941 (1970).

97. E. J. McGonigle, *Anal. Chem. 43*, 966 (1971).

98. Y. Hattori, T. Arai, T. Mori, and E. Fujihira, *Chem. Pharm. Bull. 18*, 1063 (1970).

99. H.-D. Dell and R. Kamp, *Arch. Pharm. 303*, 785 (1970) (German).

100. A. C. Mehta and S. G. Schulman, *Talanta 20*, 702 (1973).

101. J. W. Meilinck, *Pharm. Weekbl. 109*, 22 (1974).

102. N. Strojny and J. A. F. de Silva, *J. Chromatogr. Sci. 13*, 583 (1975).

103. D. G. Kaiser, W. F. Liggett, and A. A. Forist, *J. Pharm. Sci. 65*, 1767 (1976).

104. D. G. Kaiser, B. J. Bowman, and A. A. Forist, *Anal. Chem. 38*, 977 (1966).

105. D. G. Kaiser and E. M. Glenn, *J. Pharm. Sci. 61*, 1908 (1972).

106. D. G. Kaiser and A. A. Forist, *J. Pharm. Sci. 64*, 2011 (1975).

107. W. F. Bayne, T. East, and D. Dye, *J. Pharm. Sci. 70*, 458 (1981).

108a. W. Baeyens and P. De Moerloose, *J. Pharm. Sci. 66*, 1771 (1977).

108b. D. Westerlund, A. Theodorsen, and Y. Jaksch, *J. Liq. Chromatogr. 2*, 969 (1979).

109. T. Frost, *Analyst 106*, 999 (1981).

110. W. Pacha, C. Delaborde, H. P. Keller, J. Meier, and H. Rietsch, *Arzneim.-Forsch./Drug Res. 31*, 893 (1981).

111. L. Brand and S. Shaltiel, *Biochim. Biophys. Acta 75*, 145 (1963).

112. I. Weinryb and R. F. Steiner, *Biochemistry 7*, 2488 (1968).

113. M. Sun and P.-S. Song, *Photochem. Photobiol. 25*, (1977).

114. M. Roth, *Anal. Chem. 43*, 880 (1971).

115. P. Lindroth and K. Mopper, *Anal. Chem. 51*, 1667 (1979).

116. J. D. H. Cooper and D. C. Turnell, *J. Chromatogr. 227*, 158 (1982).

117. M. Maeda and A. Tsuji, *Anal. Biochem. 52*, 555 (1973).

118. E. Bayer, E. Grom, B. Kaltenegger, and R. Uhmann, *Anal. Chem. 48*, 1106 (1976).

119. P. M. Froehlich and M. Yeats, *Anal. Chim. Acta 87*, 185 (1976).

120. P. M. Froehlich and L. D. Murphy, *Anal. Chem. 49*, 1606 (1977).

121. J. C. Touchstone, J. Sherma, M. F. Dobbins, and G. R. Hansen, *J. Chromatogr. 124*, 111 (1976).

122. H. Nakamura and J. J. Pisano, *Arch. Biochem. Biophys. 177*, 334 (1976).

123. H. Nakamura, *J. Chromatogr. 131*, 215 (1977).

124. W. McHugh, R. A. Sandmann, W. G. Haney, S. P. Sood, and D. P. Wittmer, *J. Chromatogr. 124*, 376 (1976).

125. G. L. Soh, J. L. Pace, D. L. Kemper, and W. L. Ragland, *Anal. Lett. 10*, 111 (1977).

126. S. Udenfriend and S. Stein, *Pept., Proc. Am. Pept. Symp., 5th*, 14–26 (1977), M. Goodman and J. Meienhofer (eds.), Wiley, New York; in *Chem. Abstr. 88*, 148327z (1978).

127. S. Stein, Pept. Neurobiol., 9–37 (1977), H. Gainer (ed.), Plenum, New York; in *Chem. Abstr. 90*, 117300x (1979).

128. S. Udenfriend, *Proc. Int. Symp. Proteins*, 23–37 (1978), Li Choh Hao (ed.), Academic Press, New York; in *Chem. Abstr. 91*, 136335z (1979).

129. J. Barre, *Thérapie 33*, 411 (1978).

130. E. Fujimori, *FEBS Lett. 135*, 257 (1981).

131. W. M. Hadley, W. F. Bousquet, and T. S. Miya, *J. Pharm. Sci. 63*, 57 (1974).

132. K. M. Halprin and A. Ohkawara, *J. Invest. Dermatol. 48*, 149 (1967).

133. H. Steinhart and J. Sandmann, *Anal. Chem. 49*, 950 (1977).

134. S. Kirschenbaum and M. D. Glantz, *Mikrochim. Acta I*, 589 (1976).

135. M. Ahnoff, I. Grundevik, A. Arfwidsson, J. Fonselius, and B. A. Persson, *Anal. Chem. 53*, 3, 485 (1981).

136. Y. Watanabe and K. Imai, *Proceedings of the 18th International Symposium Advances in Chromatography*, no. 42, Tokyo, April 1982.

137. K. W. Kohn, *Anal. Chem. 33*, 862 (1961).

138. K. H. Ibsen, R. L. Saunders, and M. R. Urist, *Anal. Biochem. 5*, 505 (1963).

139. J. Spock and S. E. Katz, J.A.O.A.C., *46*, 434 (1963).

140. R. G. Kelly and K. D. Hoyt, *Z. Klin. Chem. Klin. Biochem.* 7, 152 (1969).

141. S. E. Katz, C. A. Fassbender, and D. Dorfman, *J. AOAC,* 54, 947 (1971).

142. T. V. Alykova, *Antibiotiki* (Moscow) 17, 353 (1972) (Russ.); in *Chem. Abstr.* 77, 28708a (1972).

143. D. M. Wilson, M. Lever, E. A. Brosnan, and A. Stillwell, *Clin. Chim. Acta* 36, 260 (1972).

144. S. E. Katz and C. A. Fassbender, *J. AOAC* 56, 17 (1973).

145. G. J. Willekens, *J. Pharm. Sci.* 66, 1419 (1977).

146. G. van Hoeck, I. Kapétanidis, and A. Mirimanoff, *Pharm. Acta Helveti.* 47, 316 (1972) (Fr.).

147. E. Ragazzi and G. Veronese, *J. Chromatogr.* 132, 105 (1977).

148. F. Schatz, *Drug Res.* 23, 426 (1973) (Germ.).

149. D. Hall, *J. Pharm. Pharmacol.* 28, 420 (1976).

150. D. Hall, *Chemother. Proc. 9th Int. Congr. Chemother.* Vol. 2 1975, pp. 111–114.

151. D. Hall, *Br. J. Clin. Pharmacol.* 4, 57 (1977).

152. S. T. Day, W. G. Crouthamel, L. C. Martinelli, and J. K. H. Ma, *J. Pharm. Sci.* 67, 1518 (1978).

153. C. van den Bogert and A. M. Kroon, *J. Pharm. Sci.* 70, 186 (1981).

154. H. Poiger and C. Schlatter, *Analyst* 101, 802 (1976).

155. A. Regosz, *Fresenius' Z. Anal. Chem.* 280, 383 (1976).

156. F. De Fabrizio, *J. Pharm. Sci.* 58, 136 (1969).

157. N. M. Alykov, *Zh. Anal. Khim.* 36, 1387 (1981) (Russ.); in *Chem. Abstr.,* 09525214702R.

158. B. Shaikh, M. Leadbetter, and E. H. Allen, *Abstract of Papers Presented at the 1981 Pittsburgh Conference on Analytical Chemistry and Applied Spectroscopy,* No. 502, Convention Hall, Atlantic City, NJ, March 1981.

159. V. A. Poltorak, *Mikrobiologiya,* 41, 341 (1972) (Russ.); in *Chem. Abstr.* 77, 150526e (1972).

160. A. M. Felix, J. W. Westley, and J. Meienhofer, *Anal. Biochem.* 73, 70 (1976).

161. K. Y. Tserng and J. G. Wagner, *Anal. Chem.* 48, 348 (1976).

162. K. Tsuji, *J. Chromatogr.* 158, 337 (1978).

163. R. J. Sturgeon and S. G. Schulman, *J. Pharm. Sci.* 66, 958 (1977).

164. A. Csiba, *J. Pharm. Pharmacol.* 31, 115 (1979).

165. R. A. A. Watson, J. Landon, E. J. Shaw, and D. S. Smith, *Clin. Chim. Acta* 73, 51 (1976).

166. J. F. Burd, R. C. Wong, J. E. Feeney, R. J. Carrico, and R. C. Bogulaski, *Clin. Chem.* 23, 1402 (1977).

167. P. Kabasakalian, S. Kalliney, and A. W. Magatti, *Anal. Chem.* 49, 953 (1977).

168. J. H. Ngui-Yen, T. Hofmann, M. Wigmore, and J. A. Smith, *Antimicrob. Agents Chemother. 20*, 821 (1981).

169. S. Matsuda, N. Kato, Y. Ban, Y. Kawada, and T. Nishiura, *Chemotherapy* (Tokyo) *29*, 1062 (1981) (Japan.); in *Chem. Abstr. 96* (7) 45772b.

170. J. E. Sinsheimer, D. D. Hong, and J. H. Burckhalter, *J. Pharm. Sci. 58*, 1041 (1969).

171. W. J. Jusko, *J. Pharm. Sci. 60*, 728 (1971).

172. K. Miyazaki, O. Ogino, and T. Arita, *Chem. Pharm. Bull.* (Tokyo) *22*, 1910 (1974).

173. R. H. Barbhaiya, P. Turner, and E. Shaw, *Clin. Chim. Acta 77*, 373 (1977).

174. K. Miyazaki, O. Ogino, H. Sato, M. Nakano, and T. Arita, *Chem. Pharm. Bull. 25*, 253 (1977).

175. A. Tsuji, E. Miyamoto, and T. Yamana, *J. Pharm. Pharmacol. 30*, 811 (1978).

176. J. Kušnír, V. Adameová, and K. Barna, *Drug Res. 28*, II, 2058 (1978).

177. A. F. Heald, C. E. Ita, and E. C. Schreiber, *J. Pharm. Sci. 65*, 768 (1976).

178. A. B. C. Yu, C. H. Nightingale, and D. R. Flanagan, *J. Pharm. Sci. 66*, 213 (1977).

179. E. Crombez, G. Van Der Weken, W. Van Den Bossche, and P. De Moerloose, *J. Chromatogr. 177*, 323 (1979).

180. E. Crombez, W. Van Den Bossche, and P. De Moerloose, in: *Recent Developments in Chromatography and Electrophoresis*, Vol. 10, A. Frigerio and M. McCamish (eds.), Elsevier, Amsterdam, 1980, pp. 261–266.

181. K. Tsuji, *J. Chromatogr 158*, 337 (1978).

182. A. Iliopoulou, R. N. Thin, and P. Turner, *Br. J. Vener. Dis. 57*, 263 (1981).

183. M. Thompson and C. R. Williams, *Anal. Lett. 9*, 179 (1976).

184. R. Clarenburg and V. R. Rao, *Drug Metab. Dispos. 5*, 246 (1977).

185. W. C. Neely and J. R. McDuffie, *J. AOAC 55*, 1300 (1972).

186. R. J. Bopp, R. E. Schirmer, and D. B. Meyers, *J. Pharm. Sci. 61*, 1750 (1972).

187. K. Hirauchi and S. Masuda, *Chem. Pharm. Bull. 25*, 1474 (1977).

188. L. A. Roberts, *J. Pharm. Sci. 58*, 1015 (1969).

189. W. Messerschmidt, *J. Chromatog. 33*, 551 (1968).

190. K. Taylor, *J. Pharm. Pharmacol. 19*, 770 (1967).

191. D. Gomes, *Rev. Port. Farm. 21*, 3, 245 (1971) (Port.); in *Chem. Abstr. 76*, 144871b (1972).

192. A. Khabarov, F. Shemyakin, and O. Fakeeva, *Novye Metody Khim. Anal. Mater. 2*, 95 (1971) (Russ.); in *Chem. Abstr. 77*, 168696v (1972).

193. S. Silvestri, *Il Farmaco-Ed. Pr. 23*, 90 (1968).

194. L. Dombrowski, A. Crain, R. Browning, and E. Pratt, *J. Pharm. Sci. 64*, 643 (1975).

195. D. Andersen, *J. Chromatogr. 41*, 491 (1969).

196a. M. Steinigen, *Deutsche Apotheker-Zeitung 112*, 51 (1972) (Germ.).

196b. P. J. Twitchett, S. M. Fletcher, A. T. Sullivan, and A. C. Moffat, *J. Chromatogr. 150*, 73 (1978).

197. G. Härtel and A. Harjanne, *Clin. Chim. Acta 23*, 289 (1969).

198. G. Cramer and B. Isaksson, *Scand. J. Clin. Lab. Invest. 15*, 553 (1963).

199. E. H. Denswil and J. B. M. Vismans, *Pharm. Weekbl. 105*, 1441 (1970) (Dutch).

200. L. Broussard, *Clin. Chem. 27*, 1929 (1981).

201. S. G. Schulman, R. Threatte, A. Capomacchia, and W. Paul, *J. Pharm. Sci. 63*, 876 (1974).

202. T. Guentert and S. Riegelman, *Clin. Chem. 24*, 2065 (1978).

203. C. Szalkowski and W. Mader, *J. Am. Pharm. Ass., Sci. Ed. 45*, 613 (1956).

204. F. Tishler, P. B. Sheth, and M. B. Giaimo, J. AOAC *46*, 448 (1963).

205. R. Haycock, P. Sheth, T. Higuchi, W. Mader, and G. Papariello, *J. Pharm. Sci. 55*, 826 (1966).

206. B. Kabadi, A. Warren, and C. Newman, *J. Pharm. Sci. 58*, 1127 (1969).

207. T. Urbanyi and H. Stober, *J. AOAC 55*, 180 (1972).

208. J. M. G. J. Frijns, *Pharm. Weekbl. 106*, 605 (1971) (Dutch).

209. B. Kabadi, *J. Pharm. Sci. 60*, 1862 (1971).

210. T. Gurkan, *Microchim. Acta, 1*, 173 (1976).

211. W. Smith and C. Clark, *J. AOAC 59*, 811 (1976).

212. W. Smith, *J. AOAC 60*, 1018 (1977).

213. G. Caille, D. Leclerc-Chevalier, and J. A. Mockle, *Can. J. Pharm. Sci. 5*, 55 (1970) (Fr.).

214. M. Szasz and G. Szasz, *Acta Pharm. Hung. 40*, 38 (1970) (Hung.).

215. V. Massa, F. Gal, and P. Susplugas, *6th Int. Symp. Chromatogr. Electrophor., Lect. Pap.*, 1970 pp. 470–475, 1971 (Fr.).

216. A. D. Thomas, *Talanta 22*, 865 (1975).

217. J. M. G. J. Frijns, *Pharm. Weekbl. 106*, 865 (1971) (Dutch).

218. C. J. G. A. Bos and J. M. G. J. Frijns, *Pharm. Weekbl. 107*, 111 (1972) (Dutch).

219. M. Gelbcke and C. Dorlet, *J. Pharm. Belg. 32*, 143 (1977) (Fr.).

220. P. O. Edlund, *J. Chromatogr. 226*, 107 (1981).

221. R. E. Jensen and R. T. Pflaum, *J. Pharm. Sci. 53*, 835 (1964).

222. R. Onishi, K. Kawamura, K. Inoue, and T. Kobayashi, *Yakugaku Zasshi 92*, 1101 (1972) (Japan.); in *Chem. Abstr. 77*, 156358y (1972).

223. S. Nakano, H. Taniguchi, and T. Furuhashi, *Yakugaku Zasshi 92*, 411 (1972) (Japan.); in *Chem. Abstr. 77*, 52426c (1972).

224. D. L. Wilson, D. R. Wirz, and G. H. Schenk, *Anal. Chem. 45*, 1447 (1973).

225. D. R. Wirz, D. L. Wilson, and G. H. Schenk, *Anal. Chem. 46*, 896 (1974).

226. R. L. DeAngelis, M. F. Kearney, and R. M. Welch, *J. Pharm. Sci. 66*, 841 (1977).

227. W. Manger, O. Steinsland, G. Nahas, K. Wakim, and S. Dufton, *Clin. Chem. 15*, 1101 (1969).

228. E. Matlina, R. Shchedrina, and E. Shirinyan, *Tr. Nov. App. Metod. 8*, 104 (1969) (Russ.).

229. C. Brunori, V. Renzini, C. Porcellati, L. Corea, C. Valori, and M. Timio, *Rass. Fisiopatol. Clin. Ter. 41*, 332 (1969) (Ital.).

230. J. Griffiths, F. Leung, and T. McDonald, *Clin. Chim. Acta 30*, 395 (1970).

231. V. Renzini, C. Brunori, and C. Valori, *Clin. Chim. Acta 30*, 587 (1970).

232. F. Geissbuhler, *Clin. Chim. Acta 30*, 143 (1970).

233. P. Antum, I. Pullar, D. Eccleston, and D. Sharman, *Clin. Chim. Acta 34*, 387 (1971).

234. H. Geissler and E. Mutschler, *J. Chromatogr. 56*, 271 (1971).

235. E. Dworzak van H. Huck, *J. Chromatogr. 61*, 162 (1971).

236. J. Sapira, T. Klaniecki, and M. Rizk, *Clin. Chem. 17*, 486 (1971).

237. A. Ardashev and P. Golikov, *Biol. Med. Issled. Dal'nem Vostoke 95* (1971) (Russ.).

238. A. Coeur, D. Cantin, and J. Alary, *Ann. Pharm. Fr. 30*, 801 (1972) (Fr.).

239. L. Grokhovskii, *Lab. Delo 9*, 530 (1972) (Russ.).

240. H. Tsuzuki, K. Tikani, K. Imai, and Z. Tamura, *Chem. Pharm. Bull. 20*, 1931 (1972).

241. B. Andersson, S. Hovmöller, C.-G. Karlsson, and S. Svensson, *Clin. Chim. Acta 51*, 13 (1974).

242. G. Schwedt, *Clin. Chim. Acta 51*, 247 (1974).

243. G. Schwedt, *Clin. Chim. Acta 57*, 291 (1974).

244. P. Waldmeier, P. De Herdt, and L. Maitre, *Clin. Chem. 20*, 81 (1974).

245. G. Schwedt, *Z. Klin. Chem. Klin. Biochem. 12*, 39 (1974) (Germ.).

246. F. Nachtmann, H. Spitzy, and R. Frei, *Anal. Chim. Acta 76*, 57 (1975).

247. E. Quek, J. Buttery, and G. De Witt, *Clin. Chim. Acta 58*, 137 (1975).

248. W. Wood and R. Mainwaring-Burton, *Clin. Chim. Acta 61*, 297 (1975).

249. T. T. Kensler, D. Brooke, and D. M. Walker, *J. Pharm. Sci. 65*, 763 (1976).

250. T. Karasawa, *Rinsho Kagaku 4*, 24 (1975) (Japan.); in *Anal. Chem. 50*, 198R (18E) (1978).

251. G. Corona, P. Frattini, B. Bonferoni, and G. Santagostino, *Il Farmaco, Ed. Pr. 32*, 53 (1976).

252. G. Schwedt, *J. Chromatogr. 143*, 463 (1977) (Germ.).

253. G. Beijersbergen Van Henegouwen, C. Kruse, and K. Gerritsma, *Pharm. Weekbl. 111*, 197 (1976) (Dutch).

254. H. Tarlin, M. Hudson, and M. Sahn, *J. Pharm. Sci. 65*, 1463 (1976).
255. Y. Miura, V. Campese, V. DeQuattro, and D. Meijer, *J. Lab. Clin. Med. 89*, 421 (1977).
256. E. J. Moerman, M. G. Bogaert, and A. F. De Schaepdrijver, *Clin. Chim. Acta 72*, 89 (1976).
257. J. De Vente, P. Bruyn, and J. Zaagsma, *J. Pharm. Pharmacol. 33*, 290 (1981).
258. D. Calverley, H. McKim, and W. Dewhurst, *J. Pharmacol. Meth. 5*, 179 (1981).
259. G. M. Anderson, J. G. Young, P. I. Jatlow, and D. J. Cohen, *Clin. Chem. 27*, 2060 (1981).
260. G. Schwedt and H. H. Bussemas, *Chromatographia 9*, 17 (1976) (Germ.).
261. G. Schwedt and H. H. Bussemas, *Anal. Chem. 283*, 23 (1977).
262. G. Schwedt and H. H. Bussemas, *Fresenius' Z. Anal. Chem. 285*, 381 (1977).
263. R. A. Kagel, S. O. Farwell, and J. D. Willett, *Abstracts of Papers Presented at the 1981 Pittsburgh Conference on Analytical Chemistry and Applied Spectroscopy*, no. 559, Convention Hall, Atlantic City, NJ, March 1981.
264. T. Hayes, *Clin. Chem. 19*, 390 (1973).
265. F. Van Hoof and A. Heyndrickx, *Anal. Chem. 46*, 286 (1974).
266. B. Klein, J. Sheehan, and E. Grunberg, *Clin. Chem. 20*, 272 (1974).
267. T. J. Hopen, R. C. Briner, H. G. Sadler and R. L. Smith, *J. Forensic Sci. 21*, 842 (1976).
268. C. I. Miles and G. H. Schenk, *Anal. Chem. 45*, 130 (1973).
269. A. Vidi and G. Bonardi, *Clin. Chim. Acta 38*, 463 (1972).
270. S. Takahashi and L. Gjessing, *Clin. Chim. Acta 36*, 369 (1972).
271. M. Hardeman, A. Den Uyl, and H. Prins, *Clin. Chim. Acta 37*, 71 (1972).
272. T. James, *J. Pharm. Sci. 62*, 669 (1973).
273. G. Schwedt, *Anal. Chim. Acta 81*, 361 (1976) (Germ.).
274. G. Schwedt and H. Bussemas, *Clin. Chim. Acta 69*, 175 (1976) (Germ.).
275. G. P. Jackman, *Clin. Chim. Acta 120*, 137 (1982).
276. T. P. Davis, C. W. Gehrke, C. W. Gehrke, Jr., T. D. Cunningham, K. C. Kuo, K. O. Gerhardt, H. D. Johnson, and C. H. Williams, *Clin. Chem. 24*, 1317 (1978).
277. H. Spiegel and A. Tonchen, *Clin. Chem. 16*, 763 (1970).
278. K. Türler and H. Käser, *Clin. Chim. Acta 32*, 41 (1971).
279. H. Wisser and D. Stamm, *Z. Klin, Chem. Klin. Biochem. 7*, 631 (1969) (Germ.).
280. Z. Oberman, R. Chayen, and M. Herzberg, *Clin. Chim. Acta 29*, 391 (1970).
281. R. Greenland and I. Michaelson, *Clin. Chem. 20*, 915 (1974).
282. A. Anton and D. Sayre, *J. Pharmacol. 138*, 360 (1962).

283. A. Anton and D. Sayre, *J. Pharmacol. 145*, 326 (1964).

284. A. Anton and D. Sayre, in: *The Thyroid and Biogenic Amines*, T. Rall and I. Kopin (eds.), North-Holland, New York, 1972, p. 398.

285. J. Kopecká and M. Jarolímková, *Biochem. Clin. Bohemoslov. 10*, 55 (1981) (Czech.).

286. H. Nohta, K. Ohtsubo, K. Zaitsu, and Y. Ohkura, *J. Chromatogr. 227*, 415 (1982).

287. S. Vickers and E. K. Stuart, *J. Pharm. Sci. 62*, 1550 (1973).

288. W. L. Pacha, *Experienta 25*, 802 (1969).

289. E. M. Philips and D. Hicks, *Eur. J. Clin. Pharmacol. 5*, 158 (1973).

290. J. A. F. de Silva, J. C. Meyer, and C. V. Puglisi, *J. Pharm. Sci. 65*, 1230 (1976).

291. C. Izard-Verchère and C. Viel, *Chim. Ther. 6*, 469 (1971) (Fr.).

292. E. R. Garrett and K. Schnelle, *J. Pharm. Sci. 60*, 833 (1971).

293. A. G. Arbab and P. Turner, *J. Pharm. Pharmacol. 23*, 719 (1971).

294. G. Bodem and C. A. Chidsey, *Clin. Chem. 18*, 363 (1972).

295. A. Roux and B. Flouvat, *J. Chromatog. 166*, 327 (1978).

296. D. G. Shand, E. M. Nuckolls, and J. A. Oates, *Clin. Pharmacol. Ther. 11*, 112 (1970).

297. D. J. Coltart and D. G. Shand, *Br. Med. J. 3*, 731 (1970).

298. P. Ambler, B. Singh, and M. Lever, *Clin. Chim. Acta 54*, 373 (1974).

299. G. J. Krol, C. A. Mannan, R. E. Pickering, D. V. Amato, and B. I. Kho, *Anal. Chem. 49*, 1836 (1977).

300. M. T. Rosseel and M. G. Bogaert, *J. Pharm. Sci. 70*, 688 (1981).

301. F. Albani, R. Riva, and A. Baruzzi, *J. Chromatogr. 228*, 362 (1982).

302. T. Kensler, D. Brooke, and D. Walker, *J. Pharm. Sci. 65*, 763 (1976).

303. M. Joseph and H. Baker, *Clin. Chim. Acta 72*, 125 (1976).

304. O. Beck, G. Palmskog, and E. Hultman, *Clin. Chim. Acta 79*, 149 (1977).

305. O. Suzuki and K. Yagi, *Clin. Chim. Acta 78*, 401 (1977).

306. T. Sourkes, G. Murphy, and B. Lara, *Br. J. Med. Pharm. Chem. 5*, 204 (1962).

307. A. Sjoerdsma, A. Vandsalu, and K. Engelman, *Circulation 28*, 492 (1963).

308. J. Meilink, *Pharm. Weekbl. 106*, 385 (1971) (Dutch).

309. B. Kim and R. Koda, *J. Pharm. Sci. 66*, 1632 (1977).

310. P. E. Grebow, J. A. Treitman, and A. K. Yeung, *J. Pharm. Sci. 67*, 1117 (1978).

311. P. E. Grebow, J. A. Treitman, A. K. Yeung, and M. M. Johnston, *J. Pharm. Sci. 70*, 306 (1981).

312. B. Oosterhuis, M. Van Den Berg, and C. J. Van Boxtel, *J. Chromatogr. 226*, 259 (1981).

313. K. Kamimura, H. Sazaki, and S. Kawamura, *J. Chromatogr. 225*, 115 (1981).

314. J. M. Finkel and S. D. Harrison, Jr., *Anal. Chem. 41*, 1854 (1969).

315. J. G. Van Damme and R. E. De Nève, *J. Pharm. Sci. 68*, 16 (1979).

316. E. Jekabsons, *J. AOAC 52*, 110 (1969).

317. C. R. Jones and L. A. King, *Biochem. Med. 2*, 251 (1968).

318. W. S. Simmons and R. L. DeAngelis, *Anal. Chem. 45*, 1538 (1973).

319. M. Osadca and E. De Ritter, *J. AOAC 53*, 1244 (1970).

320. R. S. Browning and E. L. Pratt, *J. AOAC 53*, 464 (1970).

321. W. R. Vincent, S. G. Schulman, J. M. Midgley, W. J. Van Oort, and R. H. A. Sorel, *Int. J. Pharm. 9*, 191 (1981).

322. D. E. Schwartz, B. A. Koechlin, and R. E. Weinfeld, *Chemotherapy 14*, 22 (1969).

323. C. W. Sigel and M. E. Grace, *J. Chromatogr. 80*, 111 (1973).

324. C. W. Sigel, M. E. Grace, and C. A. Nichol, *J. Infect. Dis. 128*, S580 (1973).

325. H. G. Boxenbaum and S. Riegelman, *J. Pharm. Sci. 63*, 1191 (1974).

326. G. A. Digenis, S. Shakhshir, M. Miyamoto, and H. Kostenbauder, *J. Pharm. Sci. 65*, 247 (1976).

327. J. T. Stewart and R. E. Wilkin, *J. Pharm. Sci. 61*, 432 (1972).

328. G. Ghilardelli, F. Rotilio, and R. Perego, *Il Farmaco-Ed. Pr. 30*, 398 (1975) (Ital.).

329. C. H. Hsiao, H. J. Rhodes, and M. I. Blake, *J. Pharm. Sci. 66*, 1157 (1977).

330. L. A. Roberts, *J. AOAC 53*, 117 (1970).

331. L. S. Rosenberg and S. G. Schulman, *J. Pharm. Sci. 67*, 1770 (1978).

332. L. J. Cline Love, L. M. Upton, and A. W. Ritter III, *Anal. Chem. 50*, 2059 (1978).

333. M. J. Adams, J. G. Highfield, and G. F. Kirkbright, *Anal. Chem. 49*, 1850 (1977).

334. E. Doelker, I. Kapétanidis, and A. Mirimanoff, *Pharm. Acta Helv. 44*, 647 (1969) (Fr.).

335. G. Eide-Jürgensen, I. Kapétanidis, and A. Mirimanoff, *Planta Med. Suppl.* 98 (1971) (Fr.).

336. F. Hammerstein and F. Kaiser, *Planta Med. 21*, 5 (1971) (Germ.).

337. L. F. Cullen, D. L. Packman, and G. J. Papariello, *J. Pharm. Sci. 59*, 697 (1970).

338. D. V. Naik, J. S. Groover, and S. G. Schulman, *Anal. Chim. Acta 74*, 29 (1975).

339. A. Z. Britten and E. Njau, *Anal. Chim. Acta 76*, 409 (1975).

340. R. N. Gupta, F. Eng, and D. Lewis, *Anal. Chem. 50*, 197 (1978).

341. H. S. I. Tan and C. Beiser, *J. Pharm. Sci. 64*, 1207 (1975).

342. R. N. Gupta, F. Eng, D. Lewis, and C. Kumana, *Anal. Chem. 51*, 455 (1979).

343. K. Schloszmann, *Drug Res. 22*, 60 (1972).

344. R. McAllister and S. Howell, *J. Pharm. Sci. 65*, 431 (1976).

345. J. A. F. de Silva, N. Munno, L. D'Arconte, and N. Strojny, *J. Pharm. Sci. 67*, 752 (1978).

346. N. Khalafalla and M. A. Shams-Eldeen, *Pharm. Acta Helv. 56*, 225 (1981).

347. P. J. Meffin, S. R. Harapat, and D. C. Harrison, *J. Pharm. Sci. 66*, 583 (1977).

348. R. D. Hollifield and J. D. Conklin, *Arch. Int. Pharmacodyn. Ther. 174*, 333 (1968).

349. R. D. Hollifield and J. D. Conklin, *J. Pharm. Sci. 62*, 271 (1973).

350. S. J. Saxena, I. L. Honigberg, J. T. Stewart, G. R. Keene, and J. J. Vallner, *J. Pharm. Sci. 66*, 286 (1977).

351. J. T. Stewart and C. W. Chan, *J. Pharm. Sci. 68*, 910 (1979).

352. J. T. Stewart and C. W. Chan, *Anal. Lett. B11*, 667 (1978).

353. A. M. Pastorino, *Drug Res. 28* (II), 1728 (1978).

354. U. W. Kersten, D. K. F. Meijer, and S. Agoston, *Clin. Chim. Acta 44*, 59 (1973).

355. M. J. Watson and K. McLeod, *Clin. Chim. Acta 79*, 511 (1977).

356. L. B. Wingard, Jr., E. Abouleish, D. C. West, and T. J. Goehl, *J. Pharm. Sci. 68*, 914 (1979).

357. B. A. Koechlin and L. D'Arconte, *Anal. Biochem. 5*, 195 (1963).

358. M. A. Schwartz and E. Postma, *J. Pharm. Sci. 55*, 1358 (1966).

359. J. T. Stewart and J. L. Williamson, *Anal. Chem. 48*, 1182 (1976).

360. Sy-Rong Sun, *J. Pharm. Sci. 67*, 5, 639 (1978).

361. S. Lauffer, E. Schmid and F. Weist, *Drug Res. 19*, 1965 (1969).

362. S. Lauffer and E. Schmid, *Die Pharmazie 19*, 740 (1969).

363. J. Braun, G. Caille, and E. A. Martin, *Can. J. Pharm. Sci. 3*, 65 (1968) (Fr.).

364. G. Caille, J. Braun, and J. A. Mockle, *Can. J. Pharm. Sci. 5*, 78 (1970) (Fr.).

365. J. A. F. de Silva and N. Strojny, *J. Pharm Sci. 60*, 1303 (1971).

366. J. Rieder, *Arzneim. Forsch./Drug Res. 23*, 207 (1973).

367. J. Troschütz, *Arch. Pharm. 314*, 204 (1981).

368. G. Caille, J. Braun, D. Gravel, and R. Plourde, *Can. J. Pharm. Sci. 8*, 42 (1973) (Fr.).

369. U. A. Th. Brinkman, P. L. M. Welling, G. De Vries, A. H. M. T. Scholten, and R. W. Frei, *J. Chromatogr. 217*, 463 (1981).

370. N. Strojny and J. A. F. de Silva, *Anal. Chem. 52*, 1554 (1980).

371. L. C. Bailey and A. P. Shroff, *Res. Commun. Chem. Pathol. Pharmacol.* *7*, 105 (1974).

372. J. T. Stewart, I. L. Honigberg, A. Y. Tsai, and P. Hajdu, *J. Pharm. Sci.* *68*, 494 (1979).

373. J. P. Moody, S. F. Whyte, and G. J. Naylor, *Clin. Chim. Acta 43*, 355 (1973).

374. S. G. Schulman, J. M. Rutledge, and G. Torosian, *Anal. Chim. Acta 68*, 455 (1974).

375. S. S. Brown and G. A. Smart, *J. Pharm. Pharmacol. 21*, 466 (1969).

376. J. W. Meilink, *Pharm. Weekbl.* 109, 22 (1974).

377. B. Caddy and A. H. Stead, *Analyst 103*, 937 (1978).

378. S. De Bernado, M. Weigele, V. Toome, K. Manhart, W. Leimgruber, P. Böhlen, S. Stein, and S. Udenfriend, *Arch. Biochem. Biophys.* 163, 390 (1974).

379. J. Hasegawa, A. Hurwitz, G. Krol, and R. Davies, *J. Pharm. Sci. 65*, 508 (1976).

380. G. Caille, C. Sauriol, J. A. Mockle, and J. C. Panisset, *Can. J. Pharm. Sci. 5*, 72 (1970) (Fr.).

381. K. F. Overø, *J. Chromatogr. 224*, 526 (1981).

382. K. F. Overø, *Eur. J. Clin. Pharmacol. 14*, 69 (1978).

383. E. Oeyehaug, E. T. Oestensen, and B. Salvesen, *J. Chromatogr. 227*, 129 (1982).

384. K. F. Overø, *Acta Psychiatr. Scand.* Suppl. *61*, 279, 92 (1980); in *Chem. Abstr. 93*, 60830a (1980).

385. K. Hempel, H. Ulrich, and G. Philippu, *Biol. Psychiatr. 17*, 49 (1982).

386. W. Baeyens, Ph.D. Thesis, State University of Ghent, Belgium, Faculty of Pharmaceutical Sciences, 1976 (Dutch).

387. W. Baeyens, *Analyst 102*, 525 (1977).

388. W. Baeyens, *Pharm. Weekbl. 112*, 681 (1977) (Dutch).

389. W. Baeyens, P. De Moerloose, and L. De Taeye, *J. Pharm. Sci. 66*, 12, 1787 (1977).

390. F. De Croo, G. A. Bens, and P. De Moerloose, *J. HRC & CC, 3*, 423 (1980).

391. W. Baeyens and P. De Moerloose, *Pharmazie, 32*, 764 (1977).

392. W. Baeyens, *Talanta 24*, 579 (1977).

393. T. J. Mellinger and C. E. Keeler, *Anal. Chem. 35*, 554 (1963).

394. J. B. Ragland and V. J. Kinross-Wright, *Anal. Chem. 36*, 1356 (1964).

395. T. J. Mellinger and C. E. Keeler, *Anal. Chem. 36*, 1840 (1964).

396. W. L. Pacha, *Experientia 25*, 103 (1969).

397. S. L. Tompsett, *Acta Pharmacol. Toxicol. 26*, 298 (1968).

398. D. H. Manier, J. Sekerke, J. V. Dingell, and M. Khaled El-Yousef, *Clin. Chim. Acta 57*, 225 (1974).

399. J. M. Konieczny and E. McGonigle, *J. Pharm. Sci. 63,* 1306 (1974).

400. V. R. White, C. S. Frings, J. E. Villafranca, and J. M. Fitzgerald, *Anal. Chem. 48,* 1314 (1976).

401. P. N. Kaul, M. W. Conway, and M. L. Clark, *Nature 226,* 372 (1970).

402. I. S. Forrest, S. D. Rose, L. G. Brookes, B. Malpern, V. A. Bacon, and I. A. Silberg, *Agressologie 11,* 127 (1970).

403. P. N. Kaul, M. W. Conway, M. L. Clark, and J. F. Huffine, *J. Pharm. Sci. 59,* 1745 (1970).

404. P. N. Kaul, M. W. Conway, M. K. Ticku, and M. L. Clark, *J. Pharm. Sci. 61,* 581 (1972).

405. P. N. Kaul, L. R. Whitfield, and M. L. Clark, *J. Pharm. Sci. 65,* 689 (1976).

406. W. Bauld, M. Givner, L. Engel, and J. Goldzieher, *Can. J. Biochem. Physiol. 38,* 213 (1960).

407. L. Penzes and G. Oertel, *J. Chromatog. 51,* 322, (1970).

408. L. Penzes and G. Oertel, *J. Chromatog. 51,* 325 (1970).

409. M. Numazawa and T. Nambara, *Taisha 8,* 371 (1971) (Japan.).

410. L. Lee and R. Hähnel, *Clin. Chem. 17,* 1194 (1971).

411. A. Howarth and D. Robertshaw, *Clin. Chem. 17,* 316 (1971).

412. K. Outch, P. Dennis, and A. Larsen, *Clin. Chim. Acta 40,* 377 (1972).

413. G. Adessi and M. Jayle, *Ann. Biol. Clin.* (Paris) *30,* 127 (1972) (Fr.).

414. G. Adessi and M. Jayle, *Ann. Biol. Clin.* (Paris) *30,* 219 (1972) (Fr.).

415. T. James, *J. AOAC 54,* 1192 (1971).

416. T. James, *J. AOAC, 56,* 86 (1973).

417. A. Craig, J. W. Leek, and R. F. Palmer, *Clin. Biochem. 6,* 34 (1973).

418. U. Goebelsmann, *Clin. Chim. Acta 43,* 285 (1973).

419. T. James, *J. Pharm. Sci. 61,* 1306 (1972).

420. R. Dvir and R. Chayen, *J. Chromatog. 45,* 76 (1969).

421. W. C. Foster, R. A. Donato, and J. N. McKenna, *Clin. Chem. 17,* 651 (1971).

422. M. van Bezeij and M. Bosch, *Z. Klin, Chem. Klin. Biochem. 13,* 381 (1975).

423. K. L. Oehrle and K. Vogt, *Acta Endocrinol.* (Copenhagen) Suppl. *82,* 202, 61 (1976).

424. K. Schollberg, C. Blümel, and E. Seiler, *Zbl. Pharm. 109,* 811 (1970) (Germ.).

425. D. Tenhaeff, D. Maruhn, and E. Ahl, *Clin. Chim. Acta 31,* 37 (1971).

426. S. Csevari and I. Csaba, *Lab. Delo, 1,* 8 (1982) (Russ.); in *Chem. Abstr.* 96 (11) 80135t (1983).

427. Rhys-Williams, A. T., S. A. Winfield, and R. C. Belloli, *J. Chromatogr.* 235, 461 (1982).

428. T. James, *J. Pharm. Sci. 59,* 1648 (1970).

429. A. Mariani and C. Mariani-Vicari, *Zbl. Pharm. 111,* 19 (1972) (Germ.).

430. V. Graef, *Z. Klin. Chem. Klin. Biochem. 8,* 320 (1970) (Germ.).

431. T. Kawasaki, M. Maeda, and A. Tsuji, *J. Chromatogr. 226,* 1 (1981).

432. S. Pal, *Clin. Chim. Acta 33,* 215 (1971).

433. D. Egg and H. Huck, *J. Chromatogr. 63,* 349 (1971) (Germ.).

434. W. Eechaute, G. Demeester, and I. Leusen, *Steroids 16,* 277 (1970).

435. W. Eechaute, G. Demeester, and I. Leusen, *Clin. Chim. Acta 32,* 297 (1971).

436. D. Alkalay, L. Khemani, and M. F. Bartlett, *J. Pharm. Sci. 61,* 1746 (1972).

437. E. Njau, *J. Pharm. Sci. 66,* 1478 (1977).

438. A. Britten and E. Njau, *Anal. Chim. Acta 76,* 409 (1975).

439. A. Britten and E. Njau, *J. Chem. Soc.* (Perkin I), 158 (1976).

440. W. Johnson, *J. Am. Chem. Soc. 65,* 1317 (1943).

441. L. Birkofer, S. Kim, and D. Engels, *Chem. Ber. 95,* 1495 (1962).

442. R. Segura and M. Gotto, Jr., *J. Chromatogr. 99,* 643 (1974).

443. S. Levi and R. Reisfeld, *Anal. Chim. Acta 97,* 343 (1978).

444. J. Bartos and M. Pesez, *The Analysis of Organic Materials,* Vol. 11: *Colorimetric and Fluorometric Analysis of Steroids,* Academic, London, 1977.

445. P. De Moor, P. Osinski, R. Deckx, and O. Steeno, *Clin. Chim. Acta 7,* 475 (1962).

446. E. Z. Naugol'nykh, *Lab. Delo 9,* 532 (1969) (Russ.); in *Chem. Abstr. 72,* 9533v (1970).

447. Y. Sasaki, R. Kawashima, and S. Uesaka, *Nippon Chikusan Gakkai-Ho 41,* 632 (1970) (Japan.).

448. S. Borrell and P. Vega, *Rev. Clin. Espan. 119,* 9 (1970) (Span.).

449. M. Martin and A. Martin, *Clin. Chim. Acta 27,* 379 (1970).

450. B. R. Clark, R. T. Rubin, A. Kales, and R. Poland, *Clin. Chim. Acta 27,* 364 (1970).

451. J. Bumba and A. Janicek, *Vnitr. Lek 17,* 182 (1971) (Czech.); in *Chem. Abstr. 74,* 135729t (1971).

452. T. Bezverkhaya and G. Povolotskaya, *Lab. Delo 2,* 74 (1971) (Russ.); in *Chem. Abstr. 74,* 120696f (1971).

453. K. M. Pirke and D. Stamm, *Z. Klin. Chem. Klin. Biochem. 10,* 243 (1972) (Germ.).

454. T. R. Koch, L. Edwards, and M. E. Chilcote, *Clin. Chem. 19,* 258 (1973).

455. H. Kaplan, L. Kany, and R. Abraham, *Clin. Chim. Acta 51,* 113, (1974) (Germ.).

456. T. Kawasaki, M. Maeda, and A. Tsuji, *J. Chromatogr. 163,* 143 (1979).

457. W. Funk, R. Kerler, E. Boll, and V. Dammann, *J. Chromatogr. 217,* 349 (1981).

458. E. M. Kazin, G. G. Avdeev, R. I. Andreeva, and N. G. Blinova, *Vop. Biol., Mater. Nauch. Konf. Biol. Kafedr* 64 (1970) (Russ.); in *Chem. Abstr. 77*, 161684f (1972).

459. B. Raczynska, A. Swiecicki, I. Lenartowska, and B. Wancowicz, *Pediat. Pol. 47*, 195 (1972) (Pol.); in *Chem. Abstr. 77*, 31141s (1972).

460. H. Suemori, *Nichidai Igaku Zasshi 31*, 714 (1972) (Japan.); in *Chem. Abstr. 77*, 149295x (1972).

461. P. P. Golikov, *Veg. Blokada Adrenal. Sist. 71* (1969) (Russ.); in *Chem. Abstr., 73*, 84236z (1970).

462. T. Takeda, C. Nishizawa, Y. Noto, K. Uchida, H. Nakabayashi, U. Miwa, Z. Saito, S. Morimoto, and G. Murakami, *Horumon To Rinsho 18*, 401 (1970) (Japan.); in *Chem. Abstr. 73*, 84225v (1970).

463. T. Usui, H. Kawamoto, and S. Shimao, *Clin. Chim. Acta 30*, 663 (1970).

464. P. De Moor, O. Steeno, M. Raskin, and H. Hendrikx, *Acta Endocrinol. 33*, 297 (1960).

465. V. Graef and Hj. Staudinger, *Z. Klin. Chem. Klin. Biochem. 8*, 368 (1970) (Germ.).

466. T. Usui and H. Kawamoto, *Rinsho Byori 20*, 387 (1972) (Japan.); in *Chem. Abstr. 77*, 123470a (1972).

467. L. E. Mejer and R. C. Blanchard, *Clin. Chem. 19*, 710 (1973).

468. L. E. Mejer and R. C. Blanchard, *Clin. Chem. 19*, 718 (1973).

469. H. Kawamoto, T. Usui, and T. Yamane, *Rinsho Kagaku 10*, 226 (1981); in *Chem. Abstr. 96* (13) 97828u.

470. I. S. Kokoreva and V. A. Rudenko, *Fiz.-Khim. Metody Issled. Anal. Biol. Ob'ektov Nekot. Tekh. Mater.* 51 (1971), from *Ref. Zh., Khim* 1972, Abstr. No. 3G206 (Russ.).

471. M. Gelbcke, *J. Pharm. Belg. 31*, 397 (1976) (Fr.).

472. T. Panalaks and P. M. Scott, *J. AOAC 60*, 583 (1977).

473. C. W. Thorpe, G. M. Ware, and A. E. Pohland, *Abstracts of Papers Presented at the 1981 Pittsburgh Conference on Analytical Chemistry and Applied Spectroscopy*, no. 272, Convention Hall, Atlantic City, NJ, March 1981.

474. L. A. King, J. N. Miller, D. T. Burns, and J. W. Bridges, *Anal. Chim. Acta 68*, 205 (1974).

475. L. A. Gifford, W. P. Hayes, L. A. King, J. N. Miller, D. T. Burns, and J. W. Bridges, *Anal. Chem. 46*, 94 (1974).

476. L. A. King and L. A. Gifford, *Anal. Chem. 47*, 17 (1975).

477. A. A. Khabarov, V. P. Baranov, and Sh. Sh. Shasaitov, *Deposited Doc.*, VINITI, 5413–80, 8 pp. (1980) (Russ.).

478. W. A. Dill, A. Leung, A. W. Kinkel, and A. J. Glazko, *Clin. Chem. 22*, 908 (1976).

479. A. Bowd, P. Byrom, J. B. Hudson, and J. H. Turnbull, *Talanta 18*, 697 (1971).

480. F. J. Bullock (Arthur D. Little, Inc.) U.S. Pat. 3, 656, 906 (Cl. 23/230 B; GOln), 18 Apr 1972, Appl. 28, 015, 13 Apr 1970; 8pp.

481. A. Dionyssiou-Asteriou and C. J. Miras, *J. Pharm. Pharmacol. 27*, 135 (1975).

482. R. Bourdon, *Eur. J. Toxicol. Environ. Hyg. 9*, 11 (1976) (Fr.).

483. J. A. Vinson, D. Patel, and A. Patel, *Anal. Chem. 49*, 163 (1977).

484. T. P. Ivanov, *Monatsh. Chem. 97*, 1499 (1966).

485. C. P. Ivanov and Y. Vladovska-Yukhnovska, *J. Chromatogr. 71*, 11 (1972).

486. Y. Vladovska-Yukhnovska, C. P. Ivanov, and M. Malgrand, *J. Chromatogr. 90*, 181 (1974).

487. I. Durko, Y. Vladovska-Yukhnovska, and C. P. Ivanov, *Clin. Chim. Acta 49*, 407 (1973).

488. M. Richter, *Zbl. Pharm. 115*, 611 (1976).

489. J. P. Gramond and R. R. Paris, *Plant. Med. Phytother. 5*, 315 (1971).

490. L. E. Zel'tser, Sh. T. Talipov, and L. A. Morozova, *Zh. Anal. Khim. 36*, 8, 1477 (1981) (Russ.).

491. S. G. Schulman and L. S. Rosenberg, *J. Phys. Chem. 83*, 447 (1979).

492. S. Haing Lee, L. R. Field, W. N. Howald, and W. F. Trager, *Anal. Chem. 53*, 467 (1981).

493. V. Maes, Y. Engelborghs, J. Hoebeke, Y. Maras, and A. Vercruysse, *Mol. Pharmacol. 21*, 100 (1982).

493bis. W. Voelter, R. Huber, and K. Zech, *J. Chromatogr. 217*, 491 (1981).

494. W. Baeyens and P. De Moerloose, *Anal. Chim. Acta 107*, 291 (1979).

495. A. Haussler and P. Hajdu, *Arzneim. Forsch. 14*, 710 (1964).

496. M. R. Kelly, R. E. Cutler, A. W. Forrey, and B. M. Kimpel, *Clin. Pharm. Ther. 15*, 178 (1974).

497. W. Sadée, M. Dagcioglu and S. Riegelman, *J. Pharm. Sci. 61*, 7, 1126 (1972).

498. A. Gochman and C. L. Gantt, *J. Pharmacol. Exp. Ther. 135*, 312 (1962).

499. W. Sadée, S. Riegelman, and L. B. Johnson, *Steroids 17*, 595 (1971).

500. R. B. Smith, R. V. Smith, and G. J. Yakatan, *J. Pharm. Sci. 65*, 8, 1208 (1976).

501. H. Held, B. Kaminski, and H. F. v. Oldershausen, *Drug Res. 20*, 1927 (1970).

502. R. Puech and P. Viallet, *Eur. Biophys. Congr., Proc.*, 1971, 4, 193, E. Broda (ed.), Verlag Wiener Med. Akad., Vienna, Austria, 1971.

503. G. Guebitz, R. W. Frei, and H. Bethke, *J. Chromatogr. 117*, 337 (1976).

504. J. A. F. de Silva, N. Strojny, and N. Munno, *Anal. Chim. Acta 66*, 23 (1973).

505. J. A. F. de Silva, N. Strojny, F. Rubio, J. C. Meyer, and B. A. Koechlin, *J. Pharm. Sci. 66*, 353 (1977).

506. Fardos Abdel Fattah Mohamed, Luminescence Analysis of Some Pharmaceutically Important Imidazoles; Correlation Between Fluorescence, Low temperature and Room Temperature Phosphorescence Behaviour, *Ph.D. Thesis*, State University of Ghent, Belgium, 1981.

507. Y. Takeda, A. Nishikata, T. Osawa, and T. Ukita, *Eisei Kagaku, 16*, 63 (1970) (Japan.).

508. J. D. Martucci and S. G. Schulman, *Anal. Chim. Acta 77*, 317 (1975).

509. S. G. Chakrabarti and I. A. Bernstein, *Clin. Chem. 15*, 1157 (1969).

510. W. Baeyens and P. De Moerloose, *Analyst 103*, 359 (1978).

511. W. Baeyens and P. De Moerloose, *Anal. Chim. Acta 110*, 261 (1979).

512. L. A. Tiktin, Y. P. Borisyuk, and V. V. Konstantinov, *Onkologiya* (Kiev) *1*, 165 (1970) (Russ.); in *Chem. Abstr. 75*, 97322r (1971).

513. H. Hatano, Y. Yamamoto, M. Saito, E. Mochida, and S. Watanabe, *J. Chromatogr. 83*, 373 (1973).

514. R. J. Hurtubise, T. W. Allen, and H. F. Silver, *Anal. Chim. Acta 126*, 225 (1981).

515. Y. Yang, A. P. D'Silva, and V. A. Fassel, *Anal. Chem. 53*, 894 (1981).

516. J. R. Maple and E. L. Wehry, *Anal. Chem. 53*, 266 (1981).

517. J. R. Maple and E. L. Wehry, *Anal. Chem. 53*, 1244 (1981).

518. D. Murray and R. B. Cundall, *Analyst 106*, 335 (1981).

519. S. Uchiyama, T. Kondo, and I. Kawashiro, *Yakugaku Zasshi 89*, 828 (1969).

520. S. Uchiyama, H. Tanabe, and Z. Tamura, *Chem. Pharm. Bull. 20*, 357 (1972).

521. Y. Iitaka, S. Uchiyama, and Z. Tamura, *Chem. Pharm. Bull. 20*, 1181 (1972).

522. S. Uchiyama and Z. Tamura, *Chem. Pharm. Bull. 23*, 1032 (1975).

523. L. R. Snyder and J. J. Kirkland, *Introduction to Modern Liquid Chromatography*, 2nd ed., Wiley, New York, 1979, pp. 731–742.

524. R. W. Frei, L. Michel, and W. Santi, *J. Chromatogr. 126*, 665 (1976).

525. R. W. Frei and J. F. Lawrence, *Chemical Derivatization in Analytical Chemistry*, Vol 1: Chromatography, Plenum, New York, 1981.

526. R. M. Cassidy and R. W. Frei, *J. Chromatogr. 72*, 293 (1972).

527. E. D. Pellizzari and C. M. Sparacino, *Anal. Chem. 45*, 378 (1973).

528. J. N. Miller, *Anal. Proc.* (London) *18*, 264 (1981).

529. R. W. Frei and J. F. Lawrence, *Chemical Derivatization in Analytical Chemistry, Vol. 2: Separation and Continuous Flow Techniques*, Plenum, New York, 1982.

530. P. Froehlich, *Applications of Luminescence Spectroscopy to Quantitative Analyses in Clinical and Biological Samples*, in: *Modern Fluorescence*

Spectroscopy, Vol. 2, E. L. Wehry (ed.), Plenum, New York, 1976, Ch. 2, pp. 69–76.

531. P. A. Asmus, J. W. Jorgenson, and M. Novotny, *J. Chromatogr. 126,* 317 (1976).

532. W. Baumann, *Fresenius' Z. Anal. Chem. 284,* 31 (1977).

533. W. Slavin, A. T. Williams, and R. F. Adams, *J. Chromatogr. 134,* 121 (1977).

534a. L. H. Thacker, *J. Chromatogr. 136,* 213 (1977).

534b. P. J. Harman, G. L. Blackman, and G. Phillipou, *J. Chromatogr. 225,* 131 (1981).

535. T. Jupille, *J. Chromatogr. Sci. 17,* 160 (1979).

536. J. F. Lawrence, *J. Chromatogr. Sci. 17,* 113 (1979).

537. F. T. Noggle, Jr., *J. Assoc. Off. Anal. Chem. 63,* 702 (1980).

538. G. Schwedt and E. Reh, *Chromatographia 14,* 249 (1981).

539. G. Schwedt and E. Reh, *Chromatographia 14,* 317 (1981).

540. A. H. M. T. Scholten, P. L. M. Welling, U. A. T. Brinkman, and R. W. Frei, *Adv. Chromatogr.* (Houston), 15, 239 (1980).

541. L. W. Hershberger, J. B. Callis, and G. D. Christian, *Anal. Chem. 53,* 971 (1981).

542. C. J. Appellof and E. R. Davidson, *Anal. Chem. 53,* 2053 (1981).

543. E. Voigtman, A. Jurgensen, and J. D. Winefordner, *Anal. Chem. 53,* 1921 (1981).

544. K. Ogan and E. Katz, *Anal. Chem. 53,* 160 (1981).

545. K. Ohkubo, A. Nakamoto, and H. Yamamoto, *Abstracts of Papers Presented at the 1981 Pittsburgh Conference on Analytical Chemistry and Applied Spectroscopy,* no. 162, Convention Hall, Atlantic City, NJ, March 1981.

546. M. S. Klee, R. R. Antcliff, L. A. Carreira, and L. B. Rogers, *op. cit.,* no. 171.

547. B. Cunico and S. R. Abbott, *idem.,* no. 645.

548. M. Uihlein and E. Schwab, *Chromatographia 15,* 140 (1982).

549. B. A. Berkowitz, J. H. Asling, S. M. Snider, and E. L. Way, *Clin. Pharmacol. Ther. 10,* 320 (1969).

550. B. A. Berkowitz and E.L. Way, *Clin. Pharmacol. Ther. 10,* 681 (1969).

551. A. H. Beckett, J. F. Taylor, and P. Kourounakis, *J. Pharm. Pharmacol. 22,* 123 (1970).

552. K. O. Borg and A. Mikaelsson, *Acta Pharm. Suecica 7,* 673 (1970).

553. D. Westerlund and K. O. Borg, *Acta Pharm. Suecica 7,* 267 (1970).

554. B. A. Persson, *Acta Pharm. Suecica 7,* 337 (1970).

555. D. Westerlund, K. O. Borg, and P.-O. Lagerström, *Acta Pharm. Suecica 9,* 47 (1972).

556. P.-O. Lagerström, K. O. Borg, and D. Westerlund, *Acta Pharm. Suecica* *9*, 53 (1972).

557. M. Hoshino and A. Tsuji, *Bunseki Kagaku 22*, 163 (1973) (Japan.); in *Chem. Abstr.* 79, 9958g (1973).

558a. M. D. Baker, H. Y. Mohammed, and H. Veening, *Anal. Chem. 53*, 1658 (1981).

558b. Y. Hiraga and T. Kinoshita, *J. Chromatogr. 226*, 43 (1981).

559. R. C. Simpson, H. Y. Mohammed, and H. Veening, *Abstracts of Papers Presented at the 1981 Pittsburgh Conference on Analytical Chemistry and Applied Spectroscopy*, no. 558, Convention Hall, Atlantic City, NJ, March 9–13, 1981.

560. L. L. Dent, J. T. Stewart, and I. L. Honigberg, *Anal. Lett. 14*, B13, 1031 (1981).

561. J. T. Stewart and D. M. Lotti, *J. Pharm. Sci.*, 60, 461 (1971).

562. J. T. Stewart and D. M. Lotti, *Anal. Chim. Acta 52*, 390 (1970).

563. L. J. Dombrowski and E. L. Pratt, *Anal. Chem. 43*, 1042 (1971).

564. W. Baeyens, G. A. Bens, P. De Moerloose, and L. De Taeye, *Pharmazie 35*, 87 (1980).

565. J. E. Sinsheimer, D. D. Hong, J. T. Stewart, M. L. Fink, and J. H. Burckhalter, *J. Pharm. Sci. 60*, 141 (1971).

566. A. De Leenheer, J. E. Sinsheimer, and J. H. Burckhalter, *J. Pharm. Sci. 61*, 273 (1972).

567. J. Monforte, R. J. Bath, and I. Sunshine, *Clin. Chem. 18*, 1329 (1972).

568. R. W. Frei and J. F. Lawrence, *J. Chromatogr. 83*, 321 (1973).

569. Y. Watanabe and K. Imai, *Abstracts of Papers Presented at the 1981 Pittsburgh Conference on Analytical Chemistry and Applied Spectroscopy*, no. 647, Convention Hall, Atlantic City, NJ, March 9–13, 1981.

570. K. Imai and Y. Watanabe, *Anal. Chim. Acta 130*, 377 (1981).

571. J. L. Andrews, P. Ghosh, B. Ternai, and M. W. Whitehouse, *Arch. Biochem. Biophys. 214*, 386 (1982).

572. J. R. Benson and P. E. Hare, *Proc. Nat. Acad. Sci. USA 72*, 619 (1975).

573. S. Udenfriend, S. Stein, P. Böhlen, W. Dairman, W. Leimgruber, and M. Weigele, *Science 178*, 871 (1972).

574. A. M. Felix and M. H. Jimenez, *J. Chromatogr. 89*, 361 (1974).

575. B. R. Larssen and F. G. West, *J. Chromatogr. Sci. 19*, 259 (1981).

576. H. G. Nowicki, *J. Forensic Sci. 21*, 154 (1976).

577. C. Branca, E. Gaetani, C. F. Laureri, and M. Vitto, *Farmaco*, Ed. Prat., *36*, 518 (1981) (Italian).

578. J. A. F. de Silva and N. Strojny, *Anal. Chem. 47*, 714 (1975).

579. P. De Moerloose and W. Baeyens, *Verhandelingen van de Koninklijke Academie voor Geneeskunde van België, XXXVIII*, nr. 6, 252 (1976) (Dutch).

580. T. K. Hwang, J. N. Miller, and D. T. Burns, *Anal. Chim. Acta 99*, 305 (1978).

581. R. W. Thomas and N. J. Leonard, *Heterocycles 5*, 839 (1976).

582. L. K. Tan and R. A. Patsiga, *Anal. Lett. 10*, 437 (1977).

583. R. W. Frei, J. F. Lawrence, U. A. T. Brinkman, and I. Honigberg, *J. High Resolut. Chromatogr. Chromatogr. Commun. 2*, 11 (1979).

584. H. Khalaf and M. Rimpler, *Hoppe-Seyler's Z. Physiol. Chem. 358*, 505 (1977) (Germ.).

585. G. Guebitz, R. Wintersteiger, and A. Hartinger, *J. Chromatogr. 218*, 51 (1981).

586. J. Fourche, H. Jensen, and E. Neuzil, *Anal. Chem. 48*, 155 (1976).

587. M. V. Cerda and F. C. Mongay, *Quim. Anal. 30*, 15 (1976). (Span.).

588. A. De Leenheer, J. E. Sinsheimer, and J. H. Burckhalter, *J. Pharm. Sci. 61*, 1659 (1972).

589. M. P. Wong and K. A. Connors, *Anal. Chem. 50*, 2051 (1978).

590. S. Mizutani, T. Nakajima, A. Matsumoto, and Z. Tamura, *Chem. Pharm. Bull. 12*, 850 (1964).

591. S. Mizutani, Y. Wakuri, N. Yoshida, T. Nakajima, and Z. Tamura, *Chem. Pharm. Bull. 17*, 2340 (1969).

592. R. Brandt and N. D. Cheronis, *Microchem. J. V*, 110, (1961).

593. J. E. Spikner and J. C. Towne, *Anal. Chem. 34*, 1468 (1962).

594. Y. Ohkura, K. Ohtsubo, K. Zaitsu, and K. Kohashi, *Anal. Chim. Acta 99*, 317 (1978).

595. K. Ohtsubo, Y. Okada, K. Zaitsu, and Y. Ohkura, *Anal. Chim. Acta 110*, 335 (1979).

596. H. Shanfield, F. Hsu, and A. J. P. Martin, *J. Chromatogr. 126*, 457 (1976).

597. S. Uchiyama and M. Uchiyama, *J. Chromatogr. 153*, 135 (1978).

598. S. Uchiyama and M. Uchiyama, *J. Liq. Chromatogr. 3*, 681 (1980).

599. R. Segura and X. Navarro, *Advances in Chromatography 1981, Proceedings of the 16th Int. Symp.*, A. Zlatkis (ed.), Barcelona, 1981, p. 319.

600. R. Segura and A. M. Gotto, Jr., *J. Chromatogr. 99*, 643 (1974).

601. L. Zhou, H. Shanfield, F.-S. Wang, and A. Zlatkis, *Advances in Chromatography 1981, Proceedings of the 16th Int. Symp.*, A. Zlatkis (ed.), Barcelona, 1981, p. 331.

602. H. J. C. F. Nelis and J. E. Sinsheimer, *Anal. Biochem. 115*, 151 (1981).

603. H. J. C. F. Nelis, S. C. Abry, and J. E. Sinsheimer, *Anal. Chem. 54*, 213 (1982).

604. G. Avigad, *J. Chromatogr. 139*, 343 (1977).

605. L. H. Kormos, R. L. Sandridge, and J. Keller, *Anal. Chem. 53*, 1122 (1981).

606. E. F. Ullman, M. Schwarzberg, and K. E. Rubenstein, *J. Biol. Chem. 251*, 4172 (1976).

607. L. S. Forster, *Photochem. Photobiol. 23*, 445 (1976).

608. U. E. Handschin and W. J. Ritschard, *Anal. Biochem. 71*, 143 (1976).

609. M. W. Burgett, S. J. Fairfield, and J. F. Monthony, *Clin. Chim. Acta 78*, 277 (1977).

610. S. G. Thompson, *Clin. Biochem. Anal. 10*, 107 (1981).

611. L. A. Kaplan, I. W. Chen, N. Gau, J. Fearn, H. Maxon, C. Volle, and E. A. Stein, *Clin. Biochem. 14*, 182 (1981).

612. G. C. Visor and S. G. Schulman, *J. Pharm. Sci. 70*, 469 (1981).

613. A. J. Quattrone, C. M. O'Donnell, J. McBride, P. B. Mendershausen, and R. S. Putman, *J. Anal. Toxicol. 5*, 245 (1981).

614. M. E. Jolley, *J. Anal. Toxicol. 5*, 236 (1981).

615. W. D. Hinsberg, K. H. Milby, S. D. Lidofsky, and R. N. Zare, *Proc. SPIE—Int. Soc. Opt. Eng. 286*, 132 (1981).

616. T. M. Li, J. L. Benovic, and J. F. Burd, *Anal. Biochem 118*, 102 (1981).

617a. M. M. Collins, C. H. Casavant, and D. P. Stites, *J. Clin. Microbiol. 15*, 456 (1982).

617b. J. H. Rho and J. L. Stuart, *Anal. Chem. 50*, 620 (1978).

617c. J. N. Miller, T. A. Ahmad, and A. F. Fell, *Anal. Proc.* (London) *19*, 37 (1982).

618. F. L. Vandemark and J. L. DiCesare, *Chromatogr. Newsl. 9*, 33 (1981).

619. I. K. Moules, E. K. Rooney, and A. G. Lee, *FEBS Lett. 138*, 95 (1982).

620. G. M. Gandolfo and M. V. Torresi, *Clin. Lab.* (Rome) *5*, 138 (1981).

621. J. C. Touchstone, R. E. Levitt, and S. S. Levin, *Abstracts of Papers Presented at the 1981 Pittsburgh Conference on Analytical Chemistry and Applied Spectroscopy* no. 438, Convention Hall, Atlantic City, NJ, March 1981.

622. C. L. Yuan, S. S. Kuan, and G. G. Guilbault, *Anal. Chim. Acta 134*, 47 (1982).

623. L. G. Dickson, C. B. Johnson, and D. R. Johnson, *Clin. Chem. 19*, 301 (1973).

624. R. G. Price, N. Dance, B. Richards, and W. R. Cattell, *Clin. Chim. Acta 27*, 65 (1970).

625. T. Sekine, H. Itakura, T. Namihisa, T. Takahashi, H. Nakayama, and Y. Kanaoko, *Chem. Pharm. Bull. 29*, 3286 (1981).

626. G. M. Gandolfo and M. V. Torresi, *Clin. Lab.* (Rome) *5*, 263 (1981).

627. I. Ando, Y. Horiuchi, A. Miura, and T. Harumoto, *Arerugi 22*, 435 (1973) (Japan.); in *Chem. Abstr. 79*, 134090u (1973).

628. M. Watanabe and K. Hayashi, *Clin. Chim. Acta 37*, 417 (1972).

629. P. A. Toseland, M. J. Michelin and S. A. Price, *Clin. Chim. Acta 37*, 477 (1972).

630. I. K. Genefke, *Acta Pharmacol. Toxicol 31*, 554 (1972).

631. T. G. Rosano, J. M. Meola, and T. A. Swift, *Clin. Chem. 28*, 207 (1982).

632. M. Watanabe, Y. Watanabe, and M. Okada, *Clin. Chim. Acta 27*, 461 (1970).

633. W. D. Hinsberg III, K. H. Milby, and R. N. Zare, *Anal. Chem. 53*, 1509 (1981).

634. P. J. Cashion, H. J. Notman, T. B. Cadger, and G. M. Sathe, *Anal. Biochem 80*, 654 (1977).

635. T. Hirata, M. Kai, K. Kohashi, and Y. Ohkura, *J. Chromatogr. 226*, 25 (1981).

636. L. Malina, K. Jojkova, and E. Hejlova, *Cas. Lek. Cesk. 121*, 146 (1982) (Czech.).

637. K. S. Hammond and D. S. Papermaster, *Anal. Biochem. 74*, 292 (1976).

638. S. Grimaldi, H. Edelhoch, and J. Robbins, *Biochemistry 21*, 145 (1982).

639. J. M. Lisy and B. C. Gerstein, *Anal. Chem. 45*, 1536 (1973).

640. G. M. Anderson and W. M. Purdy, *Anal. Lett. 10*, 493 (1977).

641. G. N. Lewis and M. Kasha, *J. Am. Chem. Soc. 66*, 2100 (1944).

642. R. J. Keirs, R. D. Britt, Jr., and W. E. Wentworth, *Anal. Chem. 29*, 202 (1957).

643. S. Freed and W. Salmre, *Science 128*, 1341 (1958).

644. C. A. Parker and C. G. Hatchard, *Analyst 87*, 664 (1962).

645. S. Freed and M. H. Vise, *Anal. Biochem. 5*, 338 (1963).

646. J. D. Winefordner and H. W. Latz, *Anal. Chem. 35*, 1517 (1963).

647. H. W. Latz, *Ph.D. thesis*, University of Florida, Gainesville, 1963.

648. T. C. O'Haver and J. D. Winefordner, *Anal. Chem. 38*, 602 (1966).

649. H. C. Hollifield and J. D. Winefordner, *Anal. Chem. 40*, 1759 (1968).

650. M. Zander, *Phosphorimetry: The Application of Phosphorescence to the Analysis of Organic Compounds*, Academic Press, New York, 1968.

651. J. D. Winefordner, W. J. McCarthy, and P. A. St. John, *Methods of Biochemical Analysis*, Vol. XV, D. Glick (ed.), Interscience, New York, 1967.

652. C. A. Parker, *Photoluminescence of Solutions*, Elsevier, New York, 1968.

653. J. D. Winefordner, P. A. St. John, and W. J. McCarthy, *Fluorescence Assay in Biology and Medicine*, Vol. II, S. Udenfriend (ed.), Academic Press, New York, 1969.

654. J. D. Winefordner, "Phosphorimetry", in: *Accuracy in Spectrophotometry and Luminescence Measurements*, in Proc. Conf. N.B.S. 22–24 March 1972, NBS Spec. Publ. 378, Washington, 1973.

655. W. J. McCarthy, "Phosphorescence Spectrometry", in: *Spectrochemical Methods of Analysis*, J. D. Winefordner (ed.), Wiley-Interscience, New York, 1971.

656. J. J. Aaron and J. D. Winefordner, *Talanta 22*, 707 (1975).

657. C. M. O'Donnell and J. D. Winefordner, *Clin. Chem. 21*, 285 (1975).

658. W. Baeyens, *Pharm. Weekbl.* III, *41,* (part I) (1976) (Dutch).

659. W. Baeyens, *Pharm. Weekbl.* III, *43,* 1075 (part II) (1976) (Dutch).

660. J. D. Winefordner and H. W. Latz, *Anal. Chem. 35,* 1517 (1963).

661. J. D. Winefordner and M. Tin, *Anal. Chim. Acta 31,* 239 (1964).

662. H. C. Hollifield and J. D. Winefordner, Anal. Chim. Acta 36, 351 (1966).

663. H. C. Hollifield and J. D. Winefordner, *Talanta 12,* 860 (1965).

664. L. B. Sanders, J. J. Cetorelli, and J. D. Winefordner, *Talanta 16,* 407 (1969).

665. R. Zweidinger and J. D. Winefordner, *Anal. Chem. 42,* 639 (1970).

666. J. L. Ward., G. L. Walden, and J. D. Winefordner, *Talanta 28,* 201 (1981).

667. U. P. Wild, *Chimia,* 30, 382 (1976).

668. J. J. Dekkers, G. Ph. Hoornweg, K. J. Terpstra, C. Maclean, and N. H. Velthorst, *Chem. Phys. 34,* 253 (1978).

669. J. C. Sutherland, C. P. Yarter, and S. D. Putney, *Photochem. Photobiol. 23,* 141 (1976).

670. E. Lue Yen and J. D. Winefordner, *Anal. Chem. 49,* 1262 (1977).

671. A. T. R. Williams, *Int. Lab.* 90 (November–December, 1981).

672. S. G. Schulman, "Fluorescence and Phosphorescence Spectroscopy: Physicochemical Principles and Practice", Pergamon Press, New York, 1977.

673. G. G. Guilbault, Practical Fluorescence: Theory, Methods and Techniques, Dekker, New York, 1973.

674. E. L. Wehry, *Modern Fluorescence Spectroscopy,* Vols. I, II, Plenum Press, New York, 1976.

675. J. D. Winefordner, S. G. Schulman, and T. C. O'Haver, *Luminescence Spectrometry in Analytical Chemistry,* Wiley-Interscience, New York, 1972.

676. D. M. Hercules, *Fluorescence and Phosphorescence Analysis: Principles and Applications,* Interscience, New York, 1966.

677. M. D. Lumb, *Luminescence Spectroscopy,* Academic Press, New York, 1978.

678. J. W. Bridges, L. A. Gifford, W. P. Hayes, J. N. Miller, and D. Thorburn Burns, *Anal. Chem. 46,* 1010 (1974).

679. L. A. Gifford, *Proc. Soc. Anal. Chem.* 272 (November, 1973.)

680. L. A. Gifford, J. N. Miller, D. T. Burns, and J. W. Bridges, *J. Chromatogr. 103,* 15 (1975).

681. J. N. Miller, D. L. Phillipps, D. T. Burns, and J. W. Bridges, *Anal. Chem. 50,* 613 (1978).

682. J. R. McDuffie and W. C. Neely, *Anal. Biochem. 54,* 507 (1973).

683. L. D. Morrison and C. M. O'Donnell, *Anal. Chem. 46,* 1119 (1974).

684. W. A. Dill, L. Chucot, T. Chang, and A. J. Glazko, *Clin. Chem. 17,* 1200 (1971).

685. L. A. Gifford, W. P. Hayes, L. A. King, J. N. Miller, D. Thorburn Burns, and J. W. Bridges, *Anal. Chim. Acta 62,* 214 (1972).

686. C. I. Miles and G. H. Schenk, *Anal. Chem. 45*, 130 (1973).

687. L. A. Gifford, J. N. Miller, D. L. Phillipps, D. T. Burns, and J. W. Bridges, *Anal. Chem. 47*, 1699 (1975).

688. N. Strojny and J. A. F. de Silva, *J. Chromatogr. Sci. 13*, 583 (1975).

689. J. A. F. de Silva, N. Strojny, and K. Stika, *Anal. Chem. 48*, 144 (1976).

690. N. Strojny, J. A. F. de Silva, and B. Carroll, *Abstracts of Papers Presented at the 1981 Pittsburgh Conference on Analytical Chemistry and Applied Spectroscopy*, no. 803, Convention Hall, Atlantic City, NJ, March 1981.

691. I. Hornyák and L. Székelyhidi, *Anal. Chim. Acta 120*, 415 (1980).

692. G. D. Boutilier, C. M. O'Donnell, and R. O. Rahn, *Anal. Chem. 46*, 1508 (1974).

693. R. T. Mayer, G. M. Holman, and A. C. Bridges, *J. Chromatogr. 90*, 390 (1974).

694. G. D. Boutilier and J. D. Winefordner, *Anal. Chem. 51*, 1391 (1979).

695. G. D. Boutilier and J. D. Winefordner, *Anal. Chem. 51*, 1384 (1979).

696. A. U. Acuña, A. Ceballos, and M. J. Molera, *An. Quim.*, 72, 410 (1976).

697. A. W. H. Mau and M. Puza, *Photochem. Photobiol. 25*, 601 (1977).

698. V. I. Stenberg, S. P. Singh, and N. K. Narain, *Spectrosc. Lett. 8*, 639 (1977).

699. A. I. Al-Mosawi, J. N. Miller, and J. W. Bridges, *Analyst 105*, 448 (1980).

700. M. Yamaguchi, S. Miyamoto, and K. Kohashi, *Anal. Chim. Acta 121*, 289 (1980).

701. Y. Ohkura, M. Yamaguchi, M. Kai, and K. Kohashi, *Chem. Pharm. Bull. 28*, 1913 (1980).

702. G. Sogliero, D. Eastwood, and R. Ehmer, *Abstracts of Papers Presented at the 1981 Pittsburgh Conference on Analytical Chemistry and Applied Spectroscopy*, no. 198, Convention Hall, Atlantic City, NJ, March 1981.

703. L. V. S. Hood and J. D. Winefordner, *Anal. Biochem. 27*, 523 (1969).

704. K. L. Bell and H. C. Brenner, *Biochemistry 21*, 799 (1982).

705. E. M. Schulman and C. Walling, *Science 178*, 53 (1972).

706. M. Roth, *J. Chromatogr. 30*, 276 (1967).

707. J. B. F. Lloyd and J. N. Miller, *Talanta 26*, 180 (1979).

708. D. J. Brown, *J. Chem. Soc. 1*, 1974 (1958).

709. R. A. Paynter, S. L. Wellons, and J. D. Winefordner, *Anal. Chem. 46*, 736 (1974).

710. J. J. Aaron and J. D. Winefordner, *Talanta 22*, 707 (1975).

711. P. G. Seybold and W. White, *Anal. Chem. 47*, 1199 (1975).

712. T. Vo-Dinh, E. Lue Yen, and J. D. Winefordner, *Anal. Chem. 48*, 8, 1186 (1976).

713. E. M. Jakovljevic, *Anal. Chem. 49*, 2048 (1977).

714. J. B. F. Lloyd, *Analyst 103*, 775 (1978).

715. E. M. Schulman and R. T. Parker, *J. Phys. Chem. 81,* 1932 (1977).

716. T. Vo-Dinh, G. L. Walden, and J. D. Winefordner, *Anal. Chem. 49,* 1126 (1977).

717. E. Lue Yen-Bower, J. L. Ward, G. Walden, and J. D. Winefordner, *Talanta 27,* 380 (1980).

718. R. T. Parker, R. S. Freedlander, and R. B. Dunlap, *Anal. Chim. Acta 119,* 189 (1980).

719. R. T. Parker, R. S. Freedlander, and R. B. Dunlap, *Anal. Chim. Acta 120,* 1 (1980).

720. J. N. Miller, *Trends Anal. Chem. 1,* 31 (1981).

721. R. J. Hurtubise, *Solid Surface Luminescence Analysis: Theory, Instrumentation, Applications,* Dekker, New York, 1981.

722. G. J. Niday and P. G. Seybold, *Anal. Chem. 50,* 1577 (1978).

723. E. L. Wehry, *Anal. Chem. 52,* 75R (1980).

724. D. L. McAleese, R. S. Freedlander, and R. B. Dunlap, *Anal. Chem. 52,* 2443 (1980).

725. J. L. Ward, E. Lue Yen-Bower, and J. D. Winefordner, *Talanta 28,* 119 (1981).

726. R. M. A. von Wandruszka and R. J. Hurtubise, *Anal. Chim. Acta 93,* 331 (1977).

727. R. M. A. von Wandruszka and R. J. Hurtubise, *Anal. Chem. 49,* 2164 (1977).

728. T. Vo-Dinh, E. Lue Yen, and J. D. Winefordner, *Talanta 24,* 146 (1977).

729. E. Lue-Yen Bower and J. D. Winefordner, *Anal. Chim. Acta 102,* 1 (1978).

730. T. Vo-Dinh and J. R. Hooyman, *Anal. Chem. 51,* 1915 (1979).

731. T. Vo-Dinh, R. B. Gammage, and P. R. Martinez, *Anal. Chim. Acta 118,* 313 (1980).

732. T. Vo-Dinh, R. B. Gammage, and P. R. Martinez, *Anal. Chem. 53,* 253 (1981).

733. C. D. Ford and R. J. Hurtubise, *Anal. Chem. 50,* 610 (1978).

734. M. L. Meyers and P. G. Seybold, *Anal. Chem. 51,* 1609 (1979).

735. R. T. Parker, R. S. Freedlander, E. M. Schulman, and R. B. Dunlap, *Anal. Chem. 51,* 1921 (1979).

736. C. D. Ford and R. J. Hurtubise, *Anal. Chem. 51,* 659 (1979).

737. C. D. Ford and R. J. Hurtubise, *Anal. Chem. 52,* 656 (1980).

738. R. J. Hurtubise, *Talanta 28,* 145 (1981).

739. F. Abdel Fattah, W. Baeyens, and P. De Moerloose, in: *Bioluminescence and Chemiluminescence: Basic Chemistry and Analytical Applications,* M. A. De Luca and W. D. McElroy (eds.), Academic Press, New York, 1981, pp. 335–346.

740. L. J. Cline Love, M. Skrilec, and J. G. Habarta, *Anal. Chem. 52,* 754 (1980).

741. M. Skrilec and L. J. Cline Love, *Anal. Chem. 52,* 1559 (1980).

742. L. J. Cline Love, J. G. Habarta, and M. Skrilec, *Anal. Chem. 53,* 437 (1981).

743. L. J. Cline Love and M. Skrilec, *Anal. Chem. 53,* 1872 (1981).

744. L. J. Cline Love and M. Skrilec, *Abstracts of Papers Presented at the 1981 Pittsburgh Conference on Analytical Chemistry and Applied Spectroscopy,* no. 799, Convention Hall, Atlantic City, NJ, March 1981.

745. M. Skrilec and L. J. Cline Love, *Abstracts of Papers Presented at the 1981 Pittsburgh Conference on Analytical Chemistry and Applied Spectroscopy,* no. 800, Convention Hall, Atlantic City, NJ, March 1981.

746. C. Gooijer, J. J. Donkerbroek, R. W. Frei, and N. H. Velthorst, *Abstracts of Papers Presented at the 1981 Pittsburgh Conference on Analytical Chemistry and Applied Spectroscopy,* no. 801, Convention Hall, Atlantic City, NJ, March 1981.

747. J. J. Donkerbroek, J. J. Elzas, C. Gooijer, R. W. Frei, and N. H. Velthorst, *Talanta 28,* 717 (1981).

748. J. J. Donkerbroek, N. J. R. van Eikema Hommes, C. Gooijer, N. H. Velthorst, and R. W. Frei, *Chromatographia 15,* 218 (1982).

CHAPTER

3

THE FLUORESCENCE OF ORGANIC NATURAL PRODUCTS

O. S. WOLFBEIS

Institute of Organic Chemistry
Division of Analytical Chemistry
Karl Franzens University
Graz, Austria

3.1.   INTRODUCTION

The last articles dealing with the native fluorescence of natural products in a broad sense were written in 1939 and 1950 by Dhéré (1,2). Since that time so much progress has occurred and so much knowledge has been gained because of improved instrumentation and the advances in lasers and time resolution methods that the literature has become difficult to overview. Several useful books treat parts of the subject (3–11) mainly with emphasis on biochemical aspects.

The present article covers the literature up to the end of 1983, with a few references for 1984. No attempt was made to report the literature for Sections 3.1, 3.2, and 3.8 in complete form. Rather, fundamental facts and more recent results were favored over specialized work or work done before 1960. The older literature is cited in newer articles and papers. Further, only compounds of well-established molecular structure have been considered, and the native fluorescence of "organized" systems, such as cells, micelles, or membranes, had to be omitted. Bioluminescence and chemiluminescence (12) have also been disregarded.

An incredible number of organic natural products are known to fluoresce, either at room temperature or at low temperature. In fact, only a few classes of compounds are said to possess no native fluorescence at above 250 nm, which today is the lower wavelength limit of most commercial instrumentation. The lack of fluorescence is either due to the lack of π-electron systems as in the case of fatty alcohols, some steroids, and insect pheromones or due to the presence of rather short and nonfluorescent π-electron systems, as in the case of fatty acids; aldoses and ketoses; the mono-, di-, and some trienes; vitamin C; and most of the amino acids. In addition, the presence of a transition metal ion in a fluorescent system may render it nonfluorescent (e.g., heme and related transition metal complexes).

Fluorescence methods have, in recent years, found their most widespread use in physical biochemistry and in analytical chemistry, including clinical chemistry. Combinations of steady-state and time-resolved measurements of polarization and decay time measurements have allowed deeper insight into structure and rapid molecular processes of both biomolecules and complex biosystems. More reliably than any extrinsic probe that may report its possibly altered environment, the native fluorescence of natural fluorophores does report the true static and dynamic behavior.

Aside from the presentation of spectral and photophysical data, attention was paid to the application of fluorescence in analytical and clinical chemistry. Fluorimetry is most frequently used in the determination of vitamins, alkaloids, chlorophylls, and metabolites. Many of the methods are now performed routinely, mainly with the use of HPLC instrumentation in combination with fluorescence detection.

Luminescence data and, roughly, the experimental conditions for many groups of substances are compiled at the end of most chapters. It is hoped that they may be useful in rapid estimations on the choice of analytical wavelengths, and also in the identification of unknown fluorescing products.

3.2. AMINO ACIDS

Aside from the chlorophylls, the amino acids in free and bound forms are certainly the group of natural products whose fluorescence properties have been studied most thoroughly. Several reviews deal with this subject (3–10,13,14,39).

Among the frequently occurring amino acids three are fluorescent, namely, phenylalanine (Phe), tyrosine (Tyr), and tryptophan (Trp). The fluorescence data for these three acids are compiled in Table 3.1. A very weak fluorescence, maximizing at 300 nm (pH 7.5, room temperature) has been reported for histidine (40) when excited at 230 nm. The quantum yield was estimated to be very low (10^{-4}). Unfortunately, in this study the histidine was used without previous purification. Ledvina and LaBella (41) have demonstrated that many nonfluorescent amino acids on simulated acid hydrolysis exhibit measurable emission. Phe, Tyr, and Trp possess quite different photophysical properties and shall now be discussed separately.

3.2.1. Phenylalanine (Phe)

Because of its low quantum yield, its low molar extinction, and its location in the shortwave UV, the fluorescence of Phe is weak and frequently hidden. It has therefore received much less attention than the fluorescence of the other two aromatic amino acids.

As with Tyr and Trp, the fluorescence properties of Phe are strongly affected by environmental factors. It is now accepted, that at 20°C the fluorescence lifetime (τ) and quantum yield (ϕ_f) in aqueous solution are 6.8 ns and 0.025 (16,20), respectively. ϕ_f and τ vary strongly with temperature. At 2°C the values are 0.04 and 9.6 ns, while at 70°C they are 0.08 and 2.0 ns (20). Evidence has been presented that in concentrated solution Phe can form an excited state dimer (20).

Fluorescence decreases by ~30% on protonation of the carboxylate group (at pH below 2.3), and by ~15% on deprotonation of the $-NH_3^+$ group (16).

On the basis of the singlet excited state and the triplet-state energies of Phe, it has generally been considered that its electronic energy can be transferred to Tyr, and further to Trp (8,39). On strong UV irradiation Phe suffers both photoionization (42,43) and destruction (44,45).

Methods for the fluorimetric assay of Phe in blood or urine at the ultramicro level (as the basis of mass screening of infants for phenylketonuria) are based on its native fluorescence or on its reaction with fluorogenic reagents (46). The latter are by far more sensitive.

3.2.2. Tyrosine (Tyr)

This amino acid possesses the relative highest ϕ_f value (see Table 3.1), but it has long been known that ϕ_f is reduced by 40–90% when Tyr is bound into a protein (47). Based on its fairly strong native fluorescence, Tyr may be assayed in the picomolar level in protein hydrolysates even in the presence of Trp by a HPLC method (48).

Table 3.1. Photophysical Data for the Amino Acids Phenylalanine (Phe), Tyrosine (Tyr), and Tryptophan (Trp)[a]

Amino Acid	λ_{ex} (nm)	λ_f (nm)	Molar Extinction ($cm^{-1} M^{-1}$) at Excitation Maximum	Fluorescence Quantum Yield ϕ_f	Fluorescence Lifetime τ (ns)
Phe	260	282	200	0.04 (15), 0.04 (16), 0.024 (17)	8.0 (16), 6.4 (19), 6.8 (20)
Tyr	275	303	1500	0.21 (15), 0.21 (16), 0.14 (21), 0.23 (22)	7.5 (16), 3.2 (19), 3.6 (23), 3.52 (24)
Trp	287	348	6000 / 6300 at 280 nm	0.20 (15), 0.20 (25), 0.12 (26), 0.13 (21), 0.14 (27), 0.14 (28), 0.13 (29)	3.0 (19), 2.6 (21), 2.9 (30), 2.8 (27), 3.0 (31), 2.0 (28), 2.8 (32), 3.14 and 0.5 (33), 5.4 and 2.1 (34), 3.3 and 0.43 (35), 3.32 and 0.43 (29), 3.2 and 0.6 (36), 3.13 and 1.45 (37), 3.13 and 0.53 (38)

[a] In aqueous solution at pH ~7 and room temperature; for τ and ϕ_f the apparently most reliable values are underlined.

The fluorescence properties of Tyr are influenced by many factors, among which are temperature, pH, ionic strength of the solution, presence of quenchers or energy acceptors like Trp, and peptide bonds, but not by oxygen (47).

Fluorescence intensity is diminished by approximately 0.8–1.0% per °C increase in temperature (49). Since Tyr possesses three dissociable functional groups (pK_a's 2.20, 9.11, 10.07), the spectra are highly pH dependent. The fluorescence maximum remains the same from low pH up to 10, but is shifted to 345 nm at above pH 10. The weak 345-nm emission arises from the phenolate form (50).

Fluorescence intensity iş maximal between pH 4 and 9, and drops sharply at extreme pH values. The dependence of the quantum yields of Tyr and tyrosinate does not fit a simple excited state acid–base equilibrium model, but a more complicated system where carboxylate is also capable of simultaneously quenching Tyr by a mechanism not involving proton transfer.

Rayner et al. (24) were the first to show that in some cases moderate phenolate fluorescence can be detected even in neutral solution in the presence of high buffer base concentrations (Fig. 3.1). This is consistent with the large shift in the excited state pK_a value of the phenolic hydroxy group (4.2 ± 0.4), predicted from the Förster cycle calculation, and confirms that proton transfer is a mechanism by which carboxylate ions quench Tyr fluorescence (50).

Tyr conjugate base fluorescence at 345 nm in neutral solution has also been observed by others (51,52). Again, the appearance of anion fluorescence was interpreted in terms of excited-state dissociation. Tyrosinate emission at 345 nm in neutral solution has meanwhile been identified in three peptides (53–55). Quantum yields are rather low because of rapid intersystem crossing (56).

Several attempts have been made to separate the fluorescence of Tyr from either Phe or Trp fluorescence or both. The Tyr and Phe emissions could be resolved most satisfactorily by using derivative spectroscopy (57). Detection of Tyr fluorescence in the presence of Trp is more difficult, since both amino acids have similar excitation spectra and overlapping flurorescence bands. In addition, the molar absorptivity of Trp at 280 nm is about four times greater than that of Tyr. The problem was solved elegantly by synchronous spectroscopy (58), that is, recording fluorescence intensity versus emission wavelengths, when both excitation and emission monochromators are scanned synchronously. The method has been refined using second-derivative spectroscopy (59) and is said to be as good as more complex techniques (60).

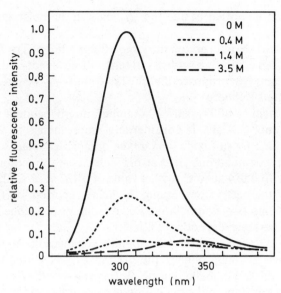

Fig. 3.1. Fluorescence emission spectra of *N*-acetyltyrosine in aqueous solvents at pH 7.5 and various concentrations of potassium phosphate. Concentration, 50 μM; excitation wavelength, 270 nm. The longwave emitting species in 3.5 *M* phosphate is the tyrosinate anion. (Redrawn from ref. 51).

Another method to resolve Tyr from Trp emission is based on phase-sensitive fluorescence detection (61). A phase fluorimeter is used, modified only by addition of a phase-sensitive detector. Choice of the detector phase angle so as to be out of phase with one component gives the emission spectrum of the second component.

3.2.3. Tryptophan (Trp)

The fluorescence of Trp in aqueous solution has been studied in great detail. This is largely due to the utility of Trp fluorescence as an intrinsic probe for the elucidation of structure and function of proteins and enzymes. Trp emission in water is maximally excited at 287 nm and peaks at 348 nm. The origins of the large Stokes' shifts in polar solvents have received much attention. Studies of the solvent effect of varying polarity indicated that the shift was linearly correlated with the solvent dielectric properties and the refractive indices (62–66). However, these correlations do not hold for hydrogen-bonding solvents. Emission occurs from the solvent relaxed ^1L_a state formed within 10 ps after excitation (67). This state has been identified beside the ^1L_b state in the polarized excitation spectrum (65).

Walker et al. (68) interpreted the large Stokes' shift of Trp fluorescence in polar hydrogen-bonding solvents in terms of a solvent–indole exciplex (69), while Lami (70) proposed that the large shift may be due to emission from a solvated Rydberg state.

The fluorescence of Trp suffers a temperature-dependent blue shift in organic solvents (71) and is considerably more intense in ethanol (66), glycol (72), dimethylsulfoxide, and mixed organic–aqueous solvents (73). The fluorescence quantum yield at pH 7 decreases monotonically from 0.201 at 10°C to 0.024 at 80°C, with a value of 0.137 at 25°C (27,29). These are probably the best values among the many reported for Trp. At least in polar solvents two nonradiative decays diminish ϕ_f, one of which is strongly temperature dependent (32).

Both ϕ_f and decay time (τ) correspond to changes in the pH value. ϕ_f and τ decrease strongly outside the pH range 3–11 with values of 0.42 for ϕ_f at pH 10, and 0.06–0.08 at pH 2, respectively (29,74).

ϕ_f remains constant in the excitation wavelength range from 290 to 360 nm, but decreases by 31% in going from 260 to 230 nm (75). A report that ϕ_f for L-Trp differs by about 10% between excitation by left and right circular polarized light was shown to be in error (76). No difference is evident.

Unlike 3-methylindole and N-acetyl tryptophanamide the fluorescence of Trp does not demonstrate a single exponential decay. Fleming et al. (34,29) examined the fluorescence decay using laser excitation, and Rayner and Szabo (33,38) used conventional photon counting to resolve the fluorescence into two exponentially decaying components, which are shown in Fig. 3.2. The one with an apparent emission maximum at 335 nm had a lifetime of 0.5 ns, while the other with a maximum at 350 nm had a lifetime of 3.1 ns (pH 7, 20°C). This was interpreted in terms of emission from different rotamers or conformers (38). Previously this emission was ascribed to the presence of different ionic species (31,33). Table 3.2 summaries the decay times and quantum yields of Trp and several of its derivatives.

In studying the Trp decay kinetic as a function of pH, buffer concentration, and temperature, the findings of Rayner and Szabo were practically confirmed (35–37) and former erronous values (34) were shown to result probably from photodecomposition (77).

Recently, in reexamining the Trp fluorescence lifetime puzzle, Gudgin et al. (36) found that at pH above 7.0 the decay becomes even triple-exponential, with the appearance of a long-lived component ($\tau = 9.1$ ns),

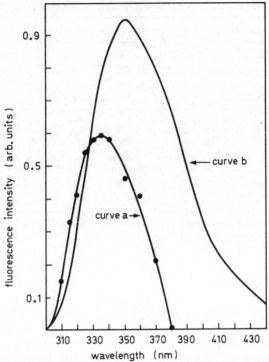

Fig. 3.2. Resolution of the steady-state fluorescence spectrum of aqueous tryptophan (pH 7, 20°C) into the two components from the fluorescence decay data. Curve a: 0.5-ns component, multiplied by 10; curve b: 3.1-ns component. Clearly, the 0.5-ns component makes a very small contribution (below 10%) to the total steady-state spectrum. (Redrawn from ref. 38.)

whose contribution to the total emission intensity increases rapidly with increasing pH, at the expense of the other two.

Jameson and Weber (78) resolved the pH-dependent heterogenous decay of Trp by phase and modulation measurements. They found decay times of 3.1 ns for the zwitterion, and 8.7 for the anion. The same technique was applied to determine the lifetime of each fluorophore in a Tyr–Trp mixture (61).

In an attempt to estimate the efficiency of various deactivation processes of Trp at room temperature and neutral pH, Fleming et al. (34) calculated the following values: fluorescence quantum yield 0.14, phosphorescence quantum yield 0.08, photoionization (electron ejection) 0.25, and nonradiative decay (intramolecular quenching) 0.53.

Table 3.2.   Kinetic Parameters of the Fluorescence Decay of Trp Derivatives[a,b]

Compound		λ_{em} (nm)	τ_1 (ns)	τ_2 (ns)	ϕ_f
Tryptophan	pH 7.0	330	3.13	0.53	0.14
	pH 7.0	350	3.09	0.65	
	pH 10.5	330	10.1	4.55	0.31
N-acetyltryptophanamide		330	3.00		0.14
N-acetyltryptophan		330	4.80		0.21
Tryptophanamide	pH 5	330	1.61		0.11
	pH 9	330	6.86		
Tryptophan ethyl ester		350	1.87	0.51	
	pH 5	330	1.53	0.47	0.029
N-acetyl tryptophan Ethyl ester		330	1.87	0.81	0.066

[a] Values reported are for measurements made on degassed buffered solutions at 20°, pH 7 (unless otherwise noted), with excitation wavelength 280 nm.
[b] Data from ref. 38.

The intramolecular quenching efficiency of the ammonium group has been discussed in terms of an electrostatic effect on the quenching efficiency of the carboxyl group, and not a direct quenching process (28,34,47). However, Nakanishi and Tsubei (79), in their stopped-flow fluorescence measurements on Trp have demonstrated that at least 83% of the fluorescence enhancement in D_2O relative to H_2O is due to the exchange of the ammonium hydrogens by deuterium. These reports indicate that the ammonium group can act as a quencher.

There are at least two quenching mechanisms to be discussed: (1) a proton transfer mechanism as suggested by some workers (80,81) and (2) an electron transfer mechanism as suggested by others (43). Beddard and co-workers (35) as well as Robbins et al. (29) applied mode-lock lasers to time-resolve the Trp fluorescence in the subnanosecond region and found strong evidence for an intramolecular process, dominated by a proton transfer reaction from the NH_3^+ group to the 2-position of the indole ring, rather than for an electron transfer. At pH below 3, the protonated carboxyl group (pK_a 2.38) becomes the center for another radiative process, which involves an intramolecular charge transfer interaction between the excited indole ring and the electrophilic carbonyl group. At pH 11 the behavior of Trp closely resembles that of 3-methylindole, where the principal nonradiative decay routes are intersystem crossing and photoionization.

The conclusions of Beddard et al. (35) are in some contradiction to those of Werner and Forster (82), who postulate that, at least in the case of small peptides, quenching by NH_3^+ does not involve proton transfer. This is concluded from the fact that peptide quenching is more effective when binding is at the amino rather than at the carboxyl end of Trp.

Aside from intramolecular quenching, the fluorescence of Trp is known to be dynamically quenched by the folloiwng ions and neutral molecules: bromide and iodide (39,83), nitrate (84), oxygen (85), hydrogen peroxide (86), acrylamide (37), NAD(P)H (87), glucose (88,89), trichloroethanol (90), histidine (91), SH groups, and S–S bonds (8), and polynucleotides (92). Other factors that are known to influence the fluorescence properties of Trp are pressure (93,94) and photodecomposition (77).

Since the fluorescence of both free and bound Trp is affected by so many external factors, particular care must be employed when interpreting the Trp fluorescence in complex systems such as proteins. On the other hand, Trp can act as a very sensitive probe for its environment.

In extension of previously reported Trp assays based on its native fluorescence, automated methods for the determination in the nanogram level in cerebrospinal fluids (95), blood (96), and serum (97,98) have been worked out. Trp was similarly assayed in food (99,100) and ingredients using HPLC separation techniques (100). The same method was applied to determine Trp (and its metabolites) in plasma after enrichment on porous polystyrene polymer (101). By applying laser excitation fluorimetry, Richardson (102) was able to improve the detection limit by a factor of thousand, when 0.25 nmol was detectible.

3.3. PROTEINS

The work that has been performed on the native fluorescence of peptides, proteins, and enzymes (''proteins'') is so extensive that it is impossible to summarize it briefly. A series of excellent reviews has been written on this topic, and so this chapter can be confined to fundamental aspects and more recent progress.

3.3.1. Phenylalanine and Tyrosine Fluorescence in Proteins

It is known that only proteins possessing Phe, Tyr, or Trp, or combinations thereof, exhibit native fluorescence. Proteins lacking native fluorescence can, however, be labeled with a fluorescent tag, with the inherent disadvantage of disturbing the system under investigation.

In the case of Phe, fluorescence is weak and can only be detected in the absence of other aromatic amino acids. However, its fluorescence has been used to study the enzyme superoxide dismutase (103) and other proteins (104).

The native fluorescence of Tyr has been used more frequently to characterize proteins, in particular those that lack Trp (class A proteins). The work up to 1975 has been reviewed by Cowgill (47). Tyr fluorescence is much weaker ($\sim$10% only) in protein-bound than in free form. Both quantum yield and decay time depend on whether substitution occurs at the amine or carboxyl end (105). The emission maximum of protein-bound Tyr does not vary as strongly as the Trp maximum and lies at 302 nm in neutral aqueous solution. Only in the last years has it been established that Tyr, in neutral solution, can also fluoresce from its phenolate species (at 345 nm) because of photodissociation in the presence of nucleophilic buffer ions. The proximity of an amino group may also favor photodissociation. A number of Trp-lacking proteins is known to fluoresce from S_1-tyrosinate at 345 nm in neutral solution (53–55,106) (see also the appendix) but the majority of class A proteins (e.g., ribonuclease A, insulin, troponine C, pancreatic trypsin inhibitor, ovomucoid) has maximal fluorescence at 302 nm (47,104,107–109).

The fluorescence of the apoenzyme part of aspartate aminotransferase is characteristic of buried Trp's. When, however, the apoenzyme binds pyridoxalphosphate to form the holoenzyme, the observed fluorescence is the one of Tyr (1386).

3.3.2. Tryptophan Fluorescence in Proteins

The fluorescence of Trp in proteins is quite heterogenous (110) in having different maxima (308–350 nm), quantum yields (0.04–0.60), and decay times (0.04–7.0 ns). In proteins containing both Tyr and Trp the latter may be excited selectively at 290–295 nm. But even when exciting at shorter wavelengths, where both Tyr and Trp are excited, protein fluorescence is dominated by Trp due to energy transfer from Tyr to Trp (111). Such an energy transfer requires a certain distance (maximally 5–10 nm) between donor (Tyr) and acceptor (Trp) and does not occur otherwise. Measurements of the efficiency of energy transfer therefore allow estimations of intramolecular distances (112). For several proteins it has been shown that the energy transfer efficiency depends highly on the state of denaturation (113), a fact that may contribute to the reconciliation of previous conflicting reports.

The location of the fluorescence maximum is a sensitive indicator of the polarity of the Trp environment. Pajot (114) has compiled the fluorescence maxima of proteins before and after denaturation. For the natural state, maxima between 322 and 342 nm were given, whereas the denatured proteins generally fluoresce maximally at around 345 nm. The most shortwave emission maximum that has apparently been reported for a Trp protein is at 308 nm in the case of azurin (115). It is assumed that the more hydrophobic the environment of Trp is, the shorter the wavelength of the emission maximum will be.

The fluorescence of Trp in proteins can be quenched either by energy acceptors, which themselves are fluorescent (such as NADH or pyridoxal phosphate), or by internal or external nonfluorescent quenchers such as iodide, cesium, or europium ion (116,117,118), acrylamide (120), or oxygen at high pressures (85). External quenchers are used to determine whether a Trp moiety is exposed and thus accessible to collisional quenching. Clearly, the probability of collisional quenching also increases with the radiative lifetime.

Phosphate ion can also act as a quencher, a fact that appears significant in studying interactions of proteins with nucleic acids (119).

For systems having more than one fluorophor (e.g., many proteins) the classical Stern–Volmer equation was modified by Lehrer (83a) to account for the different quenching constants and decay times of each chromophore (120). The mathematics for a situation in which the fluorescence of a single fluorophore experiences dynamic quenching by two or more quenchers has also been treated (83b).

Trp in neutral aqueous solution demonstrates a double exponential decay, but the peptide model N-acetyl tryptophanamide shows a single exponential decay (Table 3.2). It is therefore surprising that for several model peptides (34,35,82,121,122), for native peptides (82), and for proteins such as lysozyme (123), glucagon (35,38,122,124), LADH (122), and human and bovine serum albumin (125,126) double exponential decays have been reported. For the pair of diastereomeric dipeptides, D,D-Trp-Trp-methylester and L,L-Trp-Trp-methylester, different—and in each case double exponential—decays were measured (127).

The instrumental manifold that is available for measuring even complex decay times has been extended by a method called phase-sensitive fluorescence spectroscopy (12,61,128). It permits direct recording of the individual emission spectra of two fluorophores in a two-component mixture or in proteins, and the detection of the lifetime of each species in the mixture. For Trp in proteins resolution seems possible when the lifetimes differ by more than 1 ns (129a). Further, the method allows the recording of the fluorescence of the initially excited and solvent-relaxed state of N-acetyl trypthophanamide (the most simple peptide model) (129b).

Fluorescence decay studies with proteins are often performed in combination with polarization measurements (130–133). In a steady-state experiment the sample is excited by vertically polarized light, which produces an ensemble of preferentially aligned excited molecules. The orientation of these molecules then becomes randomized by rotational Brownian motion, so that horizontally polarized fluorescence becomes measurable. By means of the Perrin equation (130) the rotational relaxation time of the protein can be obtained, provided the lifetime is known. In a time-resolved experiment (133), pulsed excitation is applied and depolarization is measured as a function of time.

Munro et al. (134) studied the depolarization of proteins with single Trp residues and found different degrees of rotational freedom in the subnanosecond time range. In an alternative approach Lakowicz and Weber (135) applied high-pressure oxygen quenching in their steady-state as well as time-resolved fluorescence anisotropy measurements.

The technique of simultaneously measuring the perpendicular polarized time spectra allowed an estimation of the picosecond rotational diffusion of the peptide model compound, N-acetyl tryptophanamide (136). Detection of the two polarization directions by the same detector is an important feature of this method, which avoids normalization procedures.

Using a molecular dynamics simulation, Levy and Szabo (137) attempted to calculate the influence of picosecond motions of Tyr in proteins on the initial depolarization. Cautious predictions on molecular motions for the four Tyr's in the enzyme pancreatic trypsin inhibitor could be made.

Circular polarization of luminescence (CPL) is a potentially useful method in protein research (138), allowing *inter alia* the measurement of excited-state optical activity and chirality. The topic has been reviewed by Steinberg (138). Lobenstine et al. (139) reported the first fluorescence-detected circular dichroism (FDCD) of proteins. In combination with CD and absorption measurements, an equivalent of the Kuhn dissymmetry factor was introduced, which is a measure of the chirality around the fluorescent Trp and any chromophores that transfer energy to it.

As usual, new theories and methods primarily find application in the life sciences. As a result, in this chapter on proteins aside from achievements obtained by conventional spectrometry, some quite new trends and developments could be reported. Their application to the study of other groups of natural products seems likely and will provide a large field of research.

3.4. COENZYMES AND VITAMINS

The intrinsic fluorescence of coenzymes has been studied with respect to understanding their dynamic behavior, particularly in enzyme complexes. In contrast, the fluorescence of vitamins has mainly been investigated to work out fluorimetric determinations, except perhaps for the B_6 group.

This chapter deals with the fluorescence of reduced nicotinamide and flavine coenzymes, and with the vitamins A, B_1, B_2, B_6, B_{12}, D, E, and folic acid. No report on the fluorescence of other vitamins has come to my attention, though one would expect by comparison vitamins K_5 to K_7 to be fluorescent. The "coenzyme" ATP will, because of its similarity to the nucleic bases, be treated in the section on nucleic acid components. Tables 3.3 and 3.5 summarize fluorescence quantum yield data and decay parameters and Table 3.4 the spectral maxima of the coenzymes and vitamins.

3.4.1. Nicotinamide Coenzymes

The reduced forms of nicotinamide adenine dinucleotide (NADH) and nicotinamide monoucleotide (NMN) as well as the NADH phosphate (NADPH) show fairly strong fluorescence with a maximum at around 460 nm, when excited at 340 nm in neutral aqueous solution. The respective oxidized forms are nonfluorescent, except in strong alkali. In NAD(P)H, emission comes from the 1,4-dihydronicotinamide fluorophore with its absorption band at 340 nm, where the adenine part of the molecule does not absorb. When excitation is performed at the main absorption band of adenine at 260 nm, energy is transferred to the nicotinamide part of the molecule with an efficiency of about 34% (140,141).

The quantum yield of NADH in water (141) is 0.019 at 25°C. In contrast to former reports (141) the lifetimes of the dinucleotides were found to be the sum of two exponentials (142,143), but for NADH the long-lived component (τ_2 0.69 ns) makes only a minor contribution as compared to the short-lived component (0.25 ns, 82%). In Table 3.3 the decay properties of several nicotinamide species are summarized. Further data are given in Table 3.4. The average lifetimes reported from different laboratories are in sufficient agreement, being (143) 0.36 ns for NMN, 0.42 ns for NADH, and 0.41 for NADPH in aqueous solutions at 20°C. The emission heterogeneity has been interpreted in terms of an intramolecular exciplex mechanism and the assumption of an equilibrium between folded and unfolded dinucleotide conformers (143). A definite trend was observed in which both lifetime and $\emptyset_f$ change by the same factor with temperature (141).

The native fluorescence of NAD(P)H has been used for its determination (144,145) and to follow the kinetics of a manifold of hydrogenases and dehydrogenases (144). Aside from the enzymes, substrates, coenzymes (like ATP), metal ions, and inhibitors may also be assayed, de-

Table 3.3. Fluorescence Decay Times of Nicotinamide-
Derived Coenzymes in Aqueous Solution at Various
Temperatures[a,b]

Species	τ_1 (ns)[c]	τ_2 (ns)[c]	T (°C)
NMNH	0.28 (0.99)	1.7 (0.01)	20
NMNH	0.25 (0.95)	1.0 (0.05)	2
NADH	0.25 (0.82)	0.69 (0.18)	20
NADH	0.29 (0.71)	0.81 (0.29)	2
NADPH	0.31 (0.86)	0.70 (0.14)	20
NADPH	0.29 (0.73)	0.77 (0.27)	2

[a] Data from Ref. 143.
[b] Double-exponential decay with somewhat different data has also
been reported by Gafni and Brand (142) for NADH and NMNH,
and by Brochon et al. (149) for the NADPH–glutamate dehydrogenase complex.
[c] The values in brackets give the contribution to the total decay
("weighting factor").

pending on what the limiting factor is. NADH fluorescence can also serve
to control physicochemical parameters such as oxygen concentration in
bioreactors (146,147).

When NAD(P)H binds to proteins, the excitation and emission maxima
may suffer considerable changes. Extreme values of 325/440 to 365/478
nm (excitation–emission maxima) have been reported (148). When bound
to liver alcohol dehydrogenase, the fluorescence decay of NAD(P)H can
again be described best as a sum of two exponentials, with average lifetimes as high as 5.5–6.0 ns (142). The decay parameters, quantum yields,
and emission maxima depend on solvent polarity and viscosity. A similar
situation is found for NADH in the ternary complex with guanosine tri-
phosphate and beef liver glutamate dehydrogenase (149).

3.4.2. Flavin Coenzymes

The flavins are part of in vivo redox systems catalyzing one-electron and
two-electron transfer reactions. They also play an important role in flavoproteins, and probably in blue light photoreceptors. Chemically they
are isoalloxazines, existing in oxidized and reduced forms. Among the
naturally occurring derivatives riboflavin (vitamin B_2) and the coenzymes
FMN (flavin mononucleotide) and FAD (flavin adenine dinucleotide) are
the most important ones. Heelis has given a review on the photophysical
properties of flavins (150a).

The fluorescence spectra of FMN and FAD have been studied in aqueous and polar organic solvents; see ref. 150b for a compilation of data. All emission spectra are broad and structureless, with peaks around 530 nm. However, the excitation polarization spectra of flavins can be considerably different from one species to the other (151,152). In apolar solvents the spectra become structured (153,154) and the Stokes' losses are reduced to 850–1300 cm^{-1}.

Visser et al. (155) studied the blue fluorescence of various cationic flavins in an ethanolic glass. Despite the small differences in their chemical structures large differences in spectral shapes and in the ratio of ϕ_f and ϕ_p were observed. The absorption, excitation, and emission spectra of flavin molecules in the gas phase have been investigated in the same group (156). Schmidt (152) measured corrected excitation and emission spectra of isotropically and anisotropically embedded flavins. The latter were anchored to vesicles by means of aliphatic chains, thereby mimicking the specific binding to the apoprotein in flavoproteins. Large differences in the polarization were explained by different specific interactions with the matrix.

Song (157) has performed quantum-mechanical computations on the electronic spectra of flavins. The results are particularly useful in the interpretation of fluorescence spectra, fluorescence polarization and excited-state basicities. It is important to note that some of the quantities presented are not readily obtainable from experimental approaches.

Riboflavin and FMN show about the same quantum yields (0.25 at pH ~7) (158,159) in pure water solution, and the effect of pH is the same. For FAD the spectral distribution of the emission remains the same, but its ϕ_f in neutral solution is 10 times smaller than that of FMN, and the effect of pH is quite different. These differences are known to result from both static and dynamic quenching effects (160). Quenching also occurs when flavin interacts with electron-rich compounds such as tryptophan, cysteine (161,162—on a model peptide), and other amino acids (163), with bromide (164) and iodide (165), with oxygen under high pressure (85), and on binding to proteins (94,166,167).

The lifetime of FMN fluorescence (4.7 ns at room temperature) is approximately a factor of 2 longer than that of FAD (160,168). This has been attributed to intramolecular quenching of the flavin fluorophor by the adenyl residue and the existence of two interconvertible conformers (169).

The decay of flavin fluorescence follows in fluid solution a monoexponential law, while in the vapor phase a nonsingle-exponential decay is found, together with much shorter lifetimes (156). When flavins bind to proteins, fluorescence decay becomes more complex (162,170,171). The changes induced by up to 11-kbar pressures on the fluorescence properties

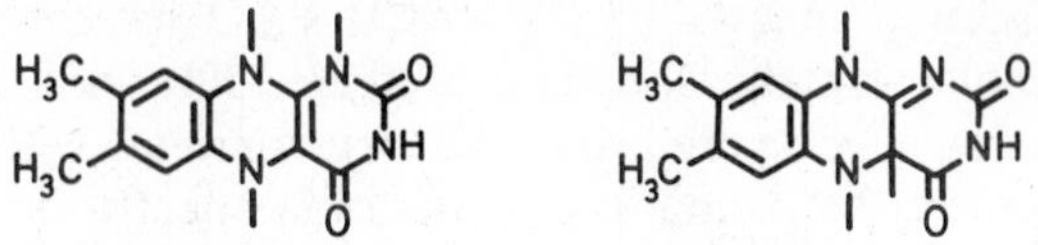

1,5 - dihydroflavin   4a,5 - dihydroflavin

Scheme 1.   Structures of the two common dihydroflavin species.

of various flavodoxins have been used to study binding regions and apparent volume changes. Some of the changes are irreversible (162).

In contrast to the oxidized flavins treated up to here, reduced flavins have a rather variable absorption between 310 and 410 nm and have been long considered to be nonfluorescent (172). Among the many possible reduced flavin forms, the 1,5-dihydro and the 4a,5-dihydro isomer are important in enzymic catalysis (Scheme 1).

1,5-Dihydroflavins are nonfluorescent in ethanol solution at room temperature, but show emission in a glassy matrix at 77 K, maximizing at around 500 nm (172) 4a,5-Dihydroflavins are also almost nonfluorescent in fluid solution at room temperature and become fluorescent at low temperature in rigid glasses (172) or when bound to apoflavoproteins (173). The absorption and fluorescence properties of oxidized and reduced forms of various flavoproteins have been compiled by Ghisla (174).

Visser et al. (175) studied the fluorescence lifetimes and polarized emission spectra of reduced flavin model compounds and flavoproteins. The fluorescence decay of most reduced forms was not single exponential. In many cases reasonable fits were obtained by assuming a decay function consisting of a sum of two exponential terms.

Frequent use has been made of the intrinsic fluorescence of flavins for their determination (3,4,176,177,178), partially in combination with HPLC techniques (179). Analytical methods have also been worked out for mixtures of FMN and FAD (176,180). Concentration changes of flavins in intact organs may be followed directly by measuring their fluorescence with the help of a surface fluorimeter (181).

3.4.3.   Vitamin A (Retinol)

For all-*trans*-retinol and its acetate-corrected excitation maxima between 325 and 328 nm have been reported, which agree with the respective absorption maxima. The corrected emission maxima range from 475 to 510 nm, depending on the solvent (182). Most quantum yield reports give values between 0.03 and 0.006 for retinol in cyclohexane or ethanol solutions at room temperature (182–184) but on cooling to 77 K ϕ_f increases to approximately 0.45–0.62 (183,184). In marked contrast to the behavior of retinal, the fluorescence quantum yield of retinol does not depend on the wavelength of excitation (185a). The primary photoprocesses in retinol after pulsed laser excitation have been shown to involve fluorescence, intersystem crossing, and remarkably enough, in polar solvent the formation of a longwave-absorbing transient, which was identified as retinylic cation (185b).

The intrinsic radiative lifetimes of retinols are significantly longer than those calculated from the oscillator strength of the observed, low-energy main absorption band systems (assigned as $^1B_u^* \leftarrow {}^1A_g$ transition) (186). This behavior has been attributed to a relatively forbidden $^1A_g^* \leftarrow {}^1A_g$ transition discovered only in 1973 (184,187,188). The transition energy lies below and borrows intensity from the strongly allowed $^1B_u^* \leftarrow {}^1A_g$ transition. Also, the high degree of fluorescence depolarization (189,190) indicates the origin of fluorescence to be from a long-axis-polarized transition and has been explained in terms of the borrowing mechanism (187,190).

The model of a $^1A_g^*$ state from which fluorescence occurs (187) can also explain the discrepancies between the observed and calculated natural fluorescence lifetimes in terms of different states involved in absorption (a B_u second excited singlet state) and emission (an A_g first excited singlet state). The forbidden $^1A_g^* \leftarrow {}^1A_g$ transition has recently been shown (191) to occur extremely weakly to 360 nm. The $^1B_u^*$ state of retinyl acetate in polymeric films is significantly perturbed by high pressures, while the $^1A_g^*$ state remains unaffected (192).

It was stated by Brey et al. (192) that fluorescence decay of retinyl acetate in toluene solution was highly nonexponential, but exponential in polymeric films. When retinol binds to β-lactoglobulin, its spectra and photophysical properties are considerably changed (193).

Various fluorimetric methods have been presented for the determination of vitamin A in biological materials (182,194), dairy products (195), and food (196). The assay in serum was automatized (197). Practically all procedures are based on the extraction of the vitamin into hexane or cyclohexane and subsequent fluorimetry, but the photolability of retinol is a potential source of error. In human blood the carotenoid phytofluorene may interfere with the assay (194).

3.4.4. Vitamin B₁ (Thiamine)

Only recently was it discovered (198,1290) that thiamine hydrochloride and its monophosphate, in ethanolic glass at 77 K, possess native fluorescence in the UV when excited at 270 nm (Table 3.4), but the emissions are absent at room temperature. The reported yields are fairly high (0.30 and 0.32, respectively) and the radiative lifetimes rather long (8.1 ns). By comparison with model compounds it was concluded (198) that emission comes from the π,π^*-excited pyrimidine portion of the molecule. There is evidence that thiamine exists in a folded conformation, where pyrimidine and thiazole are in weak contact (199). Since in the absence of a thiazolium group fluorescence intensity is much higher, one would assume that this group acts as an intramolecular quencher.

The spectral and kinetic characteristics of the low temperature luminescence of aqueous and salt-water solutions of thiamine diphosphate and also its pyrimidine and thiazole components were studied at various pH values (1290). At neutral pH the fluorescence is said to consist of two broad bands centered at 335 and 414 nm.

The native fluorescence of thiamine has found no application for its determination, since the thiochrome method is sensitive, selective, and well established.

3.4.5. Vitamin B_2 (Riboflavin)

Riboflavin has been studied quite thoroughly. Because of its photoinstability many researchers have also used either 9-methyl-isoalloxazine as a model substance, or the riboflavin photodecomposition products lumiflavin and lumichrome. The latter, however, is an alloxazine. The fluorescence properties of riboflavin have been reviewed by Udenfriend (3,4), with special reference to analysis, and by Heelis (150a).

Among the flavins, riboflavin (and FMN) shows the most intense fluorescence, the quantum yield at pH 7 being around 0.25. In neutral aqueous solution fluorescence is maximal at around 515 nm (Table 3.4). The degree of emission polarization of riboflavin in isotropic and anisotropic media is constant over the whole spectrum (200).

The dependence of ϕ_f and fluorescence maximum on environmental factors has been investigated in great detail (201,202). In mixed aqueous solutions ϕ_f increases with the organic solvent fraction up to maximally 0.37. Riboflavin fluorescence is constant in intensity between pH 3 and 8, but is quenched outside of this range. By absorptiometric titration two pK_a's can be calculated, one for the protonation (0.1), and one for the deprotonation (10.0). On the opposite the fluorimetric titration gives a pK_a^1 (for the protonation) of 1.7. The difference in the ground state pK_a results from different ground-state protonation (at N_1) and excited-state protonation (at N_{10}) (203; see, however, the comment in ref. 164).

Aside from apparent quenching at high and low pH values, the fluorescence of riboflavin is reduced by addition of purines (204), caffeine (204b), amino acids (205–207), various other nucleophiles (208), iodide, silver (I), and hydroquinone (204), and under high oxygen pressures (85).

Riboflavin tetrabutyrate undergoes characteristic spectral changes upon hydrogen bonding with trichloroacetic acid or trifluoroacetic acid (212). Hydrogen bonding is considered to occur with increasing concentration of acid, first at N_1, then at O_{12}, O_{14} and N_3–H, and finally at N_5.

The fluorimetric determination of riboflavin is a USP and AOAC standard method, for which improvements have been suggested (176,178,209–211) and for which automated procedures have been devised (213,214). The application of lasers as powerful light sources has lowered the detection limits down into the subpart per trillion range (215,216).

3.4.6. Vitamin B_6 (Pyridoxine) and Its Metabolites

Pyridoxine, its derivatives, and its metabolites are known to exist at neutral pH as a mixture of differently charged and tautomeric species, which makes the interpretation of its electronic spectra somewhat difficult. Based on the decisive work of Bridges et al. (217), Morozov and co-workers have made assignments of the absorption as well as fluorescence bands with great diligence (218) by studying also various O- and N-methylated derivatives and by using quantum-chemical methods. For most compounds the fluorescence intensity is maximal in the near neutral range (pH 6–7) (217) and varies strongly with temperature (219). Though all pyridoxine congeners exhibit appreciable native fluorescence (Tables 3.4 and 3.5), which may be used for their assay (3,4,102), the more general method involves MnO_2- or cyanide-catalyzed air oxidation to highly fluorescent pyridoxic acid lactone (3,4,212,220). Unfortunately, all vitamin B_6 compounds are photolabile (222).

The coenzyme form of vitamin B_6 is pyridoxine phosphate, which has a fluorescence maximum at 378 nm when excited at 318 nm in neutral aqueous solution. The emission maximum is invariably 378 nm down to pH 1, despite changes in the absorption spectrum due to protonation of the zwitterion. This has been interpreted in terms of excited-state dissociation of the phenolic hydroxy group (223). Churchich has also investigated the fluorescence properties of pyridoxamine phosphate (223,224) and pyridoxic acid phosphate (225). Work involves polarization and energy transfer studies, the effect of temperature and pH on the spectra, and the spectral changes on binding to proteins. For pyridoxal-5-phosphate one notes large differences in the spectral data given by Churchich (251) and Bridges et al. (217) (see Table 3.4).

In enzymes, pyridoxal is covalently bound via an azomethine link to the ϵ-amino group of a lysine residue. The resulting Schiff base shows quite different spectra. There are two major bands in absorption (maxima at 325 and 410 nm) and two in fluorescence (430 and 510 nm). The two absorption bands have been attributed to the enolimine (325 nm) and ketoenamine form (410 nm), respectively. The relative population of the two tautomers is solvent polarity dependent (226–228). The two fluorescence bands have different excitation spectra and can be assigned in a similar fashion to the enolimine form (430 nm) and the ketoenamine form (510 nm). By invoking this model, and after studying several Schiff-base models, it has been claimed (229) that it is not necessary to assume different protonation of the pyridine nucleus in pyridoxal Schiff bases to explain the dual fluorescence. The results obtained in the solvent and pH dependence of pyridoxal Schiff bases have been used to interpret the fluorescence properties of enzymes (226–229).

3.4.7. Vitamin B$_{12}$ (Cyanocobalamin)

Duggan et al. (230) reported that cyanocobalamin exhibits native fluorescence with a maximum at 305 nm, when excited at 275 nm. The emission arises from the benzimidazole residue (231). The corrin chromophore with the central cobalt(III) ion is nonfluorescent, but descobalt-cobalamine shows the usual corrin-type fluorescence (232). Despite an apparent low detection limit (0.14 ppm) (102) no use of the intrinsic fluorescence of vitamin B$_{12}$ has been made in analytical chemistry.

3.4.8. Vitamin D$_2$ (Calciferol)

It is known that vitamins of the D series become strongly fluorescent on treatment with strong acid (233). However, the formation of the fluorophore is an irreversible process, and not due to protonation. Aaron and Winefordner (234) report the fluorescence of calciferol in ethanol solution with maxima of 348 nm (excitation) and 420 nm (emission). One notes a disagreement between the reported excitation maximum and the UV-absorption maximum (265 nm) (235) and the fact that nonfluorescent steroids can be contaminated with fluorescent trienones (236).

3.4.9. Vitamin E (Tocopherols)

α-, β- and γ-tocopherols dissolved in ethanol fluoresce maximally at 340 nm, the best excitation being at 295 nm (230,237). Tocopherol acetate is said to be nonfluorescent. The low detection limits obtained by conven-

Table 3.4. Spectral Data for the Fluorescence of Coenzymes and Vitamins

Coenzyme–vitamin	λ_{ex} (nm)	λ_f (nm)	Experimental Conditions	Ref.
NADH	340	435	Water, pH 7	230
	340	460	Water, pH 8	5
	350	450	Water	244
	363	470	Water, 37°C	245
	340	470	Water, 25°C	141
NADPH	340	460	Water, pH 8	5
FMN[a]	365	520	$CHCl_3$/AcOH	5
	450	530	Water, pH 7	158
FAD	365	520	Water, pH 6	5
	267	525	Water	102
	450	530	Water, pH 7	246
	450	530	Water, pH 2.9	246
Coenzyme F_{420}	390	460	Water, pH 4.4	247
	420	470	Water, pH 9.8	247
Reduced ubiquinone	290	325	Ethanol	4
Vitamin A (retinol)	325	470	Ethanol	193
	321	513	Pentane–hexane	194
	345	490	Cyclohexane	182
	340	490	Cyclohexane	248
	324	494	3-methylpentane	184
9-*cis* retinol	317	510	Pentane–hexane	194
13-*cis* retinol	325	520	Ethanol	193
	324	529	Pentane–hexane	194
Anhydroretinol	367, 387	~562	Ethanol	193
Retinyl acetate	322	510	Pentane–hexane	194
	360	508	Ethanol	234
	337	500	Methanol	102
Retro retinylacetate	346, 367	~538	Ethanol	193
Vitamin B_1 (thiamin)	270	325	Ethanol, 77 K	198
Vitamin B_2 (riboflavin)	370	520	Water, pH 7	230
	370, 455	520	Water, pH 7	234
	375	540	Water	102
	377	526	Water, pH 7	216
	447	517	Water	203
	436	531	Water	201
	436	510	Water	214
	436	527	95% dioxane	201
Riboflavin tetrabutyrate	345	536	Water	249
	436	520	Ethanol	250

Table 3.4.   (*continued*)

Coenzyme–vitamin	λ_{ex} (nm)	λ_f (nm)	Experimental Conditions	Ref.
Vitamin B$_6$ compounds[b]				
Pyridoxine and	291	390	Water, pH 1	251
pyridoxine-5'-	324	390	Water, pH 7	251
phosphate	310	375	Water, pH 12.5	251
Pyridoxamine and	295	395	Water, pH 1	218, 223
pyridoxamine-5'-	327	395	Water, pH 7	
phosphate				
Pyridoxal-5'-phosphate	390	500	Water, pH 7.4	251
	330	410	Water, pH 6	217
	390	530	Water, pH 12.5	251
	330	400	Water, pH 7.5–8.5	217
Pyridoxic acid and	317	420	Water, pH 3	225
pyridoxic acid-	317	425	Water, pH 7	251
5'-phosphate	307	415	Water, pH 11	251
Vitamin B$_{12}$	275	305	Water	230
Vitamin E (tocopherol)	295	340	Ethanol	234, 237
Folic acid	365	450	Water, pH 7	230
	365	440	Ethanol	234
Tetrahydrofolic acid	310	360	Water, pH 3	243

[a] The luminescence data of various flavoproteins in oxidized and reduced form have been reported by S. Ghisla, V. Massey, J. M. Lhoste, and S. G. Mayhew, Biochemistry *13*, 589 (1974).
[b] The fluorescence characteristics of the various ionized forms of vitamin B$_6$ derivatives are given in more detail in ref. 4, pp. 310 ff., and in refs. 217 and 218.

tional fluorimetry (0.1 μg ml^{-1}) (234) have been used in both the extractive tocopherol assay (237,238) and in combination with chromatographic separation procedures (239,240). Thompson et al. (240) reported the fluorescence intensities of the four tocopherols in 12 kinds of solvents and found that fluorescence is maximal in dioxane solution, but not detectable in either dichloromethane, chloroform, or tetrachloromethane. The emission maxima vary from 324 to 329 nm, the excitation maxima from 291 to 300 nm.

3.4.10.   Folic Acids

Folic acid is composed of two fluorophors, the *p*-aminobenzamide system and the 2-amino-4-hydroxypteridine (pterine) system. In isolated form both exhibit fluorescence, the benzamide at 360 nm and the pterine at

Table 3.5. Fluorescence Quantum Yields (ϕ_f's and Decay Times (τ) of Coenzymes and Vitamins

Coenzyme–Vitamin	ϕ_f	τ (ns)	Experimental Conditions[a]	Ref.
NADH and NADPH	0.019	see Table 3.3	Water	141
FMN		4.5	Water, pH 7	164,[b] 252
		4.55–4.70	Water, pH 6.9	160[c]
	0.25		Water, pH 7	158, 246
FAD	0.02		Water, pH 7	158, 246
		2.7–2.8	Water, pH 6.9	160[c]
	0.04		Water	176, 102
Reduced flavins and flavoproteins				172–175
Vitamin A (retinol)	0.02	4.5	Pentane–hexane	253
	0.006	2.35	Methanol	254
	0.018	3.7	Cyclohexane	184
	0.016	3.0	EtOH–MeOH, 4:1	164
	0.008	2.8	Ethanol	193
9-*cis*-retinol	0.007	6.0	Pentane–hexane	253
13-*cis*-retinol	0.02	5.0	Pentane–hexane	253
		2.0	Ethanol	193
Retinylacetate	0.11	8.7	Polymethyl methacrylate	192
	0.09	8.5	Polystyrene	192
	0.004	1.53	Ethanol	193, 254
	0.02	6.0	Pentane–hexane	253
Vitamin B$_1$ (thiamine)	0.3	8.1	Ethanol, 77 K	198
Thiamine pyrophosphate	0.27	6.1	Ethanol, 77 K	198
Vitamin B$_2$ (riboflavin)	0.26	4.2	Water, pH 7	176, 85
	0.32[d]		Ethanol	151
Riboflavin–tetrabutyrate	0.22[e]		Water	249
Pyridoxin	0.02	0.7	Water, acidic	218
	0.03	0.7	Water, neutral	218
	0.047	0.6	Water, alkaline	218
Pyridoxin phosphate	0.024	0.56	Water, acidic	218
	0.045	0.95	Water, neutral	218
	0.032	0.68	Water, alkaline	218
Pyridoxamine	0.1	1.45	Water, acidic	218
	0.11		Water, pH 7.2	222
	0.09	1.40	Water, neutral	218
	0.07	0.8	Water, alkaline	218
Pyridoxamine phosphate	0.11		Water, pH 7.2	256

Table 3.5.   (*continued*)

Coenzyme–Vitamin	ϕ_f	τ (ns)	Experimental Conditions[a]	Ref.
Pyridoxal	0.03	1.5	Water, acidic	218
	0.05		Water, pH 7.2	222
	0.02	1.3	Water, neutral	218
	0.08		Water, pH 7	255
Pyridoxal phosphate	0.12		Water, pH 7.2	256
Pyridoxic acid	0.15	9.0	Water, pH 7.4	225

[a] Room temperature unless otherwise specified.
[b] Decay times at various temperatures also given.
[c] Decay times at lower pH's also given.
[d] Fluorescence quantum yields in various organic solvent–water mixtures are compiled in ref. 150.
[e] Quantum yields up to 0.58 are found in organic solvents; see Ref. 150.

440 nm. However, in folic acid, where the two moieties are linked covalently together, fluorescence peaks invariably at 440 nm, whatever the excitation wavelength is (230,241). It appears likely that because of the spatial proximity, the intrinsic fluorescence of the *p*-aminobenzamide system in folic acid is quenched by energy transfer to the pterine system, since the absorption maximum of pterine is close to the emission maximum of the *p*-aminobenzamide. On the other hand, equimolar mixtures of the two fluorophors show two fluorescence bands (241).

Aside from folates, nature knows dihydrofolates (242), which are characterized by violet fluorescences peaking at around 430 nm and maximal excitation between 290 and 320 nm, and tetrahydrofolates (243) with UV fluorescences peaking at around 360 nm (excitation at around 310 nm, see Tables 3.4 and 3.5).

3.5.   NUCLEIC ACID COMPOUNDS

3.5.1.   General Remarks

Unlike the amino acids, most of which are nonfluorescent, all purine and pyrimidine bases, their nucleosides and nucleotides, as well as DNA and RNA, are fluorescent. The intensity is in each case, however, very low. Room temperature photophysical data of nucleic acid components are summarized in Table 3.7.

The fluorescence of the purines and pyrimidines is particularly low in neutral aqueous solution, but increases by a factor of around 10 after acidification to pH 2, and by a factor of 100 and more on cooling the solution to 77 K. Therefore most studies up to 1972 have been done at low temperatures and/or in acidic solutions (257).

There are several unusual features in the fluorescence properties of nucleic acid components:

1. A strong dependence of the quantum yield on temperature.
2. Extremely short lifetimes.
3. A disagreement between excitation and absorption spectra.
4. A high degree of fluorescence polarization even at room temperature.
5. A more or less expressed dependence of the quantum yield on the excitation wavelength.
6. A strong dependence of the emission maximum on the temperature in acidified solutions.

Daniels and Hauswirth (258) investigated the fluorescence properties of the purine and pyrimidine bases in neutral aqueous solution at room temperature. Corrected emission and excitation spectra were presented (see Table 3.7) and were compared with low-temperature data. All spectra have maxima in the UV, and guanine is found to be the most strongly fluorescent base. Purines fluoresce at least three times more than pyrimidines.

Fluorescence lifetimes are generally very short as estimated by several methods (258–260). The *non*radiative relaxation times of the S_1-state for uracil, uridine, and thymine lie in the range 4 ± 2 ps, while for adenine and cytosine they are far below 4 ps (261).

A common finding among several investigators is that the excitation spectra have maxima at longer wavelengths than the first absorption maxima (260). Becker and Kogan (262) have shown that for thymine this does not result from variations of the quantum yield with excitation, but rather on the presence of H-bonded and non-H-bonded species. The discrepancy between excitation and absorption maximum was interpreted (263,264) in terms of (1) hidden transitions that lead to the fluorescent state, (2) of fluorescence from small quantities of tautomers being present, and (3) excitation into higher vibronic levels, which would result in more efficient nonradiative decay.

Even at 25°C the fluorescence spectra of thymine, cytosine, adenine, and guanine are highly polarized (265,266). Figure 3.3 shows the absorp-

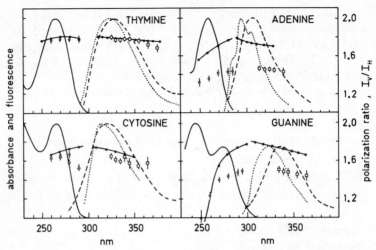

Fig. 3.3. Absorbance, fluorescence, and polarization spectra for DNA bases: (——) absorbance, neutral aqueous, 20°C; (– – – –) fluorescence, neutral aqueous, 22°C; (·····) fluorescence glycol–water (6:4), − 125°C: O, △: polarization ratio versus excitation wavelengths at 22°C; □: polarization ratio versus emission wavelength, 22°C; (–·–·–·–): polarization ratios at − 125°C. *Thymine:* O, at 340-nm fluorescence; □, at 280-nm excitation; (–·–·–·) at 325-nm fluorescence (left) and at 290-nm excitation (right). *Cytosine:* O, at 340-nm fluorescence; △, at 325-nm fluorescence; □, at 270-nm excitation; (–·–·–·) at 325-nm fluorescence (left) and 287-nm excitation (right). *Adenine:* O, at 340-nm fluorescence; □, at 280-nm excitation; (–·–·–·) at 310-nm fluorescence (left) and 280-nm excitation (right). *Guanine:* O, at 345-nm fluorescence; □, at 290-nm excitation; (–·–·–·) at 323-nm fluorescence (left) and 295-nm excitation (right). Error bars indicate standard deviation from average of three or more experiments. (Redrawn from ref. 266.)

tion, fluorescence, and polarization spectra of DNA bases. Wilson and Callis (267) made an extensive study on the absorption, fluorescence, and fluorescence polarization of nucleotides as a function of excitation and emission wavelengths. The monophosphates of cytidine, thymidine, guanosine, and adenosine, together with their respective dimers and polymers, were investigated in glycol–water glasses at around − 120°C. While the spectra of monomers and polymers are not very different (except for cytidyl-cytosine), the polarization degrees differ significantly.

3.5.2.   Pyrimidines

Thymine in neutral aqueous solution fluoresces maximally at 338 nm with ϕ_f 0.00014 at λ_{ex} 265 nm (264). The maximum is unaltered in 0.5 M H_2SO_4. The corrected excitation spectrum does not tally the absorption spectrum. It has frequently been reported that ϕ_f decreases strongly with increasing excitation energy (258,263,264,268,269). On the contrary, in a more recent work Becker and Kogan report (262) that for thymine and uracil at low temperature ϕ_f is independent of excitation energy.

Thymine, thymidine, and uracil have profoundly different properties in polar–aprotic versus polar–protic solvents with respect to the character of the nature of electronic transitions and fluorescence lifetimes. The species involved in the emission process of the pyrimidines is said to be the keto (lactam) form (262). The anisotropy of the polarized fluorescence excitation spectrum of thymine in neutral aqueous solution is essentially constant across the absorption spectrum, but shows a pronounced depolarization at the red edge of the band. It is suggested that this is due to a hidden n,π^*-transition (270a).

Morgan and Daniels (270b) studied cytidine-5'-phosphate and polymeric forms thereof. Quantum yields, corrected emission spectra, and their degree of polarization were given. The emission spectrum of cytidyl–cytidine is red-shifted to the spectrum of cytidine monophosphate. The emission spectra could be resolved into three (in the case of polycytidylic acid even into four) components. These are fluorescences from the monomer, the excimer (monomer phosphorescence), and excimer phosphorescence. For all species the quantum yields range from 1×10^{-5} to 3×10^{-4}. Excimer emission originates from the stacked fraction of the polymeric forms, and different stacking geometries were recognized from differences in excimer polarization.

Rapoport et al. (270c) have suggested that another possible cause for the difference between absorptions and luminescence excitation spectra of cytidine monophosphate may result from the change in the protonation pK value from 4.3 at 300 K to 5.9 at 77 K.

3.5.3. Purines

Fluorescence from adenine, adenosine, and AMP occurs from the π,π^* state (271). Protonation has only a small effect on the position and width of the absorption spectrum, whereas the fluorescence is greatly broadened and red-shifted over that of the corresponding neutral species (272).

In contradiction to former reports, Vigny and Duquesne (273) as well as Knighton et al. (274) have found that absorption and excitation spectra of purine and pyrimidine bases are in sufficient agreement. From a study on 11 closely related adenine derivatives it was concluded (274) that fluorescence from adenine and adenosine is primarily from a minor tautomer protonated at the 7-position (Scheme 2). This tautomer has ϕ_f of ~0.1, a lifetime of several nanoseconds, and is present at levels of 0.1–0.5% at pH 1.5. The excited state lifetimes, as estimated from the apparent ab-

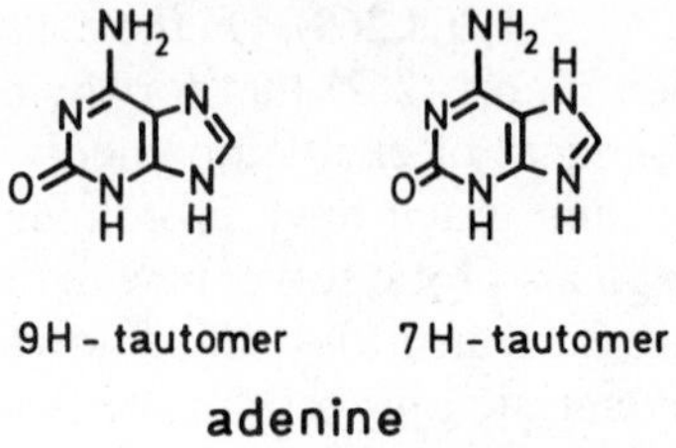

9H - tautomer   7 H - tautomer

adenine

Scheme 2.   Structures of the two adenine tautomers (7H and 9H).

sorption spectra, are only a few picoseconds (266,275). A striking point of the weak emission of adenine and other purine bases is the large anisotropy (265,266,275), which approaches that seen in rigid media. Nonradiative processes seem to be much more rapid than the process of rotational diffusion.

Morgan and Daniels (276) reported corrected emission spectra for adenosine, AMP, adenyladenosine and polyadenylic acid. Generally, the quantum yields are higher for the polymers than for the monomers. The total emission spectrum of adenyladenosine and polyadenylic acid could again be resolved into four components: monomer-like fluorescence, excimer fluorescence, monomer phosphorescence, and excimer phosphorescence. The quantum yields lie in the 5×10^{-5} to 3×10^{-4} range and agree favorably with those determined previously by Vigny (277). As for polycytidylic acid, a stacking model was used to explain excimer emission and differences in emission polarization.

Wilson and Callis (278) investigated the apparent photophysics of adenine and guanine in a neutral ethylene glycol–water mixture at temperatures between 140 and 165 K. Again the conclusion was made that emission stems from the N_7–H tautomer instead of the more abundant N_9–H form. The same is true for guanine. The model of a fluorescence of the N_7–H tautomer allowed the explanation of former discrepancies in lifetime and polarization studies.

Guanosine is known to be the base that is preferentially attacked by alkylating mutagens. Singer (279) measured the absorption, excitation, and emission spectra of various N-and O-methyl derivatives of guanine. The fluorescence spectra differ markedly, and two excitation maxima (one around 300 ± 10, the other at 350 ± 10 nm) are apparent.

Hemminki (280) gave the fluorescence excitation spectra of a number of alkylated guanines at six different pH values. Large differences in the intensities were observed, and the identification of guanosine alkylation products via their pH-dependent fluorescence spectra was suggested.

3.5.4. DNA and RNA

The fluorescence properties of DNA and RNA have been reviewed by Konev (39) and by Wang (281). Their work is of extreme interest for the performance of molecular biological studies, for example, on proteins. Unfortunately, the native fluorescences of DNA and RNA are very weak, but yeast t-RNA is an exception in containing a highly fluorescent base called the Y base. Therefore, many investigators have used intercalating fluorescent probes such as ethidium bromide or benzo[a]pyrene (282). Alternatively, highly fluorescent adenine substitutes such as 2-amino-purine or ϵ-adenosine have found use.

In contradiction to earlier reports (259,283,284) both DNA and RNA exhibit room temperature fluorescence, which becomes more intense by about three orders of magnitude at 77 K. At this temperature the fluorescence of native DNA is from excimer states of adenine and thymine, while the fluorescences of guanine and cytosine are quenched (285). In denatured DNA probably all four bases contribute to fluorescence. Lifetimes of adenine and thymine were estimated from quantum yield data to be around 10 ps (285).

Vigny and Ballini (286) made a study on the room-temperature fluorescence of the five common nucleotides and of DNA. Except for GMP, the red shift when going from rigid to fluid solutions is small (CMP, TMP) or negligible (AMP, UMP). Room-temperature quantum yields range from 0.3×10^{-4} (UMP) to 1.2×10^{-4} (CMP, TMP). It was found that addition of the ribose and phosphate group leave the quantum yields of cytosine, thymine, and uracil unchanged, whereas those of adenine and guanine are decreased by a factor of 5.

The former finding (285,287) that fluorescence of DNA is dominated by emissions from exciplexes involving adenine and thymine was corroborated (286). DNAs from several sources showed maximal emission between 330 and 335 nm, with quantum yields between 0.6 and 0.8×10^{-4}. The excitation spectrum disagrees with the absorption spectrum, but resembles that of a synthetic AMP–TMP polymer, which would indicate a favored involvement of adenine and thymine in the total DNA emission spectrum.

One notes that the question of the efficiency of energy transfer in DNA still seems unresolved. Values of exciton transfer along 50–100 base pairs have been published (288,289).

The reported fluorescence properties of native DNA are still contradictory. Data presented by Anders (290), Daniels (291), and Aoki and Callis (293) are compiled in Table 3.6. The recent study by Aoki and Callis (293) seems to be the most careful investigation performed so far. Par-

Table 3.6. Fluorescence Data for DNA as Reported by Various Authors

	Ref. 290	Ref. 292	Ref. 293	Ref. 286
DNA species	Calf thymus	Salmon sperm	Various sources	Various
Purification	Without	—	Extensive	Partially
Solvent	Distilled water	Water	π or Tris buffer	Tris buffer
Concentration	$1.4 \times 10^{-4}\ M$	$10^{-4}\ M$	$1.3 \times 10^{-3}\ M$	10^{-3}–$10^{-4}\ M$
pH value	6.8	6.4	7.0–7.3	7
Excitation (nm)	260, 300 (laser)	270 (fluorimeter)	250–280 (fluorimeter)	260 (fluorimeter)
Excitation bandwidth	0.1 nm	21 nm	13 nm	—
Emission maximum (nm)	352	349	325–330	330–335[b]
Half-width	94 nm	93 nm	—[a]	~66 nm
Quantum yield	4×10^{-5}	2×10^{-5}	3.4–11.6×10^{-5}	6–8×10^{-5c}

[a] The emission band is asymmetric with a broad shoulder at around 450 nm.
[b] A single photon-counting instrumentation was used.
[c] 3×10^{-5} for a highly purified DNA preparation.

ticular attention was paid in this work to the purification of the DNAs obtained from several sources. Fluorescences generally have maxima near 330 nm, quantum yields of about 4×10^{-5}, and excitation spectra similar to but not identical with the absorption spectra. A longwave component at around 450 nm in the emission spectrum, being absent in the spectrum of a mononucleotide mixture, is ascribed to exciplex emission from a singlet state, rather than to phosphorescence. Changes in fluorescence with temperature and pH agree favorably with well-established DNA denaturation processes, irrespective of the source or purification process. The results agree with data given by Vigny and Ballini (286), but are in some disagreement with other reports (290,291) (Table 3.7).

The possibility of a phototautomerism of guanosine–cytosine base pairs has frequently been discussed in the literature and its likelihood has been derived from theoretical considerations (294). Remarkably, no experimental evidence, for example, by applying fluorescence techniques, has been presented so far.

3.6. ALKALOIDS

Although metabolic amines of the phenylethylamine type are sometimes referred to as alkaloids (295), this chapter is confined to classical ones, that is, those that occur in the plant kingdom. The phenethylamines will be discussed later, together with other metabolic amines (Section 3.9.2).

Except for alkaloids without an aromatic or heteroaromatic ring, and the imidazole-derived compounds, practically all alkaloids are fluorescent, with nicotine being an important exception. Because of the structural manifold, however, the emission spectra can differ strongly from each other, depending on the emitting fluorophore. Scheme 3 shows the frequently encountered alkaloid fluorophores, together with the names of typical representatives.

Unfortunately, a classification following spectroscopic criteria cannot be in agreement with one made by an alkaloid chemist. The latter classifies alkaloids according to the structural skeleton, irrespective of whether they are aromatic or not. Thus the strychnos alkaloids are often treated under the indole heading, but, as 2,3-dihydroindoles, their spectral properties more clearly resemble those of acetanilide. Similarly, the fluorescence of the tetrahydroisoquinoline laudanosine is quite different from isoquinoline, while the fluorescence of the chemotaxonomically related papaverine is practically identical with isoquinoline.

A common feature of all nitrogen heterocycles, except indole and carbazole, is the increase in basicity upon photoexcitation. On the other

Table 3.7. Excitation and Emission Wavelengths (nm), Quantum Yields (ϕ_f), and Lifetimes (τ) of Nucleic Acid Components

Component	λ_{ex} (nm)	λ_f (nm)	ϕ_f	τ (ns)	Experimental Conditions[a]	Ref.
Adenine	—	307	0.06	—	Water, pH 7, 77 K	292
	270	295/310	0.02		Glycol–water, 140 K	278
	280	335	2.7×10^{-4}		Ethanol	271
	280	296, 308	1.9×10^{-1}	2.1	Ethanol, 77 K	271
	280	375			Water, pH 1	230
	265	365	2.1×10^{-3}	0.009^b	Water, pH 7.3	258
	265	365	0.7×10^{-4}		Water, pH 1.5	274
Adenosine	280	345	2.5×10^{-4}		Ethanol	271
	280	320	1.6×10^{-2}		Ethanol, 77 K	271
	270	392	7.7×10^{-4}		Water, pH 1.5	274
	248–266		$5.2–6 \times 10^{-5}$		Water	276
AMP	248	312	5×10^{-5}		Water, pH 7	286
ADP	280	365	1.1×10^{-4}		Ethanol	271
	280	360	3.4×10^{-2}	0.9	Ethanol, 77 K	271
ATP	280	360	9.5×10^{-5}		Ethanol	271
	280	370	2.6×10^{-2}	1.1	Ethanol, 77 K	271
	285	365	—		Water, pH 1	230
Cytosine	267	313	8.2×10^{-5}	0.0009^b	Water	258
	270	315	0.02		Water, pH 1, 77 K	292
	270	312	0.06		Water, pH 7, 77 K	292

CMP	248	330	1.2×10^{-4}	—	Water, pH 7	286
Guanine	275	328	3.0×10^{-4}	0.003^b	Water	258
	260	355	0.005	—	Water, pH 2	263
	245	365	—	—	Water, pH 1	230
	285	325	0.05–0.2	—	Glycol–water, 160 K	278
GMP	248	340	8×10^{-5}	—	Water, pH 7	286
	265	320	0.066	—	Water, pH 7, 77 K	284
Thymine	265	320	1×10^{-4}	0.0015^b	Water, pH 7	258
	240–290	330	0.035	—	Alcohols, 77 K	262
	270	316	—	—	Water, pH 7, 77 K	292
	270	338	0.0001	—	Water, pH 7	264
Thymidine	240–290	330	0.08–0.09	—	Alcohols, 77 K	262
	270	318	—	—	Water, 77 K	292
TMP	248	330	1.2×10^{-4}	—	Water, pH 7	286
Uracil	258	309	4.5×10^{-5}	0.0014^b	Water, pH 7	258
	260	315	8×10^{-5}	—	Water, 77 K	292
	250–285	300	$\sim 1 \times 10^{-4}$	—	Alcohols, 77 K	262
Uridine	260	318	1×10^{-4}	—	Water, 77 K	292
UMP	248	320	3×10^{-5}	—	Water, pH 7	286

[a] Room temperature unless otherwise specified.
[b] Calculated from the apparent absorption spectrum and ϕ_f.

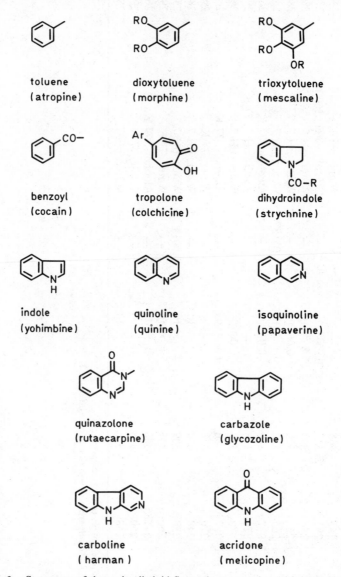

toluene
(atropine)

dioxytoluene
(morphine)

trioxytoluene
(mescaline)

benzoyl
(cocain)

tropolone
(colchicine)

dihydroindole
(strychnine)

indole
(yohimbine)

quinoline
(quinine)

isoquinoline
(papaverine)

quinazolone
(rutaecarpine)

carbazole
(glycozoline)

carboline
(harman)

acridone
(melicopine)

Scheme 3. Structures of the main alkaloid fluorophores together with typical representatives.

hand, indole and carbazole tend to become stronger acids in their excited (fluorescent) states (296,297). In practice this has no consequences for the emission spectra in inert solvents, but can give rise to unusual pH effects and fluorescence emission bands in aqueous solvent systems.

3.6.1. Alkaloids with Nonheteroaromatic Fluorophores

Because of the lack of an extended π-electron system the tropa and morphine alkaloids generally fluoresce far in the UV. It has been shown for morphine and its model compounds that the fluorescence maximum goes to longer wavelengths as they change to polar solvents (298). For cocaine, at 77 K, no such shift was observed (299). For phenethylamine-derived alkaloids a charge transfer interaction between fluorophore and a remote amino group can under certain conditions take place, which gives rise to a longwave exciplex emission (300). Unlike codeine, the fluorescence of morphine is quenched in 0.1 N sodium hydroxide (301).

The hallucinogenic alkaloid mescaline is rather weakly fluorescent (302,303). Two fluorescence bands are observed for methanol solutions, when excitation is at 260 nm, but only one is observed when excitation is at 273 nm (302).

Fluorimetric methods for the determination of the following opium alkaloids have been reported: morphine (301,304), codeine (301), narcotine (305), and normorphine (306). Chalm and Wadds have worked out a procedure for the spectrofluorimetric determination of mixtures of the principal opium alkaloids by use of a combination of extraction and selective quenching methods (307). However, most of the fluorimetric methods published in the last years that make use of the native fluorescence of opium alkaloids are based on previous TLC or HPLC separation and subsequent excitation with the 254-nm mercury line. Morphine, being the metabolite of heroin, is most frequently assayed by fluorimetry of its highly fluorescent oxidation product pseudomorphine (308).

Colchicine is an uncommon alkaloid in possessing a tropolone fluorophore. Tropolone itself has been investigated carefully (309–311), and some of the results have been used to interpret the colchicine fluorescence (311,312). Dissolved in water, colchicine fluoresces weakly ($\phi_f \sim 0.001$) with maximum intensity at 460 nm. When bound to albumin the intensity is markedly increased (and shifted to 430 nm) (313,314), a fact that is the basis of a fluorimetric assay of the tuberculine–colchicine complex (315).

3.6.2. Indole Alkaloids

The indoles form one of the largest groups of alkaloids. Detailed studies have been presented on the photophysics of the parent compound indole

(32,69), and the properties of the related tryptophan have been discussed in Section 3.2.3.

The simple, naturally occurring derivatives of indole have properties similar to those of the parent compound. Fluorescence is maximally excited at 270–290 nm and shows maximum intensity between 330 and 350 nm, depending on the polarity of the solvent. Polar solvents favor longer-wavelength emissions. Burnett and Audus have reported the solution spectra of various indole-derived alkaloids, together with pH effects thereon (316). MacNeil et al. measured the respective spectra on TLC plates, together with the effect of spraying with dilute hydrochloric acid (317). The solvent and pH dependence of the emissions of hydroxyindoles and indole-3-acetic acids has also been investigated in detail (318–321). The fluorescence of indoles is quenched by protons in both aqueous and nonaqueous solvents (164).

Introduction of a hydroxy or methoxy group into positions 4, 5, or 7 causes a small shortwave shift in the excitation and emission maxima, while the same groups in position 6 induce a small longwave shift. Many hydroxyindoles suffer complete fluorescence quenching in strongly alkaline solutions, together with oxidative decomposition.

With increasing complexity of the molecule unexpected phenomena may be encountered. Rauh et al. (322) have shown that in indole alkaloids such as reserpine, deserpine, and rescinnamine the electronic excitation energy is intramolecularly transferred to an energy acceptor—a second and independent π-electron system—whose fluorescence is finally observed. This intramolecular singlet–singlet energy transfer can occur with 100% efficiency.

Jung and Jungova (323) have presented a voluminous investigation of the fluorescence of 19 rauwolfia alkaloids. Most of them possess a tetrahydrocarboline skeleton, but in a spectroscopic sense they are indole derivatives. Indeed the fluorescence maxima (see Table 3.8) are typical for indoles rather than for carbolines. In methanol solution they cover the 340–355-nm region. In contrast to the carboline spectra they are not influenced markedly by addition of dilute hydrochloric acid, but react more to ammonia.

A very unusual and still not fully elucidated phenomenon is the appearance of a greenish-yellow emission of 5-hydroxy- and 5-methoxyindoles dissolved in, or sprayed with, 3 M mineral acid (323,324). It can be used to determine 5-oxyindoles in the presence of others (323). A more detailed discussion is presented in section 3.9.2 (see also Table 3.20).

It has been reported (325) that psilocin and bufotenine, two hallucinogenic 5-hydroxyindoles, fluoresce when irradiated at 254 nm, whereas psilocybine, the psilocine phosphate ester, does not. This observation is not supported by the findings of Gillespie (326), who gives a detection limit of 1 μg ml^{-1} for psilocybine. Gillespie has also reported the fluorescence properties of various other hallucinogens such as dimethyl tryptophanamide, MDA, STP, mescaline, and LSD.

The emission properties of the ergot alkaloids found widespread interest with respect to the fluorimetric determination of these drugs. Bowd et al. (327) have reported the luminescence characteristics of nine lysergic acid derivatives, of which LSD is probably the most well known.

A common property of all ergot alkaloids is their intense fluorescence in the visible part of the spectrum, and the large shift between excitation and emission maximum, which, for aqueous solvents, generally exceeds 8000 cm^{-1}. The unusually large Stokes' shift of LSD and its congeners (being even larger than the 5300-cm^{-1} Stokes' shift of the exciplex-forming indole fluorophore in, for example, tryptophan) has not thus far been interpreted. Since maxima in absorption are much less solvent polarity dependent than in emission, large differences in the dipole moments of ground and excited singlet state are to be expected. The fluorescence maxima are practically pH independent up to pH 12, but intensity drops at extreme values. In 1 N sodium hydroxide the emission maximum is shifted to 536 nm (326). It is interesting to note that the nonconjugated ergoline agroclavine has a fluorescence maximum much shorter (350 nm in water) than the conjugated ergolines (427–435 nm).

Fluorescence quantum yields as high as 0.7–0.8 were found (327) for the Δ^9-ergoline series in ethanol solution at 25°C. The values are somewhat higher for LSD, lysergic acid, and ergometrine in water, but the ergotamine and ergocristine epimers, on the other hand, showed rather low ϕ_f values in water (0.05–0.14). The lifetimes are about 6 ns in ethanol and are increased to around 12 ns on changing to water as a solvent, except for the two previously mentioned epimers, whose lifetimes drop to 1.3 and 2.9 ns, respectively.

Detection limits as low as 6.5 ng mL^{-1} for LSD and 1.3 ng mL^{-1} for lysergic acid have been reported (303). Various methods and conditions, which account for the manifold of samples wherein LSD is to be determined, have been worked out for LSD and other ergot alkaloids, using their native fluorescence (328–334).

The strychnos alkaloids are derivatives of 2,3-dihydroindole with an acyl group attached to the nitrogen atom (see the formula in Scheme 3). Although no reports on the native fluorescence of these alkaloids are evident, one would expect them to be fluorescent by analogy to the parent fluorophore acetanilide, which exhibits fluorescence maxima of 317 nm in cyclohexane and 356 nm in ethanol respectively (335). Ibogain, a related hallucinogen with a 2,3-dihydroindole fluorophore, indeed shows native fluorescence peaking at 355 nm in aqueous methanol, when excited at 290 nm (303).

3.6.3.   Quinoline and Isoquinoline Alkaloids

Both quinoline and isoquinoline are nitrogen heterocyclics with small singlet-to-triplet energy gaps. Therefore, the ratio of fluorescence to phosphorescence intensity is highly dependent on the polarity of the solvent. Polar solvents favor fluorescence, as do electron-donating substituents such as hydroxy-, methoxy-, or amino groups. Fluorescence and triplet formation yields decrease with increasing temperature (336).

Introducing electron-delivering substituents shifts the absorption and emission spectra to longer wavelengths, but the effect appears to be more pronounced when substitution occurs at the aromatic ring than at the pyridine ring (337).

For both quinoline and isoquinoline the fluorescence quantum yields are dependent on the excitation wavelength and are highest when excitation is at the red edge of the longest-wave absorption band (338,339). For isoquinoline, evidence has been presented that fluorescence decays double-exponentially (340). Also, there is indication that the thermally relaxed ^1L_b and ^1L_a states of the protonated forms of quinoline and isoquinoline are degenerate and that fluorescence occurs from both states (341).

The fluorescences of the protonated and quarternized forms of quinoline (342), 6-methoxyquinoline (343,344), quinine (343–345), and alkaloids of the furoquinoline type (337) are strongly quenched by halide and pseudohalide anions.

Following photoexcitation, quinolines and isoquinolines (341), just as other heterocyclics such as pyridines (346), acridines (347), acridones (348), and phenazines (349), become more basic. In other words, the pK_a value of the conjugate acid increases, frequently by several orders of magnitude (341,350). Provided the excited-state lifetime is long enough and the excited-state equilibrium is established, the following reaction may take place as a result of photoexcitation:

$$\text{base}(S_0) + H_2O \xrightarrow{h\nu} \text{base·H}^+(S_1) + OH^-$$

Fluorescence will now be observed from the protonated form, despite the excitation of the neutral molecule. Such a situation is found for 6-methoxyquinoline, but not for quinine (351).

Quinine is the alkaloid whose properties have probably been investigated most thoroughly. Dissolved in 1 N sulfuric acid it is a frequently used standard for quantum yield determinations (352,353), despite the fact that it shows the following anomalies (352):

1. In acid solution it has an excitation spectrum that fails to coincide with the absorption spectrum at long wavelengths.
2. The quantum yield is 23% higher with 390-nm excitation than it is with 313-nm excitation. There is, however, a contradictory report (354).
3. The emission spectra of both quinine and its parent compound, 6-methoxyquinoline, are shifted toward the red with increasing excitation wavelength. This effect, named *edge excitation red shift*, has been studied in detail (338). It was concluded that at least two average conformations, each with its own distinct electronic energy, fluoresce (355).
4. The fluorescence polarization of the emission band of quinine in a rigid medium varies with wavelength in a way suggesting that the emission band arises from two singlet states simultaneously.
5. The polarization of the excitation spectrum suggests the presence of a hidden electronic transition in the longwave absorption band.
6. The fluorescence of quinine cation decays double-exponentially (356).
7. The fluorescence is pH dependent even in the pH 1–2 range (357).

Schulman et al. (351) performed a detailed study on the pH dependence of spectra, decay times, and quantum yields of quinine, quinidine, and 6-methoxyquinoline. It is interesting to note that while the spectra of the pair of epimers, quinine and quinidine, are almost identical with respect to their maxima and pH dependences, the fluorescence decay times and quantum yields differ considerably. At neutral pH, quinine has a decay time of 13.4 ns and shows a ϕ_f value of 0.50, whereas quinidine has values of 19.9 and 0.43, respectively.

Montagu et al. (337) studied the absorption and fluorescence properties of several quarternary quinolinium alkaloids extracted from rutaceae plants. The furoquinolinium derivatives with no oxygen function attached to the aromatic ring fluoresce in the UV (maximally at around 345 nm), those with a methoxy group in positions 6, 7, or 8 fluoresce blue (maximally between 450 and 500 nm). Cationic quinolines with free hydroxy groups fluoresce even green (515–520 nm) which, in view of studies on hydroxyquinoline model compounds (358,359) can possibly be interpreted in terms of excited-state tautomerization prior to emission. Again, various anions act as efficient quenchers.

Introducing a carboxylic acid function gives rise to unusual fluorescences due to changes in the dissociation constants of acidic and basic functions following photoexcitation. As a result, the fluorescence behavior of cinchophen and cinchoninic acid is rather complex (360).

The isoquinolines form another large family of alkaloids. They can be divided into two main groups. One comprises compounds whose heterocyclic ring is not hydrogenated, the second the so-called tetrahydroisoquinolines. Fluorimetrically, the two groups can easily be distinguished by the location of the fluorescence maxima and by pH effects thereon. Tetrahydroisoquinolines fluoresce in a much shorter time (310–320 nm in ethanol at room temperature) than the isoquinolines (~380 nm). The latter suffer a longwave shift to around 450 nm after acidification, while the former exhibit a practically unchanged fluorescence maximum.

Smekal et al. have investigated spectra, quantum yields, and lifetimes of several berberines (361,362) and protoberberines (363–365), which are known to form complexes with DNA. Some of the data are given in Tables 3.8 and 3.9 at the end of this section. All investigated alkaloids were found to be highly fluorescent, but substituents strongly influence both the absolute value of ϕ_f and the fluorescence maximum, so that generalizations can hardly be made. The maxima of the fluorescence bands of protoberberines and pseudoprotoberberines cover the 540–570-nm region, whereas the tetrahydroberberines fluoresce maximally between 320 and 330 nm (365). Fluorescence polarization (361), quantum yields (363), and lifetimes (363) of berberines are quite sensitive toward effects of DNA binding (364), which allows the mutual interactions to be studied more precisely than with absorption methods (362). Interestingly, for some alkaloids double-exponential decay kinetics were found (see Table 3.9).

Ito et al. (366) have compiled the qualitatively measured fluorescence intensities of 50 isoquinolines, and the relationship between fluorescence and chemical structure was examined.

The fluorescences of tubocurarine chloride (367) and various other tetrahydroisoquinoline alkaloids (368) in ethanol solution do not vary strongly with structural changes. Excitation maxima range from 280 to 290 nm, emission maxima from 310 to 318 nm. Hernandezine was found to exhibit an additional longwave band, which was ascribed to an intramolecular exciplex. Quantum yields from 0.4 to 1.8% were found at room temperature, which can increase up to 35% at 77 K. The emission parameters have again been found to depend on the absolute configuration of epimers, as was previously reported for the quinine–quinidine pair of isomers (351).

3.6.4. Carboline Alkaloids

From a spectroscopic point of view only alkaloids with a carboline or dihydrocarboline structure are typical of the carboline alkaloids, whereas the spectra of tetrahydrocarbolines resemble more the indole spectra. The intense fluorescence of the harmala alkaloids has been recognized very early, but only recently the solvent and pH dependence of the fluorescence properties of harman (369), harmine, and harmole (370) were investigated in some detail. The spectra are not strongly influenced by organic solvents, but depend highly on the acidity of the aqueous solution. Unusual changes in the basicity of the photoexcited harmala derivatives give rise to the formation of unusual excited state (and fluorescent) tautomers. The results obtained with harman were corroborated by similar findings for carboline (371), which is the unsubstituted fluorophore of harmane.

The fluorescences of 3,4-dihydroharmanes (e.g., harmaline) are not very well investigated. Both harmaline and harmalol have absorptions and emissions at longer wavelengths than the harmanes. The fluorescence maxima are around 470 nm in methanol and water and are not shifted after addition of acid. However, the intensity is much higher in acidic solutions (323,530). The harmalol spectra are similar to harmaline in their pH dependence, except for a longwave shift to 520 nm at pH values above 12 (530).

Fluorodaturatine, another strongly fluorescent dihydrocarboline derivative (372), is spectrally similar to harmaline in showing an emission maximum of 487 nm in methanol solution, which only shifts to 492 nm in 1 N sulfuric acid. The respective excitation maxima remain practically the same (397 and 395 nm). Further studies are certainly desirable.

The native fluorescence of the hallucinogens harmine and harmaline has been used to study their interaction with calf thymus DNA (373) and to probe the alkali ion dependent enzyme ATPase (374). Harmole is metabolized to either harmole sulfate (375) or its glucuronide (376), whose unusual longwave fluorescence points to a zwitterionic structure, as was observed for related compounds (370).

3.6.5. Acridone Alkaloids

Acridone derivatives have been found in nature as late as 1948, when Hughes et al. (377) reported the isolation of five alkaloids from rutaceae species. Meanwhile, many acridones have been isolated and most of them were found to be fluorescent (for a review see ref. 378).

Table 3.8.   Absorption Maxima (first figure) and Fluorescence Maxima (second figure) of Acridone Derivatives in Various Kinds of Solvents at 22°C[a]

Acridone	Methanol	0.01 N NaOH	Water, pH 5	1 N H$_2$SO$_4$
N-methyl	381/427	383/435, 455	383/435, 455	385/475
2-hydroxy	421/448, 470	439/560	423/468	421/485
2-methoxy	415/452	416/468, 495	416/457	416/490
3-hydroxy	385/406, 426	400/480	389/411, 432	390/460

[a] Wavelengths in nanometers.

The acridone fluorophore has been investigated with respect to its pH (348,379,380), temperature (381), solvent (348,381,382), and alkyl substituent (383) dependence. No spectral data could be located in the literature for naturally occurring acridones, all of which bear hydroxy or alkoxy groups. We have therefore measured the absorption and fluorescence spectra of model acridones in methanol and in aqueous solutions of various acidity (Table 3.8). It is evident that the fluorescence maxima are much more acidity dependent than the absorption maxima. 1-Hydroxyacridone is nonfluorescent. Introduction of a methoxy group shifts the absorption and fluorescence maxima to longer wavelengths, except perhaps for the 3-oxy isomers. The fluorescences of the naturally occurring trioxyacridones are green throughout (377,384).

Emission intensity is highest (and most longwave) in polar protic solvents. While the introduction of methoxy groups can intensify fluorescence, a hydroxy group can have an adverse effect (383).

3.6.6.   Miscellaneous Heterocyclic Alkaloids

Unusually longwave emission maxima have been reported for the pyridine-derived alkaloid anabasine and its derivatives (385) at 77 K and various pH values. The fluorescences of the three isomeric hydroxypyridines, which may serve as model compounds, have also been investigated as a function of acidity and solvent (386).

The carbazoles form another important group of alkaloids, for which no reports on the native fluorescence are evident. The parent fluorophore is well investigated with respect to solvent (387–389) and pH dependence (390), as well as the quenching (391,392) of its fluorescence.

The quinazoline-derived alkaloids can, by analogy to the behavior of simple quinazolines (393), be expected to be fluorescent. Finally, some alkaloids are derivatives of quinazolinium (394) and indolizine (395), two systems with strong inherent fluorescence.

Table 3.9. Excitation Wavelengths, Fluorescence Maxima, and Experimental Conditions of Alkaloid Fluorescences Reported in the Literature

Alkaloid	Fluorophore Type	λ_{ex} (nm)	λ_f (nm)	Experimental Conditions[a]	Ref.
Ajmalicine	Indole	285	355[b]	Methanol	323
Ajmaline	Indole	295	355[c]	Methanol	323
Balfurodinium chloride	Quinolinium	310	448	Water	337
Berbamine	Trioxytoluene	285	318	Ethanol	368
Berberine	Isoquinolinium	432, 352	548	Ethanol	365
		380	510	DMF	397, 531
		433, 353	510	TLC plate	398
Berberubine	Isoquinolinium	455, 353	563	Ethanol	365
		521	632	Ethanol–NaOH	364
Berberubine methyl ether	Isoquinolinium	435	549	Ethanol	364
Boldine	Bisdioxytoluene	308	375[c]	0.1 N H_2SO_4	399
		330	420	pH 10	399
		312	350	Ethanol	399
Caffeine	Purine	270	303	Water	396
Canadine	Toluene	288	327	Ethanol	364
Cinchonidine	Quinoline	315	445	Water, pH 1	329
Cinchonine	Quinoline	320	420	Water, pH 1	3
Cinchoninic acid	Quinoline	315	391	Water, pH 10	360
		315	395	Water, pH 2.5	360
		326	529	Conc. H_2SO_4	360
Cinchophen	Quinoline	324	380	Water, pH 10	360
		340	407	Water, pH 2.5	360
		373	535	Conc. H_2SO_4	360

Table 3.9. *(continued)*

Alkaloid	Fluorophore Type	λ_{ex} (nm)	λ_f (nm)	Experimental Conditions[a]	Ref.
Cocaine	Benzoic ester	277	300	Organic solvents, 77 K	299
Codeine	Dioxytoluene	245, 285	350	Water, pH 1	400
Colchicine	Tropolone	370	418	EPA, 77 K	312
		355	422	EPA, 77 K	311
		360	435	Water	315
		360	500	Isoamyl alcohol	315
Coptisine	Isoquinolinium	467, 363	568	Ethanol	364
Coraline	Isoquinolinium	429	472	Ethanol	364
Coralydine	Toluene	287	320	Ethanol	364
Corydaline	Toluene	282	318	Ethanol	364
Corysamine	Isoquinolinium	455, 352	548	Ethanol	365
Dauricine	Dioxytoluene	285	316	Ethanol	368
3,4-Dehydroreserpine	3,4-Dihydrocarboline	395	490[b]	Methanol	323
Deserpidine	Indole	254	313[a]	Ether, 77 K	322
		280	360	Methanol	323
		280	365	Water, pH 1	401
Ellipticine	Carboline	301, 425	532	0.1 N H$_2$SO$_4$	407
		290, 382	437, 514	Methanol	407
Emetine	Dioxytoluene	290	320	Dilute acid	402
Ergocristine and ergocristinine	Indole	320	405[e,f]	Ethanol	327
Ergometrine and ergometrinine	Indole	320	405[e,f]	Ethanol	327
Ergotamine and ergotaminine	Indole	320	404[e,f]	Ethanol	327

212

Fangchinoline	Trioxytoluene	285	314	Ethanol	368
Fluorodaturatine	3,4-Dihydrocarboline	390	487	Methanol	—[g]
		395	540(sh), 491	Water, pH 7	—[g]
Gramine	Indole	370	543	0.1 N NaOH	—[g]
		275	340	Methanol	316
Harmaline	3,4-Dihydrocarboline	385	480[b]	Methanol	323
		313	425, 470	Acidic water	530
		380	480	Methanol–HCl	323
Harmalol	Carboline	313	420, 465	Acidic water	530
Harman	Carboline	345	375[b]	Methanol	323
		368	433[h]	1 N H_2SO_4	369
		390	491	pH 13	369
Harmine	Carboline	303	353[h]	Methanol	370
		357	419	1 N H_2SO_4	370
		359	420	Water, pH 7	370
		370	480	1 N NaOH	370
Harmole	Carboline	345	367, 415[h]	Methanol	370
		365	421	1 N H_2SO_4	370
		335	465	1 N NaOH	370
Hernandezine	Tetraoxytoluene	285	316, 398	Ethanol	368
Hydrastine	Dioxybenzoate	370	460	DMF	531
5-Hydroxyindole-3-acetic acid	Indole	306	350	Water[c]	320
		300	545	Acid of H_0 −1.0	320
		295	350[c]	Methanol	316
		290	355	TLC plate	317
		307	460	1 N NaOH	321
5-Hydroxytryptophan	Indole	293	356[i]	TLC plate	317
		295	340	Water	318

Table 3.9. *(continued)*

Alkaloid	Fluorophore Type	λ_{ex} (nm)	λ_f (nm)	Experimental Conditions[a]	Ref.
5-Hydroxytryptophol	Indole	308	351[i]	TLC plate	317
Ibogaine	2,3-Dihydroindole	290	355	Methanol–water	303
Indole-3-acetic acid	Indole	292	362	Water[c]	320
		293	356	TLC plate	317
		285	365[c]	Methanol	316
		307	420	1 N NaOH	321
Indole-3-acetonitrile	Indole	280	350	Methanol	316
Isodeserpidine	Indole	280	355[b]	Methanol	323
Isoptelefolonium perchlorate	Quinolinium	324	479, 509	Water	337
Isotetrandrine	Trioxytoluene	285	312	Ethanol	368
Jatrorrhizine	Isoquinolinium	440, 352	538	Ethanol	365
Laudanosine	Dioxytoluene	285	315	Ethanol	368
Laudanosoline	Dioxytoluene	285	321	Ethanol	368
LSD	Indole	313	438[e]	0.001 N HCl	326
		312	430	Methanol–water	303
Lysergic acid	Indole	320	402[e,f]	Ethanol	327
		320	394[f]	Ethanol	327
		311	420	Methanol–water	303
Mescaline	Trioxybenzoyl	273	315, 356	Methanol	326
Mescaline·HCl	Trioxybenzoyl	270	386	Methanol–water	303
13-Methoxyberberine	Isoquinolinium	433	562	Ethanol	364

Compound	Chromophore			Medium	
10-Methoxydeserpidine	Indole	285	355[b]	Methanol	323
5-Methoxyindole-3-acetic acid	Indole	307	430	1 N NaOH	321
5-Methoxytryptophol	Indole	305	350[i]	TLC plate	317
Morphine	Dioxytoluene	304	350[i]	TLC plate	317
		285	350	Water, pH 7	301
		285	350	Water, pH 1	400
		287	335[k]	Ether, 77 K	298
Narcotin	Trioxytoluene	335	400	Chloroform	307
		315	375	CHCl$_3$–trichloroacetic acid	307
Norcoralyne	Isoquinolinium	424	475	Ethanol	364
Nortenuipine	Trioxytoluene	285	315	Ethanol	368
Obamegine	Trioxytoluene	285	317	Ethanol	368
Oxyacanthine	Trioxytoluene	285	316	Ethanol	368
Oxyberberine	Isoquinolinone	386	427	Ethanol	364
Oxyepiberberine	Isoquinolinone	396	439	Ethanol	364
Palmatine	Isoquinolinium	433, 350	552	Ethanol	365
Papaverin	Isoquinoline	315	347	Chloroform	307
		415	452	CHCl$_3$–trichloroacetic acid	307
Phaeanthine	Trioxytoluene	285	312	Ethanol	368
Platidesminium·HClO$_4$	Quinolinium	307	345	Water	337
Pseudocoptisine	Isoquinolinium	380, 346	556	Ethanol	365
Pseudoepiberberine	Isoquinolinium	380, 341	550	Ethanol	365
Pseudopalmatine	Isoquinolinium	380, 342	546	Ethanol	365
Psilocine	Indole	276	320	Methanol–water	303
		292	314	Water, pH 7	320

Table 3.9. (*continued*)

Alkaloid	Fluorophore Type	λ_{ex} (nm)	λ_f (nm)	Experimental Conditions[a]	Ref.
Psilocybine	Indole	270	341	Methanol–water	303
Ptelefolonium picrate	Quinolinium	272	309[h]	Methanol	326
Quinidine	Quinoline	345	464	Water	337
Quinine	Quinoline	347	447	Water, pH 2	351
		352	450	0.1 N H₂SO₄	337
		347	448	Water, pH 2	351
		331(330)	382(380)	Water, pH 7 (pH 10)	351
Raubasine: see ajmalicine					
Repandinine	Trioxytoluene	285	312	Ethanol	368
Rescinnamine	Indole–trimethoxycinnamic acid	310	400	Water, pH 1	3
		254	340[d]	Ether, 77 K	322
		325	435[b]	Methanol	323
Reserpine	Indole–trimethoxybenzoic acid	254	313[d]	Ether, 77 K	322
		295	360[b]	Methanol	323
		300	375	Water, pH 1	329
Reserpinine	Indole	300	360	Methanol	323
Ribalinium perchlorate	Quinolinium	342	515	Water	337
Rutalinium perchlorate	Quinolinium	344	518	Water	337
Sarpagine	Indole	300	340[b]	Methanol	323
Serpentine	Carboline	305, 370	440[b]	Methanol	323
		365	445	0.1 N H₂SO₄	399
Sinactine	Toluene	287	322	Ethanol	365
Stylopine	Toluene	290	328	Ethanol	365

216

Tetrahydroberberine	Toluene	295	330	Dimethyl-formamide	531
Tetrahydrocorysamine	Toluene	291	327	Ethanol	365
Tetrahydroharman	Indole	285	350[b]	Methanol	323
Tetrahydropalmatine	Toluene	282	318	Ethanol	365
Tetrahydropapaverine	Dioxytoluene	285	312	Ethanol	368
Terahydropseudoepiberberine	Toluene	290	328	Ethanol	365
Tetrandrine	Trioxytoluene	285	312	Ethanol	368
Thalictricavine	Toluene	291	327	Ethanol	365
Tryptophol	Indole	280	365[c]	Methanol	316
		290	360	TLC plate	317
		307	420	1 N NaOH	321
Tubocurarine	Oxytoluene	280	317	Ethanol	368
Xanthine	Purine	315	435	Water, pH 1	230
		285	355	Methanol	323
Yohimbine	Indole	280	350[c]	0.1 N H_2SO_4	399
		282	360	Water, pH 7	399

[a] Room temperature unless otherwise specified.
[b] Effect of adding HCl and NH_3 also studied.
[c] pH dependence given in detail.
[d] O–O transition.
[e] ~380 nm at 77 K.
[f] ~435 nm in water.
[g] Unpublished results.
[h] Data for various other pH values for benzene and methanol solutions also given.
[i] Effect of spraying with acid also studied.
[j] 427 nm in water.
[k] Spectra in chloroform, acetonitrile, and dimethyl sulfoxide also given.

Table 3.10. Fluorescence Quantum yields (θ_f's) and Decay Times (τ) of Selected Alkaloids

Alkaloid	ϕ_f	τ^a (ns)	Experimental Conditions[b]	Ref.
Acridone	—	14.8	Water	380
	—	26.0	2.3 M HClO$_4$	380
Agroclavine	0.37	4.1	Ethanol	327
	0.18	2.5	Water	327
Berbamine alkaloids[c]	0.004–0.017	—	Ethanol	368
	0.08–0.35	0.5–1.3	Ethanol, 77 K	368
Berberine	0.05	1.3	Water	361
Bisbenzylisoquinoline alkaloids[c]	0.007–0.15	0.9–2.5	Ethanol	368
	0.14–0.86	0.8–8.5	Ethanol, 77 K	368
Carboline	—	22	Water, pH 3.6	371
	—	1.6	Water, pH 14	371
Colchicine	0.25	2.4 and 5.37[d]	EPA, 77 K	311
Coptisine	0.11	2.3	Water	361
Coralyne	0.63	—	Water	361
Corysamine	0.07	1.4	Water	361
Ergometrine[e]	0.74	6.2	Ethanol	327
	0.80	12.3	Water	327
Ergometrinine[e]	0.74	6.2	Ethanol	327
	0.63	11.6	Water	327
Harmane	0.89	—	1 N H$_2$SO$_4$	369
Harmine	0.51	—	1 N H$_2$SO$_4$	370
Harmol	0.41	—	1 N H$_2$SO$_4$	370
Indole-3-acetic acid	—	2.6	Water	403
Indolizine	0.72	41	Methanol	395
LSD	0.73	6.3	Ethanol	327
	0.84	12.3	Water	327
Lysergic acid	0.72	5.3	Ethanol	327
	0.92	11.5	Water	327
Norcoralyne	0.45	3.5	Water	361
Oxyberberine	0.12	—	Water	361
Protoberberine	0.40	6.4	Water	361
Pseudopalmatine	0.02	—	Water	361
Quinine	—	~20 and ~2[d]	1 N H$_2$SO$_4$	356
	—	~1.05	Ethanol	338
	0.54	—	1 N H$_2$SO$_4$	404
	0.50	13.4	Water, pH 7	351
	0.009	—	Water, pH 11	351
	0.77	—	1 N H$_2$SO$_4$	141
Quinidine	0.50	24.6	Water, pH 2	351
	0.43	19.9	Water, pH 7	351
	0.014	—	Water, pH 11	351
Tryptophol	—	7.7	Water	164

[a] Measured lifetime.
[b] Room temperature unless otherwise specified.
[c] Nine examples.
[d] Double-exponential decay.
[e] Various other ergot alkaloids also investigated.

Caffeine, theophylline, and theobromine form another group of compounds with alkaloid-type activity. Their fluorescences are rather short-wave. For example, caffeine, being nonfluorescent in cyclohexane solution, fluoresces maximally at around 303 nm when excited at 270 nm in neutral aqueous solution (396). (See Tables 3.9 and 3.10.)

3.7. OXYGEN RING COMPOUNDS

Among the variety of naturally occurring oxygen ring compounds (405) that can be expected to fluoresce, a few groups have received particular attention, namely, coumarins, furocoumarins, and flavonoids. Their intense fluorescences were described very early, since they are observable in the daylight. Studies in the last 10 years, however, have shown that many other oxygen ring compounds are fluorescent, albeit in the ultraviolet.

Practically all natural oxygen ring compounds bear a free hydroxy group, which makes their spectra pH dependent. Some possess a sugar residue, which may suffer decomposition when running spectra in more concentrated mineral acids. These compounds are partially responsible for the flavor and aroma of many plants, and they play an important role in chemotaxonomy. Because of its sensitivity, fluorimetry has been able to make important contributions to the analysis of oxygen ring compounds.

3.7.1. Chromenes and Chromones

Reports on the native fluorescence of these types of oxygen heterocycles are scarce. 2,2-Dimethyl-2H-chromene is nonfluorescent, but introduction of an 8-methoxy group gives rise to fluorescence (406). The related compounds precocene I and II (two insect hormones) fluoresce in the UV (407). The data are given in Table 3.13 at the end of this section.

Vitamin E (tocopherol), which may be regarded as a hydrogenated chromene, is also fluorescent. Its properties have been discussed in the coenzyme and vitamin section.

The cannabis constituents, being partially hydrogenated dibenzopyranes, likewise fluoresce in the UV, possessing maxima around 318 nm in ethanol solution at room temperature (408). The spectra change radically on prolonged irradiation, yielding new peaks at 366 and 383 nm.

Chromones, unlike most chromenes, fluoresce mainly in the visible. The first—and visual—examination of the fluorescence of chromones excited by daylight in sulfuric acid solution was performed by Rangas-

Table 3.11. Excitation Wavelengths (First Figure) and Fluorescence Maxima (Second Figure) of Some Oxygenated Chromones, Flavones, and Isoflavones in Methanol Solution, and the Effect of Shift Reagents Thereon (wavelengths in nanometers)

Oxygen Ring Compound	In Methanol	+ water[a]	+ Na-acetate[b]	+ H_2SO_4[c]
3-Methylchromones				
7-Hydroxy	305/464	305/460	305/460	360/504
7-Hydroxy-6-methoxy	317/393	317/448	350/450	350/486
6,7-Dihydroxy	324/426	324/454	355/454	355/597
6-Hydroxy-7-methoxy	323/416	323/414, 553	340/533	350/597
Flavones				
3-Hydroxy	340/405, 532[d]	340/412, 514[d]	340/410, 514[d]	340/412, 529[d]
5-Methoxy	330/459	330/473	330/473	330/535
7-Hydroxy	335/419, 543[d]	340/540	360/528	370/428, 542[d]
7-Methoxy	330/404	330/414	330/412	360/440
8-Methoxy[e]	330/475	330/495	330/488	340/468, 554
2'-Methoxy[e]	330/428	330/452	—	340/516
3'-Methoxy[e]	330/438	330/466	—	340/547
4'-Hydroxy	326/421	326/428	326/425	373/466
4'-Methoxy	325/412	325/424	325/424	378/462
Isoflavones				
7-Hydroxy	305/465	305/465	305–340/462	340/503
6,7-Dihydroxy	327/422	327/470, 500	355/470	350/500
6,7-Dimethoxy	320/402	320/410	320/408	355/460
7,4'-Dihydroxy	310/440	310/480	350/450	—[f]
6-Hydroxy-7,4'-dimethoxy	325/434	325/492	350/527	330/603
6,4'-dihydroxy-7-methoxy	325/416	325/426, 526	355/526	330/608
6,7-dihydroxy-4'-methoxy	325/427	330/471	360/470	330/590

Source: From Wolfbeis (407).

[a] 3 ml water to 1 ml methanol stock solution.
[b] 1 ml 5% aqueous sodium acetate to 3 ml methanol solution.
[c] 1 ml of a 50% acid to 3 ml methanol solution.
[d] Dual emission.
[e] The hydroxy derivative is very weakly fluorescent.
[f] Fluorescence is quenched.

220

wami and Seshadri (409). Another study was presented by Sreerama-Murti (410). Generally (409–411), the fluorescence color in sulfuric acid solution is described to be violet or blue.

Gallivan (412) and others (413) report that chromone, 2-methylchromone, and 3-methylchromone are not fluorescent in alcoholic solution at room temperature, but are highly so in strong sulfuric acid (413). Chromanone (2,3-dihydrochromone) exhibits weak fluorescence (λ_{max} 410 nm, $\phi_f \sim 10^{-4}$) in ethanol at room temperature (412).

All natural chromones are oxygenated in either 5-, 6-, or 7-position. As far as they have been investigated, only the 5-hydroxy derivatives are nonfluorescent (411,414,415). The latter form fluorescent beryllium chelates (414,415). Huitink (416) has measured the absorption and fluorescence spectra as well as the alkali stability of several 7-hydroxychromones, which are known (413) to undergo either photodissociation or phototautomerization.

We have measured the fluorescence properties of several oxychromones in methanol solution and the effect of shift reagents thereon (Table 3.11). Shift reagent techniques proved to be extremely useful in the identification of flavons and related natural products (417).

The results show all chromones to possess characteristic emission maxima and to suffer distinguished spectral shifts after addition of either water, acid, or base. These data, together with those obtained with absorptiometry, may serve to identify chromones. An unusually large shift in emission is observed for 6-hydroxy-3-methylchromone and 6-hydroxyisoflavone following additon of 50% sulfuric acid (1 ml) to the methanol solution (3 ml): Fluorescence turns from blue to orange, the lifetime drops sharply (from a few ns to below 300 ps), but the fluorescence excitation spectrum remains practically unchanged. We assume a similar excited-state process to be operative as was suggested for the process of the formation of a longwave-emitting species of 5-hydroxyindole (324) in strong mineral acid.

The excited-state behavior of 3-hydroxychromone is unusual in that it forms an intramolecular phototautomer (418). In 3-methylpentane solution the excitation maximum is at around 330 nm, but the emission maximum is at 490 nm. This tautomerization parallels that of flavonol (*vide infra*).

3.7.2. Coumarins

More than 300 coumarins have been found in various plants (419). Most frequently they are oxygenated in the 7-position and, in decreasing order, in the 6-, 5-, 8-, 4- and 3-position. Early qualitative reports on the fluorescence have been made with respect to solvent, pH (409,420), and structural effects (421).

Casparis and Manella (422) and Goodwin and Kavanagh (423) performed extensive studies on pH versus intensity profiles of a great number of natural coumarins, but no spectral data were given. The 366-nm line of the mercury lamp was used as the excitation light source, and therefore some compounds are said to be nonfluorescent, probably only because of the lack of absorption at this wavelength. Later, Concilius reported the R_f values and visual fluorescence color of 15 natural coumarins and furocoumarins, and the effect of NH_3 and $AlCl_3$ thereon (424). Their work is significant, when rapid estimations on the presence or absence of certain coumarins are to be made.

7-Alkoxycoumarins generally have a purple fluorescence, whereas 7-hydroxy and 5.7-dihydroxycoumarins tend to fluoresce blue. The fluorescence of 6- and 7-hydroxycoumarins is intensified with sodium carbonate or ammonia, when excited at 366 nm. Solutions in sulfuric acid fluoresce intensely in daylight. 7-Hydroxycoumarins develop a green fluorescence in strongly alkaline solutions.

Mattoo (425) and Wheelock (426) presented the first spectral data and relative intensity measurements. It became clear that an electron-donating substituent in position 7 intensifies fluorescence most efficiently and shifts its maximum most widely into the visible. An electron-accepting substituent in position 3 has the same beneficial effect.

Coumarin itself is said to be nonfluorescent in fluid solution (427), but to emit weakly (ϕ_f below 0.004, λ_{max} 348 nm) in an ethanol glass at 77 K. Addition of monovalent cations, which complex at the carbonyl group, results in a dramatic increase in intensity (428). Introduction of a hydroxy group makes all coumarins become fluorescent at room temperature, except perhaps 8-hydroxycoumarin (420,422). Table 3.13 summarizes the special properties of various oxycoumarins and, roughly, the conditions under which the spectra were obtained.

Fedorin and Georgievski (429) have studied several oxycoumarins with respect to the influence of solvent and position of the substituent on the intensity and the location of the emission band. A comparison of the fluorescence maxima in dry methanol solution shows that both 3- and 4-hydroxycoumarin fluoresce in the UV (430,431,433) while the 7-, 5-, and 6-isomers exhibit visible fluorescence (429–443). In aprotic solvents the emission maxima of alkoxy derivatives are similar to those of the hydroxy derivatives, but 4-methoxycoumarin- being nonfluorescent (430)- and 5-methoxycoumarin (432) (whose large shift in the emission maximum as compared with the hydroxy derivative has so far not been interpreted) are notable exceptions. In protic solvents, but particularly in water, the spectra of 7-hydroxy- and 7-alkoxycoumarins differ largely.

Striking differences are found in fluorescence quantum yields (ϕ_f). Thus 3-hydroxycoumarin has a ϕ_f value of only 0.006 at pH 6.44, which drops to zero at high pH (433). Similarly, 4-hydroxycoumarin is only weakly emissive. 7-Hydroxy-4-methylcoumarin ("4-MU"), on the other hand, has ϕ_f of 0.70 in both alkaline and acidic solution (434), and 5,7-dimethoxycoumarin ("citropten") has ϕ_f 0.65 (427). In more than 50% sulfuric acid all hydroxycoumarins fluoresce blue from their protonated forms.

The fundamental difference between coumarins and flavones lies first in the fact that coumarins appear to fluoresce most intensely in alkaline solutions, wherein flavones emit only weakly. Second, coumarins, being α-pyrones, do not fluoresce intensely in ~30% sulfuric acid, whereas flavones do.

The effect of pH on the fluorescence properties of various 7-hydroxycoumarins, most of which are not natural products but efficient laser dyes, has been reported (436–441). Because of the changes in intensity with pH and their favorable pK_a values, umbelliferones have been recommended as pH indicators for the near neutral pH range (434,443). Esters, ethers, and glycosides of umbelliferones are used as fluorogenic enzyme substrates (444).

The fluorescence intensity of umbelliferones is greatly enhanced by introduction of electron-delocalizing substituents such as phenyl in position 3 (421,426,436,439). The plant estrogen coumestrol, which may be considered as a rigidized 3-phenylumbelliferone, possesses fluorescence quantum yields as high as 0.64 (with decay time 3.3 ns) in ethanol at room temperature (445). 4-Phenylumbelliferone, a model for related natural coumarins, shows quite unusual fluorescences in aqueous solutions and undergoes a reversible photochemical ring-opening reaction during the lifetime of the excited singlet state (446).

Fluorescence decay times from 3 to 20 ns have been measured for umbelliferone (7-hydroxycoumarin) and its 4-methyl derivative (4-MU) (447–450). 4-MU has received particular attention as a broadband tunable laser dye with remarkable excited state properties (447–453). For instance, the fluorescence spectra of 4-MU are completely different if taken either 0–6 ns or 18–33 ns after pulsed excitation (449).

Several attempts have been made to compute the fluorescence properties of coumarins by molecular orbital methods. Thus, favorable agreement was found between the EHMO-calculated and experimental fluorescence maxima of the 4-MU photoautomer (454). Fabian (453) applied semiempirical SCF methods to compute the absorption and fluorescence maxima of coumarin and some of its hydroxy derivatives. The largest difference between calculated and experimental wavenumbers of the maxima is 1100 cm^{-1}. Similar results were obtained for hydroxycoumarins by application of the PPP or the pars-orbital method (456).

The native fluorescence of natural coumarins has greatly facilitated their identification and quantification in various natural products such as grapefruit peel and expressed lime oil (457,458), mandarin, lemon, and bergamot oils (458–461), suntan cosmetics (1397), *Hydrangea* species and related plants (462a), clover. (*Melilotus albus*) (462b), and pharmaceutical preparations (463,464).

3.7.3. Furocoumarins

Psoralens are naturally occurring furocoumarins with photosensitizing, skin-erythema-inducing, and carcinogenic properties (466). Table 3.13 gives the excitation and emission maxima of the natural furocoumarins. One notes a particular bathochromic effect of an 8-alkoxy group on the emission maxima and large differences between the room temperature and 77 K spectra.

Among the two main groups of furocoumarins the aflatoxins are highly fluorescent, while psoralens show more intense phosphorescence than fluorescence. Quantum yields have been reported for psoralen (0.019), 8-methoxypsoralen (0.013), and 5-hydroxy- and 5-methoxypsoralen (0.019 and 0.023, respectively) in ethanol at 77 K (465). The values for angelicine (isopsoralene) and the structurally related 5,7-dimethoxycoumarin are distinctly higher (0.33 and 0.65, respectively).

The fluorescent state of psoralene appears to be of the π,π^* type, with an n,π^* state lying slightly higher. At 25°C in ethanol the lifetime is 1.8 ns (466,468). The choice of solvent has been shown to affect the intersystem crossing yields and hence ϕ_f of psoralene, coumarin, furochromene, and difurobenzene derivatives (469,470). 8-Methoxypsoralene (8-MOP), which is used in the treatment of psoriasis, has short lifetimes and low quantum yields in solution (468). Fluorescence intensity decreases with decreasing solvent polarity, and the maxima shift to shorter wavelengths (469). Its fluorescence is quenched by weak acids and shifted to longer wavelengths by strong acids (471).

Analytical methods for psoralene assays are mainly based on combinations of chromatographic methods with fluorescence detection (461,472,473). Recently, an 8-MOP metabolite was detected in human interstitial fluid, whose spectral characteristics (excitation maximum 390 nm, emission maximum 470 nm) differ from 8-MOP (474).

Aflatoxins, the second large group of furocoumarins, are metabolites of *Aspergillus flavus* and are the most potent hepatocarcinogens known. Because of their occurrence in food, their fluorescence properties have been studied thoroughly with particular emphasis on analytical applications. The fluorescence of aflatoxins B_1, B_2, G_1, and G_2 in the solid state (475) as well as in solutions of various organic solvents has been described (432,477–479). Uwaito et al. (480) presented the emission maxima of aflatoxin B_3 and of the palmotoxins B_0 and G_0 in chloroform.

The fluorescence intensities of the aflatoxins are not subject to oxygen quenching and show a good relationship with concentration in calibration (linearity from ~0.2 to 600 μM). Using laser fluorimetry a detection limit of 100 pg was found for aflatoxin B_1, with linear analytical graphs extending from 10^{-6} to 10^{-13} M (481,482). Many procedures for the analysis of aflatoxins using chromatographic separation techniques have been published. The interested reader is referred to the compilation of fluorimetric methods in the biennial review in *Analytical Chemistry* (483,484).

Rubratoxin, another fungal metabolite, can be detected fluorimetrically over a wide concentration range (485), but less sensitively than aflatoxins, since its quantum yield is only 0.06, and the lifetime is 211 ps.

3.7.4. Flavonoids

Flavanones, flavones, and flavonols together form the largest group of naturally occurring oxygen ring compounds. In a chemical sense they are 4H-benzopyran derivatives, with a phenyl group in positions 2 or 3. They bear up to seven hydroxy groups, which may also be methylated or linked to a sugar residue.

The first systematic observations on their fluorescences were made by Indian scientists (409–411). Kühn and Löw (486) pointed out that polyhydroxyflavones with a 5-hydroxy group may be distinguished from their isomers having a 5-methoxy group by observing their fluorescence in acetic acid anhydride solution, only the latter being fluorescent. Addition of B_2O_3 makes the 5-hydroxyflavones become greenish-yellow fluorescent.

Later, Harborne (487) and Beckmann and Geiger (488) reported qualitatively the fluorescence colors of several flavonoids on TLC plates, while Spiegl (489) described R_f values and visual colors of flavonoids and phenolic carboxylic acids on cellulose plates before and after treatment with $AlCl_3$ and Neu's reagent. These early observations allow rapid estimations on the presence or absence of frequently encountered flavonoids to be made. Another rapid and sensitive test for flavonols is based on the appearance of a greenish-yellow fluorescence after treatment of TLC spots with 1% aqueous ammonia (490). Fifteen flavonols were examined.

Hayashi et al. (491) found a poor correlation between flavone carbonyl stretching frequencies and fluorescence emission. A Russian group presented the fluorescence spectra of flavonol (492) and other flavones (493) and correlated ground- with excited-state basicities (494). Kopach et al. (495) recorded the fluorescence properties of galangin, kaempferol, morin, quercetin, and rutin in methanol and in the solid state. Fluorescences at room temperature are sometimes weak, but increase markedly at 77 K. The colors are yellow, as is typical for flavonol derivatives. Unlike other reports, which state many flavonols to show two fluorescence bands (*vide infra*), this work gives only one fluorescence maximum for methanol solutions at room temperature, but two for temperatures of 77 K.

The most systematic investigation on the fluorescence maxima of 100 flavonoids on cellulose TLC plates and in solution was performed by Geiger et al. (496,497), who also studied the effect thereon of various acidic, basic, and complexing shift reagents. In a recent paper (498) 6-, 8-, and 2'-hydroxyflavones, which formerly had received only little attention, were specifically investigated. Some of the data are presented in Table 3.13 at the end of this section, while the effects of shift reagents on the fluorescence maxima of model oxyflavones are given in Table 3.11.

The results of our measurements (499) on all isomeric monooxyflavones (except for the photolabile 3-methoxyflavone) are summarized in Table 3.12, together with some data from the literature. From the above-mentioned studies the following generalizations on the fluorescence properties of flavonoids can be made:

1. 3-, 6-, 7-, and 4'-Hydroxyflavones fluoresce strongly, 2'-, 3'-, and 8-hydroxyflavones fluoresce weakly, 5-hydroxyflavones are practically nonfluorescent.*

2. Flavones exhibit exceptionally large Stokes' shifts (6.8×10^3 to 1.0×10^4 cm^{-1}).

3. The fluorescence of flavonols (3-hydroxyflavones) is green, whereas others fluoresce blue.

4. 5-Hydroxyflavones become increasingly fluorescent, albeit weakly, with an increasing number of oxygen functions being present in the molecule. A 3-hydroxy group has a particular beneficial effect.

---

*5-Hydroxyflavones begin to fluoresce greenish when the TLC spot is sprayed with a methanol solution of diphenylboric acid β-aminoethylester (496).

5. Many flavonols exhibit two fluorescence bands: a violet or blue and a green or yellow one.
6. Relatively, the fluorescence intensities are highest in strong sulfuric acid solution, weaker in neutral solution, and low (except for the 3- and 7-hydroxy compounds) in alkaline solution.
7. Fluorescence is higher in polar than in apolar solvents.
8. Fluorescence intensities are distinctly higher for methoxyflavones than for the respective hydroxyflavones.

The unusually large Stokes' shift of flavonols has attracted considerable attention. Thus, the fluorescence of the most simple flavonol (3-hydroxyflavone) has been investigated as a function of solvent (500), temperature (500,501), viscosity (500,501), pressure (501), pH of the solution (502), and with respect to deuteration effects (500,502,503), excitation energy dependence of ϕ_f (504), and to kinetics of the excited-state proton transfer processes in the nanosecond time range (418,503). It now appears accepted that flavonol, during the lifetime of the excited state, forms an intramolecular photoautomer that gives rise to a green fluorescence in addition to the "normal" violet fluorescence. The risetime of the green (phototautomer) emission is consistent with the decay time of the violet fluorescence from nontautomerized flavonol (504,418). (See Table 3.12)

The lack of fluorescence from 5-hydroxyflavones (and of 5-hydroxychromone and -isoflavone) may be due to a diabatic proton transfer process in the S_1 state along the hydrogen bond between the 5-hydroxy and the 4-keto group. This assumption is supported by the fact that 5-methoxyflavone as well as 5-hydroxyflavone metal complexes are fluorescent.

7-Hydroxyflavone has been shown to undergo photodissociation and to fluoresce from its anion even in weakly acidic solutions (505).

The ability of flavones to form fluorescent complexes was recognized very early. Morin is probably the most popular reagent for aluminum(III) ions, but complexes may also be formed with various other flavonols and with 5-hydroxy- and 3',4'-dihydroxyflavones. The emission from these chelates has served for the determination or identification of flavones (506–509), metal ions, and nonmetals like boron (510–512). The native fluorescence of flavonoids has similarly been used to identify them in plant products after chromatographic separation (491,513).

3.7.5. Miscellaneous Oxygen Ring Compounds

Isoflavonoids are of limited taxonomic distribution, but are of interest in chemotaxonomy and as estrogens. The few reports in the literature (416,514) and our unpublished results indicate that

Table 3.12. Longwave Absorption (Excitation) Wavelengths of the Isomeric Oxyflavones, Together with Quantum Yields (θ_f) and Measured Decay Times (τ) in Methanol Solution at 23°C

Flavone	λ_{max}^{abs} (nm)	λ_{max}^{flu} (nm)	θ_f (water)	τ (ns)
3-Hydroxy	306, 344	405, 532[a]	0.005[b,c]	2.34[f,g,h]
5-Methoxy	289, 319	459	—	3.2
6-Hydroxy	302	455	—	—
6-Methoxy	302	444	—	3.3
7-Hydroxy	308	419, 543[a]	0.004[d]	—
7-Methoxy	309	404	0.04[e]	1.4
8-Hydroxy	295 (sh)	471	—	—
8-Methoxy	264, 296 (sh)	475	—	4.9
2'-Hydroxy	306	427	—	—
2'-Methoxy	287	428	—	3.6
3'-Hydroxy	296	428	—	—
3'-Methoxy	293	438	—	6.7
4'-Hydroxy	325	421	0.02	—
4'-Methoxy	317	412	0.95	2.0

[a] Dual emission.
[b] For the 530-nm emission. The shortwave component decays with a lifetime of 1.4 ns.
[c] Reference 501 gives 0.024 for glycerol and 0.008 for a 1:1 glycerol–isobutanol mixture as solvents.
[d] 0.42 in 60% sulfuric acid.
[e] 0.52 in 60% sulfuric acid.
[f] Reference 501 gives 3.08 ns for a glycerol solution.
[g] Reference 503 gives 4.58 ns for the tautomer in methylcyclohexane.
[h] Reference 504 gives 2.0 ns for the gas phase at 316 K.

1. The fluorescence properties of isoflavones resemble more closely those of chromones than those of flavones.
2. 6-, 7-, and 4'-Hydroxyisoflavones are the most fluorescent, but 5-hydroxyisoflavones are nonfluorescent.
3. Fluorescence intensity is higher in polar than in apolar solvents and highest in strong sulfuric acid.
4. The fluorescence intensities of methoxyisoflavones are higher than those of the corresponding hydroxyisoflavones.

Table 3.11 gives the fluorescence maxima of selected isoflavones and the effect of shift reagents thereon. Depending on structural factors, hydroxyisoflavones (like hydroxychromones and hydroxyflavones) form intensely fluorescing borate and aluminum(III) complexes, which may serve for elucidation of their structure (407).

The fluorescences of the tetronic acid derivatives atromentic, pulvinic, and xerocomic acid in methanol and in hydrogen carbonate solution were reported, all of which maximize beyond 470 nm (515). When irradiated, new and shortwave bands of higher intensity are produced.

Patulin, an antimicrobial and carcinogenic furopyrone, has been assayed fluorimetrically via a TLC separation procedure (516). Dehydroacetic acid is nonfluorescent, but forms a fluorescent complex with boric acid (517). Triacetic acid lactone (6-methyl-4-hydroxy-2-pyrone) is weakly fluorescent, when excited at 280 nm in methanol, with an emission maximum at 325 nm (407).

Xanthone does not fluoresce in neutral aqueous or organic solution, but shows fluorescence in sulfuric acid at room temperature (ϕ_f 0.072) (518). Three reports on the xanthone fluorescence maxima are in agreement (Table 3.13). Hydroxy- and methoxy-substituted xanthones are also reported to fluoresce in strong sulfuric acid, with Stokes' shifts ranging from 3500 to 7000 cm^{-1}. However, no precise spectral data are given (518). Recently, the fluorescence properties of 3-hydroxyxanthone, for which an unusual excited-state species was detected, and of its methyl ether, were studied in detail (519). Leprocybin, a yellow-greenish fluorescing pyranoxanthone, was isolated from a fungus (520).

Ellagic acid, one of the most widely distributed natural yellow coloring matter, fluoresces rather weakly in methanol at room temperature (λ_{max} 412 nm), but intensely in the presence of aluminum(III) or borate (see Table 3.13).

3.7.6. Some Unusual Excited-State Proton Transfer Reactions

As mentioned briefly, the dissociation constants (or pK_a values) of phenols may change dramatically upon photoexcitation. While this has no effect on the spectra of solutions in weak dielectric solvents, it has important consequences on the spectra in aqueous solvents. Since all biological solutions are aqueous, the elucidation of the excited-state behavior of natural products is important to understand their properties when exposed to sunlight.

It has been found that a variety of plant products containing a hydroxy group in conjugation with a keto group (via two or three conjugated C–C double bonds) at no pH value is able to fluoresce from its neutral form, as they do in ether or methanol (Fig. 3.4). This means that in the excited (fluorescent) state they do not have the same structures as they do in the ground state. Many of these compounds (with pK_a values from 7 to 9) fluoresce from their anion form in both alkaline and weakly acidic solution up to pH 3. In more acidic solution fluorescence is from a tautomeric

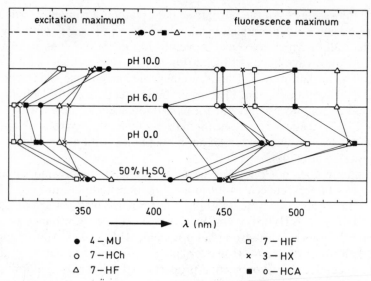

Fig. 3.4. Excitation and fluorescence maxima of 4-methylumbelliferone (7-hydroxy-4-methylcoumarin, 4-MU), 7-hydroxy-2-methylchromone (7-HCh), 7-hydroxyflavone (7-HF), 7-hydroxyisoflavone (7-HIF), 3-hydroxyxanthone (3-HX) and orthohydroxycinnamic acid (*o*-HCA) in aqueous solutions of various acidity. The upper (dotted) line gives the fluorescence maxima in dry methanol solution. Obviously, the spectral shifts in absorption are not paralleled by the shifts in fluorescence.

form and from the protonated forms (in strong mineral acid). Figure 3.4 shows the excitation and fluorescence maxima of six typical compounds at various acidities. Evidently, the shifts in excitation (or absorption) do not parallel the shifts in emission. Thus, at pH 6 anion fluorescence as a result of photodissociation is the only emission, despite the fact that the species directly excited is the neutral molecule (except, partially, orthohydroxy cinnamic acid). On going to pH ~0.0, the excitation maximum remains unchanged, but the fluorescence maximum is shifted to much longer wavelengths.

The longwave-emitting species have been considered to be phototautomers, formed through proton loss at the phenolic hydroxy group and by proton gain at the keto group. Scheme 4 shows the structures of a typical ground-state species (7-hydroxyflavone) and its longwave fluorescing phototautomer, to whose structure the zwitterionic form may contribute significantly. A proposed (505,519) exciplex structure is also shown.

Aside from 4-methylumbelliferone (4-MU) (438,447–454), phototautomerizations and dissociations have been detected for 4,7-dihydroxy-

Table 3.13. Excitation Wavelengths and Fluorescence Emission Maxima (nm) of Natural Oxygen Ring Compounds (see also Tables 11 and 12)

Compound	λ_{ex} (nm)	λ_f (nm)	Experimental Conditions[a]	Ref.
		2H-chromenes		
3-Methyl	290	340	Ethanol	406
2,2,3-Trimethyl-8-methoxy	290	360	Ethanol	406
Precocene I	315	354	Methanol	407
Precocene II	321	374	Methanol	407
		4H-chromones		
Unsubstituted	336	419	60% H_2SO_4	413
2-Methyl	334	409	60% H_2SO_4	413
3-Methyl	346	421	60% H_2SO_4	413
3-Hydroxy	ca. 320	490	Methylpentane	418
	ca. 320	455	Me-pentane, 77 K	418
7-Hydroxy	300	479	Water, pH 4.5	416
	338	471	Water, pH 10.5	416
7-Hydroxy-2-methyl	302	379	Benzene	413
	359	426	60% H_2SO_4	413
	336	445	Water, pH > 10	413
	310	460	Water, pH 3	413
	332	460	Water, pH 10.5	416
	295	485	1 N H_2SO_4	413
		Coumarins		
Unsubstituted	320	435	H_2SO_4–acetic acid	521
	322	440	Sulfuric acid, H_0 −8.92	430
3-Hydroxy	316	372	Methanol	433
	319	369	Benzene	433
	327	380	Water, pH 7.6	433
	—	390–400	Water, pH 1	522
	—	480	Water, pH 14	522
4-Hydroxy	315	374	Water, pH 6 to 11	430
	—	357	Water, pH <4	430
	—	387	H_2SO_4, H_0 −8	430
	300	357	Methanol	431
	317	335	Ethanol, 77 K	465

Table 3.13. (*continued*)

Compound	λ_{ex} (nm)	λ_f (nm)	Experimental Conditions[a]	Ref.
4-Hydroxy-3-phenyl	348	398	Methanol	407
5-Hydroxy	320	400	Water, pH 2	432
5-Methoxy	318	475	Water, pH 2	432
6-Hydroxy	341	431	Methanol	407
	413	580	Water, pH 10	407
6-Hydroxy-4-methyl	—	427	Ethanol	426
6-Methoxy-4-methyl	—	418	Ethanol	426
7-Hydroxy	343	450	Water, pH 7	523
(umbelliferone)	333	392	Methanol	431
	330	440	Methanol (aqueous?)	524
	—	441	95% ethanol	426
	—	385	Chloroform	429b
	—	386	100% ethanol	525
	—	375	Ethanol, 77 K	465
	376	454	Water, alkaline	436
	330	480	Water, pH <2	437
7-Methoxy	320	415	Water	523
(herniarin)	—	385	Ethanol	426
7-Hydroxy-4-methyl	322	476	Water, pH 1	430
(4-MU)	345	412	H_2SO_4, $H_0 - 8.9$	430
	—	450	Water, pH's above 2	430
	—	380	Chloroform	430
	320	442	95% ethanol	426
	367	449	Alkaline water	436
	320	380	100% ethanol	525
7-Methoxy-4-methyl	—	392	Ethanol, 95%	426
	—	374	100% ethanol	525
	—	381	Water	525
7-Hydroxy-4-acetic acid	305	388, 462	Ethanol–water	449
7-Hydroxy-3-phenyl	350	430	Methanol	439
	386	462	Water, pH 9	439
	340	428, 472	Water, pH 4	439
	342	433, 465	0.1 N HCl	439
7-Hydroxy-4-phenyl	342	428	Methanol	446
	378	515	Water, pH 10	446
	330	517	Water, pH 5	446
	330	524	Water, pH 1	446

Table 3.13. (*continued*)

Compound	λ_{ex} (nm)	λ_f (nm)	Experimental Conditions[a]	Ref.
4,7-Dihydroxy	326	352	Methanol	431
	315	418	Water, pH 4	431
	325	391	Water, pH 10	431
	315	465	1 N H_2SO_4	431
5,7-Dimethoxy	335	420	Ethanol	459
(limettin)	330	378	Ethanol, 77 K	465
6,7-Dihydroxy	350	440	Methanol	429b, 524
(esculetin)	—	420	Chloroform	429b
	350	455	0.1 N H_2SO_4	524
	—	460	Water, pH 1.5	527
	—	475	Water, pH 7–11	527
	390	465	Water, pH 10	524
6-Hydroxy-7-methoxy	350	440	Methanol	524
(isoscopoletin)	350	430	0.1 N H_2SO_4	524
	370	460	Water, pH 10	524
7-Hydroxy-6-methoxy	350	420	Methanol	429b, 524
(scopoletin)	350	430	0.1 N H_2SO_4	524
	390	460	Water, pH 10	524
	—	410	Chloroform	429b
6,7-Dimethoxy	350	420	Methanol	429b, 524
(scoparone)	—	410	Chloroform	429b
	350	425	0.1 N H_2SO_4	524
	350	430	Water	429b, 524
6-Hydroxy-7-glucoside	—	416	Chloroform	429b
(cichorin)	—	440	Methanol	429b
	—	456	Water	429b
7-Hydroxy-6-glucoside	—	400	Chloroform	429b
	—	416	Methanol	429b
	—	458	Water	429b
6,7-Methylenedioxy	350	415	Methanol	524
(ayapine)	365	430	Water, pH 1	524
7,8-Dihydroxy	—	414	Chloroform	429b
(daphnetin)	—	434	Methanol	429b
	—	454	Water	429b
7-Hydroxy-6,8-dimethoxy (isofraxidine)	330	433	Methanol	407
Drimatol[b]	357	435	Methanol	407
Isodrimatol[b]	361	436	Methanol	407

Table 3.13. (*continued*)

Compound	λ_{ex} (nm)	λ_f (nm)	Experimental Conditions[a]	Ref.
Coumestrol	344	388	Ether	445
	348	399	Ethanol	445
	342	435	Water, pH 5	445
	388	519	Water, pH 9	445
	354	481	40% H_2SO_4	445
		Psoralens		
Unsubstituted	330	409	Ethanol, 77 K	465
	360	440	Water–ethanol, 77 K	526
4-Methyl	360	420	Water–ethanol, 77 K	526
5-Hydroxy	345	442	Ethanol, 77 K	465
5-Methoxy (bergapten)	—	427	Ethanol, 77 K	465
	350	460	Ethanol–water, 77 K	526
	320	470	Ethanol	461
	322	450	Silica TLC plate	461
5-Geranoxy	314	474	Ethanol	459
8-Methoxy	350	554	Conc. H_2SO_4	471
	300	455	Chloroform	469
	300	475	Ethanol	469
	300	510	Water	469
	~320	463	Acetonitrile	949
	~320	470	Ethanol	949
	~320	493	Trifluoro-ethanol	949
	345	440	Ethanol, 77 K	465
	360	470	Water–ethanol, 77 K	526
8-Geranoxy	312	460	Ethanol	459
5-Geranoxy-8-methoxy	318	517	Ethanol	459
Isopsoralene (angelicine)	330	401	Ethanol, 77 K	465

Table 3.13. (*continued*)

Compound	λ_{ex} (nm)	λ_f (nm)	Experimental Conditions[a]	Ref.
		Aflatoxins		
B$_1$	363	426	Methanol	482
	365	413	Chloroform	483
	365	425	Methanol	528
	370	435	Water, pH 2	432
B$_2$	365	413	Chloroform	483
	365	425	Methanol	528
B$_3$	365	432	Chloroform	483
G$_1$	363	450	Methanol	482
	365	430	Chloroform	483
	365	450	Methanol	528
G$_2$	365	430	Chloroform	483
	365	450	Methanol	528
Citrinin	365	505	TLC plate	948
Ochratoxin A	340	475	TLC plate	948
Palmotoxin B$_0$	365	410	Chloroform	483
Palmotoxin G$_0$	365	425	Chloroform	483
Rubratoxin B	305	410	Acetonitrile	485
		Flavones[c]		
3-Hydroxy-4′methoxy	313	430, 532 435, 538[d]	methanol	496, 497
5,6-Dimethoxy	313	500	TLC plate	496
5-Methoxy-7-hydroxy	313	540/460[d]	—	497, 498
5,7-Dimethoxy	313	452/460[d]	—	496, 497
6,7-Dihydroxy	313	465	TLC plate	496
6,7-Dimethoxy	313	440	TLC plate	496
7-Hydroxy-2′-methoxy	313	415	TLC plate	498
7,4′-Dihydroxy	313	410/413[d]	—	496, 497
7,4′-Dimethoxy	313	406	Methanol	497
7,8-Dimethoxy	313	487/485[d]	—	496, 497
2′,3′-Dimethoxy	313	475	TLC plate	496
3′,4′-Dihydroxy	313	452/466[d]	—	496, 497
3′,4′-Dimethoxy	313	440/460[d]	—	496, 497
7-Hydroxy-3′, 4′-dimethoxy	313	430/440[d]	—	496–498
Apigenine-trimethyl ether	365	425/430[d]	—	496, 497
Azaleatin	365	540/480, 535[d]	—	496, 497
Azaleatin-3-galactoside	365	445	TLC plate	496
Datiscetin	313	510	TLC plate	498

235

Table 3.13. (*continued*)

Compound	λ_{ex} (nm)	λ_f (nm)	Experimental Conditions[a]	Ref.
Fisetin	365	535	TLC plate	496
Fisetin-7,3',4'-trimethyl ether	313	496	TLC plate	496
Fisetin-tetramethyl ether	313	442	TLC plate	496
Galangin	365	530/530[d]	—	496, 497
	362	535, 585 461, 535[e]	At 77 K[f]	495
Galangin-trimethyl ether	365	447	TLC plate	496
Gossypetin-8-glucoside	365	547	TLC plate	498
6-Hydroxykaempferol	365	537	TLC plate	498
Kaempferid	365	525/530[d]	—	496, 497
Kaempferol	365	525/540[d]	—	496, 497
	366	562, 585/ 450, 538[e]	77 K	495
Kaempferol-7-glucoside	365	535	TLC plate	496
Kaempferol-5, 4'-dimethyl ether	365	535	TLC plate	496
Luteolin-5-methyl ether	313	430/434[d]	—	496, 497
Luteolin-5-glucoside	313	440	TLC plate	496
Luteolin-5,3'-dimethyl ether	313	430/434[d]	—	496, 497
Luteolin-tetramethyl ether	313	427/432[d]	—	496, 497
6-Methoxykaempferol	365	540	TLC plate	498
Morin	365	500/505, 550[d]	—	496, 497
	388	575/556[e]	—	495
	388	571/529, 578[e]	77 K	495
Myricetin	365	540/494, 550[d]	—	496, 497
Myricetin-3'-methyl ether	365	540	TLC plate	496
Patuletin	365	532	TLC plate	498
Patuletin-7-glucosid	365	545	TLC plate	498
Quercetin	372	581/543[e]	—	495
	372	571/455, 538[e]	At 77 K	495
	365	530/480, 533[d]	—	496, 497

Table 3.13. (*continued*)

Compound	λ_{ex} (nm)	λ_f (nm)	Experimental Conditions[a]	Ref.
Quercetin-7-glucoside	365	540	TLC plate	496
Quercetin-5,3'-dimethyl ether	365	540	TLC plate	496
Quercetin-5,3', 4'-trimethyl ether	365	540	TLC plate	496
Quercetin-pentamethyl ether	365	445	TLC plate	496
Rhamnazin	365	522	TLC plate	496
Rhamnetin	365	525	TLC plate	496
Robinetin	365	540	TLC plate	496
Rutin	362	592/467, 535[e]	At 77 K[f]	495
Tamarixetin	365	532	TLC plate	496
Tamarixetin-7-rhamnoglucoside	365	535	TLC plate	496
Isoflavones[g]				
Unsubstituted	307	473	60% H_2SO_4	514
7-Hydroxy	302	489	Water, pH 4.5	416
	339	479	Water, pH 10.5	416
	316, 327	405, 479	Methanol	514
	351	448	60% H_2SO_4	514
	303	507	10% H_2SO_4	514
Daidzein	300	440	Methanol	1317
Formonenetin (biochanin B)	302	464	Methanol	1317
Genistein-(prunetol) afrormosin	319	463	Methanol	1317
Kakkatin	325	416	Methanol	1317
Glycitein	319	467	Methanol	1317
Texasin	326	427	Methanol	1317
Miscellaneous				
Atromentic acid	480	550	Methanol[h]	515
Cannabinols	~280	318	Ethanol	408
Ellagic acid	340	410	Methanol[i]	407
3-Hydroxyxanthone	352	453	95% H_2SO_4	519
	343, 365	465	Water, pH 7	519
	318	510	Methanol–H_2SO_4	519
3-Methoxyxanthone	349	455	95% H_2SO_4	519
	312, 337	391	Water, pH 7	519

Table 3.13. (*continued*)

Compound	λ_{ex} (nm)	λ_f (nm)	Experimental Conditions[a]	Ref.
Pulvinic acid	450	535	Methanol[h]	515
Triacetic acid lactone	280	325	Methanol	407
Xanthone	410	456	H_2SO_4	518
	330	450	H_2SO_4	529
	390	458[j]	H_2SO_4–acetic acid	521

[a] At room temperature, unless otherwise specified.

[b] A 6,7,8-trialkoxycoumarin.

[c] The following flavonoids are practically nonfluorescent: most 5-hydroxyflavones without a 3-hydroxy group; acacetin and its 7-methyl ether, apigenin, its 7- and 4'-methyl ethers, chrysin and its 7-methyl ether, chrysoeriol and its 7-glucoside, datiscetin-3-rutinosid, 7,8-dihydroxy- and 2',3'-dihydroxyflavon, diosmetin and its 7-rutinoside, galangin-3-methyl ether, genkwanin, gossypetin and its 3-sophoroside-8-glucoside, herbacetin-3-sophoroside-8-glucoside, isoëtin, kaempferol-3-glucoside, kaempferol-3-glucoside-7-rhamnoside, luteolin and its 7-, 3'-, and 4'-ethers and glucosides, myricetin-3-galactoside, patuletin-3-glucoside, primetin, quercetin-3-glucoside and its 7-rhamnoside, quercetin-3,7,3',4'-tetramethyl ether, scutellarein and its 7-glucoside, tricetin and its 7-glucoside, tricin.

[d] The first figure(s) give the maximum of a spot on a TLC cellulose plate, the second the maximum of a methanol solution.

[e] The first figure(s) gives the maximum of the compound in the solid state, the second the maximum of a methanol solution.

[f] No fluorescence at room temperature.

[g] See also Table 11.

[h] Fluorescence maximum in 0.01 M sodium bicarbonate solution also given.

[i] Fluorescence is weak in methanol solution, but greatly enhanced and shifted to 502 nm by added $AlCl_3$, and even more intensified by borax (λ_{max} 468 nm).

[j] 425 nm at 77 K.

coumarin (431), 7-hydroxy-2-methylchromone (413), 7-hydroxyflavone (505), 7-hydroxyisoflavone (514), 3-hydroxyxanthone (519), and orthohydroxycinnamic acid (407).

The λ_{max} versus pH plot is typical for these compounds and their derivatives. No such excited-state processes are found for the respective methyl ethers and for the isomeric hydroxy compounds. The assumption of a rapid excited-state proton transfer can now be used to interpret the unusual solvent dependence of the fluorescence of certain hydroxylated natural products. 4-Methylumbelliferone in nonaqueous solvents such as chloroform or dry alcohols fluoresces in a manner similar to its methyl ether, with maxima below 400 nm (Table 3.13). A few percent of admixed water suffices, however, to shift the emission maximum to around 460

ground state species   phototautomer

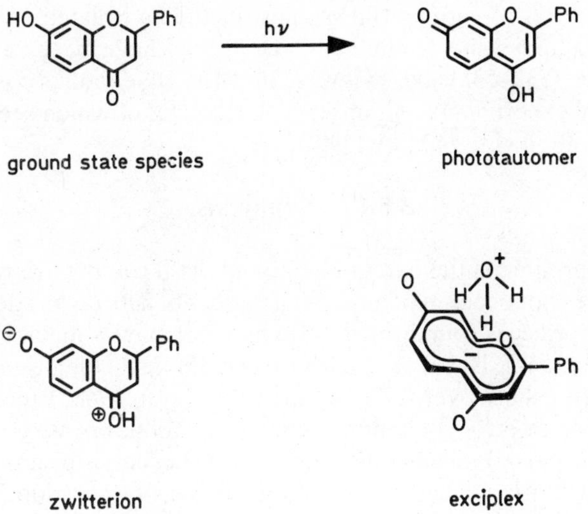

zwitterion   exciplex

Scheme 4. Structures of a typical ground-state tautomeric species and its possible excited state forms (7-hydroxyflavone).

nm (anion fluorescence), while a few percent of admixed acid shifts the maximum to even 480 nm (tautomer emission).

Similar effects were reported for the other above-mentioned compounds. Since the changes in the absorption and excitation spectra in each instance remain small, it can be concluded that the dissociation constants and the tautomeric equilibria in the photoexcited state do not correspond to the respective equilibria in the ground state.

3.8. DYES AND PIGMENTS

Living nature is an admirable manufacturer of beautiful colors, many of which are fluorescent. Among the natural dyes the chlorophylls (Chl's) are considered the most typical representatives. Since Chl's play a vital role in the natural photosynthetic process by virtue of their ability to bring about a charge separation, their excited-state properties have been investigated in much more detail than those of any other pigment. The properties of natural dyes can, however, be quite different in the aggregated state, as found frequently in nature, and in the dissolved state, as usually investigated. Because of space limitation this chapter will be confined to fundamental fluorescence properties of Chl's and other pigments in liquid or solid solution.

The properties of various Chl species, including isolated protein complexes, thylakoids, and whole cells and leaves have been reviewed in several books (532–535) and reviews (536–541). Older books cover a manifold of early experiments and observations, some of which seem worthy of reinvestigation (1,2,542,543,1399).

3.8.1. Porphyrins

Free-base porphine is the parent compound of all cyclic tetrapyrrole pigments. It does not occur in nature, but a variety of substituted derivatives, called porphyrins, is found, many of which complex a metal ion (metalloporphyrins) (544). Dhèrè (1,2), Schwartz (545), and Udenfriend (4) have written articles that cover the early literature up to 1968. Free-base porphyrins fluoresce strongly in homogeneous solution, but very little when colloidally dispersed in aqueous media. Fluorescence is best observed in dilute hydrochloric acid, but high concentrations of chloride (546), bromide, iodide, and tartrate (547) can cause fluorescence quenching.

In general, the fluorescence of porphyrins and metalloporphyrins is more efficiently excited at around 410 nm (the so-called Soret band) than at the longest-wave transition (the blue-green band). The shortwave excitability of the chlorophylls is of great advantage for practical purposes, since excitation into the longwave band, which is extremely close to the first emission maximum, frequently gives rise to interferences from Rayleigh and Raman scatter.

The fluorescence properties of important porphyrins are compiled at the end of this section in Tables 3.15 and 3.16.

The porphyrin fluorescence intensities are dependent on the pH value of the solution. Fink and Hörburger (547,548) have developed a method to recognize different porphyrins from plots of intensity versus pH, which are characteristic for even structurally similar compounds such as copro-, uro-, isouro-, etio-, meso-, hemato-, deutero-, and protoporphyrin. The acid–base properties of protoporphyrin and its derivatives are different in homogeneous and micellar solution (549). Gradyushko et al. (550) have studied the effect of acid strength on octaethylporphyrins. Fluorescence of the acidic forms is weaker than of the neutral molecules and is markedly quenched by bromide and iodide. Austin and Gouterman report the emission data for species resulting from addition of acids, bases, and metal salts to etioporphyrin in various organic solvents (551).

For the parent compound, porphine, two nondegenerate tautomers with NH–NH axes at right angles have been shown to exist at liquid helium temperature in an octane Shpolskii matrix (552–554). The two tautomers can be transformed into each other by photoexcitation. Solovev et al. (555,556) have demonstrated that the fluorescence polarization of solid porphyrin solutions at 77 K suffers changes with time of irradiation due to phototautomerizations. The photoinduced orientational anisotropy of the molecules causes similar changes in the quasi-line spectra obtained from snowy low-temperature matrices (557–563). Selective laser excitation of porphines, the four isomeric etioporphyrins (564), and of chlorins together with their deuterated derivatives (557) allowed a detailed interpretation of the vibronic structure of the emission spectra. As an alternative to low-temperature studies the spectroscopy in ultracold supersonic expansions gives detailed informations on rovibronic states in the gas phase (565).

Solid solutions of the meso-, deutero-, and coproporphyrin free bases in 1 N hydrochloric acid have been studied under selective laser excitation. At 2 K the fluorescence bands consist of well-resolved, narrow zero-phonon lines and phonon side bands (566). In an octane matrix at 5 K the chemically related pheophorbides respond to laser irradiation to give quasi-line fluorescence spectra with lines similar to but not identical with those of Chl (567).

High-resolution spectra and polarization data have also been published for octaethylporphyrin (568,569), protoporphyrin (557), and chlorin (569,570).

Porphyrins such as etioporphyrin tend to form dimers and higher aggregates with increasing concentration (553,571). Fluorescence quantum yields are between 0.04 and 0.1 in benzene or toluene (572,573) and increase by around 10% after deuteration of the two readily exchangeable protons (573). Gradyushko et al. (574) have shown that for porphyins and chlorins the sum of the quantum yields for fluorescence and intersystem crossing is approximately unity, indicating that internal conversion plays a negligible role.

The lifetimes of free-base porphyrins are in the order of a few nanoseconds (Table 3.16). The experimental data are in large disagreement with values obtained by mathematical methods from absorption and fluorescence data (572,575) with values of more than 100 ns, for example, for mesoporphyrin.

Hematoporphyrin is largely used in photodynamic systems for its high photosensitizing activity and for its efficiency in inducing the regression of different kinds of tumors. Its fluorescence is pH dependent (576) with relatively high quantum yields in acidic solution (0.15 in acetic acid) (577,578). Emission decays double exponentially with τ_1 of 3.9 ns (6% contribution) and τ_2 of 15.4 ns (94% contribution) in pH 7 solution. The two decays are ascribed to monomeric and dimeric forms (578). Struc-

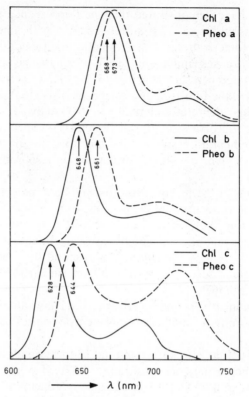

Fig. 3.5. Fluorescence spectra of chlorophylls *a*, *b*, and *c* and of the respective pheophytins in ether. (Redrawn from ref. 580.)

turally related porphyrins have also been shown to act as photosensitizers, for example, in the photo-oxidation of Trp (579).

The chlorins (e.g., the pheophytins) play an important role as energy acceptors in bacterial photosynthesis. Their fluorescence spectra are long-wave shifted (as compared with porphyrins, see Figure 3.5) to around 770 nm (580), with quantum yields around 0.1 and lifetimes from 2–6 ns. The fluorescence is polarized (583) and strongly quenched by methyl viologen, *m*-dinitrobenzene (581), polyenes (584), fucoxanthin (584), and at concentration above 0.7 mM, possibly because of the formation of pheophytin dimers (585).

At 4.2 K in an octane matrix, pheophytin-a shows two broad emission bands with maxima around 676 and 708 nm, superimposed by sharp vibronic bands (586,562,694). The pheophytin spectra are pH dependent

(587), but are scarcely influenced by the nature of the acid counterion. Protopheophytins and proto-Chl's show delayed fluorescence (588). Dvornikov et al. measured the fluorescence spectra (589), quantum yields (589), and fluorescence polarization spectra (590) of the pheophytins and calculated the probabilities of radiative and nonradiative transitions.

The chlorins, like the porphyrins, form two tautomeric species whose fluorescence excitation spectra at 4.2 K are different (591) and whose single-site-excited emission spectra were recorded (592). Interestingly, good spectral resolution is not obtained with broadband excitation.

The protopheophytins do not differ strongly from the pheophytins (Table 3.15) (593–596), but are distinctly changed as compared with the spectra of the proto-Chl's. The excitation spectra are much more complex than the respective spectra of proto-Chl's because of the loss of symmetry.

Bonellin, a chlorin found in marine animals, was characterized by Matthews et al. (597). The corrin ring in vitamin B_{12} is nonfluorescent, but metal-free corrins show emission (598). From polarization ratios in the excitation spectrum the directions of the four lowest energy transitions were estimated.

While chlorins (dihydroporphyrins) have been investigated thoroughly, there is only limited work available on further hydrogenated porphyrins. Sevchenko et al. (599) presented the quasi-line spectra of porphyrins, dihydroporphyrins, and two tetrahydroporphyrins and compared them with related substances.

The following studies on the fluorescences of porphyrins may also be of interest. Mauzerall concluded from changes in the spectra that porphins like uroporphyrins form molecular complexes with large heterocyclic molecules (adenine, caffeine, nicotinamide, and pyridinium salts) (600). Again, methyl viologen was found to act as an efficient quencher. Gouterman (601) studied amorphous thin films of octaethylporphine and found characteristic differences in comparison to crystalline films. The fluorescence properties are also highly dependent on the way the films are prepared. Copro-, uro-, and protoporphyrins (602) as well as chlorins (603) bind to albumin and other proteins, but thereby suffer almost no changes in their spectral maxima and quantum yields.

Because of the importance of precise assays of free-base porphyrins in clinical, biochemical, and forensic sciences a manifold of fluorimetric methods has been published (484). Schwartz et al. (545), Udenfriend (604), and Guilbault (605) reviewed the methods in the older literature. In addition to former work a variety of methods for the determination of porphyrins has been worked out (606–610), some in combination with chromatographic separation steps (611). Some assays are based on the fluorescence of the porphyrin–zinc complexes (612,613,892), which is measured in a hematofluorimeter.

3.8.2. Metalloporphyrins (Chlorophylls)

Porphyrins form strong complexes with metal ions among which chlorophyll (Chl) is probably the best known. Fluorescence of the complexes is generally observed only from closed-shell metal chelates (613–615), whereas transition metal complexes are not fluorescent but are phosphorescent. Some metalloporphyrins with no unpaired electrons (Mg^{2+}, Zn^{2+}, etc.) are not fluorescent in anhydrous nonpolar solvents, but give fluorescent compounds in the presence of polar ligands.

As a result of the continuous efforts to elucidate the mechanism of the photosynthetic process and owing to the need for artificial photosynthesis, a huge literature on metalloporphyrins has accumulated. Research seems to have become even more active and sophisticated in the last few years.

Early studies on the fluorescence spectra of Chl's have been presented by Dhéré (1,2) and Zscheile and Harris (616); later studies were carried out by French et al. (580) and by Livingston et al. (617). Except for Dhéré's work the reported fluorescence spectra are corrected and—in comparison to more recent data—very precise. Figure 3.5 shows the fluorescence spectra of the three Chl's and the corresponding pheophytins in ether solution.

As a general rule it can be said that introduction of a metal atom shifts the fluorescence maxima to shorter wavelengths, whereas hydrogenation of the pyrrole ring (to give chlorins) has the opposite effect. The two bands observed in the spectra of Chl's (and pheophytins) are now attributed to emissions from monomeric Chl (657 nm for Chl a in moist alkane solvent) and from oligomeric species (760 nm for Chl a). A weak fluorescence of Chl a with maximum at around 720 nm is attributed to the hydrated dimer (618). Supplementary bands in the luminescence spectrum of Chl were identified to result from a proto-Chl impurity (619). Chl c, unlike Chl's a and b, exhibit delayed fluorescence when adsorbed onto filter paper (620). The fluorescence properties of Chl's are summarized in Table 3.14. Data for other metalloporphyrins are given in Tables 3.15 and 3.16.

A variety of newer publications, which cannot be discussed in detail, describe the emissive properties of Chl's, including concentration (616,621,622), temperature (616,622), and solvent effects (616,623–625). Quantum yields are very low in dry apolar solvents (624–626), but are strongly increased by traces of water and other hydrogen bonding molecules (590,617,625–628). Similar effects of polar additives have been found for the zinc complexes of pheophytin and protoporphyrin (629).

Some of the quantum yields reported in the literature (Table 3.14) seem to have been determined in nondried solvents. For benzene solutions values around 0.1 have been given (158,572,580,589). The quantum yield of the polymeric form of Chl a dihydrate is two orders of magnitude lower than that of the monomeric form (628). The in $vivo$ quantum yield of Chl is around 0.03 (533). Likewise, the lifetimes of Chl's in solution are on the order of a few nanoseconds (618,630–632) (Table 3.14), but much shorter in vivo (633), except for delayed fluorescence (634).

Fluorescence methods have been used together with others to study the aggregation of Chl's in solution (553,621,623,635–637). Chl a dimers have been identified by their fluorescence bands at 625 and 690 nm, together with higher aggregates fluorescing at around 740 nm (618,630,632). Dimerizations have also been observed for several porphyrins and their zinc complexes at low temperatures (621,638,639).

Krooyman et al. (640) have been able to characterize the Chl–water complexes by their fluorescence spectra and by fluorescence-detected magnetic resonance. Fong et al. (641) report the corrected fluorescence excitation spectrum of Chl dimer hydrate, which is probably the photoactive center in photosystem I. The modifications of Chl in nonpolar solvents by polar substances such as pyridine, ether, or acetone have been explained in terms of the existence of equilibria between dimers, monosolvates, and polar substances (642,643).

The fluorescence of Chl's in solution is quenched with increasing concentration (585,631,644,646) as a function of the intermolecular distance (621,645,647–648) partially because of light reabsorptions (631). Quenching is consistent with a Förster-like mechanism (649) and, possibly, proceeds as well via aggregated Chl species (585,650). Indeed, an analysis of the fluorescence spectra of self-assembled species of Chl a with ethanol or a second Chl molecule indicated that the pathways for dissipation of energy in these multicomponent systems are rather complex (651,652). When Chl's are coadsorbed on a hydrophobic surface, electronic energy is transferred from Chl b to Chl a (653).

Various external quenchers such as oxygen (652,654), iodide,* p-benzoquinones (656,657), anthraquinones (658), manganese complexes (659), β-carotene (660), fucoxanthine (584), nitrobenzenes (661,662), methyl-

---

* The fluorescence of Chl a is not quenched by nitrite, but by bromide, iodide, and sulfite with increasing efficiency. 0.1 M solutions of these ions reduce the fluorescence of Chl a (0.8 μM in water containing 5% acetone) to 88, 69, and 47%, respectively, of its initial intensity (655). It is also efficiently quenched by oxygen when adsorbed on kieselgel (1406).

Table 3.14. Photophysical Data of Chlorophylls and Bacteriochlorophylls in Solution at Room Temperature Unless Otherwise Specified.

Solvent	λ_{ex} (nm)	λ_f (nm)	ϕ_f	Fluorescence Lifetime (ns)[a]	Ref.
			Chlorophyll a		
Acetone	429[b]	664[b]			757
Ether	428[b]	661[b]	0.33[d]		757
Hexane	430	663[b]			624
Ether	430	665			624
Chloroform	430	673			624
Benzene	430	671[b]			624
Methanol	430	675[b]	0.32[d]		624
Pyridine	430	677[b]	0.35[d]		624
Acetone	430	669[b]			624
Ether			0.32,[e] 0.24[f]	5.1	671
Methanol			0.23,[e] 0.24[f]	6.9	671
Benzene				7.8	671
Ethanol[c]		676, 728[c]	0.23[e]	4.4	622, 758
Ethanol 77 K		682, 740[c]			622
EPA, 77 K			0.51	8.6	759
Ether	428	666		6.1,[h,i] 5.9[h,k]	652
Pyridine		680, 730		6.4,[h,i] 7.2[h,k]	651, 652
Ethanol, 95%				5.6,[i] 5.2	670, 758
Methanol				5.4,[h,i] 5.9[h]	631, 652
Toluene	337	673, 719		2.0 (670 nm), 2.9 (725 nm)	651
Toluene, −72°C	337	679, 730		2.1 (670 nm), 4.0 (725 nm)	651
90% acetone	435	667			737
Ether	405, 436	668, 722			580
Lecithin liposomes				1.4 and 0.12[g]	673
			Chlorophyll b		
Hexane	Soret	645			624
Ether	Soret	645	0.16[l]		624
			0.12[e]		
Chloroform	Soret	653			624

246

Solvent	Soret	Red band	ϕ	τ (ns)	Emission
Benzene	Soret	651			624
Methanol	Soret	661	0.084,[l] 0.10[e]		624
Acetone	Soret	652			624
Pyridine	Soret	662			624
Ether	405, 436	648, 708	0.17	3.9	580, 671
Benzene		651		6.3	671
Methanol	470		0.06[m]	5.9	671
90% acetone	Soret	648			737
Ether	Soret	652			580
Acetone	Soret				580
		Chlorophyll c			
Ether	405, 436	629, 688			580
90% acetone	435	635			737
Filter paper	430	605			620
		Chlorophyll d			
Ether	405, 436	699, 745			580
		Bacteriochlorophyll a			
Ethanol	360, 782	795	0.14	2.50	731
Ether	360, 770	782	0.19	3.21	731
Pyridine	360, 780	793	0.26	3.53	731
		Protochlorophyll a			
Ether	405, 436	631			580
Acetone	405, 436	626			580
Ether	432	630, 685			595
Ether	622	630			596
		Vinylprotochlorophyll a (bacterial proto-Chl a)			
Ether	437	630, 685			596

[a] The lifetime of Chl's in vivo are much shorter (633).
[b] Corrected.
[c] At low concentrations; more concentrated solutions have longer-wavelength maxima (581).
[d] From ref. 580.
[e] From ref. 158.
[f] From ref. 625.
[g] Double-exponential decay.
[h] Corrected for reabsorption.
[i] In deoxygenated solvent.
[k] In aerated solvent.
[l] From ref. 760.
[m] From ref. 625.

Table 3.15. Excitation Wavelengths and Fluorescence Maxima of Pyrrole Pigments and Selected Photoreceptors

Compound	λ_{ex} (nm)	λ_f (nm)	Experimental Conditions[a]	Ref.
Allophycocyanine	405, 436	663	Water	580
Bacteriopheophytin	350	764	Ether	731
	525	761	Ether	580
Bilirubin	435	500	Phosphoric acid	765
	440	530	Water–detergent	766
	<450	525	Organic glass, 77 K	767
	390, 455	493, 517 (sh)	Ethanol, 77 K	768
Bilirubin–HSA	460	525	Water, pH 8.4	865
	455	520	Water, pH 7.4	770
	435	530	Water, pH 8	771
Bilirubin–BSA	460	540	Water, pH 7.4	770
Bilirubin–sheep albumin	460	528	Water, pH 7.4	770
Biliverdin	380	725	Ethanol, 77 K	773
	705	735	Methanol–H$^+$, 77 K	774
		710, 770[b]	Ethanol, 77 K	775
Biliverdinato–zinc				
α-form	705	752	Methanol	776
β-form	695	740	Methanol	776
γ-form	695	735	Methanol	776
δ-form	700	742	Methanol	776
Coproporphyrin	Soret	620, 651, 672, 689	Pyridine	1
	400	603, 650	Dilute HCl	609

Compound			Solvent/conditions	Ref.
Coproporphyrin[b]	401	623	Chloroform or TLC plate	611
Coproporphyrin	405	595, 618, 652[d]	2 N HCl	545
	405	597, 623, 654,[d] 671, (691)	Dioxane	545
Coproporphinato–zinc	404	595, 650	0.1 N HCl	613
	405	580, 620	0.1 N HCl	613
Etioporphyrin I				
monomer	400	624	Petrol ether, 77 K	553
dimer	420	629, 626	Petrol ether, 77 K	553
	546	628, 695	Benzene	572
	Soret	624, 651, 671,[d] 689	Pyridine	1
Etioporphinato–Mg	400	625	Chloroform	673
	~420	582	EPA, 77 K	614b
	Soret	589	Chloroform	615
	~410	589	Petrol ether, 77 K	553
Etioporphinato–Zn	546	597, 625	Benzene	572
Hematoporphyrin	~410	597, 619, 653	2 N HCl	580
	Soret	625, 654, 673,[d] 691	Pyridine	1
Isophorcabilin	405, 544	610, 675	Water, pH 7.3	577
	605	793	Water, pH 10	774
	690	730	Water, pH 4	774
	690	730	Methanol, 179 K	774
Mesoporphyrin	~410	617	EPA, 77 K	614
	Soret	621, 651, 670,[d] 689	Pyridine	1
Mesoporphyrin[b]	546	625, 695	Benzene	572
Mesoporphinato-Mg	Soret	586	Chloroform	615

Table 3.15. (*continued*)

Compound	λ_{ex} (nm)	λ_f (nm)	Experimental Conditions[a]	Ref.
Mesoporphinato-Zn	Soret	572	EPA, 77 K	614a
Pheophytin a	390	667	90% acetone	737
	415	673	Ether	580
Pheophytin b	435	657	90% acetone	737
	430	661	Ether	580
Pheophytin c	435	657	90% acetone	737
	430	648	Ether	580
Pheophytin d	440	701	Ether	580
Phorcabilin	555	715	Water, pH 10	774
	635	705	Water, pH 4	774
	645	695	Methanol, 179 K	774
Phycocyanin	406, 435	637	Water	580
Phycoerythrin	406, 435	578	Water	580
Phytochrome (P_r)	670	690	Water	761
	380, 620	680	Glycerol/water, 14 K	777
	380	674	Glycerol/water, 14 K	773
Protopheophytin a	638	653	Ether	596
	418, 638	653, 713	Ether	595
	406, 435	648, 712	Acid acetone	580
Protoporphyrin	Soret	634, 663, 683, 701[d]	Pyridine	1
	405	605, 632, 669, 704[d]	Dioxane	545
	410	504, 626, 655	2 N HCl	545
	400	608, 658	Acidified serum	609

	Soret			
Protoporphyrin IX	410	624, 687	Dilute blood, pH 6.5	613
	415	630^b	On filter paper	764
	400	638	Chloroform	763
	415	635	Tetrahydrofuran	778
	410	632^b	Chloroform or TLC plate	611
Protoporphyrin–H$^+$	415, 582	634	Water–detergent	762
Protoporphyrin–H$_2^{2+}$	408, 610	612	HCl–detergent	762
	413, 602	606	HCl–detergent	762
Protoporphinato–Mg	428	600^b	On filter paper	764, 738
Protoporphinato–Zn	424	587, 650	Dilute blood, pH 6.5	613
	420	587	On filter paper	738, 764
	424	594	Dilute blood	610
	415, 578	589	Water–detergent	762
	420	587^b	On filter paper	738, 764
Pyrromethenone	374–413	440–475	Various solvents	779
Tripyrrin	575	600	Methanol–BF$_3$	779
Uroporphyrin	405	596, 619, 653	2 N HCl^c	545
	400	605, 658	Acidified serum	609
	355	615, 680	Water, pH 9.5	600
	Soret	633, 656, 674, 692	Pyridine	1
Uroporphyrin^b	405	627	CHCl$_3$ or TLC plate	611
Uroporphyrin III	410	583, 650	1 M HCl	613

^{a} At room temperature unless otherwise specified.
^{b} Dimethyl ester.
^{c} Maxima are at 600, 626, 658, 676, and 695 nm in dioxane.
^{d} Most intense band in italics.

Table 3.16. Fluorescence Quantum Yields and Experimental Decay Times of Pyrrole Pigments, except the Chl's

Compound	ϕ_f	τ (ns)	Experimental Conditions[a]	Ref.
Bacteriopheophytin	0.13	2.54	Ether	731
Bonellin	0.07	6.3	Ethanol	597
Bilirubin–HSA		0.074–0.072	Water, pH 7.4	777
Bilirubin–BSA		0.032–0.037	Water, pH 7.4	777
Bilirubin–HSA	0.001		Water, pH 7.4	778
	0.5		Water, 77 K	778
Biliverdin	0.0001	—[g]	Ethanol	775
	0.01		Methanol, 77 K	774
	0.0005		Ethanol, 77 K	775
	0.00027		Ethanol–H[+]	775
	0.026		Ethanol/H[+], 77 K	775
Coproporphyrin I	0.054		0.1 N HCl	613
	0.07	6.3	Ethanol	613
	0.20		Ethanol, 77 K	613
Etioporphyrin	0.09		Benzene	572
Etioporphinato–Zn	0.04		Benzene	572
Hematoporphyrin	0.15		Acetic acid	577
	0.11–0.32[b]	3.94, 15.4[c]	Water, pH 7.4	578
Hematoporphyrin IX	0.005	0.2	Water, pH 7.2	779
Isophorcabilin	0.04		Water, pH 10	774
	0.024		Water, pH 4	774
Mesoporphyrin	0.08		Benzene	572
	0.12		EPA, 77 K	759
Octaethylporphin	0.08[e]		Toluene	780
Octaethyl- porphinato–Zn	0.039		Toluene	780
Pheophytin a				
monomer		6.3	Toluene–pyridine	582
dimer		0.11	Ethanol	582
Pheophytin	0.18		Benzene	158
Protoporphyrin IX		0.23	Water, pH 7.2	779
	0.018	1.14	Ethanol–H[+]	779
	0.07		Dilute serum	613
		23	Benzene	778
Protoporphinato–Zn	0.047		Benzene	778
Pyrromethenone	~0.0005	~0.006[f]	Methanol[d]	779
Tetraphenylporphin	0.13	15.7	Benzene[d]	781
Tripyrrin	0.13		Methanol–BF$_3$	779
Uroporphyrin III	0.051		1 M HCl	613

[a] Room temperature unless otherwise given.
[b] Highly concentration dependent.
[c] Double exponential.
[d] Various other solvents also studied.
[e] Deuterated derivatives also investigated.
[f] In chloroform (860).
[g] ϕ_f and lifetime data are given in Tables 1–3 in ref. 1349.

viologen (658) and other dyestuffs (663) diminish both quantum yields and lifetimes.

The quenching of Chl a by benzoquinone is solvent viscosity dependent, with strong contributions from static quenching (664). In a study on the dynamic quenching of mesoporphyrin and its zinc complex by benzoquinone, dinitrobenzene, phenylhydrazine, and methyliodide (665) the deactivation constants for the singlet and triplet states as well as the quantum yields for the formation of the two states were calculated. The singlet states are much more efficiently quenched.

Related studies were done on Chl a, etioporphyrin I, its magnesium and zinc complexes (666,667), on porphins and chlorins (668) with methyl iodide as an external quencher. The rate constants for the intramolecular transitions between singlet and triplet states were calculated from quantum yields and lifetimes (668), and the effects of molecular structure on the radiationless deactivation processes were discussed (669).

The fluorescence decay times are also dependent on solvent (630–633,670,671), concentration (631), analytical wavelengths (672), and other factors (646,673). They are, however, scarcely influenced by the temperature down to 4.2 K (674). In each instance the decay behavior is complex because of the presence of monomeric and oligomeric species, the appearance of prompt and delayed fluorescence, and the biphasic decay of the oligomers (628,653,654).

The degree of polarization of porphyrins is independent of excitation wavelength within the longest-wave absorption band. Values of ~0.18 are characteristic of a planar oscillator (675). The polarization degree increases substantially as the porphyrin symmetry is lowered from D_{4h} to D_{2h} or C_{2v} as in porphins and chlorins, respectively. The $S_0 \rightarrow S_2$ transition is polarized perpendicular to the $S_0 \rightarrow S_1$ transition, yielding a negative degree of fluorescence polarization. For Chl a the polarization between 520 and 560 nm is negative, but at any other wavelength between 360 and 660 nm it is positive (676). Thus, the transition moments of the absorption bands at 665 and 533 nm have directions approximately perpendicular to each other. Similar results were obtained by Goedheer (677) in a study on the polarized fluorescences of Chl a, bacteriochlorophyll (BChl), pheophytin, and bacteriopheophytin. The data on porphyrins were confirmed by Gouterman and Stryer (678). Chl a and Chl b have similar polarization spectra between 400 and 680 nm (646).

Further polarization studies were presented by Kaplanova and Vacek (679) on Chl a in low-temperature matrices, by Bär et al. (680) on related porphyrins and by Dvornikov et al. (589) in a study on Chl's and pheophytins.

The spectra of Chl *a* become more resolved at low temperatures. There are many lines (1–10 cm^{-1} wide), referred to as quasi-lines, seen in the low-temperature spectra in octane and ether (681,682), hexane, pyridine, toluene, ether, and butanol (681–686), in other alcohols (679), in octane (586,679,687), in methyltetrahydrofuran (688,689), dodecane, and hexadecane (687). The fine structure in the spectra of Chl *a*, proto-Chl *a*, and their pheophytins in ether at 4.2 K has been described and interpreted by Bykovskaya et al. (682).

Further low-temperature studies have been performed with Chl-related metalloporphyrins (553,690,691). At extremely low temperatures the spectral resolution of Chl (692), porphine (562,693), chlorin (591,694), and proto-Chl (695) can be increased enormously by means of photochemical hole burning. This method allows the measurement of the homogeneous linewidth of the O–O band of the $S_1 \leftarrow S_0$ transitions. Interestingly, the resolution of the fluorescence bands resulting from white (broadband) excitation is much worse than that in the spectra obtained by narrowband ("single-site") laser excitation.

The low-temperature spectra become increasingly complex with loss of the molecule's symmetry. In a study on the effect of isocyclic substitution of the pyrrole ring on the luminescence and polarization of porphyrins large differences between symmetrically and unsymmetrically substituted porphyrins (e.g., deoxyphylloerythrins and phyllerythrins) were discovered (596).

Two-photon laser excitation of Chl *a* (697) and other metalloporphyrins (698) gives rise to $S_4 \rightarrow S_0$ and $S_3 \rightarrow S_0$ fluorescences in the blue part of the spectrum. Chl a solutions are laser active (651,700–702). The maximum of stimulated emission is around 808 nm.

Mainly to mimic the natural pigment and its efficient energy transfer, the following model systems containing Chl have been investigated by fluorimetry: microcrystals (703), monolayers (647,648,704–709), bilayers (710), detergent micelles (711–713), lipid vesicles (645,656,673,714), hydrophobic–hydrophilic interfaces (715), solid solutions of cholesterol (716), lecithin (710,717), polystyrene (718–721), polyvinyl alcohol (722) and polymethyl methacrylate (723), nematic liquid crystals (724,725), covalently bound to polymers (726) and after binding to albumin (603).

The optical properties of proto-Chl pigments are different from the respective Chl's (Table 3.14). Fluorescence data (580,593,595,727) and polarization spectra (593) for various solvents, in films and in adsorbed form (593,727) have been reported. The main red and blue bands of the proto-Chl pigments show fluorescence polarizations parallel to each other (595), in contrast with the case of the Chl's where they are perpendicular.

Observation of a decrease in the fluorescence intensity of protochlorophyllide *a* in a nonpolar solvent has been used by Seliskar and Ke (728), together with spectral and light-scattering changes, as an indication of aggregate formation. The aggregates seem not to be uniform (729). Resonant hole burning into the fluorescence spectra of proto-Chl's at temperatures between 1.8 and 30 K gives highly resolved and narrow O–O transition lines (695).

The bacteriochlorophylls have received considerable attention by virtue of their ability to achieve charge separation in photosynthesis with the help of one single photosystem. The fluorescences of BChl's and their respective pheophytins are at distinctly longer wavelengths (750–850 nm) than the chlorophylls and chlorins. BChl's tend to aggregate in mixed dioxane–water solutions. The aggregates show no measurable fluorescence, but addition of more than 1% of proto-Chl, Chl *a*, or vinylproto-Chl donors to the BChl acceptor component leads to the appearance of fluorescence in BChl (730).

Connolly et al. (731) studied the effects of solvent on the fluorescence maxima, lifetimes, and quantum yields of BChl *a* and, partially, bacteriopheophytin *a*. The decay times (2.2–3.6 ns) were found to be longer when the central metal atom is hexacoordinated than when pentacoordinated, and shorter when the macrocycle is hydrogen bonded.

Site selection spectroscopy of BChl at 4.2 K allowed the estimation of the angle of reorientation of the molecule upon photoexcitation (592).

The properties of Chl's in photosynthetic material differ distinctly from the properties in homogeneous solution. The emission spectra show at least three bands, but their location changes with the species. The short-wave bands are usually associated with photosystem II, whereas the long-wave band, which is preferentially excited via Chl *a*, is attributed to photosystem I. Much research is performed at low temperatures where quantum yields and lifetimes are 4–10 times higher. In intact photosynthetic units the fluorescence of Chl is quenched through electron transfer mainly by the primary quinone acceptor of photosystem II and by the secondary plastoquinone acceptor. Detailed descriptions of the state of the art in understanding the natural photosynthetic process are given in excellent reviews (533–541,732,733). The photophysics of in vivo BChl also remains under active research (734,735).

Chl's and related metalloporphins (612,613) can be detected easily by fluorimetry, even as mixtures (736,737) and in the presence of pheophytins by selective monochromatic excitation and emission wavelengths (737). Chl *c* can be selectively assayed by virtue of its delayed fluorescence when adsorbed on filter paper (738). In the determination of Chl's in marine and freshwater phytoplankton two basic procedures are commonly used, namely, measurement *in vivo* and after extraction with 80% acetone. In *in vivo* measurements the sample is usually taken directly for fluorimetry, but it may be first concentrated or treated with the fluorescence enhancer [3-(3,4-dichlorophenyl)-1,1-dimethylurea (or a related herbicide) (739,740)]. In the extractive procedures (741–743) the phytoplankton is isolated, ground, and extracted.

Other interesting features of fluorimetric analysis involve remote sensing of Chl's in seawater (744) and spectroscopic studies of interstellar grains (745). With the use of low-temperature fluorimetry and Weber's method of matrix calculations, the pools of Chl *a* and Chl *b* in higher plants have been shown to consist of four different chromophores that could be characterized by their different excitation and fluorescence maxima (747). The unesterified proto-Chl's separated from etiolated plants are heterogeneous and contain Chl's and Chl precursors with different excitation and emission maxima (747). Various forms of natural Chl's have been identified by a combination analysis including curve shapes, spectral relation studies, and derivative spectroscopy (748). The following Chl's (absorption and fluorescence maxima in brackets) were identified:

Chl (680/691),   Chl (691/697),   Chl (698/706),   Chl (702/715),
Chl (709/726),   Chl (715/738),   Chl (720–724/750–755).

Cytochromes, hemoerythrins (749), hemoglobin (750), and hemocyanin (which in fact is not a tetrapyrrole) exhibit fluorescences from the tryptophan part of the molecule, whereas the pyrrole chromophore is nonfluorescent. Fluorescence (and phosphorescence) data, together with energy transfer studies between tryptophan or flavin mononucleotide (FMN) and the tetrapyrrole allowed estimations on the relative orientations between cytochrome *b* and flavin (751). For cytochrome *c*, an iron-containing protein, an extremely weak fluorescence from the tetrapyrrole ring has been detected (752).

The metal-free cytochrome *c*, on the other hand, is strongly fluorescent with quantum yield 0.1 and a decay time of 10 ns at room temperature in water solution (752,753). In a glassy matrix at 4.2 K highly resolved emission spectra can be observed that were used to investigate the excited-state vibrational structure of the electron transport heme protein (754). Similarly, removal of the iron atom from hemoglobin to give desoxyferrohemoglobin, makes the pyrrole pigment fluorescent (755).

c-Type cytochromes can be detected on acrylamide gels by virtue of their intrinsic fluorescence with great sensitivity (756).

3.8.3. Photoreceptors

Most chromophores of photoreceptors are of the bilindione (noncyclic tetrapyrrole) type, which is known to be fluorescent, at least in the protein-bound form. Among these, phytochrome seems to be the most widespread photoreceptor, and its function and properties have been reviewed (782,783).

The two forms of phytochrome (P_f and P_{fr}), which govern the photomorphogenetic process, contain the biliverdin chromophore, but have quite different photophysical properties. The "large" forms exhibit only weak fluorescence at room temperature (773,784), while the "small" forms have both higher quantum yields ($\sim 10^{-2}$) and longer lifetimes (785). The fluorescence of biliverdin is red; a second band in the blue found at low temperatures (773) is most likely the result of photodecomposition (786).

The fluorescence excitation polarization of "small" phytochrome in buffer–glycerol at 77 K is high (0.1) at the main absorption band (785), and the quantum yields are dependent on excitation wavelength and temperature (773,787,788). Hermann et al. (784) found the quantum yields of large phytochrome at 293 K in Tris buffer of pH 7.8 to be around 0.0015. At 77 K in 10% glycerol–water the yields are 0.015 at excitation wavelength 380 nm and 0.039 at 630 nm, respectively (787).

The primary photoprocesses of intact phytochrome excited with red and blue light at 77 K are complex because of the existence of the two interconvertible forms P_r and P_{fr} (789). The primary photoprocess in large phytochrome proceeds within the lifetime scale of a few picoseconds, but its nature is yet to be established. One likely mechanism is based on intrapyrrolic proton transfer (790).

Energy transfer occurs from flavin mononucleotide to P_r at 77 K, apparently within flavin–phytochrome complexes (789), and from tryptophan residues to the fluorophore in large phytochrome, but not in the small one (786).

Tegmo-Larsson et al. (791) studied biliverdin dimethylester in lipid vesicles. The pigment is incorporated into the lipophilic part of the liposome bilayer, wherein fluorescence is more efficient than in ethanol solution.

The phorcabilins and isophorcabilins, which may be considered as rigidified biliverdins, are about 10 times more fluorescent than the latter (774). Their photophysical properties were described by Pertrier et al. (774,776,792).

The phycobiliproteins form the second large group of photoreceptors, which have attracted considerable attention. They show relatively short

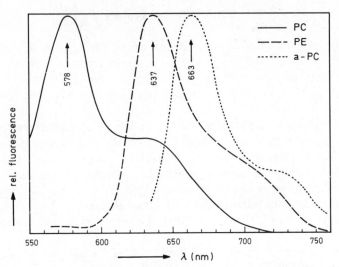

Fig. 3.6. Fluorescence spectra of phycocyanine (PC), phycoerythrine (PE), and allophycocyanine (a-PC) in water. (Redrawn from ref. 580.)

lifetimes, high quantum yields (760,793), and large fluorescence depolarizations in water (794). In contrast to the Chl's, phycobiliproteins exhibit only small Stokes' shifts between their most efficient excitation bands and their fluorescence maxima.

French et al. (580) and Dale and Teale (795,796) characterized allophycocyanins, phycoerythrins, and phycocyanins by their spectra (see Fig. 3.6). It was concluded that energy is transferred from sensitizing to fluorescing chromophores of the same molecule (795,796). Fluorescence spectra of isolated intact phycobilisomes suggested (797) that the transfer occurred first between phycobiliproteins and that light was emitted through allophycocyanine.

Gantt (797) determined lifetimes (2.0–3.2 ns), quantum yields (0.48–0.98), and polarization degrees of phycobilisomes in neutral aqueous solution at room temperature. In a further study it was shown that the mean energy transfer time from photoexcited phycoerythrin to allophycocyanin within the phycobilisome was about 280 ps (798).

The application of time-resolved emission spectroscopy has greatly benefited the understanding of primary photophysical processes of photoreceptors. To resolve the puzzle of effective energy transfer to the reaction centers, investigations have been performed on the complete antenna system in vivo (802–804), on isolated phycobilisomes (798,805), and on phycobiliproteins (806,807).

The decay of the fluorescence of six phycobilisomes from a blue-green alga was exponential at low excitation intensities (808). Lifetimes ranged from 1.55 to 2.66 ns in water of pH 7 at 23°C. Both the risetime of fluorescence and the mean energy transfer time were below 12 ps. The fluorescence from the C-phycoerythrins in the phycobilisomes decayed within 31 ps; the energy transfer time was measured to be 34 ps (808). The times of energy transfer from B-phycoerythrin to R-phycocyanine (300 ± 200 ps), from allophycocyanine to Chl a (500 ± 200 ps) (802,809), and from biliproteins to photosystem I in heterocysts (809) were also reported.

Porter et al. measured the energy transfer rates in intact, isolated phycobilisomes (805) and in the pigments of *Porphyridium cruentum* (804), on a picosecond time scale. The wavelength-resolved fluorescences from the pigment showed contributions from phycoerythrin B, phycocyanine R, allophycocyanine, and Chl due to energy transfer from the primarily excited phycoerythrin.

The results of time-resolved fluorescence spectroscopy are in agreement with a model of transfer of absorbed energy on a picosecond time scale from phycoerythrin or phycocyanine via phycocyanine and allophycocyanine to the reaction centers (810–812).

Fluorescence polarization data and effects of denaturation and renaturation on spectra were used to study energy transfer efficiency and conformations of C-phycocyanins (813).

The fluorescences of phycobilisomes and phycobiliproteins are quenched by glutaraldehyde and benzoquinone (814). Fluorescence is also lost upon denaturation with urea (815).

The following photoreceptors have been investigated in addition to the ones mentioned before: stentorin (816,817) [a protein-bound hypericin, which undergoes photodissociation (818)], phycocyanines from *Chroomonas* species (molecular topography on the basis of fluorescence polarization) (819) and from *Agmenellum quadruplicatum* (spectra of intact and denatured protein complexes) (820), phycoerythrocyanines (fluorescence spectra and their changes with time, effects of buffer concentration) (821) allophycocyanine from *Anabaena cylindrica* (polarization, deconvolution of complex emission spectra, association equilibria) (832,833), allophycocyanine B of *Porphyridium cruentum* (absorption and fluorescence spectra, energy transfer from biliproteins) (824), allophycocyanines I, II, III, and B from *Nostoc sp.* (energy transfer between phycobilisomes, phycocyanines, and thylakoids) (825,826) allophycocyanin II (salt effects on association equilibria) (827), peridinin* (polarization of fluorescence,

---

* Peridinin is a polyene photoreceptor rather than a pyrrole.

molecular topology of the photosynthetic complex with Chl a and protein) (829,830), phycocyanine C from *Gloeocapsa alpicola* (831) and phycoerythrin from *Anacystis nidulans* (831,832), phycobilisomes from Red Sea macroalgae (energy migration) (833), of *Synechoccus* 6301 and a mutant [energy transfer (834), aggregation in the absence and presence of detergents (835)], of *Anabaena variabilis* [composition of complexes, energy transfer (836), energy transfer from Chl a (837)], of various blue-green and red algae [spectra of intact phycobilisomes (838,839), temperature effects on aggregation (838)], of *Microcystis aeruginosa* (reversible photoeffect of dithionite) (840), of phycoerythrins [polarization, lifetimes, and quantum yields (841), energy transfer to phycocyanine (842,843) and chlorophyllin (843)], of phycoerythrines from *Cryptomonas maculata* (spectra of three forms) (844), and of phycocyanines (energy transfer to photosystem I in *Anabaena variabilis*) (845).

3.8.4. Noncyclic Pyrrole Pigments

Metabolic degradation of cyclic tetrapyrroles leads to products with absorption and emission maxima at distinctly shorter wavelengths. Most of them bind strongly to protein to give the so-called biliproteins, some of which were already treated in the photoreceptor section.

Free bilindiones and denatured biliprotein are only weakly fluorescent, but quantum yields up to 1 may be found in undenatured protein complexes. Freshly prepared solutions of biliverdin dimethylester in ethanol fluoresce at room temperature with maxima at 710 and 770 nm and a quantum yield of 0.00011 (846,847). Protonation shifts the maxima to longer wavelengths and decreases ϕ_f. When a neutral solution was kept standing at room temperature, an additional band at 500 nm developed, probably identical with a previously found (777) blue component in the total emission. The apparent impurity has been recognized as a dihydrobiliverdin derivative (847). Interestingly, in earlier studies no fluorescence was observed from biliverdin solutions at room temperature (777,848).

The four different biliverdin IX dimethyl esters have different fluorescence maxima in ambient temperature methanol solutions (776). Their zinc complexes have distinctly stronger fluorescence. Fluorescence is also enhanced in aqueous solutions of pH below the pK_a values of 4.6. Solutions of biliverdin dimethyl ester and its protonated form exhibit solvato- and thermochromism, which were interpreted in terms of variations of the relative populations of conformers (849).

From the polarized fluorescence excitation it was concluded that biliverdin assumes a conformation more similar to that of a nonsymmetric cyclic polyene than that of a linear polyene (850). When biliverdin is incorporated into liposome membranes, certain preferred conformations have been detected together with irreversible changes (791). The conformationally flexible bilindione fluoresced more efficiently in the membranes than in ethanol solution.

Etiobiliverdin IV (851) and its lactim isomer (852) are further examples of weakly fluorescent bilindiones. The fluorescence of etiobiliverdine in freshly prepared methanol solution decays within 68 ± 3 ps (853). For the dimethyl esters of phorcabilin, a pigment related to biliverdin, much stronger fluorescences in both alkaline and acidic solutions have been observed (774,854). The increase in fluorescence intensity is thought to result from the steric geometry of the molecule, which prevents an intramolecular proton transfer as a quenching process. Fluorescence is further increased on lowering the temperature (855). Pyrromethenes and pyrromethenones are frequently investigated as the simplest biliverdin models. Various authors described their emissive properties (775,779,856,857). Pyrromethenones are fluorescent in the solid state (857) but not in solution. They form highly fluorescent zinc and BF_3 complexes (779). Pyrromethenones fluoresce weakly ($\phi_f \sim 10^{-3}$) with maxima between 400 and 450 nm, dependent on structure, solvent, and association degree (779,858). At 77 K the ϕ_f values increase by a factor of around 20 (775,856) at the expense of the photochemical E/Z-isomerization quantum yield (859).

The fluorescence of pyrromethenones in ethanol at room temperature decays single-exponentially within 6 ± 2 ps (860). Unlike the free pyrromethenone its boron trifluoride complex exhibits strong fluorescence with a decay of about 10 ns, possibly due to the loss of a pyrrolic (excited-state quenching) proton (860).

For a fluorescent homolog of pyrromethenone an intramolecular proton transfer from a pyrrol to a pyridine nitrogen atom was invoked to explain the unusually large Stokes' shift (9560 cm^{-1} in hexane) between excitation and emission, and the decrease in ϕ_f in going from hexane to methanol solvent (858).

Tripyrroles (tripyrrins) and open-chain tetrapyrrolones are practically nonfluorescent, but form longwave fluorescing (λ_{max} around 600 nm) complexes with boron trifluoride (779).

Urobilinoids (e.g., stercobilins and urobilins) are also nonfluorescent in solution, but fluorescent in the solid state (861) or after oxidation. This is the basis for their fluorimetric assay (861–863). Alternatively, the zinc (864) or mercury complexes (861) may be used for the quantitation.

Free bilirubin (BR) shows very weak fluorescence in fluid solution at room temperature (766,865–868). ϕ_f is greatly enhanced in organic glasses at 77 K (766–768,869) and in polymer solution (870).

In normal blood plasma BR is essentially completely bound to albumin and in this state exhibits much more intense fluorescence (765,865,871,872). Binding to animal albumins (872) and to rabbit hemopexin (873) was also observed. Fluorescence titration of albumin with BR allowed the determination of the binding constants via Scatchard plots (872). Stopped-flow fluorimetry was used to follow the binding kinetics. The time required for half the BR fluorescence to appear is about 100 ms (770).

BR quenches the fluorescence of tryptophan (770,872–874), and it has been suggested that a charge transfer from Trp residues to BR may be operative (770,872,874). A distance of 2.8 nm between the first BR and the single Trp in HSA was estimated from energy transfer experiments (872).

The spectra of bound BR are dependent on pH in the near neutral range (771,875) and on buffer viscosity (870). Bilirubin in vesicles, in human serum albumin (HSA) and bovine serum albumin, and in organic solvents decays biexponentially with τ_1 of 20–90 ps (99% contribution) and τ_2 of 1–2 ns (876,877). The 525-nm fluorescence of BR in ammoniacal methanol at 77 K consists of two (or more) components with a decay time smaller than 5 ns for the major component (869).

Circular polarized fluorescence spectra also suggest a contribution of more than one species to the measured emission, and different excited-state conformations were invoked (876,877).

Because of the formation of photoproducts, the fluorescence intensity of protein-bound BR is not constant under irradiation (878). The fluorescence quantum yield of HSA-BR exhibits a reciprocal relation with the quantum yield of photoisomerization (876) and decay times of 19.3 ± 3 ps at 22°C (879) were found. At low irradiation intensities the fluorescence signal of BR-HSA solutions first decreases by 10–20% and then increases substantially. It was possible to observe a photoequilibration between Z/Z-BR and it E-isomer (photo-BR) (880) formed via a twisted common intermediate (880,881). The isomerization yield is about 0.002 at room temperature and 0.44 at 77 K (876).

A pigment different from Z/Z-BR was detected by fluorimetry in blood specimens from newborn infants undergoing phototherapy (880). There is, however, still a lot of speculation on the structure of the BR photoproducts (870,876,878). The structure of unbound photobilirubin has recently been established (882).

BR sensitizes the photomodification of biological systems such as DNA, erythrocyte membranes, and sulfur-containing amino acids (883).

Albumin-bound BR can be determined fluorimetrically (765,769,771,884,885), preferentially with the help of a specially designed filter instrument called a hematofluorimeter (886). Other assays have been described for albumin titratable BR in jaundiced neonatals (887) for total and free BR (via a fluorescence-quenching method) (888), for serum bile acids (889,890), and for BR in plasma and serum by the Roth fluorimetric method (891). The fluorescence of BR is said not to interfere with the zinc protoporphyrin detection with the help of a hematofluorimeter, provided excitation is at 410–430 nm and fluorescence is taken at above 580 nm (892).

3.8.5. Quinoid Dyes and Pigments

Unsubstituted quinones are colored, but are usually weakly fluorescent. Introduction of a hydroxy or amino group renders quinones more strongly fluorescent. All naturally occurring quinones, among which the anthraquinones (AQs) have been subjected to the most detailed investigations, bear hydroxy groups. The fluorescences of alizarin (1,2-dihydroxy-AQ), quinizarin (1,4-dihydroxy-AQ), anthrarufin (1,5-dihydroxy-AQ), purpurin (1,2,4-trihydroxy-AQ), and other hydroxy-AQ's in boric acid solution (893,894) and in the gas phase (895,896) have been reported very early. Stokes' shifts up to 90 nm (equal to around 4000 cm^{-1}) between excitation and emission maxima are observed, with fluorescence peaks ranging from 450 to 560 nm. The quantum yield of quinizarin in ambient temperature solutions is around 0.2. Table 3.17 summarizes the data for several naturally occurring quinones.

It has been noted that compounds with fluorescence maxima above 530 nm are not subject to quenching by either oxygen or nitric oxide (895,896). The effects of high pressures of gases on the emission spectra have also been described (896).

In recent years, high-resolution spectra of various natural anthraquinones have been obtained in Shpolskii matrices at 4.2 K. Among these are naphthazarin (898–900), dihydroxy-AQ's (899,900), and tetrahydroxy-AQ's (901–903). From the laser-induced fluorescence emission spectra of quinizarin and its deuterated derivatives at 12 K, in heptane, it was concluded that the geometries of the S_0 and S_1 states are similar.

There is no evidence for an intramolecular proton transfer of quinizarin after photoexcitation (904). Anthrarufin, however, shows a different behavior in showing dual fluorescence at low temperature. A structured band between 490 and 540 nm, arising from the Franck–Condon excited state, and a stronger structureless fluorescence at wavelengths greater than 550 nm, attributed to a phototautomer, was observed (897).

The fluorescence of quinizarin has also been investigated at 2 K by photochemical hole-burning techniques (905,906). It was concluded that

Table 3.17. Fluorescence Excitation Wavelengths and Fluorescence Emission Maxima of 'Naturally Occurring Quinones

Compound	λ_{ex} (nm)	λ_f (nm)	Experimental Conditions[a]	Ref.
Aclacinomycin A	441, 456	565, 517 (sh)	Methanol	916
Adriamycin	458	557 (sh), 588, 633 (sh)	Methanol	927
	495	552	Water, pH 5	933
	550	588	Water, pH 11	933
	580	596	Conc. H_2SO_4	933
Anthrapurpurin	420–460	500, 610	Ethanol–water[c]	909
Anthrarufin	440	580, 610	Ethanol–water	909
	460	560	Gas phase	895
Carminomycin	492	538	Acetonitrile–H^+	943
Cercosporin	495, 574	618, 675 (sh)	Methanol	916
Chrysomycin[b]	394	508	Methanol	916
Chlortetracycline	400	530	Water, pH <7.5	928
	345	430, 530	Water, pH >7.5	928
Chlortetracycline– aluminum complex	365	555	Water, pH 5	930
Daunomycin	490	560	Isoamyl alcohol	912
Daunorubicin	480	522, 532, 544	EPA, 77 K	911
Doxorubicin, see adriamycin				
Elsinochrome A	570	585, 625	Methanol	916
	620	646	Methanol–NaOH	916
Hypericin	355, 550	601, 615, 651	Ethanol	913
	355, 550	596, 611, 644	Ethanol, 77 K	913
	—	595, 610 (sh), 643, 700 (sh)	Ethanol	914
	—	605, 653, 710	Acetone	944
Isocercosporin	575, 484	620, 675 (sh)	Methanol	916
	650	678	Methanol–NaOH	916
Javanicin	543	570, 605	Methanol	916
Lachnanthocarpon	448	628	Methanol	916
Lampyrine	565	600, 635	Ethanol–HCl	945
	565	622, 672	Ethanol–KOH	945
Lanigerine	Broadband	611, 660	Solid state	945
	563, 587	608, 635	Ethanol	945
δ-Naphthocyclinon	405	460, 471, 487	Methanol	915
Oxytetracycline–Al^{3+}	390	460	Water, pH 5	930
Oxypennicilliopsine	560	595, 609, 642	Ethanol	946
	560	606, 621, 652	Pyridine	946
Phleichrome	544, 583	617, 674 (sh)	Methanol	916
	650	676	Methanol–NaOH	916

Table 3.17.   (*continued*)

Compound	λ_{ex} (nm)	λ_f (nm)	Experimental Conditions[a]	Ref.
Purpurin	470	560	Gas phase	896
Quinizarin	475	570	Ethanol–water	909
	480	522, 532, 544, 558	EPA, 77 K	911
	500	524, 537, 548, 560	Heptane, 12 K	897
	475	550	Gas phase	895
	458	530, 562, 600	Methanol	927
Resistomycin	457	557, 614	Methanol	916
Strobinine	Broadband	630	Ethanol	945
Tetracycline–Al^{3+}	465	555	Water, pH 5	930
Tetracycline–Ca^{2+}	405	530	Water	947
Violamycin BI	500	586	Water	934

[a] At room temperature unless otherwise specified.
[b] Actually not a quinone, but structurally related.
[c] Effects of pH changes and salt addition (Ca^{2+}, Mg^{2+}, Sr^{2+}, Zn^{2+}, Mo^{3+}, and Y^{4+}) also investigated.
[d] Fine structures observed and interpreted.

electronic excitation energy is localized in the quasi-aromatic rings containing the intramolecular hydrogen bonds (899). Vibrational studies on quinizarin in its ground and excited singlet states, and the effect of deuteration thereon, have been presented by Smulevich et al. (907,908).

Roman et al. (909) followed the changes in the absorption, excitation, and emission spectra of 1,4- and 1,5-dihydroxy-AQ's and of 1,2,7-trihydroxy-AQ with pH, togethter with the effect of concentration changes. The fluorescence intensities of the anions were found to be rather weak. A similar study was performed on 1,8-dihydroxy-AQ (910).

Capps and Vala (911) measured the absorption and emission spectra of quinizarin in hydrocarbon and alcoholic solvents at 14 K. They found no evidence for aggregation of molecules due to intermolecular hydrogen bounding. Vibrational analyses of the spectra revealed the presence of 3, 8, and 9 singly excitable sites in octane, heptane, and hexane matrices, respectively. The lifetime is 6.5 ns in hexane.

Fluorescence polarization measurements allowed the assignment of the nature of the electronic transitions. The absence of phosphorescence was attributed to the intramolecular hydrogen bonding, which displaces the parent AQ n,π^* states above the π,π^* states, thereby rendering the intersystem crossing pathway inefficient (911).

Quinizarin is the fluorophor of the drug daunorubicin, which contains an amino sugar residue. Their spectral properties are almost identical (911). Both may be detected and determined by their native fluorescences following extraction and TLC separation (912).

Various other natural antibiotics contain the AQ or a related quinoid π-electron system. As long ago as 1939, Dhéré (1) reported the fluorescence properties of hypericin and penicilliopsin at room and liquid nitrogen temperature. Fluorescence colors and, partially, fluorescence maxima of other pigments, such as lanigererine, strobinine (brilliant red), lampyrine, fagopyrine, citroroseine, and oxypenicilliopsine, were also given. As far as spectral data were given they are included in Table 3.17.

Hypericine fluoresces red and the spectral resolution shows (913,914) three peaks at 601, 615, and 650 nm, respectively, in ethanol solution. At 77 K the maxima are slightly shifted to the blue (913). Based on absorption, fluorescence lifetime, and polarization measurements, the fluorophore of *Stentor* photoreceptor has been identified as hypericin (816) linked to a protein. In this form hypericin suffers photodissociation to fluoresce from its anion form, notwithstanding that the neutral molecule is exclusively excited (818). Non-protein-bound hypericin shows no such phenomenon.

Krone and Zeeck (915) reported the fluorescence spectra of several naphthocyclinone derivatives. In methanol the longest-wave absorption bands are at around 410 nm and the fluorescence maxima between 460 and 500 nm. The emissions come from the 2,3-dihydro-1,4-naphthoquinone chromophore of the naphthocyclinones.

Wolfbeis and Fürlinger (916) measured the fluorescence spectra and detection limits of quinoid antibiotics and phototoxic pigments. Fluorescences are moderately intense in methanol solution but can increase after addition of aluminum chloride, provided a chelating hydroxy group is present. Fluorescence can also increase in sulfuric acid, although some compounds seem to decompose to form more strongly fluorescent products. The data for elsinochrome A, resistromycin, aclacinomycin, chrysomycin, solanione, and lachnanthocarpon are included in Table 3.17.

The latter two pigments are derived from hydroxyphenalenone, a widely distributed chromophore in nature (917). The parent compound, 9-hydroxyphenalenone, is known to undergo a rapid intramolecular proton tunneling along a symmetric double energy minimum in the S_0 and the S_1 state (918,919). The fluorescence spectra and their polarizations have also been reported for other phenalenone-type molecules (920).

Many hydroxyquinones change their absorption and emission spectra considerably in the presence of inorganic ions as a result of complex formation. This is the basis of assays for both ions and quinones. Thus, boron forms fluorescent chelates with hydroxyanthraquinones (921,922) and with carminic acid (the chief constituent of cochineal) in sulfuric acid, sometimes even in aqueous borate solution (924). Beryllium ion is complexed by quinizarin to give highly fluorescent chelates (925). Senna, cascara, and aloe constituents have been assayed fluorimetrically in biological matter in the presence of sodium dithionite and borate (926). Antibiotics such as adriamycin (doxorubicin) bind to DNA and thereby undergo changes in the emission spectra (927).

An extensive literature is available on the fluorescence of tetracycline antibiotics, which have also been used as probes for calcium deposition in bones and as divalent cation probes for membranes by virtue of their ability to form fluorescent metal complexes. At pH values below 7.5 excitation at 400 nm results in an emission peaking at 530 nm. Above pH 7.5 a strong fluorescence is observed at 430 nm on excitation at 345 nm, in addition to the former band at 530 nm (928). Both bands are sensitive to divalent metal binding. Excitation energy transfer from the 430-nm band to the 530-nm band is observed in metal complexes, and from tryptophan residues in proteins to the green band of the magnesium complexes.

Clearly, fluorimetry offers a sensitive method for tetracycline assays. Although tetracyclines are only weakly fluorescent themselves in 0.1 N hydrochloric acid, they may be converted thermally into their anhydro salts with increased emission (929). Increase in fluorescence is also achieved by addition of di- or trivalent cations or alkaline degradation to give isotetracyclines.

A study on the relation between tetracycline structure and the nature of the metal ion was presented by Mani and Foltran (931). In an independent study it was found that anhydrotetracyclines interact with metal ions at the C_{11} carbonyl oxygen atom (942). Alykov investigated the reaction of oxy- and chlortetracycline, morphocycline, and rubomycin with $H_3B_3O_6$ in strong sulfuric acid by fluorimetry (932). Fluorescences, when observable, are invariably red. Sturgeon and Schulman determined the ground- and excited-state dissociation constants of the related antibiotic, doxorubicin (adriamycin). The prevalence of only one absorbing species having intense, longwave absorption in the pH 4–7 region and one emitting species having intense fluorescence in the pH 1–7 region suggested that the spectrometric analysis in aqueous solutions be carried out in dilute acidic solutions. A similar study was done on certain violamycin antibiotics (934).

Tetracycline assays in biological matter are based on solvent extraction of complexes with calcium (935), magnesium (936), or beryllium ion (937) or after separation directly on TLC plates (938,939). The aluminum(III) complex allows nonextractive determinations of tetracyclines in deproteinized plasma (930). Detailed procedures have been described to determine tetracycline (935), chlortetracycline (935), and oxytetracycline (940), via the aluminum complexes of the thermally formed anhydro bases (941).

Further fluorimetric determinations of quinoid antibiotics have been described for carminomycin and its 13-dihydro derivative (943), and for minocycline (930). The native fluorescence of nonquinoid antibiotics will be discussed in section 3.9. Unexpectedly, some quinoid pigments such as lachnanthopyron, phlebiarubrone, and xylindein were found to be nonfluorescent (916).

3.8.6   Miscellaneous Dyes and Pigments

The pterins and lumazines are yellow pigments, most of which fluoresce blue with maxima around 450 nm and excitation maxima around 370 nm. Fluorescent pterins have been isolated from tuberculosis (950) and diphtheria bacilli (951), from Mississippi alligators (952), and from related animals (953,954). They have been identified by their absorption and fluorescence spectra as ichthyopterine (952–955).

Leucopterin and xanthopterin are widely distributed in nature (953). Their fluorescence spectra have been reported (956). Their fluorescences are said to be blue at any pH between zero and 13, which is in contrast to an earlier report (1409) that only leukopterin fluoresces blue, whereas xanthopterin fluoresces blue in neutral solution, green in weak acid or acidified methanol, and red in 20% sulfuric acid. Chrysopterin in 0.004 N hydrochloric acid fluoresces violet-blue.

In a systematic study, the excitation and emission maxima of more than 70 pterins, xanthopterins, isoxanthopterins, lumazines, hydroxylumazins, and hydrogenated pteridins were measured after electrophoretic or TLC separation, and the effects of pH and substituents on the spectra were described (1410).

The vitamin folic acid is another widespread pterin derivative (pteroylglutamine), but has been discussed in the vitamin section (3.4.10). Isoxanthopterine fluoresces fairly strongly and much more intensely after binding to DNA (957).

The fluorescence of 2-aminopteridin-4-one (pterin) is pH dependent (958), its quantum yield is highest (0.057) at pH 10. Fluorescence maxima in aqueous solutions are between 440 and 465 nm at room temperature and between 418 and 424 nm at 77 K. The emission of neopterin (a tumor marker) in pH 6.4 solution is maximal at 438 nm, when excited at 353 nm (1408).

The fluorescence of lumazine (2,4-dihydroxypteridin) has been thoroughly studied by Lippert et al. (959–961). The work included pH effects on fluorescences, polarization data and shifts in tautomeric equilibria, and pK_a values following photoexcitation. 7-Hydroxylumazine was studied in a similar fashion (960,961). Fluorescence spectra and their relation to the chemical structure of the pterin fluorophore were also discussed by Polonovski et al. (962) and by Bertrand (963).

When lumazines bind to proteins they become more strongly fluorescent with emission characteristics similar to the bioluminescence of certain bacteria (964). Pterin-6-carboxylic acid has tentatively been identified as a strongly fluorescent metabolite of cancer cells (965). When spotted on chromatographic paper it shows E-type delayed fluorescence at around 450 nm (966).

Biopterin is another widely spread natural pterin with blue emission. It can act as a photosensitizer and is able to quench the fluorescence of tryptophan (967). Reduced biopterins have practically no intrinsic fluorescence, but may be converted by oxidation to the fluorescent biopterin forms, which can be quantitated by fluorimetry (968).

Among the natural dyes, the anthocyanines are in particular responsible for the red and blue colors of flowers. Their fluorescences have not been studied in great detail. Mullick (969) observed the fluorescence colors of six anthocyanidines on cellulose TLC plates and found them throughout red. After exposure to ammonia vapors the intensity increased distinctly. He noted, however, that certain chalcones and aurones also fluoresced red on exposure to ammonia.

It was noted that fluorescence is not observed when a 5-hydroxy group is present in anthocyanidins. However, those devoid of oxygen in 5-position or carrying a methoxyl or sugar at this position do fluoresce, pelargonidin-5-glucosides giving a yellow or greenish-yellow color, and paeonidin-3,5-diglycosides a bright orange one (405).

The luminescences of anthocyanidines undergo characteristic time-dependent color changes after exposure to ammonia (969). Pretreatment with formic or sulfuric acid has another pronounced effect on color sequences and stabilities. The fluorescence colors following molybdate or lead acetate spraying are different for various anthocyanidines and therefore of diagnostic value.

Nasu and Nakazawa (970) measured the emission spectra of four anthocyanines at 77 K in ethanol. Two (or three in the case of pelargonidine) emission bands are observed. The maxima of the emissions depended on the excitation wavelength and the shapes of the excitation spectra depend on the emission wavelength. The agreement between absorption and excitation spectra is sometimes poor. The results may possibly be interpreted by the well-known tendency of anthocyanines to forms aggregates.

In a subsequent paper (971) a study on the light emissions of anthocyanines in strawberry at 77 K was presented. Two bands resulting from a nonidentified pigment were observed under longwave (500 nm) excitation, but only one band was seen under 430-nm excitation. No agreement between absorption and excitation spectra was found.

Pellegrino et al (972) report the fluorescence data of uncharacterized extracts from *Streptocarpus holstii* and *Anthurium andreanum*, and from the respective intact flowers. Relatively short lifetimes (10–128 ps) and high-fluorescence polarization anisotropies ranging from 0.13 to 0.35 were found.

Lipofuscin granules in age pigments emit yellow light under ultraviolet excitation, and different age-related fluorophores, derived from other intracellular sources fluoresce blue. Prodigiosin, a red pigment from *Serratia marcescens*, is said to fluoresce greenish-yellow from its dimeric form, but not from the monomer (974).

Azulene derivatives have attracted considerable theoretical interest due to their fluorescence from the second excited singlet state (975). It has been suggested that this anomaly might be caused by a large energy gap of 14,000 cm^{-1} between the S_0 and S_1 state (976). The fluorescence spectra (maxima between 370 and 400 nm) (977) and quantum yields (0.02–0.03) (978) are solvent dependent, thus allowing the determination of excited-state dipole moments (977). The effect of temperature on its photophysical properties has also been reported (979). More recently, azulene was shown to exhibit another anomaly, namely, radiative transitions from its S_2 to the S_1 state (980). The $S_2 \rightarrow S_0$ and $S_1 \rightarrow S_0$ fluorescences have been studied at low temperatures, but only the former one has a sharp zero-phonon line pattern (981). Both emissions are pressure dependent (982). Delayed fluorescence from the second excited singlet state due to homo-triplet–triplet annihilation was recently observed for a trimethyl-azulene (983).

3.9.   MISCELLANEOUS NATURAL PRODUCTS

3.9.1.   Polyenes, Retinal

Saturated compounds such as linear or cyclic alkanes (984–986) as well as ethylenes (987) generally do not fluoresce efficiently. Tetramethylethylene shows a very broad weak fluorescence (λ_{max} 265 nm) with a quantum yield ϕ_f of $\sim 10^{-4}$ and a lifetime of around 10 ps. Short lifetimes, low emission efficiencies, and large Stokes' shifts are typical of flexible molecules for which radiationles deactivation may occur via molecular stretching or twisting motions.

Butadiene (988) and hexatriene (989) in fluid solution at room temperature are nonfluorescent. In these molecules radiationless deactivation is so rapid that emission from singlet states could not be detected (989,990).

Fluorescence, however, has been observed from certain sterically fixed trienes of the steroid series. Thus, fluorescence spectra and lifetime measurements of cholesta-4,6,8(14)-triene, and cholesta-5,7,9(11)-triene at 77 K allowed the assignment of the longest-wave transitions to a ^1A_g-like and to a ^1B_u-like excited state, respectively (991). The spectrum of estra-1,3,5(10)-triene-3,17-ol was also described (992). It is apparent from these results that small changes in the molecular geometry may reverse completely the order of excited states with different symmetries.

Cholestan-5,7,9(11)-triene-3β-ol was used as a nondisturbing lipophilic fluorescent probe, and its metabolism in rat liver was studied by fluorescence (993). Spectral data of some fluorescent cholestatrienes are given in Tables 3.18 anbd 3.19.

One- and two-photon excitation spectra of isotachysterol (994) (an all-*trans*-hexatriene) were obtained at 77 K, and the relative energies of the excited ^1A_g and ^1B_u states were estimated. The fluorescence spectrum and the two-photon excitation spectrum could be fitted to the same electronic transition assigned to a $2\ ^1A_g^{-*}$ state. The same state was assigned to the two-photon excited state of hexatriene (995).

Five by-products in charges of commercial steroid preparations were recognized by their blue-green fluorescence and were identified as steroid-4,6,8(14)-trien-3-ones (236). Excitation at 360 nm gave fluorescences peaking at 460–480 nm. Fluorescence is maximal in polar solvents and is pH dependent, despite the lack of acidic or basic sites in the molecule. Quantum yields vary from 0.15 to 0.27 in water solution. The compounds are rather photolabile, with half-lifetimes of ~ 1 h under irradiation. 6,8-Dehydrotestosterone was used as a fluorescent probe to study steroid–protein interactions of human serum albumin fractions (996). For a steroid out of the ergosta-tetraenone series native fluorescence was observed visually (997), but not further characterized.

Alkyl substituted linear conjugated polyenes with four (998), five (999,1000), or six (999) double bonds are throughout fluorescent and have a ^1A_g-like symmetry of their first excited singlet states. In these species, the transition from the ground state to the ^1A_g excited state is slightly lower in energy than the transition to the ^1B_u excited state.

Table 3.18. Excitation Wavelengths and Emission Maxima (in nanometers) and, Roughly, the Experimental Conditions for Natural Polyenes

Polyene	λ_{ex} (nm)	λ_f (nm)	Experimental Conditions[a]	Ref.
Amphotericin	360	440	Glycol–water	1033
Cholesta-4,6,8-(14)triene	298	346	Isopentane, 77 K	991
Cholesta-5,7,9-(11)-trien-3-ol	327, 346	348, 370, 392	Isopentane, 77 K	991
Enanthotoxin	370	425	Chloroform[b]	1029
Filipin III	355	480	Cholesterol	1031
	355	475	Glycol–water	1033
Isomethylbixin	494, 528	572, 618, 674	Xylene, 77 K	1019
Isotachysterol	292, 305	353	EPA, 77 K	994
Nistatin	315	395	Glycol–water	1033
cis-Parinaric acid	319	432	Decane[b]	1009
trans-Parinaric acid	314	425	Decane[b]	1009
Phytofluene	349, 366	470, 487	Hexane	194
all-trans-Retinal	380	600	EPA	185a
	380	540	EPA, 77 K	185a
	405	545	Solid film	1061
	347[c]	530	Hydrocarbon	1057
	347[c]	640	Methanol	1057
all-trans-Retinal	420	535	Ethanol, 77 K	1046
	400	520	Methylpentane[d,e]	1043
9-cis-Retinal	390	555	Solid film	1061
	390–400	520–525	3-Methylpentane[d,e]	1055, 1058
11-cis-Retinal	347[c]	610	Hydrocarbon	1057
	347[c]	640	Methanol	1057
13-cis-Retinal	347[c]	550	Hydrocarbon	1057
	347[c]	640	Methanol	1057
	400	535	Solid film	1061
Retinylidene-n-butylamine	~400	505	Methylpentane[d,e]	1055, 1058
	355	480	Hexane[d]	1050

[a] At room temperature unless otherwise specified.
[b] Other solvents also studied.
[c] Laser excitation.
[d] At 77 K.
[e] Moist solvent.

Table 3.19. Fluorescence Lifetimes and Quantum Yields of Natural Polyenes

Polyene	ϕ_f	τ (ns)	Experimental Conditions[a]	Ref.
Carotene	—	<0.001[b]	Benzene	1023
Cholesta-4,6,8(14)-triene	0.13	1.7	Isopentane[c]	991
Cholesta-5,7,9(11)-trien-3-ol	0.5–1.0	2.6	Isopentane[c]	991
cis-Parinaric acid	0.017	1.3	Methanol	1009
	0.054	5.2	Decane	1009
trans-Parinaric acid	0.009	<1.0	Methanol	1009
	0.031	3.1	Decane	1009
all-trans-Retinal	0.0001		Hydrocarbon	1057
	0.005		Hydrocarbon[c]	1057
	0.004		Methanol	1057
	<0.0001[d]	1.0	Methylpentane[c]	185a
	0.064[e]	0.61[e]	Ethanol, 77 K	1045
	0.012[f]	0.67[f]	Ethanol, 77 K	1045
	0.05	1.0	Methylpentane[c,g]	1043
9-cis-Retinal	<0.0005	1.6[d]	Methylpentane[c]	185a
11-cis-Retinal	<0.0005	2.0[d]	Methylpentane[c]	185a
	0.00009	—	Hydrocarbon	1057
	0.004	—	Methanol	1057
13-cis-Retinal	<0.0001	2.5[d]	Methylpentane	185a
	0.00008	—	Hydrocarbon	1057
	0.024[c]	—	Methanol	1057
	0.004		Methanol	1057
Retinylidene-n-butylamine	—	0.0022 and 0.265[h]	Hexane	1050

[a] Room temperature unless otherwise specified.
[b] The lifetime of the excited state.
[c] At 77 K.
[d] Excitation wavelength 380 nm.
[e] Excitation wavelength 420 nm.
[f] Excitation wavelength 360 nm.
[g] Moist solvent.
[h] Double-exponential decay.

273

As already mentioned in the vitamin A section, linear polyenes have intrinsic lifetimes that are much longer than the value obtained by integrating their absorption spectra according to the method of Strickler and Berg. It is assumed that this is due to fluorescence from the low-lying ^1A_g state (187,1001). Thus, the intrinsic lifetime of decatetraene is 500 ns compared to the value of 1.8 ns obtained from its integrated absorption (998).

Hudson and Kohler (187) were the first to present evidence that a low-lying ^1A_g state rather than the ^1B_u state was the lowest in a photoexcited tetraene. Because of symmetry considerations the state cannot be populated directly by excitation. It was argued that this state ordering would also hold for all linear polyenes, except perhaps butadiene and hexatriene.

In subsequent work it was shown that linear polyenes with five and six conjugated double bonds have a symmetry-forbidden nature of the lowest S_1 states (999). The dipole-forbidden character of the lowest S_1 state in 1,3,5,7-octatetraene was experimentally confirmed by running the spectra in n-octane at 4.2 K, where the symmetry labels g and u are strictly valid (1002).

Various authors have studied the photophysics of the more readily accessible linear diphenyl polyenes. Since natural polyenes do not bear phenyl substituents, these papers have not been considered in this chapter, despite their usefulness in the elucidation of the ordering of electronically excited states.

The low-energy ^1A_g excited state of polyenes gives rise to relatively longwave fluorescence. There is also evidence that the molecules in their excited states are severely distorted. The fluorescence spectra and the energy order of the excited states of the following tetraenes have been reported: $trans$-$trans$-octatetraene (990,1001,1003,1004) mono-cis-octatetraene (1005), cis-cis-octatetraene (1006), decatetraene (1007), α-and β-parinaric acid (1008–1010) and cis-eleostearic acid (1010).

Among these, the parinaric and eleostearic acids are of particular interest as fluorescent lipid probes. Their fluorescences occur at considerably lower energies than their absorptions and the wavelength of the emission is nearly independent of the solvent. However, the quantum yields (0.02–0.05) and lifetimes (1.3–5.2 ns) of the most frequently used cis-parinaric acid are strongly affected by both solvent and temperature (1009) (Tables 3.18 and 3.19). This phenomenon can be utilized to follow phase transitions (1008,1011) and lateral phase separation processes by changes in the fluorescence polarization (1012).

Cis-parinaric acid and cis-eleostearic acid bind to bovine serum albumin at approximately five binding sites (1010). Binding to proteins, membranes, or phospholipids increases ϕ_f up to 0.3 and the lifetimes up to 30 ns (1011). Parinaric acid can act as an energy acceptor from photoexcited tryptophan (1010).

The fluorescence spectra of decapentaene (1004), dimethylundecapentaene (1013), undecapentaene carboxylic acid (1014), phytofluene (194), dimethyltridecahexaene (1015), and dodecahexaene (1004) have been described. All exhibit the characteristic gap between the origin of the strongly allowed absorption and the origin of fluorescence.

The pentaene phytofluene has fluorescence emission spectra that closely overlap those of retinol (cf. Section 3.4.3) and therefore interferes in the fluorimetric vitamin A determination (194).

Various authors have reported that carotenes with 11 or 12 conjugated C–C double bonds show visible (1016,1017) or even UV fluorescence (1018), while others claimed it to be nonfluorescent. Thus, Lyalin et al. (1016) report the observation of a green fluorescence from β-carotene in octane at 77 K. Wolf and Stevens (1018), on the other hand, found no visible emission with either β-carotene and a number of other carotenoids, but rather a fluorescence in the UV similar for all carotenoids, which was maximally excited at 280 nm. Cherry et al. (1017) observed weak green fluorescence of cyclohexane, ethanol or benzene solutions of all-*trans* and 15-*cis* β-carotene at room temperature. Kuhn and co-workers measured the fluorescence spectra of polyene carboxylic acids such as isomethylbixin at 77 K (1019). Delayed emission was reported for β-carotene (1017) and lycopene (1020).

On the other hand, it was said that carotenes lack fluorescence with quantum yields greater than 10^{-4} in both the UV and visible part of the spectrum (1021). The visible fluorescences formerly observed were attributed to impurities, since these emissions remained measurable even when the β-carotenes were photobleached. This also holds for lycopene and canthataxin.

It seems now most likely that the fluorescence of carotenes is extremely weak, if at all existent, and that the lifetimes are extremely short. Song and Moore (1022) expected β-carotene to have lifetimes of 1 ps or less. It could indeed be shown by resonance Raman experiments that the deexcitation lifetime of the β-carotene excited singlet state is not greater than 1 ps (1023). A previous report of a decay time of ~55 ps for both α- and β-carotene (1024) in chloroform at high concentrations was supposed to result from a minor impurity (1025).

The mechanism by which the carotenoids of photosynthetic organisms can act in an accessory light-harvesting role was recently examined (1026). It was concluded that, since fluorescence quantum yields of pigments such as β-carotene are vanishingly small, a dipole–dipole (long-range) energy transfer mechanism must be discounted. Rather, electron-exchange-mediated transfer of excited-state energy would be favored on the grounds of the excellent overlap between the transitions from the low-lying singlet states of carotenoids and the absorption bands of chlorophyll acceptors. Evidence has been presented that the required close approach of the two entities involved in the energy transfer process is reached prior to the photoexcitation event (1023). In elegant synthetic an spectroscopic studies it was shown that singlet energy transfer from a carotenoid to a porphyrin in a covalently coupled pair occurs in high dependence on the molecular arrangements (1027,1028).

Certain polyenes exhibit toxic activities. The fluorescence properties of the following polyene antibiotics have been described: enanthotoxin (fluorimetric determination) (1029), flavopentin (luminescence of polyene-treated tissue) (1030), filipin III (1031,1032), amphotericin B (1031–1033), nystatin and lagosin (1031,1033), etruscomycin and pimaricin (1033).

Filipin III has been recommended as a probe for aqueous suspensions and membranes of cholesterol by measuring the polarized fluorescence intensities (1031). The effect of solvent viscosity on the fluorescence polarization has likewise been described. (1033).

Addition of cholesterol to aqueous solutions of filipin decreases the fluorescence intensity, whereas with pimaricin it is more than 80-fold higher. Addition of nystatin and amphotericin B has little effect. It was concluded that filipins can exist in conformational states or bonded conditions such as dimers that do not bind sterols in a specific manner and that such forms undergo changes in aqueous solutions (1034).

Careful HPLC separation of commercial preparations of the antifungal heptaene amphotericin B showed that the compound itself does not fluoresce, but usually contains an unidentified fluorescent contaminant (1035). This casts some doubt on the results previously obtained with "amphotericin B" (1031–1034).

A variety of retinal related compounds has been investigated by fluorescence techniques in efforts to clear the photophysics of vision. Das and Becker (1036) prepared a series of retinal-related polyenals and polyenones with 2–10 double bonds (including the carbonyl group) and described their emission spectral properties. Those with short chain length (two to four double bonds) are nonemitting with a singlet n,π^* state lowest. Those with intermediate chain length (five or six C–C double bonds) exhibit fluorescence at 77 K that is strongly dependent on the exciting wavelengths and is sensitive to environmental factors. Those with long chain length (seven or eight double bonds) show fluorescence both at 298 and 77 K with an intensity essentially independent of exciting wavelengths. The nature of the emitting states of the longer polyenals is probably of mixed character.

The authoritative study of Das and Becker involves absorption, excitation, and emission spectra; effects of solvent and hydrogen bonding agents (such as alcohols, phenol, and trichloroacetic acid); concentration effects; and lifetime and quantum yield studies. In other papers the spectral and photophysical data for polyene systems ending with functional groups other than the aldehyde group were reported (1037–1039).

In a recent study (1040) the emission properties of linear polyenals of the series $CH_3-(CH=CH)_n-CHO$ with n ranging from 2 to 5 were described. The work involves the assignment of electronic transitions and the identification of the nature of the fluorescent state as a function of the chain length. Aside from the dipole-forbidden singlet-excited state of primarily ^1A_g character near the strongly allowed ^1B_u state polyene aldehydes possess an additional low-lying $^1(n,\pi^*)$ state that plays a very important role in determining the spectral and photodynamic behavior of the polyenes (1040,1041).

Extremely narrow optical spectra of 2,4,6,8,10-dodecapentaenal (a compound with the same number of double bonds as retinal but without the β-ionylidene ring) and of its Schiff base were obtained in a Shpolskii matrix at 77 K (1042). Fluorescence emission was assigned to a forbidden $\pi^* \leftarrow \pi$ transition of $^1A_g^{*-}$-like character (1043). The site-selected optical spectra of the same compound at 4.2 K are similar to retinal (where hydrogen bonding is necessary for emission) and to other linear polyenes with their progression of single- and double-bond modes (1042).

The fluorescence of retinal, the chromophore linked to opsin in rods and cones in visual pigments, has received considerable attention. In the visual pigment, 11-*cis* retinal is covalently bonded via a most likely protonated Schiff base to the ε-amino group of a lysine residue of opsin. It is assumed that in one of the first steps of the visual process the 11-*cis* form is photochemically converted with quantum yield of ~0.67 to prelumirhodopsin with an all-*trans* configurated retinal Schiff base, possibly accompanied by a proton transfer process at the protonated C–N double bond.

A manifold of publications deals with the native fluorescence of 9-, 11-, or 13-*cis*-retinal (185a,1043,1044) and the all-*trans* isomer (185a,1043–1048) of Schiff bases and their protonated forms (185a,1049–1052). The photophysics of a homolog of all-*trans*-retinal with one additional C–C double bond has also been studied in detail (1053,1054). The spectral properties of vitamin A along with a discussion of the nature of its fluorescent state are presented in Section 3.4.3.

The fluorescence of all-*trans*-retinal is very unusual because of the existence of three low-lying states [$^1B_u^{*-}$ (π,π^*), $^1A_g^{*-}$ (π,π^*), and $^1n,\pi^*$ in character] of comparable energy. The basis for the anomalous excitation wavelength dependence of the quantum yield (1043–1045,1055) has been resolved. It was shown that both hydrogen-bonded and non-hydrogen-bonded forms of retinal coexist in nondried solvents and only the hydrogen-bonded form fluoresces. The nonemitting species with a lowest lying n,π^* state absorbs a fraction of incident light, giving rise to differences in the absorbance ratio between fluorescing and nonfluorescing species, and therefore for wavelength dependence of ϕ_f (1043).

Evidence has been presented that fluorescence can come from a dimeric form of retinal, since its fluorescence can be detected in carefully dried solvents, provided the concentration is high enough (1056).

As for other long-chain polyenes, the lowest π,π^* state of 11-*cis* and all-*trans*-retinal is of $^1A_g^{*-}$ character (1047,1055), but a lowest-lying n,π^{*-} state is believed to be present in dry hydrocarbon solvent. Consequently, there is a large disagreement between the measured radiative lifetime and the value calculated from the integrated absorption (185a,1036,1043,1044).

The room temperature emission maxima of retinal and its E/Z-isomers are strongly dependent on the nature of the solvent (1057–1059) and shift from 530 nm in hydrocarbon solvent to 640 nm in methanol. In a laser photolysis study on the photoisomerization of retinals the laser-stimulated room temperature emission spectra and quantum yields of the all-*trans*, 11-*cis*, and 13-*cis* isomers were reported (1057).

Becker et al. (185a) tried to estimate the quantum yields for the various processes for all-*trans*- and 11-*cis*-retinal near 295 K. For *trans*-retinal, fluorescence and triplet-sensitized photoisomerizations have low efficiency. However, intersystem crossing and direct photoisomerization are fairly efficient in hydrocarbon solvent, but not in methanol. For 11-*cis*-retinal the situation is similar, except for a rather efficient photoisomerization via triplet sensitization.

Other emission anomalies of *trans*-retinal are the temperature dependence of the radiative lifetime (1045,1046), the enhancement of fluorescence by inorganic ions that usually act as quenchers (1046), and the nearly maximal polarization of the strongly allowed ^1B_u transition at around 370 nm with respect to the fluorescence oscillator under photoselection conditions at 77 K (1046,1060).

Solid films of all-*trans*-retinal and its 9-*cis* and 11-*cis* isomers have high positive fluorescence polarizations (0.35–0.40) at room temperature (1061). The broad shape, the absence of a 0–0 band and large Stokes' shifts (3000–3500 cm^{-1}) indicate that the absorption and fluorescence bands of the retinals are severely Franck–Condon forbidden.

Schiff bases of retinal and related polyenals seem to be the most appropriate models for the rhodpsin chromophore. Palmer et al. (1051) measured the fluorescence excitation and emission spectra of simple polyene

Schiff bases. For all three detected species (the free, hydrogen-bonded, and protonated base, respectively) the fluorescence origins are substantially red shifted from the origins of the strongly allowed absorptions, which is attributed to fluorescence from $^1A_g^{*-}$-like states.

Schiff bases with short polyene chainlength ($n = 2$–3, where n is the number of the conjugated double bonds including the C=N double bond) do not fluoresce. Polyene Schiff bases with intermediate chain length ($n = 4,5$) show fluorescence at 77 K, but scarcely at room temperature. The Schiff bases with relatively long chain length ($n = 5$–7) show strong or moderate fluorescence at 77 K (1052).

The retinal Schiff base with n-butylamine fluoresces maximally at 500 nm in 3-methylpentane and at 609 nm in EPA at 77 K, when excited at 365 and 375 nm, respectively. The quantum yields are 0.052 in methylpentane (0.144 in EPA); the experimental decay times are 5.4 (6.2) ns. Protonation shifts the maximum to 671 nm (in methylpentane, excitation at 490 nm), the quantum yield is 0.0098, and the decay time 1.0 ns now (1052). Evidently, protonation has a pronounced effect on both the emission maximum and the quantum yield, but not on the natural radiative lifetime $\tau_0 (= \tau/\phi_f)$, which for the protonated form is 5 ns.

Time-resolved fluorescence measurements have been performed on various model systems for the visual pigment (1058,1062), retinal (1062), retinylidene-n-butylamine (1050), and rhodopsin (1063). The retinal Schiff base with n-butylamine shows a biexponential decay in n-hexane at temperatures between 4 and 298 K, but not in acetonitrile (1050). At 298 K the decays are 22 and 265 ps (no weighting factors given) with a more rapid increase in the amplitude of the slow component at lower temperatures. The results were interpreted in terms of a vibrational–torsional deactivation model of the singlet excited state of two different conformers.

The photophysics of rhodopsin (1064,1065), bacteriorhodopsin (1065), and the ultrafast processes of the photoconversion of rhodopsin to prelumirhodopsin (1066) have recently been reviewed in detail and will not be discussed here.

3.9.2. Amino Acid Metabolites

The three amino acids: phenylalanine (Phe), tyrosine (Tyr), and tryptophan (Trp) are metabolized mainly in three ways: (1) by decarboxylation (e.g., the formation of tryptamine from Trp), (2) by oxidation (e.g., DOPA from Tyr), and (3) by methylation (e.g., dimethyltryptamine from tryptamine). In the case of Trp, the oxidative indole ring opening is a fourth metabolic pathway to give strongly fluorescent products. Since biogenic amines have distinct physiological activities, including blood pressure regulations and intoxications, their determination (e.g., on the basis of the native fluorescence) is of great practical importance.

It has been recognized that certain amines show strong fluorescence, despite a complete lack of a π-electron system (1067). It is, therefore, likely that biogenic amines derived from amino acids other than Phe, Tyr, or Trp show fluorescence. No such observation has been reported to date. Pitet et al. (1068) studied the native fluorescence of some aromatic amino alcohols in acidic, neutral, and basic solutions.* Fluorescences of the protonated forms are negligibly small.

There are only a few reports available on the fluorescence of the main metabolites of Phe. Hurst and Toomey (1069) determined phenethylamine (and other amines such as tyramine and serotonin) in dairy products by combination of HPLC separation and fluorescence detection (excitation 285, emission 320 nm). The method is said to be as sensitive as the one with postcolumn derivatization with o-phthalaldehyde.

Many 2-phenethylamines exhibit fluorescence because of intramolecular exciplex formation when the amino group is tertiary (1070–1072). The fluorescence maxima are at 290–320 nm in cyclohexane, which is at about 10–30 nm at longer wavelengths than the fluorescence of benzene. The maxima of the exciplex emissions are dramatically shifted to the red (λ_{max} 380–400 nm) in going to acetonitrile solvent (1071).

The metabolites derived from tyrosine have longer-wave excitation and emission maxima than the phenethylamines (Table 3.21). Duggan et al. (230) have measured the excitation and emission maxima of many metabolic amines and acids derived from Tyr and Trp. Fluorescence activation and emission maxima are mostly in the ultraviolet, except for the hydroxyanthranilic acids and hydroxykynurenines. A particular high fluorescence intensity—and therefore sensitivity in the determination—was found for adrenaline, anthranilic acid, 3-hydroxy- and 5-hydroxyanthranilic acid, hydroxykynurenines, noradrenaline, serotonin, and tryptamine. Detection limits were lower than 0.006 μg mL^{-1}.

The fluorescence of tyramine is similar to that of Tyr, with the exception that it is not quenched at low pH values (16). Enzymatic oxidation of Tyr affords 3,4-dihydroxyphenylalanine (DOPA), which shows intrinsic fluorescence at 323 nm with an excitation maximum at 285 nm (230).

To enhance the fluorescence yield of biogenic amines and related compounds several derivatization techniques have come into use, in particular for the relatively weakly fluorescent Phe and Tyr metabolites. The most frequently employed methods are the trihydroxyindole-procedure (ca-

---

* One notes, in some cases, large disagreements between the absorption and excitation maxima.

techol oxidation and rearrangement) (1073) and derivatization reagent methods with fluorescamine (1074), dansyl chloride (1075), or o-phthalaldehyde (1076), but all of them require additional reaction steps.

Catecholamines possess native fluorescences, which are usually excited at 285 nm and are maximal between 325 and 340 nm. These excitation and emission maxima are subject to changes, depending on pH and the type of solvent system used. Norepinephrine, epinephrine, DOPA, dopamine, methyldopa, deoxyepinephrine, and tyramine have been determined in this way in high sensitivity in rat brain, heart extracts, and human serum (1077). Jackman assayed urinary catecholamines and urinary metanephrines (1078) such as 3-methoxytyramine, tyramine, (nor)metanephrine, and (nor)synephrine by a combination of HPLC and fluorescence detection. Excitation light of the rather unusual wavelength 200 nm was applied.

In tissue sections the primary catecholamines can be visualized by converting them into highly fluorescent isoquinolines with the help of formaldehyde or glyoxylic acid (1079,1080).

Metabolites of Trp possess the strongest native fluorescence. Their photophysical properties in neutral aqueous solution are similar to those of Trp, which has been discussed in Sections 3.2.3 and 3.6.2. Spectral data of various biogenic amines are given in Table 3.21 at the end of this section.

Bridges and Williams (320) performed an extensive study on the pH dependence of the fluorescence spectra of indole, indol-3-acetic acid and its 5-hydroxy derivative, Trp, indoxyl acetate, physostigmine, oxindole, bufotenine, psilocin, 6-hydroxytryptamine, and some indolenines. It was found that indole and its 3-, 5-, and 6-hydroxy derivatives are strongly fluorescent, whereas 2-, 4-, and 7-hydroxyindoles are not. The fluorescence of 3-hydroxy indoles is in the red (λ_{ex} 495 nm, λ_f 570 nm) (1081), whereas N-methylindoxyl fluoresces green (λ_{ex} 430 nm, λ_f 501 nm) (1082).

Because of the lack of a quenching carboxy group, the fluorescence of tryptamine and related compounds is more intense than that of Trp at low pH values. It has been utilized along with solvent extraction to measure this amine in urine and tissue (1083). Various chromatographic separation procedures coupled with detection of the native fluorescence have been reported. These include assays for anthranilic acid, tryptamine, indole-3-acetic acid, and 5-hydroxyindoles (1084) and for anthranilic acid and its hydroxy derivatives in urine (1085). McNeil et al. (317) determined the excitation and emission maxima of 15 metabolic indoles on silica gel plates and found detection limits from 0.03 to 0.05 μg per spot. It was also found that spraying with methanolic sulfuric acid plus dimethylsulfoxide produced longwave shifts in both excitation and emission at the expense of the detection limits (1 μg). All indoles tended to photodecompose.

Other assays that are based on native fluorescences were reported for 5-hydroxyindole-3-acetic acid (1086) (by HPLC), Trp and nine of its metabolites after preconcentration and HPLC separation (101), 5-methoxy-N-methyl-tryptamine in plants by TLC fluorescence scanning (1087), eight metabolic amines from intoxicating snuffs ("epeña") (1088), serotonin in blood platelets (1089,1090), various Trp metabolites in biological matter (98,1077,1091–1094) and urine (1094,1095), and for tryptamine and serotonin (1096). Uric acids were identified in HPLC fractions of serum and plasma by using a fluorescence detector with an excitation wavelength of 325 nm and a cutoff filter at 340 nm (1097).

When tryptamin, serotonin, or 5-methoxytryptamine bind to nucleic acids, their fluorescence is quenched as a result of energy transfer (or charge transfer) from the indole to the purine and pyrimidine rings (1098).

An unusual and ambiguously interpreted phenomenon is the shift in the fluorescence emission maxima of 5-hydroxy- and 5-alkoxyindoles from about 340 nm to about 540 nm when the pH value of the aqueous solution becomes lower than zero, e.g., in 3 N hydrochloric acid (1099). There is no corresponding change in the absorption spectrum. Since the excitation spectra are also the same as in neutral or weakly acidic solution, it has been assumed that an excited state dynamic process is operative.

It has been shown that only 5-oxyindoles exhibit this phenomenon, whereas 4-hydroxytryptamine, 6-methoxyindole, and others do not (1100). The longwave fluorescence is also emitted in ethanolic 3 N sulfuric acid at room temperature, but not at 77 K (320). When this solvent is replaced by the more viscous glycerol 3 N sulfuric acid at 20°C, the fluorescence intensity falls to 25% of that in ethanolic sulfuric acid. Table 3.20 summarizes the photophysical data for typical 5-oxyindoles.

While there is general agreement that the green-emitting species of photoexcited 5-oxyindoles is formed by a protonation step in the S_1 state, there are large differences as to the nature of the process. Williams and Bridges (320) assumed protonation of the phenolic oxygen atom. Chen (1100) favored a protonation at the indole nitrogen over the former possibility on grounds of electron density shifts from oxygen to nitrogen in the excited state. Excited state pK_a calculations with the help of the Förster cycle favor N-protonation.

However, both structures seem unlikely because the indoles as well as phenols are generally stronger acids in the S_1 state than in the ground state. Therefore, it may be assumed that they would also be protonated less readily in the excited than in the ground state. Since indoles are not

Table 3.20. Photophysical Properties of 5-Oxyindoles in Neutral and 3 M Hydrochloric Acid Solution at Room Temperature, Excitation Wavelength 290 nm

5-Oxyindole	In Water, pH 7.4			In 3 M HCl		
	λ_f (nm)	ϕ_f	τ (ns)	λ_f (nm)	ϕ_f	τ (ns)
5-Hydroxytryptophan	349	0.26	3.9	545	0.046	6.0
5-Methoxytryptamine	350	0.31	5.3	547	0.047	6.8
5-Methoxyindole-3-acetic acid	350	0.23	6.5	542	0.043	7.2
Serotonin	338	0.25	4.4	554	0.048	5.7
5-Methoxyindole	333	0.23	4.6	529	0.041	3.8

Source: From Chen (1100).

protonated yet in 3 M hydrochloric acid in the 3-position, which is the protonation site in the S_0 state, both O- and N-protonations are unlikely.

Further sites of protonation that have been suggested are the 3-position (by analogy to the ground state process) (317), 4-position (after theoretical calculations on serotonine in its excited state) (324), and the 6-position (1101). The problem is not solved yet.

During their studies on the photophysics of indoles (1102,1103), Lumry and co-workers found that 5-methoxyindoles, unlike indole, do not form exciplexes with solvent molecules. Fluorescence maxima do not vary strongly with the polarity of the solvent (321 nm in cyclohexane, 331 nm in water), the quantum yield is 0.29 in water at 25°C, and the measured lifetime is 4.0 ns.

The green emission of 5-oxyindoles has served to determine very specifically the following metabolites in the presence of other indoles: serotonin (323,1087,1090,1105–1108), 5-hydroxytryptamine, 5-hydroxytryptophan, and 5-hydroxyindole-3-acetic acid (1109,1110).

The metabolic ring opening of the indole nucleus leads to products of the kynurenine and anthranilic acid type. Their excitation maxima are at distinctly longer wavelengths, and fluorescence emission is already in the visible (Table 3.21). The native fluorescence of 3-hydroxyanthranilic acid, which is maximal between pH 7.0 and pH 7.4, was seen very early (1111). It may be utilized for its quantitation in tissue (1112). Bell and Mainwaring (1113) measured the fluorescences of 3-hydroxyanthranilic acid, kynurenine and its 3-hydroxy derivative, and xanthurenine after TLC separation. In an attempt to identify fluorescent urine metabilites, Wachter et al. (1114) determined the in situ excitation and emission maxima of 15 metabolic acids such as anthranilic acid and its hydroxy derivatives, kynurenic and xanthurenic acid, indoxyl sulfate, and indolylacetic acids after paper electrophoresis.

Fukunaga et al. (1115) described the fluorescence properties of kynurenine and its amide, N'-formylkynurenine and its amide, and kynuramine in aqueous solutions of various pH values. Kynurenine and formylkynurenine were studied in great detail, including the effects of temperature, ionic strength, protein denaturants, solvent polarity, and by quenching neighboring groups.

Kynurenines differ from N'-formylkynurenine in that their fluorescence at pH values below 2 is strongly quenched by the protonated amino group. For the formyl derivative fluorescence is high at pH below 2, but low in alkaline solutions. In contrast to Trp, the fluorescence of kynurenine cannot be quenched by a carboxyl group.

Kynurenine in aqueous solution is a weak emitter. With decreasing solvent polarity the intensity increases logarithmically, and the maximum shifts to the blue. The sensitivity toward solvent polarity is much more expressed than that of Trp and can therefore better report the polarity of the microenvironment.

These findings are the basis of a new technique to probe the Trp environment in proteins by transforming it into kynurenine or its formyl derivative by mild ozonolysis (1115). The location of the emission maximum is now a better reporter for polarity estimation of the protein surrounding than the fluorescence intensity.

The large shifts between the excitation and emission maximum of formylkynurenine (325/434 nm in water, 320/510 nm in cyclohexane) is unusual. It has been suggested that intramolecular hydrogen bonding plays an important role and that excited state proton transfer may occur (1116,1117). The green fluorescence of formylkynurenine (a potent photosensitizer) (1118) in apolar solvents was assigned to the phototautomer, whose structure is shown in Scheme 5. Similar effects were observed with o-acylaminoacetophenones, provided they have an intramolecular hydrogen bond (1117).

Pirie (1119) studied the native fluorescence of N'-formylkynurenine and of proteins in the human lens. The 440-nm fluorescence in neutral or acid solutions shifts to 400 nm at pH values above 10, accompanied by a 20-fold increase in intensity. Reduction of the formylkynurenine to give (N-formylaminophenyl)-homoserine replaces the former spectrum by a peak of emission at 345 nm excited at 285 nm.

The fluorescence of proteins exposed to sunlight shows similar fluorescences, provided they contain a Trp unit.

3-Hydroxykynurenine-O,β-glucoside was identified in the water-soluble fraction in human lens and characterized (1120) by its native fluorescence. Kynuramine was used as a fluorescent probe for the catalytic site of plasma amine oxidase (1121). Binding causes spectral shifts, changes in the fluorescence polarization and intensity, thus indicating a rigid attachment of the substrate to the enzyme. It was stated that the use of fluorescent substrates offers the possibility of monitoring conformational changes occurring prior to the catalytic event.

3.9.3. Phenols, Aldehydes, Ketones, and Acids

Because of the phenolic character of ring A, the estrogens possess the strongest native fluorescence among the steroids. Like phenol itself, estriol, β-estradiol, and estrone are fairly strong fluorescers having excitation and emission peaks at 285 and 325–330 nm, respectively (230). The intensity of estradiol is about eight times that of estrone, which contains a carbonyl group that can accept energy and therefore can act as a quencher.

Equilin excited at 290 nm gives a fluorescence maximum near 345 nm at about one-tenth of the intensity of β-estradiol. Equilenin is the strongest fluorescent estrogen, with peaks at 340 nm in excitation and at 370 nm in the emission spectrum (in ethanol). Its intensity is around 10 times that of β-estradiol.

Weinreb and Werner (1127) studied the fluorescence of a series of estrogens with and without keto functions. In all ketoestrogens—with the exception of equilenin—there was a very strong transfer of energy from ring A to the keto group. Thus, the spectrum of estrone in ethanol at room temperature consists of two bands, one at 325 nm, the other at around 430 nm in the intensity ratio of 6:1. The latter band is ascribed to fluorescence from the keto group.

The difference between the luminescence of hydroxyestrogens and their keto analogs in solution and in solid films is related to the crystal structure of the steroids (1128). The spectral shifts can be correlated with the number of hydrogen bonds. The intrinsic fluorescence of estrone in alcoholic and alkaline aqueous solution was utilized to follow its radiolysis by γ-rays (1129).

Changes in the electronic spectra of eugenol and isoeugenol with pH allowed the determination of the pK_a values and dipole moments in the ground and first excited singlet states (1130).

Since estrogen determinations are essential to the proper management of pregnancy, numerous fluorimetric procedures have been developed, as, for example, the determination of estrogens in pregnancy urine (1131) and total estrogens in urine after automated hydrolysis (1132). It is said that the use of native fluorescence of estrogens gives more reliable results than the method with the Ittrich–Kober reagent, which requires careful control of the experimental conditions to provide reproducibility.

The phenol ether β-asarone has been determined in vermouths by HPLC separation and UV fluorescence detection (1133). The fluorescences of phenolic lignin model compounds (1134) and of various tannins as a function of the pH value was studied (1135). Synthetic tannins usually have a violet fluorescence, which becomes green with increased pH. Natural tannins become brown with increased pH.

Steroids with a keto group as the only π-electron system show rather poor fluorescence (1136). The excitation maxima correspond with the n,π^* transitions of low molar absorbance (18–45 M^{-1} cm^{-1} in ethanol) in the UV spectrum. The fluorescence emission peaks of all androstane ketones are around 415 nm, except for the 11-keto isomer (360 nm). The spectra of steroids with more than two conjugated double bonds have already been discussed in the polyene section (3.9.1).

A general feature of the emission of ketones possessing lowest n,π^* singlet states is the small value of the fluorescence quantum yield (0.001–0.0001). Therefore the native fluorescence has never been used to characterize or analyze natural products with a keto group as the only fluorophore. In addition, aliphatic ketones have low molar absorbances and are rather photoreactive (1137).

The fluorescence spectra of formaldehyde (1138,1139), acetaldehyde (1140), acetone (1141), 2,3-butanedione (biacetyl) (1142,1143), and certain cyclic 1,2-diketones (1143), steroid 1,2-diketones (1144), glyoxal (1145), pyruvic acid and alkyl pyruvates (1146), aliphatic diketones containing two remote fluorophores of opposite chirality (1147), rigidized β, γ-enones (1148), benzaldehyde and acetophenone (1149), naphtaldehyde and acetonaphthone (1150), and of various benzophenones (1151,1152) have been described. Usually, aromatic ketones fluoresce efficiently in strong mineral acids only (1149–1152).

Aromatic aldehydes and ketones are well known for their phosphorescences. However, some aromatic carbonyl compounds with medium-sized aromatic systems have been found to fluoresce in hydroxylic polar solvents, at room temperature (1151,1153). Weak delayed fluorescence has been observed from solutions of substituted benzophenones at room temperature (1151,1154). Fluorescence is strongly enhanced in sulfuric acid, wherein the molecules exist in protonated form, causing the low-lying n,π^* transitions to be raised in energy.

Introduction of one or more hydroxy or methoxy groups into aromatic aldehydes and ketones gives compounds, some of which are responsible

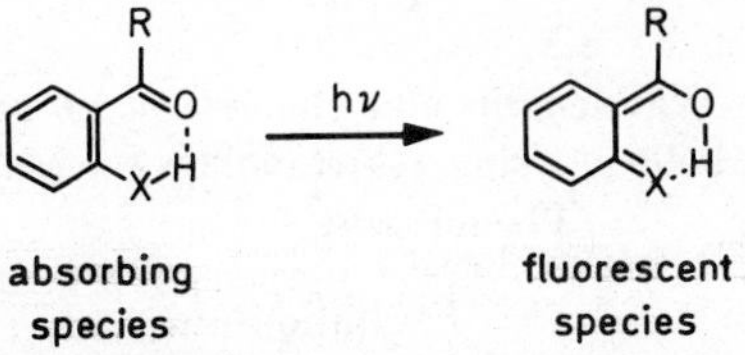

absorbing
species

fluorescent
species

Scheme 5. Structures of salicylic aldehyde (R = H, X = O), salicylic acid (R = OH, X = O), o-hydroxyacetophenone (R = CH$_3$, X = O) and N'-formylkynurenine (R = CH$_2$—CH(NH$_2$)—COOH, X = N—CHO) and their respective phototautomers. Flourescence is observed mainly from the tautomeric enol form (either cis or trans), which—in the ground state—reverts thermally to the carbonyl tautomer.

for the smell and flavor of flowers and plants. Aromatics with a strong charge transfer transition from the hydroxy to the carbonyl group are particularly fluorescent, since this new state is frequently below the n,π^* state with its tendency to facilitate intersystem crossing.

The emission spectra of the three isomeric hydroxybenzaldehydes have been described (1155). Salicylaldehyde was found to be fluorescent at room temperature and at 77 K, whereas for m-hydroxybenzaldehyde fluorescence was found at 77 K only. For p-hydroxybenzaldehyde, its methyl ether, and for o-methoxybenzaldehyde no fluorescences could be detected. In this work, all phosphorescence and fluorescence maxima are said to lie between 2.10×10^4 and 2.20×10^4 cm^{-1} except for salicylaldehyde (1.90×10^4–2.00×10^4 cm^{-1}).

In a more recent work it has been shown that salicylaldehyde after photoexcitation at 327 nm undergoes a shift of the phenolic proton along the hydrogen bond to form a fluorescent quinoid tautomer (Scheme 5). This species is responsible for the extremely longwave-shifted fluorescence, peaking at 525 nm in cyclohexane. The quantum yield is only 0.001 and drops even further when excitation is performed below 320 nm (1156).

p-Hydroxybenzaldehyde is nonfluorescent in hydrocarbon solvent at both room temperature and 77 K. Addition of the proton accepting base triethylamine gives rise to the formation of a fluorescent ion pair (1157).

The existence of an excited state proton-transferred species as for salicylaldehyde was also deduced from the unusual longwave emission of o-hydroxyacetophenone (1156). No such effects are observed with the o-methoxy derivatives of benzaldehyde and acetophenones, whose fluorescence maxima are in the UV. o-Hydroxybenzophenones are practically nonfluorescent in organic solvents (1158).

Among the natural carboxylic acids, naphthalene-1-acetic acid (1159) and naphthoxyacetic acid (1160) (two plant hormones) have been shown to exhibit native fluorescence in the ultraviolet. The amino acid meta-

Table 3.21. Excitation Wavelengths and Fluorescence Emission Maxima (in nm) of Amino Acid Metabolites, Urinary Metabolites, and Naturally Occurring Phenols and Acids

Compound	Excitation/ Emission maximum	Experimental Conditions	Ref.
N-acetylserotonine	311/352	TLC plate	317
N-acetyltryptamine	307/420	1 *N* NaOH	321
Adrenaline	285/325	Water, pH 1	230
5α-androstan-3-one	288/413	Ethanol	1136
2-Aminobenzoic acid	300/405	Water, pH 7	230
(anthranilic acid)	300/395	Water, pH 1	230
	300/405	TLC plate	1113
	320/395	Paper strip	1114
3-Aminobenzoic acid	315/399	Paper strip	1410
4-Aminobenzoic acid	295/345	Water, pH 11	230
Caffeic acid	365/450	Paper strip	1114
Deoxybilianic acid	290/415	Ethanol	1136
N,N-Diethylserotonine	295/340	Water	1088
	295/340, *540*	3 *N* HCl	1088
N,N-Diethyltryptamine	277/355	Water, pH 7	326
3,5-Dihydroxybenzoic acid	315/450	Paper strip	1114
3,4-Dihydroxymandelic acid	285/330	Water, pH 7	230
3,4-Dihydroxy- phenethylamine	285/325	Water, pH 1	230
3,4-Dihydroxyphenylacetic acid	280/330	Water, pH 7	230
3,4-Dihydroxyphenylalanine	285/325	Water, pH 7	230
3,4-Dihydroxyphenylserine	280/320	Water, pH 1	230
N,N-Dimethyl-5- methoxytryptamine	295/340	Water	1088
	295/340, *540*	3 *N* HCl	1088
N,N-Dimethyltryptamine	277/355	Water, pH 7	326
	278/350	Water	1088
Epinephrine	285/325	Water, pH 1	3
Equilenine	340/370	99% ethanol	230
Equiline	290/345, 420	99% ethanol	230
Estradiol	285/330	99% ethanol	230
Estrone	285/325	99% ethanol	230
	290/325, 430	Ethanol	1122
Eugenol	280/320	Water, pH 2	1130
	295/351	Water, pH 13	1130
Ferulic acid	350/440	Paper strip	1114

Table 3.21. (*continued*)

Compound	Excitation/ Emission maximum	Experimental Conditions	Ref.
N'-formylkynurenine	319/440	Water–methanol	1122
	330/440	Water, pH 2–8	1119
	315/400	Water, pH 11	1119
Fulvic acid	~350/450	Water, pH 10	1211
Gallic acid	300/370	Water, pH 11	1195
Gentisic acid	318/442	Water, pH 7	1192
	330/437	0.01 N NaOH	1192
Homogentisic acid	290/340	Water, pH 7	230
Homovanillic acid	270/315	Water, pH 7	230
3-Hydroxyanthranilic acid	320/415	Water, pH 7	230
	328/415	Paper strip	1410
5-Hydroxyanthranilic acid	340/430	Water, pH 7	230
2-Hydroxybenzoic acid: see salicylic acid			
3-Hydroxybenzoic acid	314/423	Water, pH 12	1191
	287/341	Water, pH 11	1190
	320/400	Paper strip	1114
	– /345	Chloroform	1190
4-Hydroxybenzoic acid	277/307	Water, pH 1	1190
	290/330	Paper strip	1114
	– /311	Chloroform	1190
2-Hydroxycinnamic acid	320/427	Methanol	1200
	311/410, *500*	Water, pH 7	1200
	360/500	0.1 N NaOH	1200
3-Hydroxycinnamic acid	314/416	Methanol	1200
	310/407	Water, pH 7	1200
	337/515	0.1 N NaOH	1200
4-Hydroxycinnamic acid	350/440	Water, pH 7	230
	340/430	Paper strip	1114
	309/418	0.1 N H_2SO_4	1200
	310/382	Methanol	1200
5-Hydroxyindole-3-acetic acid	307/420	1 N NaOH	321
	300/340	Paper strip	1114
	300/355	Water, pH 7	230
	290/355	TLC plate	317
	300/545	3 N HCl	320
	306/342	Water, pH 2	320
	306/350	Water, pH 5–10	320
3-Hydroxykynurenine	365/460[b]	Water, pH 11	230

Table 3.21. (*continued*)

Compound	Excitation/ Emission maximum	Experimental Conditions	Ref.
5-Hydroxykynurenine	375/460	Water, pH 11	230
4-Hydroxymandelic acid	300/380	Water, pH 7	230
4-Hydroxyphenylacetic acid	280/310	Water, pH 7	230
4-Hydroxyphenylpyruvic acid	290/345	Water, pH 7	230
4-Hydroxyphenylserine	290/320	Water, pH 1	230
5-Hydroxytryptamine, see serotonine			
5-Hydroxytryptophane[c]	280/360	Water, pH 4.6	1123
	293/356	TLC plate	317
	307/380	1 N NaOH	321
5-Hydroxytryptophol	308/351	TLC plate	317
Humic acid	270/460	Water, pH 10	1203
Indole-3-acetic acid[d]	290/350	Paper strip	1114
Indoxyl sulfate	290/380	Paper strip	1114
Isoeugenol	258, 294/342	Water, pH 2	1130
	281, 309/396	Water, pH 13	1130
Kynurenine	370/490	Water, pH 11	230
Kynurenic acid	335/385	Water, pH 7	1124
	335/410	Water, pH 1	1124
	340/435	14 N H_2SO_4	1125
	335, 365/380, 470	Paper strip	1114
Melatonine	300/540	3 N HCl	1212
	300/350	TLC plate	317
	307/430	1 N NaOH	321
2-Methoxycinnamic acid	322/410	Methanol	1200
	311/404	Water, pH 7	1200
3-Methoxycinnamic acid	315/414	Methanol	1200
	302/400	Water, pH 7	1200
4-Methoxycinnamic acid	309/390	Methanol	1200
	296/392	Water, pH 7	1200
5-Methoxyindole-3-acetic acid[c]	305/350	TLC plate	317
	307/430	1 N NaOH	321
5-Methoxytryptamine[c]	307/430	1 N NaOH	321
Noradrenaline	285/325	Water, pH 1	230
Physostigmine	265/315	Water	320
5α-Pregnan-3-ol-20-one	286/413	Ethanol	1136
Salicylic acid	314/410	Water, pH 5.5	1191
	314/402	Water, pH 12	1191

Table 3.21. (*continued*)

Compound	Excitation/ Emission maximum	Experimental Conditions	Ref.
	303/403	10 μ*M* in methanol	1167
	303/452	10 mM in methanol	1158
	310/435	Water, pH 10	1108
	296/408	Water, pH 7	1171
	~300/444	Chloroform	1171
	322/392	6 *M* KOH	1165
	329/405	Conc. H_2SO_4	1165
	304/350, 438	Methanol–H$^+$,	1168
	302/398	Methanol–OH$^-$,	1168
	313/415	Methylcyclohexane	1168
	300/410	Paper strip	1114
Salicylaldehyde	327/525	Cyclohexane	1156
Methyl salicylate	302/366, 448	Water, pH 1–7	1171
	~305/450	Chloroform	1171
	300/350, 450	Methanol	1165
	308/350, 450	Methylcyclohexane	1168
Salicyluric acid	330/400	Paper strip	1114
Serotonin^c	295/340	Water, pH 2	230
	295/540	Water, pH 0	230
	295/550	3 *N* HCl	1104
	312/347	TLC plate	317
	307/380	1 *N* NaOH	321
Stipitatic acid	380/430	Water, pH 2.5	1376
Stipitatonic acid	420/465	Water, pH 2.5	1376
Tryptamine	290/360	Water, pH 7	230
	307/420	1 *N* NaOH	321
Tryptophol^d	290/360	TLC plate	317
	307/420	1 *N* NaOH	321
Tyramine	275/310	Water, pH 1	230
Uric acid	325/370	Water, pH 11	230
Xanthurenic acid	350/460	Water, pH 11	230
	370/530	5 *N* NaOH	1125
	340/430	Paper strip	1114

^{a} All data for room temperature measurements.
^{b} The compound was later said to be not fluorescent (1124).
^{c} See also Table 3.18.
^{d} See also Table 3.9.

bolites, 3,4-dihydroxyphenylacetic acid, homovanillic acid, and others also have rather shortwave emission maxima. They are listed in Table 3.21.

Wachter et al. (1114) have measured the fluorescence excitation and emission maxima of a variety of hydroxy-substituted aromatic carboxylic acids, which are frequently encountered as urine constituents resulting from the metabolism of aromatic amino acids. The spectra were run directly on paper strips after electrophoretic separation of mixtures at pH 8.6. Nanogram amounts could be detected without staining or other chemical treatment of the spots. Identification was greatly facilitated with the help of the reference substance salicyluric acid.

Contrary to previous results, the energies and quantum yields of the fluorescence of benzoic acid in hydrocarbon at 77 K are independent in the 280-nm excitation region (1161). Fluorescence is exclusively from the dimeric form at 77 K, in fluid solution and in the gas phase (1161,1162), as well as in host single crystals (1163). The laser-induced fluorescence excitation spectra showed that in the gas phase electronic excitation of benzoic acid dimers is essentially localized in one monomer unit (1162).

The fluorescence of benzoic acid is drastically enhanced in strong sulfuric acid. The titration of intensity versus acidity gives a pK_a value of 1.5 for the first excited singlet state (1164).

The unusual fluorescence properties of salicylic acid and its esters and ethers have received much attention. In 1955, Weller (1165) found very large Stokes' shifts in the emission spectra of salicylic acid and its methyl ester, but not from its O-methyl ether. The unusually longwave-shifted fluorescence was said to stem from an excited-state tautomer. Its formation was explained (1165,1166) in terms of a reversal of the acidic and basic functionalities in the molecule. Photoexcitation results in both an increase in acidity of the phenol and in basicity of the carbonyl group. The shifts are so dramatic that the phenolic proton is able to protonate the basic carbonyl group. This is shown schematically in Scheme 5.

Further studies revealed that the photophysics of salicylic acid is even more complex. The fluorescence maxima depend critically on the concentration (1158) and, in aqueous solution, on the pH value (see Table 3.21). The bathochromic shifts observed on increasing the concentrations were interpreted in terms of dimerization (1158,1168,1169). Remarkably, in water solution no evidence for association was found in the spectra (1169).

In neutral aqueous solution, wherein salicylic acid exists as a single charged anion, fluorescence from the phototautomer is still observed (1170). It was concluded that the intramolecular proton transfer in the excited state is mediated by intramolecular hydrogen bonding in the ground and Franck-Condon excited states (1171).

The fluorescence intensity is highest for the monosodium salt of salicylic acid, with a ϕ_f value of 0.28 in aqueous solution (158). Sodium salicylate in solid form has ϕ_f 0.58 (1172). The decay time of salicylic acid in ethanol at room temperature is 8.8 ns (1173).

Salicylic acid esters have been used as models for salicylic acid which do not dimerize. They have two emission bands too, one in the ultraviolet, one in the visible (1165,1166). The bands were again attributed to fluorescences from nontautomerized and tautomerized states, respectively.

In protic solvents the two maxima are at 355 and 444 nm, and their relative intensities depend on the excitation wavelength. This was interpreted in terms of equilibria between two different hydrogen-bridged ground state species (1174). A third species detected in methanol solution was due to traces of the phenolate, generated by dissociation.

In recent years, studies have been published on the picosecond decay kinetics of salicylic acid and its methyl ester (1168). Also studied were: the solvent dependence of the lifetime (1175,1176) and nonradiative decay rate (1176), the high-resolution excitation spectrum of methyl salicylate in a free jet (1177), the fluorescence lifetimes of alkyl salicylates in the gas phase and in cyclohexane (1178,1179), the dispersed fluorescence of gaseous methyl salicylate, including decay time measurements (1180) and the photon energy relaxation at energies above the O-O transition (1181).

Evidence has been presented that the phototautomerism of methyl salicylate in aqueous solution proceeds biprotonic, involving diffusion-limited protonation of the carbonyl group and dissociation of the hydroxyl group by the solvent (1171). It is now generally assumed that methyl salicylate in its ground state consists of a mixture of two (1168,1174,1178,1179,1182) or even three (1183) conformers.

Salicylic acid is one of the metabolites of aspirin and can be determined by fluorimetry in serum (1184), urine (1185,1186), and related biological matter (1187). Other metabolites such as salicyluric acid (1187,1188) and other soluble conjugates (1189) are also fluorescent.

The fluorescence properties of *m*- and *p*-hydroxybenzoic acids and their methylated derivatives were studied by Paul and Schulman (1190). Except for methyl *m*-hydroxybenzoate, all were found to fluoresce in the pH 1–10 range. However, the fluorescences of all species, except those of the dianions and monoanions derived from *m*- and *p*-hydroxybenzoic acid, are very weak. There is no evidence of an excited state tautomerism, which was explained in terms of short excited state lifetimes.

The fluorimetric detection of salicylic acid and *m*-hydroxybenzoic acids in mixtures is possible, because at pH 5.5 only the *ortho*isomer is fluorescent, and at pH 12.5 both isomers are fluorescent (1191). The fluorescence of salicylic acid, under 314 nm excitation, drops sharply below pH 5, while that of the *meta*isomer drops below pH 12.

Gentisic acid (2,5-dihydroxybenzoic acid) has been studied in aqueous solutions of various acidities (1192). Fluorescence was observed at all pH values above 3, but not from the uncharged and protonated forms. The longwave emission found in neutral solution was attributed to the tautomerized form. The latter has the negative charge located at the phenolate oxygen rather than at the carboxyl group, which, in the excited state, is not dissociated. Fluorescence of gentisic acid was also observed when spotted on paper chromatograms (1193).

In a similar way, methyl 2,6-dihydroxybenzoate shows two emission bands with intensity ratios depending on the excitation wavelengths (1194).

Gallic acid shows fairly strong and pH-dependent fluorescence in aqueous solutions and chemiluminescence in oxygenated alkaline solutions (1195). The characteristic blue and red shifts in the emission spectra are in accord with the trends usually observed for these functional groups. It was reported that gallic acid forms a copper(II) complex with green fluorescence (1196).

Spiegl (489) studied the visible fluorescence of various mono-, di-, and trihydroxybenzoic acids and some of the methylethers on TLC plates. Only 2-hydroxy-, 2,5-dihydroxy-, and 3,5-dihydroxybenzoic acids showed visible emission (violet or blue). Spraying with aluminum trichloride reagent makes 2,4-dihydroxybenzoic acid fluorescent, whereas with β-aminoethyl diphenylborate ("Neu's reagent") the 3,4-dihydroxy- and 3,4,5-trihydroxy derivatives develop strong violet fluorescence.

Hydroxycinnamic acids are more widely distributed in the plant kingdom than hydroxybenzoates. According to Spiegl (489), 3-hydroxy-, 4-hydroxy-,3,4-dihydroxy-, and 4-hydroxy-3-methoxycinnamic acid exhibit a violet fluorescence on TLC plates. Wolf (1197) investigated the fluorescences of 4-hydroxycinnamic acid, chlorogenic acid, ferulic acid, and sinapic acid. Except for the first compound, the fluorescence spectra from ethanolic solutions were found quite similar with maxima between 430 and 445 nm. The *p*-hydroxy acid fluoresced in 0.1 *N* sodium hydroxide only. Ferulic acid was the most intensely fluorescent compound, its intensity being around five times as great as that of the other compounds.

The absence of appreciable fluorescence of *p*-coumaric acid in ethanol tends to confirm the observation by Bate-Smith (1198), who reported a violet fluorescence in either acid or alkaline solution, but not at neutrality, whereas other cinnamic acids found in leaves exhibited violet, blue, or green fluorescences.

Neish (1199), on the other hand, estimates a fluorescence maximum of 440 nm for *p*-coumaric acid under 350-nm excitation. Our recent investigations showed that *p*-coumaric acid is indeed fluorescent, albeit weakly so (1200). The data for the three isomeric monohydroxycinnamic acids, their respective esters and ethers are given in Table 3.21.

The fluorescence intensities of the three hydroxycinnamic acids are not strong in methanol or water, but increase after addition of base. Methoxycinnamic acids and their esters fluoresce much more strongly than the respective phenols. *o*-Hydroxycinnamic acid develops a particularly strong green fluorescence (λ_{max} 500 nm) at pH values above 8, which practically disappears at lower alkalinities. However, even at pH 4 still a weal green fluorescence can be detected, evidently as a result of photodissociation.

The fluorescence in acidic solutions is so weak that it has been stated that *o*-hydroxycinnamic acid in its uncharged or monoanion form is nonfluorescent (1201). At pH values below 4 a new and extremely longwaveshifted emission band is found, which we attribute to an excited state species formed by photo dissociation of the phenol group (see Section 3.7.6 and Fig. 3.4). No such longwave bands are found with *o*-methoxycinnamic acid. Its fluorescence is blue in 0.1 *N* acid and violet in neutral solution.

Both fluorescence excitation bands of *o*-coumaric acid are positively polarized in ethanol solution at 77 K (1202).

Humic substances are a complex group of organic compounds characterized mainly by functional groups such as carboxyl and phenolic hydroxyl. Their native fluorescence is used for routine water quality monitoring. The fluorescence excitation spectra of fulvic acid and humic acid were recorded at different pH values and sodium chloride concentrations (1204). The spectra of both compounds exhibited emission maxima at 465 nm, the intensities being highest in alkaline solution and in the absence of salts. The excitation spectra can be used to differentiate between fulvic and humic acid.

These results corroborate former studies (1205) on three humic acids, which—at nearly equal wavelengths of emission—showed distinct deviations of the exciting wavelength. Futhermore, the relative fluorescence intensities and their pH dependence were different for the three species.

The fluorescence spectrum of fulvic acid is broad and featureless with a maximum around 450 nm (1206), which shifts to longer wavelengths as the excitation wavelength is increased (1207). The fluorescence is quenched upon binding to paramagnetic metal ions (1208,1209). The fluorescence titration with copper(II) ion allowed the determination of complexing capacities and stability constants (1210).

Fluorescence polarization and lifetime measurements as a function of pH, concentration, and ionic strength were used to calculate rotational relaxation times (2.0 ns) of fulvic acid in solution (1211). Since the rotational relaxation time did not change over the pH range 5–8 over the concentration range from 33 to 330 μM, it was suggested that fulvic acid does not aggregate or change conformation.

3.9.4.   Other Natural Products

Aside from the quinone and polyene antibiotics (Sections 3.8.5 and 3.9.1) some polypeptide antibiotics are known to fluoresce, provided they do contain an aromatic amino acid. Representative examples are crotoxin (1213), gramicidin A (1214), and megacin Cx (1215), all of which exhibit typical tryptophan fluorescences. Their fundamentals have already been discussed in detail (Section 3.3.2).

Actinoxanthine, an antineoplastic protein type antibiotic with a supposed naphthalenecarboxylic acid fluorophore, shows fluorescence with maximum at 420 nm (1216). The isolated chromophore of neocarcinostatin (2-hydroxy-5-methoxy-7-methylnaphthoate) is as active as the holoprotein. Both exhibit a fluorescence maximum at 420 nm when excited at 340 nm (1217). The fungal metabolite griseofulvin has fairly strong native fluorescence (1218) with excitation peaks at 295 and 315 nm, an emission peak at 420 nm and a quantum yield of 0.108 (in methanol) (1219). Its short lifetime (0.6 ns) is consistent with the lack of oxygen quenching. The spectra of TLC spotted griseofulvin are almost the same.

The toxic fungal metabolites ochratoxin A and B show strong native fluorescence (1220,1221). Their excitation spectra are different in having maxima at 340 and 325 nm, respectively, while the emission maxima are at 475 nm in both cases. It was also found that the fluorescence maxima of ochratoxin A are concentration dependent (1219), ranging from 333 to 376 nm in excitation and from 426 to 465 nm in emission. The fluorescence spectra of ochratoxins on TLC plates have been reported (excitation maximum ~340 nm, emission maximum ~475 nm), and their fluorodensitometric determination has been described (948,1220).

Gillespie and Schenk (1221) presented fluorescence data for the mycotoxins ochratoxin A and B, their respective methyl esters, patulin and zearalenone, together with the fluorimetric detection limits (0.5–0.007 ppm in methanol or ethanol). The quantum yields for the ochratoxins are greater than 0.8, but only 0.02 for zearalenone in ethanol at room temperature. Plots of pH versus intensity showed that maximum intensity of ochratoxins is reached at pH values above 9.

The potent estrogenic mycotoxin zearalenone can be assayed with high sensitivity by laser fluorimetry (1222). For sterigmatocystin a fluorodensitometric method with low sensitivity has been published (1223). A more sensitive assay is based on its conversion to the more fluorescent monoacetate (1124).

Lasalocid A, a poultry coccicidal, has a salicylic-acid-type fluorophore with maximal excitation at 310 nm and maximal emission at 420 nm in water or methanol. The structural aspects of this ionophore have been studied by various fluorescence techniques (1225). From energy transfer and circular polarization measurements it was concluded that lasalocid binds trivalent lanthanides stronger than calcium(II) (1225) and that the conformation of the complexes depend on the metal ion and solvent (1226). It can be determined in biological matter with high sensitivity by HPLC separation combined with fluorescence detection (1227).

Antimycin A, another salicylate-type antibiotic, has an intrinsic fluorescence that is maximal in the range of pH 7–9 and that is influenced by the amount of ethanol in the aqueous solution (1228). Fluorescence was utilized for its quantitation (1229).

The fluorescence of the bleomycins with their bithiazole and aminopyrimidine fluorophores has received considerable attention. Bleomycin A in aqueous solution has an excitation maximum at 300 nm and an emission peak positioned at 355–360 nm (1230). The fluorescence spectrum is asymmetric and has a weak shoulder around 400 nm. There is an indication that the 360-nm band comes from the bithiazole rings and that the 4-aminopyrimidine is responsible for the 400-nm band. A hydrolytic fragment of bleomycin containing the aminopyrimidine moiety showed indeed a pronounced 400-nm band (1230).

Fluorescence quenching of bleomycins after binding to metal ions such as gallium(III) (1231), trivalent lanthanides (1232), iron(II), and copper(II) (1233) has been investigated. The binding specificity of copper-chelated bleomycin A_5 to various nucleic acids was determined by a fluorescence quenching method (1234).

In tallysomycin A an extra sugar unit is present, but the fluorescence properties are identical with those of bleomycin A_2 (1233). Phleomycin lacks the aromatic conjugation system shared by bleomycins because of hydrogenation of one of the double bonds. As a result, the fluorescence spectrum shows a markedly reduced intensity at 360 nm (one-tenth of bleomycin) and the 400-nm shoulder becomes more prominent (1233). A mode-locked krypton ion laser was used as an excitation light source to measure the fluorescence decay times of chromomycin with its dihydroxynaphthalene fluorophore (1235).

Table 3.22. Excitation Wavelengths and Emission Maxima of Natural Products Treated in Section 9.4.

Compound	Excitation/ Emission (nm)	Experimental Conditions[a]	Ref.
Actinoxanthine	335/420	Water	1216
Antimycin A	350/420	Water, pH 8	1229
Griseofulvin	295, 320/420	Methanol	1219
	295/420	TLC plate	1219
Lasalocid	310/420	Methanol[e]	1225
Luciferin	384/537	Water, pH 11	1236
	328/537	Water, pH 5	1236
	330/415	Ethanol	1236
Luciferyl adenylate	336/548	Water, pH 5	1236
Bleomycin A_2	300/360, 400[b]	Water	1233
Neocarcinostatin	340/420	Methanol	1217
Ochratoxin A[c]	380/428[d]	Methanol	1221
	333/465	Ethanol, 0.1 mM	948
	376/426	Ethanol, 1 μM	948
Ochratoxin B[c]	366/418[d]	Methanol	1221
Patulin	278/340	Ethanol	1221
Phleomycin	300/360, 400[b]	Water	1233
Tallysomycin	300/355, 400[b]	Water	1233
Zearalenone	275, 365/465	Methanol	1221

[a] At room temperature.
[b] Shoulder.
[c] The respective ethyl esters have very similar spectra (1221).
[d] Corrected.
[e] Fluorescence 8.7 times stronger in benzene (1417).

Although luciferin most likely is not the emitting species, the emission characteristics of its monoanion are similar to the spectrum of firefly bioluminescence. The fluorescent state of luciferin is characterized by a strong charge transfer from the hydroxybenzothiazole to the thiazoline ring, resulting in large lowering of the ground state pK_a value of the phenolic hydroxy group (from 8.7 to $-$ 0.8), an increase in the excited state dipole moment and a strong solvent and temperature dependence of the Stokes' shift (1236,1237).

In aqueous solution of pH between 4 and 8 luciferin undergoes photodissociation and fluoresces from the anion form (λ_{max} 560 nm) (1236). In pure ethanol, fluorescence is from the undissociated form (λ_{max} 435 nm). The fluorescence quantum yields of luciferin are 0.62 at pH 11 and 0.25 at pH 5. They are much smaller in organic solvents such as ethanol

(0.03). Divalent cations (e.g., zinc) cause longwave shifts in both excitation and emission. The fluorescence polarization of luciferin phenolate anion is significantly higher than that of the phenol form under identical conditions in the absence of rotational polarization (1237).

A supposed polypyrrole-type luciferin isolated from a dinoflagellate displayed strong blue fluorescence (λ_{max} 474 nm) when excited at 390 nm (1238). The emission spectrum closely matches the bioluminescence spectrum. Table 3.22 summarizes the spectral properties of some of the natural products treated in this section.

3.10. APPENDIX

During and after the preparation of this manuscript important contributions to the knowledge of the fluorescence of natural products have been published. They are briefly summarized in this chapter, thus completing the relevant literature up to the end of 1983.

Amino Acids.   The fluorescence properties of the aromatic amino acids remain under active research. Quenchings of tyrosine (Tyr) fluorescence by copper(II) ion (1210), glycine, alanine, and related amino acids (1239), and -in contrast to former reports (47)-, by molecular oxygen (1240) were observed. The Tyr emission maximum shifts to longer wavelengths by 1–5 nm, when the concentration of urea or polyethylene glycol in the aqueous solution is increased (1241). Simultaneously, the quantum yield of fluorescence of Tyr is increased, being 0.29 in 60% polyethylene glycol.

The nature of the fluorescent state of indole (the fluorophore of tryptophan) is still under investigation. In an excellent study, Meech et al. (1242) have explained the unusually large Stokes' shift in polar solvents by the existence of an ^1L_a state with increased charge transfer character, relative to the initially excited state. Complex formation between indoles and protic solvents has also been studied by fluorescence techniques (1243,1244a).

The biexponential fluorescence decay of tryptophan (Trp) in pH 7 solution at 20°C was confirmed. The two components have lifetimes of 3.15 and 0.54 ns, respectively (1244b), and are independent of emission wavelength. The authors prefer the Szabo–Rayner interpretation with the assumption of two conformers fluorescing (38) over the model of partial quenching of the indole by a protonated amino group (29). The photophysics of Trp in water, deuterated water, and in nonaqueous solvents was studied as a function of acidity (1245). As previously reported (36), the decay at pH values above the pK_a of the amino group is triple exponential, while at near neutral pH the decay is biexponential. The lifetimes become longer in deuterated water. The steady-state fluorescence quenching by hydrogen ion at pH values below 3 shows a definite wavelength effect, consistent with less pronounced quenching of the subnanosecond component, whose emission maximum is at 330 nm. The Stern–Volmer plots show a marked curvature in the direction of decreasing Stern–Volmer constant as the acidity increases.

The multiple exponential decay of Trp and its derivatives was discussed in terms of a model of different conformers (1246) that was previously invoked by others. The strictly single-exponential decay of the model peptide, N-acetyl tryptophanamide (NATA) was confirmed (1246), and a lifetime of 3.05 ns was reported for solutions of pH 7 and 20°C (1244). The collisional quenching of NATA by iodide was studied on a picosecond time scale, and the interaction radius was determined (1247). The decay of NATA deviates from single exponentiality at high iodide concentrations, but does not become double exponential. Leismann et al. (1248) determined the subnanosecond fluorescence decay of Trp using a nanosecond flashlamp.

Excited singlet–triplet crossing of Trp in dimethylsulfoxide is enhanced in the presence of trivalent lanthanide ions, leading to fluorescence quenching (1249). The photophysics of Trp has been reviewed (1383).

Evidence has been presented that electron ejection, one of the major competitive pathways for the deactivation of excited tryptophan, occurs in a time that is several orders of magnitude shorter than the fluorescence lifetime (1250).

Second derivative fluorescence spectroscopy has been applied in a method for an easy, accurate, and rapid determination of tyrosine and tryptophan (1251). It is based on the addition of small amounts of a standard solution to the samples, followed by measurement of the increase in the distance between a selected minimum and an adjacent maximum in the second derivative spectrum.

Peptides. Another example of tyrosinate fluorescence (λ_{max} 340 nm) of a protein at neutral pH was observed with the H_1 histones from calf thymus and those from a fruitfly (1252). In bovine testes calmodulin both Tyr and tyrosinate fluorescence were observed (1384). In the β-subunit of the bovine brain S 100-b protein, however, Tyr residues show the typical 303-nm fluorescence (1253). Permyakov and co-workers (1254) continued their studies on the fluorescence of class A proteins and have recently studied the tyrosyl and phenylalanine fluorescence of two pike albumins.

Acrylamide, a useful neutral quencher for Trp, has been applied to study the Tyr fluorescence of calmodulin and its Ca(II) complex, taking into account the inner filter effect of acrylamide at the excitation wavelength of Tyr (1255). The structure of the histone H2A-H2B dimer in solutions of different ionic strength was investigated by the same method (1256).

Burstein has written a review on intrinsic protein luminescence as a tool for studying fast structural dynamics (1257) and a book edited by Steiner (1258) contains two chapters on protein fluorescence, one dealing with nanosecond pulse fluorimetry, the other with the use of fluorescence anisotropy decay in the study of macromolecules. However, many of the methods described there make use of an added fluorescent probe rather than of the native fluorescence of the protein. The intrinsic fluorescence of proteins has recently been reviewed (1385).

The model peptides, tryptophylglycine, tryptophylalanine, glycyltryptophan, alanyltryptophan, glycyltryptophylglycine, and Trp itself have been investigated by using steady-state and subnanosecond spectroscopy (1250). For Trp, and the peptides where the tryptophyl residue is the N-terminal, the involvement of the state of protonation on the excited state dynamics and the absorption and emission spectra was demonstrated. It was shown, for the first time, that the deprotonated amino group gives rise to red-shifted absorption and emission spectra and to longer fluorescence decay time compared with the protonated form. It is suggested that the role of the protonated amino group in the fluorescence quenching of the N-terminal tryptophyl compounds is not proton transfer to the indole ring, but an enhancement of charge transfer from the indole ring to the adjacent carbonyl group.

More than 40 papers have appeared in 1983 that describe the fluorescence of proteins and its use to study their properties. Only a few can be cited here. Mention should be made of studies on the molecular dynamics of lysozymes (1260) and 14 other proteins (1261), conformational changes by using acrylamide as a dynamic quencher (various proteins) (1262), microenvironments within proteins such as bacteriorhodopsin (1263), intramolecular distances within proteins such as alkaline bacterial proteinases (1264), effects of metal ion binding [Ca^{2+}–parvalbumin (1265), Ca^{2+}–melittin and calmodulin (1266)], protein–protein interactions (thrombin–antithrombin) (1267), and interconversion of two forms of human prostatic acid phosphatase (1267).

The usefulness of quenching studies on proteins to elucidate their tertiary and quaternary structures as well as their motilities in solution becomes more and more evident. Thus, the quenching of Tyr fluorescence by molecular oxygen and time resolution methods have served to study the accessibility and motional freedom of tyrosines in small peptides and random coil proteins (1240).

The quenching of the Trp residues by oxygen in an alcohol dehydrogenase and an alkaline phosphatase was used to estimate the rate of reaching the buried Trp's in these enzymes (1268). Cinnamic acid amide, mesityl oxide, acrylic acid, and certain ketones were introduced as new quenchers of Trp fluorescence of NATA, alcohol dehydrogenase, and alkaline phosphatase, and their quenching constants were given (1269). The utility of the established dynamic quenchers acrylamide and molecular oxygen was demonstrated in an investigation on the fluorescence quenching of a liver alcohol dehydrogenase (1270).

Quenching by molecular oxygen and fluorescence polarization of Trp residues was also used to study the effect of ligand binding and conformational changes in various proteins (1387).

A model was presented for the quenching of fluorophors (such as Trp) in a protein interior (1388). At low quencher concentration the quenching process is determined by the acquisition rate of quenchers by the protein, the migration rate of quenchers in the interior, and the exit rate of quenchers from the protein. In cases where the fluorescence observed in the absence of a quencher could be described by a single exponential decay, the presence of quenchers led to double exponential decay times, and the aforementioned exit rates of quenchers could be determined from experimental data.

A fluorescent compound called pyridinoline was identified as a component of collagen, which is present in addition to tyrosine and another unidentified third fluorophore (1271). The fluorescence of pyridinoline is maximal at 395 nm and depends largely on the pH value of the solution.

Phalloidin and α-amanitin two extremely poisonous cyclic peptides from a mushroom, possess two unusual Trp-derived fluorophores. In comparison with Trp their excitation and emission maxima are shifted to the red, and the quantum yields are much lower (1272).

Fluorescence emission after two-photon excitation at 532 nm by means of a laser was observed in apohemoglobin, hemoglobin, albumin, and Trp at room temperature (1273). Fluorescence originates from Trp residues in proteins. In contrast, no fluorescence of a biphotonic nature could be detected from lysozyme and Tyr.

The total UV fluorescence of rat serum was presented as a contour plot ("fluorescence topogram"). It consists almost exclusively of contributions resulting from Trp, partially due to energy transfer from Tyr, and suffers a significant decrease in intensity with increasing tumor burden (1274). The UV fluorescence topogram of human serum is similar (Fig. 3.7) (1275).

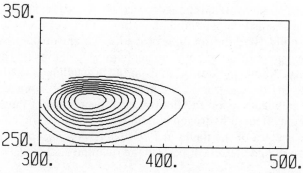

Fig. 3.7. Ultraviolet part of the total fluorescence of human serum. Dilution 1:500, pH 7.40; abscissa: emission wavelength (in nm), ordinate: excitation wavelength (in nm); contour lines are from 10 to 90% of the peak fluorescence intensity in 10% increments. The peak at 388/342 nm is assigned to fluorescence from tryptophan, whereas tyrosine does not form a peak of its own (at 275/303 nm), but rather transfers its energy to tryptophan. (From ref. 1275.)

The fluorescence of Trp-containing proteins adsorbed at solid–liquid solution interfaces can be excited by evanescent waves of totally reflected light. This is the basis for a new technique called *total internal reflection intrinsic fluorescence* that allows in situ studies on adsorbed proteins with 0.1-s time resolution (1276).

Medium-sized peptides have been separated by HPLC methods and were sensitively detected by their native fluorescence resulting from the presence of Trp moieties (1277).

Coenzymes and Vitamins. Fluorescence lifetime data obtained by the phase modulation technique were used to resolve the heterogeneous decay of horse liver alcohol dehydrogenase. The two lifetimes found (3.95 and 7.45 ns at 20°C) were in good agreement with data obtained by other methods (1278).

Salmon et al. (1279) studied the ratio of free and bound NAD(P)H [reduced nicotinamide adenine dinucleotide (phosphate)] as a metabolic indicator in various cell types. The fluorescence maxima of free NAD(P)H (475 nm) and protein-bound NAD(P)H (450 nm) were discerned, and the effect of microinjection of enzyme substrates such as glucose-6-phosphate, malate, or ADP under 365-nm excitation was followed. Under 436-nm excitation, microinjections of these enzyme substrates result in a redox change of the flavins, indicated by a decrease in fluorescence at 510 nm.

Visser (1280) has studied the rotational motion of two flavins with octadecyl side chains in dipalmitoyl phospholipid bilayers. At temperatures between 10 and 50°C the rotation proved to be anisotropic, which indicated composite motion of both the aliphatic side chain and the isoalloxazine moiety. Above the transition from the crystalline to the liquid crystalline state the depolarization is complete within the average lifetime.

Further publications on the fluorescence properties of flavins involve a study on the effect of hydrogen bonding on the polarization of riboflavin (1281), the spectra of the flavin in *Photobacterium phosphoreum* (1282), and the picosecond lifetime of FAD in the D-amino acid oxidase–benzoate complex (1283).

A most interesting article on the enigma of the spectral properties of old yellow enzyme (NADPH oxidoreductase) has been published (1284). Spectra, lifetimes, and quantum yields were reported for free and enzyme-bound FMN. Free FMN fluoresces much more intensely than enzyme-bound FMN and shows a single exponential decay (5.7 ns). The enzyme exhibits an excitation wavelength-dependent decay, possibly because of contributions from nucleic acid excimers.

Formation of singlet charge transfer states in the course of the fluorescence quenching of lumiflavin, riboflavin tetrabutyrate, and D-amino acid oxidase by indole have been directly observed by means of time-resolved absorption measurements in the 100–3000-ps time regime (1285). Photoexcited lumiflavin was also shown to undergo an electron transfer reaction with a variety of electron donors, which thereby act as quenchers (1286).

The fluorescence spectra of flavin mononucleotide, encapsulated in surfactant-entrapped water pools, in apolar solvents show unresolved bands, which—in comparison to those in water solution—are red-shifted (1287). The decay is double-exponential, with $\tau_1 = 3.7–4.6$ ns (weighting factors-0.57–0.76), and $\tau_2 = 0.2–1.6$ ns (weighting factor 0.24–0.43). Lifetimes are much shorter in the presence of bromide ion. The emission properties of riboflavin-5-phosphate, riboflavin, and 10-formylmethyl-isoalloxazine in hydrophilic matrices such as polyvinyl alcohol (PVA) or glycerol show some differences when compared with those of aqueous solutions (1288). In PVA the fluorescence maxima are blue-shifted (-16 to -18 nm), and show smaller Stokes losses ($\sim 3130–3200$ cm^{-1}). Lifetimes are much smaller, and the angles between the transition moments of absorption and emission are bigger in glycerol than in PVA.

Two new metabolites of riboflavin, namely, its 7α-hydroxy and 8α-hydroxy derivatives, have been identified in human urine (1289). Their absorption and fluorescence properties are very similar to those of riboflavin.

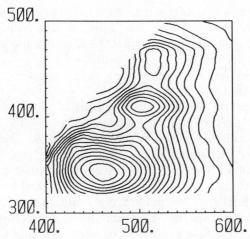

Fig. 3.8. Visible fluorescence of human serum. Dilution 1:20, pH 8.11; abscissa: emission wavelength (in nm), ordinate: excitation wavelength (in nm); contour lines from 10 to 90% of the intensity of the highest peak (at 345/460 nm) in 5% increments. The peaks are assigned to fluorescences from reduced nicotinamide adenine dinucleotide (phosphate) (maximum at 345/460 nm), pyridoxal phosphate Schiff base (410/505 nm), and bilirubin (~460/515 nm), respectively. Shoulders at 365/425 nm and 370/500 nm are tentatively assigned to pyridoxic acid lactone and flavin mononucleotide, respectively. Note that the bilirubin fluorescence consists of a double peak. (From ref. 1275.)

The fluorescence of thiamine pyrophosphate at 77 K in aqueous solutions was found to vary with pH, which is attributed to protonation of the pyrimidine fluorophore in the pH 6.5–7.5 range (1290).

Four tocopherols and two tocotrienols were determined in feedstuffs (1291), plasma and erythrocytes (1292) by HPLC with fluorescence detection (excitation at 294 nm, emission at 325 nm). α-Tocopherol was recommended as a lipoid fluorescent probe for liposomes and was spectrally characterized (1293).

NAD(P)H, flavins, and enzyme-bound pyridoxal phosphate are mainly responsible for the longwave part of the total fluorescence of rat sera (1294). In human serum the fluorescence of enzyme-bound bilirubin forms an additional peak (Fig. 3.8) (1275).

Nucleic Acid Components.   Gibson and Turnbull (1295) reported the singlet excited-state parameters of adenine, adenosine, and adenosine phosphates in ethanol. The fluorescence quantum yield of adenine was found to be extremely low at room temperature (0.00024), but much higher at 77 K (0.19). The experimental lifetime was 2.1 ns at 77 K.

Whereas the absorption spectrum of adenosine is identical with that of adenine, its fluorescence is shifted to the red. The 9-ribose substituent seems to have a quenching effect, since the lifetimes and quantum yields are reduced. In adenosine-5'-monophosphate the fluorescence is further shifted to the red, and the lifetimes are slightly shorter. 5'-ADP and 5'-ATP have similar characteristics.

The effect of pH on the fluorescence spectra of purines and pyrimidines has been investigated over a wide acidity range, and the pK_a values of the first excited singlet states were determined (1389). The effects of pH were also studied by another group (1390) which utilized the differences in the spectra for a selective HPLC detection of certain purines, pyrimidines, their methylated derivatives, nucleic acids, and bases.

Murray and Becker (1296) published a study on the fluorescence excitation and emission spectra, quantum yields, and lifetimes of the pyrimidine bases orotic acid and iso-orotic acid. The data were compared with the parent compounds thymine and uracil, and an assignment of the nature of the lowest excited singlet states was made. In no case was any excitation wavelength dependence of the quantum yields observed.

Smith and Melhuish (1297) reported the fluorescence of the dinucleotide guanyl(3-5')cytidine at low temperatures, at pH 6. There is indication that the singlet state of an exciplex is formed in 1:1 ethylene glycol–water, where the bases are close together. The latter is concluded from the fact that the dinucleotide exhibits a marked hypochromism in emission.

The fluorescence emission spectra and lifetimes of adenine, polyadenylic acid, cytidine, and thymidine were determined at room temperature, making use of synchrotron radiation as a powerful excitation energy source (1298). Fluorescence quantum yields have been determined for adenine ($2.6 \times 10^{-4}\ M$ in phosphate buffer of pH 7 at room temperature) and polyadenylic acid ($3 \times 10^{-4}\ M$), unfortunately, under conditions of very high optical densities of the solutions. No evidence of a lifetime of more than 0.1 ns was found for the four DNA nucleosides.

The decay profiles of polyadenylic acid are dependent on the excitation and emission wavelengths and are complex. However, at an emission wavelength of 460 nm only one decaying species is apparent, with a lifetime of 4 ns. Despite this lifetime and the fact that the measurements were carried out at room temperature in the presence of oxygen, this decay was attributed to an excimer phosphorescence. At shorter wavelengths the emission profiles could be fitted to double-exponential decays in the 0.1–1.7-ns time range.

The first lifetime measurements of DNA fluorescence were performed with natural and synthetic material at 270-nm excitation. An exponentially decaying emission with a lifetime of 29 ± 0.4 ns for calf-thymus DNA, and of 3.0 ± 0.3 ns for poly(dA-T) was observed. This is most likely an excimer fluorescence (1299).

Selective laser excitation on nucleic acids has been performed taking advantage of small differences in the absorption spectra, absorption cross sections from the excited states, or lifetimes of the excited states (1391,1392). A review of electronic states and luminescence of nucleic acid systems has appeared (1300).

Alkaloids. The emission parameters of atropine, cocaine, hyoscine (scopolamine), homatropine, and tropic acid in ethanol at room temperature were determined (1301). The fluorescence maxima range from 280 to 307 nm, with quantum yields between 0.04 and 0.002. Microgram quantities of these drugs may be determined by fluorimetry.

The fluorescence spectrum of yohimbine, reserpine, rescinnamine, corynanthine, ajmalicine, reserpic acid, and 6-methoxyindole were measured in solution at 298 K and 77 K (1301). The data obtained provide a basis for the analytical determination of yohimbine and reserpine at low concentrations. Insight is afforded into quenching processes that degrade emission in reserpine and rescinnamine molecules.

Berberine derivatives have been used as fluorescent probes for the investigation of the energized state of mitochondria (1303), and the pH dependence of the absorption and fluorescence of sanguinarine (1304) and β-carbolines (1393) was described (1304). The strong intrinsic fluorescence of harmala alkaloids has been utilized to improve the detection limits by a factor of 100 over the photometric sensitivity (1305). Spectrofluorimetry, liquid chromatography, and fluorescence scanning TLC as methods for the determination of quinidine in serum were compared, and their precision was evaluated (1306).

The binding of colchicine analogs to purified tubulin is accompanied by an increase in fluorescence polarization (1307). Scatchard analysis of the titration curve gave values for the stochiometry and affinity constants. The mode of binding of various ellipticine-derived alkaloids to DNA was studied by fluorescence and polarization techniques (1394).

Oxygen Ring Compounds. The kinetics and mechanism of the excited state tautomerization of 7-hydroxy-4-methylcoumarin was reinvestigated by means of steady-state and time-resolved fluorescence spectroscopy. The rate constants of the excited state process were interpreted in terms of an intramolecular proton transfer, promoted by water molecules with a hydrogen bond surrounding the excited molecule (1308).

The excited state proton transfer of 3-hydroxyflavone ("flavonol," see Section 3.7.4) has been studied in more detail. Salman and Drickamer (1309) measured the solvent polarity and pressure dependence of the quantum yields and lifetimes. The radiative lifetime was found to depend strongly on the low-frequency dielectric constant of the solvent.

The time- and wavelength-resolved fluorescence data of flavonol were interpreted by assuming a photodynamic process that is well described by a model of two pathways of excited-state proton transfer (1310). The deuterium effects on the kinetics of the proton transfer were studied and were found to be small and not significantly temperature dependent (1311).

The tautomerization can be described by a kinetic model that includes two pathways (1311). The data revealed an irreversible excited state isomerization that occurs at a rate that is temperature and solvent dependent.

McMorrow and Kasha (1312) found that the low-temperature spectroscopic properties of flavonol and 3-hydroxy-2-methylchromone are controlled by traces of water in hydrocarbon solvents. In the absence of water, alcohol, or other proton-accepting molecules only the green (tautomer) fluorescence is found, apparently independent of both temperature and viscosity. This is in agreement with similar findings that the ratio of the intensity of the violet and green fluorescence bands at room temperature is highly dependent on the nature of the organic solvent (502).

Following the addition of water, the emission of flavonol in quickly frozen methylcyclohexane glass at 77 K can be resolved into three bands with maxima around 410, 490, and 520 nm, respectively. They were attributed, in that order, to the "normal" molecule's fluorescence, perturbed tautomer fluorescence, and tautomer fluorescence (1312). The tautomerism may even be observed in an isolated-site Shpolskii matrix (1395).

A single exponential risetime is observed for the tautomer fluorescence of 3-hydroxyflavone in the absence of hydrogen-bonding impurities in the temperature range of 298–77 K. The observed risetime in hydrocarbon solvents is less than 8 ps at 298 K, while at 77 K it is 37 $\pm$ 6 ps. The discrepancy with other reports is attributed to hydrogen-bonding solvent impurities that inhibit the intramolecular proton transfer process (1313).

Itoh and co-workers performed additional studies on the transient absorption of the phototautomers of 3-hydroxyflavones and 3-hydroxychromone (1314). Their photoexcitation by a laser flash into the transient absorption band ("two-step laser excitation") affords exclusively the green fluorescence of the excited state tautomer (1315a). The lifetime of this tautomer in its ground state (λ_{max}^{abs} 440 nm) is unexpectedly long (several microseconds). In addition, the triplet–triplet absorption band [λ_{max}

400 nm (1314), 405 nm (502)] as well as the $S_i \rightarrow S_n$ absorption of the tautomer (λ_{max} 460 nm) have been identified (1314).

The excited-state proton transfer of 7-hydroxyflavone (505) (see Section 3.7.6) has also been investigated by the two-step laser excitation fluorescence technique (1315b). It was demonstrated that in methanol, at 200 K, proton transfer takes place, forming two transient phototautomers with lifetimes of ~0.7 and less than 0.2 ns, respectively. Their fluorescence maxima are different (518 and 530 nm). Thus, three species contribute to the overall fluorescence of 7-hydroxyflavone, namely, the unchanged ("neutral") molecule (λ_{max} 410 nm) and the two excited state tautomers.

The fluorescence properties of the aluminum(III) chelates of naturally occurring flavones, such as kaempferol, quercetin, myricetin, apigenin, luteolin, and naringenin, were reported (1316). Fluorescence is due to chelating hydroxy groups in ortho- or peri-position to the carbonyl group. Chelates formed with 3-hydroxyflavones ("flavonols") show the shortest emission wavelength (1316), since complexation suppresses the formation of longwave fluorescent tautomers (497).

The fluorescence properties of various hydroxy- and methoxyisoflavones and the effect of complexing reagents was studied (1317). The results are similar to those obtained for flavones (Section 3.7.4) in that 5-hydroxyisoflavones are nonfluorescent, whereas all others fluoresce. Methoxyisoflavones exhibit stronger fluorescence than the respective hydroxy derivatives, and fluorescence efficiency increases rapidly with solvent polarity.

The fluorescence spectra of biflavones in typical solvents have been published (1318). All compounds under investigation are fully eitherified so that no excited state proton transfer reactions of phenolic protons can be expected. As a result, no unusual longwave emissions were observed. Fluorescence maxima are between 450 and 476 nm in water, and between 440 and 472 nm in isoamyl alcohol.

A laser flash photolysis and fluorescence study was performed on aminomethyltrimethylpsoralen in the presence and absence of DNA (1319). The strong temperature and solvent dependence of the photophysical properties of psoralens are due to efficient internal conversion that arises from the proximity of the lowest-energy n,π^* and π,π^* states ("proximity effect") rather than from changes in the intersystem crossing rates (949). The internal conversion rates increase with temperature and decreasing solvent polarity. Decay times and quantum yields of 8-methoxypsoralen in methylpentane, acetonitrile, ethanol, and trifluoroethanol were determined.

5-Methoxypsoralen was detected by fluorimetry in plasma (1320), perfumes and suntan cosmetics (1396), and the luminescent photoproducts of 8-methoxypsoralen were characterized (474,1321). A multianalytical method based on the native fluorescence of aflatoxins B and G, citrinin, and ochratoxin in molded foods was described (907), and the effect of solvent composition on the emission spectra of aflatoxins B and G was reported (1397). Sterigmatocystin can be quantitated via its yellow-fluorescent complex with aluminum trichloride (907).

Urine samples collected in Kenya were shown (1322) to contain a compound whose fluorescence spectrum was identical to 2,3-dihydro-2-(7'-guanyl)-3-hydroxyaflatoxin B_1, indicating the interaction between the ultimate carcinogenic form of aflatoxin B_1 (administered by food) and cellular nuclear acids in vivo.

Porphyrins.   Selective laser-excited fluorescence spectra of coproporphyrin were reported (1323), and quantitation methods for coproporphyrins (1324,1325) and uroporphyrin (1324) in urine, utilizing the native fluorescence at 610 nm, were described. Detection limits are as low as 70 pg.

The emission anisotropy of porphin and chlorin in a single crystal matrix at 77 K allowed the attribution of vibration modes to symmetry types (1325). The vibrations in the quasi-line spectra of the two compounds are rather similar. The dependence of the excitation selectivity on the excitation frequency was investigated for protoporphyrin IX at 3.8 K (1326). Vibrational frequencies were obtained for the ground and excited states when the excitation energy was scanned in the region of pure electronic and vibronic bands, respectively.

Hematoporphyrin displays a double-exponential decay in aqueous solution (τ_1 14.3 ns, τ_2 3.0 ns). When solutions are stored at 37°C for 48 h, a blue-shifted emission is observed, which was interpreted in terms of aggregation (1327). The dimerization and tautomerization of etioporphyrin I in hydrocarbon solvents was also detected by fluorescence (1328).

A new band in the fluorescence spectrum of pheophytin at low temperature was found after irradiation into its 0–0 absorption band at 670 nm, which was assigned to a tautomeric form (1329). The tautomer is stable in the dark at 4.2 K but decomposes at higher temperatures.

Kapinus et al. (1330) report the deuterium isotopic effect on the fluorescence quenching of pheophytin and porphyrin by oxidizing agents. Lamola et al. (1331) showed by fluorescence studies that more than 90% of added protoporphyrin is bound to albumin in a solution of human plasma, while less than 1% is bound to hemopexin.

Metalloporphyrins. The submicrosecond oscillatory delayed fluorescence observed in nonpolar solutions of hydrated chlorophyll is attributed to the dihydrate, in which the water molecule is bound to the magnesium atom on each side of the porphyrin skeleton. The observed quantum oscillations are described in terms of symmetry selection rules for electron tunneling matrix elements and electron shuttling between bonding and antibonding states (1332).

Chlorophyll b was submitted to site selection fluorescence spectroscopy in membranes of lecithin vesicles (1333). At 10 K, the spectra displayed zero-phonon lines as pronounced as those in the best organic glasses. It was found that the chromophore of chlorophyll b is close to the surface of the membrane but not in direct contact with the aqueous phase.

The fluorescence spectra of chlorophyll a in benzene, carbon tetrachloride, and hexane were reported (1334). Selective excitation of fluorescence from either the monomer or the dimer was achieved using a pulsed laser tuned into resonance with the Soret band of the corresponding species.

Stelmakh and Tsvirko (1400) studied the effect of vibrational energy storage on the fluorescence spectra and quantum yields of metalloporphyrins in solution. The fluorescence decay times of Chl a (conc. ~ 10 μM) was found to depend on both the excitation and emission wavelengths (1401), and the laser-induced emission of the Chl a dimer in solution was analyzed (1402).

The emission maxima of chlorophyll a in acetone shift to longer wavelengths when the solvent is deuterated (1335). Simultaneously, the quantum yield and the stability at room temperature decrease, suggesting the formation of more aggregates in deuterated than in nondeuterated solvents.

For zinc(II) porphyrin in irradiated micellar solutions an oscillatory behavior in fluorescence intensity was observed (1336). Six fluorescent protochlorophylls were identified in an aqueous micellar solution of an extract of pumpkin seeds by computer analysis of the absorption and fluorescence bands (1337).

The temperature effects on the solubility and the efficiency of energy transfer of pheophytin a, Chl a or b in dimyristoyllecithin vesicles were reported (1338). At lower temperatures phase separations were observed. The photochemical electron transfer and charge separation process in porphyrin-quinone systems as studied on a picosecond time scale were reviewed (1403).

The fluorescence decay characteristics of the isolated light-harvesting chlorophyll a/b protein have been investigated using low-intensity, subnanosecond single-photon counting (1339). In the monomeric state, in detergent micelles, the Chl a/b protein exhibits biexponential decay (τ_1 1.2 ns, τ_2 3.3 ns), with the two components having very similar weights.

A new model system for Chl photochemistry was introduced, which consists of polyethylene-squalene particles, to which Chl and N-(3-pyridyl)-myristamide were adsorbed (1340). In spite of evidence for the presence of associated Chl species along with the monomer, the lifetimes were found undiminished.

Because of their strong light absorption and large Stokes shifts (when excitation is into the Soret bands), porphyrins and chlorophylls have been used as labels in fluorescence immunoassays (1341).

Substitution of Fe(II) for Zn(II) in *Hansenula anomala* cytochrome *c* provides a luminescent derivative suitable as a probe for the determination of the interaction of cytochrome *c* with flavocytochrome *b* (1342). It was used to investigate the reversible association with proteolytic fragments of oxidized and reduced flavocytochrome b_2 by fluorimetry (1343).

Photoreceptors.   The single-photon timing method was applied to follow the fluorescence kinetics of energy transfer in phycocyanin 645 from the cryptomonad *Chroomonas sp.* as a function of both the excitation and emission wavelength, using low-intensity excitation (1344). The kinetics were found to be dominated by a fast (15 ps) and a slow (1.44 ns) decay component, the amplitudes of which depended strongly on both the excitation and emission wavelengths. A third component with a small relative amplitude and a lifetime in the range 360–680 ps has been found as well. The fast decay component is attributed to intramolecular energy transfer from sensitizing to fluorescing chromophores.

The solution and single crystal fluorescence for C-phycocyanin isolated from the cyanophyte *Agmenellum quadrupl.* were compared (1345). Although the absorbance peak of a crystal suspension is red-shifted by only ~7 nm relative to the solution forms, the single-crystal fluorescence is red-shifted by around 60 nm. Depending on the excitation wavelength, the crystal fluorescence spectrum exhibits one or two peaks. Similar results were obtained with B-phycoerythrin from a red alga.

The thermal denaturation of monomeric and trimeric phycocyanins from *Spirulina platensis* was studied by means of static and polarized time-resolved fluorescence spectroscopy under low excitation photon fluxes (1346). The complex decay can be fitted satisfactorily with a biexponential (τ_1 70–400 ps, τ_2 1–3 ns) for both the isotropic and polarized part. The results are discussed in terms of sensitizing and fluorescing chromophores in an extended model formerly proposed by Teale and Dale (795). In the isotropic emission the short lifetime is attributed to sensor–

fluorescer (donor–acceptor) energy transfer and the longer one to the decay of the final acceptor. The polarized part is dominated by an extremely short decay time, which is related to the sensor–fluorescer transfer, as well as to resonance transfer between the longer-lifetime fluorescent chromophores.

Despite difficulties in the reproducible preparation of phycobiliproteins, they have been recommended for use in immunoassay techniques because of their high absorptivities, high quantum yields, and excitation and emission bands across the visible spectrum (1347). Covalent binding to antibodies allowed fluorescence immunoassays in the picomolar range by both the solid-phase sandwich and fluorescence excitation transfer method. Phycoerythrin was covalently coupled via an S–S link to allophycocyanin and the efficiency of the energy transfer (90%) was measured (1348). Because of large Stokes' shift between absorption (545 nm) and emission (660 nm) the system was said to be useful in fluorescence-activated cell sorting, fluorescence microscopy, and fluorescence immunoassay.

Noncyclic Pyrrole Pigments. Braslavsky et al. (1349) have written a review on conformation, photophysics, and photochemistry of bile pigments, including some recent unpublished work. Bilirubin dimethyl ester exhibits two fluorescence emission bands in frozen ethanol at 77 K, having maxima of 448 nm and 408 nm, and decay times of 1.9 and 11.9 ns, respectively (1350). The former component is much stronger in intensity. Its lifetime drops to around 20 ps at 295 K (876,879).

Dipyrromethenones, including xanthobilirubinic acid, its methylester, and a vinyl analog of neoxanthobilirubinic acid, bind to human serum albumin and thereby exhibit a more than 20-fold increase in fluorescence (1351). Configurational isomerization on a ps time scale is the major deactivation pathway for decay of the S_1 states of the HSA-bound dipyrromethenones. The results of these studies of "half bilirubins" allow a clear evaluation of the effects of protein binding on the photophysics and photochemistry of bilirubin.

Biliverdin dimethyl ester was investigated by means of laser-induced optoacoustic spectroscopy (1352). The previously published scheme (791,1353), which attributes the fluorescence of biliverdin ester to two families of monomeric conformers, is extended to include the associated species, which accounts for the concentration dependence of the optoacoustic data (1352). The results were confirmed and the scheme was further refined after a study on the picosecond kinetics of the relaxation (1404).

Phorcarubin and isophorcarubin, two tetrapyrroles with fixed conformation, have been reinvestigated, and the corrected fluorescence excitation and emission spectra of their esters were reported (1354). Both spectra depend on the emission and excitation wavelengths, thereby indicating the presence of more than one emitting species.

The room temperature fluorescence maxima and quantum yields of dihydrobilatrienes *a*, *b*, *c* have been measured for various *E*/*Z* isomers in free-base and protonated form (1355). Förster-cycle calculations predict shifts in the tautomeric equilibria of the tetrapyrrole system, but no spectral evidence was found for such a process, possibly because of the short lifetime of the excited state.

Quinones.   High spectral resolution and picosecond time-resolved studies on the dye naphthazarin in solid neon showed very extensive contributions of vibrationally unrelaxed fluorescence (1356). The highly resolved excitation and emission spectra allowed a vibration analysis and the conclusion that free naphthazarine exists in the 1,4-naphthoquinone rather than in the 1,5-isomeric form.

1,8-Dihydroxyanthraquinone has a weak native fluorescence, which is greatly enhanced by addition of magnesium(II), thus allowing a sensitive determination of the ion in aqueous ethanol (1357). Fluorescence intensity in ethanol dropped sharply as increasing amounts of water were added.

Anoshin et al. have presented low-temperature studies on thin layers of 5,8-dihydroxy-1,4-naphthoquinone (1358) and 1,4,5,8-tetradeuteroxy-9,10-anthraquinone (1359). The vibrations observed in the fluorescence spectra were interpreted. The anthraquinone showed emissions from two centers with different O–O transitions and vibronic structures.

The fluorescence of the antibiotics tetracycline and oxytetracycline is quenched by purins, pyrimidines, and amino acids (1407). A simple method was developed for the routine monitoring of daunorubicin or adriamycin and of their chief fluorescent metabolites in plasma of cancer patients. It is based on HPLC separation and fluorescence detection at 585 nm after excitation at 470 nm (1360).

Miscellaneous Dyes.   A new class of anthocyanidins was discovered, and their fluorescence maxima were reported (1361). Fluorescence is said to be blue (λ_{max} 410–423 nm), whereas the absorption maxima lie between 535 and 547 nm. This would imply a fluorescence from an excited state higher in energy than the S_1 state, which, however, seems unlikely in view of the "normal" red fluorescences of other anthocyanines (969–972). The excitation maxima of these compounds are said to be at 360–370 nm, a spectral region close to an absorption minimum.* The effect of flavone sulfonic acid and quercetin-5'-sulfonate on the spectra of eight flavylium salts (anthocyanins) has been determined and was found to be quite different in that the former enhances luminescence, whereas the latter has the opposite effect (1398).

Polyenes, Retinal. The marked solvent dependence of the fluorescence quantum yield of cholesta-4,6,8(14)-trien-3-ones (ϕ_f ~0.3 in water, almost zero in aprotic solvents) is interpreted in terms of level crossing between the energies of the n,π^* and π,π^* states (1362). Fluorescence is only observed when the π,π^* state is lowest in energy (i.e., in protic solvents).

Highly resolved emission and one-photon excitation spectra and fluorescence lifetimes for all-*trans*-1,3,5,7-octatetraene in a series of *n*-alkane crystals, maintained at liquid helium temperatures, have been measured (1363). In the *n*-octane host, the $1\,^1A_g$ to $2\,^1A_g$ transition is strictly symmetry forbidden. In other hosts the spectra exhibit both allowed and forbidden components. The vibronic development, fluorescence lifetime, and degrees of transition probabilities were analyzed to elucidate the connections between local environment, molecular conformation, and electronic energy. A similar study, which includes two-photon excitation spectra, was performed on *cis-cis*-octatetraene (1364) and cis-trans-octatetraene (1413,1414).

Laser-induced fluorescence spectra have been recorded for β-carotene in carbon disulfide (1365), chloroform, ethanol, and lipid–water mixtures (1366). The lifetime of β-carotene is extremely short (265 fs) (1365). In lipid–water mixtures the profiles of laser excitation and absorption spectra are closely corresponding (1366). The observations are explained in terms of a low-lying ^1A_g excited state. A related study was performed on 15,15'-*cis*-β-carotene and 15,15'-*trans*-β-carotene (1415).

The fluorescence relaxation kinetics from all-*trans* retinal in polar and nonpolar solvents have been investigated, and an activation energy of ~1 kcal mol^{-1} has been measured (1367). The results, in conjunction with previous picosecond absorption measurements by others, strongly suggest that initially excited molecules relax to three singlet excited states.

A streak camera detection technique was used to record the fluorescence of bacteriorhodopsin at room temperature induced by single subpicosecond light pulses. The lifetime of the bacteriorhodopsin fluorescence was <2 ps (1368).

---

* A reexamination of the dyes showed that they have an additional weak fluorescence in the red (typically 590 nm), when excitation is into the longest-wave absorption band (1412).

Amino Acid Metabolites.   The technique to convert Trp residues into strongly and longwave fluorescent kynurenine by ozonolysis was applied to follow the binding between thermolysin and talopeptin (1369). Replacement of the Trp residue did not seriously affect the apparent rate constant. There was an indication of energy transfer from Trp (the donor) to kynurenine (the acceptor), and the distance between these two was estimated to be 180 pm.

5-Hydroxytryptamine, its amino acid precursors, and its acid metabolite were simultaneously determined in discrete brain regions by HPLC with fluorescence detection (1370). The photophysics of 5-methoxytryptophan has been studied as a function of pH (1371). The fluorescence spectrum, which maximizes at distinctly shorter wavelengths than tryptophan, undergoes a pronounced red shift as the pH is increased, which is ascribed to the deprotonation of the amino group.

Phenols, Aldehydes, Ketones, and Acids.   Single rotational level S_1 emission has been observed for jet-cooled acetaldehyde (1372). A molecular beam study was performed on the photochemistry and fluorescence of glyoxal (1373), and the fluorescence properties of various norcamphors were described (1374). The latter show the typical weak fluorescence of ketones with emission maxima between 383 and 413 nm (in acetonitrile).

The fluorescence excitation and dispersed fluorescence spectra of tropolone in ultracold gas phase generated by a supersonic expansion have been observed (1375). Free tropolone exhibits bands of doublet structure arising from proton tunneling of the intramolecular hydrogen bridge, which disappear by formation of intermolecular hydrogen-bridged complexes. These results provide conclusive evidence for proton tunneling in both the S_0 and S_1 states.

Stipitatonic acid and stipitatic acid, two natural tropolonecarboxylic acids, have quite different fluorescence spectra and fluorescence efficiencies, despite a closely related chemical structure (Table 3.21). This allowed the fluorimetric assay of the former in the presence of the latter (1376). The spectra of various hydroxy- and aminobenzoic acids (most of which are urinary metabolites) have been reported, after they had been separated by electrophoresis (1410).

The structures and dynamic processes of the excited states of salicylaldehyde and o-hydroxyacetophenone were investigated by means of steady-state and picosecond time-resolved emission spectroscopy (1377). The unusual red-shifted fluorescence is assigned to an enol tautomer with two conformations and two decay times at room temperature. At 77 K, in nonpolar solvents, the closed conformer of salicylaldehyde is converted to the open conformer by UV irradiation, changing the fluorescence into phosphorescence.

In a study of solvent effects on the fluorescences of coumaric acids, it was found (1378) that their native fluorescences are weak and short-lived. Except in hydroxylic solvents, the fluorescence maximum could be correlated with solvent polarity parameters. Cinnamic acid, methyl cinnamate, and p-methoxycinnamic acids were found to be nonfluorescent in all solvents studied. The emission maxima of ortho- and meta-coumaric acid and their methyl ethers suffer considerable red shifts with increasing solvent polarity.

The fluorescence spectra of five humic acids showed maximum excitation at 360 nm and developed maximum emission in the range 430–455 nm, but the humic acids from sediments showed an additional peak at 410 nm (1379). Fluorescence from peat humic acid was broad, and a secondary emission was observed with a maximum at 520 nm, which was attributed to an excimer.

Fluorescence has been used to monitor the interaction of cations with humic acids in solutions of different ionic strengths. This shows promise for distinguishing between metal–humate complex formation and the coagulation of a colloid (1379). Corrected fluorescence spectra were reported for fulvic acids isolated from soil and water (1380).

Other Natural Products. The interactions of six analogs of bleomycin type antitumor antibiotics with poly (deoxyadenylylthymidylic acid) (a DNA model) was monitored by the quenching of the fluorescences of these derivatives in the presence of the polynucleotide at moderate ionic strength of the solution (1381). Two classes of binding sites were observed, one of which appears to be primarily intercalative in nature and one of which appears to be governed primarily by ionic interactions. It is demonstrated that bleomycin A_2 is extremely sensitive to the nucleic acid structure (which, *inter alia*, is affected by the ionic strength of the solution), and that the substituent determines to a large extent the mode of binding.

The fluorescence and phosphorescence spectrum of firefly oxyluciferin at 77 K are identical, raising the question of the multiplicity of the excited state resulting in bioluminescence (1382). Similar findings were reported for various oxyluciferins from *Cypridina* at temperatures between 77 and 300 K.

REFERENCES

1. Ch. Dhéré, La spectrochimie de fluorescence dans l'etude des produites naturel, *Prog. Chem. Org. Nat. Prod. 2*, 301 (1939).

2. Ch. Dhéré, Progres recents en spectrochimie de fluorescence des produits biologiques, *Prog. Chem. Org. Nat. Prod. 6*, 311 (1950).

3. S. Udenfriend, *Fluorescence Assay in Biology and Medicine*, Vol. I, Academic Press, New York, 1962.

4. S. Udenfriend, *Fluorescence Assay in Biology and Medicine*, Vol. II, Academic Press, New York, 1969.

5. G. G. Guilbault, *Practical Fluorescence: Theory, Methods, and Techniques*, Dekker, New York, 1973.

6. J. B. Birks (ed.), *Excited States of Biological Molecules*, Wiley, New York, 1976.

7. A. J. Pesce, C. G. Rosen, and T. L. Pasby (eds.), *Fluorescence Spectroscopy: An Introduction for Biology and Medicine*, Dekker, New York, 1971.

8. R. F. Steiner, and J. Wynryb (eds.), *Excited States of Proteins and Nucleic Acids*, Plenum Press, New York, 1971.

9. R. F. Chen and H. Edelhoch (eds.), *Biochemical Fluorescence: Concepts*, Vol. I, Dekker, New York, 1975.

10. R. F. Chen and H. Edelhoch (eds.), *Biochemical Fluorescence: Concepts*, Vol.II, Dekker, New York, 1976.

11. J. R. Lakowicz, *Principles of Fluorescence Spectroscopy*, Plenum Press, New York, 1983.

12. M. A. DeLuca (ed.), *Bioluminescence and Chemiluminescence, Methods in Enzymology*, Vol.57, Academic Press, New York, 1978.

13. C. Helene and J. Maurizot, Interactions of oligopeptides with nucleic acids, *CRC Crit. Rev. Biochem. 10*, 213–258 (1981).

14. H. Morawetz and I. Z. Steinberg (eds.), Luminescence from biological and synthetic macromolecules, *Ann. N.Y. Acad. Sci. 366*, (1981).

15. F. W. J. Teale and G. Weber, *Biochem. J. 65*, 476 (1956).

16. J. Feitelson, *J. Phys. Chem. 68*, 391 (1964).

17. R. F. Chen, *Anal. Lett. 1*, 35 (1967).

18. I., Tatischeff, P. Vigny, R. Klein, and M. Duquesne, *C. R. Acad. Sci. Fr. 276D*, 1217 (1973).

19. M. Y. Gladchenko, L. G. Kostko, L. G. Pikulik, and A. N. Sevchenko, *Dokl. Akad. Nauk. Beloruss. SSR 9*, 647 (1965).

20. E. Leroy, H. Lami, and G. Laustriat, *Photochem. Photobiol. 13*, 411 (1971).

21. R. F. Chen, G. G. Vurrek, and M. Alexander, *Science 156*, 949 (1967).

22. J. W. Longworth and R. O. Rahn, *Biochim. Biophys. Acta 147*, 526 (1967).

23. M. Fayet and Ph. Wahl, *Biochem. Biophys. Acta 229*, 102 (1971).

24. D. M. Rayner, D. Th. Krajcarski, and A. G. Szabo, *Can. J. Chem. 56*, 1238 (1978).

25. R. W. Cowgill, *Arch. Biochem. Biophys. 100*, 36 (1963).

26. H. C. Börresen, *Acta Chem. Scand. 21*, 920 (1967).

27. J. Eisinger, and G. Navon, *J. Chem. Phys. 50*, 2069 (1969).

28. J. Feitelson, *Isr. J. Chem. 8*, 241 (1970).

29. R. J. Robbins, G. R. Fleming, G. S. Beddard, G. W. Robinson, P. J. Thistlethwaite, and G. J. Woolfe, *J. Am. Chem. Soc. 102*, 6271 (1980).

30. I. Weinryb and R. F. Steiner, *Biochemistry 7*, 2488 (1968).

31. W. D. DeLauder and Ph. Wahl, *Biochemistry 9*, 2750 (1970).

32. E. P. Kirby and R. F. Steiner, *J. Phys. Chem. 74*, 4480 (1970).

33. D. M. Rayner and A. G. Szabo, *Can. J. Chem. 56*, 743 (1978).

34. G. R. Fleming, J. M. Morris, R. J. Robbins, G. J. Woolfe, P. J. Thistlethwaite, and G. W. Robinson, *Proc. Natl. Acad. Sci. USA 75*, 4652 (1978).

35. G. S. Beddard, G. R. Fleming, G. Porter, and R. J. Robbins, *Phil. Trans. R. Soc. Lond. 298A*, 321 (1980).

36. E. Gudgin, R. Lopez-Delgado, and W. R. Ware, *Can. J. Chem. 59*, 1037 (1981).

37. M. R. Eftink and C. A. Ghiron, *Photochem. Photobiol. 33*, 749 (1981).

38. A. G. Szabo and D. M. Rayner, *J. Am. Chem. Soc. 102*, 554 (1980).

39. S. V. Konev, *Fluorescence and Phosphorescence of Proteins and Nucleic Acids*, Plenum Press, New York, 1967.

40. I. Tatischeff, P. Vigny, R. Klein, and M. Duquesne *C. R. Acad. Sci. Fr. 276D*, 1217 (1973).

41. M. Ledvina and F. S. LaBella, *Anal. Biochem. 36*, 174 (1970).

42. J. Feitelson and E. Hayon, *J. Phys. Chem. 77*, 10 (1973).

43. D. V. Bent and E. Hayon, *J. Am. Chem. Soc. 97*, 2606 (1975).

44. D. R. Cooper and J. Feitelson, *Biochem. J. 97*, 139 (1965).

45. E. Fujimori, *Biochemistry 5*, 1034 (1966).

46. J. A. Ambrose, *Ann. N.Y. Acad. Sci. 196*, 295 (1972).

47. R. W. Cowgill, *Tyrosyl fluorescence in proteins and model peptides*, in Ref. 10.

48. S. W. Bailey and J. E. Aylin, *Anal. Biochem. 107*, 156 (1980).

49. J. A. Galley and G. M. Edelman, *Biochim. Biophys. Acta 60*, 499 (1962).

50. J. L. Cornog and W. R. Adams, *Biochim. Biophys. Acta 66*, 356 (1963).

51. O. Shimizu, J. Watanabe, and K. Imakubo, *Photochem. Photobiol. 29*, 915 (1979).

52. T. Alev-Behmoaras, J. J. Toulme, and C. Helene, *Photochem. Photobiol. 30*, 533 (1979).

53. A. G. Szabo, D. Krajcarski, and D. M. Rayner, *J. Lumin. 18/19*, 582 (1979).

54. J. W. Longworth, *Ann. N.Y. Acad. Sci. 366*, 237 (1981).

55. B. T. Lim and T. Kimura, *J. Biol. Chem. 255*, 2440 (1980).
56. M. Schnarr and C. Helene, *Photochem. Photobiol. 36*, 91 (1982).
57. D. A. Terhaar and J. L. DiCesare, Perkin Elmer Fluorescence Report 1979, No. F2-71.
58. J. N. Miller, Proc. Anal. Div. Chem. Soc. *16*, 203 (1979).
59. J. N. Miller, T. A. Ahmad, and A. F. Fell, *Anal. Proc. 19*, 37 (1982).
60. J. N. Miller and L. A. King, *Biochim. Biophys. Acta 393*, 433 (1975), and references cited therein.
61. J. R. Lakowicz and H. Cherek, *J. Biol. Chem. 256*, 6348 (1981).
62. L. J. Andrews and L. S. Forster *Photochem. Photobiol. 19*, 353 (1974).
63. I. Gryczinski and A. Kawski, *Z. Naturforsch. 30A*, 287 (1975).
64. M. Sun and P. S. Song, *Photochem. Photobiol. 25*, 3 (1977).
65. B. Valeur and G. Weber, *Photochem. Photobiol. 25*, 441 (1977).
66. L. C. Pereiro, I. C. Ferreira, M. P. F. Thomasz, and M. I. B. M. Jorge, *J. Photochem. 9*, 425 (1978).
67. G. Porter, E. S. Reid, and C. J. Tredwell, *Chem. Phys. Lett. 29*, 469 (1974).
68. M. S. Walker, T. W. Bednar, and R. Lumry, *J. Chem. Phys. 47*, 1020 (1967).
69. R. Lumry and M. Hershberger, *Photochem. Photobiol. 27*, 819 (1978).
70. H. Lami, *J. Mol. Struct. 45*, 437 (1978).
71. I. Gonzalo and J. L. Escudero, *J. Phys. Chem. 86*, 2896 (1982).
72. R. MacGuire and I. Feldman, *Photochem. Photobiol. 18*, 119 (1973).
73. P. M. Froehlich and M. Yeats, *Anal. Chim. Acta 87*, 185 (1976).
74. R. W. Ricci, *Photochem. Photobiol. 12*, 67 (1970).
75. H. B. Steen, *J. Chem. Phys. 61*, 3997 (1974).
76. E. W. Lobenstine and D. H. Turner, *J. Am. Chem. Soc. 102*, 7787 (1980).
77. K. P. Ghiggino, G. R. Mant, D. Phillips, and A. J. Roberts, *J. Photochem. 11*, 297 (1979).
78. D. M. Jameson and G. Weber, *J. Phys. Chem. 85*, 953 (1981).
79. M. Nakanishi and M. Tsuboi, *Chem. Phys. Lett. 57*, 262 (1978).
80. S. S. Lehrer, *J. Am. Chem. Soc. 92*, 3459 (1970).
81. T. L. Bushueva, L. P. Busel, and E. A. Burstein, *Stud. Biophys. 52*, 41 (1978).
82. T. C. Werner and L. S. Forster, *Photochem. Photobiol. 29*, 905 (1979).
83. (a) S. S. Lehrer, *Biochemistry 10*, 3254 (1971); (b) O. S. Wolfbeis and E. Urbano, *Anal. Chem. 55*, 1904 (1983).
84. M. N. Ivkova, M. S. Vedekina, and N. A. Burshtein, Mol. Biol. *5*, 214 (1971).
85. J. R. Lakowicz and G. Weber, *Biochemistry 12*, 4161, 4171 (1973).
86. P. Cavatorta, R. Tavilla, and A. Mazzini, *Biochim. Biophys. Acta 578*, 541 (1979).

87. F. Torikata, L. S. Forster, and R. E. Johnson, *J. Biol. Chem. 254*, 3516 (1979).

88. D. C. Kramp and I. Feldman, *Biochim. Biophys. Acta 537*, 406 (1978).

89. I. Feldman and D. C. Kramp *Biochemistry 17*, 1542 (1978).

90. M. R. Eftink, J. L. Zajicek, and C. A. Ghiron, *Biochim. Biophys. Acta 491*, 473 (1977).

91. M. Shinitzky and I. Goldman, *Eur. J. Biochem. 3*, 139 (1967).

92. T. LeDoan, J. J. Toulme, R. Santus, and C. Helene, *Photochem. Photobiol. 34*, 309 (1981).

93. T. G. Politis and H. G. Drickamer, *J. Chem. Phys. 75*, 3203 (1981).

94. T. M. Li, J. W. Hook, H. G. Drickamer, and G. Weber, *Biochemistry 15*, 3205 (1976).

95. F. Flentge, K. Venema, and J. Korf, *Anal. Biochem. 11*, 234 (1974).

96. I. M. Kuznetsova, I. I. Kirik, and K. K. Turoverov, Mol. Biol. *15*, 989 (1981).

97. G. G. Guilbault and P. M. Froehlich, *Clin. Chem. 19*, 1112 (1974).

98. A. M. Krstulovic, P. R. Brown, D. M. Rosie, and P. B. Champlin, *Clin. Chem. 23*, 1984 (1977).

99. R. Oste, B. M. Nair, and A. Dahlqvist, *J. Agric. Food Chem. 24*, 1141 (1976).

100. A. D. Jones, C. H. S. Hitchcock, and G. H. Jones, *Analyst 106*, 968 (1981).

101. I. Morita, T. Masujima, H. Yoshida, and H. Imai, *Anal. Biochem. 118*, 142 (1981).

102. J. K. Richardson, *Anal. Biochem. 83*, 754 (1977).

103. E. A. Permyakov, E. A. Burstein, Y. Sawada, and I. Yamazaki, *Biochim. Biophys. Acta 491*, 149 (1977).

104. E. A. Permyakov, V. V. Yarmolenko, L. P. Kalinichenko, and E. A. Burstein, *Bioorgan. Khim. 7*, 1660 (1981).

105. P. Gauduchon and Ph. Wahl, *Biophys. Chem. 8*, 87 (1978).

106. C. E. Svenberg, H. C. Pant, S. Shapiro, F. Wong, and R. Cavanagh, *Biophys. J. 37*, 248A (1982).

107. R. P. Homer and S. R. Allsopp, *Biochim. Biophys. Acta 434*, 297 (1976).

108. A. Rehage and F. X. Schmidt, *Biochemistry 21*, 1499 (1982).

109. J. D. Johnson, J. H. Collins, and J. D. Plotter, *J. Biol. Chem. 253*, 6451 (1978).

110. W. C. Galley, Heterogeneity in protein emission spectra, in Ref. 10.

111. L. Stryer, Fluorescence energy transfer as a spectroscopic ruler, *Ann. Rev. Biochem. 47*, 819 (1978).

112. P. W. Schiller, The measurement of intramolecular distances by energy transfer, in Ref. 10.

113. Y. Saito, H. Tachibana, H. Hayashi, and A. Wada, *Photochem. Photobiol. 33*, 289 (1981).

114. P. Pajot, *Eur. J. Biochem. 63*, 263 (1976).

115. E. A. Burstein, E. A. Permyakov, V. A. Yaskin, S. A. Burkhanov, and A. Finazzi-Agro, *Biochem. Biophys. Acta 491*, 155 (1977).

116. S. S. Lehrer, Perturbation of intrinsic protein fluorescence, in ref. 10.

117. R. F. Chen, The effect of metal cations on intrinsic protein fluorescence, in ref. 10.

118. W. Altekaar, *Biopolymers 16*, 341, 369 (1977).

119. C. Helene, *Jerusalem Symp. Quantum Chem. Biochem.: Excited States in Chemistry and Biochemistry, 10*, 65–78 (1977).

120. M. R. Eftink and C. A. Ghiron, *Anal. Biochem. 114*, 199 (1981), and refs. therein.

121. A. G. Szabo and D. M. Rayner, *Biochem. Biophys. Res. Comm. 94*, 909 (1980).

122. L. Brand, L. B. A. Ross, and W. R. Laws, *Ann. N.Y. Acad. Sci. 366*, 197 (1981).

123. C. Formoso and L. S. Forster, *J. Biol. Chem. 250*, 3738 (1975).

124. A. Grinvald and I. Z. Steinberg, *Biochim. Biophys. Acta 427*, 663 (1976).

125. W. B. De Lauder and Ph. Wahl, *Biochim. Biophys. Res. Comm. 42*, 398 (1971).

126. G. Hazan, E. Haas, and I. Z. Steinberg, *Biochem. Biophys. Acta 434*, 144 (1976).

127. C. D. Tran, G. S. Beddard, R. McConnell, C. F. Hoyng, and J. H. Fendler, *J. Am. Chem. Soc. 104*, 3002 (1982).

128. T. V. Veselova, A. S. Cherkasov, and V. I. Shirokov, *Opt. Spectrosc. 29*, 617 (1970).

129. (a) J. R. Lakowicz and H. Cherek, *J. Biochem. Biophys. Meth. 5*, 19 (1981) and *Biophys. J. 37*, 148 (1982); (b) J. R. Lakowicz and A. Balter, *Photochem. Photobiol. 36*, 125 (1982).

130. I. Z. Steinberg, Fluorescence polarization: Some trends and problems, in Ref. 9.

131. R. E. Dale and J. Eisinger, Polarization energy transfer, in ref. 9.

132. S. A. Levison, Fluorescence polarization kinetic studies on macromolecular reactions, in ref. 9.

133. J. R. Lakowicz, *J. Biochem. Biophys. Meth. 2*, 91–119 (1980).

134. I. Munro, I. Pecht, and L. Stryer, *Proc. Natl. Acad. Sci. USA 76*, 56 (1979).

135. J. R. Lakowicz and G. Weber *Biophys. J. 10*, 591 (1980).

136. R. W. Wijnandts van Resandt and L. DeMaeyer, *Chem. Phys. Lett. 78*, 219 (1981).

137. R. M. Levy and A. Szabo, *J. Am. Chem. Soc. 104*, 2073 (1982).

138. I. Z. Steinberg, *Ann. Rev. Biophys. Bioeng. 7*, 113 (1978).

139. E. W. Lobenstine, W. C. Schaefer, and D. H. Turner, *J. Am. Chem. Soc.* *103*, 4936 (1981).

140. R. T. Spencer and G. Weber, *Ann. N.Y. Acad. Sci.* *158*, 361 (1969).

141. T. G. Scott, R. T. Spencer, N. J. Leonhard, and G. Weber, *J. Am. Chem. Soc.* *92*, 687 (1970).

142. A. Gafni and L. Brand, Biochemistry *15*, 3165 (1976).

143. A. J. W. G. Visser and A. van Hoek, *Photochem. Photobiol. 33*, 35 (1981).

144. G. G. Guilbault, *Handbook of Enzymatic Methods of Analysis* Dekker, New York, 1976.

145. H. B. Burch, *Meth. Enzymol. 18B*, 11 (1971).

146. D. E. F. Harrison and B. Chance, *Appl. Microbiol. 19*, 446 (1970).

147. W. Beyeler, A. Eisele, and A. Fiechter, *Eur. J. Appl. Microbiol. Biotechnol. 13*, 10 (1981).

148. R. A. Morton, *Biochemical Spectroscopy*, Adam Hilger, London, 1975, p. 425.

149. J. C. Brochon, Ph. Wahl, J. M. Jallon, and M. Iwatsubo, *Biochemistry 15*, 3259 (1976).

150. (a) P. F. Heelis, *Chem. Soc. Rev. 11*, 15 (1982); (b) G. D. Fasman (ed.), *Handbook of Biochemistry and Molecular Biology*, 3rd ed., CRC Press, Cleveland, 1976, Vol. 2, p. 208.

151. M. Sun, T. A. Moore, and P. S. Song, *J. Am. Chem. Soc. 94*, 1730 (1972).

152. W. Schmidt, *Photochem. Photobiol. 34*, 7 (1981).

153. J. K. Eweg, F. Müller, A. J. W. G. Visser, C. Veeger, D. Bebelaar, and J. D. W. van Voorst, *Photochem. Photobiol. 30*, 463 (1979).

154. A. J. W. G. Visser and F. Müller, *Helv. Chim. Acta 62*, 593 (1979).

155. A. J. W. G. Visser, A. van Hoek, and F. Müller, *Helv. Chim. Acta 63*, 1296 (1980).

156. J. K. Eweg, F. Müller, D. Bebelaar, and J. D. W. van Voorst, *Photochem. Photobiol. 31*, 435 (1980).

157. P. S. Song, *Ann. N.Y. Acad. Sci. 158*, 410 (1969).

158. G. Weber and F. W. J. Teale, *Trans. Faraday Soc. 53*, 646 (1957).

159. J. Koziol, *Photochem. Photobiol. 9*, 45 (1969).

160. Ph. Wahl, J. C. Auchet, A. J. W. G. Visser, and F. Müller, *FEBS Lett. 44*, 67 (1974).

161. M. C. Falk and D. B. McCormick, *Biochemistry 15*, 646 (1976).

162. A. J. W. G. Visser, T. M. Li, H. G. Drickamer, and G. Weber, *Biochemistry 16*, 4883 (1977).

163. N. A. Garcia and C. M. Previtali, *Anal. Asoc. Quim. Argent. 70*, 19 (1982); *Chem. Abstr. 96*, 176441 (1982).

164. N. Lasser and J. Feitelson, *J. Phys. Chem. 77*, 1011 (1973).

165. G. Weber, F. Tanaka, B. Y. Okamoto, and H. G. Drickamer, *Proc. Nat. Acad. Sci. USA 71*, 1264 (1974).

166. F. Müller and S. G. Mayhew, *Meth. Enzymol. 66*, 350 (1980).

167. A. J. W. G. Visser, T. M. Li, H. G. Drickamer, and G. Weber, *Biochemistry 16*, 4879 (1977).

168. R. D. Spencer and G. Weber, in: *Structure and Function of Oxidation-Reduction Enzymes*, A. Akeson and H. Ehrenberg (eds.), Pergamon Press, Oxford, 1972, p. 393.

169. A. J. W. G. Visser and F. Müller, *Meth. Enzymol. 66*, 373 (1980).

170. Ph. Wahl, J. C. Auchet, A. J. W. G. Visser, and C. Veeger, *Eur. J. Biochem. 50*, 413 (1975).

171. H. K. Sarkar, P. S. Song, T. Y. Leong, and W. R. Briggs, *Photochem. Photobiol. 35*, 593 (1982).

172. S. Ghisla, V. Massey, J. M. Lhoste, and S. G. Mayhew, *Biochemistry 13*, 589 (1974), and ref. therein.

173. S. Ghisla, B. Entsch, V. Massey, and M. Husein, *Eur. J. Biochem. 76*, 139 (1977).

174. S. Ghisla, *Meth. Enzymol. 66*, 360 (1980).

175. A. J. W. G. Visser, S. Ghisla, V. Massey, F. Müller, and C. Veeger, *Eur. J. Biochem. 101*, 13 (1979).

176. J. Koziol, *Meth. Enzymol. 18B*, 253 (1971).

177. J. H. Wassink and S. G. Mayhew, *Anal. Biochem. 86*, 609 (1975).

178. K. Yagi, *Meth. Enzymol. 18B*, 290 (1971).

179. A. D. Lumley and R. A. Wiggins, *Analyst 106*, 1103 (1981), and ref. therein.

180. E. J. Faeder and L. M. Siegel, *Anal. Biochem. 53*, 332 (1973).

181. I. Hassinen and T. Jämsä, *Anal. Biochem. 120*, 365 (1982).

182. J. Kahan, *Meth. Enzymol. 18C*, 574 (1971).

183. P. S. Song, Q. Chae, M. Fujita, and H. Baba, *J. Am. Chem. Soc. 98*, 819 (1976).

184. K. Chihara, T. Takemura, T. Yamaoka, N. Yamamoto, A. Schaffer, and R. S. Becker, *Photochem. Photobiol. 29*, 1001 (1979).

185. (a) R. S. Becker, G. Hug, P. K. Das, A. M. Schaffer, T. Takemura, N. Yamamoto, and W. Waddell, *J. Phys. Chem. 80*, 2265 (1976); (b) T. Rosenfeld, A. Alchalal, and M. Ottolenghi, *Chem. Phys. Lett. 20*, 291 (1973).

186. J. P. Dalle and B. Rosenberg, *Photochem. Photobiol. 12*, 151 (1970).

187. B. S. Hudson and B. E. Kohler, *J. Chem. Phys. 59*, 4984 (1973).

188. J. R. Andrews and B. S. Hudson, *J. Chem. Phys. 68*, 4587 (1978).

189. T. A. Moore and P. S. Song, *Chem. Phys. Lett. 19*, 128 (1973).

190. S. Hotchandani, P. Paquin, and R. M. LeBlanc, *J. Lumin. 20*, 59 (1979).

191. S. L. Bondarev, and M. V. Belkov, Spectrosc. Lett. *14*, 617 (1981).

192. L. A. Brey, G. B. Schuster, and H. G. Drickamer, *J. Chem. Phys. 71*, 2765 (1979).

193. R. D. Fugate and P. S. Song, *Biochim. Biophys. Acta 125*, 28 (1980).

194. J. M. Thompson, P. Erdody, R. Brien, and T. K. Murray, *Biochem. Med. 5*, 67 (1971).

195. J. N. Thompson, P. Erdody, W. B. Maxwell, and T. K. Murray, *J. Dairy Sci. 55*, 1077 (1972).

196. J. W. Erdman, S. F. Hou, and P. A. LaChance, *J. Food Sci. 38*, 447 (1973).

197. J. Kahan, *Int. J. Vitam. Nutr. Res. 43*, 127 (1973); *Acta Chem. Scand. 21*, 2515 (1967).

198. G. P. Gibson and J. H. Turnbull, *J. Chem. Soc.*, Perkin II, 1288 (1980).

199. A. A. Gallo, J. J. Mieyal, and H. Z. Sable, in: *Bioorganic Chemistry*, E. E. van Tamelen (ed.), Academic Press, New York, 1978, p. 147.

200. B. Siodmiak and R. Drabent, *Acta Phys. Pol. 44A*, 659 (1973).

201. J. Koziol and E. Knobloch, *Biochem. Biophys. Acta 102*, 289 (1965).

202. G. D. Fasman (ed.), *Handbook of Biochemistry and Molecular Biology*, 3rd ed., CRC Press, Cleveland, 1976, p. 206 f.

203. S. G. Schulman, *J. Pharm. Sci. 60*, 628 (1971).

204. (a) G. Weber, *Biochem. J. 47*, 114 (1950); (b) G. Weber, *Trans. Faraday Soc. 44,* 85 (1948).

205. M. A. Slifkin, *Nature 197*, 275 (1963).

206. J. F. Pereira and G. Tollin, *Biochem. Biophys. Acta 143*, 79 (1967).

207. N. A. Garcia, J. Silber, and C. Previtali, *Tetrahedron Lett. 24*, 2073 (1977).

208. D. G. Whitten, J. W. Happ, G. L. B. Carlson, and M. T. McCall, *J. Am. Chem. Soc. 92*, 3499 (1970).

209. E. De Ritter, *J. Ass. Off. Anal. Chem. 53*, 542 (1970).

210. I. D. Lumley and R. A. Wiggins, *Analyst 106*, 1103 (1981).

211. J. H. Wassink and S. G. Mayhew, *Anal. Biochem. 68*, 609 (1975).

212. K. Yagi, N. Ohishi, K. Nishimoto, J. D. Choi, and P. S. Song, *Biochemistry 19*, 1553 (1980).

213. O. Pelletier and R. Brassard, *J. Pharm. Sci. 63*, 1138 (1974).

214. J. R. Kirk, *J. Ass. Off. Anal. Chem. 57*, 1085 (1974).

215. J. H. Richardson, B. W. Walling, D. C. Johnson, and L. W. Hrubesh, *Anal. Chim. Acta 86*, 263 (1976).

216. N. Ishibashi, T. Ogawa, T. Imasaka, and M. Kunitake, *Anal. Chem. 51*, 2096 (1979).

217. J. W. Bridges, D. Davies, and R. T. Williams, *Biochem. J. 98*, 451 (1966).

218. Y. V. Morozov and L. P. Cherkashina, *Mol. Photochem. 8*, 45 (1977) (review).

219. J. F. Gregory and J. R. Kirk, *J. Food Sci. 42*, 1073 (1977).

220. Z. Tamura and S. Takanashi, *Meth. Enzymol. 18A*, 471 (1970).

221. D. Mikac-Devic and C. Tomanic, *Clin. Chim. Acta 38*, 235 (1972).

222. R. F. Chen, *Science 150*, 1593 (1965).

223. J. E. Churchich, *Biochim. Biophys. Acta 102*, 280 (1965).

224. J. E. Churchich, in: *Modern Fluorescence Spectroscopy*, Vol. 2E. L. Wehry (ed.), Plenum Press, New York, 1976, p. 217.

225. J. E. Churchich, *J. Biol. Chem. 247*, 6953 (1972); and ref. therein on former work.

226. M. Arrio-Dupont, *Photochem. Photobiol. 12*, 297 (1970).

227. M. Arrio-Dupont, *Biochem. Biophys. Res. Comm. 44*, 653 (1971).

228. S. Shaltiel and M. Cortijo, *Biochem. Biophys. Res. Comm. 41*, 594 (1970).

229. S. Veinberg, S. Shaltiel, and I. Z. Steinberg, *Isr. J. Chem. 12*, 421 (1974).

230. D. F. Duggan, R. L. Bowman, B. B. Brodie, and S. Udenfriend, *Arch. Biochem. Biophys. 68*, 1 (1957).

231. J. W. Longworth, R. C. Rahn, and R. G. Shulman, *J. Chem. Phys. 45*, 2930 (1966).

232. R. D. Fugate, Ch. A. Chin, and P. S. Song, *Biochim. Biophys. Acta 421*, 1 (1976).

233. Guilbault, G. G. *Practical Fluorescence: Theory Methods, and Techniques*, Dekker, New York, 1973, p. 335.

234. J. J. Aaron and J. D. Winefordner, *Talanta 19*, 21 (1972).

235. R. C. Weast (ed.), *Handbook of Chemistry and Physics*, 55th ed., CRC Press, Cleveland, 1971.

236. R. Müller, R. Palluk, and M. Kempfle, *Z. Anal. Chem. 311*, 369 (1982).

237. D. Duggan, *Arch. Biochem. Biophys. 84*, 116 (1959).

238. S. L. Taylor, M. P. Lamden, and A. L. Tappel, *Lipids 11*, 530 (1976).

239. L. G. Hansen and W. J. Warwick, *Clin. Biochem. 3*, 225 (1970).

240. J. N. Thompson, P. Erdody, and W. P. Maxwell, *Anal. Biochem. 50*, 267 (1972).

241. C. Thiery, *Eur. J. Biochem. 37*, 100 (1973).

242. F. Otting and F. M. Huennekens, *Arch. Biochem. Biophys. 152*, 429 (1972).

243. K. Uyeda and J. C. Rabinowitz, *Anal. Biochem. 6*, 100 (1963).

244. H. J. Leese and J. R. Bronk, *Anal. Biochem. 45*, 211 (1972).

245. A. L. Milkovsky and H. A. Larody, *Arch. Biochem. Biophys. 181*, 270 (1977).

246. O. A. Bessey, O. H. Lowry, and R. H. Love, *J. Biol. Chem. 180*, 755 (1949).

247. A. Pol, C. van der Drift, G. D. Vogels, T. J. H. M. Cuppen, and W. H. Laarhoven, *Biophys. Res. Comm. 92*, 255 (1980).

248. B. D. Drujan, R. Castillon, and E. Guerrero, *Anal. Biochem. 23*, 44 (1968).

249. T. Kotaki and K. Yagi, *J. Biochem. 68*, 509 (1970).

250. J. Koziol, *Photochem. Photobiol. 5*, 41 (1966).

251. G. D. Fasman (ed.), *Handbook of Biochemistry and Molecular Biology*, 3rd ed., CRC Press, Cleveland, 1976 vol. 2, p. 215.

252. A. Bowd, D. Byron, J. B. Hudson, and J. H. Turnbull, *Photochem. Photobiol. 11*, 445 (1970).

253. G. D. Fasman (ed.), *Handbook of Biochemistry and Molecular Biology*, 3rd ed., CRC Press, Cleveland, 1976 vol. 2, p. 211.

254. A. J. Thomson, *J. Chem. Phys. 51*, 4106 (1969).

255. J. E. Wampler and J. E. Churchich, *J. Biol. Chem. 244*, 1477 (1969).

256. J. E. Churchich, *Biochim. Biophys. Acta 79*, 643 (1964).

257. J. Eisinger and R. G. Shulman, *Science 161*, 1311 (1968).

258. M. Daniels and W. Hauswirth, *Science 171*, 675 (1971).

259. A. A. Lamola and J. Eisinger, *Biochim. Biophys. Acta 240*, 313 (1971), and refs. therein.

260. R. W. Wilson, J. Morgan, and P. R. Callis, *Chem. Phys. Lett. 36*, 618 (1975), and refs. therein.

261. A. A. Oraevsky, A. V. Sharkov, and D. N. Nikogosyan, *Chem. Phys. Lett. 83*, 276 (1981).

262. R. S. Becker and G. Kogan, *Photochem. Photobiol. 31*, 5 (1980).

263. H. C. Börresen, *Acta Chem. Scand. 17*, 921 (1963); *21*, 2463 (1967).

264. W. Hauswirth and M. Daniels, *Photochem. Photobiol. 13*, 157 (1971).

265. J. P. Morgan and M. Daniels, *Photochem. Photobiol. 27*, 73 (1978).

266. P. R. Callis, *Chem. Phys. Lett. 61*, 563 (1979).

267. R. W. Wilson and P. R. Callis, *J. Phys. Chem. 80*, 2280 (1976).

268. P. Vigny, *C. R. Acad. Sci. Fr. 272D*, 3206 (1971).

269. M. Daniels, *Proc. Nat. Acad. Sci. USA 69*, 2488 (1972).

270. (a) J. P. Morgan and M. Daniels, *J. Phys. Chem. 86*, 4004 (1982); (b) idem, *Photochem. Photobiol. 31*, 207 (1981); (c) V. L. Rapoport, V. M. Bakulov, *Vestn. Leningr. Univ., Fiz. Khim. 1982*, 90; from *Chem. Abstr. 97*, 171814 (1982).

271. E. P. Gibson and J. H. Turnbull, *J. Photochem. 11*, 313 (1979).

272. H. C. Börresen, *Acta Chem. Scand. 21*, 11 (1967).

273. P. Vigny and M. Duquesne, *Excited States of Biological Molecules*, J. B. Birks (ed.), Wiley-Interscience, New York, 1976, p. 166.

274. W. B. Knighton, G. O. Giskas, and P. R. Callis, *J. Phys. Chem. 86*, 49 (1982).

275. B. E. Anderson and P. R. Callis, *Photochem. Photobiol. 32*, 1 (1980).

276. J. P. Morgan and M. Daniels, *Photochem. Photobiol. 31*, 101 (1980).

277. P. Vigny, *C. R. Acad. Sci. Fr., 277D*, 1941 (1973).

278. R. W. Wilson and P. R. Callis, *Photochem. Photobiol. 31*, 323 (1980).

279. B. Singer, *Biochemistry 11*, 3939 (1972).

280. K. Hemminki, *Acta Chem. Scand. 34*, 603 (1980).
281. S. Y. Wang (ed.), *Photochemistry and Photobiology of Nucleic Acids*, Academic Press, New York, 1976.
282. S. Georghiou, in: *Modern Fluorescence Spectroscopy*, Vol. 3, E. L. Wehry (ed.), Plenum Press, New York, 1981, p. 193.
283. S. Udenfriend and P. Zaltzman, *Anal. Biochem. 3*, 49 (1962).
284. M. Gueron, J. Eisinger, and R. G. Shulman, *J. Chem. Phys. 47*, 4077 (1967).
285. J. Eisinger, M. Gueron, R. G. Shulman, and T. Yamane, *Proc. Nat. Acad. Sci. USA 55*, 1015 (1966).
286. P. Vigny and J. P. Ballini, in: *Excited States in Organic Chemistry and Biochemistry*, B. Pullman and M. Goldblum (eds.), Reidel, Dordrecht, Holland, 1977, p. 1.
287. C. Helene, M. Ptak, and R. Santus, *J. Chim. Phys. 65*, 160 (1968).
288. S. I. Shapiro, A. J. Campillo, V. H. Kollman, and W. B. Goad, *Opt. Comm. 15*, 308 (1975).
289. A. Anders, *Opt. Comm. 26*, 339 (1978).
290. A. Anders, *Chem. Phys. Lett. 81*, 270 (1981).
291. M. Daniels, in: *Physicochemical Properties of Nucleic Acids*, Vol. I, J. Duquesne (ed.), Academic Press, New York, 1973, p. 99.
292. J. W. Longworth, R. O. Rahn, and R. G. Shulman, *J. Chem. Phys. 45*, 2930 (1960).
293. T. I. Aoki and P. R. Callis, *Chem. Phys. Lett. 92*, 327 (1982).
294. S. K. Srivastava and P. C. Mishra, *J. Theor. Biol. 93*, 655 (1981), and refs. therein.
295. R. H. F. Manske (ed.), *The Alkaloids: Chemistry and Physiology*, Vols. 1–17, Academic Press, New York, 1950–1979.
296. J. F. Ireland and A. H. Wyatt, *Adv. Phys. Org. Chem. 12*, 131 (1976).
297. S. G. Schulman, in: *Modern Fluorescence Spectroscopy*, Vol. 2, E. L. Wehry (ed.), Plenum Press, New York, 1976, p. 239.
298. S. P. Singh and V. I. Stenberg, *Spectrosc. Lett. 11*, 845 (1978).
299. V. I. Stenberg, S. P. Singh, and N. K. Narain, *Spectrosc. Lett. 10*, 639 (1977).
300. R. S. Davidson, R. Bonneau, J. Joussot-Dubien, and K. J. Toyne, *Chem. Phys. Lett. 63*, 269 (1979).
301. R. Brandt, S. Ehrlich-Rogozinski, and N. D. Cheronis, *Microchem. J. 5*, 215 (1961).
302. A. M. Gillespie, *Anal. Lett. 2*, 609 (1969).
303. J. J. Aaron, L. B. Sanders, and J. D. Winefordner, *Clin. Chim. Acta 45*, 375 (1973).
304. R. Brandt, M. J. Olsen, and N. D. Cheronis, *Science 139*, 1063 (1963).

305. G. A. Wadds, *Proc. Soc. Anal. Chem. 5*, 209 (1968).

306. P. Balatre, M. Traisnel, and J. Delcambre, *Ann. Pharm. Fr. 19*, 171 (1961).

307. R. A. Chalmers and G. A. Wadds, *Analyst 95*, 234 (1970).

308. W. D. Darwin and E. J. Cone, *J. Pharm. Sci. 69*, 253 (1980).

309. E. F. Breheret and M. M. Martin, *J. Lumin. 17*, 49 (1978).

310. R. Rosetti and L. E. Brus, *J. Chem. Phys. 73*, 1546 (1980).

311. R. Croteau and R. M. Leblanc, *Photochem. Photobiol. 28*, 33 (1978); and *J. Luminesc. 15*, 353 (1977).

312. H. Roigt and R. M. Leblanc, *Can. J. Chem. 50*, 1959 (1972).

313. T. Arai and T. Okuyama, *Seikagaku 45*, 19 (1973); in *Chem. Abstr. 79*, 88525 (1973).

314. B. Bhattacharyya and J. Wolf, *Proc. Nat. Acad. Sci. USA 71*, 2627 (1974).

315. T. Arai and T. Okuyama, *Anal. Biochem. 69*, 443 (1975).

316. D. Burnett and L. J. Audus, *Phytochemistry 3*, 395 (1964).

317. J. D. MacNeil, M. Häusler, R. W. Frei, and O. Hutzinger, *Anal. Biochem. 45*, 100 (1972), and refs. therein.

318. H. Sprince, G. R. Rowley, and D. Jameson, *Science 125*, 442 (1957).

319. B. L. Van Duuren, *J. Org. Chem. 26*, 2954 (1961).

320. J. W. Bridges and R. T. Williams, *Biochem. J. 107*, 225 (1968).

321. M. G. M. Balemans and F. C. G. van de Veerdonk, *Experientia 23*, 906 (1967).

322. R. D. Rauh, T. R. Evans, and P. A. Leermakers, *J. Am. Chem. Soc. 90*, 6897 (1968).

323. Z. Jung and M. Jungova, *Ceskoslov. Farm. 22*, 195 (1973); *Anal. Abstr. 25*, 4031 (1973).

324. T. Kishi, M. Tanaka, and J. Tanaka, *Bull. Chem. Soc. Jap., 50*, 1267 (1977), and refs. therein.

325. E. G. C. Clarke, *J. Forensic Sci. 7*, 46 (1967).

326. A. M. Gillespie, *Anal. Lett. 2*, 609 (1969).

327. A. Bowd, J. B. Hudson, and J. H. Turnbull, *J. Chem. Soc.*, Perkin II, 1312 (1973).

328. J. Axelrod, *Ann. N.Y. Acad. Sci. 66*, 435 (1957).

329. S. Udenfriend, D. E. Duggan, B. M. Vasta, and B. B. Brodie, *J. Pharm. Exp. Ther. 120*, 26 (1957).

330. T. Niwaguchi and T. Inoue, *J. Chromatogr. 59*, 127 (1971).

331. J. M. Weber and T. S. Ma, *Mikrochim. Acta 1976*, I, 581.

332. J. Christie, M. White, and J. Wiles, *J. Chromatogr. 120*, 496 (1976).

333. P. J. Twitchett, S. M. Fletcher, A. T. Sullivan, and A. C. Moffat, *J. Chromatogr. 150*, 73 (1978).

334. A. H. M. T. Scholten and R. W. Frei, *J. Chromatogr. 176*, 349 (1979).

335. H. Shizuko, *Bull. Chem. Soc. Jap.* 42, 52 (1969).

336. T. Lai and E. C. Lim, *Chem. Phys. Lett.* 62, 507 (1969).

337. M. Montagu, P. Levillain, M. Rideau, and J. C. Chenieux, *Talanta 28*, 709 (1981).

338. K. Itoh and I. Azumi, *J. Chem. Phys.* 62, 3431 (1975).

339. S. Okajima and E. C. Lim, *J. Chem. Phys.* 69, 1929 (1978).

340. A. K. Jameson, S. Okajima, H. Saigusa, and E. C. Lim, *J. Chem. Phys.* 75, 4729 (1981).

341. S. G. Schulman and A. C. Capomacchia, *J. Am. Chem. Soc.* 95, 2763 (1973).

342. S. C. Chao, J. Tretzel, and F. W. Schneider, *J. Am. Chem. Soc.* 101, 134 (1979).

343. W. West, R. H. Müller, and E. Jette, *Proc. Roy. Soc. Lond. Ser. A., 121*, 229, 294 and 313 (1928).

344. O. S. Wolfbeis and E. Urbano, *Z. Anal. Chem. 314*, 577 (1983).

345. J. Eisenbrand and M. Reisch, *Z. Anal. Chem. 179*, 352 (1961).

346. J. F. Ireland and A. H. Wyatt, *Adv. Phys. Org. Chem. 12*, 131 (1976), p. 174.

347. A. Weller, *Z. Elektrochem. 61*, 956 (1957).

348. H. Kokubun, *Z. Elektrochem. 62*, 599 (1958).

349. A. Grabowska and B. Pakula, *Photochem. Photobiol. 9*, 339 (1969).

350. W. R. Moomaw and M. F. Anton, *J. Phys. Chem. 80*, 2243 (1976).

351. S. G. Schulman, R. M. Threatte, A. C. Cappomacchia, and W. L. Paul, *J. Pharm. Sci. 63*, 876 (1974).

352. R. F. Chen, *Anal. Biochem. 19*, 374 (1967), *Proc. Nat. Acad. Sci. USA 60*, 598 (1968).

353. J. N. Demas and G. A. Crosby, *J. Phys. Chem. 75*, 991 (1971).

354. J. E. Gill, *Photochem. Photobiol. 9*, 313 (1969).

355. A. N. Fletcher, *J. Phys. Chem. 72*, 2742 (1968).

356. D. V. O'Connor, S. R. Meech, and D. Phillips, *Chem. Phys. Lett. 88*, 22 (1982).

357. W. R. Dawson and M. W. Windsor, *J. Phys. Chem. 72*, 3251 (1968).

358. M. Goldman and E. L. Wehry, *Anal. Chem. 42*, 1178 (1970).

359. S. G. Schulman, *Anal. Chem. 43*, 285 (1971).

360. B. Zalis, A. C. Cappomacchia, D. Jackman, and S. G. Schulman, *Talanta 20*, 33 (1973).

361. E. Smekal, *Stud. Biophys. 87*, 213 (1982).

362. E. Smekal, J. Koudelka, and M. A. Hung, *Stud. Biophys. 81*, 89 (1980).

363. E. Smekal, *Stud. Biophys. 87*, 211 (1982).

364. E. Smekal and S. Pavelka, *Stud. Biophys. 64*, 183 (1977).

365. E. Smekal and S. Pavelka, *Coll. Czech. Chem. Comm. 41*, 3157 (1976).

366. H. Ito, T. Kita, and T. Tomimatsu, *Tokushima Daigaku Yakugaku Kenkyu Nempo 16*, 5 (1967); in *Chem. Abstr. 70*, 37952h (1969).

367. E. P. Gibson and J. H. Turnbull, *Analyst 104*, 582 (1979).

368. E. P. Gibson and J. H. Turnbull, *J. Chem. Soc.*, Perkin II, 1696 (1980).

369. O. S. Wolfbeis, E. Fürlinger, and R. Wintersteiger, *Monatsh. Chem. 113*, 509 (1982).

370. O. S. Wolfbeis and E. Fürlinger, *Z. Phys. Chem.* (N. F.) *129*, 171 (1982).

371. R. Sakurovs and K. P. Ghiggino, *J. Photochem. 18*, 1 (1982).

372. J. Maier, J. Jurenitsch, F. Heresch, E. Haslinger, G. Schulz, M. Pöhm, and K. Jentzsch, *Monatsh. Chem. 112*, 1425 (1981).

373. G. Duportail and H. Lami, *Biochim. Biophys. Acta 402*, 20 (1975).

374. J. S. Charnock, C. L. Bashford, and J. C. Ellory, *Biochim. Biophys. Acta 436*, 413 (1976).

375. K. P. Wong, *Anal. Biochem. 62*, 149 (1974).

376. K. P. Wong and T. L. Sourkes, *Anal. Biochem. 21*, 444 (1967).

377. G. K. Hughes, F. N. Lahey, J. R. Price, and L. J. Webb, *Nature 162*, 223 (1948).

378. S. Johne and D. Gröger, *Pharmazie 27*, 195 (1972).

379. S. G. Schulman and W. J. M. Underberg, *Anal. Chim. Acta 107*, 411 (1979).

380. S. G. Schulman, B. S. Vogt, and M. W. Lovell, *Chem. Phys. Lett. 75*, 224 (1980).

381. M. Siegmund and J. Bendig, *Ber. Bunsenges. Phys. Chem. 82*, 1061 (1978), and *Z. Chem. 21*, 227 (1981).

382. M. Siegmund and J. Bendig, *Z. Naturforsch. 35A*, 1076 (1980).

383. L. Villemey, *Ann. Chim. Phys. 12*, Ser. 5, 642 and 779 (1950).

384. G. K. Hughes, K. G. Neill, and E. Ritchie, *Aust. J. Sci. Res. Ser. A 3*, 497 (1950).

385. L. E. Zeltser and S. T. Talipov, *Zhurn. Anal. Khim. 33*, 1260 (1978).

386. A. Weisstuch, P. Neidig, and A. C. Testa, *J. Lumin. 10*, 137 (1975).

387. D. F. Bender, E. Sawicki, and R. Wilson, *Anal. Chem. 36*, 1011 (1964).

388. A. Perry, P. Tidwell, J. Cetorelli, and J. D. Winefordner, *Anal. Chem. 43*, 781 (1971).

389. A. Ahmad and G. Durocher, *Photochem. Photobiol. 34*, 573 (1981).

390. A. C. Capomacchia and S. G. Schulman, *Anal. Chim. Acta 59*, 471 (1972).

391. J. R. Lakowicz and D. Hogen, *Chem. Phys. Lipids 26*, 1 (1980).

392. G. E. Johnson, *J. Phys. Chem. 84*, 2940 (1980).

393. B. V. Golomolzin, L. D. Shcherak, and I. Y. Postovskii, *Khim. Geterotsikl. Soedin. 1969*, 1131.

394. C. Parkanyi, G. M. Sanders, and M. Van Dijk, *Recl. Trav. Chim. Pay-Bas 100*, 161 (1981).

395. D. A. Lerner, P. M. Horowitz, and E. M. Evleth, *J. Phys. Chem. 81*, 12 (1977).

396. J. N. Kikkert, G. R. Kelly, and T. Kurucsev, *Biopolymers 12*, 1459 (1973).

397. J. A. Mockle, *Can. J. Pharm. Sci. 5*, 55 (1970).

398. F. Korte and H. Weitkamp, *Angew. Chem. 70*, 434 (1958).

399. T. Gürkan, *Mikrochim. Acta 1976*, I, 173.

400. R. Bowman, P. Caulfield, and S. Udenfriend, *Science 122*, 32 (1955).

401. J. A. Gordon and D. J. Campbell, *Anal. Chem. 29*, 488 (1957).

402. B. Davies, M. G. Dodds, and E. G. Tomich, *J. Pharm. Pharmacol. 14*, 249 (1962).

403. R. F. Chen, G. G. Vurek, and N. Alexander, *Science 156*, 949 (1967).

404. W. H. Melhuish, *J. Phys. Chem. 65*, 229 (1961); for a discussion of the quinine quantum yields, see J. N. Demas and G. A. Crosby, *J. Phys. Chem. 75*, 1007 (1971).

405. F. M. Dean, *Naturally Occurring Oxygen Ring Compounds*, Butterworths, London, 1963.

406. E. Davin, C. Balny, and R. Guglielmetti, *C. R. Acad. Sci. Fr., Ser. C. 275*, 79 (1972).

407. O. S. Wolfbeis, unpublished results.

408. A. Bowd, P. Byron, J. B. Hudson, and J. H. Turnbull, *Talanta 18*, 697 (1971).

409. S. Rangaswami and T. R. Seshadri, *Proc. Indian Acad. Sci. 12A*, 375 (1940).

410. V. V. Sreerama Murti, S. Rajagopalan, and L. Ramachandra Row, *Proc. Indian Acad. Sci. 34A*, 319 (1951).

411. S. K. K. Jatkar and B. N. Mattoo, *J. Indian Chem. Soc. 33*, 559 (1956).

412. J. B. Gallivan, *Can. J. Chem. 48*, 3928 (1970).

413. O. S. Wolfbeis and A. Knierzinger, *Z. Naturforsch. 34A*, 510 (1979).

414. A. Murata, T. Suzuki, and T. Ito, *Bunseki Kagaku 14*, 630 (1965).

415. T. Ito and A. Murata, *Bunseki Kagaku 20*, 1422 (1971).

416. G. Huitink, *Talanta 27*, 977 (1981).

417. T. J. Mabry, K. R. Markham, and M. B. Thomas, *The Systematic Identification of Flavonoids*, Springer-Verlag, New York, 1979.

418. M. Itoh, K. Tokomura, Y. Tanimoto, Y. Okada, H. Takeuchi, K. Obi, and I. Tanaka, *J. Am. Chem. Soc. 104*, 4146 (1982).

419. R. D. H. Murray, J. Mendez, and S. A. Brown, *The Natural Coumarins*, Wiley-Interscience, Chichester, 1982.

420. S. Rangaswami, T. R. Seshadri, and V. Venkateswarlu, *Proc. Indian Acad. Sci. 13A*, 375 (1941).

421. V. Balaiah, T. R. Seshadri, and V. Venkatesvarlu, *Proc. Indian Acad. Sci. 16A*, 68 (1942).

422. P. Casparis and E. Manella, *Centen. Soc. Suisse Pharm.* (1843–1943), 347; in ref. 423.

423. R. H. Goodwin and F. Kavanagh, *Arch. Biochem. Biophys. 36*, 442 (1952).

424. F. Corcilius, *Arch. Pharm.* (Weinheim) *289*, 81 (1955).

425. B. N. Mattoo, *Trans. Faraday Soc. 52*, 1184 (1956).

426. C. E. Wheelock, *J. Am. Chem. Soc. 81*, 1348 (1958).

427. P. S. Song and W. H. Gordon, *J. Phys. Chem. 74*, 4234 (1970).

428. P. S. Song and Q. Chae, *J. Lumin. 12/13*, 831 (1976).

429. G. F. Fedorin and V. P. Georgievski, *Zhurn. Prikl. Spektrosk. 20*, 153 (1974), and *21*, 165 (1974).

430. G. J. Yakatan, R. J. Juneau, and S. G. Schulman, *J. Pharm. Sci. 61*, 749 (1972).

431. O. S. Wolfbeis and G. Uray, *Monatsh. Chem. 109*, 123 (1978).

432. E. A. Bababunmi, M. R. French, R. J. Rutman, O. Bassir, and L. G. Dring, *Biochem. Soc. Trans. 3*, 940 (1975).

433. O. S. Wolfbeis, *Z. Phys. Chem.* (N. F.) *125*, 15 (1981).

434. R. F. Chen, *Anal. Lett. 1*, 423 (1968).

435. C. N. Ou, P. S. Song, M. L. Harter, and I. C. Felkner, *Photochem. Photobiol. 24*, 487 (1976).

436. W. R. Sherman and E. Robins *Anal. Chem. 41*, 1180 (1969); fluorescences at pH's between 8.5 and 10.

437. D. W. Fink and W. R. Koehler *Anal. Chem. 42*, 990 (1970).

438. G. J. Yakatan, R. J. Juneau, and S. G. Schulman, *Anal. Chem. 44*, 1044 (1972).

439. O. S. Wolfbeis, *Z. Naturforsch. 32A*, 1065 (1977).

440. O. S. Wolfbeis, W. Rapp, and E. Lippert, *Monatsh. Chem. 109*, 899 (1978).

441. S. C. Haydon, *Spectrosc. Lett. 8*, 815 (1975).

442. G. G. Guilbault, *Practical Fluorescence: Theory, Methods, and Techniques*, Dekker, New York, 1973. p. 598.

443. O. S. Wolfbeis, E. Fürlinger, H. Kroneis, and H. Marsoner, *Z. Anal. Chem., 314*, 577 (1983).

444. G. G. Guilbault, *Enzymatic Methods of Analysis*, Pergamon Press, New York, 1970.

445. O. S. Wolfbeis and K. Schaffner, *Photochem. Photobiol. 32*, 143 (1980).

446. O. S. Wolfbeis, E. Lippert, and H. Schwarz, *Ber. Bunsenges. Phys. Chem. 84*, 115 (1980).

447. P. E. Zinsli, *Z. Angew. Math. Phys.* (Basel) *23*, 1003 (1972), and J. Photochem. *3*, 55 (1974).

448. M. Nakashima, J. A. Sousa, and R. C. Clapp, *Nature 235*, 16 (1972).

449. G. S. Beddard, S. Carlin, and R. S. Davidson, *J. Chem. Soc.*, Perkin II, 262 (1977).

450. R. K. Bauer and A. Kowalczyk, *Z. Naturforsch. 35A,* 946 (1980).

451. C. V. Shank, A. Dienes, A. M. Trozzolo, and J. A. Myer, *Appl. Phys. Lett. 16,* 409 (1970).

452. T. Kindt, E. Lippert, and W. Rapp, *Z. Naturforsch. 27A,* 1371 (1972).

453. A. Bergman and J. Jortner, *J. Lumin. 6,* 390 (1973).

454. J. R. Huber, M. Nakashima, and J. A. Sousa, *J. Phys. Chem. 77,* 860 (1973).

455. W. Fabian, *Z. Naturforsch. 33B,* 628 (1978).

456. W. Fabian, *J. Mol. Struct. 69,* 227 (1980), and *85,* 77 (1981).

457. J. F. Fisher, H. E. Nordby, A. C. Weiss, and W. L. Stanley, *Tetrahedron 23,* 2523 (1967).

458. E. Martin and Ch. Berner, *Mitt. Geb. Lebensmittel-Untersuch. Hygi. 64,* 251 (1973).

459. H. W. Latz and B. C. Madsen, *Anal. Chem. 41,* 1180 (1969), and refs. therein.

460. H. W. Latz and A. D. Ernes, *J. Chromatogr. 166,* 189 (1978).

461. G. Vernin, J. P. Bianchini, and A. Siouffi, *Parfum. Cosmet. Arômes 30,* 49 (1979).

462. (a) R. K. Ibrahim, *J. Chromatogr. 42,* 544 (1969); (b) P. Schwarze and R. Hoffschmidt, *Naturwiss. 45,* 518 (1958).

463. J. P. Gramond and R. R. Paris, *Plant Med. Phytother. 5,* 315 (1971); in *Chem. Abstr. 76,* 131561 (1972).

464. S. M. Khafagy, N. N. Sabri, and A. H. Abou Ouia, *Egypt. J. Pharm. Sci. 19,* 293 (1978).

465. W. Mantulin and P. S. Song, *J. Am. Chem. Soc. 95,* 5122 (1973).

466. P. S. Song and K. J. Tapley, *Photochem. Photobiol. 29,* 1177 (1979) and *32,* 813 (1980); reviews.

467. C. N. Ou, P. S. Song, M. L. Harter, and I. C. Felkner, *Photochem. Photobiol. 24,* 487 (1976).

468. W. Poppe and L. I. Grossweiner, *Photochem. Photobiol. 22,* 217 (1975).

469. M. Sasaki, T. Sakata, and M. Sukigara, *Chem. Lett.* 701 (1977).

470. R. V. Bensasson, E. J. Land, and C. Salet, *Photochem. Photobiol. 27,* 273 (1978).

471. H. R. Battacharjee, E. L. Menger, and G. S. Hammond, *J. Am. Chem. Soc. 102,* 1977 (1980).

472. H. H. Wisneski, *J. Assoc. Off. Anal. Chem. 59,* 547 (1976).

473. P. Chambon, B. Huit, and R. Chambon-Mougenot, *Ann. Pharm. Fr. 27,* 635 (1969).

474. J. Blais, L. Dubertret, F. Gaborian, and P. Vigny, *Photochem. Photobiol. 35,* 423 (1982).

475. J. A. Robertson and W. A. Pons, *J. Assoc. Off. Anal. Chem. 51,* 2024 (1968).

476. R. B. Carnaghan, R. D. Hartley, and J. O'Kelly, *Nature* **200**, 1101 (1963).
477. J. A. Robertson, W. A. Pons, and L. A. Goldblatt, *J. Agric. Food Chem.* **15**, 798 (1967).
478. J. Chelkowski, *Photochem. Photobiol.* **20**, 279 (1974).
479. B. L. Van Duuren, T. L. Chan, and F. M. Irani, *Anal. Chem.* **40**, 2024 (1968).
480. A. O. Uwaifo, G. O. Emerole, and O. Bassir, *J. Agric. Food Chem.* **25**, 1218 (1977).
481. M. R. Berman and R. N. Zare, *J. Am. Chem. Soc.* **98**, 620 (1976).
482. M. R. Berman and R. N. Zare, *Anal. Chem.* **47**, 1200 (1975).
483. O. J. Francis, L. J. Lipinski, J. A. Gaul, and A. D. Campbell, *J. Assoc. Off. Anal. Chem.* **65**, 672 (1982).
484. C. E. White and A. Weissler, *Anal. Chem.* **42**, 57R (1970), and **44**, 182R (1972); A. Weissler, *Anal. Chem.* **46**, 500R (1974); C. M. O'Donnell and T. N. Solie, *Anal. Chem.* **48**, 175R (1976), and **50**, 189R (1978); E. L. Wehry, *Anal. Chem.* **52**, 75R (1980), **54**, 131R (1982), and **56**, 156R (1984).
485. W. C. Neeley, M. Y. Siraj, L. B. Smith, and A. W. Hayes, *J. Ass. Off. Anal. Chem.* **61**, 601 (1978).
486. R. Kühn and I. Löw, *Ber. Dtsch. Chem. Ges.* **77**, 211 (1944).
487. J. B. Harborne, *J. Chromatogr.* **2**, 581 (1959).
488. S. Beckmann and H. Geiger *Phytochemistry* **2**, 281 (1963).
489. P. Spiegl, *J. Chromatogr.* **39**, 93 (1969).
490. M. Katyal and D. E. Ryan, *Anal. Lett.* **2**, 499 (1969).
491. T. Kayashi, S. Kawai, and T. Ohno, *Chem. Pharm. Bull.* **19**, 792 (1971).
492. N. A. Tyukavkina, N. N. Pogodaeva, E. I. Brodskaya, and Y. M. Saposhnikov, *Khim. Prirod. Soedin.* (Engl. ed.) 613 (1975).
493. Y. L. Frolov, Y. N. Saposhnikov, S. S. Barer, N. N. Pogodaeva, and N. A. Tyukavkina, *Izv. Akad. Nauk SSSR, Ser. Khim.* (Engl. ed.) 2279 (1974).
494. Y. L. Frolov, Y. N. Saposhnikov, N. N. Chipanina, V. F. Sidorkin, and N. A. Tyukavkina, *Izv. Akad. Nauk SSSR, Ser. Khim.* (Engl. ed.) 258 (1978).
495. M. Kopach, S. Kopach, and V. Klechek, *Zhurn. Org. Khim.* (Engl. ed.) **16**, 1467 (1980), and refs. on previous work performed by Russian scientists.
496. H. Homberg and H. Geiger, *Phytochemistry* **19**, 2443 (1980).
497. O. S. Wolfbeis, M. Begum, and H. Geiger, *Z. Naturforsch.* **39B**, 231 (1984).
498. H. Geiger and H. Homberg, *Z. Naturforsch.* **38B**, 253 (1983).
499. O. S. Wolfbeis, M. Leiner, P. Hochmuth, and H. Geiger, *Ber. Bunsenges. Phys. Chem.* **88**, 759 (1984).
500. P. K. Sengupta and M. Kasha, *Chem. Phys. Lett.* **68**, 382 (1979).

501. O. A. Salman and H. G. Drickamer, *J. Chem. Phys.* 75, 572 (1981).

502. O. S. Wolfbeis, A. Knierzinger, and R. Schipfer, *J. Photochemistry 21,* 67 (1983).

503. G. J. Woolfe and P. J. Thistlethwaite, *J. Am. Chem. Soc. 103,* 6916 (1981).

504. M. Itoh and H. Kurokawa, *Chem. Phys. Lett. 91,* 487 (1982).

505. R. Schipfer, O. S. Wolfbeis, and A. Knierzinger, *J. Chem. Soc.,* Perkin II, 1443 (1981).

506. T. B. Gage and S. H .Wender, *Anal. Chem. 22,* 708 (1950).

507. P. Hagedorn and R. Neu, *Arch. Pharm.* (Weinheim) *286,* 486 (1953).

508. R. Neu, *Mikrochim. Acta* 1169 (1956).

509. A. S. Zaprjanova, *Mikrochim. Acta I,* 561 (1976).

510. E. M. Nevskaya and V. A. Nazarenko, *Zhurn. Anal. Khim. 27,* 1699 (1972) (review).

511. M. Katyal and S. Prakash, *Talanta 24,* 367 (1977) (review).

512. A. Gomez-Hens and M. Valcarcel, *Analyst 107,* 465 (1982) (review).

513. J. Hubach and M. Trojna, *Coll. Czech. Chem. Comm. 35,* 1575 (1970).

514. O. S. Wolfbeis and R. Schipfer, *Photochem. Photobiol. 34,* 567 (1981).

515. M. C. Gaylord and L. R. Brady, *J. Pharm. Sci. 60,* 1503 (1971).

516. T. F. Salem and B. G. Swanson, *J. Food Sci. 41,* 1237 (1976).

517. T. Shibazaki, *Yakugaku Zasshi 88,* 601 (1968); in *Chem. Abstr. 69,* 64463 (1968).

518. K. Mizutani, H. Miyazaki, K. Ishigaki, and H. Hosoya, *Bull. Chem. Soc. Jap. 47,* 1596 (1974).

519. O. S. Wolfbeis and E. Fürlinger, *J. Am. Chem. Soc. 104,* 4069 (1982).

520. L. Kopanski, M. Klaar, and W. Steglich, *Liebigs Ann. Chem.* 1280 (1982).

521. N. Filipescu, S. K. Chakrabarti, and P. G. Tarassoff, *J. Phys. Chem. 77,* 2276 (1973).

522. R. T. Williams, *J. Roy. Inst. Chem. 83,* 611 (1959).

523. R. W. Thomas and N. J. Leonhard, *Heterocycles 5,* 839 (1976).

524. D. G. Crosby and R. V. Berthold, *Anal. Biochem. 4,* 349 (1962).

525. T. Hinohara, K. Amano, and K. Matsui, *Nippon Kagaku Kaishi 247* (1976).

526. M. A. Pathak, B. Allen, D. J. E. Ingram, and J. H. Fellman, *Biochim. Biophys. Acta 54,* 506 (1961).

527. J. Grzywacz and S. Taszner, *Z. Naturforsch. 37A,* 262 (1982).

528. R. B. Hartley and V. O'Kelly, *Nature 200,* 1101 (1963).

529. E. Sawicki, T. W. Stanley, W. C. Elbert, and M. Morgan, *Talanta 12,* 605 (1965).

530. S. G. Hadley, A. S. Muraki, and K. Spitzer, *J. Forensic Sci. 19,* 657 (1974). (The legends of Figs. 3 and 4 in this paper should be reversed.)

531. G. Caille, D. Leclerc-Chevalier, and J. A. Mockle, *Can. J. Pharm. Sci.* *5*, 55 (1970).

532. M. Gouterman, in: *The Porphyrins,* Vol. III, D. Dolphin (ed.), Academic Press, New York, 1978, p. 1.

533. Govindjee (ed.), *Photosynthesis,* Academic Press, New York, 1982.

534. H. T. Witt, in: *Excited States of Biological Molecules,* Wiley, New York, 1976.

535. F. K. Fong (ed.), *Light Reaction Path of Photosynthesis,* Springer-Verlag, New York, 1982.

536. Govindjee and R. Govindjee, *Biochem. Rev. 48,* 25 (1977).

537. H. Levanon and J. R. Norris, *Chem. Rev. 78,* 185 (1978).

538. Govindjee and P. A. Jursinic, *Photochem. Photobiol. Rev. 4,* 125 (1979).

539. A. Harriman and J. Barber, *Top. Photosynth. 3,* 243 (1979).

540. G. S. Beddard (ed.), *NATO Adv. Stud. Ser. 34A,* 225 (1980).

541. J. Lavorel and A. L. Etienne, in: *Primary Processes in Photosynthesis,* J. Barber (ed.), Elsevier, Amsterdam, 1977, p. 203.

542. J. A. Radley and J. Grant, *Fluorescence Analysis in Ultraviolet Light,* Chapman and Hall, London, 1959.

543. P. W. Danckwortt and J. Eisenbrand, *Lumineszenzanalyse im filtrierten ultraviolettem Licht,* Akad. Verlagsges. Geest & Portig, Leipzig, 1964.

544. J. E. Falk, *Porphyrins and Metalloporphyrins,* Elsevier, Amsterdam, 1964.

545. S. Schwartz, M. Berg, I. Bossenmaier, and H. Dinsmore, *Meth. Biochem. Anal. 8,* 221 (1960).

546. S. Schwartz, L. Zieve, and C. J. Watson, *J. Lab. Clin. Med. 37,* 843 (1951).

547. H. Fink and W. Hoerburger, *Z. Physiol. Chem. 218,* 181 (1933).

548. H. Fink and W. Hoerburger, *Z. Physiol. Chem. 220,* 123 (1933) and *225,* 49 (1934).

549. A. A. Savitskii, E. V. Vorobeva, I. V. Berezin, and N. N. Ugarova, *J. Colloid Interface Sci. 84,* 175 (1981).

550. A. T. Gradyushko, V. N. Knyukshto, K. N. Solevev, and M. P. Tsvirko, *Zh. Prikl. Spektrosk. 23,* 444 (1975).

551. A. Everett and M. Gouterman, *Bioinorg. Chem. 9,* 281 (1978).

552. W. G. Van Dorp, T. J. Schaafsma, M. Soma, and J. H. Van der Waals, *Chem. Phys. Lett. 21,* 221 (1973); Mol. Phys. *30,* 1701 (1975); and *Mol. Phys. 32,* 1703 (1976).

553. M. P. Tsvirko and K. N. Solovev, *Zh. Prikl. Spektrosk. 20,* 115 (1974).

554. R. M. Macfarlane and S. Völker, *Chem. Phys. Lett. 69,* 151 (1980).

555. I. E. Zalesski, V. N. Kotlo, A. N. Sevchenko, K. N. Solovev, and S. F. Shkirman, *Dokl. Akad. Nauk SSSR 207,* 1314 (1972).

556. G. A. Zagusta, V. N. Kotlo, S. R. Shkirman, and K. N. Solovev, *Zh. Prikl. Spektrosk. 27,* 164 (1977).

557. L. A. Bykovskaya, A. D. Gradyushko, R. I. Personov, Y. V. Romanov-skii, K. N. Solovev, A. S. Starukhin, and A. M. Shulga, *Zh. Prikl. Spektrosk. 29*, 1088 (1978) (Engl. ed.), p. 1510.

558. A. T. Gradyushko, K. N. Solovev, A. S. Starukhin, and A. M. Shulga, *Opt. Spektrosk. 43*, 70 (1977).

559. A. T. Gradyushko, K. N. Solovev, and A. S. Starukhin, *Opt. Spektrosk. 40*, 267 (1976).

560. A. T. Gradyushko, K. N. Solovev, A. S. Starukhin, and A. M. Shulga, *Izv. Akad. Nauk SSSR, Ser. Fiz. 39*, 1938 (1975).

561. S. Völker, R. M. Macfarlane, A. Z. Genack, H. P. Trommsdorff, and J. H. Van der Waals, *J. Chem. Phys. 67*, 1759 (1977).

562. S. Völker and R. M. Macfarlane, *Chem. Phys. Lett. 53*, 8 (1978), and *61*, 421 (1979).

563. H. P. H. Thijssen, A. I. M. Dicker, and S. Völker, *Chem. Phys. Lett. 92*, 7 (1982).

564. O. S. Yudina and R. I. Personov, *Biofizika 19*, 42 (1974) (Engl. ed.), p. 39, and refs. therein.

565. U. Even and J. Jortner, *J. Chem. Phys. 77*, 4391 (1982).

566. M. N. Saposhnikov and V. I. Alekseev, *Chem. Phys. Lett. 87*, 487 (1982).

567. R. Avarmaa and R. Tamkivi, *Esti NSV Tead. Akad. Toim., Funs. Mat. 29*, 442 (1980); in *Chem. Abstr. 94*, 60331 (1981).

568. L. L. Gladkov, A. T. Gradyushko, K. N. Solovev, A. S. Starukhin, and A. M. Shulga, *Zh. Prikl. Spektrosk. 29*, 304 (1978), and refs. therein.

569. K. N. Solovev, S. P. Shkirman, and G. A. Zagusta, *Zh. Prikl. Spektrosk. 14*, 1055 (1971).

570. A. N. Sevchenko, K. N. Solovev, S. F. Shkifman, and A. G. Sarzhev-skaya, *Dokl. Akad. Nauk SSSR 153*, 1391 (1963).

571. K. A. Zachariasse and D. G. Whitten, *Chem. Phys. Lett. 22*, 527 (1973), and refs. therein.

572. P. G. Seybold and M. Gouterman, *J. Mol. Spectrosc. 31*, 1 (1969).

573. K. N. Solovev, V. N. Knyukshto, M. P. Tsvirko, and A. T. Gradyushko, *Opt. Spektrosk. 41*, 569 (1976).

574. A. T. Gradyushko, A. N. Sevchenko, K. N. Solovev, and M. P. Tsvirko, *Photochem. Photobiol. 11*, 387 (1970).

575. A. N. Sevchenko, *Izv. Akad. Nauk SSSR, Ser. Fiz. 26*, 53 (1962) (Engl. ed.), p. 54.

576. R. C. Srivastava, V. D. Anand, and W. R. Carper, *Appl. Spectrosc. 27*, 444 (1973).

577. G. Cauzzo, G. Gennari, G. Jori, and J. D. Spikes, *Photochem. Photobiol. 25*, 389 (1977).

578. A. Andreoni, R. Cubeddu, S. DeSilvestre, G. Jori, P. Laporta, and E. Reddi, *Z. Naturforsch. 38C*, 83 (1983).

579. D. Kessel and E. Rossi, *Photochem. Photobiol. 35*, 37 (1982), and refs. therein.

580. C. S. French, J. H. C. Smith, H. I. Virgin, and R. L. Airth, *Plant Physiol. 31*, 369 (1956).

581. D. Holten, M. W. Windsor, W. W. Parson, and M. Gouterman, *Photochem. Photobiol. 28*, 951 (1978).

582. R. R. Bucks, T. L. Netzel, I. Fujita, and S. G. Boxer, *J. Phys. Chem. 86*, 1947 (1982).

583. J. V. Knop and L. Knop, *Z. Naturforsch. 27A*, 1663 (1972).

584. L. I. Paramonova, J. Naus, V. D. Kreslavskii, and Y. M. Stolovitskii, *Biofizika 27*, 197 (1982).

585. M. J. Yuen, L. L. Shipman, J. J. Katz, and J. C. Hindman, *Photochem. Photobiol. 32*, 281 (1980).

586. R. J. Platenkamp, H. J. den Blanken, and A. J. Hoff, *Chem. Phys. Lett. 76*, 35 (1980).

587. M. N. Usacheva, *Biofizika 6*, 983 (1971).

588. A. A. Krasnovskii, V. A. Shuvalov, F. F. Litvin, and A. A. Krasnovskii, *Dokl. Akad. Nauk SSSR 199*, 1181 (1971).

589. S. S. Dvornikov, V. N. Knyukshto, K. N. Solovev, and M. P. Tsvirko, *Opt. Spektrosk. 46*, 689 (1979).

590. S. S. Dvornikov, V. N. Knyukshto, A. N. Sevchenko, K. N. Solovev, and M. P. Tsvirko, *Dokl. Akad. Nauk SSSR 240*, 1457 (1978).

591. A. M. Dicker, M. Noort, H. P. H. Thijssen, S. Völker, and J. H. van der Waals, *Chem. Phys. Lett. 78*, 212 (1981).

592. H. P. H. Thijssen and S. Völker, *Chem. Phys. Lett. 82*, 478 (1981).

593. A. A. Krasnovskii and M. I. Bystrova, *Biokhimiya 27*, 812 (1962).

594. M. I. Bystrova and A. A. Krasnovskii, *Biofizika 10*, 812 (1966).

595. C. Houssier and K. Sauer, *Biochim. Biophys. Acta 172*, 492 (1969).

596. G. P. Gurinovich, A. N. Sevchenko, and K. N. Solovev, *Opt. Spektrosk. 10*, 750 (1961).

597. J. I. Matthews, S. E. Braslavsky, and P. Camillieri, *Photochem. Photobiol. 32*, 733 (1980).

598. A. J. Thomson, *J. Am. Chem. Soc. 91*, 2780 (1969).

599. A. N. Sevchenko, K. N. Solovev, and S. F. Shkirman, *Izv. Akad. Nauk SSSR, Ser. Fiz. 34*, 527 (1970).

600. D. Mauzerall, *Biochemistry 4*, 1801 (1965).

601. F. J. Kampas and M. Gouterman, *J. Lumin. 14*, 121 (1976), and *17*, 439 (1978).

602. W. T. Morgan, A. Smith, and P. Koskelo, *Biochim. Biophys. Acta 624*, 271 (1980).

603. E. I. Zenkevich, G. A. Kochubeev, and A. M. Kulba, *Biofizika 26*, 389 (1981).

604. S. Udenfriend, *Fluorescence Assay in Biology and Medicine*, Vol. II, Academic Press, New York, 1969 p. 394.

605. G. G. Guilbault, *Practical Fluorescence: Theory, Methods, and Techniques*, Dekker, New York, 1973 p. 327.

606. T. L. Hanna, D. N. Dietzler, C. H. Smith, S. Guptan, and H. S. Zarkowsky, *Clin. Chem. 22*, 161 (1976).

607. R. E. Ford, *Clin. Chem. 26*, 964 (1980).

608. R. B. Schifman and P. R. Finley, *Clin. Chem. 27*, 153 (1981).

609. S. Granick, S. Sassa, J. L. Granick, R. D. Levere, and A. Kappas, *Proc. Nat. Acad. Sci. USA 69*, 2381 (1972), and *Biochem. Med. 8*, 135 (1973).

610. A. A. Lamola, M. Joselow, and T. Yamane, *Clin. Chem. 21*, 93 (1975).

611. M. Doss, *Z. Anal. Chem. 252*, 104 (1970); *J. Chromatogr. 63*, 113 (1971); and *J. Clin. Chem. Clin. Biochem. 8*, 208 (1970).

612. A. A. Lamola, J. Eisinger, W. E. Blumberg, T. Kometani, and B. F. Burnham, *J. Lab. Clin. Med. 89*, 881 (1977).

613. D. K. Lavallee and J. T. Norelius, *Clin Chem. 23*, 282 (1977).

614. (a) J. B. Allison and R. S. Becker, *J. Chem. Phys. 32*, 1410 (1960); (b) R. S. Becker and J. B. Allison, *J. Phys. Chem. 67*, 2669 (1963).

615. F. Haurowitz, *Chem. Ber. 68*, 1795 (1935).

616. F. P. Zscheile and D. G. Harris, *J. Chem. Phys. 47*, 623 (1943).

617. R. Livingston, W. F. Watson, and J. McArdle, *J. Am. Chem. Soc. 71*, 1542 (1949).

618. F. K. Fong, M. Kusunaki, L. Galloway, T. G. Matthews, F. E. Lytle, A. J. Hoff, and F. A. Brinkman, *J. Am. Chem. Soc. 104*, 2759 (1982).

619. R. Avarmaa and R. Tamkivi, *Zh. Prikl. Spektrosk. 27*, 259 (1977).

620. Y. Onoue, K. Hiraki, and Y. Nishikawa, *Bull. Chem. Soc. Jap. 54*, 2633 (1981).

621. J. B. Birks (ed.), *Excited States of Biological Molecules*, Wiley, New York, 1976, p. 15.

622. P. S. Stensby and J. L. Rosenberg, *J. Phys. Chem. 65*, 906 (1961).

623. M. Mazurek, B. D. Nadolski, A. M. North, M. J. Park, and R. A. Pethrick, *J. Photochem. 19*, 151 (1982).

624. A. M. North, R. A. Pethrick, M. Kryszewsky, and B. Nadolski, *Acta Phys. Polon. 54A*, 797 (1978).

625. L. S. Forster and R. Livingston, *J. Chem. Phys. 20*, 1315 (1952).

626. V. B. Estigneev, V. L. Gavrilova, and A. A. Krasnovskii, *Dokl. Akad. Nauk SSSR 71*, 1542 (1949).

627. S. Freed and K. M. Sancier, *J. Am. Chem. Soc. 76*, 198 (1954).

628. V. B. Estigneev, V. A. Gavrilova, and A. A. Krasnovskii, *Dokl. Akad. Nauk SSSR 70*, 261 (1950).

629. R. Livingston and S. Weil, *Nature 170*, 750 (1952).

630. J. S. Connolly, A. F. Janzen, and E. B. Samuel, *Photochem. Photobiol.* *36*, 559 (1982), and refs. therein.

631. M. Kaplanova and K. Cermak, *J. Photochem.* *15*, 313 (1981).

632. A. C. DeWilton and J. A. Koningstein, *J. Phys. Chem.* *87*, 185 (1983).

633. I. Moya and R. Garcia, *Biochim. Biophys. Acta 722*, 480 (1983), and values compiled in their Table 1.

634. A. R. Crofts, C. A. Wraight, and D. E. Fleischmann, *FEBS Lett.* *15*, 89 (1971).

635. J. S. Brown and C. S. French, *Plant Physiol.* *34*, 305 (1959).

636. R. P. H. Krooyman, T. J. Schaafsma, G. Jansen, R. H. Clarke, D. R. Hobart, and W. R. Leenstra, *Chem. Phys. Lett.* *68*, 65 (1979).

637. L. J. Shipman, T. M. Cotton, J. R. Norris, and J. J. Katz, *J. Am. Chem. Soc.* *98*, 8222 (1976).

638. K. A. Zachariasse and D. G. Whitten, *Chem. Phys. Lett.* *22*, 527 (1973).

639. S. S. Brody and M. Brody, Nature *189*, 547 (1961).

640. R. P. H. Krooyman, T. J. Schaafsma, and J. E. Kleibeuker, *Photochem. Photobiol.* *26*, 235 (1977).

641. F. K. Fong, V. J. Koester, and J. C. Polles, *J. Am. Chem. Soc.* *98*, 6406 (1976).

642. R. Journeaux, G. Chene, and R. Viovy, *J. Chim. Phys., Phys. Chim. Biol.* *74*, 1203 (1977).

643. G. R. Seely, *Photochem. Photobiol.* *27*, 639 (1978).

644. G. S. Beddard and G. Porter, *Nature 260*, 366 (1976), and refs. therein.

645. C. Bojarski, *Z. Naturforsch. 37A*, 150 (1981).

646. D. Wong, K. Vacek, H. Merkelo, and Govindjee, *Z. Naturforsch. 33C*, 863 (1978).

647. S. M. B. de B. Costa, J. R. Froines, J. M. Harris, R. M. Leblanc, B. H. Oger, and G. Porter, *Proc. Roy. Soc. Lond., Ser. A 326*, 503 (1972).

648. A. G. Tweet, W. D. Bellamy, and G. L. Gaines, *J. Chem. Phys. 41*, 2068 (1964).

649. L. L. Shipman and D. L. Housman, *Photochem. Photobiol. 29*, 1163 (1979).

650. A. J. Alfano and F. K. Fong, *J. Am. Chem. Soc. 104*, 2767 (1982), and refs. therein.

651. M. J. Yuen, L. L. Shipman, J. J. Katz, and J. C. Hindman, *Photochem. Photobiol. 36*, 211 (1982).

652. J. S. Connolly, A. F. Janzen, and E. B. Samuel, *Photochem. Photobiol. 36*, 559 (1982), and refs. therein.

653. R. J. Eissler and M. J. Dutton, *Photochem. Photobiol. 33*, 385 (1981).

654. W. Vidaver, R. Popovic, D. Bruce, and K. Colbow, *Photochem. Photobiol. 34*, 633 (1981).

655. O. S. Wolfbeis, unpublished results.

656. J. P. Dodelet, M. F. Lawrence, M. Ringuet, and R. M. Leblanc, *Photochem. Photobiol. 33*, 713 (1981), and refs. therein.

657. L. V. Natarajan and R. E. Blankenship, *Photochem. Photobiol. 37*, 329 (1983).

658. K. Kano, T. Sato, S. Yamada, and T. Ogawa, *J. Chem. Phys. 87*, 566 (1983).

659. R. G. Brown, A. Harriman, and G. Porter, *J. Chem. Soc., Faraday II 73*, 113 (1976).

660. G. S. Beddard, R. S. Davidson, and K. R. Threthewey, *Nature 267*, 373 (1977).

661. S. L. Bondarev and G. P. Gurinovich, *Opt. Spektrosk. 36*, 687 (1974).

662. G. S. Beddard, S. Carlin, L. Harris, G. Porter, and C. J. Tredwell, *Photochem. Photobiol. 27*, 433 (1978), and refs. therein.

663. D. Frackowiak and D. Baumann, *Acta Phys. Polon. 62A*, 157 (1982).

664. E. I. Kapinus and I. I. Dilung, *Biofizika 25*, 181 (1980).

665. S. L. Bondarev, G. P. Gurinovich, and V. S. Chernikov, *Izv. Akad. Nauk SSSR, Ser. Fiz. 34*, 641 (1970).

666. A. T. Gradyushko, V. A. Mashenkov, K. N. Solovev, and M. P. Tsvirko, *Zh. Prikl. Spektrosk. 9*, 514 (1968).

667. S. S. Dvornikov, K. N. Solovev, M. P. Tsvirko, and A. T. Gradyushko, *Zh. Prikl. Spektrosk. 25*, 1011 (1976).

668. A. T. Gradyushko, A. N. Sevchenko, K. N. Solovev, and M. P. Tsvirko, *Photochem. Photobiol. 11*, 387 (1970).

669. B. M. Dzhagarov, A. I. Sagun, S. L. Bondarev, and G. P. Gurinovich, *Biofizika 22*, 565 (1977).

670. A. Müller, R. Lumry, and H. Kokubun, *Rev. Sci. Instrum. 36*, 1214 (1965).

671. S. S. Brody and E. Rabinowitch, *Science 125*, 555 (1957).

672. R. Avarmaa and R. Tamkivi, *Opt. Spektrosk. 45*, 247 (1978).

673. S. S. Brody, *Z. Naturforsch. 37C*, 260 (1982).

674. R. Avarmaa, T. Soovik, R. Tamkivi, and R. Tonissoo, *Stud. Biophys. 65*, 213 (1977).

675. G. D. Yegorova, V. A. Mashenkov, K. N. Solovev, and N. A. Yushkevich, *Biofizika 18*, 40 (1973).

676. R. Stupp and H. Kuhn, *Helv. Chim. Acta 35*, 2469 (1952).

677. J. C. Goedheer, *Nature 176*, 928 (1955).

678. M. Gouterman and L. Stryer, *J. Chem. Phys. 37*, 2260 (1962).

679. M. Kaplanova and K. Vacek, *Photochem. Photobiol. 20*, 371 (1974).

680. F. Bär, H. Lang, H. Schnabel, and H. Kuh, *Z. Electrochem. 66*, 346 (1962).

681. F. F. Litvin, R. I. Personov, and R. I. Korotayev, *Dokl. Akad. Nauk SSSR 188*, 1169 (1969).

682. L. A. Bykovskaya, F. F. Litvin, R. I. Personov, and J. V. Romanovskii, *Biofizika 25*, 13 (1980).

683. R. A. Avarmaa and K. K. Rebane, *Stud. Biophys. 48*, 209 (1975).

684. R. A. Avarmaa and K. K. Mauring, *Zh. Prikl. Spektrosk. 28*, 658 (1978).

685. R. A. Avarmaa, R. Tamkivi, S. Kiisler, and V. Nömm, *Izv. Akad. Nauk ESSR 29*, 39 (1980); in *Chem. Abstr. 92*, 206340 (1980).

686. R. H. Clarke and R. H. Hofeldt, *J. Chem. Phys. 61*, 4582 (1974).

687. J. Hala, I. Pelant, L. Parma, and K. Vacek, *J. Lumin. 26*, 117 (1981), and *24/25*, 303 (1981).

688. J. Fünfschilling and D. F. Williams, *Photochem. Photobiol. 22*, 151 (1975).

689. J. Fünfschilling and D. F. Williams, *Photochem. Photobiol. 26*, 109 (1977).

690. G. W. Cauters, J. van Egmond, T. J. Schaafsma, I. J. Chan, W. G. van Dorp, and J. H. van der Waals, *Ann. N.Y. Acad. Sci. 206*, 711 (1973).

691. J. Bohandy and B. F. Kim, *J. Chem. Phys. 76*, 1180 (1982).

692. K. Rebane and R. Avarmaa, *J. Photochem. 17*, 311 (1981).

693. S. Völker, R. M. Macfarlane, A. Z. Genack, H. P. Trommsdorff, and J. H. van der Waals, *J. Chem. Phys. 67*, 1759 (1977).

694. S. Völker and R. M. Macfarlane, *J. Chem. Phys. 73*, 4476 (1980).

695. R. Avarmaa, K. Mauring, and A. Suisala, *Chem. Phys. Lett. 77*, 88 (1981).

696. S. S. Dvornikov, V. N. Knyukshto, K. N. Solovev, and A. M. Shulga, *Zh. Prikl. Spektrosk. 33*, 653 (1980).

697. M. Asano and J. A. Koningstein, *Chem. Phys. 57*, 1 (1981).

698. G. F. Stelmakh and M. P. Tsvirko, *Opt. Spektrosk. 48*, 185 (1980).

699. M. R. Wasielewski, R. L. Smith, and A. G. Kostka, *J. Am. Chem. Soc. 102*, 6923 (1980).

700. J. C. Hindman, R. Kugel, A. Svirmickas, and J. J. Katz, *Proc. Nat. Acad. Sci. USA 74*, 55 (1977).

701. D. Leupold, S. Mory, R. König, P. Hoffman, and B. Hieke, *Chem. Phys. Lett. 45*, 567 (1977).

702. N. E. Geacintov, D. Husiak, T. Kolubaev, J. Breton, A. J. Campillo, S. L. Shapiro, K. R. Winn, and P. K. Woodbridge, *Chem. Phys. Lett. 66*, 154 (1979).

703. J. Nakata, T. Imura, and K. Kawaba, *J. Phys. Soc. Jap. 42*, 146 (1977).

704. T. Trosper, R. B. Park, and K. Sauer, *Photochem. Photobiol. 7*, 451 (1968).

705. O. Gonem, H. Levanon, and L. K. Patterson, *Isr. J. Chem. 21*, 271 (1981).

706. B. Ke, in: *The Chlorophylls*, L. P. Vernon and G. R. Seely (eds.), Academic Press, New York, 1966, p. 253 (review on monolayers).

707. A. G. Tweet, G. L. Gaines, and W. D. Bellamy, *J. Chem. Phys. 41*, 1008 (1964).

708. D. Moebius, in: *Topics in Surface Chemistry*, E. Kay and P. S. Bagus (eds.), Plenum Press, New York, 1978, p. 75 (review).

709. T. Trosper, R. B. Park, and K. Sauer, *Photochem. Photobiol. 7*, 451 (1968).

710. A. Kelly and G. Porter, *Proc. Roy. Soc. Lond. 315A*, 149 (1970).

711. E. I. Zenkevich, A. P. Losev, and G. P. Gurinovich, *Molek. Biol.* (Engl. ed.) *6*, 666 (1972).

712. K. Csatorday, E. Lehoczki, and L. Szalay, *Biochim. Biophys. Acta 376*, 268 (1975).

713. E. Lehoczki and K. Csatorday, *Biochim. Biophys. Acta 396*, 86 (1975).

714. K. Colbow, *Biochim. Biophys. Acta 314*, 320 (1973).

715. G. R. Seely and V. Senthilathipan *J. Phys. Chem. 87*, 373 (1983).

716. G. Porter and G. Strauss, *Proc. Roy. Soc. Lond. 295A*, 1 (1966).

717. A. Kelly and L. K. Patterson, *Proc. Roy. Soc. Lond. 324A*, 117 (1971).

718. V. K. Gorshkov, *Biofizika* (Engl. ed.) *14*, 25 (1969).

719. K. Vacek J. Naus, M. Svabova, E. Vavrinec, M. Kaplanova, and J. Hala, *Stud. Biophys. 62*, 201 (1977).

720. K. Vacek, D. Wong, and Govindjee, *Photochem. Photobiol. 26*, 269 (1977).

721. V. I. Godik, M. Urbanova, A. Y. Borisov, and K. Vacek, *Stud. Biophys. 82*, 179 (1981).

722. D. Monkhor and K. Vacek, *Stud. Biophys. 72*, 117 (1978), and refs. therein on former work.

723. W. Hägele, D. Schmid, F. Drissler, J. Naus, and H. C. Wolf, *Z. Naturforsch. 33A*, 1197 (1978).

724. D. Frackowiak, D. Bauman, and I. Betkowska, *Photochem. Photobiol. 33*, 779 (1981).

725. D. Frackowiak, D. Bauman, H. Manikowski, and T. Martynski, *Biophys. Chem. 6*, 369 (1977).

726. E. Hasegawa, T. Kanayama, and E. Tsuchida, *J. Polym. Sci., Polym. Chem. Ed. 15*, 3039 (1977).

727. A. A. Krasnovskii and M. I. Bystrova, *Biofizika 10*, 480 (1965).

728. C. J. Seliskar and B. Ke, *Biochim. Biophys. Acta 153*, 685 (1968).

729. M. Brouers, *Photosynthetica 13*, 9 (1979).

730. E. I. Zenkevich, G. Kochubeev, A. P. Losev, and G. P. Gurinovich, *Molek. Biol. 13*, 888 (1979).

731. J. S. Connolly, E. B. Samuel, and A. F. Janzen, *Photochem. Photobiol. 36*, 565 (1982), and refs. therein.

732. J. P. Thornber, J. P. Markwell, and S. Reinmann, *Photochem. Photobiol. 29*, 1205 (1979).

733. G. G. Guilbault, *Practical Fluorescence: Theory, Methods, and Techniques*, Dekker, New York, 1973, p.543.

734. H. J. M. Kramer, H. Kingma, T. Swarthoff, and J. Amesz, *Biochim. Biophys. Acta 681*, 359 (1982).

735. J. D. Bolt, C. N. Hunter, R. A. Niederman, and K. Sauer, *Photochem. Photobiol. 34*, 653 (1981).

736. E. I. Zenkevich and A. P. Losev, *Zh. Prikl. Spektrosk. 13*, 1032 (1970).

737. K. G. Boto and J. S. Bunt, *Anal. Chem. 50*, 392 (1978), and refs. therein.

738. Y. Onoue, M. Kotani, K. Hiraki, T. Shigematsu, and J. Nishikawa, *Bunseki Kagaku 31E*, 45 (1982).

739. S. Roy and L. Legendre, *Marine Biol. 55*, 93 (1979).

740. L. V. Warwick, *J. Phycol. 15*, 429 (1979).

741. C. S. Yentsch and O. W. Menzel, *Deep Sea Res. 10*, 221 (1963).

742. C. S. Yentsch and M. Clarence, *J. Marine Res. 37*, 471 (1979).

743. E. J. Fee, *Limnolog. Oceanogr. 21*, 767 (1976).

744. F. E. Hoge and R. N. Swift, *Appl. Opt. 20*, 3197 (1981).

745. F. M. Johnson, NASA Contract Rep. 1970, NASA CR-1667; in *Chem. Abstr. 74*, 92945 (1970).

746. C. A. Rebeiz, F. C. Belanger, G. Freyssinet, and D. G. Saab, *Biochim. Biophys. Acta 590*, 234 (1980).

747. F. C. Belanger and C. A. Rebeiz, *Biochem. Biophys. Res. Comm. 88*, 365 (1979).

748. F. F. Litvin, I. N. Stadnichuk, and V. P. Kruglov, *Biofizika 23*, 450 (1978).

749. J. B. R. Dunn, A. W. Addison, R. E. Bruce, J. S. Loehr, and T. M. Loehr, *Biochemistry 16*, 1743 (1977).

750. B. Alpert, D. Jameson, and G. Weber, *Photochem. Photobiol. 31*, 1 (1980).

751. J. L. Risler, *Biochemistry 10*, 2664 (1971).

752. P. M. Champion and G. J. Perreault, *J. Chem. Phys. 75*, 490 (1981).

753. J. M. Vanderkooi and M. Erecinska, *Eur. J. Biochem. 60*, 199 (1975).

754. P. J. Angiolillo, J. S. Leigh, and J. M. Vanderkooi, *Photochem. Photobiol. 36*, 133 (1982).

755. P. Sebban, M. Coppey, B. Alpert, L. Lindqvist, and D. M. Jameson, *Photochem. Photobiol. 32*, 727 (1980).

756. M. B. Katan, *Anal. Biochem. 74*, 132 (1976).

757. D. K. Lavallee, T. J. McDonough, and L. Cioffi, *Appl. Spectrosc. 36*, 430 (1982).

758. O. D. Dmitrievskii, V. L. Ermolaev, and A. N. Terenin, *Dokl. Akad. Nauk SSSR 114*, 751 (1957).

759. B. M. Dzhagarov and G. P. Gurinovich, *Opt. Spektrosk. 30*, 425 (1971).

760. P. Latimer, P. T. Bannister, and E. Rabinowitch, *Science 124*, 585 (1956).

761. S. B. Hendricks, W. L. Butler, and H. Siegelmann, *J. Phys. Chem. 66*, 2250 (1962).

762. J. E. Falk, *Porphyrins and Metalloporphyrins*, Elsevier, Amsterdam, 1964. p. 87.

763. J. M. Goldstein, W. M. McNabb, and J. F. Hazel, *J. Chem. Educ. 34*, 604 (1957).

764. O. Yoshiaki, K. Hiraki, and Y. Nishikawa, *Bull. Chem. Soc. Jap. 103*, 7370 (1981).

765. M. Roth, *Clin. Chim. Acta 17*, 487 (1967).

766. A. Cu, G. Bellak, and D. A. Lightner, *J. Am. Chem. Soc. 97*, 2579 (1975).

767. R. Bonnett, J. Dalton, and D. E. Hamilton, *J. Chem. Soc., Chem. Comm., 1975*, 639.

768. I. B. C. Matheson, G. J. Faini, and J. Lee, *Photochem. Photobiol. 21*, 135 (1975).

769. A. A. Lamola, J. Eisinger, W. E. Blumberg, S. C. Patel, and J. Flores, *Anal. Biochem. 100*, 25 (1979).

770. R. F. Chen, *Arch. Biochem. Biophys. 160*, 106 (1974).

771. J. Krasner, *Biochem. Med. 7*, 135 (1973).

772. A. A. Lamola, J. Eisinger, W. E. Blumberg, S. C. Patel, and J. Flores, *Eur. J. Biochem. 33*, 500 (1973).

773. P. S. Song, Q. Chae, and D. A. Lightner, *J. Am. Chem. Soc. 95*, 7892 (1973).

774. C. Pertrier, P. Jardon, C. Dupuy, and R. Gautron, *J. Chim. Phys. 78*, 519 (1981).

775. A. R. Holzwarth, H. Lehner, and E. E. Braslavsky, *Liebigs Ann. Chem. 1978*, 2002.

776. C. Pertrier, C. Dupuy, P. Jardon, and R. Gautron, *Photochem. Photobiol. 29*, 389 (1979).

777. P. S. Song, Q. Chae, D. A. Lightner, W. R. Briggs, and D. Hopkins, *J. Am. Chem. Soc. 95*, 7893 (1973).

778. S. J. Chantrell, C. A. MacAuliffe, R. W. Munn, and A. C. Pratt, *J. Chem. Soc., Chem. Comm. 1975*, 470.

779. H. Falk and F. Neufingerl, *Monatsh. Chem. 110*, 987 (1979).

780. U. Even, J. Magen, and J. Jortner, *Chem. Phys. Lett. 88*, 131 (1982).

781. A. Harriman and R. J. Hosie, *J. Chem. Soc., Faraday Trans. II 77*, 1695 (1981).

782. L. H. Pratt, *Photochem. Photobiol. Rev. 4*, 59 (1979).

783. H. Scheer, *Angew. Chem. 93*, 230 (1981) (Engl. ed., p. 241).

784. G. Hermann, B. Kirchhoff, K. J. Appenroth, and E. Müller, *Biochem. Biophysiol. Pflanzen 178*, 177 (1983).

785. P. S. Song, Q. Chae, and J. D. Gardner, *Biochim. Biophys. Acta 576*, 479 (1979).

786. A. R. Holzwarth, S. E. Braslavsky, S. Culshaw, and K. Schaffner, *Photochem. Photobiol. 36*, 581 (1982).

787. P. S. Song, Q. Chae, and W. R. Briggs, *Photochem. Photobiol. 22*, 75 (1975).

788. P. S. Song and Q. Chae, *J. Lumin. 12/13*, 831 (1976).

789. P. S. Song, H. K. Sarkar, I. S. Kim, and K. L. Poff, *Biochim. Biophys. Acta 635*, 369 (1981).

790. P. S. Song, Q. Chae, and J. Gardner, *Biochim. Biophys. Acta 576*, 479 (1979).

791. I. M. Tegmo-Larsson, S. E. Braslavsky, S. Culshaw, R. M. Ellul, C. Nicolau, and K. Schaffner, *J. Am. Chem. Soc. 103*, 7152 (1981).

792. C. Pertrier, C. Jardon, C. Dupuy, and R. Gautron, *J. Chim. Phys. 76*, 97 (1979).

793. F. Macdowall and M. Walker, *Photochem. Photobiol. 7*, 109 (1968).

794. E. A. Chernitskii and S. V. Konev, *Biofizika 8*, 561 (1963).

795. F. W. J. Teale and R. E. Dale, *Biochem. J. 116*, 161 (1970).

796. R. E. Dale and F. W. J. Teale, *Photochem. Photobiol. 12*, 99 (1970).

797. E. Gantt and C. A. Lipschultz, *Biochim. Biophys. Acta 292*. 858 (1973).

798. J. Grabowski and E. Gantt, *Photochem. Photobiol. 28*, 47 (1978).

799. D. Frackowiak, K. Fiksinski, and H. Pienkowska, *Photobiochem. Photobiophys. 2*, 21 (1981).

800. C. Vernotte, *Photochem. Photobiol. 14*, 163 (1971).

801. C. Vernotte and I. Moya, *Photochem. Photobiol. 17*, 245 (1973).

802. G. Tomita and E. Rabinowitch, *Biophys. J. 2*, 483 (1962).

803. S. S. Brody, G. Porter, C. J. Tredwell, and J. Barber, *Photobiochem. Photobiophys. 2*, 11 (1981).

804. G. Porter, C. J. Tredwell, G. F. W. Searle, and J. Barber, *Biochim. Biophys. Acta 501*, 232 (1978).

805. G. F. W. Searle, J. Barber, G. Porter, and C. J. Tredwell, *Biochim. Biophys. Acta 501*, 246 (1978).

806. D. Wong, F. Pellegrino, R. R. Alfano, and B. A. Zilinskas, *Photochem. Photobiol. 33*, 651 (1981).

807. J. Grabowski and E. Gantt, *Photochem. Photobiol. 28*, 39 (1977).

808. F. Pellegrino, D. Wong, R. R. Alfano, and B. A. Ziliskas, *Photochem. Photobiol. 34*, 691 (1981).

809. R. B. Petterson, E. Dolan, A. E. Calvert, and B. Ke, *Biochim. Biophys. Acta 634*, 237 (1981).

810. E. Gantt, *Photochem. Photobiol. 26*, 685 (1977).

811. E. Gantt, *Ann. Rev. Plant Physiol. 32*, 327 (1981).

812. P. Hefferle, M. Nies, W. Wehrmeyer, and S. Schneider, *Photobiochem. Photobiophys. 5*, 41 (1983).

813. B. Zickendraht-Wendelstadt, J. Friedrich, and W. Rüdiger, *Photochem. Photobiol. 31*, 367 (1980).

814. H. P. Köst, K. Rath, G. Wanner, and H. Scheer, *Photochem. Photobiol. 34*, 139 (1981).

815. H. Scheer and W. Kufer, *Z. Naturforsch. 32C*, 513 (1977).

816. E. B. Walker, T. Y. Lee, and P. S. Song, *Biochim. Biophys. Acta 587*, 129 (1979).

817. P. S. Song, *Biochim. Biophys. Acta 639*, 1 (1981).

818. P. S. Song, E. B. Walker, R. A. Auerbach, and G. W. Robinson, *Biophys. J. 35*, 551 (1981).

819. J. Jung, P. S. Song, R. J. Paxton, M. S. Edelstein, R. Swanson, and E. E. Hazen, *Biochemistry 19*, 24 (1980).

820. E. E. Gardner, S. E. Stevens, and J. L. Fox, *Biochim. Biophys. Acta 624*, 187 (1980).

821. R. MacColl, G. O'Connor, G. Crofton, and K. Csatorday, *Photochem. Photobiol. 34*, 719 (1981).

822. R. MacColl, K. Csatorday, D. S. Berns, and E. Traeger, *Biochemistry 19*, 2817 (1980).

823. M. Mimuro, A. Murakami, and Y. Fujita, *Arch. Biochem. Biophys. 215*, 266 (1982).

824. A. C. Ley, W. L. Butler, D. A. Bryant, and A. N. Glazer, *Plant Physiol. 59*, 974 (1977).

825. B. A. Zilinskas, L. S. Greenwald, C. L. Bailey, and P. Kahn, *Biochim. Biophys. Acta 592*, 267 (1980).

826. O. D. Canaani and E. Gantt, *Biochemistry 19*, 2950 (1980).

827. R. MacColl, K. Csatorday, D. S. Berns, and E. Traeger, *Arch. Biochem. Biophys. 208*, 42 (1981).

828. P. S. Song, in: *The Blue Light Syndrome*, H. Senger, (ed.), Springer-Verlag, New York, 1980, p. 157.

829. P. S. Song, P. Koka, B. B. Prezelin, and F. T. Haxo, *Biochemistry 15*, 4422 (1976).

830. P. Koka and P. S. Song, *Biochim. Biophys. Acta 495*, 220 (1977).

831. I. N. Stadnichuk and L. A. Mineeva, *Biofizika 25*, 727 (1980).

832. K. Csatorday, *Biochim. Biophys. Acta 504*, 341 (1978).

833. O. D. Bekasova, V. A. Romanyuk, and V. I. Zvalinski, *Biofizika 26*, 74 (1981).

834. G. Yamanaka, A. Glazer, and R. C. Williams, *J. Biol. Chem. 255*, 11004 (1980).

835. A. Glazer, R. C. Williams, G. Yamanaka, and H. K. Schachman, *Proc. Nat. Acad. Sci. USA 76*, 6162 (1979).

836. M. H. Yu, A. Glazer, and R. C. Williams, *J. Biol. Chem. 256*, 13130 (1981).

837. N. Murata, *Plant Cell. Physiol. 1977*, 9; in *Chem. Abstr. 87*, 50276 (1977).

838. E. Gantt, C. A. Lipschultz, J. Grabowski, and B. K. Zimmerman, *Plant Physiol. 63*, 615 (1981).

839. O. D. Bekasova, L. M. Shubin, and V. B. Evstigneev, *Izv. Akad. Nauk SSSR, Ser. Biol. 1979*, 198.

840. O. D. Bekasova, N. O. Bukov, and N. V. Karapetyan, *Biokhimiya 46,* 287 (1981).

841. D. Frackowiak, J. Dudkiewicz, J. Grabowski, K. Fiksinski, and H. Manikowski, *Photosynthetica 13,* 21 (1979).

842. D. Frackowiak, *Bull. Acad. Polon. Sci., Ser. Sci. Biol. 27,* 162 (1979).

843. D. Frackowiak, H. Pienkowska, and J. Szurkowski, *Photosynthetica 16,* 496 (1982).

844. E. Mörschel and W. Wehrmeyer, *Arch. Microbiol. 113,* 83 (1977).

845. K. Ohki and T. Kath, *Plant Cell Physiol. 1977,* 15; in *Chem. Abstr. 87,* 50277 (1977).

846. S. E. Braslavsky, A. R. Holzwarth, H. Lehner, and K. Schaffner, *Helv. Chim. Acta 61,* 2219 (1978).

847. A. R. Holzwarth, H. Lehner, S. E. Braslavsky, and K. Schaffner, *Liebigs Ann. Chem. 1978,* 2002.

848. E. N. Tobin and W. R. Briggs, *Photochem. Photobiol. 18,* 487 (1973).

849. S. E. Braslavsky, A. R. Holzwarth, E. Langer, H. Lehner, J. I. Matthews, and K. Schaffner, *Isr. J. Chem. 20,* 196 (1980).

850. Q. Chae and P. S. Song, *J. Am. Chem. Soc., 97,* 4176 (1975).

851. H. Falk and F. Neufingerl, *Monatsh. Chem. 110,* 907 (1979).

852. H. Falk and T. Schlederer, *Liebigs Ann. Chem. 1979,* 1560.

853. M. E. Lippitsch, A. Leitner, M. Riegler, and F. R. Aussenegg, *Springer Ser. Chem. Phys. 14,* 327 (1981).

854. C. Petrier, W. Kufer, H. Scheer, and R. Gautron, unpublished results (from ref. 783).

855. R. Gautron, P. Jardon, C. Pertrier, M. Choussy, M. Barbier, and M. Vuillaume, *Experientia 32,* 1100 (1976).

856. J. Kossanyi, S. Sabbah, A. Queguiner, and J. Duflos, *J. Photochem. 9,* 282 (1978).

857. A. Stern and H. Molvig, *Z. Phys. Chem. 176A,* 209 (1936).

858. H. Falk, K. Grubmayr, and F. Neufingerl, *Monatsh. Chem. 108,* 1185 (1977).

859. H. Falk and F. Neufingerl, *Monatsch. Chem. 110,* 1243 (1979).

860. M. E. Lippitsch, A. Leitner, M. Riegler, and F. R. Aussenegg, *Kvantov. Elektron. 9,* 1034 (1982); in *Chem. Abstr. 97,* 153384 (1982).

861. Ch. Dhéré, La spectrochimie de fluorescence dans l'etude des produites naturel, *Prog. Chem. Org. Nat. Prod. 6,* 301 (1939). p. 324 f, and references therein.

862. S. Udenfriend, *Fluorescence Assay in Biology and Medicine,* Vol. I, Academic Press, New York, 1962 p. 309.

863. S. Udenfriend, *Fluorescence Assay in Biology and Medicine,* Vol. II, Academic Press, New York, 1969, p. 399.

864. J. B. F. Lloyd and N. T. Weston, *J. Forensic Sci. 27*, 352 (1982), and references therein.

865. G. H. Beaven, A. d'Albis, and W. B. Glatzer, *Eur. J. Biochem. 33*, 500 (1972).

866. E. J. Land, *Photochem. Photobiol. 24*, 475 (1976).

867. R. W. Sloper and T. G. Truscott, *Photochem. Photobiol. 31*, 445 (1980).

868. Some commercial BR preparations may contain an impurity with a pink-red fluorescence: A. J. Fatiadi and R. Schaffer *Experientia 27*, 1139 (1971).

869. J. Dalton, L. R. Milgrom, and R. Bonnett, *Chem. Phys. Lett. 61*, 242 (1979).

870. A. A. Lamola and J. Flores, *J. Am. Chem. Soc. 104*, 2530 (1982).

871. D. A. Lightner and A. Cu, *Photochem. Photobiol. 26*, 427 (1977).

872. R. F. Chen, in: *Fluorescence Techniques in Cell Biology*, A. A. Thaer and M. Sernetz (ed.), Springer-Verlag, New York, 1973, p. 273.

873. W. T. Morgan, W. Muller-Eberhard, and A. A. Lamola, *Biochim. Biophys. Acta 532*, 57 (1978).

874. C. B. Berge, B. S. Hudson, R. D. Simoni, and L. A. Sklar, *J. Biol. Chem. 254*, 391 (1979).

875. J. J. Lee and G. D. Gillispie, *Photochem. Photobiol. 33*, 757 (1981).

876. C. D. Tran and G. S. Beddard, *Biochim. Biophys. Acta 678*, 497 (1981).

877. C. D. Tran and G. S. Beddard, *J. Am. Chem. Soc. 104*, 6741 (1982).

878. A. A. Lamola, J. Flores, and F. H. Doleiden, *Photochem. Photobiol. 35*, 649 (1982), and refs. cited therein.

879. B. I. Greene, A. A. Lamola, and C. V. Shank, *Proc. Nat. Acad. Sci. USA 78*, 2008 (1981).

880. A. A. Lamola, W. E. Blumberg, R. MacClead, and A. Fanaroff, *Proc. Nat. Acad. Sci. USA 78*, 1882 (1981).

881. A. K. Brown, J. Eisinger, W. E. Blumberg, J. Flores, G. Boyle, and A. A. Lamola, *Pediatrics 65*, 767 (1980).

882. H. Falk, N. Müller, M. Ratzenhofer, and K. Winsauer, *Monatsh. Chem. 113*, 1421 (1982).

883. G. Jori, F. Rubaltelli, and E. Rossi, *Springer Ser. Opt. Sci. 22*, 145 (1978).

884. G. Navon and R. Panigel, *Clin. Chim. Acta 91*, 221 (1979).

885. J. Eisinger and J. Flores, *Anal. Biochem. 94*, 15 (1979).

886. W. E. Blumberg, J. Eisinger, A. A. Lamola, and D. Zuckerman, *J. Lab Clin. Med. 89*, 712 (1977).

887. S. B. McCluskey, G. N. B. Storey, G. K. Brown, D. G. More, and W. J. O'Sullivan, *Clin. Chem. 21*, 1638 (1975).

888. S. Nagaoka and M. L. Cowger, *Anal. Biochem. 96*, 364 (1979).

889. R. Beke, G. A. DeWeerdt, J. Parijs, and F. Barbier, *Clin. Chim. Acta 71*, 27 (1976).

890. T. Osuga, K. Mitamura, F. Mashige, and K. Imai, *Clin. Chim. Acta 75*, 81 (1977).

891. T. Uete, *Clin. Chim. Acta 59*, 259 (1975).

892. A. A. Lamola, J. Eisinger, and W. E. Blumberg, *J. Lab. Clin. Med. 93*, 345 (1979).

893. K. Neelankatam, L. R. Row, and V. Venketeswarlu, *Proc. Indian Acad. Sci. 18A*, 364 (1943).

894. K. Neelankantam and M. V. Sitaraman, *Proc. Indian Acad. Sci. 21A*, 45 (1945).

895. A. V. Karyakin and A. N. Terenin, *Izv. Akad. Nauk SSSR, Ser. Fiz. 13*, 9 (1949).

896. A. V. Karyakin, A. N. Terenin, and Y. I. Kalenichenko, *Dokl. Akad. Nauk SSSR 67*, 305 (1949).

897. M. H. Van Benthem, G. D. Gillispie, and R. C. Haddon, *J. Phys. Chem. 86*, 4281 (1982).

898. A. N. Anoshin, D. N. Shigorin, and M. V. Gorelik, *Zh. Fiz. Khim. 53*, 761 (1979).

899. A. N. Anoshin, E. A. Gastilovich, and D. N. Shigorin, *Zh. Fiz. Khim. 54*, 2474 (1980).

900. E. A. Gastilovich and A. N. Anoshin, *Opt. Spektrosk. 51*, 255 (1981).

901. A. N. Anoshin, E. A. Gastilovich, G. T. Kryushkova, B. K. Sokolov, D. N. Shigorin, and Y. Y. Frolov, *Opt. Spektrosk. 44*, 107 (1978).

902. E. A. Gastilovich, A. N. Anoshin, and V. M. Ryaboi, *Zh. Fiz. Khim. 54*, 2446 (1980).

903. E. A. Gastilovich, G. T. Kryuchkova, and D. N. Shigorin, *Opt. Spektrosk. 38*, 500 (1975); *39*, 235 (1975).

904. T. P. Carter, G. D. Gillispie, and M. A. Connolly, *J. Phys. Chem. 86*, 192 (1982).

905. F. Drissler, F. Graf, and D. Haarer, *J. Chem. Phys. 72*, 4996 (1980).

906. R. Friedrich and D. Haarer, *J. Chem. Phys. 76*, 61 (1982).

907. G. Smulevich, L. Angeloni, S. Giovannarchi, and M. P. Marzocchi, *Chem. Phys. 65*, 313 (1982).

908. G. Smulevich, A. Amirav, U. Even, and J. Jortner, *Chem. Phys. 73*, 1 (1982).

909. M. Roman, A. Fernandez-Gutierrez, and C. Marin, *Afinidad 34*, 481 (1977).

910. M. Roman, A. Fernandez-Gutierrez, and M. C. Mahedero, *An. Quim. 74*, 439 (1978).

911. R. N. Capps and M. Vala, *Photochem. Photobiol. 33*, 673 (1981).

912. H. S. Schwartz, *Anal. Biochem. 7*, 396 (1973).

913. G. Scheibe and A. Schöntag, *Chem. Ber. 75*, 2019 (1942).

914. C. S. French, J. H. C. Smith, H. I. Virgin, and R. L. Airth, *Plant Physiol.* *31*, 369 (1956).

915. B. Krone, and A. Zeeck, *Liebigs Ann. Chem. 1983*, 471.

916. O. S. Wolfbeis, and E. Fürlinger, *Mikrochim. Acta 1983*, III, 385.

917. R. G. Cooke and J. M. Edwards, *Prog. Chem. Org. Nat. Prod. 41*, 153 (1981).

918. V. E. Bondybey, R. C. Haddon, and J. H. English, *J. Chem. Phys. 80*, 5432 (1984).

919. R. Rosetti, S. Rayford, R. C. Haddon, and L. E. Brus, *J. Am. Chem. Soc. 103*, 4303 (1981).

920. M. Nepras, M. Titz, and V. Slavik, *Coll. Czech. Chem. Comm. 40*, 2120 (1975).

921. J. A. Radley, *Analyst 69*, 47 (1944).

922. A. Holme, *Acta Chem. Scand. 21*, 1679 (1967).

923. L. Szebelledy and F. J. Gaal, *Z. Anal. Chem. 98*, 255 (1934).

924. R. Joachimovits, *Monatschr. Geburtshilfe Gynäkol. 83*, 42 (1929).

925. M. H. Fletcher, C. C. White, and M. S. Sheftel, *Anal. Chem. 18*, 179 (1946).

926. A. C. Lane, *Anal. Chem. 45*, 1911 (1973).

927. L. Angeloni, G. Smulevich, and M. P. Marzocchi, *Spectrochim. Acta 38A*, 213 (1982).

928. M. K. Matthew and P. Balaram, *J. Inorg. Biochem. 13*, 339 (1980).

929. J. H. Boothe, J. Morton, J. P. Petisi, R. G. Wilkinson, and J. H. Williams, *J. Am. Chem. Soc. 75*, 4621 (1953).

930. D. Hall, *J. Pharm. Pharmacol. 28*, 420 (1976).

931. J. C. Mani and G. Foltran, *Bull. Soc. Chim. Fr. 1971*, 4141.

932. N. M. Alykov, *Izv. Vyssh. Uchebn. Zaved., Khim. Khim. Tehnol. 20*, 827 (1977); *Chem. Abstr. 87*, 189354 (1977).

933. R. J. Sturgeon and S. G. Schulman, *J. Pharm. Sci. 66*, 958 (1977).

934. G. Loeber, V. Kleinwächter, Z. Balcarova, H. Fritzsche, and D. G. Strauss, *Stud. Biophys. 71*, 203 (1978).

935. H. Poiger and C. Schlatter, *Analyst 101*, 808 (1976), and ref. therein.

936. M. Lever, *Biochem. Med. 6*, 216 (1972).

937. T. V. Alykova, *Antibiotika 17*, 353 (1972).

938. E. Ragazzi and G. Veronese, *J. Chromatogr. 32*, 105 (1977).

939. G. Van Hoek, I. Kapetanidis, and A. Mirimanoff, *Pharm. Acta Helv. 47*, 316 (1972).

940. B. Scales and D. A. Assinder, *J. Pharm. Sci. 62*, 913 (1973).

941. R. G. Kelly, L. M. Peets, and K. D. Hoyt, *Anal. Biochem. 28*, 222 (1969).

942. L. J. Stoel, E. C. Newman, G. L. Asleson, and C. W. Frank, *J. Pharm. Sci. 65*, 1794 (1976).

943. A. V. Poteshnik and V. E. Sevchenko, *Antibiotiki 27*, 820 (1980).

944. N. Pace and G. Mackinney, *J. Am. Chem. Soc. 63*, 2570 (1941).

945. Ch. Dhèrè, Progress recents en spectrochimie de fluorescence des produits biologiques, *Prog. Chem. Org. Nat. Prod. 6*, 311 (1950) p. 341ff.

946. Ch. Dhèrè, La spectrochimie de fluorescence dans l'etude des produites naturel, *Prog. Chem. Org. Nat. Prod. 2*, 301 (1939) p. 333f.

947. K. W. Kohn, *Anal. Chem. 33*, 863 (1961).

948. H. Johann and K. Dose, *Z. Anal. Chem. 314*, 139 (1983).

949. T. Lai, B. T. Lim, and E. C. Lim, *J. Am. Chem. Soc. 104*, 7631 (1982).

950. Ch. Dhèrè, Progress recents en spectrochimie de fluorescence des produits biologiques, *Prog. Chem. Org. Nat. Prod. 6*, 311 (1950) p. 330.

951. O'L. M. Crowe and A. Walker, *J. Opt. Soc. Am. 34*, 135 (1944).

952. A. Pirie and D. M. Simpson, *Biochem. J. 40*, 14 (1946).

953. D. M. Simpson, *Analyst 72*, 382 (1947).

954. R. Hüttel and G. Sprengling, *Liebigs Ann. Chem. 554*, 69 (1943).

955. M. Polonovski, S. Guinard, M. Besson, and R. Vieillefosse, *Bull. Soc. Chim. Fr. 12*, 924 (1945).

956. W. Jacobson and D. M. Simpson, *Biochem. J. 40*, 3 and 9 (1946).

957. J. M. Lagowski and H. S. Forrest, *Proc. Nat. Acad. Sci. USA 58*, 1541 (1967).

958. C. Chahidi, M. Aubailly, A. Momzikoff, M. Bazin, and R. Santus, *Photochem. Photobiol. 33*, 641 (1981).

959. E. Lippert and H. Prigge, *Z. Elektrochem. 64*, 662 (1960).

960. H. Prigge and E. Lippert, *Ber. Bunsenges. Phys. Chem. 69*, 458 (1969).

961. E. Lippert, *Acc. Chem. Res. 3*, 74 (1970).

962. M. Polonovski, R. Vieillefosse, and M. Besson, *Bull. Soc. Chim. Fr. 12*, 78 (1945).

963. D. Bertrand, *Bull. Soc. Chim. Fr. 12*, 1010 (1945).

964. A. J. W. G. Visser and G. Lee, *Biochemistry 21*, 2218 (1982), and refs. therein.

965. T. M. Rossi, J. M. Quarles, and I. M. Warner, *Anal. Lett. 15*, 1083 (1982).

966. R. T. Parker, R. S. Freedlander, E. M. Schulman, and R. B. Dunlap, *Dev. Biochem. 4*, 61 (1978); in *Chem. Abstr. 91*, 35175 (1979).

967. A. Momzikoff and R. Santus, *C. R. Acad. Sci. Fr. 293*, 15 (1981).

968. K. Yazawa and Z. Tamura, *J. Chromatogr. 254*, 327 (1983).

969. D. B. Mullick, *Phytochemistry 8*, 2003 (1969).

970. Y. Nasu and T. Nakazawa, *Nippon Nogeikagaku Kaishi 55*, 935 (1981).

971. Y. Nasu and T. Nakazawa, *Nippon Nogeikagaku Kaishi 55*, 1113 (1981).

972. F. Pellegrino, P. Sekuler, and R. R. Alfano, *Photobiochem. Photobiophys. 2*, 15 (1981).

973. G. E. Eldred, G. V. Miller, W. S. Stark, and L. Feeney-Burns, *Science* *216*, 757 (1982), and refs. therein.

974. D. P. Dimitrov, *Z. Naturforsch. 25B*, 762 (1970).

975. H. Beer and H. C. Longuet-Higgins, *J. Chem. Phys. 23*, 1390 (1955).

976. H. J. Griesser and U. P. Wild, *Chem. Phys. 52*, 117 (1980).

977. H. Yamaguchi, T. Ikeda, and H. Mametsuka, *Bull. Chem. Soc. Jap. 49*, 1762 (1976).

978. S. Murata, C. Iwanaga, T. Toda, and H. Kokubun, *Ber. Bunsenges. Phys. Chem. 76*, 1176 (1972).

979. H. J. Grisser and U. P. Wild, *J. Photochem. 12*, 115 (1980).

980. G. D. Gillispie and E. C. Lim, *J. Chem. Phys. 65*, 4314 (1976).

981. H. J. Griesser and U. P. Wild, *J. Chem. Phys. 73*, 4715 (1980).

982. D. J. Mitchell, H. G. Drickamer, and G. B. Schuster, *J. Am. Chem. Soc. 99*, 7489 (1977).

983. D. Kemp and B. Nickel, *Chem. Phys. Lett. 97*, 66 (1983).

984. M. S. Henry and W. P. Helman, *J. Chem. Phys. 56*, 5734 (1972).

985. F. P. Schwarz, D. Smith, S. G. Lias, and P. Ausloos, *J. Chem. Phys. 75*, 3800 (1981).

986. H. T. Choi and S. Lipsky, *J. Phys. Chem. 85*, 4089 (1981).

987. F. Hirayama and S. J. Lipsky, *J. Chem. Phys. 62*, 576 (1975).

988. V. Vaida and G. M. McClelland, *Chem. Phys. Lett. 71*, 436 (1980).

989. R. M. Gavin, G. Risemberg, and S. A. Rice, *J. Chem. Phys. 85*, 3160 (1970).

990. M. F. Granville, G. R. Holtom, and B. E. Kohler, *J. Chem. Phys. 72*, 4671 (1980).

991. J. R. Andrews and B. S. Hudson, *Chem. Phys. Lett 60*, 380 (1979).

992. A. Weinreb and A. Werner, *Chem. Phys. Lett 3*, 231 (1969).

993. D. C. Wilton, *J. Biochem. 208*, 521 (1982).

994. B. M. Pierce, J. A. Bennett and R. R. Birge, *J. Chem. Phys. 77*, 6343 (1982).

995. A. C. Lasaga, R. J. Aerni and M. Karplus, *J. Chem. Phys. 73*, 5230 (1980).

996. R. Palluk, R. Müller, and M. Kempfle, *Z. Anal. Chem. 31*, 370 (1982).

997. H. Morimoto, I. Imada, T. Murata, and N. Matsumoto, *Liebigs Ann. Chem. 708*, 230 (1967).

998. J. R. Andrews and B. S. Hudson, *Chem. Phys. Lett. 57*, 600 (1978).

999. R. L. Christensen and B. E. Kohler, *J. Phys. Chem. 80*, 2197 (1976).

1000. R. L. Christensen and B. E. Kohler, *J. Chem. Phys. 63*, 1837 (1975).

1001. R. M. Gavin, C. Weisman, J. K. MacVey, and S. A. Rice, *J. Chem. Phys. 68*, 522 (1978).

1002. M. F. Granville, G. R. Holtom, B. E. Kohler, R. L. Christensen, and K. L. D'Amico, *J. Chem. Phys. 70*, 593 (1979).

1003. L. A. Heimbrook, J. E. Kenny, B. E. Kohler, and G. W. Scott, *J. Chem. Phys.* 75, 4338 (1981).

1004. K. L. D'Amico, C. Manos, and R. L. Christensen, *J. Am. Chem. Soc.* 102, 1777 (1980).

1005. M. Hossain, Ph.D. thesis, Wesleyan University, 1982; in ref. 1006.

1006. M. Hossain, B. E. Kohler, and P. West, *J. Phys. Chem.* 86, 4918 (1982).

1007. J. R. Andrews and B. S. Hudson, *Chem. Phys. Lett.* 57, 600 (1978).

1008. L. A. Sklar, B. S. Hudson, and R. D. Simoni, *Proc. Nat. Acad. Sci. USA* 72, 1649 (1975).

1009. L. A. Sklar, B. S. Hudson, M. Petersen, and J. Diamond, *Biochemistry* 16, 813 (1977).

1010. L. A. Sklar, B. S. Hudson, and R. D. Simoni, *Biochemistry 16*, 5100 (1977).

1011. L. A. Sklar, B. S. Hudson, and R. D. Simoni, *Biochemistry 16*, 819 (1977).

1012. L. A. Sklar, G. P. Miljanich, and E. A. Dratz, *J. Biol. Chem.* 254, 9592 (1979).

1013. R. L. Christensen and B. E. Kohler, *J. Chem. Phys.* 65, 63 (1975).

1014. P. W. Danckwortt, and J. Eisenbrand, *Lumineszenzanalyse im filtrierten ultraviolettem Licht*, Akad. Verlagsges. Geest & Portig, Leipzig, 1964 p. 129.

1015. R. A. Auerbach, R. L. Christensen, M. F. Granville, and B. E. Kohler, *J. Chem. Phys.* 74, 4 (1980).

1016. G. N. Lyalin, G. I. Kobyshev, and A. N. Terenin, *Dokl. Akad. Nauk. SSSR 150*, 407 (1963).

1017. R. J. Cherry, D. Chapman, and J. Langelaar, *Trans. Faraday Soc. 64*, 2304 (1968).

1018. F. T. Wolf and M. V. Stevens, *Photochem. Photobiol. 6*, 597 (1967).

1019. K. W. Hausser, R. Kuhn, and E. Kuhn, *Z. Phys. Chem. 29B*, 417 (1935).

1020. G. N. Lewis and M. Kasha, *J. Am. Chem. Soc. 66*, 2100 (1944).

1021. C. Tric and V. Lejeune, *Photochem. Photobiol. 12*, 339 (1970).

1022. P. S. Song and T. A. Moore, *Photochem. Photobiol. 19*, 435 (1974).

1023. R. F. Dallinger, W. H. Woodruff, and M. A. J. Rodgers, *Photochem. Photobiol. 33*, 275 (1981).

1024. A. J. Campillo, R. C. Hyer, V. H. Kollman, S. L. Shapiro, and H. D. Sutphin, *Biochim. Biophys. Acta 387*, 533 (1975).

1025. R. F. Dallinger, W. H. Woodruff and M. A. J. Rodgers, *Photochem. Photobiol. 33*, 273 (1981), footnote p. 277.

1026. K. R. Naqvi, *Photochem. Photobiol. 31*, 523 (1980).

1027. G. Dirks, A. L. Moore, T. A. Moore, and D. Gust, *Photochem. Photobiol. 32*, 277 (1980).

1028. A. L. Moore, G. Dirks, D. Gust, and T. A. Moore, *Photochem. Photobiol. 32*, 691 (1980).

1029. B. Del Castillo, A. Garcia de Marina, and M. P. Martinez-Honduvilla, *Ital. J. Biochem.* 29, 233 (1980).

1030. M. A. Schneider, E. Stilbans, and G. A. Gavrilov, *Antibiotiki* (Moscow) 27, 693 (1982).

1031. R. Bittman and F. A. Fischkoff, *Proc. Nat. Acad. Sci. USA* 69, 3795 (1972).

1032. R. Bittman, W. C. Chen, and O. R. Anderson, *Biochemistry 13*, 1364 (1974).

1033. J. R. Chantrez-Antoranz, C. Sainz Villanueva, and E. Otero Aenlle, *An. Real Acad. Farm. 1978*, 77.

1034. F. Schroeder, J. F. Holland, and L. L. Bieber, *Biochemistry 11*, 3105 (1972).

1035. N. O. Petersen and P. F. Henshaw, *Can. J. Chem. 59*, 3377 (1981).

1036. P. K. Das and R. S. Becker, *J. Phys. Chem. 82*, 2093 (1978).

1037. P. K. Das, G. Kogan, and R. S. Becker, *Photochem. Photobiol. 30*, 689 (1979).

1038. R. S. Becker and P. K. Das, *J. Phys. Chem. 84*, 2300 (1980), and ref therein.

1039. R. S. Becker and P. K. Das, *Photochem. Photobiol. 32*, 739 (1980).

1040. P. K. Das and R. S. Becker, *J. Phys. Chem. 86*, 921 (1982).

1041. P. K. Das and R. S. Becker, *J. Phys. Chem. 86*, 921 (1982), refs. 1–7; therein.

1042. B. S. Hudson and R. T. Loda *Chem. Phys. Lett. 81*, 591 (1981).

1043. R. S. Becker, P. K. Das, and G. Kogan, *Chem. Phys. Lett. 67*, 463 (1979).

1044. R. L. Christensen and B. E. Kohler *Photochem. Photobiol. 19*, 401 (1974).

1045. R. D. Fugate and P. S. Song *J. Am. Chem. Soc. 101*, 3408 (1979).

1046. P. S. Song, Q. Chae, M. Fujita, and H. Baba, *J. Am. Chem. Soc. 98*, 819 (1976).

1047. R. R. Birge, J. A. Bennett, L. M. Hubbard, H. L. Fang, B. M. Pierce, D. S. Kligler, and G. E. Leroi, *J. Am. Chem. Soc. 104*, 2519 (1982).

1048. R. S. Becker, K. Inuzuka, J. King, and D. E. Balke, *J. Am. Chem. Soc. 93*, 43 (1971).

1049. T. Takemura, P. K. Das, G. Hug, and R. S. Becker, *J. Am. Chem. Soc. 100*, 2626 (1978).

1050. J. Everaert and P. M. Rentzepis, *Photochem. Photobiol. 36*, 543 (1982).

1051. B. Palmer, B. Jumper, W. Hagan, J. C. Baum, and R. L. Christensen, *J. Am. Chem. Soc. 104*, 6907 (1982).

1052. P. K. Das, G. Kogan, and R. S. Becker, *Photochem. Photobiol. 30*, 689 (1979).

1053. P. K. Das and G. L. Hug, *J. Phys. Chem. 87*, 49 (1983), and refs. therein.

1054. N. Selvarajan, P. K. Das, and G. L. Hug, *J. Photochem. 20*, 355 (1982).

1055. R. S. Becker, K. Inuzuka, J. King, and D. E. Balke, *J. Am. Chem. Soc.* *93*, 43 (1971).

1056. T. Takemura, G. Hug, P. K. Das, and R. S. Becker, *J. Am. Chem. Soc.* *100*, 2631 (1978).

1057. B. Veyret, S. G. Davies, M. Yoshida, and K. Weiss, *J. Am. Chem. Soc.* *100*, 3283 (1978).

1058. W. H. Waddell, A. M. Schaffer, and R. S. Becker, *J. Am. Chem. Soc.* *95*, 8223 (1973).

1059. T. Takemura, P. K. Das, G. Hug, and R. S. Becker, *J. Am. Chem. Soc.* *98*, 7099 (1976).

1060. T. A. Moore and P. S. Song, *Nature (New Biol.)242*, 30 (1973).

1061. S. Hotchandani, P. Paquin, and R. M. Leblanc, *Can. J. Chem. 56*, 1985 (1978).

1062. A. Kropf and R. Hubbard, *Photochem. Photobiol. 12*, 249 (1970).

1063. A. G. Doukas, P. Y. Lu, and R. R. Alfano, *Biophys. J. 35*, 547 (1981).

1064. M. Montal, *Biochim. Biophys. Acta 559*, 231 (1979).

1065. R. R. Birge, *Ann. Rev. Biophys. Bioeng. 10*, 315 (1981).

1066. M. Ottolenghi, *Adv. Photochem. 12*, 97 (1980).

1067. M. S. Henry and W. P. Helman, *J. Chem. Phys. 56*, 5734 (1972).

1068. G. Pitet, A. Verdier, L. S. Nang, and A. Lattes, *Ann. Pharm. Fr. 26*, 135 (1968).

1069. W. J. Hurst and P. B. Toomey, *Analyst 106*, 394 (1981).

1070. D. Bryce-Smith, A. Gilbert, and G. Klunklin, *Chem. Comm. 1973*, 330.

1071. R. A. Beecroft and R. S. Davidson, *Chem. Phys. Lett. 77*, 77 (1981).

1072. A. M. Swinnen, M. van der Auweraer, F. C. De Schruyver, C. Windels, R. Goedeweek, A. Vannerem, and F. Meeus, *Chem. Phys. Lett. 95*, 467 (1983).

1073. E. Ueda, N. Yoshida, K. Nishimura, T. Ioh, S. Autoku, S. Ganno, and T. Kokubu, *Clin. Chim. Acta 80*, 447 (1977).

1074. K. Imai, *J. Chromatogr. 105*, 135 (1975).

1075. M. S. F. Ross, *J. Chromatogr. 141*, 107 (1977).

1076. S. S. Simons and D. F. Johnson, *J. Am. Chem. Soc. 98*, 7098 (1976).

1077. A. M. Krstulovic and A. M. Powell, *J. Chromatogr. 171*, 345 (1979).

1078. G. P. Jackman, *Clin. Chem. 27*, 1202 (1981), and *Clin. Chim. Acta 120*, 137 (1982).

1079. O. Lindvall and A. Björklund, *Histochemistry 39*, 97 (1974).

1080. A. Björklund, A. Nobin, and U. Stenevi, *J. Histochem. Cytochem. 19*, 286 (1971).

1081. G. G. Guilbault and D. N. Kramer, *Anal. Chem. 37*, 120 (1965).

1082. G. G. Guilbault, M. H. Sadar, R. Glazer, and C. Skou, *Anal. Lett. 1*, 365 (1968).

1083. A. Sjoerdsma, J. A. Oates, P. Zaltzman, and S. Udenfriend, *J. Pharmacol. Exp. Ther. 126*, 217 (1959).

1084. N. Seiler, G. Werner, and M. Wiechmann, *Naturwiss, 20*, 643 (1963).

1085. K. F. Nakken, *Scand. J. Lab. Invest. 15*(Suppl. 76), 78 (1963).

1086. O. Beck, G. Palmskog, and E. Hultman, *Clin. Chim. Acta 79*, 149 (1977).

1087. W. Majak and R. J. Bose, *Phytochemistry 16*, 749 (1977).

1088. B. Holmstedt, *Arch. Intern. Pharmacodyn. 156*, 285 (1965).

1089. M. R. Hardemann, A. DenUyl, and H. K. Prins, *Clin. Chim. Acta 37*, 71 (1972).

1090. J. Goldman and R. J. Thibert, *Microchem. J. 22*, 85 (1977).

1091. Y. Yamada, Y. Sugimoto, and K. Horisaka, *Anal. Biochem. 129*, 460 (1983).

1092. J. S. Swan, E. Y. Kragten, and H. Veening, *Clin. Chem. 29*, 1082 (1983).

1093. A. J. Lewis, P. J. Holden, R. C. Ewan, and D. R. Zimmerman, *J. Agric. Food Chem. 24*, 1081 (1976).

1094. G. M. Anderson and W. C. Purdy, *Anal. Chem. 51*, 283 (1979).

1095. A. P. Graffeo and B. L. Karger, *Clin. Chem. 22*, 184 (1976).

1096. D. D. Chilcote, *Clin. Chem. 20*, 421 (1974).

1097. A. M. Krstulovic, P. R. Brown, and D. M. Rosie, *Anal. Chem. 49*, 2237 (1977).

1098. C. Helene, J. L. Dimicoli, and F. Brun. *Biochemistry 10*, 3802 (1971).

1099. S. Udenfriend, D. F. Bogdanski, and H. Weissbach, *Science 122*, 972 (1955).

1100. R. F. Chen, *Proc. Nat. Acad. Sci. USA 60*, 598 (1968).

1101. *Ibid.*, note added in proof, p. 605.

1102. T. R. Hopkins and R. Lumry, *Photochem. Photobiol. 15*, 555 (1972).

1103. M. V. Hershberger and R. W. Lumry, *Photochem. Photobiol. 23*, 391 (1976).

1104. D. F. Bogdanski, A. Pletscher, B. B. Brodie, and S. Udenfriend, *J. Pharmacol. Exp. Ther. 117*, 82 (1956).

1105. J. H. Thompson, C. A. Spezia, and M. Angulo, *Anal. Biochem. 31*, 321 (1969).

1106. A. Yuwiler, S. Plotkin, E. Geller, and E. R. Ritvo, *Biochem. Med. 3*,426 (1970).

1107. M. L. Das, *Biochem. Med. 6*, 299 (1972).

1108. H. Weissbach, T. P. Waalkes, and S. Udenfriend, *J. Biol. Chem. 230*, 865 (1958).

1109. C. A. Fischer and M. H. Aprison, *Anal. Biochem. 46*, 67 (1972).

1110. G. G. Guilbault and P. M. Froehlich, *Clin. Chem. 20*, 812 (1974).

1111. A. H. Bockman and B. S. Schweigert, *J. Biol. Chem. 186*, 153 (1950).

1112. P. A. Toseland, M. J. Michelin, and S. A. Price, *Clin. Chim. Acta 37*, 477 (1972).

1113. E. B. Bell and W. I. P. Mainwaring, *Clin. Chim. Acta 35*, 83 (1971).

1114. H. Wachter, K. Grassmayr, W. Gütter, A. Hausen, and G. Sallaberger, *Mikrochim. Acta 1972*, 861.

1115. Y. Fukunaga, Y. Katsuragi, T. Izumi, and F. Sakiyama, *J. Biochem. 92*, 129 (1982).

1116. M. P. Pileni, P. Walrant, and R. Santus, *Chem. Phys. Lett. 42*, 89 (1976).

1117. M. P. Pileni, *Bull. Soc. Chim. Fr. 1980*, II, 546.

1118. P. Walrant and R. Santus, *Photochem. Photobiol. 20*, 455 (1974).

1119. A. Pirie, *Biochem. J. 128*, 1365 (1972).

1120. M. Bando, A. Nakajima, and K. Satoh, *J. Biochem. 89*, 103 (1981).

1121. J. B. Massey and J. E. Churchich, *J. Biol. Chem. 252*, 8081 (1977).

1122. T. Buxton and G. G. Guilbault, *Clin. Chem. 20*, 765 (1974).

1123. F. Geeraerts, L. Schimpfesse, and R. Crokaert, *Chromatographia 15*, 449 (1982).

1124. S. Udenfriend, *Fluorescence Assay in Biology and Medicine*, Vol. II, Academic Press, New York, 1973, p. 236.

1125. K. Satoh and J. M. Price, *J. Biol. Chem. 230*, 781 (1958).

1126. G. Cohen, R. A. Fishman, and A. L. Jenkins, *J. Lab. Clin. 67*, 520 (1966).

1127. A. Weinreb and A. Werner, *Photochem. Photobiol. 20*, 313 (1974).

1128. A. Weinreb and A. Werner, *Photochem. Photobiol. 26*, 567 (1977).

1129. A. E. Mantaka-Marketon, *Z. Phys. Chem.* (Wiesbaden) *122*, 153 (1980).

1130. I. Janic, A. Kawski, and I. Gryczynski, *Bull. Acad. Pol. Sci., Ser. Sci. Math., Astr., Phys. 23*, 635 (1975).

1131. J. Bramhall and A. Z. Britten, *Clin. Chim. Acta 68*, 203 (1976).

1132. J. T. Taylor, J. G. Knotts, and G. J. Schmidt, *Clin. Chem. 26*, 130 (1980).

1133. E. J. Wojtowicz, *J. Agric. Food Chem. 24*, 526 (1976).

1134. H. Konshin, F. Sundholm, and G. Sundholm, *Acta Chem. Scand. 30B*, 262 (1976).

1135. M. Popovici, A. D. Petrescu, P. Baetoniu, and A. Redes, *Ind. Usoara 16*, 205 (1969); in *Chem. Abstr. 71*, 51253 (1969).

1136. A. Weinreb and A. Werner, *Photochem. Photobiol. 15*, 443 (1972).

1137. N. J. Turro, *Modern Molecular Photochemistry*, Benjamin/Cummings Publ. Co., Menlo Park, Calif., 1978.

1138. P. W. Fairchild, K. Shibuya, and E. K. C. Lee, *J. Chem. Phys. 75*, 3407 (1981), and refs. therein.

1139. H. Orita, H. Morita, and S. Nagakura, *Chem. Phys. Lett. 81*, 409 (1981).

1140. M. Noble, E. C. Apel, and E. K. C. Lee, *J. Chem. Phys. 76*, 2219 (1983), and ref. 11 therein.

1141. D. A. Hansen and E. K. C. Lee, *J. Phys. Chem. 62*, 183 (1975).

1142. M. Almgren, *Photochem. Photobiol. 6*, 829 (1967).

1143. L. A. Sarphatie, P. L. Verheijdt, and H. Cerfontain, *Recl. Trav. Chim. Pay-Bas 102*, 9 (1983).

1144. A. Weinreb and A. Werner, *Photochem. Photobiol. 29*, 755 (1979).

1145. C. Michel and A. Tramer, *Chem. Phys. 42*, 315 (1979), and refs. therein.

1146. M. V. Encina and E. A. Lissi, *J. Photochem. 15*, 177 (1981), and refs. therein.

1147. P. H. Schippers and H. P. J. M. Dekkers, *J. Am. Chem. Soc. 105*, 145 (1983).

1148. P. H. Schippers, J. P. M. van der Ploeg, and H. P. J. M. Dekkers, *J. Am. Chem. Soc. 105*, 84 (1983).

1149. W. M. Paul, P. J. Kovi, and S. G. Schulman, *Spectrosc. Lett. 6*, 1 (1973).

1150. P. J. Kovi, A. C. Capomacchia, and S. G. Schulman, *Spectrosc. Lett. 6*, 7 (1973).

1151. R. E. Brown, L. A. Singer, and J. H. Parks, *Chem. Phys. Lett. 14*, 193 (1972).

1152. J. F. Ireland and P. A. H. Wyatt, *J. Chem. Soc., Faraday Trans.* II, 161 (1973).

1153. K. Bredereck, T. Förster, and H. G. Oesterlin, in *Luminescence of Organic and Inorganic Materials*, H. P. Kallman and G. M. Spruch (eds.), Wiley, New York, 1962, p. 161.

1154. J. Saltiel, H. C. Curtis, L. Metts, J. W. Miley, J. Winterle, and M. Wrighton, *J. Am. Chem. Soc. 92*, 410 (1970).

1155. A. E. Lyubarskaya, V. I. Minkin, and M. I. Knyashanskii, *Teoret. Eksp. Khim. 8*, 71 (1972).

1156. J. Catalan, F. Toribio, and A. U. Acuna, *J. Phys. Chem. 86*, 303 (1982).

1157. H. Baba, *J. Phys. Chem. 49*, 1763 (1968).

1158. W. Klöpffer, *Adv. Photochem. 10*, 311 (1977).

1159. V. A. Jolliffe and C. W. Coggins, *J. Agric. Food Chem. 18*, 394 (1970).

1160. A. W. Davidson, *J. Ass. Off. Anal. Chem. 53*, 179 (1970).

1161. J. C. Baum, *J. Am. Chem. Soc. 102*, 716 (1980).

1162. D. E. Poeltl and J. K. McVey, *J. Chem. Phys. 78*, 4349 (1983).

1163. J. C. Baum and D. S. McClure, *J. Am. Chem. Soc. 102*, 720 (1980).

1164. A. C. Hopkinson and P. A. H. Wyatt, *J. Chem. Soc. B*, 1333 (1967).

1165. A. Weller, *Naturwiss. 42*, 175 (1955).

1166. A. Weller, *Z. Elektrochem. 60*, 1144 (1956).

1167. A. Weller, *Prog. React. Kinet. 1*, 188 (1961).

1168. K. K. Smith and K. J. Kaufmann, *J. Phys. Chem. 82*, 2286 (1978).

1169. V. M. Belyakova and V. L. Rapoport, *Zh. Fiz. Khim. 52*, 2706 (1978).

1170. S. G. Schulman and H. Gershon, *J. Phys. Chem.* 72, 3297 (1968).

1171. P. J. Kovi, C. L. Miller, and S. G. Schulman, *Anal. Chim. Acta 61*, 7 (1972).

1172. M. J. Adams, J. G. Highfield, and G. F. Kirkbright, *Analyst 106*, 850 (1981).

1173. S. F. Kilin and I. M. Rozman, *Opt. Spectrosc. 15*, 266 (1963).

1174. W. Klöpffer and G. Naundorf, *J. Lumin. 8*, 457 (1974).

1175. W. Klöpffer and G. Kaufmann, *J. Lumin. 20*, 283 (1979).

1176. K. K. Smith and K. J. Kaufmann, *J. Phys. Chem. 85*, 2895 (1981).

1177. L. A. Heimbrook, J. E. Kenny, B. E. Kohler, and G. W. Scott, *J. Chem. Phys. 75*, 5201 (1981).

1178. F. Toribio, J. Catalan, F. Amat, and A. U. Acuna, *J. Phys. Chem. 87*, 817 (1983), and refs. therein.

1179. A. U. Acuna, F. Amat-Guerri, J. Catalan, and F. Gonzales-Tablas, *J. Phys. Chem. 84*, 629 (1980).

1180. P. M. Felker, W. R. Lambert, and A. H. Zewail, *J. Chem. Phys. 77*, 1603 (1982), and ref. 2–13 therein.

1181. A. U. Acuna, J. Catalan, and F. Toribio, *J. Phys. Chem. 85*, 241 (1981).

1182. K. Sandros, *Acta Chem. Scand. 30A*, 761 (1976).

1183. D. Ford, P. J. Thistlethwaite, and C. J. Woolfe, *Chem. Phys. Lett. 69*, 246 (1980).

1184. D. Blair, B. H. Rumack, and R. G. Peterson, *Clin. Chem. 24*, 1543 (1978).

1185. G. Pitet, J. Cros, and R. Caujolle, *C. R. Acad. Sci. Fr. 268D*, 868 (1969).

1186. M. Kleinerman, *Anal. Lett. 10*, 205 (1972).

1187. S. Udenfriend, *Fluorescence Assay in Biology and Medicine*, Vol I, Academic Press, New York, 1962 p. 425; and S. Udenfriend, *Fluorescence Assay in Biology and Medicine*, Vol II, Academic Press, New York, 1969 p. 518.

1188. E. B. Truitt, A. M. Morgan, and J. M. Little, *J. Am. Pharm. Ass. 44*, 142 (1955).

1189. D. Schachter and J. G. Manis, *J. Clin. Invest. 37*, 800 (1958).

1190. W. L. Paul and S. G. Schulman, *Anal. Chim. Acta 69*, 195 (1974).

1191. G. A. Thommes and E. Leininger, *Anal. Chem. 30*, 1361 (1958).

1192. M. S. Stallings and S. G. Schulman, *Anal. Chim. Acta 78*, 483 (1975).

1193. L. Nanninga and B. Bink, *Nature 168*, 389 (1951).

1194. E. M. Kosower and H. Dodiuk, *J. Lumin. 11*, 248 (1975/76).

1195. D. W. Fink and J. D. Strong, *Spectrochim. Acta 38A*, 1295 (1982).

1196. G. Kisilevich, *Anal. Abstr. 7*, 3337 (1960).

1197. F. T. Wolf, *Adv. Front. Plant Sci. 21*, 169 (1968).

1198. E. C. Bate-Smith, *Chem. Ind. 1954*, 1457.

1199. A. C. Neish, *Phytochemistry 1*, 1 (1961).

1200. O. S. Wolfbeis and M. Begum, unpublished results (1982).

1201. S. Udenfriend, *Fluorescence Assay in Biology and Medicine*, Vol I, Academic Press, New York, 1962 p. 29.

1202. P. S. Song, S. C. Shim, and W. W. Mantulin, *Bull. Chem. Soc. Jap. 54*, 315 (1981).

1203. G. L. Brun and D. L. D. Milburn, *Anal. Lett. 10*, 1209 (1977).

1204. K. Ghosh and M. Schnitzer, *Can. J. Soil Sci. 60*, 373 (1980).

1205. U. Müller-Wegener, *Z. Pflanzen-Ernähr. Bodenk. 140*, 563 (1977); *Chem. Abstr. 87*, 191537 (1977).

1206. P. K. Seal, K. B. Roy, and S. K. Mukherjee, *J. Indian Chem. Soc. 41*, 212 (1964).

1207. M. Levesque, *Soil Sci. 113*, 346 (1972).

1208. R. A. Saar and J. H. Weber, *Anal. Chem. 52*, 2095 (1980).

1209. M. Schnitzer and K. Ghosh, *Soil Sci. Soc. Am. J. 45*, 25 (1981).

1210. D. K. Ryan and J. H. Weber, *Anal. Chem. 54*, 986 (1982).

1211. A. J. Lapen and W. R. Seitz, *Anal. Chim. Acta 134*, 31 (1982).

1212. J. Axelrod and H. Weissbach, *J. Biol. Chem. 236*, 211 (1961).

1213. M. R. Hanley, *Biochemistry 18*, 1681 (1979).

1214. S. V. Sychev, N. A. Nevskaya, S. Iordanov, E. N. Shepel, A. I. Miroshnikov, and V. T. Ivanov, *Bioorg. Chem. 9*, 121 (1980).

1215. W. S. A. Brusilow and D. L. Nelson, *J. Biol. Chem. 256*, 159 (1981).

1216. V. Z. Pletnev, N. V. Starovoitova, L. A. Chupova, and E. S. Efremov, *Bioorg. Khim. 8*, 169 (1982).

1217. M. A. Napier, B. Holmquist, D. J. Strydom, and I. H. Goldberg, *Biochemistry 20*, 5602 (1982).

1218. K. Eisenbrandt, *Pharmazie 19*, 406 (1964).

1219. W. C. Neely and J. R. McDuffie, *J. Ass. Off. Anal. Chem. 55*, 1300 (1972), and refs. 4, 10–12 therein.

1220. F. S. Chu, *J. Ass. Off. Anal. Chem. 53*, 696 (1970).

1221. A. M. Gillespie and G. M. Schenk, *Anal. Lett. 10*, 161 (1977).

1222. G. J. Diebold, N. Karny, and R. N. Zare, *Anal. Chem. 51*, 67 (1979).

1223. C. W. Holzapfel, I. F. H. Purchase, and P. S. Steyn, *South African Med. J. 40*, 1100 (1966).

1224. J. L. Vorster and I. F. H. Purchase, *Analyst 93*, 694 (1968).

1225. F. S. Richardson and A. D. Gupta, *J. Am. Chem. Soc. 103*, 5716 (1981), and refs. 9–13 therein.

1226. I. Z. Steinberg, R. Panigel, and G. Navon, *Biophys. Chem. 7*, 217 (1977).

1227. M. Kaykaty and G. Weiss, *J. Agric. Food Chem. 31*, 81 (1983).

1228. S. N. Sehgal, K. Singh, and C. Vezina, *Anal. Biochem. 12*, 191 (1965).

1229. S. N. Seghal and C. Vezina, *Anal. Biochem. 21,* 266 (1967).

1230. M. Chien, A. P. Grollman, and S. P. Horwitz, *Biochemistry 16,* 3641 (1977).

1231. R. E. Lenkinski, B. E. Peerce, J. E. Dallas, and J. D. Glickson, *J. Am. Chem. Soc. 102,* 131 (1980).

1232. R. E. Lenkinski, B. E. Peerce, R. P. Pillai, and J. D. Glickson, *J. Am. Chem. Soc. 102,* 7088 (1980).

1233. C. H. Huang, L. Galvan, and S. T. Crooke, *Biochemistry 18,* 2880 (1979).

1234. H. Kasai, H. Naganawa, T. Takita, and H. Umezawa, *J. Antibiot. 31,* 1316 (1978).

1235. L. L. Steinmetz, J. H. Richardson, B. W. Wallin, and W. A. Bookless, *Chem. Abstr. 90,* 95072 (1979).

1236. R. A. Morton, T. A. Hopkins, and H. H. Seliger, *Biochemistry 8,* 1598 (1969).

1237. J. Jung, C. A. Chin, and P. S. Song, *J. Am. Chem. Soc. 98,* 3949 (1976).

1238. J. C. Dunlap and J. W. Hastings, *Biochemistry 20,* 983 (1981).

1239. C. M. Previtali and N. A. Garcia, *An. Asoc. Quim. Argent. 71,* 31 (1983).

1240. J. R. Lakowicz and B. P. Maliwal, *J. Biol. Chem. 258,* 4794 (1983).

1241. A. I. Dragan and S. N. Khrapunov, *Stud. Biophys. 96,* 127 (1983).

1242. S. R. Meech, D. Phillips, and A. G. Lee, *Chem. Phys. 80,* 317 (1983); and refs. therein.

1243. T. Tamaki, *J. Phys. Chem. 87,* 2383 (1983).

1244. (a) T. Montoro, C. Jouvet, A. Lopez-Camillo, and B. Soep, *J. Phys. Chem. 87,* 3582 (1983); (b) R. W. Wijnandts van Resandt, R. H. Vogel, and S. W. Provencher, *Rev. Sci. Instrum. 53,* 1392 (1982).

1245. E. Gudgin, R. Lopez-Delgado, and W. R. Ware, *J. Phys. Chem. 87,* 1559 (1983).

1246. J. W. Petrich, M. C. Chang, D. B. McDonald, and G. R. Fleming, *J. Am. Chem. Soc. 105,* 3824 (1983).

1247. R. W. Wijnandts van Resandt, *Chem. Phys. Lett. 95,* 205 (1983).

1248. H. Leismann, H. D. Scharf, W. Strassburger, and A. Wollmer, *J. Photochem. 21,* 275 (1983).

1249. V. Anantharaman and J. Chrysochoos, *J. Less-Common Metals 93,* 59 (1983).

1250. J. C. Mialocq, E. Amouyal, A. Bernas, and D. Grand, *J. Phys. Chem. 86,* 3173 (1982).

1251. J. C. Garcia-Borron, J. Escribano, M. Jiminez, and J. L. Iborra, *Anal. Biochem. 125,* 277 (1982).

1252. J. Jordana, J. L. Barbero, F. Montero, and L. Franco, *J. Biol. Chem. 258,* 315 (1983).

1253. J. Baudier and D. Gerard, *J. Neurochem. 40,* 1765 (1983).

1254.   E. A. Permyakov, V. N. Medvekin, L. P. Kalinichenko, and E. A. Burstein, *Arch. Biochem. Biophys. 227,* 9 (1983).

1255.   A. Follenius and D. Gerard, *Photochem. Photobiol. 38,* 373 (1983).

1256.   (a) S. N. Khrapunov, A. I. Dragan, A. F. Protas, and G. D. Berdyshev, *Mol. Biol.* (Moscow) *17,* 992 (1983); (b) A. I. Dragan, S. N. Khrapunov, A. F. Protas, and G. D. Berdyshev, *Stud. Biophys. 96,* 187 (1983).

1257.   E. A. Burstein, *Mol. Biol.* (Moscow) *17,* 455 (1983).

1258.   R. F. Steiner (ed.), *Excited States of Biopolymers,* Plenum Press, New York, 1983, chaps. 3, 4.

1259.   M. C. Chang, J. W. Petrich, D. B. McDonald, and G. R. Fleming, *J. Am. Chem. Soc. 105,* 3819 (1983).

1260.   T. Ichiye and M. Karplus, *Biochemistry 22,* 2884 (1983).

1261.   I. M. Kuznetsova and K. K. Turoverov, *Mol. Biol.* (Moscow) *17,* 741 (1983).

1262.   M. R. Eftink, *Biophys. J. 43,* 323 (1983).

1263.   E. A. Permyakov and V. L. Shnyrov, *Biophys. Chem. 18,* 145 (1983).

1264.   M. C. Genov, M. Shopova, R. Boteva, F. Richelli, and G. Jori, *Biochem. J. 215,* 413 (1983).

1265.   E. A. Permyakov, E. A. Burstein, V. I. Emelyanenko, Y. M. Aleksandrov, K. V. Glagolev, V. N. Makhov, and M. N. Yakimenko, *Biofizika 28,* 393 (1983).

1266.   Y. Maulet and J. A. Cox, *Biochemistry 22,* 5680 (1983).

1267.   J. J. McTigue and R. L. van Etteln, *Prostate 3,* 165 (1982).

1268.   D. B. Calhoun, J. M. Vanderkooi, G. V. Woodrow, and S. W. Englander, *Biochemistry 22,* 1526 (1983).

1269.   D. B. Calhoun, J. M. Vanderkooi, and S. W. Englander, *Biochemistry 22,* 1533 (1983).

1270.   M. R. Eftink and D. M. Jameson, *Biochemistry 21,* 4443 (1982).

1271.   M. Tsuda, T. Ono, T. Ogawa, and J. Kawanishi, *Biochem. Biophys. Res. Comm. 104,* 1407 (1982).

1272.   E. Fürlinger and O. S. Wolfbeis, *Biochim. Biophys. Acta 760,* 411 (1983).

1273.   S. P. Jiang, S. H. Lian, and K. C. Ruan, *Chem. Phys. Lett. 104,* 109 (1984).

1274.   M. Leiner, O. S. Wolfbeis, R. J. Schaur, and H. M. Tillian, *IRCS Med. Sci. 11,* 675 (1983).

1275.   O. S. Wolfbeis, and M. Leiner, *Anal. Chim. Acta,* submitted.

1276.   R. A. Van Wagenen, S. Rockhold, and J. D. Andrade, in: *Biomaterials: Interfacial Phenomena and Applications,* S. L. Cooper and N. A. Peppas (eds.), ACS Adv. Chem. Ser. No. 199, 1982, chap. 23.

1277.   T. D. Schlabach and T. C. Wehr, *Anal. Biochem. 127,* 222 (1982).

1278.   J. M. Beecham, J. R. Knutson, J. B. A. Ross, B. W. Turner, and L. Brand, *Biochemistry 22,* 6054 (1983).

1279. J. M. Salmon, E. Kohen, P. Viallet, J. G. Hirschberg, A. W. Wouters, C. Kohen, and B. Thorell, *Photochem. Photobiol. 36,* 585 (1982).

1280. A. J. W. G. Visser, *Biochim. Biophys. Acta 692,* 244 (1982).

1281. M. K. Machwe, *Indian J. Pure Appl. Phys. 20,* 590 (1982).

1282. S. Kasai, K. Matsui, and T. Nakamura, *Dev. Biochem. 21,* 459 (1982); in *Chem. Abstr. 97,* 194704 (1982).

1283. K. Yagi, I. Tanaka, N. Nakashima, and K. Yoshihara, *Dev. Biochem. 21,* 546 (1982); in *Chem. Abstr. 97,* 195008 (1982).

1284. J. K. Eweg, F. Müller, and J. H. van Berkel, *Eur. J. Biochem. 129,* 303 (1982).

1285. A. Karen, N. Ikeda, N. Mataga, and F. Tanaka, *Photochem. Photobiol. 37,* 495 (1983).

1286. K. Maruyama, and T. Otsuki, *Chem. Lett.* 687 (1983).

1287. A. J. W. G. Visser, J. S. Santema, and A. van Hoek, *Photochem. Photobiol. 39,* 11 (1984).

1288. K. Bystra, R. Drabent, L. Szubiajowska, and B. Smyk, *Spectrosc. Lett. 16,* 513 (1983).

1289. H. Ohkawa, N. Ohishi, and K. Yagi, *J. Biol. Chem. 258,* 5623 (1983).

1290. G. A. Gachko, L. N. Kivach, V. T. Koyava, S. A. Maskevich, Y. M. Ostrovskii, and A. M. Sarzhevskii, *Zh. Prikl. Spektrosk. 38,* 396 (1983) (Engl. ed.), p. 294.

1291. W. M. Cort, T. S. Vicente, E. H. Waysek, and B. D. Williams, *J. Agric. Food Chem. 31,* 1330 (1983).

1292. J. Lehmann and H. L. Martin, *Clin. Chem. 28,* 1748 (1982); and *29,* 1840 (1983).

1293. J. Jung and D. H. Kim, *Chem. Abstr. 99,* 49862 (1983).

1294. M. Leiner, H. J. Schaur, O. S. Wolfbeis, and H. M. Tillian, *IRCS Med. Sci. 11,* 841 (1983).

1295. E. P. Gibson and J. H. Turnbull, *J. Photochem. 11,* 313 (1979).

1296. D. Murray and R. S. Becker, *J. Phys. Chem. 87,* 625 (1983).

1297. G. J. Smith and W. H. Melhuish, *J. Photochem. 21,* 183 (1983).

1298. J. P. Ballini, M. Daniels, and P. Vigny, *J. Lumin. 27,* 389 (1982).

1299. J. P. Ballini, P. Vigny, and M. Daniels, *Biophys. Chem. 18,* 61 (1983).

1300. P. R. Callis, *Ann. Rev. Phys. Chem. 34,* 329 (1983).

1301. J. Hetherington, B. Savory, and J. H. Turnbull, *J. Photochem. 20,* 367 (1982).

1302. B. Savory and J. H. Turnbull, *J. Photochem. 23,* 171 (1983).

1303. V. Mikes and V. Dadak, *Biochim. Biophys. Acta 723,* 231 (1983).

1304. M. Maiti, R. Nandi, and K. Chaudhuri, *Photochem. Photobiol. 38,* 245 (1983).

1305. F. Sasse, J. Hammer, and J. Berlin, *J. Chromatogr. 194,* 234 (1980).

1306. J. Vasiliades and J. M. Finkel, *J. Chromatogr. 278*, 117 (1983).

1307. G. G. Chaudhury, A. Banerjee, B. Bhattacharyya, and B. B. Biswas, *FEBS Lett. 161*, 55 (1983).

1308. T. Moriya, *Bull. Chem. Soc. Jap. 56*, 6 (1983).

1309. O. A. Salmon and H. G. Drickamer, *J. Chem. Phys. 77*, 3329 (1982).

1310. A. J. G. Strandjord, S. H. Courtney, D. M. Friedrich, and P. F. Barbara, *J. Phys. Chem. 87*, 1125 (1983).

1311. A. J. G. Strandjord and P. F. Barbara, *Chem. Phys. Lett. 98*, 21 (1983).

1312. D. McMorrow and M. Kasha, *J. Phys. Chem., 106*, 4320 (1984).

1313. D. McMorrow, T. Dzugan, and T. J. Aartsma, *Chem. Phys. Lett. 103*, 492 (1984).

1314. M. Itoh, Y. Tanimoto, and K. Tokumura, *J. Am. Chem. Soc. 105*, 3339, (1983).

1315. (a) M. Itoh and Y. Fujiwara, *J. Phys. Chem. 87*, 4558 (1983); (b) M. Itoh and T. Adachi, *J. Am. Chem. Soc., 106*, 4320 (1984).

1316. A. I. Rybachenko and V. P. Georgievskii, *Dopov. Akad. Nauk Ukr. RSR, Ser. B*, 1009 (1975); in *Chem. Abstr. 84*, 97313 (1976).

1317. O. S. Wolfbeis, E. Fürlinger, H. Jha, and F. Zilliken, *Z. Naturforsch. 39B*, 238 (1984).

1318. A. Sharma, M. K. Machwe, and V. V. S. Murti, *Curr. Sci. 52*, 858 (1983).

1319. P. C. Beaumont, B. J. Parsons, S. Navaratnam, and G. O. Phillips, *Photobiochem. Photobiophys. 5*, 359 (1983).

1320. P. Progron, C. Simon, and G. Mahuzier, *J. Chromatogr. 272*, 193 (1983).

1321. A. Y. Potapenko and V. L. Sukhorukov, *Stud. Biophys. 92*, 15 (1982).

1322. H. Autrup, K. A. Bradley, A. K. M. Shamsuddin, J. Wakhisi, and A. Wasunna, *Carcinogenesis 4*, 1193 (1983).

1323. L. A. Piruzyan, V. I. Alekseev, V. I. Rakhovski, and M. N. Sapozhnikov, *Dokl. Akad. Nauk SSSR 268*, 76 (1983).

1324. C. Sobel, C. Cano, and R. E. Thiers, *Clin. Chem. 20*, 1397 (1974).

1325. (a) S. F. Shkirman, K. N. Solovev, S. M. Arabei, and G. D. Egorova, *Izv. Akad. Nauk SSSR, Ser. Fiz. 42*, 658 (1978); (b) K. N. Solovev, I. V. Stanishevskii, A. S. Starukhin, and A. M. Shulga, *ibid., 47*, 1399 (1983).

1326. V. I. Rakhovski, M. N. Sapozhnikov, and A. L. Shubin, *J. Lumin. 28*, 301 (1983).

1327. A. Andreoni and R. Cubeddu, *Chem. Phys. Lett. 100*, 503 (1983).

1328. S. S. Dvornikov, K. N. Solovev, and N. P. Tsvirko, *Zh. Prikl. Spektrosk. 38*, 798 (1983).

1329. K. Mauring and R. Avarmaa, *Chem. Phys. Lett. 81*, 446 (1981); see also refs. 522–563.

1330. E. I. Karpinus, I. Y. Kucherova, and I. I. Dilung, *Dokl. Akad. Nauk SSSR 267*, 1485 (1982).

1331. A. A. Lamola, I. Asher, U. Muller-Eberhard, and M. Poh-Fitzpatrick, *Biochem. J. 196*, 693 (1981).

1332. M. Kusunoki and F. K. Fong, *Chem. Phys. Lett. 102*, 244 (1983).

1333. J. Fünfschilling and D. Walz, *Photochem. Photobiol. 38*, 389 (1983).

1334. A. C. De Wilton and J. A. Koningstein, *Spectrosc. Int. J. 2*, 144 (1983).

1335. G. E. Bialek-Bilka, *Stud. Biophys. 95*, 65 (1983).

1336. S. Tsuchiya, H. Kanai, and M. Seno, *J. Am. Chem. Soc. 103*, 7370 (1981).

1337. B. Böddi, K. Kovacz, and F. Lang, *Biochim. Biophys. Acta 722*, 320 (1983).

1338. J. Luisetti, H. J. Galla, and H. Möhrwald, *Ber. Bunsenges. Phys. Chem. 82*, 911 (1978).

1339. W. T. Lotshaw, R. S. Alberte, and G. R. Fleming, *Biochim. Biophys. Acta 682*, 75 (1982).

1340. (a) G. R. Seely and V. Senthilathipan, *J. Phys. Chem. 87*, 373 (1983); (b) Y. Kusumoto, G. R. Seely, and V. Senthilathipan, *Bull. Chem. Soc. Jpn. 56*, 1598 (1983).

1341. J. L. Hendrix, *Clin. Chem. 29*, 1003 (1983).

1342. M. A. Thomas, V. Favaudon, and F. Pochon, *Eur. J. Biochem. 135*, 569 (1983).

1343. M. A. Thomas, M. Gervais, M. Favaudon, and P. Valat, *Eur. J. Biochem. 135*, 577 (1983).

1344. A. R. Holzwarth, J. Wendler, and W. Wehrmeyer, *Biochim. Biophys. Acta 724*, 388 (1983).

1345. J. P. Priestle, R. H. Rhyne, J. B. Salmon, and M. L. Hackert, *Photochem. Photobiol. 35*, 827 (1982).

1346. P. Hefferle, W. John, H. Scheer, and S. Schneider, *Photochem. Photobiol. 39*, 221 (1984).

1347. M. N. Kronick and P. D. Grossman, *Clin. Chem. 29*, 1582 (1983).

1348. A. N. Glazer and L. Stryer, *Biophys. J. 43*, 383 (1983).

1349. S. E. Braslavsky, A. R. Holzwarth, and K. Schaffner, *Angew. Chem. Int. Ed. Engl. 22*, 656 (1983).

1350. A. R. Holzwarth, E. Langer, H. Lehner, and K. Schaffner, *J. Mol. Struct. 60*, 367 (1980); and *Photochem. Photobiol. 32*, 17 (1980).

1351. A. A. Lamola, S. E. Braslavsky, K. Schaffner, and D. A. Lightner, *Photochem. Photobiol. 37*, 263 (1983).

1352. S. E. Braslavsky, R. M. Ellul, K. G. Weiss, H. Al-Ekabi, and K. Schaffner, *Tetrahedron 39*, 1909 (1983).

1353. S. E. Braslavsky, A. R. Holzwarth, E. Langer, H. Lehner, J. I. Matthews, and K. Schaffner, *Isr. J. Chem. 20*, 196 (1980).

1354. W. Kufer, H. Scheer, and A. R. Holzwarth, *Isr. J. Chem. 23*, 233 (1983).

1355. H. Falk and U. Zrunek, *Monatsh. Chem. 114*, 1104 (1983) and *115*, 243 (1984).

1356. V. E. Bondybey, S. V. Milton, J. H. English, and P. M. Rentzepis, *Chem. Phys. Lett. 97*, 130 (1983).

1357. M. Roman Ceba, A. Fernandez-Guttierrez, and M. C. Mahedero, *Mikrochim. Acta 1983* II, 85,

1358. A. N. Anoshin, E. A. Gastilovich, and K. A. Mishenina, *Zh. Fiz. Khim. 57*, 1435 (1983).

1359. A. K. Anoshin, E. A. Gastilovich, K. V. Mikhailova, and D. N. Shigorin, *Zh. Fiz. Khim. 57*, 1505 (1983).

1360. W. Bolanowska, T. Gessner, and H. Preisler, *Cancer Chemother. Pharmacol. 10*, 187 (1983).

1361. R. Mentlein and E. Vohwinkel, *J. Chromatogr. 268*, 138 (1983).

1362. M. Kempfle, R. Müller, R. Palluk, and K. Zachariasse, *Hoppe-Seyler's Z. Physiol. Chem. 364*, 1154 (1983).

1363. B. E. Kohler and J. B. Snow, *J. Chem. Phys. 79*, 2134 (1983).

1364. B. E. Kohler and P. West, *J. Chem. Phys. 79*, 583 (1983).

1365. L. V. Haley and J. A. Koningstein, *Chem. Phys. 77*, 1 (1983).

1366. M. Van Riel, J. Kleinen-Hammans, M. Van de Ven, W. Verwer, and Y. K. Levine, *Biochem. Biophys. Res. Comm. 113*, 102 (1983).

1367. A. G. Doukas, M. R. Junnarkar, D. Chandra, R. R. Alfano, and R. H. Callender, *Chem. Phys. Lett. 100*, 420 (1983).

1368. A. V. Sharkov, Y. A. Matveetz, S. V. Chekalin, A. V. Konyashchenko, O. M. Brekhov, and B. Y. Rootskoy, *Photochem. Photobiol. 38*, 108 (1983).

1369. K. Kitagishi, K. Hiromi, and M. Tokushige, *J. Biochem. 93*, 1045 (1983).

1370. W. A. Wolf and D. M. Kuhn, *J. Chromatogr. 275*, 1 (1983).

1371. E. F. Gudgin-Templeton and W. R. Ware, *Chem. Phys. Lett. 101*, 345 (1983).

1372. M. Noble and E. K. C. Lee, *J. Phys. Chem. 87*, 4360 (1983).

1373. J. W. Hepburn, R. J. Buss, L. J. Butler, and Y. T. Lee, *J. Phys. Chem. 87*, 3638 (1983).

1374. R. R. Sauers, M. Zampino, M. Stockl, J. Ferentz, and H. Shams, *J. Org. Chem. 48*, 1862 (1983).

1375. Y. Tomioka, M. Ito, and N. Mikami, *J. Phys. Chem. 87*, 4401 (1983).

1376. R. Bentley and C. P. Thiessen, *J. Biol. Chem. 238*, 1880 (1963).

1377. S. Nagaoka, N. Hirota, M. Sumitani, and K. Yoshihara, *J. Am. Chem. Soc. 105*, 4220 (1983).

1378. R. Sakurovs and K. P. Ghiggino, *J. Photochem. 22*, 373 (1983).

1379. T. I. Balkas, Ö. Bastürk, A. F. Gaines, I. Silahoglu, and A. Yilmaz, *Fuel 62*, 373 (1983).

1380. M. Ewald, C. Belin, P. Berger, and J. H. Weber, *Environ. Sci. Technol. 17*, 501 (1983).

1381. T. T. Sakai, J. M. Rioadan, and J. D. Glickson, *Biochemistry 21*, 805 (1982).

1382. N. Suzuki, T. Ueyama, Y. Izawa, Y. Toya, and T. Goto, *Heterocycles 20*, 1027 (1983).

1383. G. S. Beddard, *NATO Adv. Sci. Inst. Ser., Ser. A 69*, 629 (1983).

1384. S. Pundak and R. S. Roche, *Biochemistry 23*, 1549 (1984).

1385. J. W. Longworth, *NATO Adv. Sci. Inst. Ser., Ser. A 69*, 651 (1983).

1386. M. Arrio-Dupont, *Eur. J. Biochem. 91*, 396 (1978).

1387. B. P. Maliwal and R. J. Lakowicz, *Biophys. J. 19*, 337 (1984).

1388. E. Gratton, D. M. Jameson, G. Weber, and B. Alpert, *Biophys. J. 45*, 789 (1984).

1389. C. Parkyani, D. Bovin, D.-C. Shieh, S. Tunbrant, J. J. Aaron, and A. Tine, *J. Chim. Phys. 81*, 3099 (1984).

1390. S. P. Assenza and P. R. Brown, *J. Chromatogr. 289*, 355 (1984).

1391. A. Anders, *Appl. Phys. 20*, 257 (1979).

1392. A. Andreoni, R. Cubeddu, S. DeSilvestri, R. Laporta, and O. Svelto, *Phys. Rev. Lett. 45*, 431 (1980).

1393. F. T. Vert, I. Zabala-Sanchez, and A. Olba Torrent, *J. Photochem. 23*, 355 (1983).

1394. M. Caprasse and C. Housier, *Biochimie 65*, 157 (1983) and *66*, 31 (1984).

1395. D. McMorrow and M. Kasha, *Proc. Nat. Acad. Sci. USA 81*, 3357 (1984).

1396. A. Bettero and C. A. Benassi, *J. Chromatogr. 280*, 167 (1983).

1397. I. Chang-Yen, V. A. Stoute, and J. B. Felmine, *J. Ass. Off. Anal. Chem. 67*, 306 (1984).

1398. M. Santhanam, R. R. Hautala, J. G. Sweeny, and G. A. Iacobucci, *Photochem. Photobiol. 38*, 477 (1983).

1399. F. E. Lloyd, *Science 59*, 241 (1924).

1400. G. F. Stelmakh and M. P. Tsvirko, *Opt. Spektrosk. 55*, 858 (1983).

1401. M. Kaplanova and L. Parma, *Gen. Physiol. Biophys. 3*, 127 (1984).

1402. A. C. DeWilton, L. V. Haley, and J. A. Koningstein, *J. Phys. Chem. 88*, 1077 (1984).

1403. N. Mataga, *Pure Appl. Chem. 56*, 1255 (1984).

1404. A. R. Holzwarth, J. Wendler, K. Schaffner, V. Sundström, A. Sandström, and T. Gillbro, *Isr. J. Chem. 23*, 223 (1983).

1405. S. Braslavsky, *Pure Appl. Chem. 56*, 1153 (1984).

1406. O. Kautsky, *Trans. Faraday Soc. 35*, 216 (1939), and earlier work, e.g., in *Chem. Ber. 64*, 2682 (1931).

1407. J. Lahiri and R. Basu, *Indian J. Chem. 22A*, 972 (1983).

1408. A. Hausen, D. Fuchs, K. König, and H. Wachter, *J. Chromatogr. 227*, 61 (1982).

1409. E. Becker and C. Schöpf, *Liebig's Ann. Chem. 524*, 124 (1936).

1410. W. Koller, PhD thesis, University of Innsbruck, Austria, 1979.
1411. M. Viscontini, A. Kühn, and A. Engelhaaf, *Z. Naturforsch. 11B,* 501 (1956) and refs. 1–6 therein.
1412. R. Mentlein, private communication, March 1984.
1413. J. R. Ackerman, S. A. Former, M. Hossain, and B. E. Kohler, *J. Chem. Phys. 80,* 39 (1984).
1414. B. E. Kohler and T. A. Spiglanin, *J. Chem. Phys. 80,* 3091 (1984).
1415. I. W. Wylie and J. A. Koningstein, *J. Phys. Chem. 88,* 2950 (1984).
1416. A. Iwatani and H. Nakamura, *J. Chromatogr. 309,* 145 (1984).
1417. A. K. Mitra and M. M. Narurkar, *J. Org. Chem. 49,* 1293 (1984).

CHAPTER

4

DETERMINATION OF INORGANIC SUBSTANCES BY LUMINESCENCE METHODS

ALBERTO FERNANDEZ-GUTIERREZ

Department of General Chemistry
Faculty of Sciences
University of Granada
Granada, SPAIN

ARSENIO MUÑOZ DE LA PEÑA

Department of Analytical Chemistry
Faculty of Sciences
University of Extremadura
Badajoz, SPAIN

4.1. INTRODUCTION

The observation of luminescence from inorganic materials such as barite dates back to the sixteenth century. The first recorded observation of fluorescence (1) seems to have been made in the sixteenth century by the Spanish physician and botanist Nicolas Monardes. He observed a blue tint in water contained in cups made from a kind of wood called *lignum nephriticum*. However, it was Stokes who introduced the concept of fluorescence as a light emission and established the relationship between fluorescence intensity and concentration in 1852 (2). He also appears to have been the first to propose the use of fluorescence as an analytical tool, in 1854 (3). Goppelsröder (4) introduced the term *fluoreszenzanalyse* (fluorimetry) in 1867 and proposed the first analytical application of fluorimetry to the determination of Al (III) by the fluorescence of its morin chelate.

Luminescence analysis of inorganic substances consists of various methods of determining these compounds with the help of the luminescence phenomenon. The various types of molecular luminescence observed can be classified by the mode of excitation to the excited state capable of emission.

Photoluminescence refers to emission from excited energy levels produced by the absorption of ultraviolet or visible radiation. Photoluminescence, depending on the type of molecular excited state deactivated, can be classified as fluorescence or phosphorescence. Fluorescence is the spin-allowed radiative transition from the first excited singlet state, while phosphorescence is the result of a spin-forbidden radiative transition from the lowest triplet state to the ground state.

Chemiluminescence refers to emission from excited molecules generated as products of a chemical reaction. If this reaction is derived from a biological system, the luminescence is called bioluminescence. Other types of luminescence include thermoluminescence (excitation arising from thermally activated ion recombination), electroluminescence (excitation by injection of charge), radioluminescence (high-energy particle or radiation excitation), triboluminescence (friction excitation), and sonoluminescence (sound wave excitation).

This chapter will be devoted exclusively to fluorescence, phosphorescence, and chemiluminescence because the other types of luminescence have minor analytical significance. Methods of luminescence analysis of inorganic substances are based on chemical reactions that usually proceed in solution, although they also can be carried out in the solid state. Generally, the emission of luminescence is used as the analytical signal (related to the concentration of the analyte), although in some cases the quenching of this emission has been used.

Both inorganic and organic reagents can be used for the luminescence determination of inorganic substances. The luminescence measurements are generally made at room temperature, although there are a number of methods that use a lower temperature (usually liquid nitrogen temperature). The number of methods proposed for the determination of cations is greater than that for the determination of anions.

These methods are generally very sensitive and have detection limits (100–0.1 ppb) from one to four orders of magnitude lower than corresponding methods based on absorption.

From the calibration curve [the plot of luminescence intensity (I) versus analyte concentration (C)] the sensitivity (H) is expressed as the first derivative of the calibration function at a given concentration. If the calibration curve is linear over the whole range of analyte concentration, the sensitivity is equal to $\Delta I/\Delta C$, and it remains constant in this concentration range. The detection limit $C_{min,p}$ is the smallest analyte concentration that can be detected by a given method with a reasonable certainty level P. This concentration is given by $C_{min,p} = (I_a - \bar{I}_0)/H$, where I_a is the limiting detectable luminescence that can still be recorded and $\bar{I}_0$ is the average luminescence intensity of the blank. It has been suggested that a signal (I_a) be considered different from the blank value when it is at least as great as the mean blank value ($\bar{I}_0$) plus three standard deviations of the blank value ($I_a = \bar{I}_0 + 3\sigma_0$). Then $C_{min,p} = 3\sigma_0/H$. If 3σ is chosen, then, for a Gaussian distribution, the confidence level is $P = 0.997$. However, the distributions at low analyte concentrations are often broader and asymmetrical, so that 3σ corresponds at a confidence level P of about 0.9.

The sensitivity of the luminescence methods depends on the efficiency of luminescence as well as on the noise level arising from photomultiplier shot noise and from variations in source intensity. If the fluctuation of the blank appreciably exceeds the instrumental system noise level, it is the blank that restricts the limit of detection.

Luminescence methods, in general, have wider linear dynamic ranges than absorption methods. The analytical curve is usually linear up to the point where absorption of excitation radiation becomes significant. Also, these methods generally show a greater selectivity than absorption methods because, on the one hand, there are more types of molecules absorbing light than reemitting it and, on the other hand, it is possible to vary the wavelength of excitation and the wavelength of emission to maximize selectivity for a particular analyte. However, several of the methods based on the formation of complexes show low selectivity because the same reagent can react with various ions. This is usually avoided by using preliminary separation techniques, such as solvent extraction, chromatography, and ionic exchange, or by using adequate masking agents.

In luminescence analysis, the effects of contamination from reagents, laboratory air, furnishings, apparatus, and containers are important when trace levels of elements need to be determined. Parallel to the analysis of the sample, a blank test is performed. To be able to determine lower levels, it is necessary to reduce the blank luminescence. This can be done by purification of reagents and the use of quartz, polyethylene, and Teflon rather than glass vessels.

The reagents are usually the greatest source of contamination in the blank test. The water used should be distilled in a quartz apparatus after its demineralization in quartz vessels. Highest-purity solvents must be used and should not be stored in plastic containers because of the possibility of fluorescent materials being bleached out of the containers. All bottle caps should have Teflon inserts. It is necessary to keep the cells scrupulously clean and free of defects (to avoid scattered light). The use of fluorescent detergents as routine laboratory cleaners should be avoided. The remaining factors that can affect the luminescence of a system (variations of the intensity of the source, temperature, photodecomposition processes, pH, solvent composition, quenching phenomena, and presence of interfering substances) must be controlled according to the indications of the proposed procedure to avoid errors in the determination.

The precision of the luminescence methods (the degree to which the measured values of luminescence intensity or analyte concentration agree with each other) is usually good. Precision is characterized by the relative standard deviation. Accuracy is a measure of the agreement between the average and actual analyte concentration. For very low analyte concentrations inferior accuracy of determination must be accepted. In the determination of trace levels of about 10^{-3} and $10^{-4}\%$, an error of up to $\pm 10\%$ is usually obtained.

Rapid advances in luminescence methods of analysis have been occurring since approximately 1945, with the development of the fluorescence spectrometer with its photoelectric detector and high-intensity arc lamp. At present, luminescence methods are available for the determination of almost all elements of the periodic table. These methods are widely used in the analysis of very different materials such as pure metals, alloys, semiconductors, mineral samples, technological products, soil, water, air, and biological, agricultural and clinical samples. An important factor in the acceptance of luminescence methods as analytical techniques is the relatively low cost of the equipment needed compared with other techniques having similar sensitivities and detection limits.

Numerous books dealing with the study of fundamentals of theory, instrumentation, and application of luminescence have been written (5–32). Biennial reviews on the topic have appeared in *Analytical Chemistry* (33–37) and from time to time in *Zabodskaya Laboratoriya* (*Industrial Laboratory*, English translation) (38,39). Passwater has compiled references dealing with fluorescence up through 1972 (40).

4.2. INORGANIC SUBSTANCES THAT LUMINESCE IN SOLUTION

Only rare earths and uranyl compounds fluoresce in solution and therefore these ions may be determined by direct analysis using native fluorescence.

4.2.1. Rare Earth Elements

Luminescence spectra of Ce(III), Pr(III), and Nd(III) salts are broad, diffuse bands associated with transitions of electrons from the 5d to the 4f shell. In contrast, Sm(III), Eu(III), Tb(III), and Dy(III) salts are characterized by line luminescence caused by transitions of 4f electrons. Luminescence of inorganic salts of lanthanide(III) ions in organic solvents containing CH, OH, NH, and other groups with high-frequency vibrations is strongly quenched by conversion of electronic excitation energy into vibrational energy in collisions with solvent molecules. This phenomenon has not yet been used in analytical chemistry.

The fluorescence spectra of cerium salts consist of a broad band extending from about 330 to 402 nm (41). Several authors used the characteristic fluorescence of Ce(III) (λ_{ex} about 260 nm, λ_{em} 350 nm) in dilute acid solutions to develop fluorimetric methods of determination. The sensitivity is 1 ppb and the fluorescence is linear with concentration up to about 10 ppm (42). Usually, Ce(IV) is reduced to Ce(III) with Ti(III) sulfate or hydroxylamine hydrochloride. The method has been applied to the determination of cerium in rare earth preparations (43,44), in yttrium oxide (45), in the presence of Ce(IV) (46), in sea water (47), and in low-alloy steels (48).

Because the fluorescence of the rare earth ions involves the transition of f electrons that are well shielded from external influences, the spectra are usually sharp. The fluorescence spectrum of gadolinium consists of a single, bright, narrow band at 311 nm, and the fluorescence spectra of europium and terbium consists of several characteristic bands (41). Terbium salts fluoresce yellow-green and concentrations of ppb to 1 ppm can be determined. Europium fluoresces red and its minimum detectable concentration is 10 ppm.

Fassel and Heidel (49,50) have developed sensitive fluorometric methods of analysis of Pr, Nd, Sm, Eu, Gd, Tb, Dy, Er, and Tm ions with detection limits of 2.5×10^{-4}, 0.5, 5×10^{-5}, 2.5×10^{-4}, 1, 2.5×10^{-3}, 2.5×10^{-3}, 5×10^{-3}, and 5×10^{-3} μg, respectively.

Eu and Tb were determined fluorimetrically (51,52) at 620 and 550 nm, respectively, after dissolving the samples in potassium carbonate. The ranges of concentrations are 0.3–70 ppm of Tb and 4–800 ppm of Eu. Cerium and large amounts of lanthanum and scandium interfere. Tb can be determined (53) in oxides of rare earths by treatment with potassium oxalate, sodium tetraborate, and hydrochloric acid. The fluorescence is measured at 545 nm and excited at 255 nm. The potassium oxalate supresses interference by dysprosium and prevents that by praseodymium but does not affect that by terbium. Several papers appeared on the great enhancement of fluorescence intensity attainable in $POCl_3$–$SnCl_4$ solvent, as compared to aqueous solution, for Eu (54), Sm (55), and Nd (56); for example, an enhancement factor of 220 was found for Eu and a factor of 300 was found for Sm in the related "heavy atom" solvent $POCl_3$–$ZrCl_4$ (57). However, these solvents have not been employed in the luminescence analysis of lanthanide(III) ions.

4.2.2. Uranium Salts

Among uranium compounds only those with the uranyl ion (UO_2^{2+}) are fluorescent. Fluorescence spectra of U(VI) salts are observed in the 520–

620-nm range and have several bands. For the determination of uranium, the fluorescence of U(VI) is used in aqueous solutions of $Na_3P_3O_9$,HF, H_3PO_4, and H_2SO_4. Fluorescence intensity depends on the nature of the ligand and its concentration, acidity of the medium, ionic state of U(VI), and presence of impurities. To separate U(VI) ions from most elements, they are usually extracted by tributylphosphate solutions in the presence of EDTA and then reextracted by $Na_3P_3O_9$ solution. Many organic compounds quench the fluorescence of U(VI) salts in aqueous solution because of intramolecular energy or electron transfer (58).

White (59–61) has suggested the use of fluorescence of uranium in concentrated acids for its determination. For the determination of uranium in biological material, the uranium should be first precipitated as a uranium–protein complex (59).

Dobrolyubskaya determined uranium in solution by the fluorescence of uranyl nitrate at pH 5.8. Excitation at 253.7 is preferable to that at 365 nm (62). Up to 30 g of sodium carbonate per liter do not interfere with the determination of hexavalent uranium by fluorescence at 214–340 nm, nor does phosphoric acid interfere (63). For the fluorimetric determination of hexavalent uranium as sulfate in a solution containing up to 20% sodium sulfate, addition of 0.1% sodium metaphosphate enhances the intensity. The same effect is obtained with a solution of uranyl fluoride containing up to 1.5% of sodium fluoride (64).

Phosphoric acid also enhances the fluorescence of a solution containing uranyl ion. A method (65) was described for the fluorimetric determination of uranium in natural waters, in a sulfuric and phosphoric acid medium. The limit of detection is 0.3 ppm.

The direct measurement of the fluorescence of uranyl nitrate in 20% tributylphosphate, 80% synthine (hydrogenated kerosene) gave a linear response for 0.01–1 ppm in liquid N_2 with front surface reading (66).

Tri-n-octyl and tri-n-decylphosphine oxides are excellent extractants for uranium (67). Using tri-n-octylphospine oxide in ethyl acetate, uranium oxide at 1 ppm to 0.2% is extracted from a N hydrochloric acid solution of the ores. Ascorbic acid is added to mask iron and uranium is estimated by fluorescence (68). Also, uranium can be extracted from saturated calcium nitrate solution with methyl isobutyl ketone or acid-deficient aluminum nitrate solution for fluorescence reading (69).

Uranium can be determined in the presence of thorium by extracting at pH 2.5 from saturated calcium nitrate solution with tributylphosphate in hydrogenated kerosene and then back-extracting the uranium with ammonium carbonate solution (70).

The UO_2^{2+} is co-precipitated with $Zn_3(PO_4)_3$ at pH 5. This gives a fluorescent precipitate that is dispersed and stabilized with agar for reading (71). The method has been applied to the determination of uranium in sea water at 520 nm with excitation at 365 nm. The calibration graph is linear for 2–100 μg of uranium (72).

4.3. LUMINESCENCE OF CRYSTALLOPHOSPHORS

The multicomponent inorganic compounds, crystallophosphors or phosphors, are polycrystalline substances containing traces of some ionic activators of luminescence. The nature of the fluorescence and properties of these inorganic substances are discussed in several books (73–76) and reviews (77–79).

Chemical impurities and crystalline imperfections are the dominant sites for luminescence in inorganic crystals. The luminescent sites or activators are usually monatomic impurities or imperfections, but in some cases they can be diatomic. The activator ion, with a somewhat distorted arrangement of lattice ions around it, is often called a *center*. Host lattices are usually vanadates, metallic oxides, metallic oxyhalides, halides, phosphates, silicates, and so on, where the large size of the oxygen ion almost dominates the lattice structure. Also, certain sulfides and alkalis or alkaline earth halides are used.

The characteristics of the luminescence depend on the chemical potential for electrons of the crystal, on the structure of the activator system, and on the conditions under which the phosphors are prepared. These crystallophosphors are most often excited by UV light of mercury and xenon lamps, sparks, lasers, cathode rays, and xrays.

The greatest number of analytical applications based on the use of phosphors is in the determination of rare earth ions. However, other ions, such as Mn, Cu, Re, Rh, Sb, Sn, Tl, Pb, and Bi, have also been determined.

4.3.1. Rare Earth Elements

Many substances become fluorescent when doped with lanthanides or uranium. In crystallophosphors, line luminescence of ions is caused by transitions from the partially filled 4f shell of the lanthanides from Ce(III) to Yb(III). The crystallophosphor luminescence spectra of Ce(III) are in the far infrared, those of Gd(III) in the ultraviolet, and those of the other lanthanides in the visible and near infrared regions. The lanthanide activators, according to the excitation spectra, can be divided into three groups: elements in which the luminescence excitation occurs (1) in a broad 4f–5d absorption band (Ce^{3+}, Pr^{3+}, Tb^{3+}, Y^{3+}), (2) in a charge transfer band (Sm^{3+}, Eu^{3+}), and (3) in narrow 4f–4f bands (Nd^{3+}, Dy^{3+}, Ho^{3+}, Er^{3+}, and Tm^{3+}).

The electronic configuration of U(IV) is $1s^2 \ldots 5s^2 5p^6 5d^{10} 6s^2 6p^6$ and that of O^{2-}, $1s^2 2s^2 2p^6$. Superposition of these atomic orbitals results in the molecular orbitals of the UO_2^{2+} ion. The nature of the orbitals, and particularly the contribution of the unfilled 5f, 6d, and 7s orbitals of U(VI) to the formation of molecular orbitals is not well understood. Inorganic salts of U(IV) absorb intensely in the 200–300-nm region and much more weakly with a pronounced structure, in the 330–550-nm region. The nature of the absorption is still in doubt. Some authors believe that the short wavelength absorption is caused by the transition of electrons from the 5f or 6d orbitals of U(VI) (80,81), others attribute this absorption to transitions to energy levels of ligands coordinated with the UO_2^{2+} (82). The absorption in the 330–350-nm range is attributed to a single electronic state by some authors (83–86) and to several excited states by others (87–93). According to the type of matrix activated by these ions to form the crystallophosphors, we can cite the following methods of determination.

Vanadates as Matrices

Crystal phosphors based on vanadates are the most widely employed. The phosphors are obtained usually by the treatment of ammonium vanadate with oxides of yttrium, gadolinium, or rare earths.

A method (94) has been devised for the determination of six of the rare earths as impurities in yttrium oxide in which the sample is treated with NH_4VO_3 to form the crystal phosphors YVO_4 and then compared to standards containing these elements. The maxima of emission (nm) and detection limits (%) of the determination are, respectively, Sm 606, 4.5×10^{-5}; Eu 619, 1.5×10^{-5}; Dy 579, 2.4×10^{-5}; Ho 571, 8.2×10^{-5}, Er 553, 2.2×10^{-5}, and Tm 479, 7.2×10^{-5}. Other authors determined also Gd and Tb in Y_2O_3 with detection limits of 10^{-6} and 10^{-4}%, respectively; Gd in Eu compounds; Gd, Eu, and Sm in metallic uranium (95); and Eu, Sm, and Dy in rare earth mixtures (96). The same method has been applied to rare earth determinations in Sc (97) and Lu or Pr oxide (98), forming $ScVO_4$, $LuVO_4$, or $Pr_yY_{1-y}VO_4$ as phosphors. The range of application of the method is 10^{-6}–10^{-4}% of Nd, Sm, Eu, Dy, Er, Tm, and Yb in Sc_2O_3, 10^{-5}–10^{-4}% of Nd, Sm, Eu, Dy, Ho, Er, Tm, Yb in Lu_2O_3 and 10^{-3}–10^{-2}% of Nd, Sm, or Eu in Pr_6O_{11}.

Nd, Tm, and Yb can be determined by this method by their luminescence emissions in the IR region, in YVO_4 (99).

To determine Sm and Eu in neodymium oxide (100), the sample is dissolved in HNO_3, and Y_2O_3 and HN_4VO_3 are added to form the YVO_4

phosphor. The luminescence of the powder product after illumination of 303 and 313 nm is measured at 606 and 619 nm to determine Sm and Eu, respectively. A luminescence method (101) was suggested for Gd and Eu determination in geochemical samples based on the preliminary isolation of the sum of rare earth elements and the preparation of crystal phosphors (based on Y_2O_3 for Gd and on YVO_4 for Eu). The detection limit of the determination is $10^{-6}-10^{-5}\%$ for Gd and Eu. The analytical lines used are 313 and 619 nm, respectively.

The luminescence determinations of Sm, Eu, Dy, and Tb (102) in Gd_2O_3 using NH_4VO_3 and of Nd (103) in Eu_2O_3 using $EuVO_4$ as phosphors were described as well.

Oxyhalides (MOX) as Matrices

GdOX.   Novikova, Efrynshina, et al. have developed crystal phosphors based on gadolinium halogen oxide to determine several ions. Using gadolinium chloride oxide (104) as a phosphor, they determined Sm and Eu in gadolinium preparations. The mixtures were irradiated by a mercury lamp and the intensity of luminescence was measured at 565 nm for Sm and at 618 nm for Eu. The ranges of application are 7–26 ppm of Sm and 20–44 ppm of Eu.

A crystal phosphor of gadolinium fluoride oxide permits determination down to 10 ppb of Pr (105), Nd and Dy in the ranges of concentration between 2.5 and 7 ppm and 30 ppm in gadolinium oxide, respectively (106). Also, with this phosphor Tb can be determined in thulium (107) and gadolinium (108) oxides. The ranges of application are 0.20–250 ppm and 0.1–54 ppm, respectively. The luminescence of Tb at 544.5 nm can be enhanced by doping the phosphor with Ho (109), and the limit of determination of Tb in gadolinium oxide can, thereby, be lowered to 0.3 ppb. Another phosphor, $Yb_{0.5}Gd_{0.5}OF$, was used to determine this ion in the range $1 \times 10^{-5}-8 \times 10^{-2}\%$ in ytterbium oxide (110).

LaOCl and YOF   The determination of lanthanide impurities in highly pure oxides of La and Y (111) is based on the luminescence of crystal phosphors of LaOCl and YOF. La_2O_3 was dissolved in HCl, the solution was evaporated to dryness, and the residue was calcinated to 750–800°, and Y_2O_3 was treated with NH_4F and HF, the mixture was evaporated, and the residue was calcined 45–60 min at 850°. Pr (112,113) Tb, Dy, Ho (114), Nd, Sm, Eu (115), Er, Tm, and Yb (116) were determined with a detection limit between 10^{-6} and $10^{-4}\%$.

Calcium Fluoride as a Matrix

Some ions can be determined by their activation of the luminescence of the CaF_2 crystallophosphor. The phosphor is synthesized by the mixture of CaF_2 and the sample or by the precipitation of CaF_2 from a $Ca(NO_3)_2$ solution with HF solution to which the sample has been added. This is followed by annealing the product obtained at an adequate temperature. This method permits the determination of up to 200 μg of Tm_2O_3 in 200 mg of CaF_2 (117) between 1 ng and 50 μg of Nd (118) in 250 mg CaF_2 and between 0.5 ppm and 4.5% Eu (119) in fluorite crystals.

A spectral system has been described that permits selective excitation of lanthanide(III) ion luminescence in absorption bands associated with transitions inside the 4f shell. The dependence of lanthanide(III) ion absorption spectra on the immediate environment of the ion has led to methods of determination of Y(III), La(III), Gd(III), and other lanthanide(III) ions (up to $1 \times 10^{-6}\%$) by luminescence of the CaF_2 crystallophosphors activated by erbium (120).

Oxides as Matrices

CeO_2. The luminescences of Nd, Eu (121), Ho, and Er (122) are measured at 898, 590, 541, and 547 nm, respectively. The method of determination is based on the recording of the luminescence of a CeO_2 crystal phosphor activated by these ions. The sensitivities are $6 \times 10^{-5}\%$, $5 \times 10^{-6}\%$, $1 \times 10^{-4}\%$, and $4 \times 10^{-5}\%$ of Nd, Eu, Ho, and Er, respectively, in pure cerium dioxide.

Y_2O_3. Optimal conditions (123,124) were established for determination of rare earths in Y_2O_3 by measuring the luminescence of crystal phosphors into which the test material is incorporated and which are excited by adequate radiation. Rare earths detected in the visible and near ultraviolet spectral region were Pr, Sm, Eu, Gd, Tb, Dy, Ho, Er, and Tm; other rare earths had no luminescence. The different emission intensities increased linearly with rare earth (RE) concentrations up to about 1×10^{-3} mole $(RE)_2O_3$ per mole Y_2O_3, except for Gd (up to about 2×10^{-2} mole). The lower limit for emission detection was in the region between 10^{-9} and 10^{-8} mole, depending on the rare earth element; the lower limit for Gd and Sm was 10^{-4} mole because of poor emission. Gd in metallic uranium (95) and Pr and Tb (96) in a mixture of rare earth elements were also determined using the same phosphors.

Other methods using mixtures of Y_2O_3 with several substances as phosphors were reported. Tb, Dy, Ho, and Er by the use of $Y_2O_3 + CaF_2$

luminophosphors, and Eu and Sm, by the use of Y_2O_3 + $CaWO_4$, were determined in Y_2O_3 with a limit of detection of 1 ppm (125,126). Also, it is possible to determine Gd in Eu_2O_3 using a Y_2O_3 + Eu_2O_3 phosphor with a sensitivity of $1 \times 10^{-5}\%$ (127).

Sodium Fluoride as a Matrix

Uranium can be determined by the intense yellow-green fluorescence of a bead, pellet, or disk that is fused with sodium fluoride (128).

For fusion, either NaF or mixtures of NaF with Na_2CO_3, K_2CO_3, Na_2CO_3 + K_2CO_3, or LiF are used. High sensitivity and precision are obtained for uranium levels between 0.1 ng and 10 μg. This technique (129) has been applied to samples too low in uranium content for determination spectrophotometrically. The method is widely used under both laboratory and field conditions (130).

The crystal-phosphors formed in this way allow the determination of uranium in the following materials: natural water (131–135), ores (131,136–140), geochemical and geological samples (134,135,141), sodium iodide (142), Zr or Hf (143), Pu (144), phosphorites (145), high zirconium alloys (146), marine sediments (134), soils (147), ash of plants (148), blood and organs (149), bovine feces (150), biosubstrates (151), and urine (152,153).

Other matrices used are phosphates and carbonates of alkali and alkaline earth elements. Since transition metal ions are strong quenchers of U(VI) fluorescence in crystallophosphors, they are previously separated and the determination is carried out by the method of constant addition.

Other Matrices

The determination of rare earths in different materials using the luminescence of other phosphors is shown in Table 4.1.

4.3.2. Non-Rare Earth Elements

A variety of inorganic crystals becomes luminescent when certain transition metal ions are dissolved in them. The electronic transitions involved are intercombination transitions within the nd shell. The mechanism of this luminescence is explained in terms of the band theory of solids. Analytical applications were reported for Cu, Mn, Re, and Rh.

There are also several matrices that are activated by ions of the elements of the groups IIIB–VB. These ions Tl(I), Pb(II), Sn(II), Sb(III), and

Table 4.1. Other Matrices Used in the Determination of Rare Earth Elements

Matrix	Preparation	Element	Application Interval or Detection Limit	Determination in	Ref.
$TbPO_4$	$(NH_4)_2HPO_4$ + $Na_4P_2O_7$ + Tb_4O_7	Sm	10^{-4}–0.37%	Tb_4O_7	154
Y_2O_2S	—	Eu	—	Y_2O_2S	155
SiO_2–$NaSO_4$	SiO_2 + $NaSO_4$ (100:10)	Gd U	$5 \times 10^{-6}\%$ $3 \times 10^{-6}\%$	SiO_2 SiO_2	156 157
$CaWO_4$	$CaWO_4$ + 1% Gd	Sm	25–50 µg	Gd and Nd oxides	158 159
$CaWO_4$–Eu_2O_3	$CaWO_4$ + Eu_2O_3 (100:2)	Sm	$2 \times 10^{-5}\%$	Eu_2O_3	160
$PbSO_4 \cdot Ce$	$PbSO_4$ + CeO_2 + LiF (100:2:0.2)	Sm	—	CeO_2	161
$PbMO_4$	$PbMO_4$	Np	5 ppb	—	162
$Cs_2NaTbCl_6$	Tb_4O_7 + CsCl + NaCl	Eu	5×10^{-5}–$1 \times 10^{-4}\%$	Tb_4O_7	163
$SrCl_2$–NaCl	$SrCl_2$ + NaCl	Eu	0.1 ppm	Minerals and industrial products	164

Bi(III)) are known as mercurylike ions because they have the electronic shell of a mercury ion, $1s^2 \ldots np^6nd^{10}(n + 1)s^2$. The luminescence center is a mercury-like ion whose energy levels are deformed by interactions with the environment.

In Table 4.2 are shown the characteristics of the methods of determination proposed for these ions. An approach (183) using selective excitation of probe ion luminescence may be utilized for nonfluorescent ions such as PO_4^{3-}, which perturb the emission of $BaSO_4 \cdot Eu$ crystallophosphors. Although the limit of detection of PO_4^{3-} was not very good, the approach may yet prove useful with further optimization of the lattice and for application to other ions.

4.4. USE OF INORGANIC REAGENTS

Some inorganic compounds produce characteristic luminescent solutions with particular metal ions that allow their determination. The most widely used inorganic reagents are HCl and HBr, and the ions determined usually are those of the IIIB–VB groups of the periodic table. Among them, Tl(I), Sn(II), Pb(II), As(III), Sb(III), Bi(III), Se(IV), Te(IV) are characterized by an external electronic shell of the type $nd^{10}(n + 1)s^2$ (mercurylike ions). Halide complexes of Tl(I), Sn(II), and Pb(II) luminesce weakly at room temperature and strongly at low temperature. Luminescence of solutions of halide complexes of As(III), Sb(III), Bi(III), Se(IV), and Te(IV) is observed only at low temperature. Luminescence studies of halide complexes of these ions showed that the nature of the emission is similar to that of their alkali halide crystallophosphors. Under the influence of the environment the energy levels of the metal ions (luminescence center) in halide complexes are deformed, becoming closer together than those in crystals. The result is a small bathochromic shift of absorption spectra and, to a larger extent, luminescence. With excitation in the 248–253-nm range, luminescence occurs only in chloride complexes of Tl(I), in the 265–270-nm range in chloride complexes of Pb(II) and Te(IV) and at 313 nm in chloride complexes of Bi(III) and Te(IV). So mercurylike ions are determined without preliminary separation by changing the conditions for excitation (184–188).

On the other hand, Ga(III), In(III), Tl(III), Ge(IV), Sn(IV), As(V), and Sb(V) are characterized by an external electronic shell of the type nd^{10}. This is the main cause of the differences between the spectral properties of solutions with these ions as impurities. The luminescence of chloro complexes of Ga(III) is excited in a broad spectral region, 200–400 nm. For Ge(IV) and In(III) it is excited essentially in the bands with maxima

Table 4.2. Methods of Determination of Non-Rare Earth Elements by the Use of Crystallophosphors

Element	Matrix	Application Interval or Detection Limit	Determination in	Ref.
Mn	Li–Mg tungstate $Sb_2O_4 + B_2O_3$	1×10^{-8}–$1 \times 10^{-6}\%$ 0.01 ng–1 μg	Water and HCl —	165, 166 167
Cu	Ag + SZn	0.01–500 ppm	K_2SiO_3	169
Re	Cs_2SnCl_6	0.1–5000 ppm	Aqueous effluents and technological solutions	169
Rh	Al_2O_3	$10^{-5}\%$	Preconcentrated samples after coprecipitation with $Al(OH)_3$	170
Sb	CaO	1×10^{-6}–$1 \times 10^{-4}\%$	H_2SO_4 HF	171
Sn	KI	0.01 μg	Alloys HCl	172
Tl	NaI	0.05 ppm	—	173
Pb	Na Cl CaO	20 ppb 5–2000 ppb	NaCl pellets Leaves and synthetic blood	174 175–178
Bi	CaO	0.1–30 μg	$NaPO_2$, granite and sulfide ores	175, 177, 179–182

385

Table 4.3. Determination of Inorganic Ions Using HCl or HBr as reagents

Ion	Conditions	$\lambda_{ex}/\lambda_{em}$	Detection Limit (ppm)	Ref.
As(III)	7.6 M HCl	312/475	0.15	193
	7.6 M HBr	350/690	0.008	193
Sb(III)	HCl	314/625	0.06	186, 194
	7.0 M HCl	306/582	0.2	195
	HBr	360/640	0.001	194
	6.0 M HBr	360/586	0.01	196
	HBr ($-70°C$)	UFS-2 filter/555	$5 \times 10^{-7}\%$	197
	HBr	UFS-filter/visual	0.1	198
Bi(III)	4.5 M HCl	340/420	0.002	186, 194, 199
	6.0 M HCl	330/410	0.002	195
	HCl	UFS filter/visual	1.0	198
	0.5 M HCl (extraction with TBP)	313/480	0.04	200
	0.6 M HCl (extraction with BuOH)	313/460	0.1	200
	HBr	380/487	0.01	194
	0.5 M HBr (extraction with TBP)	313/510	0.07	200
Se(IV)	HCl	305, 330/390	0.4	194
	HBr	352/550	0.8	194
Te(IV)	9–10 M HCl	380/586	0.02	195, 201, 202
	4.5 M HCl	270, 380/640	0.1	199
	7.6 M HCl	262, 313/705	0.001	192
	11.0 M HCl (extraction with octylalcohol)	UFS-3 filter/660	0.03	200
	7.6 M HBr	UFS-3 filter/695	0.01	200
	2–4 M HBr (extraction with TBP)	UFS-3 filter/690	0.001	200
	2–4 M HBr (extraction with hexylalcohol)	UFS-3 filter/695	0.005	200
	2–4 M HBr (extraction with ethylacetate)	UFS-3 filter/695	0.005	200

Species	Medium	Wavelength	Value	Reference
Sn(II)	7.6 M HCl	220, 268/585	1	192
	7.6 M HBr	265, 304/574, 612	1	192
Pb(II)	7.0 M HCl	276/390	0.002	186, 195
	4.5 M HCl	270/495	0.1	199
	4.0 M HCl ($-70°C$)	272/490	0.005	203–205
	HCl ($-70°C$)	UFS filter/visual	1	198
	8.0 M HCl	272/423	0.01	206
	HCl–KCl sat. (room temp)	270/480	0.1	207
	4.0 M HCl–2 M NaCl	272/410	0.01	208
	8.0 M HBr	302/424	0.01	206
Tl(I)	7.0 M HCl (room temp)	—	0.15	209
	7.0 M HCl	—	0.02	210
	10.0 M HCl	256/380	0.2	195
	8.0 M HCl	242/396	0.05	206
	3 M HCl–NaCl sat. (room temp.)	240/420	0.01	211, 212
	3.3 M HCl—0.8 M KCl (room temp.)	250/430	0.01	213, 214
	8.0 M HBr	223, 263/428	0.05	206
	7.0 M HBr		0.025	210
As(V)	7.6 M HCl	—	37	193
	7.6 M HBr	—	3.7	193
Sb(V)	HBr	—/640	$3 \times 10^{-7}\%$	215
Ge(IV)	7.6 M HCl	275/500, 550	70	192
	7.6 M HBr	—	70	192
Ga(III)	7.6 M HCl	200–400/380, 475	1.4	192
	7.6 M HBr	230, 350/—	1.4	192
In(III)	7.6 M HCl	365/385, 440	11	192
	7.6 M HBr	—	11	192
Tl(III)	3 M HCl–NaCl sat. (room temp.)	—	0.01	211

387

at 275 nm and 265 nm, respectively, and for Sn(IV) at 272 and 302 nm. The solutions activated with ions of this second group are characterized by the recombination mechanism of radiation and thus luminesce only at high concentrations of the activator. According to this mechanism, the dependence of the intensity of emission on concentration of Ga(III), In(III), Tl(III), Ge(IV), Sn(IV), As(V), and Sb(V) has a complex nonlinear character (192).

Some reviews (79,189–191) have been published on the determination of these ions with hydrochloric or hydrobromic acids at liquid-nitrogen temperature. In Table 4.3 are summarized the methods of determination using these acids as reagents at $-196°C$, except in a few cases in which other temperatures are indicated. These methods were developed for the analysis of highly pure substances and semiconducting materials.

Pb(II) also can be determined using the intense fluorescence of $(PbCl_4)^{2-}$ on cellulose (216). A fluorescent ring-oven technique was developed based on the selective solubility of lead in ammonium acetate. This approach provided a simple procedure adaptable for air pollution studies. The analytical range was 0.1–5 μg.

Also, by using the formation of chloride complexes, the determination of other ions is possible. Thus, Ce(III) was determined in $7 M$ HCl solution at $-196°C$, in the range of concentrations between $10^{-7}–10^{-6} M$. The excitation wavelength was 252 nm and the emission was measured at 348 nm (195).

Sulfide–sulfur (217) was determined in rare earth materials by dissolving the sample in a solution of Sb(V) in $6 M$ HCl. An aliquot was placed in a platinum cell and frozen at $-196°C$ before measurement of its luminescence at 620 nm. A calibration graph (for 0–40 ppm of sulfide–sulfur in aqueous 1% NaOH) was used to convert intensities into concentrations. None of the rare earth elements present interfered.

As little as 2 ppm of Cl^- can be determined by treating the test solution with Pb acetate to contain 10 ppm of Pb. The method is based on the formation of $PbCl_6^{4-}$ in excess of Pb^{2+}. The excitation spectrum of $PbCl_6^{4-}$ is recorded in the range of 250–300 nm by measuring the intensity of luminescence at 410 nm. The height of the excitation band at 272 nm is related to the concentration of Cl^- (208). An interesting ion to consider is Cr(III). This ion presents an electronic configuration $1s^2 \ldots 3p^63d^3$. Depending on the ligand field strengths in their octahedral complexes, the lower excited states may be different, and fluorescence or phosphorescence can be observed. However, the phosphorimetric method of determining Cr(III) is better than the fluorimetric method because the structure of phosphorescence spectra is more regular than that of the fluorescence spectra. So chromium may be determined (218,219) by measurement of the luminescence of its thiocyanate complex at 764 nm, which is extracted into tributylphosphate from pH 1.0–3.0 and cooled to liquid nitrogen temperature. The detection limit of the method is 0.1 ppb, and it has been applied to the determination of Cr(III) in sewage (220) and in silica (after extracting into acetone) (221).

Other inorganic reagents used in the determination of ions in solution are sodium tungstate, tetracyanoplatinate, cerium, and uranyl sulfate.

Sodium tungstate (222) acts as a specific reagent for enhancing the fluorescence intensity of samarium, europium, terbium, and dysprosium, exciting at 265–270 nm at pH = 9 and 20°C. The maxima emission (nm) and ranges of application (ppm) are 595, 590, 545, 480, and 0.5–1, 0.005–1, 0.1–1, and 0.01–1, respectively. Working at pH 4.5 (52) the methods are more sensitive, allowing the determination of 90, 1, 3, and 7 ppb of Sm, Eu, Tb, and Dy. In these conditions Eu, Tb, and Dy form 1:1 complexes with $H_3W_6O_{21}^{3-}$.

The tetracyanoplatinate(II) ion forms insoluble fluorescent compounds with many metal ions. Fluorescent precipitates useful for detection of metal ions were obtained with Y(III), Zr(IV), Ag(I), Zn(II), Cd(II), Hg(I), Hg(II), Al(III), Pb(II), La(III), and Th(IV). Limits of detection ranged from 5 to 200 ppm (223).

By utilizing the osmium catalyzed redox reaction between cerium(IV) and arsenic(III), microgram amounts of arsenic (0.075–0.375 ppm) may be determined by measurement of the fluorescence of the cerium(III) produced. The principle may be applied to the determination of several other ions. So, Fe(II), (0.05–0.28 ppm), oxalate (0.088–0.44 ppm), osmium(VIII) (0.5–2 ppb), and iodide (0.006–0.025 ppm) may be determined by their catalytic action (224). Similarly, iridium catalyzes the reduction of cerium(IV) to cerium(III) by arsenic(III) or antimony(III). The increase in fluorescence with arsenic(III) is directly proportional to the concentration of iridium for 4–32 ppb at 20°C or 0.4–3.2 ppb at 35°C. There are numerous interferences with the latter determination (225).

Finally, the quenching of the fluorescence of uranyl sulfate in phosphoric acid by thallous ion allows the determination of this ion between 0.1 and 80 ppm. There is interference by halides, silver, ferrous ion, mercurous ion, and stannous ion (226).

4.5. USE OF ORGANIC REAGENTS

The use of organic reagents for the luminescence determination of inorganic substances is based on several kinds of reactions—binary or ternary complex formation, substitution, redox, catalytic, and enzymatic. From this point of view we will study the determination methods in the corresponding sections. Such methods can be based on the appearance of fluorescence or on the quenching of the fluorescence of an organic reagent in the presence of the analyte. The analytical signal is measured generally by using equilibrium methods and sometimes by kinetics methods.

In Table 4.4 data related to the fluorimetric determination of inorganic substances using organic reagents are compiled. Phosphorescence methods are not included on the table but are summarized in Section 4.5.9. Only those reagents proposed for the quantitative determination of these substances are listed. Some reagents, mentioned in the bibliography as fluorescent reagents for various ions, have not been considered, because there has not yet been a method of determination proposed with them. The table includes the substance determined, the reagent used, the kind of reaction on which the method is based, the interval of application or the detection limit reported, and the related references for each method. When several methods for determination of a substance using the same reagent are proposed, we give the data of the more sensitive method. In some cases, in which extraction accompanies the fluorimetric method, it is also possible to carry out the determination without the use of the extraction. The original papers should be consulted for information about procedure, applicability, and potential interferences. The values given for the detection limits should only be considered as indicators of the potential usefulness of each reagent, because of the lack of a generally agreed-on system for its estimation.

A statistical study of the proposed luminescent methods shows that around 89% are based on an appearance of fluorescence, while around 11% are based on quenching reactions. Methods based on quenching of fluorescence are less selective than those using the appearance of fluorescence. Attending to the fundamental reaction, we have to remark that most of the proposed methods (75%) are based on the formation of a fluorescent complex (including ion-association complexes); 71% of them are binary systems and 29% are ternary systems. Among the other kinds of reaction used, 4% correspond to substitution reactions, 4% to reaction in which the ion to be determined produces a catalytic effect, 3.5% to enzymatic reactions, and 6% to redox reactions. A large number of methods can be considered as equilibrium methods, 93% against only 7% based on kinetic measurements (analysis carried out before the reaction reaches equilibrium). The combination of luminescence methods with the extraction includes 30% of the total.

Most of the methods have been proposed for the determination of cations. Among them, gallium, aluminum, and beryllium are the ones that have been identified by the greatest number of methods. Zinc, terbium,

Table 4.4. Fluorimetric Determination of Inorganic Substances Using Organic Reagents

Inorganic Substance	Reagent	Method	Application Interval or Detection Limit (ppm)	Ref.
Ag	Butylrhodamine B + Br^-	TC (E)	10^{-1}–1	230,231
	Eosin	C(Q)	3.7×10^{-1}	232
	Eosin + pyridine	TC(E)	8×10^{-2}	233
	Fluorescein	C(Q)	—	234
	8-Hydroxyquinoline-5-sulfonic acid + per-sulfate	Cat, Cat(K)	1.2×10^{-2}–5	235,236
	8-Hydroxybenzaldehydethiosemicarbazone + H_2O_2 + Fe (III)	Cat(K)	5×10^{-2}–10^{-1}	234a
	2,3-Naphthotriazole	C(Q)	2.5×10^{-2}–10^{-1}	237
	1,10-Phenanthroline + eosin	TC(Q)	4×10^{-3}–4×10^{-2}	238
		TC(E)	10^{-2}–10^{-1}	239–241
	1,10-Phenanthroline + 3,4,5,6-tetrachloro-fluorescein	TC(Q)	5×10^{-2}–6×10^{-1}	242
	Phenosafranine + Br^-	TC(E)	5×10^{-3}–5×10^{-1}	242a
	Phloxin	C(Q)	—	234
	Rhodamine 6G + Br^-	TC(E)	5×10^{-2}–5×10^{-1}	243
Al	Alizarin Red S	C	8×10^{-2}–8×10^{-1}	244
	1-Amino-4-hydroxyanthraquinone	C	—	245
	3-Amino-5-sulfosalicylic acid	C	5×10^{-3}	246
	2,2'-Dihydroxyazobenzene	C(E)	5×10^{-4}–5×10^{-1}	247
	2,4-Dihydroxybenzylidenehydrazone isoni-cotinaldehyde	C	10^{-1}	248,249
	2,4-Dihydroxybenzaldehyde semicarbazone	C	8×10^{-5}–8×10^{-2}	250,251
	2,4-Dioxo-4-(4-hydroxy-6-methyl-2-pyrone-3-yl) butyric acid ethyl ester	C(E)	5×10^{-1}–1.5	252

391

Table 4.4. *(Continued)*

Inorganic Substance	Reagent	Method	Application Interval or Detection Limit (ppm)	Ref.
	Eriochrome Blue Black B	C	$0-4 \times 10^{-1}$	253, 254
	Eriochrome Red B	C	—	255
	Flazo Orange	C	10^{-3}	256
	4-Hydroxy-3-(2-hydroxy-4-methoxypheny-lazo) benzenesulfonic acid	C	$2 \times 10^{-3}-8 \times 10^{-2}$	257
	2-Hydroxy-1-naphthaldehydebenzoyl hydrazone	C	$10^{-1}-1$	258
	4-Hydroxybenzoic acid-2-hydroxy-1-na-phthylmethylene hydrazide	C	2×10^{-4}	259, 260
	2-Hydroxy-3-naphthoic acid	C	$2 \times 10^{-3}-2.5 \times 10^{-2}$	261, 262
	4-(2-Hydroxyphenylazo)resorcinol	C	$2 \times 10^{-3}-4 \times 10^{-2}$	257
	3-Hydroxypyridine-2-aldehyde-2-pyridylhy-drazone	C	$2 \times 10^{-3}-3 \times 10^{-1}$	263
	8-Hydroxyquinoline	C(E)	5×10^{-3}	264–272
	8-Hydroxyquinoline-5-sulfonic acid	C(K)	$4 \times 10^{-4}-10^{-2}$	273
	8-Hydroxyquinoline-5-sulfonic acid + methyltrioctyl ammonium	TC(E)	$0.1-2 \ \mu g$	274
	3-Hydroxy-4′,5,7-trimethoxyflavone	C	$2 \times 10^{-2}-10^{-1}$	275
	Lumogallion	C	$3 \times 10^{-2}-3.2 \times 10^{-1}$	276–280
	Lumomagneson	C	$10^{-3}-1.2 \times 10^{-1}$	281
	4-Methoxybenzoic acid-2-hydroxy-1-na-phthylmethylene hydrazide	C	$4 \times 10^{-4}-2 \times 10^{-1}$	282
	4-(Methylumbelliferone)	C	$2 \times 10^{-1}-10$	283
	6-(4-Methylsalicylideneamino)-m-cresol	C	$2 \times 10^{-4}-1.2 \times 10^{-1}$	284
	Mordant Blue 9	C	5×10^{-4}	285
	Mordant Blue 31	C	$4 \times 10^{-3}-1.2 \times 10^{-1}$	286

Reagent	Method	Sensitivity	Ref.
Morin	C	2.5×10^{-4}–5×10^{-3}	287–290
Nitromagneson	C	10^{-2}–2×10^{-1}	291
Pontachrome BBR	C	4×10^{-3}–3.6×10^{-1}	292–295
Pontachrome VSW	C	2×10^{-1}–2	292, 296
Purpurin sulfonate	C	2×10^{-3}–10^{-1}	297
1-(2-Pyridylazo)-2-naphthol	C	10^{-3}–3×10^{-1}	298, 299
Quercetin	C	5×10^{-2}	300
Quercimeritrin	C	10^{-3}–2×10^{-2}	301
2-Quinizarinsulfonic acid	C	1.1×10^{-2}–5.4×10^{-2}	302
2-(Resorcylideneamino)phenol	C	4×10^{-4}–8×10^{-1}	303
Salicylaldehyde	C	2.7×10^{-1}–2.7	304
Salicylaldehyde formylhydrazone	C	2.8×10^{-2}–8.8×10^{-1}	305
Salicylidene-o-aminophenol	C(E)	2.7×10^{-4}–2.7×10^{-7}	306–313
o-(Salicylideneamino)benzenearsonic acid	C	10^{-1}	314
o-(Salicylideneamino-2-hydroxybenzenesulfonic acid	C	4×10^{-4}–2×10^{-1}	250
o-(Salicylideneamino)-3-hydroxyfluorene	C	2.7×10^{-4}–10^{-1}	315-317
o-(Salicylideneamino)-5-methoxyphenol	C		318
N-Salicylidene-2-hydroxy-4-carboxyaniline	C	2×10^{-4}–1.2×10^{-1}	319
N-salicylidene-2-hydroxy-4-chloroaniline	C	2×10^{-4}–1.2×10^{-1}	319
N-salicylidene-2-hydroxy-5-sulfoaniline	C	8×10^{-5}–8×10^{-2}	319
Salicyloylhydrazone pyridine-2-aldehyde	C	4×10^{-3}–8×10^{-2}	320
Sulfophenylazochromotropic acid	C	2×10^{-2}	321
Sulfonaphtholazoresorcinol	C	5×10^{-4}–1.6×10^{-1}	322
Superchrome Garnet Y	C	8×10^{-4}–1.6×10^{-1}	323
As			
Butylrhodamine B + Cl⁻	TC(E)	—	324
D-glyceraldehyde-3-phosphate + glyceraldehyde-3-phosphate dehydrogenase	EnZ(K)	2×10^{-2}–2	325, 226
Rhodamine B + Cl⁻	TC(E)	6×10^{-2}–3×10^{-1}	327

Table 4.4. *(Continued)*

Inorganic Substance	Reagent	Method	Application Interval or Detection Limit (ppm)	Ref.
Au	Acridine Orange + Cl⁻	TC(E)	5×10^{-3}	328
	Acridine Yellow + Cl⁻	TC(E)	10^{-2}–4×10^{-1}	329
	Acriflavine + Cl⁻	TC(E)	10^{-2}–5×10^{-1}	330
	Bipyridylglioxaldiphenylhydrazone	Cat	5×10^{-1}–2	331
	Butylrhodamine B + Cl⁻	TC(E)	2×10^{-2}–2×10^{-1}	332, 333
	Butylrhodamine B + Cl⁻ (Crystal Violet)	TC(E)	2×10^{-3}–3.3×10^{-1}	334
	p-Dimethylaminobenzilidenerhodamine + Cl⁻	TC(E)	7×10^{-2}–7×10^{-1}	335, 336
	2,2'-Dipyridylketone azine	Cat(K)	5×10^{-2}–2.5×10^{-1}	337
	5-Hydroxy-2-hydroxymethyl-1,4-pyrone	C	10^{-2}–1	338, 339
	2-Phenylbenzo[8,9]quinolizino[4,4,6,7-fed]phenanthridinylium perchlorate	TC(E)	4×10^{-2}–3.5×10^{-1}	340
	Rhodamine B + Br⁻	TC(E)	5×10^{-2}–5	341
	Rhodamine B + Cl⁻	TC(E)	1.7×10^{-2}–1.7×10^{-1}	342
B	Acetylsalicylic acid	C	10^{-2}	343
	Alizarin Red S	C	1	344
	Benzoin	C	4×10^{-2}–2×10^{-1}	345–357
	Butylrhodamine B + F⁻	TC(E)	10^{-3}–5×10^{-2}	358, 359
	Carminic acid	C	2.5	360
	Chromotropic acid	C	10^{-2}–5×10^{-1}	361
	Dibenzoylmethane	C(E)	5×10^{-4}–5×10^{-3}	362–364
	2,4-Dihydroxybenzophenone	C	3.6×10^{-3}–1.8×10^{-1}	365-367
	1,4-Dihydroxyanthraquinone	C	4×10^{-2}–8×10^{-2}	368
	Hydroxy-2-methoxy-4-chloro-4'-benzophenone	C	3.6×10^{-4}–3.6×10^{-2}	369–377

	Morin + oxalic acid	TC	2×10^{-4}–8×10^{-4}	378–384
	Phenylfluorone	C	1	385–387
	Protocatechuic acid	C	5×10^{-1}–10 μg	388
	Quercetin + oxalic acid	TC	3×10^{-4}	383, 389, 390
	Quinalizarin	C	10^{-2}–2×10^{-1}	368
	Resacetophenone	C	10^{-1}–2.8×10^{-1}	366, 391–394
	Rhodamine 6 G + salicylic acid	TC(E)	10^{-4}–10^{-1}	395, 396
	Salicylic acid	C	10^{-3}	397, 398
	Thoron I	C	6×10^{-2}–7×10^{-1}	399, 400
Ba	Calcein	C	80	401
Be	Acetylacetonate + Trylon B	TC(E)	5×10^{-1}	402
	1-Amino-4-hydroxyanthraquinone	C	2×10^{-1}	403–405
	3-Amino-5-sulfosalicylic acid	C	5×10^{-3}–1×10^{-1}	406
	Arsenazo I	C	4×10^{-3}–3×10^{-2}	407
	Benzoin	C	1×10^{-1}	347
	Chlorophosphonazo I	C	4×10^{-3}–3×10^{-2}	407
	1(-Dicarboxymethylaminomethyl)-2-hydroxy-3-naphthoic acid	C	3.6×10^{-3}–7.2×10^{-2}	408
	5,8-Dichloroquinizarin	C	0.1–100 μg	405
	Dinaphthoylmethane	C	2×10^{-2}–2×10^{-1}	409
	2,4-Dioxo-4-(4-hydroxy-6-methyl-2-pyrone-3-yl) butyric acid ethyl ester	C(E)	5×10^{-1}–2	252
	2-Ethyl-5-hydroxy-7-methoxyisoflavone	C(E)	4×10^{-4}–1.2×10^{-2}	410
	2-Ethyl-3-methyl-5-hydroxychromone	C(E)	1×10^{-3}–2.5×10^{-2}	411
	1-Hydroxy-2-carboxyanthraquinone	C	3×10^{-2}–1.3×10^{-1}	412
	4-Hydroxy-3-(2-hydroxysalicylideneamino)benzene sulfonic acid	C(E)	5	412a
	2-Hydroxy-4-methylaniline-N-2-hydroxy-5-methylbenzylidine	C	1×10^{-3}	413

395

Table 4.4. *(Continued)*

Inorganic Substance	Reagent	Method	Application Interval or Detection Limit (ppm)	Ref.
	2-Hydroxy-5-methylaniline-N-2-hydroxy-5-methylbenzylidene	C	4×10^{-3}–2×10^{-1}	413
	o-(β-Hydroxynaphthylideneimino)benzene arsonic acid	C	4×10^{-3}–2×10^{-1}	414
	2-Hydroxy-3-naphthoic acid	C(E)	1.8×10^{-4}–1.8×10^{-3}	262, 415–419
	2-(2′-Hydroxyphenyl)benzothiazole	C	1×10^{-1}	420
	8-Hydroxyquinaldine	C(E)	8×10^{-3}–8×10^{-2}	421
	6-Methyl-1-hydroxyxanthrone	C(E)	1×10^{-3}	422
	6-(4-Methylsalicylideneamino)-m-cresol	C	8×10^{-4}–2.8×10^{-1}	284
	Morin	C	4×10^{-4}–1.6×10^{-3}	419, 423–433
	Naphthazarin	C	0.1–100 μg	405
	o-Pyridinophenol	C	1.6×10^{-3}–5.4×10^{-1}	434
	Quinalizarin	C	0.1–100 μg	405
	Quinizarin	C	2×10^{-1}	404, 405, 423, 435, 436
	2-Quinizarin sulfonic acid	C	1×10^{-3}–7×10^{-3}	437
	Resorcylidene-o-aminophenylarsonic acid	C	4×10^{-4}	438
	Resorcylidene-cysteine	C	8×10^{-3}	439
	o-(Salicylideneamino)benzene arsonic acid	C(E)	9×10^{-4}	440, 441
	o-(Salicylideneamino)-3,5-dimethylbenzene arsonic acid	C(E)	1×10^{-3}	442
	o-(Salicylideneamino)-2-hydroxybenzene sulfonic acid	C	4×10^{-4}–2×10^{-1}	250
	o-(Salicylideneamino)benzoic acid	C	2×10^{-3}–2×10^{-1}	442a

Element	Reagent/System	Method	Range	Ref.
	o-(Salicylideneamino)-5-methoxyphenol	C	–	318
	N-salicylidene-2-hydroxy-5-sulfoaniline	C	4×10^{-3}–2×10^{-1}	413
	2-(3,4,5-Trihydroxybenzylideneamino)benzene arsonic acid	C	9.7×10^{-5}	443
	2-(2,4,6-Trihydroxybenzylidene amino)benzene arsonic acid	C	7×10^{-5}	443
	Tetracycline + 5,5-diethyl-2-thiobarbituric acid	TC	1×10^{-1}–3×10^{-1}	444
Bi	Umbelliferone phosphate + phosphatase	Enz(K)	10^{-2}–3×10^{-1}	445
Br$^-$	Umbelliferone phosphate + phosphatase	Enz(K)	1–70	445
	Fluorescein + H_2O_2	R(Q)	2×10^{-3}	446, 447
	10-(3-Sulfonatopropyl)-acridinium	CF	6.5×10^{-1}–1.5	448
BrO$_3^-$	Benzyl-2-pyridylketone-2-pyridylhydrazone	R(K)	2.6×10^{-1}–1.3	449
Ca	1,5-Bis(dicarboxymethylaminomethyl)-2,6-dihydroxynaphthalene	C	10^{-4}–2×10^{-2}	450
	Calcein	C	1.6×10^{-2}–1.4×10^{-1}	401, 451–466
	1,4-Diaminoanthraquinone	C	1.5×10^{-1}–4×10^{-1}	467
	1-Dicarboxymethylaminomethyl-2-hydroxy-3-naphthoic acid	C	–	468
	1,8-Dihydroxyanthraquinone	C	5×10^{-2}–5×10^{-1}	469
	5,7-Dinitro-8-hydroxyquinoline + rhodamine B	TC(E)	5×10^{-2}	470
	3-Hydroxy-2-napthoic acid	C	10^{-2}	471
	8-Hydroxyquinoline	C	1	472, 473
	8-Hydroxyquinaldehyde-8-quinolylhydrazone	C	10^{-3}	474–480
	Isocein	C	—	481
	N-(4-Methylumbelliferone-8-ylmethyl)-1,10-diaza-18-crown-6	C(E)	—	482

Table 4.4. (*Continued*)

Inorganic Substance	Reagent	Method	Application Interval or Detection Limit (ppm)	Ref.
Cd	Tb(III)-Pyridine-2,6-dicarboxylic acid	S(Q)	10^{-4}	483
	2,2'-Azodiphenol	C	—	483a
	Benzyl-2-pyridylketone-2-pyridylhydrazone	C	1.1×10^{-2}–5.5×10^{-1}	484
	3,5'-Bis-(dicarboxymethylaminomethyl)-4,4'-dihydroxy-trans-stilbene	C	2×10^{-2}–1	485
	Calcein	C	2×10^{-3}–5.6×10^{-1}	486
	2-(3,5-Dimethylpyrazol-1-yl)-8-hydroxyquinoline	C(E)	5×10^{-2}–5×10^{-1}	487
	2-(2-Hydroxy-1-naphthyl)-dithiocarbazic acid 4-chlorobenzyl ester	C(Q)	10^{-1}–1.1	488
	2-(2'-Hydroxyphenyl)benzoxazole	C	2–40	489
	8-Quinolyl dihydrogen phosphate	C	1.1×10^{-1}–1.1	490
	8-Hydroxyquinoline-5-sulfonic acid	C	10^{-3}–2×10^{-1}	491, 492
	8-Hydroxyquinoline-5-sulfonic acid + methyltrioctylammonium	TC(E)	1–10µg	274
	1,10-Phenanthroline + dibromofluorescein	TC(E)	10^{-2}–2×10^{-1}	493
	1,10-Phenanthroline + eosin	TC(E)	2.5×10^{-2}–6×10^{-1}	494
	1,10-Phenanthroline + erythrosin	TC(E)	3×10^{-2}–8×10^{-1}	494
	1,10-Phenanthroline + iodoeosin	TC(E)	2.5×10^{-2}–8×10^{-1}	494
	8-Quinolinethiol	C(E)	5×10^{-2}–1.5	495
	3-Salicylidenedithiocarbazoic acid methyl ester	C	0–7	496
	8-(Toluene-p-sulfonamide)-quinoline	C(E)	4.5×10^{-1}–45	497, 498

Ce	1-Amino-4-hydroxyanthraquinone.	R(K)	$10^{-1}-9 \times 10^{-1}$	499
	4,8-Diamino-1,5-dihydroxyanthraquinone-2,6-disulfonate	R(K)	$2 \times 10^{-2}-3.7 \times 10^{-1}$	500
Cl⁻	8-Hydroxyquinoline-5-sulfonic acid	R	$10^{-2}-3$	501
	Sulfonaphtholazoresorcinol	C	$4 \times 10^{-2}-5$	502
	6-Methoxy-1-(3-sulfonatopropyl)-quinolinium	CF	$4.5 \times 10^{-1}-2.3$	448
	2-(5-Nitro-2-furyl)benzothiazole	CF	$10^{-1}-10$	503
	Sodium fluorescein + Ag(I)	Q	$10^{-2}-5 \times 10^{-2}$	504
ClO₄⁻	Amiloride hydrochloride	C(E)	$3.7 \times 10^{-2}-2.5$	505
	Rhodamine 6G	C(E)	$4 \times 10^{-2}-1$	506
CN⁻	p-Benzoquinone	R	$2 \times 10^{-1}-50$	507, 508
	Chloramine T + nicotinamide	R	$3 \times 10^{-1}-6$	509
	Chloramine T + pyridine barbituric acid	R	$10^{-3}-6.5 \times 10^{-2}$	510
	Cu(II)-calcein	S	$10^{2}-2.5 \times 10^{-1}$	511
	Cu(II)-leucofluorescein	R	$0-10^{-1}$	511
	Cu(II)-2-(o-hydroxyphenyl)benzoxazole	S	$0-2 \times 10^{-1}$	512
	Di(acetoxymercuri)fluorescein + KI	S(Q)	$1.4 \times 10^{-1}-3.4 \times 10^{-1}$	513
	Homovanillic acid + H₂O₂ + peroxidase	Enz(K)	$1-100$	514
	Indoxylacetate + hyaluronidase	Enz(K)	$10^{-1}-4$	515
	Pd(II)-8-hydroxyquinoline-5-sulfonic acid	S	$2 \times 10^{-2}-8 \times 10^{-1}$	516
	Pd(II)-piazselenol	S(E)	$3.3 \times 10^{-1}-3.3$	517
	Pyridoxal	Cat	2×10^{-2}	518–520
	Quinone monoxime benzensulfonate ester	R	2×10^{-2}	521
	Tetra(acetoxymercuri)fluorescein	S(Q)	$8.6 \times 10^{-2}-8.6 \times 10^{-1}$	522
C₂O₄²⁻	Resorcinol	R	8×10^{-2}	523
	Zr(IV) + flavonol	S(Q)	$5 \times 10^{-3}-10^{-1}$	524

Table 4.4. (*Continued*)

Inorganic Substance	Reagent	Method	Application Interval or Detection Limit (ppm)	Ref.
Co	Al(III)-Pontachrome BBR	C(Q)	10^{-3}	525, 526
	Al(III)-Superchrome Blue Black Extra	C(Q)	10^{-3}–2×10^{-2}	527
	Benzyl-2-pyridilketone-2-pyridilhydrazone + BrO_3^-	R,C(K)	2×10^{-2}–2.6×10^{-1}	449
	2,4-Dichlorobenzyltriphenyl phosphonium + SCN^-	TC(Q)(E)	0–4	528
	Dipyridylglioxal hydrazone	Cat(K)	8×10^{-2}–2.4×10^{-1}	529
	2-[2-Hydroxy-1-naphthyl]-dithiocarbazic acid 4-chlorobenzyl ester	C(Q)	10^{-1}–1.1	488
	Homovanillic acid + H_2O_2 + peroxidase	Enz(K)	5×10^{-1}–20	514
	Phenyl-2-pyridylketone hydrazone	Cat.	10^{-2}–8×10^{-2}	529
		Cat(K)	7×10^{-1}–2.4	529
	1,1-Phenanthroline + eosin	TC(E)	10^{-2}–10^{-1}	241
	Protiptylinium + SCN^-	TC(Q)(E)	0–6	530
	Pyridine-2-aldehyde-2-pyridylhydrazone + eosin	TC(E)	8×10^{-3}–4×10^{-1}	531
	Pyridine-2-aldehyde-2-pyridylhydrazone + BrO_3^-	R,C(K)	2×10^{-2}–1.5×10^{-1}	531a
	1-(2-Pyridylazo)-2-naphthol	C	5.9×10^{-2}	532
	Salicylfluorone + H_2O_2	Cat(K)	10^{-4}–2×10^{-3}	533
Cr	EDTA	C	5×10^{-3}	534
	Ga(III)-Morin	C(Q)	2×10^{-2}–2	535
	(*o*-Hydroxy-phenyl)benzothiazole	R(Q)	10^{-1}–10 μg	536
	Homovanillic acid + H_2O_2 + peroxidase	Enz(K)	10–10^2	514

400

Cu

Safranine T + Cl⁻	TC(E)	10^{-2}–8×10^{-2}	537
Triazinylstilbexone	C(Q)	4×10^{-3}	538
Zn(II)-Salicylaldehydeacetylhydrazone	C(Q)	4×10^{-1}–25 μg	536
ε-ADP	C(Q)	6×10^{-2}–6×10^{-1}	539
Bathocuproine	C(Q)	10^{-2}–10^{-1}	540
Benzamido(p-diethylaminobenzylidene)acetic acid	Cat(K)	2×10^{-4}–4×10^{-4}	541
2,2'-Dipyridylketone hydrazone	Cat	4×10^{-4}–10^{-3}	542, 543
	Cat(K)	10^{-1}–1	544
2,2'-Dipyridylketone azine	Cat(K)	2×10^{-1}–5×10^{-1}	544
Calcein	C(Q)	5×10^{-1}–2.5	545
Homovanillic acid + H₂O₂ + peroxidase	Enz(K)	5–1.2×10^{-2}	514
2-Hydroxybenzaldehyde azine	Cat(K)	10^{-1}–5×10^{-1}	546
N,β-Hydroxyethylanabasine	C(Q)	8×10^{-3}–1	547
2-Hydroxy-1-naphthaldehyde benzoic acid hydrazone	C(Q)	2×10^{-4}–1.1	548
2(2-Hydroxy-1-naphthyl)-dithiocarbazic acid 4-chlorobenzyl ester	C(Q)	10^{-1}–1.1	488
2(2'-Hydroxyphenyl)benzoxazole	C(Q)	10^{-1}	549
N,β-Hydroxypropylanabasine	C(Q)	10^{-3}–5×10^{-1}	547, 550
Indoxylacetate + hyaluronidase	Enz(K)	2×10^{-1}–6	515
Lumocupferron	Cat	10^{-4}	551–556
Morin-Be + H₂O₂	Cat(K)	10^{-3}–10^{-1}	557
1,10-Phenanthroline + 4,5-dibromofluorescein	TC(E)	2×10^{-3}	557a
1,10-Phenanthroline + eosin	TC(E)	10^{-2}–10^{-1}	241
1,10-Phenanthroline + rose bengal extra	TC(E)	10^{-3}–6×10^{-3}	558
Phenyl-2-pryidylketone hydrazone	Cat(K)	1.2×10^{-1}–6×10^{-1}	544
Pseudopurpurine	C	5×10^{-2}–2.5×10^{-1}	559

Table 4.4. *(Continued)*

Inorganic Substance	Reagent	Method	Application Interval or Detection Limit (ppm)	Ref.
	Salicylaldazine	C	5×10^{-2}–1	549
	$\alpha,\beta,\partial,\delta$-Tetraphenylporphine-trisulfonate	C(Q)	6×10^{-3}–6×10^{-2}	560
	Thiamine	C	10^{-1}–30	561, 562
	Thiamine	R(E)	6.5×10^{-3}–6.5×10^{-1}	563
	TRIAP	C	10^{-1}–5×10^{-1}	564
Dy	Antipyrine + salicylate	TC(E)	10^{-2}–5	565
	Bis-(l-phenylmethyl)-5-pyrazolone	C	5×10^{-5}	566
	4,4'-Methylenedi-[3-methyl-l-(2-pyridyl)-pyrazol-5-ol]	C	10^{-1}–10^{-2}%	567
	3-Methyl-1-phenyl-5-pyrazolone or tolyl derivative	C	5×10^{-3}%	568
	3-Methyl-l-(4-sulfophenyl)-5-pyrazolone	C	10^{-1}–1	569, 570
	1,10 Phenanthroline + salicylate	TC(E)	10^{-1}–1	570
	Pyridine-2,6-dicarboxylic acid	C	8×10^{-3}–8×10^3	483
	Rhodamine B + 3,5-diiodosalicylic acid	TC(E)	4×10^{-3}	571
	Tiron + EDTA	TC	10^{-5}–1	572–574
	Tiron + iminodiacetic acid	TC	8×10^{-4}–1.6×10^2	575, 576
Eu	Benzoyltrifluoroacetone + trioctylphosphine oxide	TC(E)	7×10^{-2}	577
	4-Benzoyl-3-methyl-l-phenyl-5-pyrazolone + phosphine oxides or triisobutylphosphate ester	TC(E)	1.5×10^{-2}	578, 579
	Bis-(8-hydroxy-2-quinolyl)methylamine	C	2×10^{-5}	580–582
	Dibenzoylmethane	C	10^{-5}%	583

Compound	Type	Value	Ref.
Dibenzoylmethane + diethylamine	TC	2×10^{-3}	584
Dimethylformamide or acetonitrile	C	$10-1.5 \times 10^3$	585
5,5,6,7,7,7-Heptafluoro-heptan-2,4-dione + 1,10-phenanthroline	TC	0–1	586
Hexafluoroacetylacetone + trioctylphosphine oxide	TC(E)	$3 \times 10^{-4}-1.5$	587
1,1,1,5,5,5-Hexafluoro-2,4-pentanedione	C	$10^{-3}-8 \times 10^{-2}$	588
2-Hydroxy-4-methoxybenzophenone-5-sulfonic acid	C	$2 \times 10^{-2}-4$	589
2-Naphthoyltrifluoroacetone + trioctylphosphine oxide	TC(E)	$2 \times 10^{-5}-10^{-4}$	590–592
1,10-Phenanthroline + 2-phenylcinchonic acid	TC	$10^{-3}-2 \times 10^{-3}$ µg	593, 594
1,10-Phenanthroline + 2-phenylcinchonic acid methyl ester	TC	2	164
1,10-Phenanthroline + 1-naphthoic acid	TC(E)	9×10^{-5}	595
1,10-Phenanthroline + salicylate	TC(E)	$5 \times 10^{-3}-2.5$	596
Propionylindan-1,3-dione	C	$2.8-2.3 \times 10^{-2}$	597
Pyridine-2,6-dicarboxylic acid	C	$7.5 \times 10^{-4}-7.5 \times 10^{-3}$	483
2-Thenoyltrifluoroacetone	C	$8 \times 10^{-4}-8 \times 10^{-1}$	585
2-Thenoyltrifluoroacetone + collidine or diphenylguanidine	TC(E)	$2 \times 10^{-5}-2 \times 10^{-1}$	598, 599
2-Thenoyltrifluoroacetone + hydroxyethylethylenediaminetriacetic acid	TC	$4.5 \times 10^{-5}-3 \times 10^{-1}$	600–603
2-Thenoyltrifluoroacetone + 1,10-phenanthroline	TC(E)	$2 \times 10^{-4}-10^{-1}$	594, 604–609
2-Thenoyltrifluooacetone + trioctylphosphine oxide	TC(E)	$2 \times 10^{-5}-7.6 \times 10^{-1}$	609–611
1,1,1-Trifluoro-5,5-dimethylhexane-2,4-dione + 1,10-phenanthroline	TC	2×10^{-6}	612

Table 4.4. *(Continued)*

Inorganic Substance	Reagent	Method	Application Interval or Detection Limit (ppm)	Ref.
F⁻	1,1,1-Trifluoro-5,5-dimethylhexane-2,4-dione + trioctylphosphine oxide	TC	1.5×10^{-3}–1.5	613
	Al(III)-Eriochrome Red B	S(Q)	4×10^{-3}–2	614, 615
	Al(III)-8-hydroxyquinoline	S(Q)(E)	10^{-2}–4×10^{-1}	616
	Al(III)-Morin	S(Q)	2×10^{-3}–3.5×10^{-1}	616–618
	Al(III)-Pyridylazo-2-naphthol	S(Q)	2×10^{-4}–2×10^{-2}	619
	Al(III)-Superchrome Garnet Y	S(Q)	4×10^{-3}–2	614
	Mg(II)-8-hydroxyquinoline	S(Q)	0.2–1.6 µg	620–622
	Th(IV)-Morin	S(Q)	2.5×10^{-3}–1.5×10^{-1}	623
	Zr(IV) + Calcein Blue	TC	9×10^{-3}–6×10^{-2}	624
	Zr(IV)-3-hydroxyflavone	S(Q)	2×10^{-3}–10^{-1}	625
Fe	Al(III)-Pontachrome BBR	C(Q)	2×10^{-2}–2×10^{-1}	626
	Benzyl-2-pyridylketone-2-pyridilhydrazone + BrO₃⁻	R,C(K)	7×10^{-2}–3×10^{-1}	449
	1,4-Diamino-2,3-dihydroxyanthraquinone	R	2.8×10^{-1}–6×10^{-1}	627
		R(K)	5×10^{-2}–6×10^{-1}	628
	Homovanillic acid + H₂O₂ + peroxidase	Enz(K)	2–60	514
	2-Hydroxybenzaldehyde + H₂O₂	Cat(K)	–	628a
	2-Hydroxybenzladehyde + H₂O₂ + Ag(I)	Cat(K)	5×10^{-2}–1.7×10^{-1}	234a
	Indoxylacetate + hyaluronidase	Enz(K)	2×10^{-1}–12	515
	4'-(4-Methoxyphenyl)-2,2',2''-terpyridyl	C(Q)	10^{-2}–10^{-1}	629, 630
	1,10-Phenanthroline + eosin	TC(E)	10^{-1}–1	241
	Phenylenediamine	R(E)	1×10^{-2}	630a

		R		
	Potassium hydrogen phthalate + Zn amal-gam			631
	Rhodamine B	C(Q)	1	632
	Stilbexone + H_2O_2	Cat(K)	2×10^{-3}–2×10^{-1}	633–636
	$2,2',2''$-Terpyridyl	C(Q)	10^{-2}–5×10^{-1}	637
Ga	Acridine Orange + Cl^-	TC(E)	5×10^{-3}–10^{-1}	638
	Alizarin Red S	C	2×10^{-1}–2	244
	4-Amino-2,4-dihydroxybenzyl -2,3-di-methyl-1-phenyl-5-pyrazolone	C	10^{-3}	639
	Antipyrine + Quercetin	TC(E)	10^{-2}–10^{-1}	640
	Benzyl-2-pyridylketone-2-pyridyl hydrazone	C	1.4×10^{-3}–3.5×10^{-1}	641
	4-[5-Chloro-2-hydroxyphenylazol]resorcinol	C(E)	5×10^{-4}–4.6×10^{-3}	642, 643
	5,7-Dibromo-8-hydroxyquinoline	C(E)	2×10^{-1}	644–647
	5,7-Dichloro-8-hydroxyquinoline	C(E)	2×10^{-1}	645
	2,4-Dihydroxybenzaldehyde semicarbazone	C	4×10^{-3}–4×10^{-1}	250
	Dodecyllumogallion	C	10^{-3}	648
	Eriochrome Red B	C	0–2×10^{-1}	649
	Hexyllumogallion	C(E)	10^{-3}	650
	2-Hydroxy-1-benzaldehyde-thiosemicarba-zone	C(E)	1–2×10^{-2}	651
	2-Hydroxy-5-chlorobenzaldehyde thiosemi-carbazone	C	4×10^{-3}–2.8×10^{-1}	319
	2-Hydroxy-5-methylbenzaldehye-4-ami-noantipyrine	C	10^{-4}	652
	2-Hydroxy-5-methylbenzaldehyde-thiosemi-carbazone	C	8×10^{-3}–2.8×10^{-1}	319
	2-Hydroxy-1-naphthaldehyde-thiosemicar-bazone	C	0–7×10^{-1}	653
	4-Hydroxy-3-(2-hydroxy-4-methoxypheny-lazo)-benzenesulfonic acid	C	8×10^{-2}–8×10^{-1}	257

Table 4.4. *(Continued)*

Inorganic Substance	Reagent	Method	Application Interval or Detection Limit (ppm)	Ref.
	4-(2-Hydroxyphenylazo)resorcinol	C	2×10^{-2}–4.8×10^{-1}	257
	o-(Hydroxyphenyl)benzoxazole		6×10^{-1} µg	654
	8-Hydroxyquinaldine	C(E)	10^{-3}–10^{-1}	655–657
	8-Hydroxyquinoline	C(E)	2×10^{-2}–6×10^{-1}	658–666
	8-Hydroxyquinoline-5-sulfonic acid	C	2×10^{-2}–	492
	Lumogallion	C(E)	10^{-3}–8×10^{-2}	279, 667–675
	Lumomagneson	C	8×10^{-3}–3.9×10^{-1}	676
	Mordant Blue 31	C	4×10^{-2}–4.8×10^{-1}	286
	Morin	C	10^{-1}	677–680
	Pyronine G + Cl⁻	TC(E)	—	681
	2(2'-Pyridyl)benzimidazole	C	7×10^{-2}–7×10^{-1}	682
	Pyridine-2-aldehyde 2-fuorylhydrazone	C	1×10^{-3}–5×10^{-2}	682a
	Quercetin-3'-glucoside	C	5.6×10^{-3}–1.7×10^{-1}	683
	8-Quinolinethiol	C	2×10^{-2}–6×10^{-1}	684
	Resorcylidene-4-aminoantipyrine	C	10^{-3}	685
	Resorcylidene-cysteine	C	7×10^{-3}	438
	Rhodamine B + Cl⁻	TC(E)	1.5×10^{-3}–3.3×10^{-1}	686–694
	Rhodamine 3B + Cl⁻	TC(E)	5×10^{-4}%	694
Ga	Rhodamine 3 GO + Cl⁻, Br⁻	TC(E)	1.9×10^{-2}	695, 696
	Rhodamine 4G + Cl⁻, Br⁻	TC(E)	2.5×10^{-3}	695, 696
	Rhodamine 6G + Cl⁻	TC(E)	2×10^{-3}–1.6×10^{-1}	689, 697–699
	Salicylaldehyde thiosemicarbazone	C	—	700, 700a
	4-Salicylideneaminoantipyrine	C	10^{-3}	701
	Salicylidene-o-aminophenol	C(E)	3×10^{-2}–6×10^{-1}	702–704

	Reagent	Type	Range	Ref.
	o-(Salicylideneamino)-2-hydroxy-benzene sulfonic acid	C	2×10^{-3}–4×10^{-1}	250
	N-Salicylidene-2-hydroxy-4-carboxyaniline	C	4×10^{-3}–2.4×10^{-1}	319
	N-Salicylidene-2-hydroxy-5-carboxyaniline	C	4×10^{-3}–2.8×10^{-1}	319
	N-Salicylidene-2-hydroxy-5-sulfoaniline	C	2×10^{-3}–2×10^{-1}	319
	6-(4-Methylsalicylideneamino)-m-cresol	C	4×10^{-3}–2.8×10^{-1}	284
	o-(Salicylideneamino)-5-methoxyphenol	·C	—	318
	Solochrome Black AS	C(E)	10^{-2}	705
	Solochrome Red ERS or Black WEFA	C	10^{-2}–8×10^{-2}	706
	Sulfonaphtholazoresorcinol	C	3×10^{-3}–1.5×10^{-2}	707–709
	Superchrome Garnet Y	C	2×10^{-2}–1.2	323
	3,4',5,7-Tetrahydroxyflavone	C	2×10^{-3}–2.4×10^{-2}	710
Gd	Rhodamine B + 3,5-di-iodo salicylic acid	TC(E)	4×10^{-3}	571
	Rhodamine B + salicylic acid	TC(E)	1.2×10^{-1}–1.2	711
	Thenoyltrifluoroacetone	C	10	712
Ge	Alizarincomplexone + Rhodamine 6 G	TC	2×10^{-3}–10^{-1}	713
	Quercetin	C	10^{-2}–7×10^{-2}	714
	Resacetophenone	C	10^{2}	392
	Rezarson	C	4×10^{-3}	715–717
Hf	Flavonol	C	10^{-1}	718
	3-Hydroxychromone	C	1.6×10^{-2}–3.6×10^{-1}	719
	Morin	C	4×10^{-3}–2.4×10^{-2}	720–722
	Myricetin	C	2×10^{-2}–2×10^{-1}	723
	Cannabiscitrin	C	10^{-3}–3×10^{-3}	275
	Quercetin	C	10^{-2}–1.2×10^{-1}	724–728
	Quercetin-7-glucoside	C	2×10^{-4}–5×10^{-4}	729
	Quercetin-sulfonic acid	C	2×10^{-2}–1.4×10^{-1}	730

Table 4.4. (Continued)

Inorganic Substance	Reagent	Method	Application Interval or Detection Limit (ppm)	Ref.
Hg	Butylrhodamine B + Br⁻ (crystal violet)	TC(E)	$3 \times 10^{-3}–2 \times 10^{-1}$	731
	2,2'-Dipyridylketone hydrazone	Cat	$8 \times 10^{-2}–3.2 \times 10^{-1}$	543
	NADH	C(Q)	$3 \times 10^{-2}–2 \times 10^{-1}$	732
	1,10-Phenanthroline + eosin	TC(E)	$5 \times 10^{-2}–1$	241
	Pyronine G + Br⁻	TC(E)	$2 \times 10^{-2}–2 \times 10^{-1}$	733
	Rhodamine B + Cl⁻	TC(E)	$2.5 \times 10^{-2}–1$	734
	Rhodamine B + Br⁻	TC(E)	$1.6 \times 10^{-3}–1.6 \times 10^{-2}$	735
	Rhodamine B + I⁻	TC(Q)	$1–10$	736
	Rhodamine 6 G + I⁻	TC(E)	$4 \times 10^{-1}–1.6$	737
	Thiamine	R(E)	$10^{-2}–5 \times 10^{-1}$	738, 739
H₂O₂	4-Amino-1H-1,5 benzodiazepine-3-carbonitrile hydrochloride + peroxidase	Enz	$2 \times 10^{-2}–4 \times 10^{-1}$	740
	Diacetyldichlorofluorescein + peroxidase	Enz	$10^{-3}–10^{-2}$	741, 742
	p-Hydroxyphenylacetic acid + peroxidase	Enz	10^{-3}	743
	3-(4-Hydroxyphenyl)propionic acid + peroxidase	Enz	$10^{-3}–10^{-1}$	744
	Scopoletin + peroxidase	Enz(Q)	10^{-3}	745, 746
Ho	Thenoyltrifluoroacetone	C	10	712
I⁻	Di-(acetoxymercuri)-fluorescein	S(Q)	$3.5 \times 10^{-5}–4 \times 10^{-4}$	513
	10-(3-Sulfonatopropyl)-acridinium	CF	$4.8 \times 10^{-1}–1$	448
	Uranylacetate	S(Q)	$2 \times 10^{-2}–2 \times 10^{-1}$	747

408

In	Cannabiscitrin	C	5×10^{-3}–1.1×10^{-1}	748
	8-Hydroxyquinaldine	C	2×10^{-2}	749
	8-Hydroxyquinoline	C	4×10^{-2}	749–752
	8-Hydroxyquinoline-5-sulfonic acid	C	2×10^{-2}	492
	8-Hydroxyquinoline + tetraphenyl borate	C(E)	2×10^{-2}–2	753
	Lumogallion	C(E)	4×10^{-3}–4×10^{-2}	279, 754
	Lumomagneson	C	8×10^{-2}–4.6×10^{-1}	676
	Morin + antipyrine + ClO₄⁻	TC(E)	3×10^{-3}	755, 756
	2-(2'-Pyrydyl)benzimidazole	C(E)	10^{-2}–1	682
	Pyronine G + Br⁻	TC(E)	5–40	757
	8-Quinolinethiol + CH₃COO⁻	TC(E)	5×10^{-2}	758
	Quercetin	C	10^{-2}–9.2×10^{-1}	759
	Rhodamine B + Br⁻	TC(E)	3×10^{-3}–4×10^{-1}	760–763
	Rhodamine 3 B + Br⁻	TC(E)	3×10^{-3}–3.3×10^{-1}	692, 764, 765
	Rhodamine 3 GO + Cl⁻	TC(E)	7×10^{-1}	696
	Rhodamine 4 G + Cl⁻	TC(E)	2×10^{-1}	696
	Rhodamine 6 G + BR⁻	TC(E)	1.6×10^{-3}	766–769
	Rhodamine 6 J + Br⁻	TC(E)	4×10^{-1}–1.6×10^{-1}	770
	Sulfonaphtholazoresorcinol	C(E)	2×10^{-2}–1.6×10^{-1}	771
	3,4',5,7-Tetrahydroxyflavone	C	3×10^{-3}–1.7	772
	Salicylaldehyde thiosemicarbazone	C(E)	0–4×10^{-1}	773
	Superchrome Garnet Y	C	2×10^{-1}–1.2	323
IO₃⁻	Pseudopurpurine	R	2×10^{-1}–2	774
Ir	2,2',2''-Terpyridyl	C	2–20	775
K	Anilinonaphthalene sulfonate + dibenzo-18-crown-6	TC(E)	1	776
	Eosin + 18-crown-6	TC(E)	10^{-2}–1	777

Table 4.4. (*Continued*)

Inorganic Substance	Reagent	Method	Application Interval or Detection Limit (ppm)	Ref.
La	8-Hydroxyquinoline-5-sulfonic acid + hexadecylmethylammonium	TC	1–20	777a
	1-(Dicarboxymethylaminomethyl)-2-hydroxy-3-naphthoic acid	C	3×10^{-1}–1.1	408
	Rhodamine B + 3,5-di-iodosalicylic acid	TC(E)	2×10^{-3}–3	571
Li	Dibenzothiazolyl methane	C	5×10^{-2}–2	778, 779
	5,7-Dibromo-8-hydroxyquinoline	C	2.5×10^{-1}–25	780
	1,4-Dihydroxyanthraquinone	C	5×10^{-2}–5×10^{-1}	781
	1,5-Dihydroxyanthraquinone	C	5×10^{-2}–6.5×10^{-1}	782
	1,8-Dihydroxyanthraquinone	C	5×10^{-2}–7×10^{-1}	783
	8-Hydroxyquinoline	C	4×10^{-2}–1	784, 785, 785a
	N-(4-Methylumbelliferone-8-ylmethyl)-monoaza-15-crown-5)	C(E)	—	482
Lu	Bis(8-hydroxy-2-quinolyl)methylamine	C	—	581
	1-(Dicarboxymethylaminomethyl)-2-hydroxy-3-naphthoic acid	C	3.5×10^{-1}–1.4	408
Mg	Morin + diantipyrylmethane	TC(E)	2×10^{-2}–10	786
	Morin	C(E)	1.5–10	787
	Thenoyltrifluoroacetone	C	10	712
	4-[4-(3-Arsono-5-chloro-2-hydroxyphenylazo)]-3-hydroxy-5-methylpyrazol-2-ylbenzene sulfonic acid	C	10^{-3}–5×10^{-3}	788

410

Bis(salicylideneamino)benzofuran	C(E)	2×10^{-3}–10^{-1}	789, 790
Bissalicylideneethylenediamine	C	2.5×10^{-2}	791–797
Calcein	C	10^{-2}	798–800
1-Dicarboxymethylaminomethyl-2-hydroxy-3-naphthoic acid	C	—	468
1,5-Dihydroxyanthraquinone	C	10^{-2}–10^{-1}	801
1,8-Dihydroxyanthraquinone	C	5×10^{-2}–3.5×10^{-1}	802
2,2′-Dihydroxyazobenzene	C	—	803–805
Fluoren-2-aldehyde-2-pyridylhydrazone	C(K)	4×10^{-1}–1.2	806
1-Hydroxy-2-carboxyanthraquinone	C	9×10^{-3}–4.5×10^{-2}	807
3-Hydroxy-3′,4′-dimethoxyflavone	C	10^{-1}	808
3-Hydroxy-2-naphthoic acid	C	10^{-2}	471
8-Hydroxyquinoline	C	10^{-2}	473, 795, 809, 810
1-(8-Hydroxyquinoline-7-azo)-2-naphthol-4-sulfonic acid	C	10^{-2}	811
8-Hydroxyquinoline-5-sulfonic acid	C	10^{-3}	492, 812–818
1(2-Hydroxy-3-sulfo-5-chloro-phenylazo)-2′-hydroxynaphthalene	C	2×10^{-3}	819
Isocitric dehidrogenase	Enz(K)	5	820, 821
Lumomagneson	C	4×10^{-3}	822–824
Morin	C	10^{-1}–9×10^{-1}	825
5-Pyrazolone-(4-azo-2-)l-naphthol-4-sulfonic acid	C	5×10^{-3}	826
Pyridoxal nicotinylhydrazone + amines	TC(E)	2×10^{-2}–10^{-1}	827
5,10,15,20-Tetrakis-(l-methylpyridinium-4-yl) porphine	C	9.6×10^{-3}–7.8×10^{-2}	828
3,3′,4′-Trihydroxyflavone	C	0–5 µg	829
Mn 2,3-Dioxogulonate + peroxidase	Enz	4.4×10^{-1}–2.7	830
2-Hydroxybenzaldehyde thiosemicarbazone + H_2O_2	Cat(K)	2×10^{-3}–9×10^{-3}	831

Table 4.4. (Continued)

Inorganic Substance	Reagent	Method	Application Interval or Detection Limit (ppm)	Ref.
	8-Hydroxyquinoline-5-sulfonic acid	R	2.5×10^{-3}–2	832
	Isocitric dehidrogenase	Enz(K)	10^{-1}	820, 821
	1,10-Phenanthroline + eosin	TC(E)	10^{-1}–1	241
	Homovanillic acid + H_2O_2 + peroxidase	Enz(K)	3×10^{-1}–12.5	514
Mo	Carminic acid	C(E)	10^{-1}–9×10^{-1}	834, 835
	Morin	C	10^{-1}–1.6	836
	8-Hydroxyquinoline, H_2O_2, tetraphenylborate	C(E)	2×10^{-2}–4	837
	Rhodamine B + SCN^-	TC(Q)	2×10^{-3}–2×10^{-1}	838
Na	o-Phenylphenol	C	4×10^{-1}	839
Nb	Lumogallion + H_2O_2 or F^-	TC	10^{-1}–2.5	840
	Lumogallion + oxalate	TC	10^{-1}–2.5×10^{-1}	841
	Morin + H_2O_2	TC	4×10^{-3}–4×10^{-1}	842
	Quercetin + H_2O_2	TC	10^{-1}–9×10^{-1}	843
	Sulfonaphtholazoresorcinol + H_2O_2,	TC(E)	4×10^{-2}–4×10^{-1}	844
	Sulfonaphtholazoresorcinol + F^-, oxalate or tartrate	TC(E)	4×10^{-2}–4×10^{-1}	845
NH_3, NH_4^+	Formaldehyde + diketone	Hantzsh reaction	10^{-2}–2.5×10^{-1}	846, 847
	Fluorescamine	CF	10^{-2}–1	848
	2-Ketoglutarate + NADH + glutamate dehydrogenase	Enz(Q)	1.7×10^{-5}–8.5×10^{-4}	849, 850
	o-Phtaldialdehyde	CF	4×10^{-3}–4	850a
	Phthalaldehyde	CF	3.4×10^{-3}–1.7	851, 852

Ni	Al-1-(2-pyridylazo)-2-naphthol	C(Q)	6×10^{-5}–6×10^{-3}	532
	Benzyl-2-pyridylketone-2-pyridylhydrazone + BrO_3^-	R,C(K)	5×10^{-2}–3.5×10^{-1}	449
	1,10-Phenanthroline + eosin	TC(E)	10^{-1}–1	241
NO_2^-	5-Aminofluorescein	R	4.6×10^{-5}–4.6×10^{-1}	853
	Benzidine	R(Q)	1.1×10^{-2}–5.7×10^{-2}	854
	p-Chloroaniline + 2,6-diaminopyridine + ammoniacal copper (II)	Diazotization (E)	2×10^{-3}–2×10^{-2}	855
	2,3-Diaminonaphthalene	Diazotization (E)	10^{-4}–1.7×10^{-3}	856
	Resorcinol	Nitrosation	10^{-2}–1.3×10^{-1}	857
	Rhodamine 3 B	C(E)	5×10^{-1} μg	858
NO_3^-	2,3-Diaminonaphthalene	Diazotization (E)	10^{-2}–1.1	859
	2,2'-Dihydroxy-4,4'-dimethoxybenzophenone	R	5×10^{-3}–1	860
	Fluorescein	R(Q)	10^{-2}–1.2×10^{-1}	861
	Resorcinol	Nitrosation	1.2×10^{-3}–1.6×10^{-2}	857
O_2	9,10-Dihydroacridine	R	2×10^{-1}	862
	Epinephrine	R	1	863
	Leucofluorescein	R	1	864
O_3	9,10-Dihydroacridine	R	10^{-2}	865
	2-Diphenylacetyl-1,3-indandione-1-hydrazone	R	2×10^{-2}	866
Os	4,6-Bis-(methylthio)-5-aminopyrimidine	C(Q)	5×10^{-1}–2.5	867
Pb	Eosin	C(Q)	1.5×10^{-1}	232
	Fluorescein	C(Q)	—	234
	Homovanillic acid + H_2O_2 + peroxidase	Enz(K)	10–1.85×10^2	514

413

Table 4.4. *(Continued)*

Inorganic Substance	Reagent	Method	Application Interval or Detection Limit (ppm)	Ref.
	Lumogen Light Blue	C(Q)	5–80	868
	1,10-Phenanthroline + eosin	TC(E)	10^{-1}–1	241
	Phloxin	C(Q)	—	234
Pd	Benzyl-2-pyridylketone-2-pyridylhydrazone + BrO_3^-	R,C(K)	5×10^{-2}–5×10^{-1}	449
	1,10-Phenanthroline + eosin	TC(E)	2×10^{-2}–1.4×10^{-1}	869,870
Pr	Thenoyltrifluoroacetone	C	10	712
PO_4^{3-}	Al(III)-Morin	S(Q)	5×10^{-2}–1	871, 872
	Mg(II)-3-hydroxy-3',4'-dimethoxyflavone	S(Q)	10^{-1}–1.2	873
	Molybdate + quinine	TC	10^{-2}–6×10^{-1}	874
	Molybdate + rhodamine B	TC(E)	2×10^{-2}–3×10^{-1}	875
	Molybdate + thiamine	R	5×10^{-3}–10^{-1}	876–878
	NADP + glycogen	Enz	2×10^{-3}–5×10^{-2}	879
Pt	2,2'-Dipyridylketonehydrazone	Cat(K)	2×10^{-1}–6×10^{-1}	880
Re	Acridine Orange	C(E)	5×10^{-3}–4	881, 882
	Acriflavine	C(E)	8×10^{-3}	882, 883
	Phenosafranine	C(E)	8×10^{-2}	884
	Rhodamine 3 B	C(E)	1.5×10^{-2}–3.3×10^{-1}	885
	Rhodamine 6 G	C(E)	2×10^{-2}–9×10^{-1}	886
	Rhodamine 6 J	C(E)	1.5×10^{-1}–5	887
	Rhodamine B	C(E)	3×10^{-3}–8×10^{-1}	888
	Safranine T	C(E)	2×10^{-3}–10^{-1}	889, 890

414

Ru	5-Methyl-1,10-phenanthroline	C	3×10^{-1}–2	891
	1,10-Phenanthroline	C	5×10^{-3}–3×10^{-1}	892, 893
S^{2-}	Cu(II)-2-(o-hydroxyphenyl)-benzoxazole	S	10^{-3}–10^{-1}	512
	Di(acetoxymercuri)fluorescein	S(Q)	10^{-3}–10^{-2}	894–901
	Hg(II)-2,2'-pyridylbenzimidazole	S	3×10^{-4}–3×10^{-1}	902
	Homovanillic acid + H_2O_2 + peroxidase	Enz(K)	3×10^{-1}–3.2	514
	Pd(II)-8-hydroxyquinoline-5-sulfonic acid	S	2×10^{-1}–5	516
	Tetra(acetoxymercuri) fluorescein	S(Q)	10^{-5}	201, 903–910
	3,4,5,6-Tetra(chloromercuri) fluorescein	S(Q)	0–4	911
	Thiamine + MnO_4^-	R(Q)	3×10^{-3}–2×10^{-2}	912
SO_3^{2-}	5-Aminofluorescein + formaldehyde	R(Q)	2.5×10^{-2}–6.2	913
	Homovanillic acid + H_2O_2 + peroxidase	Enz(K)	5–3.7×10^2	514
SO_4^{2-}	Th(IV)-flavonol	S(Q)	2×10^{-1}–8×10^{-1}	914
	Th(IV)-morin	S(Q)	2×10^{-1}–8×10^{-1}	915–920
	Th(IV) + quercetin	TC	8×10^{-2}–1.7	921
	Th(IV) - salicylfluorone	S	5×10^{-2}–4×10^{-1}	922, 923
	Zr(IV) + calcein blue	TC	2–12×10^3	924
$S_2O_3^{2-}$	Tetra(acetoxymercuri) fluorescein	S(Q)	3.3×10^{-3}–3.3×10^{-2}	522
$S_2O_8^{2-}$	Leucofluorescein	R	5–10	925
Sb	Acridine orange + Cl^-	TC(E)	6×10^{-3}–8×10^{-1}	926
	Benzoin	C	10^{-7}%	927
	3,7-Dihydroxyflavone	C	10^{-1}–2 µg	928
	Morin	C(E)	10^{-2}–4×10^{-1}	929
	Rhodamine 3 B + Cl^- (crystal violet)	TC(E)	8×10^{-3}–4×10^{-2}	930, 931
	Rhodamine 6 G + Cl^-	TC(E)	10^{-1}–2	932
	Safranine T + Cl^-	TC(E)	1.5×10^{-2}–1	933
	2,4',7-Trihydroxyflavone	C	4×10^{-2}–2×10^{-1}	934

Table 4.4. (Continued)

Inorganic Substance	Reagent	Method	Application Interval or Detection Limit (ppm)	Ref.
Sc	5,7-Dichloro-8-hydroxyquinoline	C	10^{-1}–4×10^{-1}	935
	2,4-Dihyroxybenzaldehydeacetylhydrazone	C	1–18 µg	936
	2,4-Dihydroxybenzaldehyde semicarbazone	C	8×10^{-4}–4×10^{-1}	250, 413, 937
	2-Ethyl-3-methyl-5-hydroxychromone	C(E)	10^{-2}–4×10^{-1}	938
	2-Ethyl-5-hydroxy-7-methoxyisoflavone	C(E)	0–2.4×10^{-1}	939
	2-Hydroxybenzaldehyde semicarbazone	C	2×10^{-3}–4×10^{-1}	413, 940–942
	5-Hydroxychromone	C(E)	0–1.6×10^{-1}	938
	2-Hydroxy-5-methylbenzaldehyde semicarbazone	C	2×10^{-3}–4×10^{-1}	413
	2-Hydroxy-3-naphthoic acid	C	2×10^{-2}–4×10^{-1}	943
	Hydroxystilbene complexon	C	10^{-2}	944
	4-Methoxybenzoic acid-2-hydroxynaphthyl-methilene hydrazide	C	2×10^{-4}–6×10^{-2}	945, 946
	Mordant Blue 31	C	4×10^{-2}–4.8×10^{-1}	286
	Morin + antipyrine	TC(E)	10^{-2}–2×10^{-1}	947
	Myricetin	C	7×10^{-3}	948
	Cannabiscitrin	C	7×10^{-4}	275, 948
	5,7-Dinitro-8-hydroxyquinoline + rhodamine B	TC(E)	2×10^{-2}–4	948a, 948b
	Quercetin-7-glucoside	C	4×10^{-3}–10^{-2}	948
	1,10-Phenanthroline + eosin	TC(E)	3.3×10^{-2}–6.6×10^{-1}	949, 950
	2-Phenylcinchonic acid + rhodamine B	TC(E)	7×10^{-3}	951, 952
	2-Phenylcinchonic acid	C	2×10^{-3}–9	953
	Salicylic acid + rhodamine B	TC(E)	7×10^{-3}	952
	Superchrome Garnet Y	C	2×10^{-1}–1.6	323

416

		R		
SCN⁻	Chloramine T + pyridine-barbituric acid	R	3×10^{-3}–1.4×10^{-1}	510
Se	3,3'-Diaminobenzidine	C(E)	10^{-2}–1	954–962
	2,3-Diaminonaphthalene	C(E)	10^{-3}–5×10^{-1}	201, 962–995
	Dithizone	C(E)	8×10^{-3}–4×10^{-1}	996
SiO₃²⁻	Ammonium molybdate + carminic acid	S	3×10^{3}	997
	Benzoin	C	8×10^{-2}–4×10^{-1}	998
Sm	Bis(8-hydroxy-2-quinolyl)methylamine	C(E)	—	580
	1,1,1,5,5,5-Hexafluoro-2,4-pentanedione	C	5×10^{-3}–10^{-1}	999
	Hexafluoroacetylacetone + trioctylphosphine oxide	TC(E)	7×10^{-3}–1.5	587
	2-Naphthoyltrifluoroacetone + trioctylphosphine oxide	TC(E)	10^{-1}–5	591
	2-Phenylcinchonic acid + 1,10-phenanthroline	TC	5×10^{-1}	594
	Rhodamine B + 3,5-di-iodosalicylic acid	TC(E)	4×10^{-3}	571
	Thenoyltrifluoroacetone	C	2	712
	Thenoyltrifluoroacetone + collidine or diphenylguanidine	TC(E)	5×10^{-2}–1	598
	Thenoyltrifluoroacetone + 1,10-phenanthroline	TC(E)	10^{-2}–1	594, 604, 605, 607–609
	Thenoyltrifluoroacetone + trioctylphosphine oxide	TC(E)	2×10^{-3}–1.5	609–611
	1,1,1-Trifluoro-5,5-dimethylhexane-2,4-dione + trioctylphosphine oxide	TC	1.5×10^{-1}–1.5	613
Sn	3,7-Dihydroxyflavone	C	8×10^{-4}–2×10^{-2}	1000
	Flavonol	C	10^{-1}–1	1001
	3-Hydroxychromone	C	1.6×10^{-2}–3.2×10^{-1}	1002

Table 4.4. (Continued)

Inorganic Substance	Reagent	Method	Application Interval or Detection Limit (ppm)	Ref.
	o-Hydroxyhydroquinonephthalein	C	$0-5.5 \times 10^{-2}$	1003
	8-Hydroxyquinoline	C(E)	$2 \times 10^{-1}-8 \times 10^{-1}$	1004
	8-Hydroxyquinoline-5-sulfonic acid	C	$5 \times 10^{-3}-2.5 \times 10^{-1}$	1005
	Morin	C(E)	$2 \times 10^{-3}-10^{-1}$	679, 929, 1004
	Morin + antipyrine	TC(E)	5×10^{-1}	1006
	6-Nitro-2-naphthylamine-8-sulfonate	R	$10^{-1}-2$	1007
	Rhodamine B + Br$^-$	TC(E)	$10^{-1}-2$	1008, 1009
	Rhodamine B + cupferron (crystal violet)	TC(E)	4×10^{-2}	1010
	o-(Salicylideneamino)-2-hydroxybenzene sulfonic acid	C	$8 \times 10^{-4}-4 \times 10^{-1}$	250
	3,4',7-Trihydroxyflavone	C(E)	$7 \times 10^{-3}-6$	1011
Sr	Calcein	C	80	401
	8-Hydroxyquinoline	C	2×10^{-1} µg	473
Ta	Butylrhodamine B + F$^-$ (brilliant green)	TC(E)	$2 \times 10^{-2}-4 \times 10^{-1}$	1012
	Morin + H$_2$O$_2$	TC(E)	$6 \times 10^{-4}-7 \times 10^{-2}$	1013
	Rhodamine 6 G + F$^-$	TC(E)	$5 \times 10^{-4}-5 \times 10^{-3}$	1014
	Rhodamine 6 J + F$^-$	TC(E)	4×10^{-4}	1015
Tb	Acetylacetone	C	$4 \times 10^{-3}-2$	1016, 1017
	Antipyrine + sodium tetraphenylborate	TC	$5 \times 10^{-2}-1$	1018
	Antipyrine + salicylate	TC(E)	$10^{-2}-5$	565, 570, 1019, 1020
	4-Benzoyl-3-methyl-1-phenyl-5-pyrazolone + phosphine oxides or triisobutilphos-	TC(E)	3.6×10^{-3}	578, 579

Reagent	Type	Range	Ref.
Diantipirylmethane + salicylic acid	TC	1.6×10^{-3}–1.6×10^2	1021
Dibenzoylmethane	C	$3 \times 10^{-4}\%$	583
Dimethylformamide or acetonitrile	C	1.73×10^{-2}–1.73×10^3	585
Dipicolinic acid	C	1.7×10^{-3}–13	1022
EDTA + acetylacetone	TC	1.6×10^{-3}–1.6	602, 1023, 1024
EDTA + 2,3-dihydroxynaphthalene	TC	10^{-3}–4×10^{-1}	1025
EDTA + salicylate	TC	7×10^{-4}–5×10^{-1}	1026
EDTA + sulfosalicylic acid	TC	6.4×10^{-3}–3.2	570, 1027, 1027a
EDTA + tiron	TC	9×10^{-5}–9×10^{-1}	570, 572–574
α-α′-Ethylenediiminodi-(o-hydroxyphenyla-cetic acid)	C	1.5×10^{-3}–1.5×10^{-2}	1028
Hexafluoroacetylacetone + trioctylphos-phine oxide	TC(E)	1.6×10^{-3}–1.5	587
Hexafluoro-2,4-pentanedione	C	10^{-3}–8×10^{-2}	588, 999
8-Hydroxyphenyliminodiacetic acid	C	2×10^{-2}–17.3	1029
3-Methyl-1-phenyl-5-pyrazolone + sodium tetraphenylborate	TC	$10^{-4}\%$	570, 1030
3-Methyl-1-tolyl-5-pyrazolone + sodium te-traphenylborate	TC	$10^{-4}\%$	570, 1030
4,4′-Methylenedi-[3-methyl-1-(2-pyridyl)-pyrazol-5-ol]	C	10^{-3}–$10^{-2}\%$	567
3-Methyl-1-(4-sulfophenyl)-5-pyrazolone	C	10^{-2}–10^{-1}	569, 570, 606
1,10-Phenanthroline + salicylate	TC(E)	5×10^{-2}–2.5	570, 596
Phenyl salicylate	C	6×10^{-2}	570, 1031
Pyridine-2,6-dicarboxylic acid	C	8×10^{-4}–8×10^{-3}	483
Pyrogallolsulfonic acid	C	1.8×10^{-2}–7.2×10^{-1}	1032
Rhodamine B + 3,5-iodosalicylic acid	TC(E)	10^{-2}	571
Tiron	C	4×10^{-6}–10^{-1}	1033

419

Table 4.4. *(Continued)*

Inorganic Substance	Reagent	Method	Application Interval or Detection Limit (ppm)	Ref.
	Tiron + iminodiacetic acid	TC	4×10^{-5}–40	570, 575, 576
	Thenoyltrifluoroacetone	C	10	712
	1,1,1-Trifluoro-5,5-dimethylhexane-2,4-dione + trioctylphosphine oxide	TC	1.5×10^{-3}–1.5×10^{-1}	613
Te	Butylrhodamine B + Br$^-$	TC(E)	4×10^{-3}–2×10^{-1}	1034
	Rhodamine B + Cl$^-$	TC(E)	3×10^{-2}–7×10^{-2}	1035
	Rhodamine 3 GO + Cl$^-$	TC(E)	7.5×10^{-1}–10^2	1036
	Rhodamine 4 G + Cl$^-$	TC(E)	1.7×10^{-1}–13.3	1036
	Rhodamine 6 J + Br$^-$	TC(E)	10^{-1}–1	1037
	2,4,4-Trihydroxybenzophenone	C	9×10^{-3}–6.5×10^{-2}	1038
Th	5,7-Dibromo-8-hydroxyquinoline + rhodamine B	TC(Q)(E)	4×10^{-2}–1.2	1039
	5,7-Dinitro-8-hydroxyquinoline + rhodamine B	TC(E)	7×10^{-3}–1.5	1040, 1041
	Flavonol	C	10^{-2}–1	1042
	Morin	C	2×10^{-2}	1043–1048
	Myricetin	C	4×10^{-2}	1049
	Quercetin	C	2×10^{-2}	1050
	3,4',7-Trihydroxyflavone	C	1.2×10^{-3}–4×10^{-1}	1051
Ti	Benzal-2-amino-3-nitrilo-4,5-diphenylfurane	C	6×10^{-2}–4×10^{-1}	1052
	Dimethoxybenzal-2-amino-3-nitrilo-4,5-diphenylfurane	C	6×10^{-2}–4×10^{-1}	1052
	3-Ethyl-5-hydroxy-2-methylbenzopyran-4-one	C(E)	10^{-2}–10^{-1}	1053

	8-Hydroxyquinoline,H_2O_2,tetraphenylborate	TC(E)	2×10^{-2}–2	1054
	Picolinaldehyde nicotinoylhydrazone	Cat	6×10^{-2}–3×10^{-1}	1055
		Cat(K)	6×10^{-2}–4×10^{-1}	1055
Tl	Acridine Orange + Cl^-	TC(E)	3×10^{-3}–5×10^{-1}	1056
	Acridine Yellow + Cl^-	TC(E)	5×10^{-3}	1057
	Acriflavine + Cl^-	TC(E)	2.5×10^{-3}–5×10^{-1}	1058
	Butylrhodamine B + Cl^- (Crystal violet)	TC(E)	3×10^{-3}–2.7×10^{-1}	1059
	1,4-Diamino-2,3-dihydroxyanthraquinone	R	10^{-1}–3	627
		R(K)	5×10^{-2}–4×10^{-1}	628
	2-(4-Methoxyphenyl)-5,7-diphenyl-4H-1,3,4-thiadiazolo-[3,2-a]pyridinium chloride + Cl^-	TC(E)	4–80	1060
	Rhodamine B + Cl^- or Br^-	TC(E)	2×10^{-4}–2×10^{-1}	1061–1065
	Rhodamine 6 G + Br^-	TC(E)	5×10^{-2}–10^{-1}	1066, 1067
	Rhodamine 6 J + Br^-	TC(E)	5×10^{-3}–2.5×10^{-1}	1068, 1069
	2,4,6-Triphenylpyrylium chloride + Cl^-	TC(E)	1.5×10^{-2}–4×10^{-1}	1070
U	Hydroxynaphthol blue	C(Q)	8×10^{-1}–3.2	1071
	N-β-Hydroxyethylanabasine	C(Q)	6×10^{-3}–1	547, 1072
	N-β-Hydroxypropylanabasine	C(Q)	6×10^{-3}–1	547
	Morin	C(Q)	5×10^{-2}	1073
	Rhodamine B + benzilic acid	TC(E)	2×10^{-1}–10	1074
	Rhodamine B + benzoic acid	TC(E)	2×10^{-3}–4	1075–1078
	Rhodamine B + 5,7-Dinitro-8-hydroxyquinoline	TC(E)	7×10^{-3}	1040
	Thenoyltrifluoroacetone	C	10^{-2}–1	1079
V	1-Amino-4-hydroxyanthraquinone	R	6×10^{-2}–3.6×10^{-1}	1080
		R(K)	10^{-1}–5×10^{-1}	1081
	2-Aminophenol + ClO_3^-	Cat	10^{-1}–5	1082

Table 4.4. (*Continued*)

Inorganic Substance	Reagent	Method	Application Interval or Detection Limit (ppm)	Ref.
	Benzoic acid	R	$5 \times 10^{-4} – 4 \times 10^{-1}$	1083
	4,8-Diamino-1,5-dihydroxyanthraquinone-2,6-disulfonate	Cat(K)	$4 \times 10^{-2} – 5 \times 10^{-1}$	1084
	4,8-Diamino-1,5-dihydroxyanthraquinone-2,6-disulfonate + Fe(III)	Cat(K)	$10^{-3} – 10^{-2}$	1085
	1,4-Diamino-5-nitroanthraquinone	R	$10^{-1} – 8 \times 10^{-1}$	1086
	Ga-Lumogallion	C(Q)	$2 \times 10^{-2} – 3 \times 10^{-1}$	1087
	4-(2-Pyridylazo)resorcinol + safranine T	TC(E)	1.4×10^{-3}	1088, 1089
	Bromate-bromide-ascorbic acid	Cat(K)	$2 \times 10^{-2} – 20$	1090
	1,3,5-Triphenyl-Δ^2-pyrazoline	R(K)	$3 \times 10^{-2} – 1.5 \times 10^{-1}$	1091
W	Alizarin red S	C	$5 \times 10^{-1} – 4$	1092
	Carminic acid	C	$4 \times 10^{-2} – 3.6 \times 10^{-1}$	834, 835
	Flavonol	C	$6 \times 10^{-2} – 4.2 \times 10^{-1}$	1093–1095
	Lumogallion	C	$7 \times 10^{-2} – 1.1$	1096
	Morin	C	$3 – 48$	722
	Rhodamine B	C(Q)	$1 – 23$	1097
Y	Bis(8-hydroxy-2-quinolyl)methylamine	C(E)	2×10^{-3}	580, 581, 1098, 1098a
	5,7-Dibromo-8-hydroxyquinoline	C(E)	1.7×10^{-2}	1099
	2,4-Dihydroxybenzaldehydesemicarbazone	C	$4 \times 10^{-2} – 1.2$	250
	5,7-Dinitro-8-hydroxyquinoline + rhodamine B	TC(E)	2×10^{-2}	1040

422

8-Hydroxyquinoline	C(E)	2×10^{-2}	749
Rhodamine B + 3,5-di-iodosalicylic acid	TC(E)	4×10^{-3}	571
Zn Benzimidazole-2-aldehyde-2-quinolylhydrazone	C	$0-10^{-3}$	1100
Benzyl-2-pyridylketone-2-quinolylhydrazone	C	$10^{-2}-3.5 \times 10^{-1}$	1101
Benzyl-2-pyridylketone-2-pyridylhydrazone	C	$1.5 \times 10^{-2}-6.5 \times 10^{-2}$	1102
8-Carboxymethyl-8-hydroxyquinoline	C	$6.5 \times 10^{-2}-6.5 \times 10^{-1}$	490
2,2'-Azodiphenol	C	—	483a
Dibenzothiazoylmethane	C	$5 \times 10^{-2}-50$	779, 1103
2,4-Dihydroxybenzaldehyde semicarbazone	C	$4 \times 10^{-2}-1.2$	250
8-Hydroxyquinoline	C	$4 \times 10^{-1}-12$	1104-1106
8-Hydroxyquinoline + zephiramine	C(E)	$0-20$	1107
8-Hydroxyquinoline-5-sulfonic acid + methyltrioctylammonium chloride	TC(E)	$3.2 \times 10^{-1}-16.2$ µg	1108
2-Hydroxybenzaldehyde thiosemicarbazone + H_2O_2 + Mn(II)	Cat(K)	$5 \times 10^{-2}-4 \times 10^{-1}$	1109
Lumocupferron	C	2×10^{-1}	1110
2-(4-Methyl-2-pyridyl)-5(6)-phenylbenzimidazole	C	$10^{-2}-2 \times 10^{-2}$	1111
1,10-Phenanthroline + dibromofluorescein	TC(E)	$10^{-2}-2 \times 10^{-1}$	493
1,10-Phenanthroline + eosin	TC(E)	$10^{-2}-10^{-1}$	241
Picolinaldehyde-2-quinolylhydrazone	C	$2.6 \times 10^{-2}-3.1 \times 10^{-1}$	1112
Pyridoxal nicotinylhydrazone + ammonia, ethylenediamine or pyridine	TC	$5 \times 10^{-2}-5.5 \times 10^{-1}$	1113
2-(2-Pyridyl)benzimidazole	C	$1.5 \times 10^{-2}-8 \times 10^{-1}$	682
8-Quinolinethiol	C(E)	$0-6 \times 10^{-1}$	1114, 1115
8-Quinolyl dihydrogen phosphate	C	1	495
Resorcylalmethionine	C	$4 \times 10^{-4}-4 \times 10^{-1}$	1116

Table 4.4. (*Continued*)

Inorganic Substance	Reagent	Method	Application Interval or Detection Limit (ppm)	Ref.
	Rhodamine B + SCN$^-$	TC(Q),TC(E)	8×10^{-2}–1.7	1117, 1118
	Salicylaldehyde formylhydrazone	C	10^{-1}–2%	1119
	Salicylaldithiocarbazinic acid esters	C	1–6	496, 1120
	4-Salicylideneaminoantipyrine	C	8×10^{-3}	1120a
	4-Salicylideaminobenzothiazole	C(E)	3×10^{-4}–10^{-1}	1121
	β-(Salicylideneamino)ethanol	C	3×10^{-1}	1122
	Thiophene-2-aldehyde-2-quinolylhydrazone	C	2×10^{-2}–2×10^{-1}	1123
	8-(Toluene-p-sulfonamide)quinoline	C(E)	5.2×10^{-1}—6.5	497, 1124, 1125
Zr	2,2'-Azodiphenol	C	10^{-2}	1145a
	Calcein Blue	C	10^{-4}–10^{-3}	1126
	Datiscin	C	5×10^{-3}–3	1127
	Flavonol	C	1.5×10^{-1}–1.5	718, 1128
	3-Hydroxychromone + SO$_4^{2-}$	TC	0–3.2×10^{-1}	1129
	8-Hydroxyquinoline	C(E)	1.1×10^{-1}–1.1	1130
	β-Hydroxy-1-naphthyl-4-aminoantipyrine	C	10^{-1}	1131
	Morin	C	2×10^{-2}–4×10^{-1}	1132–1135
	Myricetin	C	5×10^{-2}–6×10^{-1}	723
	Quercetin	C(E)	2×10^{-3}–3	1136, 1137
	Quercetin-7-glucoside	C	3.3×10^{-4}–3.3×10^{-2}	275, 729, 1138

Quercetin sulfonic acid	C	4×10^{-2}–1.6×10^{-1}	730, 1139
2-Quinolylfluorone	C	3.6×10^{-5}–9×10^{-4}	1140, 1141
2(Resorcylideneamino)phenol	C	—	1142
4-Salicylideneaminoantipyrine	C	8×10^{-3}	1143, 1144
Salicylaldehyde formylhydrazone	C	—	1119
Salicylic acid + rhodamine B	TC(E)	—	1144a
2,4',7-Trihydroxyflavone	C	5×10^{-2}	1145

Abbreviations: C, binary complex formation; TC, ternary complex formation; S, substitution reaction; R, redox reaction; Cat, catalytic reaction; Enz, enzymatic reaction; CF, fluorescent compound formation; Q, quenching; E, extraction.

europium, scandium, magnesium, indium, zirconium, and cadmium have also been determined, often using organic reagents, mainly by complex formation. Such reactions are generally applied to nontransition metal ions. Transition metal ions rarely form fluorescent chelates; however, it is possible to determine them by luminescence methods by using other types of reaction. An example of this is the paramagnetic Cu(II) ion (226a), which can be determined by a considerable number of methods. In practice, the determination of selenium in a large number of materials, with 2,3-diaminonaphthalene, has been the most applied method.

An interesting review has been published recently devoted to the fluorimetric methods of determination of anions (227). It is remarkable that approximately 70% of the methods are based on reactions of appearance of fluorescence. In general, the methods of determination of inorganic anions are based on the same kinds of reactions as the determination of cations. Borate is the anion that has been determined by the greatest number of methods, usually by complex formation. Next in order of frequency are fluoride and sulfide, which have been determined mainly by methods based on substitution reactions. Information on the luminescence determination of inorganic substances by the use of organic reagents can be found in several books and reviews (7,9,21,25,29,30,33–40,79,227–229).

4.5.1. Relationship Between Luminescence and Electronic Structure of Organic Molecules

The luminescence spectral properties of organic molecules depend on their electronic structures, on the relative positions of their lower electronically excited states, and on any change in the multiplicity and nature of their orbitals under the influence of various structural factors and intermolecular interactions. As a consequence, these influences can be considered cumulatively because any change in the molecular structure of a reagent can involve several of these factors.

Usually, all the absorbed photons lead to the lowest excited singlet state. So the nature of the lowest-lying singlet is one of the most important factors affecting the luminescent properties in the organic molecules. Only aromatic compounds and a few highly unsaturated compounds that have extended π-electron systems are highly fluorescent. This is because when the extent of the π-electron system is increased, and the energy difference between the ground state (S_0) and the first excited state (S_1) is lowered so that the rates of absorption and fluorescence increase. For these compounds the lowest-lying transition always involves a transition from a bonding π orbital to an antibonding π orbital. These transitions show high molar absorptivities and high rates of fluorescence.

When the reagent contains a heteroatom, nonbonding electrons of the heteroatom take part and $n \rightarrow \pi^*$ transitions occur. These transitions are distinguished by a relatively low molar absorptivity and low rates of fluorescence, because of the low degree of overlap between the n and π^* orbitals. So compounds with an $n \rightarrow \pi^*$ lowest lying singlet rarely show efficient fluorescence. However, the rate of intersystem crossing is increased because the energy difference between the singlet and triplet states is lower, and in the solid phase, many compounds having lowest excited singlet states of the $n \rightarrow \pi^*$ type phosphoresce efficiently.

Other factors affecting the luminescence behavior are structural rigidity, steric impediments in the molecules, and the heavy-atom, paramagnetic and substituent effects.

An increase in the structural rigidity has the effect of reducing the degree of interaction of a molecule with the medium, which tends to diminish the rate of internal conversion, resulting in an enhancement of the fluorescence. In general, steric impediments lead to nonfluorescence deactivation of the excited molecule by interaction among its groups and by preventing coplanarity of the π-system, thereby reducing the extent of the conjugation.

The presence of atoms with a high atomic number causes the excited states to have a certain degree of triplet character because of the perturbation of the electronic spins. The paramagnetic species generate a magnetic field, which also perturbs electronic spins. In both cases, the forbiddenness of intersystem crossing is lowered, allowing, consequently, the transition of the excited molecules to the triplet state, resulting in a decrease or annulment of the fluorescence intensity.

Substitution of a single functional group in an aromatic structure leads to changes in the intensity of the emission that are relatively predictable. As a general rule, substituents such as $-NH_2$, $-NHR$, $-NR_2$, $-OH$, and $-OR$ enhance the efficiency of fluorescence, varying it toward longer wavelengths, while electron-withdrawing substituents such as $-COOH$, $-COH$, $-COR$, $-NO_2$, and NO that introduce $n \rightarrow \pi^*$ transitions in the molecule lead to a decrease in the fluorescence. Other substituents such as $-SO_3H$ and alkyl groups lead to relatively slight effects in the fluorescence, while the heavy-atom effect is observed with the halogen substituents.

4.5.2. Metal Organic Complex Formation

The fluorescence resulting from the formation of a complex is a consequence of a number of competitive factors, the most important of which

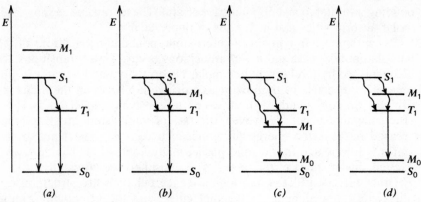

Fig. 4.1. Relative positions of lowest excited singlet state (S_1), triplet state (T_1), and metal ion (M_1, M_0), electronic levels and transitions between them in metal ion–organic reagent complexes.

is the nature and relative positions of the lower electronically excited states of the organic ligand and those of metallic ion. Other important factors are the paramagnetic character of the ion, the heavy-atom effect, and the structural rigidity.

When an organic compound, non- or weakly fluorescent, forms a complex with a metallic ion, the nonbonding electrons are associated with the metal and their energy level is lowered. As a result, formation of a metallic complex tends to increase the energies of the $n \rightarrow \pi^*$ transitions, which can change the lowest singlet excited state from $n \rightarrow \pi^*$ to $\pi \rightarrow \pi^*$. Also, an increase in the rigidity of the molecule is produced, both effects leading to a more efficient fluorescence.

If the reagent is fluorescent itself, the formation of a complex with a metallic ion can lead to a nonfluorescent structure because of the character of the ion (transition, paramagenetic, or heavy metal). Also, the complexation can lead to a fluorescent structure, with characteristics different from the original reagent. In this case the formation of a new ring, and, therefore, the enhancement of the structural rigidity, is the reason for the variations of the luminescent characteristics of the reagent.

Depending on the relative position of the lowest excited singlet state, triplet state, and electronic levels of the metallic ions, four different situations are found (Fig. 4.1):

1. If the lowest electronic level of a metallic ion (M_1) is above the S_1 level of the complex (Fig. 4.1a), then fluorescence—or more rarely, phosphorescence—is possible. This is characteristic of complexes of non-

transition metals. The metal ion acts as an inert atom being responsible for the formation of a new ring. The fluorescence spectra of these complexes are broad bands, generally located at longer wavelengths than the spectra of the reagents.

2. Transition metals use their vacant d orbitals to form complexes. In these complexes the electronic level of the metallic ion is located between the S_1 and T_1 levels of the complex (Fig. 4.1b). Hence, additional couplings can occur and thereby promote a high degree of intersystem crossing to the triplet states, because of radiationless $S_1 \rightarrow T_1$ and $S_1 \rightarrow M_1$ transitions. Moreover, the presence of new states between the usually excited states and ground state may increase the probability of internal conversion. For these reasons, usually, those complexes are nonfluorescent. Moreover, if the transition metal ion presents the internal heavy-atom effect and/or the paramagnetic effect, there is also an increase in the rate of intersystem crossing and consequently a decrease in the fluorescence. However, with a decrease in temperature, these complexes can display phosphorescence.

3. The third case is known as sensitized luminescence and is characterized when the metal ion levels are located below T_1 levels (Fig. 4.1c). Radiationless energy transfer of electronic excitation can occur: $S_1 \rightarrow M_1$ and $S_1 \rightarrow T_1 \rightarrow M_1$, and ionic luminescence $M_1 \rightarrow M_0$ is observed. This luminescence is exhibited by complexes of rare earths near the middle of the lanthanide group [Sm(III), Eu(III), Tb(III), and Dy(III)], Cr(III), and some platinum elements. The band-width characteristics of d-d* and f-f* luminescences are interesting in that they are very narrow, almost linelike. This is a consequence of the transitions arising from states localized on the metal ion so that they are not affected very much by the vibrational structure of the whole molecule.

4. Finally, with strong field ligands, low-energy metal–ligand charge transfer absorption and emission bands may be observed with suitable metals between T_1 and M_0 (Fig. 4.1d). Assignments of the transitions responsible for the absorption bands of these chelates indicate that the luminescence results from radiative transitions that are of the ligand-to-metal charge transfer kind: $\pi^* \rightarrow d$. The complexes show long lifetimes akin to those of phosphorescence. This charge-transfer phosphorescence has been observed in chelates of 1,10-phenanthroline, 2,2'-dipyridine and 2,2',2''-terpyridine with Ru(II), Os(II), and Ir(III) (78).

Binary Systems

In these methods the ion reacts with an organic reagent, to form a fluorescent binary complex or ion-association binary complex. *Binary com-*

plex refers to a complex containing one type of central ion and one type of ligand, regardless of their stoichiometric ratio. At present, there is a considerable number of reagents used for this purpose. The most important families are summarized in the following paragraphs. The observation of the data presented in Table 4.4 shows that the reagents grouped in these families are the basis of about 60% of the proposed methods for the determination of ions by formation of binary fluorescent complexes. Most of the ligands form fluorescent chelates with the same set of ions. The less heavy ions of columns IIIB and IIA of the periodic table (A1, Ga, In, Be, Mg, and Ca) and to a lesser extent, the surrounding ions, stand out among the most often determined ions.

8-Hydroxyquinoline and Derivatives.   8-Hydroxyquinoline and its derivatives are, in general, not very specific reagents. 8-Hydroxyquinoline, also known as oxine, forms fluorescent chelates with a large number of metals; among them, the chelates of A1, Ga, Li, Mg, In, Sn, Sr, Y, Zn, and Zr are of analytical use. The uncharged chelates, Mox_n are very slightly soluble in water but dissolve in organic solvents, and they can be extracted by chloroform and other insert solvents, but only if they are unhydrated. Hydrated oxinates require an oxygenated organic solvent for extraction.

In chloroform and other water-immiscible solvents, oxinates of the nontransition metals fluoresce strongly. Oxine itself shows only an extremely weak fluorescence. Compared to absorptiometric determination, the fluorimetric has the advantages of greater selectivity and sensitivity and the disadvantage of lower precision and accuracy, arising in part from the instability of oxinates exposed to UV radiation. The reaction of complex formation is represented as follows:

$$\text{(4.1)}$$

The derivatives of 8-hydroxyquinoline that have been proposed as fluorimetric reagents are summarized in the following paragraphs, together with the ions determined: 8-hydroxyquinoline-5-sulfonic acid (A1, Cd, Ga, In, Mg, Sn), 8-hydroxyquinaldine (Be, Ga, In), 8-quinolyldihydrogen phosphate (Cd, Zn), 8-quinolinethiol (Cd, Ga, Zn), 5,7-dibromo-8-hydroxyquinoline (Li, Y), 8-carboxymethyl-8-hydroxyquinoline (Zn), and 5,7-dichloro-8-hydroxyquinoline (Sc).

Azo Reagents. Among the azo reagents are 2,2'-dihydroxyazobenzene and compounds including arsonic and phosphonic acid derivatives. These azocompounds form fluorescent complexes, mainly with aluminium, gallium, indium, and magnesium. The structure of the reacting group is:

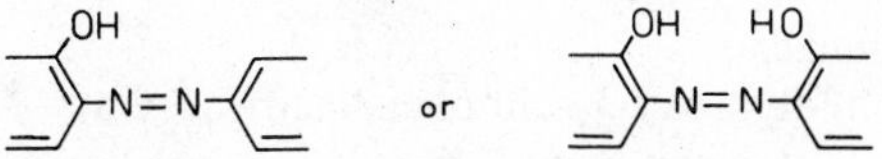

The chelate compounds formed can be of the following types:

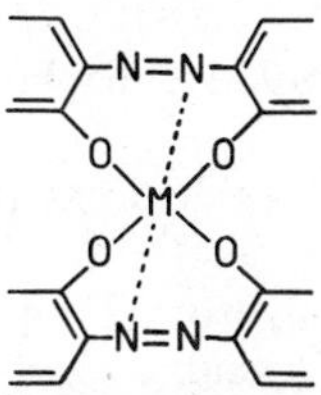

depending on the metal and its ionic charge, the pH, the excess of reagent, and other factors. In Table 4.5 are summarized the azo reagents used as fluorimetric reagents and the ions determined with each of them.

Heterocyclic Azo Reagents. The characteristic group of the heterocyclic azo reagents is represented by:

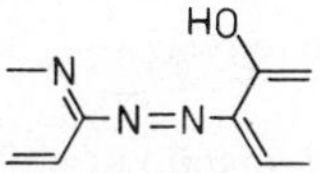

With these reagents determination methods for Al, Ga, In, Mg, and Co(III) have been proposed. The structure of the complexes formed is

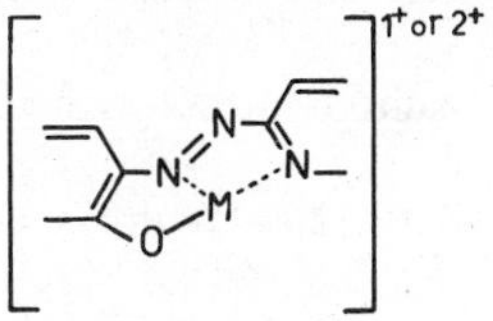

M (II or III)

Table 4.5. Azo Reagents Used in the Fluorimetric Determination of Inorganic Ions

Reagent	Determined Ion
2,2'-Azodiphenol	Cd, Zn, Zr
2'2-Dihydroxyazobenzene	Al, Mg
Mordant Blue 31 (2(2-hydroxy-5-sulfophenylazo)-1,8-dihydroxynaphthalene-3,6-disulfonic acid)	Al, Ga, Sc
Lumogallion (5-chloro-3-(2,4-dihydroxyphenylazo)-2-hydroxybenzene acid)	Al, Ga, In
Hexyllumogallion	Ga
Dodecyllumogallion	Ga
Flazo Orange (2-hydroxy-5-chlorophenylazo-2-hydroxynaphthalene)	Al
Superchrome Garnet Y (3-(2,4-dihydroxyphenylazo)-4-hydroxybenzene sulfonic acid)	Al, Ga, In, Sc
Pontachrome Blue Black R (2,2'-dihydroxy-1,1'-azonaphthalene-4-sulfonic acid)	Al
Pontachrome VSW (2,2'-dihydroxy-1,1'-naphthaleneazobenzene-5-sodium sulfonate)	Al
Sulfonaphtholazoresorcinol (4-(2,4-dihydroxyphenylazo)-3-hydroxynaphthalene-1-sulfonic acid)	Al, Ga, In, Ce
Eriochrome Blue Black B (1-(2-hydroxy-1-naphthylazo)-2-hydroxynaphthalene-4-sulfonic acid)	Al
Mordant Blue 9 (6-(5-chloro-2-hydroxy-3-sulfophenylazo)-5-hydroxynaphthalene-1-sulfonic acid)	Al
Nitromagneson (2-hydroxy-3-(2-hydroxy-1-naphthylazo)-5-nitrobenzene sulfonic acid)	Al
4-Hydroxy-3-(2-hydroxy-4-methoxy-phenylazo)benzene sulfonic acid)	Al, Ga
4-(2-Hydroxyphenylazo)resorcinol	Al, Ga
4-(5-Chloro-2-hydroxyphenylazo)resorcinol	Ga
1-(2-Hydroxy)-3-sulfo-5-chloro-phenylazo)-2'-hydroxynaphthalene	Mg
Arsenazo I (o-(1,8-dihydroxy-3,6-disulfo-2-naphthylazo)benzenearsonic acid)	Be
Chlorophosphonazo I (3-(4-chloro-2-phosphonophenylazo)chromotropic acid)	Be
Solochrome Black AS (4-sulfo-5-nitro-2-hydroxy-α-naphthalene azo-β-naphthol)	Ga
Solochrome Black WEFA (4-sulfo-6-nitro-2-hydroxy-α-naphthalene azo-β-naphthol)	Ga
Rezarson (arsonic acid(5-chloro-3-[2,4-dihydroxyphenylazo]-2-hydroxyphenyl))	Ge

Among the reagents of this group the following have been used for the fluorimetric determination of some inorganic ions: 1-(2-pyridylazo)-2-naphthol to determine Al and Co(III); lumomagneson ((2-hydroxy-3-sulfo-5-chloro-phenylazo)barbituric acid) for Al, Ga, In, and Mg; eriochrome red B (3-hydroxy-4[5-hydroxy-3-methyl-1-phenyl-4-pyrazolyl)-azo]-naphthalene-1-sulfonic acid) for Al; solochrome red ERS (4-sulfonic acid-β-naphthol-α-azo-1-phenyl-3-methyl-5-hydroxypyrazole) for Ga; and 1-(8-hydroxyquinoline-7-azo)-2-naphthol-4-sulfonic acid, 5-pyrazolone-(4-azo-2)-1-naphthol-4-sulfonic acid and 4-[4-(3-arsono-5-chloro-2-hydroxy-phenylazo)]-3-hydroxy-5-methylpryazol-2-ylbenzene sulfonic acid for Mg.

Schiff's Bases. It is possible to consider salicylidene-*o*-amino-phenol as the parent compound of this type of reagent. Its structure and the complexes that tend to form are

$$\tag{4.2}$$

The reagents summarized in Table 4.6 are well known as fluorimetric reagents and have been widely used to determine Al, Ga, Be, Mg, and Zn.

Hydrazones. Hydrazones are characterized by the grouping $\begin{matrix} \diagdown \\ C = \\ \diagup \end{matrix}$

$N - N \begin{matrix} \diagup \\ \diagdown \end{matrix}$. Several substituted hydrazones have been proposed as fluorometric reagents for metals. Among them, the compounds with the structure

form complexes with many metals, functioning as terdentate ligands. The H attached to —N— is metal replaceable and many divalent metals are extractable into organic solvents.

Table 4.6.   Schiff's Bases Used in the Fluorimetric Determination of Inorganic Ions

Reagent	Determined Ion
Salicylidene-o-aminophenol	Al, Ga
o-(Salicylideneamino)benzene arsonic acid	Al, Be
o-(Salicylideneamino)2-hydroxybenzene sulfonic acid	Al, Be, Ga, Sn
o-(Salicylideneamino)benzoic acid	Be
o-(Salicylideneamino)3,5-dimethylbenzene arsonic acid	Be
o-(Salicylideneamino)-5-methoxyphenol	Al, Be, Ga
o-(Salicylideneamino)-3-hydroxyfluorene	Al
4-Salicylideneaminoantipyrine	Ga, Zn, Zr
N-Salicylidene-4-amino-benzothiazole	Zn
N-Salicylidene-2-hydroxy-5-sulfoaniline	Al, Be, Ga
N-Salicylidene-2-hydroxy-5-carboxyaniline	Ga
N-Salicylidene-2-hydroxy-4-carboxyaniline	Al, Ga
N-Salicylidene-2-hydroxy-4-chloroaniline	Al
6-(4-Methylsalicylidenamino)-m-cresol	Al, Be, Ga
Bis(salicylideneamino)benzofuran	Mg
Bissalicylideneethylenediamine	Mg
β-(Salicylideneamino)ethanol	Zn
2-(Resorcylideneamino)phenol	Al, Zr
Resorcylidene-o-aminophenylarsonic acid	Be
Resorcylidene cysteine	Be, Ga
Resorcylideneaminoantipyrine	Ga
2-Hydroxy-4-methylaniline-N-2-hydroxy-5-methylbenzyli-dene	Be
2-Hydroxy-5-methylaniline-N-2-hydroxy-5-methylbenzyli-dene	Be
2-(3,4,5-Trihydroxybenzylideneamino)benzene arsonic acid	Be
2-(3,4,6-Trihydroxybenzylideneamino)benzene arsonic acid	Be
o-(β-Hydroxynaphthilideneimino)benzene arsonic acid	Be

$$\left[\begin{array}{c} \overset{R}{\underset{}{C}}\!=\!\overset{H}{\underset{}{N}}\!-\!N \\ N \cdots M_{\frac{1}{2}} \cdots N \end{array} \right]^{+} \xrightarrow{-H^{+}} \begin{array}{c} \overset{R}{\underset{}{C}}\!=\!N\!-\!N \\ N \cdots M_{\frac{1}{2}}\!-\!N \end{array} \tag{4.3}$$

Generally, N-heterocyclic hydrazones are fluorescent reagents for zinc. Among them, thiophene-2-aldehyde-2-quinolyl-, benzimidazole-2-aldehyde-2-quinolyl-, and picolinaldehyde-2-quinolylhydrazone have been proposed for the fluorimetric determination of this ion. Salicylaldehydeformylhydrazone has been used to determine Zn, Zr, and Al, and benzyl-2-pyridylketone-2-pyridylhydrazone for Zn, Cd, and Ga. Other ions determined with hydrazones are Ca (with 8-hydroxyquinaldehyde-8-quinolyl-); Mg (with fluoren-2-aldehyde-2-pyridyl-); Sc (with 2,4-dihydroxybenzaldehydeacetyl-); Ga (with pyridine-2-aldehyde-2-furoyl-), and Al (with isonicotinaldehyde-2,4-dihydroxybenzyliden-; 2-hydroxy-1-naphthaldehydebenzoyl-; and 3-hydroxypyridine-2-aldehyde-2-pyridyl-).

Hydroxyflavones and Chromone Derivatives. The flavone has the structure

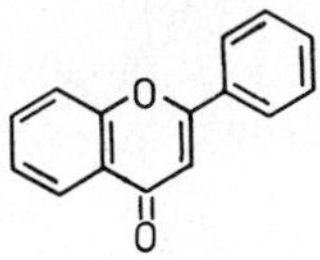

The hydroxyflavones having —OH in the 3 or 5 position or two OH groups in ortho positions form fluorescent chelates with several ions. 5-Hydroxyflavones form six-membered rings:

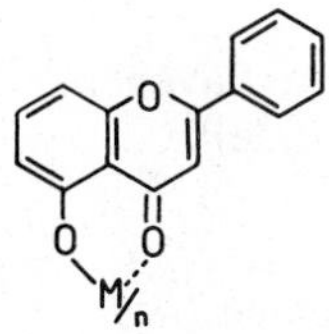

and 3-hydroxyflavones form five-membered rings:

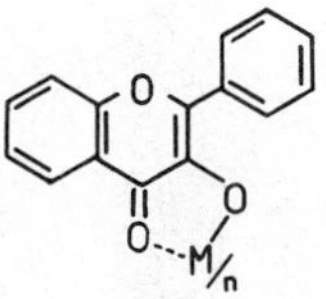

Other related compounds, with the same structural characteristics, are the chromones and the isoflavones. In Table 4.7 are summarized the proposed reagents for the fluorimetric determination of the mentioned ions.

Anthraquinones. The structure of the anthraquinone is

Table 4.7.   Hydroxyflavones and Chromone Derivatives Used in the Fluorimetric Determination of Inorganic Ions

Reagent	Determined Ion
Flavonol(3-hydroxyflavone)	Hf, Sn, Th, W, Zr
3,7-Dihydroxyflavone	Sb, Sn
2,4′,7-Trihydroxyflavone	Sb, Zr
3,3′,4′-Trihydroxyflavone	Mg
3,4′,7-Trihydroxyflavone	Sn, Th
3-Hydroxy-4′,5,7-trimethoxyflavone	Al
2-Ethyl-5-hydroxy-7-methoxyisoflavone	Be
Kaempferol(3,4′,5,7-tetrahydroxyflavone)	Ga, In
Datiscin(2′,3,5,7-tetrahydroxyflavone-3′-rutinoside)	Zr
Morin(2′,3,4′,5,7-pentahydroxyflavone)	Mg, Mo, Sb, Sn, Al, Be, Th, W, Zr, Ga, Hf, Lu
Quercetin(3,3′,4′,5,7-pentahydroxyflavone)	Al, Ge, Hf, In, Th, Zr
Quercetin 3′-glucoside	Ga
Quercetin 7-glucoside	Hf, Sc, Zr
Quercetin sulfonic acid	Hf, Zr
Quercimeritrin(3,3′,4′,5,7-pentahydroxyflavone 4-β-D-glucopyranoside)	Al
Myricetin(3,3′,4′5,5′,7-hexahydroxyflavone)	Hf, Sc, Th, Zr
Cannabiscitrin(myricetin-3′-glucoside)	Hf, In, Sc
3-Hydroxychromone	Hf, Sn
5-Hydroxychromone	Sc
2-Ethyl-3-hydroxy-5-methylchromone	Sc, Be
2-Ethyl-5-hydroxy-7-methoxyisoflavone	Sc
Isocein(8-[bis[carboxymethyl]aminomethyl]-7-hydroxy-2-methylisoflavone)	Ca

Its derivatives have been used for a long time as photometric reagents, and as fluorimetric reagents of a series of ions as well. The most widely used derivatives in fluorimetry are the hydroxyanthraquinones. They can form chelates with metallic ions to give such structures as where n generally has a value from 1 to 3. The charge of the complexes depends on the value of n, on the charge of the metallic ion, on the value of the pH of the solution, and on the nature of the medium. The fluorescence reactionability of some anthraquinonic derivatives with inorganic ions, in different media, have been reported recently (1146–1151).

The anthraquinonic derivatives proposed as fluorimetric reagents and the ions determined are summarized below: alizarin red S (1,2-dihydroxyanthraquinone-3-sulfonate) to determine Al, B, Ga, and W; anthrarrufin(1,5-dihydroxyanthraquinone) for Li and Mg; crisazin(1,8-dihydroxyanthraquinone) for Ca, Li, and Mg; quinizarin(1,4-dihydroxyanthraquinone) for B, Be, and Li; 2-quinizarin sulfonic acid for Al and Be; 5,8-dichloroquinizarin for Be; purpurin sulfonate (1,2,4-trihydroxyanthraquinone-3-sulfonate) for Al; pseudopurpurin (1,2,4-trihydroxyanthraquinone-3-carboxylic acid) for Cu; 1-hydroxy-2-carboxyanthraquinone for Be and Mg; quinalizarin (1,2,5,8-tetrahydroxyanthraquinone) for B and Be; 1-amino-4-hydroxyanthraquinone for Al and Be; 1,4-diaminoanthraquinone for Ca and carminic acid for B, Mo, and W.

Other Reagents. As we can deduce from the observation of Table 4.4, a large number of other reagents have been used for the fluorimetric determination of inorganic substances. A few of the more frequently used are mentioned here.

Several semicarbazones have been used as fluorimetric reagents; outstanding among them is the 2,4-dihydroxybenzaldehyde-semicarbazone, which has been proposed to determine Al(III), Ga(III), Sc(III), and Zn(II). The 2-hydroxybenzaldehyde semicarbazone and its 5-methyl derivative have been proposed as reagents for Sc(III). Several thiosemicarbazones have been also proposed as reagents for Ga(III); among these are 2-hydroxy-5-chlorobenzaldehyde-, 2-hydroxy-1-benzaldehyde-, 2-hydroxy-5-methylbenzaldehyde, and 2-hydroxy-1-naphthaldehyde-thiosemicarbazone.

Benzoin (α-hydroxybenzaldehydephenylacetone), in spite of being easily air-oxidized in alkaline solution and readily photodecomposed, has been widely studied as a fluorescent reagent. Deoxygenation of its solutions with nitrogen gas and brief exposure to the UV radiation are usually required for its utilization. This reagent gives fluorescent reactions with the elements Zn, B, Be, Sb, Si, Cu, and Ge. Of all these reactions the best known is the one with boron, which has allowed development of methods of detection and determination of borate. Such methods have been applied to several materials, such as soil, sea water, silicon, and iron (345–357). The reaction with antimony allows its detection (347) and determination in waters and nitric acid (927) and the reaction with silicon allows also its quantitative determination, using mannitol as a masking agent to avoid interference from boron (998). The reactions with copper (1152) and Zn (347,927) allow the detection and semiquantitative determination of these ions and the reaction of beryllium (347) and germanates (1153) have been proposed for their detections only. The reaction with germanates is given in several books for the quantitative determination of this ion. Recently, however, it has been reported that this reaction is not adequate for its detection or its quantitative determination (1154).

Calcein (2,4-bis(N,N-di(carboxymethyl)aminomethyl)fluorescein) has been widely used to determine calcium and magnesium in biological materials as well as barium, strontium, and copper (the last, by its effect of quenching on the fluorescence of the reagent).

2-Hydroxy-3-naphthoic acid has been used for the determination of Al(III), Be(II), and Sc(III). The introduction of a group complexan (dicarboxymethylaminomethyl) in position 1 in this reagent has allowed the determination of Be(II), Ca(II), Mg(II), La(III), and Lu(III).

Ternary Systems

In the ternary systems there is a central ion reacting with two different ligands. Ternary complexes have been classified by West (1155), as coordination unsaturated ternary complexes and ion association ternary complexes. Almost always ternary complexes are of interest in trace analysis because of their possibility of extraction into organic solvents. An excellent review devoted to the application of ternary complexes to spectrofluorometric analysis is the one by Haddad (229).

Coordination Unsaturated Ternary Complexes. In coordination unsaturated ternary complexes a binary complex reacts with another ligand through the free coordination positions available on the central ion. These complexes are characterized by their selectivity of formation because of the complexity of the required conditions for two ligands to react with a central ion.

Some elements of high valence, such as niobium and tantalum, often form ternary complexes of this kind. Examples are the complexes formed by these two elements with quercetin, morin or sulfonaphtholazoresorcinol, and some auxiliary complexant agents such as H_2O_2, oxalate, tartrate, or fluoride. The action of the complexant agent is to avoid the formation of polymeric-hydroxo complexes, which have a tendency to be formed with niobium and tantalum. In the complex of niobium with oxalate and sulfonaphtholazoresorcinol, the latter ligand replaces the hydroxyl groups present in the coordination sphere of the binary complex niobium–oxalate to form the ternary complex. Other examples of ternary complexes of this type are those formed by scandium with morin and antipyrine and by lutetium with morin and diantipyrylmethane. Ternary complexes of nonmetals have also been reported with the same type of ligands. The complexes of borate with morin and quercetin in presence of oxalic acid have been widely studied and applied to its determination. The proposed structure for the complex of boron with morin is

Coordination unsaturated ternary complexes have been used to determine samarium, europium, terbium, and dysoprosium, as well. The binary complexes of these elements with some organic reagents show emission known as sensitized luminescence (see Section 4.5.2). This fluorescence takes the form of a linelike spectrum that is characteristic of the metal ion itself and is easily quenched by collision with molecules of the solvent, resulting in a reduction in the sensitivity of methods based on this type of complex. To avoid this the formation of ternary complexes with a second ligand has been used. The addition of another ligand to the binary complex fills the coordination sphere of the metallic ion, preventing the interaction with the molecules of the solvent; consequently, the fluorescence is enhanced.

To improve sensitivity, use has been made of the luminescence of precipitates or suspensions of these complexes, which avoid the collisional deactivation too. However, in this case, the complexes decompose when exposed to UV light. Moreover, the luminescence strongly depends on the presence of impurities that reduce selectivity. To enhance the selectivity of these determinations, extraction of these complexes into organic solvents has been used successfully.

Examples of this type of complex, which are among those summarized in Table 4.4, are the ternary complexes formed between Tiron and EDTA or iminodiacetic acid to determine Tb and Dy; the complexes of 1,10-phenanthroline with salicylate to determine Tb, Dy, and Eu and with 2-phenylcinchoninic acid or thenoyltrifluoroacetone to determine Sm and Eu; the complexes formed between trioctylphosphine oxide and 1,1-trifluoro-5,5-dimethyl-2,4-dione or hexafluoracetylacetone to determine Tb, Sm, and Eu and 2-thenoyltrifluoroacetone with collidine or diphenylguanide to determine Sm and Eu.

Ion Association Ternary Complexes.   Ion association ternary complexes are formed between a binary-charged complex and an ion of the opposite charge. Depending on the charge of the binary complex, the association will be with a cationic or an anionic reagent.

Cationic Reagents.   Ion association ternary complexes of cationic reagents are formed by a binary complex associated with a basic dye cation (counter-ion) that generally belongs to the family of the rhodamines. These complexes may be classified into two groups: those containing an anionic binary complex formed between a metal ion and a negatively charged ligand and those formed from an anionic chelate binary complex. The negatively charged ligands used to form the metal complexes are chloride, bromide, fluoride, and thiocyanate, and the ligands used to form the anionic chelate binary complexes include 8-hydroxyquinoline derivatives, salicylic acid, phenylcinchonic acid, and benzoic acid. These methods are unselective, generally necessitating the use of masking agents or prior separation of the analyte.

The efficient fluorescence shown by rhodamine dyes is due to the oxygen bridge between the benzene rings, which promotes fluorescence in several ways. First, it serves as an electron-donating substituent. It also increases structural rigidity and forces the π systems to stay in the same plane, thereby enhancing the conjugation between them. The structure of the rhodamine dye is shown as

Table 4.8 summarizes the elements determined and the ligands used in the formation of the binary complex of each one of the rhodamine derivatives.

Table 4.8. Fluorimetric Methods Using Complexes of Rhodamine Derivatives

Counter-Ion	Ligand	Determined Ion
Rhodamine B		
$R_1, R_2, R_3, R_4 = C_2H_5$	Cl^-	As, Au, Ga
$R_5, R_6, R_7 = H$		Hg, Te, Tl
	Br^-	Au, Hg, Sn, Tl
	I^-	Hg
	SCN^-	Mo, Zn
	Molybdate	PO_4^{3-}
	Benzilic acid	U
	Benzoic acid	U
	Cupferron	Sn^a
	5,7-Dibromo-8-hydroxyquinoline	Th
	3,5-Diiodosalicylic acid	Dy, Gd, La,
		Sm, Tb, Y
	5,7-Dinitro-8-hydroxyquinoline	Ca, Th, U, Y, Sc
	Phenylcinchonic acid	Sc
	Salicylic acid	Gd, Sc, Zr
Rhodamine 3B		
$R_1, R_2, R_3, R_4, R_7 = C_2H_5$		
$R_5, R_6 = H$	Cl^-	Ga, Sb^a
	Br^-	In
Butylrhodamine B		
$R_1, R_2, R_3, R_4 = C_2H_5$	F^-	B, Ta^b
$R_5, R_6 = H$	Cl^-	As, Au, Au^a,
		Tl^a
$R_7 = C_4H_9$	Br^-	Ag, Hg^a, Tl
Rhodamine 4G		
$R_1, R_2, R_3, R_7 = C_2H_5$		
$R_4, R_5, R_6 = H$	Cl^-	Ga, In, Te
	Br^-	Ga
Rhodamine 6G		
$R_2, R_4, R_7 = C_2H_5$	F^-	Ta
$R_1, R_3 = H$	Cl^-	Ga, Sb
$R_5, R_6 = CH_3$	Br^-	Ag, In, Tl
	I^-	Hg
	Alizarin complexone	Ge
	Salicylic acid	B
Rhodamine 3GO		
$R_1, R_2, R_7 = C_2H_5$	Cl^-	Ga, In, Te
$R_3, R_4 = H$	Br^-	Ga
$R_5 = CH_3$		
Rhodamine 6J		
$R_2, R_4, R_7 = C_2H_5$	F^-	Ta
$R_1, R_3, R_5, R_6 = H$	Br^-	In, Te, Tl

[a] Organic reagent substitution technique with crystal violet.
[b] Organic reagent substitution technique with brilliant green.

441

For the determination of some ions (Tl, Sn, Ta, Au, Hg, and Sb) a new approach, known as the organic reagent substitution technique, has been reported (1059).

This technique is suggested for analyzing multicomponent samples and it involves extracting a halide complex with a triphenylmethane dye (i.e., crystal violet or brilliant green) followed by substitution of the counterion of the extracted complex by a rhodamine derivative. This combination makes it possible to take advantage of the high selectivity of the triphenylmethane group of reagents and the high sensitivity of the extraction-fluorimetric determination of the element by means of the rhodamine group of reagents.

Other dyes used as counter-ions in ion association ternary complexes are acridine orange, acridine yellow, acriflavine, safranine T, phenosafranine, pyronine G, and 3-hydroxychromone. Acridine orange allows the extraction fluorimetric determination of Ga and Sb; acridine orange, acridine yellow, and acriflavine that of Au and Tl as chloro complexes. Also, determination procedures for Cr and Sb with safranine T, for Ag with phenosafranine for Ga, Hg, and In with pyronine G, generally as chloro or bromo complexes, and the system 3-hydroxychromone-SO_4^{2-} for Zr have been proposed.

Anionic Reagents.   The ion association ternary complexes can also be formed from a cationic binary complex that is associated with an anionic reagent. This type of complex is often used to determine inorganic ions, using generally the fluorescein derivatives as the anionic reagents. The structure and luminescent properties of these reagents are similar to the rhodamine dyes. The structure of the fluorescein is

The quantum efficiency of fluorescence of the halogenated fluoresceins decreases markedly with increase in the degree of halogenation. Furthermore, the order of the fluorescence efficiency for substituted fluoresceins is Cl > Br > I because of the relative strengths of their nuclear fields, which promote increased triplet-state degeneration of the singlet excited state. On the other hand, their efficiencies of extraction into chloroform decrease with a decrease in halogenation and with a decrease in the atomic number of the halogen substituent. The derivatives used are eosin (2',4', 5',7'-tetrabromofluorescein), erythrosin (2',4',5,7'-tetraiodofluorescein), iodoeosin, 4,5-dibromofluorescein, rose bengal extra

(3,4,5,6-tetrachloro-2',4',5',7'-tetraiodofluorescein), and 3,4,5,6-tetrach-lorofluorescein. Cationic binary complexes suitable for extraction fluorometric analysis of Ag, Cd, Co, Cu, Fe, Hg, Mn, Ni, Pb, Sc, and Zn have been reported with 1,10-phenanthroline.

The complexes with these metals are non- or weakly fluorescent. The approach to determination entails the extraction, into chloroform, of the ternary complex system and the addition of ammoniacal acetone to the extract. This addition promotes an enhancement or appearance of fluorescence because of the increase in the dissociation of the ternary complex with increasing dielectric constant of the solvent mixture. If the complex is completely dissociated in the organic phase, the resulting fluorescence is entirely due to free (ionized) fluorescein dye. The fluorescent reagents are themselves only very slightly extractable into chloroform.

4.5.3. Methods Based on Substitution Reactions

Among methods based on substitution reactions are those based on the reaction of an anion with the cation that constitutes the central atom of a complex with an organic reagent. As a consequence, the organic ligand is set free, giving rise to two possibilities: (1) Methods of quenching when the organic reagent–cation complex is fluorescent and the anion liberates the reagent, which can be nonfluorescent or fluorescent with an emission spectrum different from that of the initial complex. (2) Methods of appearance of fluorescence when the organic reagent cation complex is nonfluorescent and the free reagent is fluorescent or reacts with another substance to form a fluorescent product.

All the substitution methods proposed in the references lead to the determination of anions, F^-, CN^-, and S^{2-} being the most often determined.

Since F^- forms stable and colorless complexes with Al(III), Mg(II), Th(IV), and Zr(IV), it can be determined by the quenching of the fluorescence of the complexes of Al(III) with eriochrome red B, 1-pyridylazo-2-naphthol, and superchrome Garnet Y; of Al(III) and Mg(II) with 8-hydroxyquinoline; of Al(III) and Th(IV) with morin, and of Zr(IV) with 3-hydroxyflavone. The reaction with the complex Al-1-pyridylazo-2-naphthol can be represented as

$$Al - PAN + xF^- \rightarrow AL\ F_x^{3-x} + PAN \qquad (4.4)$$

Other anions determined by quenching methods are PO_4^{3-} using the complexes Al–morin and Mg-3-hydroxy-3,4'-dimethoxyflavone; $C_2O_4^{2-}$ with Zr–flavonol; SO_4^{2-} with Th–flavonol or Th–morin; I^- with uranylacetate or di(acetoxymercuri) fluorescein; and $S_2O_3^{2-}$ with tetra(acetoxymercuri) fluorescein.

Di- and tetra(acetoxymercuri) fluorescein, in alkaline solution, show a green fluorescence, which is quenched by compounds containing the sulfhydryl group. This reaction, called the *Wronski reaction*, allows the determination of CN^- and S^{2-} and can be considered as a complex formation reaction. It has not been studied extensively, but substitution of the acetoxy group by the sulfhydryl group can occur.

The aforementioned anions can also be determined by making use of the reactions with Cu-2-(*o*-hydroxyphenyl)benzoxazole and with Pd-8-hydroxyquinoline-5-sulfonic acid. The last reaction depends on the demasking of 8-hydroxyquinoline-5-sulfonic acid by cyanide or sulfide from the nonfluorescent complex with palladium. The free reagent then coordinates with magnesium ion to form a fluorescent chelate. The reactions are shown below:

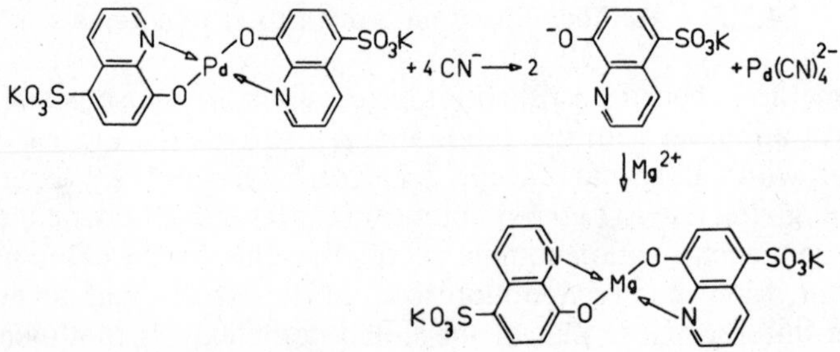

$$(4.5)$$

2,3-Diaminonaphthalene (DAN) reacts with selenite to form an orange precipitate that is soluble in ethanol and reacts with palladium(II) chloride to precipitate $Pd_2 (DAN Se)_2Cl_4$. This precipitate, known as piazselenol, is a reagent used in the determination of cyanide. On addition of the reagent to a cyanide solution, the fluorescent DAN–Se species is released as illustrated by the reaction

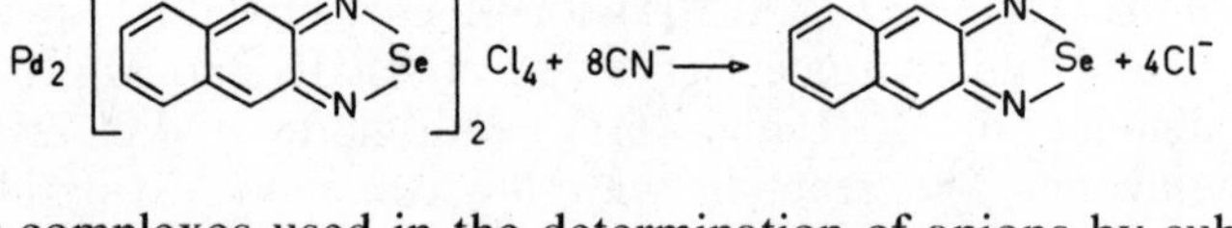

$$(4.6)$$

Other complexes used in the determination of anions by substitution methods leading to an appearance of fluorescence are Cu–calcein to determine CN^-; Hg-2-2'-pyridylbenzimidazole for S^{2-}; and Th–salicylfluorone for SO_4^{2-}.

Finally, silicon has been determined by an indirect method based on the formation of molybdosilicic acid, which is extracted with isoamyl alcohol and back-extracted with ammonia. The molybdenum is then determined by the formation of its fluorescent complex with carminic acid.

4.5.4. Methods Based on Redox Reactions

There are a number of fluorimetric methods based on redox reaction between an organic reagent and the inorganic species to be determined. Some of them have been studied from a kinetic point of view and will be discussed in Section 4.5.7. Others make use of the appearance, variation, or disappearance of fluorescence when the redox reaction reaches equilibrium.

Usually, these methods have been applied to the determination of anions, although it is also possible to determine those cations that show oxidizing or reducing characteristics. Among the inorganic substances determined by using this kind of reaction are the anions Br^-, BrO_3^-, CN^-, CrO_4^{2-}, $C_2O_4^{2-}$, IO_3^-, NO_3^-, NO_2^-, PO_4^{3-}, S^{2-}, SO_3^{2-}, $S_2O_8^{2-}$, and SCN^-; the cations Ce(IV), Cu(II), Fe(III), Hg(II), Mn(II), Sn(II), Tl(III), and V(V); and the species O_2 and O_3.

A method has been proposed to determine cyanide based on its reaction with chloroamine-T and nicotinamide. Cyanide is converted into cyanogen chloride with chloroamine-T, and the latter cleaves the pyridine ring of nicotinamide, giving a highly blue fluorescent product in alkaline medium. A similar reaction between chloramine-T and pyridine barbituric acid has been proposed to determine CN^- and SCN^-.

It is known that the reaction of cyanide with several quinone derivatives leads to fluorescent products. Methods of determination have been proposed using p-benzoquinone and quinone monoxine benzenesulfonate ester. The proposed reaction of cyanide with p-benzoquinone leads to the cyanohydroquinone derivative

$$O=\langle\ \rangle=O + 2CN^- \longrightarrow HO-\langle\ \rangle-OH \qquad (4.7)$$

Another method to determine CN^- makes use of the oxidation of leucofluorescein to fluorescein in the presence of Cu(II). The method is based on the enhancement of the oxidizing power of Cu(II) when a complex with cyanide is formed. This reagent is also oxidized by $S_2O_8^{2-}$, which allows its determination.

A method of determination of phosphate, based on the preceding effect of enhancement of the oxidizing property of a metal ion resulting from complex formation by anions, has been proposed. This anion is converted into hexamolybdatophosphate, which then reacts with nonfluorescent thiamine to produce the highly fluorescent thiochrome.

$$(4.8)$$

This reagent is also used for the determination of sulfur. The fluorescence of thiochrome, obtained by the reaction between thiamine and MnO_4^-, is quenched by the presence of S^{2-}.

Another method based on the quenching of the fluorescence of an organic reagent is the determination of sulfide by reaction with formaldehyde and 5-aminofluorescein. The aldehyde–bisulfite complex formed oxidizes 5-aminofluorescein to a nonfluorescent product.

$$(4.9)$$

Bromide has been determined by the quenching of the fluorescence of fluorescein in glacial acetic acid. Bromine substitutes for four hydrogen atoms in the fluorescein molecule forming eosin. On the other hand, nitrates oxidize fluorescein in a concentrated sulfuric acid medium to form a nonfluorescent product. So NO_3^- can be determined by the measurement of the fluorescence suppression. Another anion determined by its quenching effect is NO_2^-. This anion oxidizes benzidine, giving rise to a less fluorescent product that shows a different emission maximum. Nitrate and nitrite can also be determined by their oxidation reactions of 2,2'-dihydroxy-4,4'-dimethoxybenzophenone and 5-aminofluorescein, respectively.

The first method of spectrofluorimetric determination of IO_3^- is based on its oxidation reaction with pseudopurpurine (1,2,4-trihydroxyanthraquinone-3-carboxylic acid). The following mechanism of reaction has been proposed:

$$(4.10)$$

Other inorganic species whose fluorimetric determination is based on oxidation are oxygen and ozone. The reagents acriflavine, 9,10-dihydroacridine, epinephrine, and leucofluorescein have been used to determine oxygen and 9,10-dihydroacridine and 2-diphenylacetyl-1,3-indandione-1-hydrazone to determine ozone.

Only some cations with redox properties have been determined by using fluorimetric methods based on the oxidation or reduction of reagents to form fluorescent products. So manganese and cerium have been determined through their oxidative reactions with 8-hydroxyquinoline-5-sulfonic acid to produce highly fluorescent products, copper and mercury by the oxidation of thiamine to thiochrome and Fe(III) by the oxidation of 1,2-phenylenediamine to 2,3,-diaminophenazine. Stannous chloride, in acid solution, acts on 6-nitro-2-naphthylamine-8-sulfonic acid by reducing this compound in the presence of ammonia to the ammonium salt of 2,6-diaminonaphthalene-8-sulfonic acid, which serves for its determination. The fluorescent reaction product obtained on reacting Fe(III) with potassium hydrogen phthalate in the presence of zinc amalgam is used for the determination of this ion. The fluorescence is the result of an electron-transfer process involving Fe(II), oxygen, and phthalate at the surface of the zinc amalgam.

Finally, the oxidation reactions of several anthraquinone derivatives have been used as the basis for the determination of Fe(III), Tl(III), and V(V). The reagents used were 1,4-diamino-2,3-dihydroxyanthraquinone for Fe(III) and Tl(III) and 1,4-diamino-5-nitroanthraquinone or 1-amino-4-hydroxyanthraquinone for V(V). The oxidation process proposed in the last reaction is indicated by the following scheme:

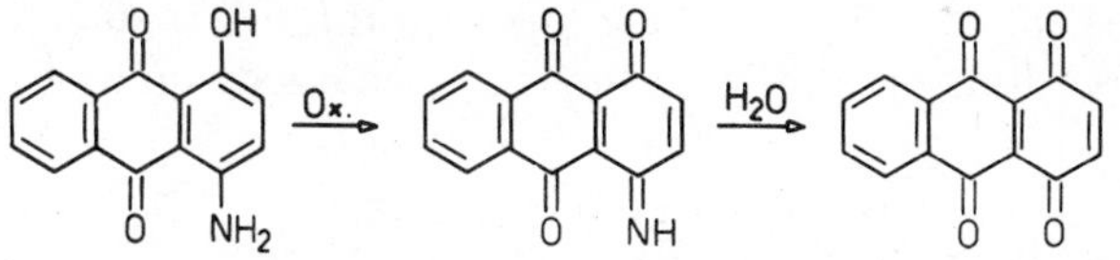

$$(4.11)$$

4.5.5. Methods Based on the Use of Catalytic Reactions

Usually, determinations based on catalysis involve oxidation-reduction systems. An oxidation-reduction reaction

$$Ox_1 + Red_1 \rightleftharpoons Ox_2 + Red_2 \qquad (4.12)$$

is likely to be slow, even though the oxidation potentials are favorable, if an unequal number of electrons are involved in the half reactions. When this is so, it has been observed that some ions effectively catalyze the reaction. A reduced (oxidized) ionic form of the catalyst must be capable of rapid oxidation (reduction) by Ox_1 (Red_1), and this form then must be rapidly reduced (oxidized) by Red_1 (Ox_1). The repetition of this cycle many times brings the system to equilibrium. The reaction rate is a function of the catalyst concentration, often a simple proportionality. The catalyzed reaction must be rapid at very low concentrations of catalyst, and Ox_1 and Red_1 when alone must show very little interaction to be useful in trace analysis.

Among the catalytic reactions used for determination of traces two groups can be distinguished. The first group comprises those reactions in which the rate of reaction is related to the concentration of catalyst. These reactions will be discussed within the kinetic methods (see Section 4.5.7). The second group comprises the nonkinetic reactions. Within this group the fluorescence measurement can be realized (1) when the reaction has been practically completed (the fluorescence does not change with the time) and (2) when the reaction is stopped before reaching equilibrium by rapid cooling, pH changes, etc. In the first group, Au(III), Co(II), and Ti(IV) have been determined by the catalysis of the aerobic oxidation of bipyridylglyoxaldiphenylhydrazone, phenyl-2-pyridylketone hydrazone, and picolinaldehyde nicotinoylhydrazone, respectively. Also, Ag(I) and V(V) have been determined by their catalytic actions on the oxidation of 8-hydroxyquinoline with persulfate and of 2-aminophenol with chlorate, respectively.

A very sensitive reaction has been proposed for determining copper; it is based on the formation of a fluorescent dimer of lumocupferron that is catalyzed by traces of copper. It has been postulated that formation of the luminescent form of the reagent passes through a stage at which divalent copper ions interact with the singly charged anion of the reagent:

$$Cu^{2+} + R^- \rightleftharpoons CuR^+ \tag{4.13}$$

The intermediate product of CuR^+ reacts with the R^- anion:

$$CuR^+ + R^- \rightleftharpoons CuR_2 \tag{4.14}$$

The CuR_2 product may break down to regenerate copper and the fluorescent dimer R_2^{2-}:

$$CuR_2 \rightleftharpoons Cu^{2+} + R_2^{2-} \tag{4.15}$$

The proposed structure for the dimer is

Using the second approach, Cu(II) and Hg(II) have been determined by their catalytic effects on the autoxidation of 2,2'-dipyridylketone hydrazone in neutral or slightly basic media, giving rise to a product that, in an acidic medium, shows an intense blue fluorescence. Acidification increases the fluorescence and stops the reaction.

Cyanide can be determined by the catalytic conversion of 4-pyridoxal to the fluorophore 4-pyridoxylactone in the reaction

$$(4.16)$$

The method consists of treating cyanide with pyridoxal at pH 7.5, heating at 50° for 60 min. The reaction was stopped by changing the pH to 10 by adding Na_2CO_3 and the fluorescence was measured.

The ions of transition metals are frequently involved in catalytic reactions. These ions are not easily determined by fluorimetry because of their structural characteristics (see Section 4.5.2). So the methods based on catalytic reactions can be an important future way for the fluorimetric determination of these ions.

4.5.6. Methods Based on the Use of Enzymatic Reactions

Enzymatic methods of analysis for inorganic species are usually based on the action of the latter on the activity of an enzyme. The rate of an enzyme-catalyzed reaction may increase (activation) or decrease (inhibition) in the presence of trace amounts of an ion allowing its determination. In other cases, the inorganic substance to be determined is one of the substrates. Enzymatic methods, however, are not very selective. Preparation and purification of the enzyme and the pH and temperature of the reaction must be carefully controlled. The ions can be determined by monitoring the reaction fluorimetrically. Information on this topic can be found in several books (1156–1158). A unique enzymic method for the determination of arsenic has been described. The enzyme glyceraldehyde-

3-phosphate dehydrogenase is used to perform an oxidation arsenolysis of 8-glyceraldehyde-3-phosphate. The rate of reaction is first order in As(V).

The conversion of nonfluorescent homovanillic acid into the highly fluorescent 2,2'-dihydroxy-3,3'-dimethoxybiphenyl-5,5'-diacetic acid is catalyzed by the oxidative enzymes peroxidase, glucose oxidase, and xanthine oxidase. Various inorganic substances inhibit these enzyme systems. This is the basis of the determination of cyanide, sulfide, dichromate, sulfite, Cu(II), Fe(II), Fe(III), Mn(II), Pb(II), and Co(II).

In a similar way, Cu(II), Fe(II), and cyanide can be determined using their inhibitory effect on the enzymic activity of hyaluronidase. The method is based on the hydrolysis of the nonfluorescent indoxyl acetate by the enzyme to give the highly fluorescent indigo white:

$$(4.17)$$

In these two reactions, the rate of change in fluorescence with time is determined and is compared to a blank containing no inhibitor. A method for the fluorimetric determination of Mn(II) in aqueous media has been developed. It is based on the acceleration by Mn(II) of the rate of enzymatic oxidation of 2,3-diketogulonate. This produces H_2O_2 and the signal observed in the fluorimetric H_2O_2 "fixed time" assay procedure is related to the Mn(II) concentration using a modified standard addition method.

The activation of isocitric dehydrogenase by Mn(II) and Mg(II) allows the determination of these ions; the inhibition of the activity of alkaline phosphate on the hydrolysis of unbelliferone phosphate to the highly fluorescent unbelliferone, causing a decrease in the slopes of the fluorescence versus time curves, is used to determine Bi(III) and Be(II).

Peroxidase catalyzes the oxidation of a wide variety of hydrogen-donating substrates with hydrogen peroxide. In the fluorimetric assay of hydrogen peroxide, based on this reaction, scopoletin, diacetyldichlorofluorescein, p-hydroxyphenylacetic acid, 3-(4-hydroxyphenyl) propionic acid, and 4-amino-1H-1,5-benzodiazepine-3-carbonitrile have been used as substrates. The reaction with diacetyldichlorofluorescein leads to diacetyldichlorofluorescein:

$$(4.18)$$

Ammonia and α-ketoglutarate react in the presence of glutamate dehydrogenase to form glutamate and oxidize a stoichiometric quantity of NADH to NAD. By suitable choice of conditions the reaction was made dependent on the NH_4^+ concentration and was followed by disappearance of NADH fluorescence. The anion phosphate can be determined through an enzymatic pathway consisting of the following three reactions. The fluorescence of $NADPH_2$ is measured.

$$\text{Glycogen} + PO_4^{3-} \xrightarrow{\text{phosphorylase}} \text{glucose-1-phosphate} \quad (4.19)$$

$$\text{Glucose-1-phosphate} \xrightarrow{\text{phosphoglucoglutase}} \text{glucose-6-phosphate} \quad (4.20)$$

$$\text{Glucose-6-phosphate} + \text{NADP} \xrightarrow{\text{glucose-6-phosphate dehydrogenase}}$$

$$\text{6-phosphogluconolactone} + NADPH_2 \quad (4.21)$$

4.5.7. Kinetic Methods

Most of the fluorimetric methods of determination of inorganic species are equilibrium methods. Such methods are based on reactions that are allowed to go to completion and the final fluorescence signal is proportional to the initial analyte concentration. However, in some cases, these reactions are relatively slow and require heating, the product is unstable, and there are interferences because of other species reacting with the reagent to form products with spectral properties similar to the product formed with the analyte. Under suitable conditions, the initial rate of the reaction, measured as a change in fluorescence signal per unit time, is linearly related to the initial analyte concentration. This is the basis of the kinetic fluorimetric methods of determination.

Until recently, because analytical methodology is dominated by measurements of systems in equilibrium or in which time is not a variable of importance, kinetic phenomena have been somewhat ignored in chemical analysis or considered mainly in the context of undesirable effects. However, in many cases, the kinetic approach may be advantageous in terms of selectivity, the reason being that the importance of numerous effects that cause nonlinear relationships between the fluorescence intensity and the analyte concentration are significantly reduced because initial rate measurements are made in the first moments of the reaction, and the concentration of the fluorescent species is much smaller than that at the completion of the reaction. Other advantages of kinetic methods are:

1. *Saving of time:* These reactions are faster than conventional techniques and are easily automated.

2. *Possible utilization of reactions that have unfavorable equilibrium constants:* In the reactions that attain equilibrium too slowly, side reactions may occur as the reaction proceeds to completion or the reactions may not be sufficiently quantitative to be adaptable to equilibrium methods. However, kinetic methods often can be used in these cases.

Kinetic methods of determination have received increased attention since the early 1960s. Only recently these methods have been combined with fluorescence. This is in part because of an aversion on the part of most analysts to measurements performed on dynamic systems (because of the obvious problem of adding time as an experimental variable) and the assumption that kinetic determinations tend to become too mathematical.

Recently, kinetic fluorimetric methods of determination of inorganic substances have been reviewed (1159). These methods are based mainly on catalytic reactions and on enzymatic, redox, and complex formation reactions. The enzymatic kinetic methods are discussed, together with the other enzymatic nonkinetic methods, in Section 4.5.6.

Independent of the kind of reaction, all the kinetic methods are based on the measurement of the rate of a chemical reaction in which at least one of the reagents present will exhibit fluorescence.

In these kinetic methods, the "initial rate," "fixed time," "fixed fluorescence," and "induction period" methods are used. In the "initial rate" method the tangent of the initial portion of the curve of fluorescence intensity versus time is measured. The "fixed time" method involves the measurement of the fluorescence after a predetermined lapse of time; the "fixed intensity" method involves measuring the time during which the fluorescence reaches a predetermined value. In the "period of induction" method, the appearance of fluorescence is preceded by a period of induction that is proportional to the concentration of analyte.

In Table 4.9 are summarized the kinetic fluorimetric methods of determination of several inorganic substances, according to the type of reaction on which they are based.

Within the methods based on catalytic reactions, the catalytic effect of the ion is carried out on reactions of autoxidation of several azines and hydrazones, of 4,8-diamino-1,5-dihydroxyanthraquinone-2,6-disulfate, and of pyridoxal. The rest are based on the catalysis of redox reaction, except the reaction of Cu(II) with benzamine (p-diethylaminobenzylidene)-acetic acid, leading to the formation of a fluorescent dimer of the reagent.

It should be pointed out that the method of determination of vanadium, making use of the autoxidation of 4,8-dihydroxyanthraquinone-2,6-disulfonate, was also modified by incorporation of Fe(III) as an activator. In the same way, the determination of Zn(II) by means of hydrogen peroxide + 2-hydroxybenzaldehyde thiosemicarbazone is realized in the presence of manganese using its modified catalytic effect on this reaction.

The induction period of the bromate–bromide–ascorbic acid system is shortened by the catalytic action of vanadium. The bromide set free is indicated by rhodamine B or acridine red, whose fluorescence is extinguished by traces of bromine. A linear relation exists between the vanadium concentration and the reciprocal of the reaction time.

Within the group of methods based on redox reactions, an interesting procedure for the kinetic determination of BrO_3^- has been proposed. The method is based on the uncatalyzed bromate oxidation of benzyl-2-pyridylketone-2-pyridylhydrazone to yield a fluorescent product. This reagent forms complexes with Fe(II), Co(II), Ni(II), and Pd(II). Addition of these ions to a solution containing the chelating agent changes the rate of oxidation of the reagent by bromate by decreasing its effective concentration. The reaction rate decreases linearly with metal ion concentration. This is the basis for sensitive methods for the determination of these ions. Also, the bromate oxidation of pyridyn-2-aldehyde-2-pyridylhydrazone has been used for the kinetic determination of Co(II).

In the references, only two kinetic fluorimetric methods of determination of inorganic ions based on complex formation reactions have been proposed. Al(III) can be determined by its slow reaction with 8-hydroxyquinoline-5-sulfonic acid to form a fluorescent product, the initial rate being proportional to aluminum concentration. The reaction of fluoren-2-aldehyde-2-pyridylhydrazone with Mg(II) produces an intense green fluorescence that decreases with time. The initial-rate method is used to carry out the determination.

Table 4.9. Kinetic-Fluorimetric Methods of Determination of Inorganic Substances

Reaction	Reagent	Determined Ion
Catalytic	Benzamido(p-diethylaminobenzylidene) acetic acid	Cu(II)
	4,8-diamino-1,5-dihydroxyanthraquinone-2,6-disulfonate	V(V)[a]
	2,2'-Dipyridylketone azine	Au(III), Cu(II)
	2,2'-Dipyridylketone hydrazone	Cu(II), Pt(IV)
	Dipyridylglyoxal hydrazone	Co(II)
	2-Hydroxybenzaldehyde azine	Cu(II)
	2-Hydroxybenzaldehyde thiosemicarbazone + H_2O_2	Mn(II), Zn(II)[b], Fe(III)[c], Ag(I)[d]
	8-Hydroxyquinoline-5-sulfonic acid + persulfate	Ag(I)
	Lumomagneson + H_2O_2	Mn(II)
	Morin-Be + H_2O_2	Cu(II)
	Phenyl-2-pyridylketone hydrazone	Co(II), Cu(II)
	Picolinaldehyde nicotinoylhydrazone	Ti(IV)
	Salicylfluorone + H_2O_2	Co(II)
	Stilbexone + H_2O_2	Fe(III)
	Bromate–bromide–ascorbic acid	V(V)
Redox	1-Amino-4-hydroxyanthraquinone	Ce(IV), V(V)
	Benzyl-2-pyridylketone-2-pyridylhydrazone	BrO_3^-, Co(II), Pd(II), Fe(II), Ni(II)
	1,4-Diamino-2,3-dihydroanthraquinone	Fe(III), Tl(III)
	4,8-Diamino-1,5-dihydroxyanthraquinone-2,6-disulfonate	Ce(IV)
	1,3,5-Triphenyl-Δ^2-pyrazoline	V(V)
Complex formation	Fluoren-2-aldehyde-2-pyridyl hydrazone	Mg(II)
	8-Hydroxyquinoline-5-sulfonic acid	Al(III)
Enzymatic	D-glyceraldehyde-3-phosphate	As(III)
	Homovanillic acid	CN^-, $Cr_2O_7^{2-}$, S^{2-}, SO_3^{2-}, Cu(II), Fe(II), Pb(II), Co(II), Mn(II), Fe(III)
	Indoxylacetate	Cu(II), Fe(II), CN^-
	Umbelliferase phosphate	Be(II), Bi(III)
	Isocitric dehydrogenase	Mg(II), Mn(II)

[a] V(V) can be also determined using this reaction in the presence of Fe(II) as activator.
[b] In the presence of Mn(II).
[c] Fe(III) can also be determined using this reaction in the presence of Ag(I) as inhibitor.
[d] In the presence of Fe(III).

4.5.8. Extraction–Luminescence Methods

Solvent extraction, like other separation and preconcentration techniques, plays a very important role in analytical chemistry. In many cases, direct methods of analysis do not allow one to determine an element without interference from other components present in the sample. Among the separation and preconcentration methods, solvent extraction holds a very important place.

At present, there are no elements that cannot be extracted from an aqueous solution into an organic solvent. Extraction is used not only for the separation and preconcentration of elements but also as an integral part of methods of analysis that combine separation and subsequent determination methods.

In fluorimetric techniques, the separation of elements is often of significant importance in order to avoid their mutual effects on the measurement of the analytical signal. To apply the extraction it is necessary to know how to convert an ion into a compound that can be completely and selectively extracted. It is necessary to select this compound and then to find optimal conditions of its formation and to choose a suitable solvent. Extraction is possible if the solubility of the compound of interest in an organic solvent is higher than that in water. If the compound is present in the form of ions, they must be changed first into an uncharged complex or an ion associated with a counter-ion. High stability of the complexes to be extracted favors efficient extraction.

Oxygen-containing extractants are hard bases, and sulfur- or phosphorus-containing compounds are soft bases. Extractants with an active nitrogen atom have intermediate behavior. Thus, oxygen-containing extractants are used for hard acid metal ions and sulfur- or phosphorus-containing compounds for extracting soft ones. To extract ions with an intermediate character, compounds containing oxygen or nitrogen are often used.

Extraction is also considered a preconcentration method. In the concentrate the relative concentration of trace elements is much higher than that in the parent sample. As a result, the effective detection limits of trace elements may be lowered.

For the determination of small amounts of metals one usually employs batch extraction in separatory funnels. Usually, when combined with luminescence methods, the trace elements are determined directly in the extract. Other methods are also available in which extraction is used only as a separation or preliminary preconcentration technique and fluorimetric determinations are carried out after back-extraction or mineralization of the extract.

Around 30% of fluorimetric methods of determination entail extraction. Some elements, such as Au, Cd, In, Re, Sb, Ta, Te, and Th, are often determined in this way.

Most often, extraction fluorimetric methods use multicomponent compounds, particularly ion association and ternary coordination complexes. When none of the ion partners is capable of fluorescence, ligand exchange reactions are used. So to determine Tl(III), the metal is first extracted into benzene as an ion associated with crystal violet and then the cation of the extracted complex is substituted by butylrhodamine B. Similar procedures were used to determine Au, Sb, Ta, Sn, and Hg. There is no need to carry out exchange reactions if the extractable ionic associate has a fluorescent component. Examples were cited in Section 4.5.2, "Ternary Systems," dealing with ternary complexes.

There is also a large number of examples of extraction fluorimetric methods in which a binary complex is extracted (see Table 4.4). In some of these cases the approaches have served to enhance the sensitivity and selectivity of previous methods proposed in aqueous solution.

Sometimes what is measured is not the fluorescence of the complex extracted but another complex formed later. A special case is the determination of molybdenum with 8-hydroxyquinoline and tetraphenylborate. The approach is based on the formation of a complex of molybdenum with 8-hydroxyquinoline and subsequent extraction with chloroform. H_2O_2 is added to the extract. This decomposes the molybdenum 8-quinolinol complex, forming a molybdenum peroxide complex, leaving free 8-quinolinol. Then tetraphenylborate is added to form the 8,8-diphenylboroquinoline complex, whose fluorescence is measured and related to the concentration of molybdenum.

Titanium and indium have been determined by using similar approaches. In some cases the extraction has been a useful tool in the fluorimetric determination of several anions. Methods of determination have been based on the formation of ion association ternary complexes with anions. For example, borate can be determined by extracting the complexes with butylrhodamine B and F^- and with rhodamine 6 G and salicylic acid into benzene, dichromate by extraction of the complex with safranine T and Cl^- into a mixture of dichloroethane, methanol and acetone, and phosphate by extraction of the complex with rhodamine B and molybdate into a mixture of chloroform and butanol. Based on the formation of binary systems, we can point out the only two methods for the fluorimetric determination of ClO_4^- by extraction of the ion association binary complexes with amiloride hypochlorite into isobutylmethylketone and with rhodamine G into benzene, and the method of determination of NO_2^- by extracting its complex with rhodamine 3 B into benzene. Recently, the method of determination of borate with dibenzoylmethane has been modified by extraction with 2-methylpentate-2,4-diol in isobutyl-methylketone. Also, the methods of determination of CN^- with the system Pd(II) and piazselenol, of F^- with aluminum oxinate and of NO_3^- and NO_2^- through reactions of diazotization, make use of extraction.

4.5.9. Phosphorimetric Methods

Phosphorimetry has been applied less than fluorimetry in the determination of inorganic substances. The number of experimental problems and limitations associated with traditional methods in phosphorimetry has been the primary factor in preventing broader application of this technique. In liquid solutions, at room temperature, molecular quenching mechanisms lead to fast radiationless deactivation of the triplet state, providing only extremely weak or nondetectable phosphorescence emission. Oxygen also contributes efficiently to the radiationless deactivation of the triplet state.

In order to observe phosphorescence, generally, it is necessary to stabilize the triplet state. This is possible in four different ways. (1) by using very carefully degassed solutions in order to remove most of the oxygen present in the solution, (2) by inserting the analyte molecule into a polymer matrix, (3) by using rigid matrices of organic solvents at low temperature ($-196°C$ usually), (4) by adsorbing the analyte on a variety of supports of room temperature.

The two first methods involve laborious and time-consuming preparations, and the third is the most widely used, its major disadvantage being the use of cryogenic equipment that does not always lend itself to quantitatively reproducible measurements. However, phosphorimetry has been used during the last 20 years for several practical qualitative and quantitative analytical methods, including the analysis of polynuclear aromatic compounds, coal tar fractions, air pollutants, impurities in petroleum fractions, detection of pesticides and fungicides in foods, and analysis of amino acids and pharmaceutical compounds in biological fluids (1160–1162).

Recently, the phenomenon of intense phosphorescence at room temperature (RTP) has been observed from ionic compounds adsorbed on a variety of supports, such as alumina, silica gel, sodium acetate, asbestos, and most often paper treated with different types of substances [NaOH, NaI, CH_3COONa, $Ph(CH_3COO)_4$, $AgNO_3$, etc.] that produce ionization, heavy-atom effects, or other phenomena favoring population of the triplet state.

These papers generally provide more intense room temperature phosphorescence from certain compounds than untreated paper. RTP has been proposed as a new analytical technique and a large variety of ionic compounds of biological and clinical interest have been investigated by this new method (1163–1165).

Little study has been devoted to the application of phosphorescence of complexes with organic ligands to inorganic trace analysis (79,190).

Phosphorescence spectra have been obtained for a few complexes at very low temperature. These make possible the determination of the central inorganic ions. When cooling is used, the phosphorescence lasts for a long time in rigid solution, and its spectrum occurs at longer wavelengths than the fluorescence spectrum.

Introduction into the organic molecule of certain inorganic ions removes the intercombination restriction and the molecule can pass from an excited state to its ground state via a triplet level. This results in enhancement of the phosphorescence and attenuation of the fluorescence. When complexes between inorganic ions and certain organic reagents are formed, the perturbing effect of the metal ion can be so strong that the phosphorescence is only observed when the ion is present. This effect is of interest from an analytical point of view.

The use of phosphoroscopes to record the phosphorescence spectra has made it possible to exclude almost completely the radiation of the blank caused by the fluorescence of the test solution and the cell and by the reflected light from the excitation source. The fluorescence of the reagent usually does not interfere with determination, even when a phosphoroscope is not used, since the phosphorescence spectrum of the complex is strongly shifted toward the long-wavelength side. These factors affect the increase in sensitivity and, at the same time, the specificity of the method, since only a limited number of ions can cause the complex to phosphoresce.

The phosphorescence of complexes has opened up new possibilities for extending the number of elements that can be determined by luminescence spectroscopy. Such elements as Fe, Cu, Co, Ni, and Cr are regarded as luminescence quenchers in fluorescence analysis. However, it is possible to find organic reagents with which these elements form brightly phosphorescent complexes at very low temperatures. In Table

Table 4.10. Phosphorimetric Determination of Inorganic Ions

Reagent	Determined Ion	Detection Limit	Ref.
Etioporphyrin II	Cu(II)	0.001 μg	1166, 1167
	Zn(II)	1 μg	1166
Dibenzoylmethane	Be(II)	—	1168–1170
	B(III)	0.1 ppb	362, 1171
Benzoylacetone	B(III)	0.4 ppb	1171–1173
2-(2′-Hydroxyphenyl)benzoxazole	Be(II)	10 ppb	1174
8-Hydroxyquinoline	Nb(V)	0.18 ppb	1175
Bis(8-hydroxy-2-quinolyl)-methy-lamine	Gd(III)	—	580, 581
Dibenzoylmethane + pyridine	Be(II)	0.0004 μg	1176

4.10 are shown several phosphorimetric methods of determination of inorganic ions.

Etioporphyrin II (1,4,5,8-tetramethyl-2,3,6,7-tetraethylporphyrin) was proposed for the phosphorimetric determination of Cu(II) in 1969 (1166). Zn(II) also forms a phosphorescent complex, making it possible to determine both Cu(II) and Zn(II) simultaneously. Later the sensitivity of the method of determination of Cu(II) with this reagent was enhanced by adopting a microtechnique, and the method was used to determine Cu(II) in silica and hydrofluoric acid (1167).

Be(II) forms a phosphorescent complex with dibenzoylmethane. This has made possible a method for the determination of this ion by using its green phosphorescence. The complex is extracted with carbon tetrachloride and the determination is possible in the presence of 1000-fold excesses of Ag, K, Na, Ca, Sr, Ba, Zn, Cu, Sn, Cd, Pb, Mo, Bi, Cr, Ni, and Sb. The method has been used to analyze plant streams and Be(II) was determined without preliminary concentration or removal from other impurities.

The same reagent gives with boric acid a highly sensitive phosphorescence reaction that has been proposed for the determination of submicrotraces of this substance (362). The method has been applied to the direct determination of boron in sea water and to solutions containing molybdenum (1171). A 2:1 dibenzoylmethane–boric acid chelate is formed and the most probable structure is

Benzoylacetone, under similar conditions, is capable of forming a phosphorescent complex with boric acid (1172,1173), which can be used for its determination (1171).

The phosphorescence characteristics of the reagents 2-(2′-hydroxyphenyl)benzoxazole and 2-(2′-hydroxyphenyl)benzothiazole and their metal complexes were studied in aqueous alcoholic medium. This study allowed the development of a procedure for the selective phosphorimetric determination of Be(II) with 2-(2′-hydroxyphenyl)benzoxazole, which can be applied to the direct analysis of mineral water (1174).

During the course of an investigation of the low-temperature fluorescence and phosphorescence characteristics of the metal complexes of 8-hydroxyquinoline, an intense phosphorescence was observed from Nb(V)-oxinate in the rigid clear glass formed from a mixture of diethyl ether–isopentane–ethyl alcohol and chloroform at − 196°C. Measurement of this phosphorescence allows the sensitive detection of niobium without the problems that may be encountered from the fluorescence of the reagent blank or from that of oxinates of other elements present if the total low-temperature emission intensity is measured. The Nb(V)-oxinate was extracted into chloroform from an aqueous citrate solution at pH 9.4 (1175).

The absorption and fluorescence spectra of bis(8-hydroxy-2-quinolyl) methylamine and its complexes with lanthanides at room temperature as well as its phosphorescence spectra at − 196°C in solvents of various polarities were studied in detail. It was possible to conclude that Y and Lu complexes show fluorescence; Sm, Eu, and Dy show sensitized luminescence; and the complex of Gd gives in frozen solution a broad phosphorescence band with a maximum at 615 nm. A phosphorimetric method of determination of Gd using this reagent was proposed (580,581).

The only method of determination of inorganic ions by room temperature phosphorescence has been proposed recently. Be(II) shows phosphorescence in a complex with dibenzoylmethane and pyridine, at room temperature, when adsorbed on filter paper. The phosphorescence spectrum of the complex is identical with the spectrum in the solid state and is characterized by a maximum at 527 nm. The method was used to determine Be in alkali and alkaline earth metal salts. Li, Na, K, Ag, Au, Mg, Ca, Sr, Ba, and Zn salts in 100-fold excess do not interfere (1176).

For a long time a phosphorimetric quenching method for the determination of oxygen has been known. The method uses the phosphorescence of acriflavine adsorbed on silica gel (or other solids). The green fluorescence is extremely sensitive to O_2 and is quenched by it. Less than 10^{-11} moles of O_2 can be detected (1172–1180).

4.6. TIME-RESOLVED FLUORESCENCE ANALYSIS

At present, with the use of fast time-resolved spectroscopic instrumentation, the measurement of the delay times of the lowest excited states of metal chelates is possible (1181). When the molecules pass from the excited state to the ground state the decay curves can be recorded. The fluorescences of many chelates show lifetimes ranging from a few to several tens of nanoseconds (1182).

There are several methods for the measurement of the lifetimes. One consists of the modulation of the excitation source and measurement of the difference in phase angle between the excitation and emission radiation (1183). Phase measurements are possible for phosphorescence and for much shorter fluorescence lifetimes (1184). In the second method, the luminescence is measured as a function of time after the source is turned off (1185,1186). In the third method, known as time-correlated single-photon counting, the source is repetitively pulsed. The time-to-amplitude converter is initiated by a single photon for each pulse, which produces a voltage proportional to the time between the source photon and the luminescence photon. The resulting counts versus voltage plot correspond to an intensity versus time plot, because the magnitude of the time interval between the two photons depends on the fluorescence lifetime (1187).

The determination of a mixture of metal ions in solution by steady-state fluorescence spectroscopy is a potentially difficult problem. The determination of mixtures of metals exhibiting sensitized luminescence is not very complicated because the linelike emission spectra are often well resolved. However, in those chelates exhibiting ligand luminescence, the wavelength distributions of the absorption and fluorescence spectra are essentially those of the ligand, perturbed slightly by the presence of the metal ions. The fact that the fluorescence spectra of metal chelates with the same organic ligand tend to be extremely similar has hampered applications of fluorimetry to metal-ion determination in mixtures using such reagents. Consequently, methodology designed for mixtures usually involves prior separation or the use of masking agents.

To avoid the problem, Lytle et al. (1188) suggested that the chelate fluorescence decay could be utilized in a manner similar to simultaneous kinetic schemes based on reaction rates. Also, they suggested that chelates of the same ligand with different metals should have different fluorescence decay times because of the heavy-atom effect, which produces a systematic variation in the rate constants for the excited states. If the fluorescence decay times are sufficiently different, timed-resolved analysis of mixtures may be practical. It was suggested that when the fluorescence lifetimes of metal chelates exceeded about 4 nsec, each compound in the mixed system could be determined simultaneously by analyzing the fluorescence decay curve (492).

To date, few applications of time-resolved fluorimetry to strictly analytical problems have been reported. Lytle et al. (1188) studied the characteristic lifetimes of the complexes of 8-hydroxyquinoline with Ca(II), Sc(III), Zn(II), Ga(III), and Ge(IV) in dimethylformamide (23, 18, 9, 5, and 3 nsec, respectively). In this paper, analytical procedures were suggested but not demonstrated. Later, Nishikawa et al. (1189) studied the complexes of this reagent with Al(III), Ga(III), and In(III) in $CHCl_3$, finding values of 17.8, 10.1, and 8.4 nsec, respectively, for the lifetimes. Between the Al(III) complex and the In(III) complex, there was a significant difference in the lifetimes (9.4 nsec). Aluminum and indium were determined by comparing the fluorescence intensity of the sample with that of the standards at a definite time. By using the composite decay curve, mixtures of aluminum and indium can be determined in $CHCl_3$.

The metal ion complexes of an 8-hydroxyquinoline derivative, 5-sulfo-8-hydroxyquinoline have also been studied by time-resolved fluorescence spectroscopy (492). The fluorescence lifetimes of the complexes of Cd(II), Ga(III), In(III), Zn(II), Al(III), and Mg(II) were determined and their possible use in differential kinetic analysis was discussed. As a consequence, the simultaneous determination of two compounds based on the difference in their fluorescence lifetimes has been reported (1190). The analysis of the decay curve can be applied to the simultaneous determination of aluminum and gallium, on the one hand, and of magnesium and cadmium on the other, for which the differences in lifetimes are about 7 nsec and about 4 nsec, respectively. Also, a method was reported for the determination of aluminum in magnesium alloy and for magnesium in aluminum alloy by taking advantage of the differences in the optimum pH value for aluminium (pH 4.5) and for magnesium (pH 10) oxinates formation.

A similar study on the fluorescence lifetimes of the complexes of Al(III), Ga(III), In(III), Y(III), Sc(III), Lu(III), and La(III) with 5,7-dichloro-8-hydroxyquinoline in $CHCl_3$ allowed the development of a method for the simultaneous determination of two components. The method was applied to the determination of Y(III) and Sc(III) in gadolinite, zenotine, and thortveitite (1191).

The emission properties of Al(III), Ga(III), and In(III) complexes with salycilidene-o-aminophenol, including the luminescence lifetimes, have been reported. From these results it was concluded that a systematic heavy atom effect was present to an extent suitable for a simultaneous kinetic analysis of mixtures of these metals (1192).

Finally, tetrachloro complexes of Al(III), Ga(III), In(III), and Tl(III) with rhodamine B have been studied to investigate the magnitude of the heavy-atom effect for ion pair ternary complexes and to see if a variation in the lifetimes of rhodamine B complexes occurred, enabling a selective fluorescence mixture analysis. It was found that time-resolved analysis by fluorescence for the rhodamine B complexes is not possible (1193).

A different case of using time-resolved fluorimetry is the determination of UO_2^{2+} in aquatic samples. The reason for using time resolution is to eliminate interferences from fluorescent organic constituents of the samples. Resolution is based on the long decay time of UO_2^{2+} luminescence (microsecond) compared with the nanosecond duration decays of the fluorescent organic interferences (1194,1195).

4.7. CHEMILUMINESCENCE ANALYSIS

Chemiluminescence occurs when a chemical reaction produces an electronically excited state that emits light on its return to the ground state. If chemiluminescence occurs in a biological system or is derived from one, it is known as bioluminescence. The main difference between chemiluminescence and fluorescence is the manner of excitation of the emitting molecule.

This phenomenon has been known for a long time. In 1877 Radziszewski found that lophine (2,4,5-triphenylimidazole) emitted green light in its reaction with oxygen in alkaline solution. Today many chemiluminescent systems are known (biological and nonbiological). The analytical applications have been confined to the second part of this century.

In the chemiluminescent reactions there is a two-step process involving excitation and emission.

$$A + B \rightarrow C^* + \text{other products}$$
$$\downarrow$$
$$C + h\nu_{(light\ measured)}$$

The chemiluminescence intensity I_{CL} is equal to $I_{CL} = \Phi_{CL} \times v$, where v is the rate of the reaction and Φ_{CL} is the chemiluminescence quantum yield of the product C^* defined as

$$\Phi_{CL} = \frac{\text{number of emitted photons}}{\text{number of reacting molecules}}$$

The chemiluminescence analysis of inorganic substances makes use of the ability of many ions to produce or to change the emission intensity. This change may be an enhancement of the emission (catalysis) or attenuation of the emission (quenching or inhibition). The analytical signal is the light emitted while the reaction is taking place.

Chemiluminescence is recorded by a fast photographic film or by highly sensitive photoelectric multipliers that can register quantum yields as low as 10^{-15}. I_{CL} can be measured as a function of time and the maximum intensity is usually proportional to the concentration of analyte. Alternatively, the area under the intensity versus time curve can be integrated for a known time period.

An alternative method of performing chemiluminescence analysis is to mix the reactants continuously in a flow system. The advantages of this approach are that steady-state chemiluminescence is easily measured and the value does not depend on mixing. The disadvantage of a flow system is that it consumes more reagent per measurement and it contains less information about reaction kinetics than the normal intensity-time curve.

In practice, the minimum detectable concentration in chemiluminescence assays usually depend on reactant purity rather than on light-measuring capability. Since chemiluminescence is an emission process, the response is often proportional to concentration up to the point where it is no longer possible to maintain an excess of other reactants relative to the analyte.

For most elements the minimum detectable concentration is about 10^{-3} μg/ml, and in many cases it is about 10^{-5} μg/ml, which makes chemiluminescence one of the most sensitive instrumental methods. Apart from the sensitivity, chemiluminescence analysis has other advantages. The instrumentation requirements are minimal. The only instrumental components required are a system for mixing the reactants, a light detector, and sometimes a filter to isolate a particular emission wavelength. The light detector may be taken from an instrument designed for some other purpose.

The mixing method can involve simply adding reactants to a vial and shaking. Often if small volumes are used, the reactants can be injected with sufficient force to induce mixing. Its disadvantage is low selectivity.

This is improved by preliminary separation or by using the differences between the catalytic or kinetic activity of complexes of the elements. Some extraction-chemiluminescence methods of determination have also been reported (1217,1346).

Chemiluminescence measurements can be realized in the gas phase or in the liquid phase. A gas mixture may be analyzed using a reaction taking place in the gas phase or by means of a reagent in solution or adsorbed on a solid carrier. The development of methods for determining gas components was created for the need to determine air pollutants. Several methods have been reported to determine sulfur compounds, ozone, nitrogen oxides, carbon monoxide, arsenic, antimony, tin, and selenium. These methods are summarized in Table 4.12.

For analysis in solution, a few different systems have been used. The most frequently employed are the alkaline oxidations of luminol (5-amino-2,3-dihydrophthalazine-1,4,-dione) and lucigenin (N,N'-dimethyl-9,9'-biacridinium nitrate).

Luminol shows a blue chemiluminescense at 425 nm when it oxidizes in an alkaline medium.

$$(4.22)$$

The most frequently used oxidant is hydrogen peroxide, although other oxidants such as hypochlorite, periodate, perborate, ferricyanide, and persulfate are used. Most oxidizing systems have an optimum pH of 10–11 that is usually obtained with metal hydroxides or carbonates.

Like luminol, lucigenin chemiluminesces on oxidation by hydrogen peroxide in basic solution.

$$(4.23)$$

Table 4.11.   Inorganic Substances Determined by Chemiluminescence with Luminol and Lucigenin

Luminol	
Catalyst	Ag(I), Co(II), Cr(III), Cu(II), Fe(II, III), Hg(II), Ir(III, IV), Mn(II), Mo(VI), Ni(II), Os(III, IV, VI, VIII), Pb(II), Pt(IV), Rh(III), Ru(III, IV, VI, VII), Sb(V), V(IV)
Inhibitor	Ag(I), Cd(II), Ce(IV), CN^-, Hf(IV), Hg(II), S^{2-}, Th(IV), Ti(IV), V(V), Zn(II)
Oxidant	As(V), Au(III), ClO^-, Cl^-, $Fe(CN)_6^{3-}$, Ge(IV), H_2O_2, I_2, Ir(IV), Ni(II), NO_2, O_2, O_3, PO_4^{3-}, SiO_3^{2-}, Tl(III), V(IV)
Lucigenin	
Catalyst	Ag(I), Bi(III), Ce(III), Co(II), Cr(III), Cu(II), Fe(III), Mn(II), Ni(II), Os(IV, VI, VIII), Pb(II), Pd(II), Tl(III)
Inhibitor	Cu(II)
Oxidant	H_2O_2, O_3
Reductant	Cr(II), Fe(II), Mo(V), V(IV)

The blue-green chemiluminescence at 470 nm is strongly influenced by catalysts. In Table 4.11 are shown the inorganic substances that have been determined with luminol and lucigenin, indicating its action on the chemiluminescent system.

Other classical reagents used for chemiluminescence analysis have been lophine (2,4,5-triphenylimidazole) (I), siloxene (II), and gallic acid (3,4,5-trihydroxybenzoic acid) (III).

(I)   (II)   (III)

Lophine chemiluminesces yellow (λ_{em} 525 nm) in basic ethanolic solutions when it is exposed to air. Siloxene and its derivatives are highly polymerized solids, which give chemiluminescence in their oxidations by hydrogen peroxide or permanganate in acidic solution. The color of the light changes from yellow to red, depending on the oxidizing agent, the degree of oxidation, and the color of the siloxene derivative. Gallic acid emits orange chemiluminescence in alkaline hydrogen peroxide solutions.

Recently, the following reagents have been proposed for the chemiluminescence determination of several inorganic substances: 4-chlorobenzoic acid-5-bromosalicylidenehydrazide for the determination of ClO^-, Cr(III), Ir(IV), Mn(II), Os(VIII), and Ru(IV); 2,3-dihydro-5-hydroxyphthalazine-1,4-dione for Co(II) and Fe(III); pyrogallol for Co(II); bis(2,4,6-trichlorophenyl)oxalate for H_2O_2, Cr(VI), Mo(VI), and V(V); tetrakis(dimethylamino)ethylene for oxygen; peroxidase for sulfur; and flavine mononucleotide for Cu(II).

The characteristics and descriptions of the various mechanisms of chemiluminescence and their analytical applications have been discussed in several review articles and monographs (79,1195–1205a). The practical development of these methods was mainly realized by Babko, Lukovskaya, and Dubovenko and more recently by Seitz and Hercules.

The chemiluminescent methods of determination of inorganic substances are now used in forensic studies, measurement of air pollutants, and analysis of trace metals in water and other materials. These methods are summarized in Table 4.12.

4.8. LUMINESCENT INDICATORS

The luminescence methods of titration are based on the changes produced in the emission of light of the indicator during the titration. These changes are observed visually or instrumentally. The luminescence indicators can be classified as fluorescent or chemiluminescent indicators. Generally, both are organic compounds, the main difference being that fluorescent indicators require excitation radiation while chemiluminescent indicators do not. As the colorimetric indicators, these can be used also in acid–base, precipitation, redox, and complexometric titrations.

Luminescent indicators have many applications and are generally used in the cases in which the colorimetric indicators lack sensitivty or are difficult to observe, as in turbid or colored solutions. In these cases they have a detection limit nearly one thousand-fold lower than the colorimetric indicators. Titrations using luminescent indicators must be carried out in a dark room or using a viewing box. Luminescence titration has been described extensively in several monographs and reviews (9,11, 13,17,26,79,1199,1370,1371).

Table 4.12. Determination of Inorganic Substance by Chemiluminescence

Inorganic Substance	Action	System	Detection Limit (ppm)	Registration	Ref.
Ag(I)	Cat.	Lucigenin–H_2O_2	0.1	PG	1211
	Cat.	Luminol–$K_2S_2O_8$–2,2′-dipyridyl	0.01	PG	1212
	Cat.	Luminol–$K_2S_2O_8$–NH_3	0.2	PG	1213
	Inh.	Luminol–$CuSO_4$–KCN–H_2O_2	0.001	PG	1214
	Cat.	Gallic acid–H_2O_2	0.3	PE	1215, 1216
As(V)	Ox.	Luminol–$(NH_4)_2MoO_4$–NH_4VO_3	0.006	PG	1217
	Ox.	Luminol–$(NH_4)_2MoO_4$–NH_4VO_3–2-propanol	0.003	PG	1217
	Ox.	O_3	0.15 ng	PE	1218
Au(III)	Ox.	Luminol–Cl^-	0.01	PG	1212, 1219–1221
Bi(III)	Cat.	Lucigenin–H_2O_2	0.2	PG	1222, 1223
Cd(II)	Cat.	Gallic acid–H_2O_2	0.8	PE	1215
	Inh.	Luminol–H_2O_2–Co^{2+}	0.02	PE	1224
Ce(III)	Cat.	Lucigenin–H_2O_2	0.8	PG	1225
Ce(IV)	Inh.	Luminol–H_2O_2–Cu^{2+}	0.1	PG	1226
	Ox.	Siloxene	7	PE	1227
Cl_2–ClO^-	Ox.	Luminol–H_2O_2	0.05	PG	1228, 1229
	Ox.	H_2O_2	4	PE	1230
	Ox.	4-Chlorobenzoic acid–5-bromo-salicylidene-hydrazide	0.05	PE	1231

ClO⁻	Cat.	Lophine–H₂O₂	0.005	PE	1232
Cl⁻	Ox.	Luminol–Na₃PO₄–NaBO₃	0.01	PE	1233
CN⁻	Inh.	Luminol–H₂O₂–Cu²⁺	3	PE	1234
Co(II)	Cat.	Lucigenin–H₂O₂	0.0005	PG	1235, 1236
	Cat.	Lucigenin–H₂O₂	0.00002	PE	1237, 1238
	Cat.	Luminol–H₂O₂	0.04	PG	1239, 1240
	Cat.	Luminol–H₂O₂	0.5	PE	1241
	Cat.	Luminol–H₂O₂–EDTA	0.000006	PE	1242
	Cat.	Luminol–H₂O₂–2-hydroxyethylethylenediaminetriacetic acid	0.0004	PE	1243
	Cat.	Luminol–H₂O₂–Na salicylate	0.0003	PG	1244
	Cat.	Luminol–H₂O₂–fluorescein–hexadecyltrimethylammonium bromide	0.00066	PE	1245
	Cat.	Luminol–H₂O₂–EDTA–Na₂B₄O₇–perdecanoic acid	0.005	PE	1246
	Cat.	Luminol–dimethylglyoxime	0.004	PE	1247
	Cat.	Gallic acid–H₂O₂	0.0004	PE	1215, 1216
	Cat.	Gallic acid–H₂O₂–methanol	0.00004	PE	1248
	Cat.	2,3-Dihydro-5-hydroxyphthalazine–1,4-dione–H₂O₂	6×10^{-7}	PG	1249
	Cat.	Lophine–H₂O₂	0.0001	PE	1232, 1250, 1251
	Cat.	Pyrogallol–H₂O₂	0.0005	PE	1252
CO₂	Ox.	O	100	PE	1253
C₂O₄²⁻	Red.	Ru(2,2'-bipyridine)₃²⁺	0.1	PE	1253a
Cr(II)	Red.	Lucigenin	0.003	PE	1254

Table 4.12. (*continued*)

Inorganic Substance	Action	System	Detection Limit (ppm)	Registration	Ref.
Cr(III)	Cat.	Luminol–H_2O_2–EDTA	0.00001	PE	1255–1258
	Cat.	Luminol–H_2O_2–EDTA–Br^-	7×10^{-6}	PE	1259
	Cat.	Lucigenin–H_2O_2–MeOH	0.005	PE	1260
	Cat.	Lucigenin–H_2O_2–EDTA	0.05	PG	1261
	Cat.	4-Chlorobenzoic acid 5-bromosalicylidene hy-drazide–KIO_4	0.002	PG, PE	1262
	Cat.	Lophine–H_2O_2	0.005	PE	1232, 1251
Cr(VI)	Cat.	Bis(2-4,6-trichlorophenyl) oxalate–H_2O_2–per-ylene	5	PE	1263
Cu(II)	Cat.	Luminol–H_2O_2	0.06	PG	1264–1270
	Cat.	Luminol–H_2O_2	0.5	PE	1241
	Cat.	Luminol–KCN	0.0001	PE	1271
	Cat.	Luminol–tellurate	0.001	PG	1272
	Cat.	Luminol–fluorescein–hexadecyltrimethylam-monium bromide	0.06	PE	1245
	Cat.	Lucigenin–H_2O_2	0.0002	PG	1273, 1274
	Inh.	Lucigenin–H_2O_2–$CoSO_4$	0.002	PG	1275
	Inh.	Lucigenin–H_2O_2–$MnSO_4$	0.0003	PG	1276
	Cat.	Lophine–H_2O_2	0.003	PE	1232, 1251
	Cat.	Flavine mononucleotide–H_2O_2	0.003	PE	1277
	Cat.	4-Aminonaphthalhydrazide–H_2O_2	0.0008	PE	1277a
Fe(II)	Red.	Lucigenin	0.003	PE	1278
	Cat.	Luminol–H_2O_2–Ac_2CH_2	$10^{-6}\%$	PG	1279
	Cat.	Luminol–O_2	5×10^{-6}	PE	1280

470

Fe(III)	Cat.	Lucigenin–H$_2$O$_2$	0.05	PG	1273
	Cat.	Luminol–H$_2$O$_2$–diethylenetriamine	0.0002	PE	1281
	Cat.	Luminol–H$_2$O$_2$–EDTA	0.003	PE	1243
	Cat.	2,3-Dihydro-5-hydroxyphthalazine–1,4-dione–H$_2$O$_2$	0.0003	PG	1249
Fe(CN)$_6^{3-}$	Ox.	Luminol	0.015	PG	1282, 1283
Ge(IV)	Ox.	Luminol–VO$_3^-$ –MoO$_4^{2-}$	0.005	PG	1284
Hf(IV)	Inh.	Luminol–H$_2$O$_2$–Cu^{2+}	0.02	PG	1285
Hg(II)	Inh.	Luminol–CNK–Cu^{2+}	0.002	PG	1214
	Cat.	Luminol–H$_2$O$_2$	0.2	PG	1286
H$_2$O$_2$	Ox.	Luminol–Cu^{2+}	0.5	PG	1267
	Ox.	Luminol–Cu^{2+}	0.0007	PE	1287–1289
	Ox.	Luminol–SbCl$_6^-$	0.01	PG	1290
	Ox.	Luminol–K$_3$ Fe(CN)$_6$	0.02	PG	1291
	Ox.	Luminol–hemin	0.002	PG	1292, 1293
	Ox.	Luminol–hemin–Cu^{2+} –triethanolamine	0.0003	PE	1294
	Ox.	Luminol–[Co(NH$_3$)$_4$(NO$_2$)$_2$]Cl	0.002	PG	1295
	Ox.	Luminol–Cu^{2+} –fluorescein–decyltrimethy-lammonium bromide	0.4	PE	1245
	Ox.	Lucigenin–Co(NO$_3$)$_2$	0.5 µg	PG	1296
	Ox.	Bis(2,4,6-trichlorophenyl)oxalate–perylene	0.002	PE	1297
I$_2$	Ox.	Luminol	1 µg	PG	1298
	Ox.	H$_2$O$_2$–NaClO	0.0003	PE	1298a
Ir(III,IV)	Cat.	Luminol–KIO$_4$	0.01	PG	1299–1300a
Ir(IV)	Cat.	Luminol–KIO$_4$–EDTA	0.001	PG	1301
	Ox.	Luminol	0.04	PG	1299–1300a

Table 4.12. (continued)

Inorganic Substance	Action	System	Detection Limit (ppm)	Registration	Ref.
	Cat.	Luminol–H_2O_2	0.01	PG	1302
	Cat.	4-Chlorobenzoic acid 5-bromosalicylidene hydrazine–KIO_4	0.0002	PG, PE	1262
Mn(II)	Cat.	Luminol–H_2O_2	0.01	PE	1241, 1303
	Cat.	Luminol–H_2O_2–1,10-phenanthroline–citrates	0.0005	PG	1304
	Cat.	Luminol–H_2O_2–1,10-phenanthroline–citrates	8×10^{-5} µg	PE	1305
	Cat.	Lucigenin–H_2O_2	0.0005	PG	1273, 1306
	Cat.	Lucigenin–H_2O_2–org. amine	0.1 µg	PG	1307
	Cat.	Gallic acid–H_2O_2	0.4	PE	1215
	Cat.	Lophine–H_2O_2	0.1	PE	1251
	Cat.	4-Chlorobenzoic acid 5-bromosalicylidene hydrazide	0.004	PG, PE	1262
MnO_4^-	Ox.	Siloxene	0.1 µg	PG	1308
Mo(V)	Cat.	Luminol–H_2O_2	1	PE	1309
	Red.	Lucigenin	0.03	PE	1310
	Cat.	Lophine	0.1	PE	1251
MoO_4^{2-}	Cat.	Lophine	0.5	PE	1251
	Cat.	Bis(2,4,6-trichlorophenyl)oxalate–H_2O_2–perylene	9.6	PE	1263
Ni(II)	Cat.	Luminol–H_2O_2	0.0006	—	1311
	Cat.	Luminol–$K_2S_2O_8$	0.02	PG	1312
	Ox.	Luminol–$K_2S_2O_8$–KIO_4	0.0005	PG	1313
	Cat.	Luminol–percaprinic acid	0.0002	PG	1314
	Cat.	Luminol–percaprinic acid	0.001	PE	1315
	Cat.	Lucigenin–H_2O_2	0.3	PG	1316, 1317

472

Analyte		Reagent			Ref.
NH$_3$	Ox.	O$_3$ (prev. transformation in NO)	0.001	PE	1318
NO	Ox.	O$_3$ or O	0.001	PE	1318–1321
	Ox.	H	0.1	PE	1322
NO$_2$	Ox.	O$_3$ (prev. transformation in NO)	0.001	PE	1318–1320
	Ox.	O	0.001	PE	1319, 1320
	Ox.	H	0.1	PE	1322
	Ox.	Siloxene	10^{-6}%	—	1323
	Ox.	Luminol	5 × 10^{-5}	PE	1324
NO$_2^-$, NO$_3^-$	Ox.	O$_3$ (prev. transformation in NO)	5 × 10^{-5}	PE	1325, 1326
O	Ox.	Luminol–(CH$_2$)$_2$SO-tert-butyl alcohol	1.6 × 10^{-7}	PG	1327
	Ox.	Tetrakis(dimethylamino)ethylene	—	PE	1328
O$_3$	Ox.	Ethylene	0.003	PE	1329–1331
	Ox.	Rhodamine B (silica gel)	0.001	PE	1332–1335
	Ox.	Rhodamine B–gallic acid	0.0003%	PE	1336
	Ox.	Luminol	—	—	1337, 1338
	Ox.	Lucigenin	—	—	1339
	Ox.	NO	0.004	PE	1320
Os(VIII, VI)	Cat.	Luminol–H$_2$O$_2$	0.0001	PG, PE	1212, 1340, 1341
Os(IV)	Cat.	Luminol–H$_2$O$_2$	0.001	PG, PE	1230, 1241
Os(III)	Cat.	Luminol–H$_2$O$_2$	0.01	PE	1341
Os(VIII, VI, IV)	Cat.	Lucigenin–H$_2$O$_2$	2 × 10^{-5}	PG	1342–1344, 1300a
Os(VIII)	Cat.	Lophine–H$_2$O$_2$	0.003	PE	1251
	Cat.	4-Chlorobenzoic acid–5-bromosalicylidene-hydrazide	0.005	PG, PE	1262
PO$_4^{3-}$	Ox.	Luminol–(NH$_4$)$_2$MoO$_4$–NH$_4$VO$_3$	0.003	PG	1345
	Ox.	Luminol–(NH$_4$)$_2$MoO$_4$–NH$_4$VO$_3$–butyl alcohol	0.0007	PG	1346

Table 4.12. (*continued*)

Inorganic Substance	Action	System	Detection Limit (ppm)	Registration	Ref.
Pb(II)	Cat.	Luminol–H_2O_2	1	PE	1241
	Cat.	Lucigenin–H_2O_2	0.2	PG	1347, 1348
	Cat.	Gallic acid–H_2O_2	1	PE	1215
Pd(II)	Cat.	Lucigenin–N_2H_4–O_2	—	PG	1349
Pt(IV)	Cat.	Luminol–H_2O_2	1×10^{-5}	PE	1350
	Cat.	Luminol–H_2O_2–I^-, SCN^-	0.5	PG	1302
Rh(III)	Cat.	Luminol–KIO_4	0.04	PG	1298, 1300, 1351
Ru(III, IV, VI, VIII)	Cat.	Luminol–KIO_4	0.003	PG	1212, 1352–1354
Ru(III, IV)	Cat.	Luminol–H_2O_2	0.01	PE	1355
	Cat.	Luminol–H_2O_2–EDTA	0.0003	PG	1212, 1252
Ru(IV)	Cat.	Luminol–$K_2S_2O_8$	0.06	PG	1300a, 1352
	Cat.	Luminol–H_2O_2	0.001	PG	1302
	Cat.	4-Chlorobenzoic acid–5-bromosalicylidene-hydrazide	0.01	PG, PE	1262
S^{2-}	Inh.	Luminol–I_2	0.005 µg	PG	1356
	Cat.	Peroxidase–H_2O_2	0.0001	PE	1357
	Cat.	N_3^-–Br_2–Br^-–methylene blue	0.003	PE	1358
Sb(V)	Cat.	Luminol–H_2O_2	0.05	PG	1359
	Cat.	Luminol–H_2O_2–dibuthyl ether	0.005	PG	1360
	Ox.	O_3	10 ng	PE	1218
Se(IV)	Ox.	O_3	35 ng	PE	1218

SiO_3^{2-}	Ox.	Luminol–$(NH_4)_2MoO_4$	0.01	PG	1345
	Ox.	Luminol–$(NH_4)_2MoO_4$–butyl alcohol	0.002	PG	1346
Sn	Ox.	O_3	110 ng	PE	1218
SO_3^{2-}	Red.	MnO_4^--riboflavin phosphate or brilliant ribof-lavin	0.9 ng	PE	1360a
SO_2	Ox.	$O^{\cdot}$	0.001	PG	1361
Th(IV)	Inh.	Luminol–H_2O_2–Cu^{2+}	1	PG	1362
Ti(IV)	Inh.	Luminol–H_2O_2–Cu^{2+}	0.0005	PE	1285
Tl(III)	Cat.	Lucigenin–H_2O_2	0.4	PG	1273,1363
	Ox.	Luminol–Na_3PO_4	0.05	PE	1364
U(IV)	Ox.	O_3–H_2SO_4	0.2	—	1365
U(VI)	Red.	Eu^{2+}–$HClO_4$	0.002	—	1365
V(IV)	Cat.	Luminol–O_2	0.01	PG	1366
	Cat.	Luminol–O_2–$P_2O_7^{4-}$	0.002	PG	1367
	Cat.	Luminol–H_2O_2	0.001	PG	1366–1368
	Cat.	Luminol–KIO_4	0.01	PG	1368
	Ox.	Luminol–$(NH_4)_2MoO_4$–$(NH_4)_3PO_4$	0.005	PG	1368
	Red.	Lucigenin	0.005	PG	1310
V(V)	Inh.	Luminol–H_2O_2–$[Co(NH_3)_4(NO_2)_2]Cl$	0.04	PG	1295
	Ox.	Siloxene	0.5×10^{-5}	PG	1269
	Cat.	Bis(2,4,6-trichlorophenyl)oxalate–H_2O_2–per-ylene	5	PE	1263
WO_4^{2-}	Cat.	Lophine	8	PE	1251
Zn(II)	Inh.	Luminol–H_2O_2–Co^{2+}	0.01	PE	1224

Abbreviations: Cat., catalyst; Inh., inhibitor; Ox., oxidant; Red., reductor; PG, photographic; PE, photoelectric.

475

4.8.1.   Fluorescent Indicators

Acid-Base Titrations

Indicators for acid–base titrations are based on changes of the quantum yield of fluorescence or the spectrum of fluorescence when changes in the pH are produced. At present, over two hundred fluorescence pH indicators are known. Listed in Table 4.13 (17,79) are indicators covering the pH ranges between 10 and 14.

Precipitation Titrations

Fluorescent adsorption indicators make evident the end point of the titration because of the difference in the quantum yield of fluorescence of the indicator when it is in solution and when is adsorbed onto the precipitate.

Shown in Table 4.14 are data related to the color of fluorescence, the ion determined, and the precipitation ion in several precipitation titrations using fluorescence adsorption indicators (9,17,1372,1373). Other adsorption indicators are 1-naphthol-5-sulfonic acid, coumaric acid, acridine orange, riboflavin, aurzin G, Vat blue, and 1,10-phenanthroline.

Redox Titrations

Generally, the reagents proposed as fluorescent redox indicators are few. These compounds change their fluorescent emission according to the state of oxidation of the indicator. Among these compounds we can include rhodamine B and 6G (1374) for the determination of U(IV) and U(V) with Ce(IV); derivatives of hydroxyphthalic acid (1375) for MnO_4^-, Ce(IV), $Cr_2O_7^{2-}$, and BrO_3^- titrations; complexes of Ru(II) with 2,2′-dipyridyl or 1,10-phenanthroline (1376) for titrations with Ce(IV); nitroferroin, ferroin, and complexes of Ru(II) with 1,10-phenanthroline, 2,2′-dipyridine, tris(4,4′-dimethyl-2,2′-bipyridine), 5-methyl-1,10-phenanthroline and 5,6-dimethyl-1,10-phenanthroline (1377) for iodimetric titrations; o-dianisidine (1378) for the determination of I^- or Fe(II) with Ce(IV); plasmochin (1379) for iodometric titrations; 5-hydroxyindol-2-carboxylic acid (1380) for the iodometric determination of As(III) and $S_2O_3^{2-}$, azine, and hydrazone of 2-pyridylcetone (1381) for the titration of As(III) and ascorbic acid with BrO_3^- and 1-amino-4-hydroxyanthraquinone (1382) for $Na_3A_3O_3$ and ascorbic acid with $NaBrO_3$ as titrimetric reagent.

Table 4.13. Fluorescent pH Indicators

Indicator	Acid Form Fluorescence	Basic Form Fluorescence	pH Transition Interval
Benzoflavine	Yellow	Green	−0.3–1.7
4-Methylumbelliferone (1st transition)	Green	Weak blue	0–2.0
Eosin	Nonfluorescent	Green	0–3.0
4-Ethoxyacridone	Green	Blue	1.2–3.2
Anthranilic acid (1st transition)	Nonfluorescent	Light blue	1.5–3.0
3-Amino-1-naphthoic acid (1st transition)	Nonfluorescent	Green	1.5–3.0
Copper-di-2-piridylcetone hydrazone	Blue	Nonfluorescent	2.0–4.0
1-Naphthoic acid	Nonfluorescent	Blue	2.5–3.5
Salicylic acid	Nonfluorescent	Dark blue	2.5–4.0
2-Naphthylamine	Nonfluorescent	Violet	2.8–4.4
Erythrosin, disodium	Nonfluorescent	Orange	3.0–4.0
Di-2-piridylcetone azine	Blue	Nonfluorescent	3.0–4.5
Quinine sulfate (1st transition)	Blue	Weak violet	3.0–5.0
Morin (1st transition)	Nonfluorescent	Green	3.1–4.4
Chromotropic acid	Nonfluorescent	Blue	3.5–4.5
3-Amino-1-naphthoic acid (2nd transition)	Green	Blue	4.0–6.0
Fluorescein	Pink-green	Green	4.0–6.0
Dichlorofluorescein	Blue-green	Green	4.0–6.6
Resorufin	Yellow	Orange	4.4–6.4
Anthranilic acid (2nd transition)	Light blue	Dark blue	4.5–6.0
Acridine	Green	Violet	5.2–6.6
3,6-Dihydroxyxanthone	Nonfluorescent	Blue green	5.4–7.6
2-Naphthol-6-sulfonic acid	Nonfluorescent	Blue	5.7–8.9
Quinoline	Blue	Nonfluorescent	6.2–7.2
Umbelliferone	Nonfluorescent	Blue	6.5–8.0
1-Naphthol	Nonfluorescent	Green blue	7.0–9.0
Coumaric acid	Nonfluorescent	Green	7.2–9.0
2-Naphthol-6,8-disulfonic acid	Blue	Light blue	7.5–9.1
1-Naphthol-2-sulfonic acid	Dark blue	Light blue	8.0–9.0
Morin (2nd transition)	Green	Yellow green	8.0–9.8
Quinine sulfate (2nd transition)	Weak violet	Nonfluorescent	9.5–10.0
Coumarin	Nonfluorescent	Light green	9.5–10.5
6,7-Dimethoxyquinoline-1-carboxylic acid	Yellow	Weak blue	9.5–11.0
3-Amino-1-naphthoic acid (3rd transition)	Blue	Nonfluorescent	11.6–13.0
2-Naphthoic acid	Blue	Violet	12.0–13.0
2-Naphthylamine-6,8-disulfonic acid	Blue	Yellow pink	12.0–14.0
Anthranilic acid (3rd transition)	Dark blue	Nonfluorescent	12.5–14.0

Table 4.14. Fluorescent Adsorption Indicators

Indicator	Fluorescence Color	Determined Ion	Precipitating Ion
Fluorescein	Pink	Cl^-	$Ag(I)$
Calcein	Yellow blue	$Ag(I)$	X^-, CN^-
Calcein blue	Blue	$Ag(I)$	X^-
Eosin	Green	X^-, CrO_4^{2-}	$Ag(I)$
		$Pb(II)$	PO_4^{3-}
Erythrosin	Green	$Pb(II)$	Oxalate
		X^-, CrO_4^{2-}	$Ag(I)$
		$Ag(I)$	X^-, CrO_4^{2-}
Phloxin	Yellow green	X^-	$Ag(I)$
		$Ag(I)$	X^-
Umbelliferone	Blue	X^-	$Ag(I)$
		$Ag(I)$	X^-
		$Pb(II)$	Oxalate, CrO_4^{2-}
		$Hg(II)$	Cl^-, SCN^-
		WO_4^{2-}	$Pb(II)$
4-Methylumbelliferone	Blue	$Pb(II)$	Oxalate, CrO_4^{2-}
		X^-	$Ag(I)$
		$Ag(I)$	X^-
		$Hg(I)$	Cl^-

Indicator	Color		
2-Naphtholsulfonic acid	Blue violet	Cl⁻	Ag(I)
Thioflavine S	Blue	Pb(II)	PO₄H²⁻
		X⁻	Ag(I)
Primulin	Blue	Ag(I)	X⁻
		Pb(II)	PO₄H²⁻
		X⁻	Ag(I)
Quinine sulfate	Blue	Ag(I)	X⁻
		Hg(I)	CrO₄²⁻ , Fe(CN)₆⁴⁻
		Cl⁻	Ag(I)
		WO₄²⁻	Pb(II)
Trypoflavine	Yellow green	X⁻, CrO₄²⁻ , Fe(CN)₆⁴⁻	Ag(I)
Morin	Green	Ag(I)	X⁻
Rhodamine C	Yellow green	Pb(II)	Fe(CN)₆⁴⁻
		X⁻ , CrO₄²⁻	Ag(I)
		Ag(I)	X⁻
Dichlorofluorescein	Green	Cl⁻	Ag(I)
3,6-Diamino-4-ethoxyacridine lactate	Green	X⁻	Ag(I)
	$\lambda_{ex} = 360$ nm $\lambda_{em} = 420$ nm	Zn(II)	Fe(CN)₆⁴⁻
8-Hydroxyquinoline-5-sulfonic acid	$\lambda_{ex} = 360$ nm $\lambda_{em} = 520$ nm	PO₄³⁻	La(III)

Table 4.15. Metallofluorescent Indicators

Determined Ion	Indicator	Titrimetric Reagent	Ref.
Al(III)	Methyl calcein	HEDTA	1383
	Methyl calcein blue	HEDTA	1383
	Morin	$H_2C_2O_4$	1384
Ba(II)	Calcein blue	EDTA	1385
Be(II)	Morin	Na sulfosalicylate	1384
Bi(III)	(DCM)A-2HN-4S	EDTA	1386, 1387
	TCM-(N-4S)D	EDTA	1386, 1387
	TCM-(2HN-4S)D	EDTA	1387
	TCM-(8NH-4,6DS)D	EDTA	1387
Ca(II)	4-[Bis(carboxymethyl)aminomethyl]-3-hydroxy-2-naphthoic acid	EDTA	1388, 1389
	Calcein	EDTA	1390, 1391
	Calcein blue	EGTA	1392
	Calcein blue	EDTA	1385, 1391, 1393
	Xanthoncomplexone	EDTA	1393
	Zincon–Hg(II)	EDTA	1394
	Zincon–Zn(II)	EDTA	1394
Cd(II)	8-Hydroxyquinoline 5-sulfonic acid	Co(II) oxalate	1395
	Zincon–Hg(II)	EDTA	1394
	Zincon–Zn(II)	EDTA	1394
Co(II)	Bis-2′-7′-N,N-glycinemethylene-4′,5′-dichlorofluorescein	EDTA	1396–1398
	Calcein	EDTA	1399
	(DCM)A-2HN-4S	EDTA	1386, 1387
	TCM-(N-4S)D	EDTA	1386, 1397

480

481

Table 4.15. (*continued*)

Determined Ion	Indicator	Titrimetric Reagent	Ref.
F⁻	Morin	Al(III)	1406
Ga(III)	Calcein blue	EDTA	1407
	Lumogallion	EDTA	1408
	Morin	EDTA	1384
Hg(II)	(DCM)-2HN-4S	EDTA	1386, 1387
	TCM-(N-4S)D	EDTA	1386, 1387
	TCM-(2HN-4S)D	EDTA	1387
	TCM-(8HN-4,6DS)D	EDTA	1387
	Zincon	EDTA	1394
In(III)	Morin	EDTA	1384
Lu(II)	4-[Bis(carboxymethyl)aminomethyl]-3-hydroxy-2-naphthoic acid	EDTA	1409, 1410
	Bis-(4-[methyl(carboxymethyl)amino-methyl]-3-hydroxy-2-napththoic acid)	EDTA	1409
Mg(II)	4-[Bis(carboxymethyl)aminomethyl]-3-hydroxy-2-naphthoic acid	EDTA	1388, 1389
	Calcein blue	EDTA	1392
	8-Hydroxyquinoline 5-sulfonic acid	Co(II) oxalate	1395
	7-(2-Hydroxy-4-sulfonaphthyalzo-8-quinolinol	EDTA	1411
Mn(II)	Methyl calcein	HEDTA	1385
	Methyl calcein blue	HEDTA	1383
Ni(II)	Bis-2',7'-*N*,*N*-glycinemethylene-4'-5'-dichlorofluorescein	EDTA	1396–1398
	Calcein	EDTA	1399
	Calcein blue	EDTA	1385

	(DCM)-2HN-4S	EDTA	1386, 1387
	Methyl calcein	HEDTA	1383
	Methyl calcein blue	HEDTA	1383
	TCM-(N-4S)D	EDTA	1386, 1387
	TCM-(2HN-4S)D	EDTA	1387
	TCM-(8HN-4,6DS)D	EDTA	1387
Pb(II)	Zincon–Hg(II)	EDTA	1394
	Zincon–Zn(II)	EDTA	1394
S^{2-}	Bis-2',7'-N,N-glycinemethylene-4',5'-dichlorofluorescein	EDTA^a	1412
Sc(II)	Morin	EDTA	1384
Sr(II)	Calcein blue	EDTA	1385
Th(IV)	Morin	EDTA	1384, 1413
Ti(IV)	Calcein	EDTA	1399
Yb(III)	4-[Bis(carboxymethyl)aminomethyl]-3-hydroxy-2-naphthoic acid	EDTA	1409, 1410
	Bis(4-[methyl(carboxymethyl)aminomethyl]-3-hydroxy-2-naphthoic acid)	EDTA	1409
Zn(IV)	8-Hydroxyquinoline-5-sulfonic acid	Co(II) oxalate	1395
	8-Hydroxyquinoline-5-sulfonic acid	EDTA	1414
Zr(IV)	Morin	NaF	1384
	Morin	EDTA	1415

^{a} Indirect method adding Cu(II) and titrating the excess with EDTA.

Abbreviations: (DCM)A-N-4S = 1-[N,N-(Dicarboxymethyl)amino]-naphthalene-4-sulfonic acid; (DCM)A-2HN-4S = 1-[N,N-(Dicarboxymethyl)amino]-2-naphthol-4-sulfonic acid; (DCM)A-8HN-4,6DS = 1-[N,N-(Dicarboxymethyl)amino]-8-naphthol-4,6-disulfonic acid; TCM-(N-4S)D = N,N',N'-Tricarboxymethyl-N-1-(naphthyl-4-sulfonic acid)diethylamine; TCM-(2HN-4S)D = N,N',N'-Tricarboxymethyl-N-1-(2-hydroxynaphthyl-4-sulfonic acid)diethylamine; TCM-(8HN-4,6DS)D = N,N',N'-Tricarboxymethyl-N-1-(8-hydroxynaphthyl-4,6-disulfonic acid)diethylamine; HEDTA = Hydroxyethylethylenediamine triacetic acid; EGTA = Ethyleneglycol-bis(aminoethyl)ether tetraacetic acid.

Complexometric Titrations

The indicators for complex formation titrations are often called metallofluorescent indicators; they change their fluorescence properties because of the increase or decrease of the ion concentration that is involved in a complex formation reaction. These changes result from:

1. Formation of a fluorescent product from a nonfluorescent reagent.
2. Formation of a fluorescent product whose color is different from the fluorescence of the reagent.
3. Quenching of the fluorescence of the reagent because of the formation of a nonfluorescent chelate.

Listed in Table 4.15 are metallofluorescent indicators used in complexometric titrations of several inorganic substances.

4.8.2   Chemiluminescent Indicators

Acid–Base Titrations

The use of chemiluminescent indicators in acid–base titrations is based on the fact that such indicators only emit light in an alkaline medium. Thus, bases as well as acids may be titrated to a point where light is quenched or where light appears. In these titrations all indicator systems employed contain hydrogen peroxide.

The indicators lucigenin (1416,1417), luminol (1418,1419), lophine (1416,1420), and pyrogallol (1421,1422) have been used. Lucigenin is the most satisfactory since no catalyst is required and the indicating reaction is reversible. The intensity of chemiluminescence is less than that obtained from catalyzed luminol oxidation but luminol is irreversibly destroyed in alkaline solutions and accordingly is recommended only for the titration of acids with bases.

The intensity of lucigenin and luminol chemiluminescence can be increased markedly by adding fluorescein. These mixtures have proved to be useful as reversible indicators for acid–base titrations.

Examples of applications of these indicators in the titration of dark-colored or turbid liquids are the determination of the acidity of methyl violet, cresol red, methyl orange, fluorescein, V_2O_5 solutions, milk, red wine, fruit juices, and molasses with lucigenin and turbid solid extracts with lucigenin, luminol, and pyrogallol.

Precipitation Titrations

Lucigenin in the presence of hydrogen peroxide is suitable as an adsorption indicator for the argentometric determination of I^- in the presence of Cl^- and Br^- (1426). The indicator, as a positive ion, is adsorbed onto the precipitated silver iodide, which is negatively charged with I^- and does not luminesce. At the end point, I^- is desorbed from silver iodide and the solution luminesces.

A titrimetric method for the determination of potassium has been worked out (1427). Potassium is precipitated with a known quantity of a standard sodium tetraphenyl borate solution and the excess of the latter reagent is back-titrated with a standard solution of lucigenin by observing the chemiluminescence. Aluminum in colored solutions can be determined by titrating with 8-hydroxyquinoline and the system luminol–H_2O_2–Co(II) as indicator. At the endpoint hydroxyquinoline removes Co(II) from its association with luminol, decreasing the luminescence (1428).

Siloxene has been employed to determine lead (II) (1429) by precipitating with potassium chromate solution. With the first excess of chromate, siloxene luminesces by oxidation. Using this method, sulfate can be determined by precipitating lead sulfate and titrating the excess of lead(II) as earlier. The same indicator was used to determine cadmium(II) by precipitation with potassium ferricyanide (1431).

Redox Titrations

Only luminol, lucigenin, and siloxene have been used as chemiluminescent redox indicators. The titration is carried out until light is emitted if an oxidizing agent is used. Occasionally, a reducing agent can be used as a titrant. The light emission ceases at the endpoint in this case. In Table 4.16 the inorganic substances that have been titrated using these indicators are summarized.

In redox titrations using luminol as an indicator, hypothalites in alkaline solution are the most widely used oxidizing agents. The titration is carried out to the point where chemiluminescence by oxidized luminol starts. Ammonia or ammonium salts must be absent because they react with hypohalites. Hypochlorite does not react with reducing agents as quickly as hypobromite and the titrations with this reagent must be carried out at a temperature of 80°C. Recently, I_2 has been also used as an oxidizing agent in these reactions.

The emission of light when lucigenin is oxidized by hydrogen peroxide in alkaline solution allows the use of this reagent as a redox indicator. In

Table 4.16. Redox Titrations with Chemiluminescent Indicators

Indicator	Ion	Titrating Reagent	Reference
Luminol	OCl^-, OBr^-	As_2O_3	1432, 1433
	As(III), Sb(III), $S_2O_3^{2-}$ SO_3^{2-}, S^{2-}, CN^- SCN^-	NaOBr	1425, 1432–1434
	As(III), Sb(III), $S_2O_3^{2-}$	NaOCl	1425, 1432, 1433
	As(III), SO_3^{2-}	I_2	1435, 1436
Lucigenin	$Fe(CN)_6^{4-}$, As(III) OCl^-, OBr^- Cr(III)	H_2O_2	1425, 1432, 1433
	OCl^-, OBr^-	$N_2H_4 \cdot H_2SO_4$	1432, 1433
Siloxene	I^-, Fe(II), Sn(II), Mo(III), Ti(III), As(III), Pb(II), $C_2O_4^{2-}$, H_2O_2 I_2^a, IO_3^{-a}, Ag^{+a} VO_3^-	MnO_4^-	1425, 1437
	Fe(II), As(III), I^- $C_2O_4^{2-}$, H_2O_2	Ce(IV)	1438, 1439
	Fe(II), Tl(I)	$Cr_2O_7^{2-}$	1439, 1440
	Fe(II)	VO_3^-	1441

^{a} Indirect methods.

titrations with hydrogen peroxide, chemiluminescence is produced only when peroxide is in excess. Hypohalites can be determined in the presence of chromates because lucigenin does not react with any other oxidizing agent but hydrogen peroxide. This is the base of a method of determination of Cr(III). To the sample, standardized NaOBr is added in excess to oxidize Cr(III) to Cr(VI). Subsequently, lucigenin is added and the excess NaOBr is titrated with H_2O_2.

Various oxidizing agents such as MnO_4^-, Ce(IV), $Cr_2O_7^{2-}$ and VO_3^- have been used in the titration of several inorganic substances using siloxene as an indicator. The siloxene emits light when a potential of about $+1.17$ V is obtained (1438). The reaction is instantaneous with a small excess of oxidant and no catalyst is required. The indicator must be used in acidic solutions and acts irreversibly.

The indirect methods using siloxene as an indicator listed in Table 4.16 involve the reduction of I_2 or IO_3^-, with zinc, to iodide, the precipitation of Ag(I) as AgI by adding an excess of iodide, which is then titrated in the presence of the precipitate and the addition of iron(II) in excess to determine VO_3^-.

Complexometric Titrations

Two chemiluminescent indicators have been used for end-point detection in chelatometric titrations: luminol and lucigenin. The emission reaction in alkaline medium in the presence of oxidizing agents is catalyzed by heavy metal ions. The mechanism of chemiluminescence is, however, different for the two indicators. So if heavy metal ions are titrated with EDTA, the endpoint of the titration is indicated by the disappearance of the light of luminol or by the appearance of the steady chemiluminescence of lucigenin. Using a direct method Ca(II), Sr(II), and Ba(II) (1442) can be determined complexometrically with luminol as an indicator and Cu(II) (1443), Cd(II), Zn(II), and Ni(II) (1444) can be determined with luminol or lucigenin.

When, in turn, an EDTA solution is titrated with copper(II), the first excess of titrant is indicated by the appearance of the chemiluminescent light of luminol or by the disappearance of the luminescence of lucigenin. Pb(II) and Hg(II) (1443) can be determined, by an indirect method, by adding EDTA in excess and back-titrating with copper(II) solution.

ACKNOWLEDGMENT

This chapter was prepared at the College of Pharmacy, University of Florida, Gainesville, 32610, while the authors were visiting professors with grants from the U.S.A.–Spanish Joint Committee for Scientific and Technological Cooperation and the Fulbright–MEC program, respectively. The authors wish to thank Dr. S. G. Schulman for his insight and helpful discussions concerning this manuscript and Ms. Vada Taylor for typing it.

REFERENCES

1. N. Monardes, *Joyfull News out of the New Founde Worlde*, Vol. 1, London, 1577; reprinted by Knopf, New York, 1925.

2. G. G. Stokes, *Phil. Trans. 142*, 463 (1852).

3. G. G. Stokes, "On the Application of the optical properties to detection and discrimination of organic substances," lecture given before the Chemical Society and the Royal Institution, 1864.

4. F. Goppelsröder, *J. Prakt. Chem. 101,* 408 (1867).

5. H. Kallman and G. M. Spruch (eds.), *Luminescence of Organic and Inorganic Materials,* Wiley, New York, 1962.

6. S. Udenfriend, *Fluorescence Assay in Biology and Medicine,* Academic Press, New York, Vol. 1, 1962; Vol. II, 1969.

7. A. Weissler and C. E. White, in *Handbook of Analytical Chemistry,* L. Meites (ed.), McGraw-Hill, New York, 1963, chap. 6.

8. W. A. Noyes, Jr., G. S. Hammond, and J. N. Pitts, Jr., *Advances in Photochemistry,* Interscience, New York, Vol. 1, 1963; Vol. 2 1964; Vol. 3, 1964; Vol. 4, 1966.

9. M. A. Konstantinova-Shlezinger, *Fluorimetry in Chemical Analysis of Mineral Raw Material,* Nedra, Moscow, 1965.

10. D. M. Hercules (ed.), *Fluorescence and Phosphorescence Analysis: Principles and Applications,* Wiley-Interscience, New York, 1966.

11. E. A. Bozhevolnov, *Luminescence Analysis of Inorganic Substances,* Khimia, Moscow, 1966.

12. G. G. Guilbault (ed.), *Fluorescence: Theory, Instrumentation, and Practice,* Dekker, New York, 1967.

13. K. P. Stolyarov and N. N. Grigorjev, *Introduction to Luminescence Analysis of Inorganic Substances,* Khimia, Leningrad, 1967.

14. E. J. Bowen (ed.), *Luminescence in Chemistry,* Van Nostrand, Princeton, 1968.

15. C. A. Parker, *Photoluminescence of Solutions with Applications to Photochemistry and Analytical Chemistry,* Elsevier, New York, 1968.

16. R. S. Becker, *Theory and Interpretation of Fluorescence and Phosphorescence,* Wiley-Interscience, New York, 1969.

17. C. E. White and R. J. Argauer, *Fluorescence Analysis: A Practical Approach,* Dekker, New York, 1970.

18. J. D. Winefordner (ed.), *Spectrochemical Methods of Analysis,* Advances in Analytical Chemistry and Instrumentation, Vol. 9, Wiley-Interscience, New York, 1970.

19. A. J. Pesce, G. G. Rosen, and T. L. Pasby, *Fluorescence Spectroscopy,* Dekker, New York, 1971.

20. J. D. Winefordner, S. G. Schulman, and T. C. O'Haver, *Luminescence Spectrometry in Analytical Chemistry,* Wiley-Interscience, New York, 1972.

21. G. G. Guilbault, *Practical Fluorescence: Theory, Methods and Techniques,* Dekker, New York, 1973.

22. G. H. Schenk, *Absorption of Light and Ultraviolet Radiation: Fluorescence and Phosphorescence Emission,* Allyn & Bacon, Boston, 1973.

23. E. L. Wehry (ed.), *Modern Fluorescence Spectroscopy*, Plenum, New York, Vols. 1 and 2, 1976; Vols. 3 and 4, 1981.

24. M. Pesez and J. Bartos, *Colorimetric and Fluorometric Analysis of Organic Compounds*, Dekker, New York, 1976.

25. S. G. Schulman, *Fluorescence and Phosphorescence Spectroscopy: Physicochemical Principles and Practice*, Pergamon, Oxford, 1977.

26. A. P. Golovina and L. V. Levshin, *Chemical Luminescence Analysis of Organic Substances*, Khimia, Moscow, 1978.

27. M. D. Lumb (ed.), *Luminescence Spectroscopy*, Academic Press, London, 1978.

28. N. J. Turro, *Modern Molecular Photochemistry*, The Benjamin Cumming Publishing Company, Menlo Park, Calif., 1978.

29. F. D. Snell, *Photometric and Fluorometric Methods of Analysis: Metals*, Parts 1 and 2, Wiley-Interscience, New York, 1978.

30. F. D. Snell, *Photometric and Fluorometric Methods of Analysis: Non-Metals*, Wiley-Interscience, New York, 1981.

31. W. R. Seitz, "Luminescence Spectrometry (Fluorimetry and Phosphorimetry)," in *Treatise on Analytical Chemistry*, P. J. Elving, E. J. Meehan, and I. M. Kolthoff (ed.), Part 1, Vol. 7; Wiley, New York, 1981, p. 159.

32. M. Zander, *Fluorimetrie*, Springer, Berlin, 1981.

33. C. E. White, *Ind. Eng. Chem. Anal. Ed. 21*, 104 (1949); *24*, 1965 (1952); *26*, 129 (1954); *28*, 621 (1956); *30*, 729 (1958); *32*, 47R (1960).

34. C. E. White and A. Weissler, *Anal. Chem. 34*, 91R (1962); *36*, 116R (1964); *38*, 115 R (1966); *40*, 116R (1968); *42*, 57R (1970); *44*, 182R (1972).

35. A. Weissler, *Anal. Chem. 46, 500R* (1974).

36. C. M. O'Donnell and T. N. Solie, *Anal. Chem. 48*, 175R (1976); *50*, 189R (1978).

37. E. L. Wehry, *Anal. Chem. 52*, 75R (1980); *54*, 198R (1982).

38. D. P. Shcherbov, *Zavod. Lab. 34*, 641 (1968).

39. D. P. Shcherbov and R. N. Plotnikova, *Zavod. Lab. 41*, 129 (1975).

40. R. A. Passwater, *Guide to Fluorescence Literature*, Plenum, New York, Vol. I, (1967); Vol. II, (1970); and Vol. III, (1974).

41. A. N. Zaidel and Y. Larionov, *Dokl. Akad Nauk. SSSR 16*, 443 (1937).

42. D. P. Shcherbov, O. D. Inyutina, R. N. Plotnikova, and V. I. Kosylanenko, *Zh. Anal. Khim.*, *27*, 1407 (1972); *Chem. Abstr. 78*, 11108c.

43. M. Furukawa, S. Sasaki, R. Nakashima, and S. Shibata, *Nagoya Kogyo Gijutsu Shikensho Hogoku 17*, 251 (1968), *Chem. Abstr. 70* 120780w.

44. N. S. Poluektov, A. I. Kirillov, M. A. Tishchenko, and Y. V. Zelyukova, *Zh. Anal. Khim. 27*, 707 (1967); *Chem. Abstr. 67*, 104844c.

45. P. Cuker and R. P. Weberling, *Anal. Chim. Acta 41*, 404 (1968).

46. W. A. Armstrong, D. W. Grant, and W. G. Humphreys, *Anal. Chem. 35*, 1300 (163).

47. T. Shigematsu, Y. Nishikawa, K. Hiraki, Sh. Goda, and Y. Tsujimoto, *Japan Anal. 20,* 575 (1971); *Chem. Abstr. 75,* 80124e.

48. D. T. Burns and M. Y. Qureshi, *Proc.* Roy Ir. Acad, Sec. B, *77,* 353 (1977).

49. V. Fassel, R. Heidel, and R. Huke, *Anal. Chem. 24,* 606 (1952).

50. V. Fassel and R. Heidel, *Anal. Chem. 26,* 1134 (1954).

51. T. Taketatsu, M. A. Carey, and C. V. Banks, *Talanta 13,* 1081 (1966).

52. I. M. Lomniczi de Upton and J. F. Possidoni de Albinati, *An. Assoc. Quim. Argent. 63,* 173 (1975).

53. G. Alberti and M. A. Massuci, *Anal. Chim. Acta 35,* 303 (1966).

54. B. C. Bhatt, G. C. Joshi, and D. D. Pant, *Indian J. Pure Appl. Phys. 11,* 226 (1973); *Chem. Abstr. 79,* 59563v.

55. P. Tokousbalides and T. Chrysochoos, *J. Phys. Chem. 76,* 3397 (1972).

56. B. M. Antipenko, I. M. Batyaev, and E. I. Lyubimov, *Opt. Spektrosk. 33,* 938 (1972); *Chem. Abstr. 78,* 77669t.

57. J. Chrysochoos and P. Toakousbalides, *Spectrosc. Lett. 6,* 435 (1973).

58. G. I. Sergeeva, *Khim. Vys. Energ. 8,* 38 (1974).

59. C. E. White, *Anal. Chem. 26,* 129 (1954).

60. C. E. White, *Anal. Chem. 24,* 87 (1952).

61. C. E. White, *Anal. Chem. 30,* 729 (1958).

62. T. S. Dobrolyubskaya, *Zh. Anal. Khim. 17,* 486 (1962).

63. T. S. Dobrolyubskaya, *Zh. Anal. Khim. 20,* 470 (1965).

64. T. S. Dobrolyubskaya, *Zh. Anal. Khim. 18,* 486 (1963).

65. A. Danielson, B. Ronnholm, L. E. Kjellstrom, and F. Ingman, *Talanta 20,* 185 (1973).

66. T. S. Dobrolyubskaya and L. I. Anikina, *Zh. Anal. Khim. 22,* 1841 (1967).

67. J. C. White, Rep. USAEC ORNL-2161, 30 pp. (1956).

68. J. Korkisch, F. Abascal, and C. Aguilar, *Rev. Lat. Quim. 4,* 185 (1973).

69. P. Pakalns and L. E. Ismay, *Mikrochim. Acta I,* 297; *II,* 217 (1976).

70. V. T. Athavale, L. M. Mahajan, N. R. Thakoor, and M. S. Varde, *Anal. Chim. Acta 21,* 353 (1959).

71. G. Alberti and A. Saini, *Anal. Chim. Acta 28,* 536 (1963).

72. M. Fujinami, K. Nagashima, and Y. Sugitani, *Jap. Anal. 28,* 616 (1979).

73. G. H. Dieke, *Spectra and Energy Levels of Rare Earth Ions in Crystals,* Interscience, New York, 1968.

74. D. T. Sviridov, R. K. Sviridova, and Yu. F. Smirnov, *Optical Spectra of Transition Metal Ions in Crystals,* Nauka, Moscow, 1976.

75. N. S. Poluektov, N. P. Efryushina, and S. A. Gava, *Determination of Microamounts of Lanthanides by Luminescence of Crystalophosphors,* Nankova Dumka, Kiev, 1976.

76. G. F. Imbush in *Luminescence Spectroscopy*, M. D. Lumb (ed.), Academic Press, London, 1978.

77. N. P. Efryushina and N. S. Poluektov, *Zh. Anal. Khim. 31*, 2418 (1976).

78. P. D. Fleischauer and P. Fleischauer, *Chem. Rev. 70*, 199 (1970).

79. A. P. Golovina, V. K. Runov, and N. B. Zorov, *Structure and Bonding 47*, 53 (1981).

80. S. P. McGlynn and J. K. Smith, *J. Mol. Spectr. 6*, 164 (1961).

81. S. P. McGlynn, J. K. Smith, and W. C. Neely, *J. Chem. Phys. 35* 105 (1961).

82. L. V. Volodko, A. I. Komyak, and L. E. Sleptsov, *Opt. Spectr. 23*, 730 (1967).

83. A. Bartecki and B. Jezowska-Trzebiatowska, *Nucleonika 6*, 267 (1961).

84. A. Bartecki and B. Jezowska-Trzebiatowska, *Nucleonika 6*, 277 (1961).

85. A. Bartecki and B. Jezowska-Trzebiatowska, *Nucleonika 6*, 287 (1961).

86. B. Jezowska-Trzebiatowska and A. Bartecki, *Spectrochim. Acta 18*, 799 (1962).

87. J. T. Bell and R. E. Biggers, *J. Mol. Spectr. 18*, 247 (1965).

88. J. B. Newman, *J. Chem. Phys. 43*, 1691 (1965).

89. S. De Jaegere and T. Govers, *Nature 205*, 900 (1965).

90. V. Baran, *Coll. Czech. Chem. Comm. 31*, 2093 (1966).

91. J. T. Bell and R. E. Biggers, *J. Mol. Spectr. 25*, 312 (1968).

92. D. S. Umrejko and G. N. Larkin, *Zh. Prikl. Spectr. 9*, 447 (1968).

93. S. De Jaegere and C. Gorller-Walrand, *Spectrochim. Acta 25A*, 559 (1969).

94. N. S. Poluektov, R. A. Vitkun, and S. A. Gava, *Zh. Anal. Khim. 24*, 693 (1969).

95. L. I. Anikina, V. V. Bagreev, T. S. Dobrolyubskaya, Yu. A. Zolotov, A. V. Karyakin, A. Z. Miklishanskii, N. G. Nikitina, P. N. Palei, and Yu. V. Yakovlev, *ibid. 24*, 1014 (1969).

96. Le Viet Binh and Tran Kim Anh, *Tap San Vat Ly 1*, 81 (1973).

97. N. S. Poluektov and S. A. Gava, *Zh. Anal. Khim. 25*, 1735 (1970).

98. N. S. Poluektov, N. I. Smirdova, N. P. Efryushina, S. A. Gava, and S. M. Gorshkova, *Zavod. Lab. 37*, 266 (1971).

99. N. S. Poluektov and S. A. Gava, *Zh. Prikl. Spektr. 9*, 268 (1968); Chem. Abstr., *70*, 15720d.

100. N. S. Poluektov, N. P. Efryushina, and N. I. Smirdova, *Ukr, Khim. Zh. 38*, 365 (1972).

101. L. I. Anikina, Yu. A. Balashov, E. K. Vul'fson, and A. V. Karyakin, *Zh. Anal. Khim. 27*, 787 (1972).

102. N. S. Poluektov and S. A. Gava, *Zavod. Lab. 35*, 1458 (1969).

103. L. I. Anikina, A. V. Karyakin, and Le Viet Binh, *Zh. Anal. Khim. 25*, 1731 (1970).

104. N. I. Smirdova, G. K. Novikova, and N. P. Efryushina, *Zh. Anal. Khim.* ibid. *35*, 471 (1980).

105. E. A. Zhikhareva, G. K. Novikova, and N. P. Efryushina, *J. Chem. Phys. 36*, 1079 (1981).

106. N. P. Efryushina, G. K. Novikova, E. A. Zhikhareva, and V. P. Dotsenko, *J. Chem. Phys. 37*, 244 (1982).

107. G. K. Novikova, E. A. Zhikhareva, and N. P. Efryushina, *Zavod. Lab. 47*, 19 (1981); *Chem. Abstr. 95*, 17593e.

108. G. K. Novikova, E. A. Zhikhareva, S. A. Gava, N. P. Efryushina, and N. S. Poluektov, *Zh. Anal. Khim. 35*, 1288 (1980); *Anal. Abstr. 40*, 4B83.

109. G. K. Novikova, E. A. Zhikhareva, N. P. Efryushina, and N. N. Zlotnikova, *Zh. Anal. Khim. 36*, 275 (1981); *Anal. Abstr. 41*, 4B72.

110. G. K. Novikova, N. P. Efryushina, E. A. Zhikhareva, and V. P. Dotsenko, *Zh. Anal. Khim. 36*, 663 (1981); *Chem. Abstr. 95*, 72556v.

111. N. S. Poluektov, N. P. Efryushina, and N. I. Smirdova, *Tr. Khim. Tekhnol. 3*, 60 (1973).

112. N. I. Smirdova, N. P. Efryushina, and N. S. Poluektov, *Zavod. Lab. 36*, 1183 (1970); *Chem. Abstr. 74*, 60605p.

113. N. S. Poluektov, N. I. Smirdova, and N. P. Efryushina, *Zh. Anal. Khim. 27*, 1616 (1972); *Chem. Abstr. 78*, 23542c.

114. N. P. Efryushina, N. S. Poluektov, N. I. Smirdova, and V. A. Kudryashov, *Zh. Anal. Chim. 28*, 1213 (1973); *Chem. Abstr. 79*, 111417s.

115. N. S. Poluektov, N. I. Smirdova, and N. P. Efryushina, *Zh. Anal. Khim. 25*, 715 (1970).

116. N. P. Efryushina, N. I. Smirdova, and V. A. Kudryashov, *Zavod. Lab. 39*, 129 (1973); *Chem. Abstr. 78*, 154510p.

117. A. I. Kirillov, R. A. Vitkun, and N. S. Poluektov, *Sovrem. Metody. Khim. Spektral. Anal. Mater.* 211 (1967); *Chem. Abstr., 68*, 26676d.

118. Yu. P. Novikov, L. I. Anikina, S. A. Ivanova, B. F. Myasoedov, and A. V. Karyakin, *Radiokhimiya 20*, 878 (1978); *Chem. Abstr. 90*, 161626g.

119. V. I. Aleshin and Yu. N. Orlov, *Vestn. Mosk. Univ. Khim. 20*, 355 (1979); *Anal. Abstr. 38*, 6B36.

120. M. V. Johnston and J. C. Wright, *Anal. Chem. 51*, 1774 (1979).

121. N. S. Poluektov, N. I. Smirdova, and N. P. Efryushina, *Zh. Anal. Khim. 25*, 1902 (1970); *Chem. Abstr. 74*, 49291n.

122. N. I. Smirdova, N. P. Efryushina, and N. S. Poluektov, *Zh. Anal. Khim. 29*, 1719 (1974).

123. L. Ozawa and T. Toryu, *Anal. Chem. 40*, 187 (1968).

124. S. A. Gava and N. S. Poluektov, *Zavod. Lab. 35*, 20 (1969); *Chem. Abstr. 70*, 84021p.

125. A. V. Karyakin, L. I. Anikina, and L. A. Filatkina, *Reins. Wiss. Techn. Int. Symp. II, Dresden, 1965*, 309 (1966).

126. A. V. Karyakin, L. I. Anikina, and L. A. Filatkina, *Zh. Anal. Khim. 21,* 1196 (1966).

127. A. V. Karyakin, L. I. Anikina, and Le Viet Binh, *Zh. Anal. Khim. 24,* 1156 (1969).

128. E. Nichols and M. K. Slattery, *J. Opt. Soc. Am. 12,* 449 (1926).

129. J. W. Owens, M. S. Energi Res. Dev. Adm. Rep. LA-6338-MS, 12 (1976).

130. C. E. White, *Anal. Chem. 24,* 1952 (1965).

131. P. Pakalns, Aust. At. Energy Comm., AAEC/TM [Rep.], AAEC/TM-552, 6, 5 (1970); *Chem. Abstr. 75,* 44598d.

132. R. Viswanathan, S. Ramaswami, and C. K. Anni, *Bull. Natl. Inst. Sci. India 38,* 284 (1968).

133. U. Croatto, S. Degetto, and L. Baracco, *Ann. Chim.* (Rome) *68,* 659 (1978).

134. A. Hazan, J. Korkisch, and G. Arrhenius, *Z. Anal. Chem. 213,* 182 (1965).

135. A. V. Karyakin, L. I. Anikina, and R. S. Melyantseva, *Geokhimiya 6,* 934 (1980); *Chem. Abstr. 93,* 135167a.

136. Z. Samsony, *Magy. Kem. Foly. 72,* 398 (1966).

137. N. I. Udal'tsova, *Izv. Sib. Otdel. Akad. Nauk. SSSR 7* (1971).

138. P. Pfeffer, *Notizbl. Hess. Landesamtes Bodenforsch. Wiesb. 85,* 425 (1957).

139. F. I. Nagi, M. Naeem, Z. Umar, J. Anwar, and A. Ali, *Rep. Pak. AEC.,* AEMC/Chem-132 (1975).

140. J. C. Veselsky and A. Woefl, *Anal. Chim. Acta 85,* 135 (1976).

141. D. T. Kaschani and A. Brauns, *Chromatography* (Suppl.) *29* (1981).

142. Z. V. Malyasova, *Metody Anal. Galogeniodv Shchelochn. Shcheloch. Metal. Vys. Chist. 2,* 108 (1971); *Chem. Abstr. 78,* 37623s.

143. P. A. Vozzella, A. S. Powell, R. H. Jale, and J. E. Kelly, *Anal. Chem. 32,* 1430 (1960).

144. R. A. Jaroszeski and C. C. Gregg, *Anal. Chem. 37,* 766 (1965).

145. M. I. Tarantsova and Yu. P. Nikol'skaya, *Izv. Sib. Otdel. Akad. Nauk. SSSR 12* (1968).

146. R. F. Rolf, *Anal. Chem. 28,* 1651 (1956).

147. A. Y. Smith and J. J. Lynch, *Can. Dep. Energy, Mines Resour. Geol. Surv. Can. Pap. 69,* 40 (1969).

148. C. Huffman, Jr., and L. B. Riley, *U.S. Geol. Surv. Prof. Pap.,* No. 700-B, 181 (1970).

149. C. Boirle, D. Bose, G. Hugot, and R. Platzer, *Acta Chim. Acad. Sci. Hung. 33,* 281 (1962).

150. U. K. At. Energy Auth. Prod. Group. Rep. PG510(s), 10 pp (1963).

151. N. M. Sadikova, E. K. Polonskaya, and M. M. Golutvina, *Med. Radiol. 15,* 65 (1970); *Chem. Abstr. 72,* 129296s.

152. L. Widua, H. Schieferdecker, and U. Hezel, *Z. Anal. Chem. 270*, 12 (1974).

153. K. Irlweck and S. Streit, *Mikrochim. Acta II*, 63 (1979).

154. S. A. Gava, G. K. Novikova, and N. P. Efryushina, *Zh. Anal. Khim. 34*, 117 (1979); *Chem. Abstr. 90*, 197046g.

155. L. Ozawa and H. Forest, *Anal. Chem. 45*, 978 (1973).

156. L. I. Anikina, T. S. Dobrolyubskaya, A. V. Karyakin, and Le Viet Binh, *Zh. Prikl. Spektr. 10*, 518 (1969).

157. T. S. Dobrolyubskaya, *Zh. Anal. Khim. 26*, 1835 (1971); *Chem. Abstr. 76*, 41689u.

158. D. V. Jayawant, N. S. Iyer, and T. K. S. Murthy, *Indian J. Technol. 14*, 151 (1976).

159. C. G. Peattie and L. B. Rogers, *Anal. Chem. 25*, 518 (1953).

160. L. I. Anikina, A. V. Karvakin, and Le Viet Binh, *Zh. Anal. Khim. 26*, 511 (1971).

161. E. A. Bozhevol'nov and O. A. Fakeeva, *Tr. Kom. Anal. Khim., Akad. Nauk. SSSR, Inst. Geokhim. Anal. Khim. 16*, 67 (1968); *Chem. Abstr. 69*, 24270p.

162. Yu. P. Novikov, L. I. Anikina, S. A. Ivanova, A. V. Karyakin, and B. F. Myasoedov, *Radiokhimiya 17*, 830 (1975); *Chem. Abstr. 84*, 53515n.

163. N. P. Efryushina, G. K. Novikova, and N. S. Poluektov, *Zh. Anal. Khim. 32*, 1731 (1977); *Chem. Abstr. 88*, 145545h.

164. M. Fabian-Mohai, E. Upor, and G. Nagy, *Acta Chim.* (Budapest) *68*, 1 (1971).

165. E. A. Bozhevol'nov and O. A. Fakeeva, *Prom. Khim. Reaktivov Osobo Chist. Veshchestv. 8*, 218 (1967); *Chem. Abstr. 69*, 64356c.

166. E. A. Bozhevol'nov and O. A. Fakeeva, *Metod. Anal. Khim. Reakt. Preparat.* No. 11, 62 (1965).

167. M. Allsalu, I. Kilk, and M. Kerikmae, *Tartu Ulikooli Toim. 219*, 168 (1968).

168. R. C. Ropp and N. W. Shearer, *Anal. Chem. 33*, 1240 (1961).

169. A. V. Karyakin, L. I. Anikina, E. N. Vasil'ev, and L. V. Borisova, *Zh. Anal. Khim. 36*, 498 (1981); *Chem. Abstr. 95*, 17535n.

170. L. I. Anikina, A. V. Karyakin, G. I. Malofeeva, N. V. Soboleva, E. V. Illarionova, and E. N. Vadil'ev, *Zh. Anal. Khim. 34*, 519 (1979); *Chem. Abstr. 91*, 48844j.

171. E. A. Bozhevol'nov, *Metod. Anal. Khim. Reakt. Preparat.* No. 4, 110 (1962).

172. E. N. Vasil'ev, V. P. Subbotin, E. A. Solov'ev, and E. A. Bozhevol'nov, *Zavod. Lab. 42*, 513 (1976); *Chem. Abstr. 85*, 171170g.

173. K. Schmidt and H. Staude, *Fresenius' Z. Anal. Chem. 234*, 241 (1968).

174. R. G. Delumyea and G. H. Schenk, *Anal. Chem. 48*, 95 (1976).

175. D. E. Ryan, R. J. Prime, and T. Holzbecher, *Anal. Lett.* 6, 721 (1973).

176. D. E. Ryan, J. Holzbecher, and H. Rollier, *Anal. Chim. Acta 73,* 49 (1974).

177. E. N. Vail'ev, O. A. Fakeeva, E. A. Solov'ev, and E. A. Bozhevol'nov, *Zh. Anal. Khim.* 28, 688 (1973); *Chem. Abstr.* 79, 38223u.

178. I. Kilk and M. Allsalu, *Tartu Ulikooli Toim.* 289, 112 (1971); *Chem. Abstr.* 78, 52264u.

179. O. A. Fakeeva, E. A. Solov'ev, and E. A. Bozhevol'nov, *Tr. Vses. Nauch-Issled. Inst. Khim. Reaktivov. Osobo, Chist. Khim. Veshchestv* No. 32, 189 (1970); *Chem. Abstr.* 77, 28479b.

180. I. Kilk, M. Allsalu, and M. Kerikmae, *Tartu Ulikooli Toim.* 289, 117 (1971); *Chem. Abstr.* 78, 52201w.

181. I. N. Astaf'eva, D. P. Shcherbov, and R. N. Plotnikova, *Zh. Anal. Khim.* 33, 1068 (1978).

182. D. E. Ryan, H. Rollier, and T. Holzbecher, *Can. J. Chem.,* 52, 1942 (1974).

183. J. C. Wright, *Anal. Chem.* 49, 1690 (1977).

184. E. A. Bozhevol'nov, E. A. Solov'ev, and O. A. Fakeeva, *Int. Conf. on Luminescence, Special Probl. Lumin., 11, Appl. Lumin. Preprint,* Budapest, (1966), p. 41.

185. E. A. Bozhevol'nov and E. A. Solov'ev, *Tr. IREA, Khim. Reakt. Preparat.* No. 30, 202 (1967), Moskow IREA.

186. A. Solov'ev and E. A. Bozhevol'nov, *Tr. Vses. Nauch.-Issled. Inst. Khim. Reakt. Osobo Chist. Khim. Veshchestv.* No. 26, 194 (1964); *Chem. Abstr.* 66, 82105n.

187. L. P. Savvina, A. P. Golovina, and E. A. Solov'ev, *Vestn. Mosk. Univ. Ser. Khim.* 4, 451 (1975).

188. E. A. Solov'ev, L. P. Savvina, A. P. Golovina, and Yu. Granovskii, *Zh. Anal. Khim.* 30, 103 (1975).

189. M. W. Belyi, Proc. of the Intern. *Conf. on Luminescence, Budapest, 1966,* Akad. Kiado, Budapest (1968), p. 807.

190. E. A. Solov'ev and E. A. Bozhevol'nov, *Zh. Anal. Khim.* 27, 1817 (1972).

191. D. P. Shchervov, I. N. Astaf'eva, and R. N. Plotnikova, *Issled. Obl. Khim. Fiz. Metod. Anal. Miner. Syr'ya,* 18 (1971); *Chem. Abstr.* 78, 66448u.

192. M. U. Belyi and I. Ya. Kushnirenko, *Zh. Prikl. Spektr. 10,* 810 (1969); *Chem. Abstr. 71,* 45419a.

193. M. U. Belyi and I. Ya. Kushnirenko, *Zh. Prikl. Spektr. 10,* 84 (1969).

194. N. M. Belyi and I. Ya. Kushnirenko, *Zh. Prikl. Spektr. 9,* 442 (1968).

195. G. F. Kirkbrigt, C. G. Saw, and T. S. West, *Talanta 16,* 65 (1969).

196. G. F. Kirkbrigt, C. G. Saw, J. V. Thompson, and T. S. West, *Talanta 16,* 1081 (1969).

197. E. A. Bozhevol'nov and E. A. Solov'ev, *Metod. Anal. Khim. Reakt. Preparat.* No. 11, 87 (1965); *Chem. Abstr. 65* 1364d.

198. A. P. Golovina, V. K. Runov, and L. P. Savvina, *Vestn. Mosk. Univ. Ser. Khim. 18,* 370 (1977).

199. I. N. Astaf'eva, D. P. Shcherbov, and R. N. Plotrikov, *Issled. Obl. Khim. Fiz. Metod. Anal. Miner. Syr'ya 30* (1971); *Chem. Abstr. 78,* 66554a.

200. L. P. Savvina, A. P. Golovina, E. A. Solov'ev, and V. K. Runov, *Zh. Anal. Khim. 31,* 1268 (1976); *Chem. Abstr. 86,* 37214z.

201. G. F. Kirkbright, C. W. Saw, and T. S. West, *Analyst 94,* 457 (1969).

202. V. V. Shemet, B. D. Luft, L. B. Khusid, and T. N. Ovseenko, *Tr. Khim. Tekhnol. 4,* 76 (1973); *Chem. Abstr. 81,* 57902x.

203. E. A. Bozhevol'nov, E. A. Solov'ev, and N. A. Lebedeva, *Zh. Anal. Khim. 26,* 1117 (1971).

204. E. A. Bozhevol'nov and E. A. Solov'ev, *Zh. Anal. Khim. 20,* 1330 (1965); *Chem. Abstr. 64,* 11850c.

205. E. A. Bozhevol'nov and E. A. Solov'ev, *Metod. Anal. Khim. Reakt. Preparat.* No. 11, 74 (1965); *Chem. Abstr. 65,* 1372b.

206. M. U. Belyi and I. Kushnirenko, *Zh. Prikl. Spectr. 9,* 272 (1968).

207. G. F. Kirkbright and C. C. Saw, *Talanta 15,* 570 (1968).

208. E. A. Solov'ev and N. A. Lebedeva, *Zh. Anal. Khim. 34,* 2167 (1979).

209. S. J. Mule and P. L. Hushin, *Anal. Chem. 43,* 708 (1971).

210. F. Barankay, *Kiserl. Orvostud. 21,* 646 (1969); *Chem. Abstr. 72,* 107239y.

211. D. P. Shchervov and A. I. Ivankova, *Prom. Khim. Reakt. Osobo Chist. Veshchestv. 8,* 191 (1967).

212. R. Bock and E. Zimmer, *Z. Anal. Chem. 198,* 170 (1963).

213. G. F. Kirkbright, T. S. West, and C. Woodward, *Talanta 12,* 517 (1965).

214. E. Y. Pesina, J. Balodis, and L. E. Tarasova, *Zh. Prikl. Spektr. 9,* 277 (1968).

215. E. A. Solov'ev, N. I. Sofina, and E. A. Bozhevol'nov, *Method. Anal. Galogenidov Shchelochn. Shclelochn. Metal. Vys. Chist. 2,* 99 (1971); *Chem. Abstr. 78,* 37621q.

216. Z. Skuric, F. Valic, and J. Propic-Marecic. *Anal. Chim. Acta 73,* 213 (1974).

217. I. M. Ivanova, E. A. Solov'ev, E. N. Stasyunas, and A. P. Golovina, *Vestn. Mosk. Univ. Khim. 20,* 583 (1979).

218. G. P. Tikhonov, A. I. Sukhanovskaya, I. V. Sinitsina, E. A. Solov'ev, and E. A. Bozhevol'nov, *Zh. Anal. Khim., 27,* 2191 (1972); *Chem. Abstr. 78,* 66396a.

219. E. A. Bozhevol'nov, E. A. Solov'ev, N. I. Sofina, V. I. Titov, G. P. Tikhonov, and Yu. V. Golubev, U.S.S.R. Patent 257, 844 (Cl. G 01 n), 20 Nov. 1969; *Chem. Abstr. 72,* 128459k.

220. E. A. Solov'ev, E. A. Bozhevol'nov, A. I. Sukhanovskaya, G. P. Tikhonov, and Yu. V. Golubev, *Zh. Anal. Khim., 25,* 1342 (1970); *Chem. Abstr. 73,* 137050t.

221. E. A. Solov'ev, G. P. Tikhonov, N. I. Sofina, and E. A. Bozhevol'nov, *Zavod. Lab. 39*, 669 (1973).

222. G. Alberti and M. A. Massuci, *Anal. Chem. 38*, 214 (1968).

223. B. C. Capelin and G. Ingram, *Talanta 17*, 187 (1970).

224. G. F. Kirkbright, T. S. West, and C. Woodward, *Anal. Chim. Acta 36*, 298 (1966).

225. D. P. Shcherbov, O. D. Inyutina, and A. I. Ivankova, *Zh. Anal. Khim. 28*, 1372 (1973); *Chem. Abstr. 79*, 142608j.

226. M. Moriyasu, Y. Yokoyama, and S. Ikeda, *Jap. Anal. 24*, 257 (1975).

226a. A. Fernandez-Gutierrez, A. Muñoz de la Peña, and M. Roman, *Ann. Chim. 74*, 1 (1984).

227. A. Gomez-Hens and M. Valcarcel, *Analyst 107*, 1274 (1982).

228. P. A. St. John, *"Fluorometric Methods for Traces of Elements,"* in J. Winefordner (ed.), "Trace Anal.: Spectrosc, Methods Elem." Chem. Anal. (NY), *46*, 1976, p. 213.

229. P. R. Haddad, *Talanta 24*, 1 (1977).

230. D. P. Shchervov and D. V. Perminova, *Issled. Razrab. Fotometrich. Metod. Opred. Mikrokolichestv. Elem. Miner.*, 149 (1967); *Chem. Abstr. 71*, 18574r.

231. D. N. Perminova and D. P. Shchervov, *Prom. Khim. Reaktivov. Osobo. Chist. Veshchestv. 8*, 1981 (1967); *Chem. Abstr. 69*, 92683a.

232. Sh. T. Talipov, A. T. Tashkhodzhaev, and L. E. Zel'tser, *Uzb. Khim. Zh. 16*, 20 (1972); *Chem. Abstr. 78*, 66574g.

233. D. P. Shcherbov and D. N. Lisitsina, *Issled. Obl. Khim. Fiz. Metod. Anal. Miner. Syr'ya* 62 (1975).

234. Sh. T. Talipov, L. E. Zel'tser and A. T. Tashkhadzhaev, *Izv. Vyssh. Ucheb. Zaved., Khim. Tekhnol. 16*, 299 (1973); *Chem. Abstr. 78*, 168179z.

234a. A. Moreno, M. Silva, and D. Perez-Bendito, *Anal. Lett. 16*, 747 (1983).

235. D. E. Ryan and B. K. Pal, *Anal. Chim. Acta 44*, 385 (1969).

236. R. L. Wilson and J. D. Ingle, Jr., *Anal. Chem. 49*, 1066 (1977).

237. G. L. Wheeler, J. Andrejack, J. H. Wiersma, and P. F. Lott, *Anal. Chim. Acta 46*, 239 (1969).

238. M. T. El-Ghamry, R. W. Frei, and G. W. Higgs, *Anal. Chim. Acta 47*, 41 (1969).

239. D. N. Lisitsyna and D. P. Shcherbov, *Zh. Anal. Khim. 25*, 2310 (1970).

240. D. P. Shcherbov and D. N. Lisitsyna, *Zavod. Lab. 39*, 656 (1973).

241. D. N. Lesitsyna and D. P. Shcherbov, *Zh. Anal. Khim. 28*, 1203 (1973).

242. I. Mori, T. Enoki, and T. Mano, *Jap. Anal. 22*, 1209 (1973).

242a. C. Sanchez-Pedreno, M. Hernandez-Cordoba, and I. Lopez-Garcia, *Quim. Anal. 1*, 97 (1982).

243. B. T. Taskarin and D. P. Shcherbov, *Issled. Razrab. Fotometrich. Metod. Opred. Mikrokolichestv. Elem. Miner.* 85 (1967); *Chem. Abstr. 71*, 18572p.

244. M. K. Akhmedli, D. A. Efendiev, and F. I. Ruvinova, *Uch. Zap. Azerb. Gos. Univ., Ser. Khim. Nauk. 4*, 10 (1973).

245. G. H. Ellis, E. G. Zook, and O. Bandesh, *Anal. Chem. 21*, 1345 (1949).

246. N. M. Alykov and A. V. Brunin, *Izv. Vyss. Ucheb. Zaved. Khim. Tethnol. 17*, 1254 (1974).

247. K. Watanabe, H. Yoshizawa, and K. Kawagaki, *Jap. Analy. 30*, 640 (1981).

248. C. Apostolopoulou, A. Voulgaropoulos, V. Simeonov, and G. Vasilikiotis, *Fresenius' Z. Anal. Chem. 309*, 3651 (1981).

249. V. Simeonov, A. Voulgaropoulos, C. Apostolopoulou, and G. Vasilikiotis, *Fresenius' Z. Anal. Chem. 311*, 16 (1982).

250. K. Morishige, *J. Inorg. Nucl. Chem. 40*, 843 (1978).

251. M. Deguchi, T. Masumoto, K. Morishige, and I. Okumura, *Jap. Anal. 28*, 127 (1979).

252. V. Drevenkar, Z. Stefanac, and A. Brbot, *Microchem. J. 21*, 402 (1976).

253. V. A. Nazarenko, M. B. Shustova, R. V. Ravitskaya, and M. P. Nikonova, *Zavod. Lab. 28*, 537 (1962).

254. V. A. Nazarenko, M. B. Shustova, G. G. Shitareva, G. Ya. Yagnyatinskaya and R. V. Ravitskaya, *Zavod. Lab. 28*, 645 (1962).

255. G. Hocman, *Acta Fac. Rerum Natur. Univ. Comeniane Chim. 13*, 75 (1968); *Chem. Abstr. 71*, 35648k.

256. C. E. White, in *Fluorescence: Theory, Instrumentation, and Practice*, G. G. Guilbault, (ed.), Dekker, New York, 1967, p. 281.

257. K. Hiraki, *Bull. Chem. Soc. Jap. 46*, 2438 (1973).

258. T. Uno, H. Taniguchi, *Jap. Anal. 20*, 1123 (1971); *Chem. Abstr. 75*, 147470k.

259. A. V. Dolgorev, Yu. A. Serikov, and V. L. Zolotavin, *Zh. Anal. Khim. 33*, 2357 (1978).

260. A. V. Dolgorev, Yu. A. Serikov, and N. A. Chesnokva, *Zavod. Lab. 45*, 691 (1979).

261. I. Cherkesov and V. Zhegalkena, *Dokl. Akad. Nauk. SSSR 118*, 309 (1958).

262. G. F. Kirkbright, T. S. West, and C. Woodward, *Anal. Chem. 37*, 137 (1965).

263. J. M. Cano, M. L. Trujillo, and A. Garcia de Torres, *Anal. Chim. Acta 117*, 319 (1980).

264. C. Gentry and L. Scherrington, *Analyst 71*, 432 (1946).

265. F. Grimaldi and G. Levine, *U.S. Geol. Survey Trace Elem. Inv. Rest. 60* (1950).

266. E. Goon, J. Petley, W. McMullen, and S. Wilberley, *Anal. Chem. 25*, 608 (1953).

267. A. Fioletova, *Zh. Anal. Khim. 14*, 739 (1959).

268. M. G. Cook, *Soil Sci. Soc. An. Proc. 32,* 292 (1968).

269. W. Mische, E. Schroeder, and K. Thinius, *Plaste Kaustsch 16,* 23 (1969).

270. T. D. Verbitskaya and Z. G. Chernorukova, *Tr. Khim. Khim. Tekhnol. 1,* 236 (1970); *Chem. Abstr. 75,* 71027b.

271. H. Rollier and D. E. Ryan, *Anal. Chim. Acta 74,* 23 (1975).

272. L. A. Gomiero and M. R. Lopes, *Publ. IEA 428,* 5 (1976).

273. R. L. Wilson and J. D. Ingle, Jr., *Anal. Chim. Acta 92,* 417 (1977).

274. Y. Kondoh, M. Kataoka, and T. Kambara, *Bull. Chem. Soc. Jap. 55,* 434 (1982).

275. L. E. Zeltser, L. A. Morozova, and Sh. T. Talipov, *Zh. Anal. Khim. 35,* 97 (1980); *Anal. Abstr. 39*(3), 3H43.

276. Y. Nishikawa, K. Hiraki, K. Morishige, and T. Shigematsu, *Jap. Anal. 16,* 692 (1967).

277. Y. Nishikawa, K. Hiraki, K. Morishige, A. Tsuchiva, and T. Shigematsu, *Jap. Anal. 17,* 1092 (1968).

278. T. Shigematsu, Y. Nishikawa, K. Hiraki, and N. Nagano, *Jap. Anal. 19,* 551 (1970); *Chem. Abstr. 73,* 48442e.

279. A. T. Pilipenko, A. I. Volkova, G. N. Pshenko, and V. P. Denisenko, *Ukr. Khim. Zh. 46,* 200 (1980); *Anal. Abstr. 39*(4), 4B60.

280. R. Gabriels, W. Van Keirsbulck, and H. Engels, *Lab. Pract. 30,* 122 (1981); *Anal. Abstr. 41*(4), 4G2.

281. M. K. Akhmedli, D. Efendiev and F. I. Ruvinova, *Zh. Anal. Khim. 29,* 450 (1974); *Chem. Abstr. 81,* 32879g.

282. A. V. Dolgorev, Yu. A. Serikov, and L. A. Rzhevina, *Zh. Anal. Khim. 33,* 1313 (1978); *Anal. Abstr. 36,* 2B58.

283. J. F. Aguila, *Talanta 14,* 1195 (1967).

284. K. Morishige, *Anal. Chim. Acta 121,* 301 (1980).

285. J. F. Possidoni de Albinati, *An. Asoc. Quim. Argent. 53,* 61 (1965).

286. K. Hiraki, *Bull. Chem. Soc. Jap. 45,* 789 (1972); *Chem. Abstr. 76,* 161875y.

287. C. E. White and C. S. Lowe, *Ind. Eng. Chem., Anal. Ed. 12,* 229 (1940).

288. D. Will, *Anal. Chem. 33,* 1360 (1961).

289. J. T. Baker, *Chemical News Leaflet,* Phillipsburg, N.J, Spring, 1970.

290. D. A. Eggert, *Stain Technol. 45,* 301 (1970).

291. S. B. Sarvin, E. S. Danilin, and T. M. Malyutina, *Zh. Anal. Khim. 16,* 1945 (1981); *Anal. Abstr. 42,* 5B43.

292. A. Weissler and C. E. White, *Anal. Chem. 18,* 530 (1946).

293. A. Danneil, *Tech. Wiss. Abhandl. Osram-Ges. 7,* 350 (1958).

294. J. F. Possidoni de Albinati, *An. Asoc. Quim. Argent. 51,* 96 (1963).

295. A. Puech, G. Kister, and J. Chanal, *J. Zentralbl. Pharm. Pharmakother. Laboratoriumsdiagn. 111,* 7 (1972).

296. J. Bognar and M. P. Szabo, *Mikrochim. Acta 221* (1969).

297. F. Capitan, M. Roman, and E. Alvarez Manzaneda, *Afinidad 30,* 623 (1973).

298. J. G. Surak, M. F. Herman, and D. T. Haworth, *Anal. Chem. 37,* 428 (1965).

299. P. R. Haddad, P. W. Alexander, and L. E. Smythe, *Talanta 21,* 123 (1974).

300. A. Davydov and A. Devekki, *Zavod. Lab. 10,* 134 (1941).

301. Z. T. Maksimycheva, Sh. T. Talipov, A. P. Pakudina, and A. S. Sadykov, *Dokl. Akad. Nauk. Uzb. SSR 2,* 36 (1973).

302. F. Capitan, M. Roman, and A. Guiraum, *An. Quim. 70,* 508 (1974).

303. T. A. Shakirova, Sh. T. Talipov, G. S. Andrushko, and A. T. Tashkhodzhaev, *Nauch. Tr. Tashkt. Gos. Univ. 435,* 68 (1973).

304. L. Ben-Dor and E. Jungreis, *Isr. J. Chem. 8,* 951 (1970).

305. Z. Holzbecker and P. Pulkrab, *Coll. Czech. Chem. Comm. 27,* 1142 (1963).

306. R. M. Dagnall, R. Smith and T. S. West, *Talanta 13,* 609 (1966).

307. I. I. Kurbatova, *Zavod. Lab. 32,* 1064 (1966); *Anal. Abstr. 14,* 7501.

308. L. A. Demina, O. M. Petrukhin, Yu. A. Zolotov, and G. V. Serebryakova, *Zh. Anal. Khim. 27,* 1731 (1972); *Chem. Abstr. 78,* 37587h.

309. D. A. Fakeeva, V. I. Konleva, and G. F. Minchenkova, *Metod. Anal. Galogenidov. Shchelochn. Shchelochn. Metal. Vys. Chist.* No. 2, *45,* (1971); Chem. Abstr., *78,* 66610r.

310. E. A. Bozhevol'nov and V. N. Yanishevskay, *Zh. Vses. Khim. Obshch. D. I. Mendeleeva 5,* 536 (1960).

311. N. B. Lebed and R. P. Pantaler, *Zavod. Lab. 31,* 163 (1965).

312. V. V. Khimor, O. S. Didkovskaya, and V. N. Kozachenko, *Zavod. Lab. 28,* 652 (1962).

313. I. I. Kurbatova, *Zavod. Lab. 32,* 1064 (1966).

314. R. Kh. Dzhiyanbaeva, A. T. Tashkhodzhaev, L. E. Zeltzer, and Kh. Khikmatov, *Nauch. Tr. Tashkt. Gos. Univ. 419,* 84 (1972).

315. C. E. White, H. McFarlane, F. Fogt, and R. Fuchs, *Anal. Chem. 39,* 367 (1967).

316. Z. Holzbecher, *Chem. Listy 47,* 680, 1023 (1953).

317. A. Dale, P. Jones, and J. Radley, Inst. Dept. U.S. Dept. of Army contract DA91-591-3309 (1965).

318. K. Morishige, *Kinki Daigaku Rikogakubu Kenkyu Hokoku 16,* 265 (1981); *Anal. Abstr. 43,* 3B61.

319. K. Morishige, *Anal. Chim. Acta 72,* 295 (1974).

320. M. Gallego, M. Valcarcel, and M. Garcia-Vargas, *Analyst 108,* 92 (1983).

321. R. K. Chernova, L. M. Kudryavtseva, I. K. Petrova, and K. I. Gur'ev, *Zh. Anal. Khim. 31,* 137 (1976).

322. A. T. Pilipenko and A. I. Zhebentyaev, *Zh. Anal. Khim. 33,* 2119 (1978); *Anal. Abstr. 36,* 6B49.

323. K. Hiraki, *Bull. Chem. Soc. Jap. 45,* 1395 (1972); *Chem. Abstr. 77I,* 42778e.

324. A. K. Babko, Z. I. Chalaya, and V. F. Mikitchenko, *Zavod. Lab. 32* 270 (1966).

325. S. R. Goode and R. J. Matthews, *Anal. Chem. 50,* 1608 (1978).

326. R. J. Matthews, S. R. Goode, and S. L. Morgan, *Anal. Chim. Acta 133,* 169 (1981).

327. S. Yamamoto and K. Kisu, *Jap. Anal. 23,* 638 (1974); *Chem. Abstr. 81,* 145243.

328. L. A. Grigoryan, D. A. Mikaelyan, and V. M. Tarayan, *Zavod. Lab. 48,* 19 (1982); *Anal. Abstr. 44,* 1B32.

329. L. A. Grigoryan, D. A. Mikaelyan, and V. M. Tarayan, *Zh. Anal. Khim. 35,* 45 (1980); *Chem. Abstr. 92,* 190669v.

330. L. A. Grigoryan, D. A. Mikaelyan, and V. M. Tarayan, *Arm. Khim. Zh. 33,* 545 (1980); *Chem. Abstr. 93,* 178979j.

331. F. Grases, F. Garcia-Sanchez, and M. Valcarcel, *Anal. Lett. 12,* 803 (1979).

332. N. K. Podberezskaya and V. A. Sushkova, *Zavod. Lab. 36,* 1048 (1970).

333. N. K. Podberezskaya, V. A. Sushkova, and E. A. Shilenko, *Zavod. Lab. 33,* 152 (1967).

334. I. A. Blyum, N. N. Pavlova, and F. P. Kalupina, *Zh. Anal. Khim. 26,* 55 (1971).

335. N. K. Podberezskaya, E. A. Shilenko, and D. P. Shcherbov, *Zavod. Lab. 36,* 661 (1970).

336. E. A. Shilenko, N. K. Podberezskaya, and D. P. Shcherbov, *Issled. Tsvet. Fluorestsent. Reakts. Opred. Blagorod. Met.* 101 (1969); *Chem. Abstr., 75,* 14542w.

337. F. Grases, J. M. Estela, F. Garcia-Sanchez, and M. Valcarcel, *Analusis 9,* 66 (1981).

338. A. Murata and T. Ujaihara, *Jap. Anal. 10,* 497 (1961).

339. N. K. Podberezskaya, D. P. Shcherbov, E. A. Shilenko, and V. S. Sushkova, *Issled. Tsvet. Fluorestsent. Reakts. Opred. Blagorod. Met.* 108 (1969).

340. T. Perez-Ruiz, C. Sanchez-Pedreno, J. A. Ortuno, and P. Molina-Buendia, *Analyst 108,* 733 (1983).

341. J. Marinenko and I. May, *Anal. Chem. 40,* 1137 (1968).

342. B. T. Taskarin and D. P. Shcherbov, *Tr. Khim. Khim. Tekhnol. 4,* 208 (1965).

343. V. N. Podchainova, L. V. Skornyakova, and B. L. Dvinyanivov, *Izv. Vyssh. Ucheb. Zaved., Khim. Tekhnol. 11,* 241 (1968).

344. I. M. Korenman, *Zh. Anal. Khim. 2,* 158 (1947).

345. C. E. White and M. H. Neustadt, *Ind. Eng. Chem., Anal. Ed. 15*, 599 (1943).

346. C. E. White, A. Weissler, and D. Busker, *Anal. Chem. 19*, 802 (1947).

347. C. E. White, *J. Chem. Educ. 28*, 369 (1951).

348. L. Sommer, *Chem. Listy 51*, 2032 (1957).

349. C. E. White and D. E. Hoffman, *Anal. Chem. 29*, 1105 (1957).

350. C. A. Parker and W. J. Barnes, *Analyst 82*, 606 (1957).

351. L. Sommer, *Coll. Czech. Chem. Comm. 24*, 99 (1959).

352. C. A. Parker and W. J. Barnes, *Analyst, 85*, 828 (1960).

353. G. Elliot and J. A. Radley, *Analyst 86*, 62 (1961).

354. D. P. Shcherbov and N. V. Kagarlitskaya, *Tr. Kaz. Nauch. Issled Inst. Miner. Syr'ya* No. 5, 255 (1961).

355. G. I. Kuchmistaya, *Metod. Anal. Khim. Reakt. Preparat.* No. 4, 69 (1962).

356. V. N. Podchainova and L. V. Skornyakova, *Tr. Ural. Politekh. Inst.* No. 163, 60 (1967).

357. Yu. L. Lel'chuk and V. A. Ivashina, *Izv. Tomsk. Politekh. Inst. 148 (1967); Chem. Abstr. 70*, 92875k.

358. A. K. Babko and Z. I. Chalaya, *Ukr. Khim. Zh. 30*, 268 (1964).

359. A. K. Babko, Z. I. Chalaya, and E. D. Voronova, *Zavod. Lab. 31*, 157 (1965).

360. T. Bruce and R. W. Ashley, Report, 1973, AECL-446, National Technical Information Service, Washington DC, 14 pp.

361. J. Lapid, S. Farhi, and Y. Koresh, *Anal. Lett. 9*, 355 (1976).

362. M. Marcantonatos, G. Gamba, and D. Monnier, *Helv. Chim. Acta 52*, 538 (1969).

363. Z. Skorko-Trybula and Z. Boguszewska, *Mikrochim. Acta 2*, 335 (1976).

364. J. Aznares, A. Bonilla, and J. C. Vidal, *Analyst 108*, 368 (1983).

365. D. Monnier, A. Marcantonatos, and M. Marcantonatos, *Helv. Chim. Acta 47*, 1980 (1964).

366. P. V. Kristalev and Ya. F. Shevchenko, *Sb. Nauch. Tr. Perm. Politekn. Inst.* No. 71, 38 (1979).

367. P. V. Kristalev and M. N. Chelnokova, *Zh. Anal. Khim. 29*, 1650 (1974); *Chem. Abstr. 82*, 51055x.

368. A. Holme, *Acta Chem. Scand. 21*, 1679 (1967).

369. M. Marcantonatos, A. Marcantonatos, and D. Monnier, *Helv. Chim. Acta 48*, 194 (1965).

370. M. Marcantonatos, D. Monnier, and J. Daniel, *Anal. Chim. Acta 35*, 309 (1966).

371. D. Monnier and M. Marcantonatos, *Anal. Chim. Acta 36*, 360 (1966).

372. M. Marcantonatos and D. Monnier, *Helv. Chim. Acta 50*, 2068 (1967).

373. D. Monnier, B. Liebich, and M. Marcantonatos, *Fresenius' Z. Anal. Chem. 247*, 188 (1969).

374. B. Liebich, D. Monnier, and M. Marcantonatos, *Anal. Chim. Acta 52*, 305 (1970).

375. D. Monnier, C. A. Menzinger, and M. Marcantonatos, *Anal. Chim. Acta 60*, 233 (1972).

376. D. Monnier and M. Marcantonatos, *Mitt. Geb. Lebens. Hyg. 63*, 212 (1972).

377. B. K. Afghan, P. D. Goulden, and J. F. Ryan, *Water Res. 6*, 1475 (1972).

378. K. Tauboeck, *Naturwissenschaften 30*, 439 (1942).

379. A. Murata and F. Yamauchi, *J. Chem. Soc. Jap., Pure Chem. Sect. 79*, 231 (1958).

380. L. Pszonicki and W. Tkacz, *Chem. Anal.* (Warsaw) *15*, 809 (1970).

381. L. Pszonicki and W. Tkacz, *Chem. Anal.* (Warsaw) *15*, 1097 (1970).

382. J. Dabrowski and L. Pszonicki, *Chem. Anal.* (Warsaw) *16*, 51 (1971).

383. W. Tracz and L. Pszonicki, *Anal. Chim. Acta 87*, 177 (1976).

384. W. Tracz and L. Pszonicki, *Anal. Chim. Acta 90*, 339 (1977).

385. D. P. Shcherbov and R. N. Korzheva, *Tezisy Dokladov Sovesh. Lyuminest.* 65 (1958).

386. D. P. Shcherbov and R. N. Korzheva, *Tr. Kaz. Nauchn. Issled. Inst. Miner. Syr'ya* No. 2, 217 (1960).

387. V. A. Nazarenko and S. Vinkovetskaya, *Zh. Anal. Khim. 26*, 802 (1971).

388. K. Hiraki, T. Isoda, T. Kitano, H. Hirayama, and Y. Nishikawa, *Kinki Daigaku Rikogakubu Kenkyu Hokoku 10*, 23 (1975); *Chem. Abstr. 84*, 25491h.

389. L. Kommenda, *Chem. Listy 47*, 531 (1953).

390. W. Tkacz and L. Pszonicki, *Chem. Anal.* (Warsaw) *16*, 535 (1971).

391. K. Neelakantam and L. Rao, *Proc. Indian Acad. Sci. 16a*, 349 (1942).

392. N. Raju and G. Rao, *Nature, 174*, 400 (1954).

393. N. A. Raju and K. Neelakantum, *Curr. Sci. 27*, 432 (1958).

394. G. G. Rao and N. A. Raju, *Z. Anal. Chem. 167*, 325 (1969).

395. A. K. Babko and A. E. Vasilevskaya, *Ukr. Khim. Zh. 33*, 314 (1967).

396. A. E. Vasilevskaya, *Nauch. Tr. Vses. Inst. Miner. Resur. 5*, 22 (1971).

397. T. Shibazaki, *Jap. Anal. 12*, 385 (1963).

398. T. Shibazaki, *Yakugaku Zasshi 88*, 1393 (1968).

399. M. Marcantonatos, D. Monnier, and A. Marcantonatos, *Helv. Chim. Acta 47*, 705 (1964).

400. V. Rigin and N. Meinichenko, *Zavod. Lab. 33*, 3 (1967).

401. J. Koerbl and F. Vyrda, *Chem. Listy 51*, 1457 (1957).

402. E. A. Morgen, E. A. Vlasov, and A. Z. Serykh, *Zh. Pribl. Khim.* (Lenograd) *43*, 2744 (1970); *Chem. Abstr. 75*, 146179q.

403. C. E. White and C. S. Lowe, *Ind. Eng. Chem. Anal. Ed. 13*, 809 (1941).

404. F. Hyslop, E. D. Palmes, W. C. Alford, A. R. Monaco, and L. T. Fairhall, *Natl. Inst. Health Bull. 181*, 49 (1943); *Chem. Abstr.* 38, 4681.

405. E. S. Przheval'skii, T. A. Belyavskaya, and A. P. Golovina, *Vestn. Mosk. Univ. 11, Ser. Mat. Mekh. Phiz. Khim.* No. 1, 191 (1956); *Chem. Abstr. 51*, 11916i.

406. N. M. Alykov and A. I. Cherkesov, *Zh. Anal. Khim. 31*, 1104 (1976); *Chem. Abstr. 86*, 255097.

407. S. B. Savvin, E. S. Danilin, and T. M. Malyutina, *Zh. Anal. Khim. 35*, 457 (1980), *Chem. Abstr. 93*, 35339b.

408. B. Budesinsky and T. S. West, *Anal. Chim. Acta 42*, 455 (1968).

409. D. E. Ryan, M. Granda, and M. Janmohammed, *Anal. Chim. Acta 76*, 467 (1975).

410. A. Murata, M. Tominaga, and T. Suzuki, *Jap. Anal.* 23, 1349 (1974).

411. T. Ito and A. Murata, *Jap. Anal. 20*, 1422 (1971).

412. F. Capitan, F. Salinas, and L. M. Franquelo, *Anal. Lett. 8*, 753 (1975).

412a. M. Deguchi, K. Nanba, K. Morisige, and I. Okumura, *Jap. Anal. 29*, 91 (1980); *Anal. Abstr., 39*, 3B44.

413. K. Morisige, *Anal. Chim. Acta 73*, 245 (1974).

414. Sh. T. Talipov, A. T. Tashkhodzhaev, L. E. Zel'tser, and Kh. Khikmatov, *Dokl. Akad. Nauk. Uzb. SSR 31*, 36 (1974).

415. A. I. Cherkesov and T. S. Zhigalkina, *Zavod. Lab. 27*, 658 (1961).

416. A. I. Cherkesov and T. S. Zhigalkina, *Trudy Astrak. Tekh. Inst. Rybn. Prom. Khoz. 8*, 25 (1962).

417. S. A. Wicks and R. W. Burke, *Natl. Bur. Stand. (U.S.) Spec. Publ. 492*, 85 (1977).

418. D. V. Gladilovich, N. N. Grigor'ev, and K. P. Stolyarov, *Zh. Anal. Khim. 33*, 2113 (1978); *Chem. Abstr. 90*, 47932d.

419. V. Maly and L. Sommer, *Chem. Listy 73*, 538 (1979); *Anal. Abstr. 38* 2B61.

420. Z. Holzbecker, *Coll. Czech. Chem. Comm. 20*, 193 (1955).

421. K. Motojinia, *Bull. Chem. Soc. Jap. 29*, 75 (1956).

422. A. Murata and M. Nakamura, *Jap. Anal. 21*, 1365 (1972); *Chem. Abstr. 78*, 52133a.

423. M. H. Fletcher, C. E. White, and M. S. Sheftel, *Ind. Eng. Chem. Anal. Ed. 18*, 179 (1946).

424. F. W. Klempeter and A. Martin, *Anal. Chem. 22*, 828 (1950).

425. H. Laitinen and P. Kivalo, *Anal. Chem. 24*, 1467 (1952).

426. G. Welford and J. Harley, *J. Am. Ind. Hyg. Assoc. Quartz 13*, 332 (1952).

427. C. W. Sill and C. P. Willis, *Anal. Chem. 31*, 598 (1959).

428. L. L. Galkina, *Byull. Nauch-Tekh. Inform. Min. Geol. SSSR Ser. Izuch. Vesh. Sost. Min. Syr'ya Takh. Obogas. Rud.* No. (3), 19 (1967); *Anal. Abstr. 15*, 5873.

429. R. N. Plotnikova, R. P. Ashaeva, and D. P. Shcherbov, *Issled. Razrab. Fotometrich. Metod. Opred. Mikro. Elem. Min. 56* (1967); *Chem. Abstr. 71*, 18495x.

430. R. F. Chen, A. N. Schechter, and R. L. Berger, *Anal. Biochem. 29*, 68 (1969).

431. P. W. West and E. Jungreis, *Anal. Chim. Acta, 45*, 188 (1969).

432. T. Takata, M. Hitosugi, T. Kadowaki, Y. Inoue, and K. Seki, *Sangyo Igaku 20*, 114 (1978); *Anal. Abstr. 38*, 2029.

433. I. Kubicek, *Prac. Lek. 34*, 49 (1982); *Anal. Abstr. 44*, 1H11.

434. L. Kabrt and Z. Holzbecher, *Coll. Czech. Chem. Comm. 41*, 540 (1976); *Chem. Abstr. 85*, 201541.

435. M. J. Bosa, *J. Indian Chem. Soc. Ind. News Ed. 14*, 61 (1951); *Chem. Abstr. 46*, 4948c.

436. W. M. Dressel and R. A. Ritchery, *U.S. Bur. Min., Inform. Cir. 7946* (1960); *Chem. Abstr. 54*, 13964d.

437. A. Guiraum and J. L. Vilchez, *Quim. Anal. 29*, 265 (1975).

438. A. T. Tashkhodzhaev, L. E. Zel'tser, Sh. T. Talipov, and Kh. Khikmatov, *Zavod. Lab. 41*, 280 (1975).

439. A. T. Tashkhodzhaev, L. E. Zel'tser, Sh. T. Talipov, and Kh. Khikmatov, *Uzb. Khim. Zh.* (6), 9 (1975).

440. Sh. T. Talipov, A. T. Tashkhodzhaev, L. E. Zel'tser, and Kh. Khikmatov, *Zh. Anal. Khim. 28*, 807 (1973); *Chem. Abstr. 79*, 48949z.

441. Sh. T. Talipov, A. T. Tashkhodzhaev, L. E. Zel'tser, and Kh. Khikmatov, *Nauch. Tr. Tashk. Gos. Univ.* No. 419, 89 (1972); *Chem. Abstr. 79*, 121538w.

442. Z. Holzbecher and K. Volka, *Coll. Czech. Chem. Comm. 35*, 2925 (1970).

442a. K. Morishige, *Kinki Daigaku, Rikogakubu Kenkuy Hakaku*, No. 17, 411 (1982); *Anal. Abstr. 45*, 1B47.

443. T. Sabirova, A. T. Tashkhodzhaev, L. E. Zel'tser, and Sh. T. Talipov, *Dokl. Akad. Nauk. Uzb. SSR*, No. 8, 41 (1978).

444. T. Naito, H. Nagano, and T. Yosui, *Jap. Anal. 18*, 1068 (1969); *Chem. Abstr. 72*, 35830.

445. G. G. Guilbault, M. H. Sadar, and M. Zimmer, *Anal. Chim. Acta, 44*, 361 (1969).

446. H. Axelrod, J. Bonelli, and J. Lodge, *Environ. Sci. Technol. 5*, 420 (1971).

447. D. P. Shcherbov and D. N. Lisitsyna, *Issled Obl. Khim. Fiz. Metod. Anal. Miner. Syr'ya* No. 4, 74 (1975); *Chem. Abstr. 87*, 94892p.

448. O. S. Wolfbeis and E. Urbano, *Fresenius' Z. Anal. Chem. 314*, 577 (1983).

449. F. Garcia Sanchez, A. Navas Diaz, and J. J. Laserna, *Anal. Chem. 56*, 253 (1983).

450. B. Budesinsky and T. S. West, *Talanta 16*, 399 (1969).

451. Y. Wilkins, *Talanta 4*, 80 (1960).

452. B. L. Kepner and D. M. Hercules, *Anal. Chem. 35*, 1238 (1963).

453. E. A. Bozhevol'nov and S. U. Kreingol'd, *Metod. Anal. Khim. Reakt. Preparat.* No. 4, 85 (1962).

454. R. O. Hurst, *Can. J. Biochem. 42*, 287 (1964).

455. H. M. Hattingberg, W. Klaus, H. Lullman, and S. Zepf, *Experientia 22*, 553 (1966).

456. B. Kleim, J. H. Kaufman, and J. Isaacs, *Clin. Chem. 13*, 1071 (1967).

457. G. G. Rudolph, J. J. Holler, and W. J. Ford, *Clin. Chim. Acta 18*, 187 (1967).

458. C. Robinson and J. A. Weatherall, *Analyst 93*, 722 (1968).

459. A. B. Borle and F. Briggs, *Anal. Chem. 40*, 339 (1968).

460. H. G. Classen, P. Marquardt, and M. Spath, *Arzneim. Forsch.* 18, 211 (1968).

461. T. Uemura, *Sci. Rep. Tohoku Imp. Univ. Ser.* 4, *34*, 31 (1968).

462. B. Fingerhut, A. Poock, and H. Miller, *Clin. Chem. 15*, 870 (1969).

463. M. Lewin, M. Wills, and D. Baron, *J. Clin. Pathol. 22*, 222 (1969).

464. W. Mische, E. Schroeder, and K. Thinius, *Plaste Kautsch. 16*, 23 (1969).

465. G. B. Moser and H. W. Gerarde, *Clin. Chem. 15*, 376 (1969).

466. J. F. Bandrowski and C. L. Benson, *Clin. Chem. 18*, 1411 (1972).

467. M. Roman, A. Fernandez-Guitierrez, and F. Cardenas, *Microchem. J. 25*, 576 (1980).

468. C. Nagasaki and K. Toei, *Jap. Anal. 21*, 87 (1972); *Chem. Abstr. 76*, 107582n.

469. M. Roman, A. Fernandez-Gutierrez, and M. C. Mahedero, *Ind. Chem. Soc.* (in press).

470. S. V. Beltyujova, V. N. Drobyazko, L. I. Kononenko, and N. S. Poluektov, *Ukr. Khim. Zh. 42*, 83 (1976).

471. K. Kasiura, *Chem. Anal. 20*, 389 (1975).

472. C. Miller and R. Magee, *J. Chem. Soc.* 3183 (1951).,

473. V. T. Lieu, C. A. Handy, *Anal. Lett.* 7, 267 (1974).

474. E. A. Bozhevol'nov, S. U. Kreingol'd, and L. I. Sosenkova, *Tr. Vses. Nauch. Issled. Inst. Khim. Reakt. Osobo, Chist. Khim. Veshchestv.* No. 30, 176 (1967).

475. E. A. Bozhevol'nov, V. M. Dziomko, L. F. Federova, and I. A. Krasavin, USSR Patent 210458 (1968); *Chem. Abstr. 68*, 1599d.

476. E. A. Bozhevol'nov and L. F. Federova, *Metod. Anal. Khim. Reakt. Preparat.* No. 15, 48 (1968).

477. E. A. Bozhevol'nov and L. F. Federova, *Metod. Anal. Khim. Reakt. Preparat.* No. 15, 44 (1968); *Chem. Abstr. 70*, 86125.

478. E. A. Bozhevol'nov, L. F. Federova, I. A. Krasavin, and V. M. Dziomko, *Zh. Anal. Khim. 24*, 531 (1969).

479. E. A. Bozhevol'nov, O. A. Fakeeva, and V. I. Komleva, *Metod. Anal. Galogenidov. Shchelochn. Shchelochn. Metal. Vys. Chist.* No. 2, 34 (1971); *Chem. Abstr. 78,* 66597s.

480. L. F. Federova and E. A. Bozhevol'nov, *Metod. Anal. Khim. Reakt. Preparat.* No. 18, 7 (1971); *Chem. Abstr. 77,* 159848u.

481. G. H. Huitink, *Anal. Chim. Acta, 70,* 311 (1974).

482. H. Nishida, Y., Katayama, H. Katsuki, H. Nakamura, M. Takagi, and K. Ueno, *Chem. Lett.* No. 11, 1853 (1982).

483. T. L. Miller and S. I. Seukfor, *Anal. Chem. 54,* 2022 (1982).

483a. A. Cabrera, T. S. Durand, R. Gallego, and T. L. Peral, *Rev. Roy. Acad. Cienc. Exactas, Fis. Nat. Madrid 77,* 389 (1983); *Chem. Abstr. 99,* 151277r.

484. J. J. Laserna, A. Navas, and F. Garcia Sanchez, *Anal. Lett. 14,* 833 (1981).

485. B. Budesinsky and T. S. West, *Analyst 94,* 182 (1969).

486. A. J. Hefley and B. Jaselskis, *Anal. Chem. 46,* 2036 (1974).

487. E. A. Bozhevol'nov, L. F. Federova, L. A. Krasavin, and V. M. Dziomko, *Zh. Anal. Khim. 35,* 1722 (1970).

488. I. Kasa and G. Bajnoczy, *Period. Polytech. Chem. Eng. 18,* 289 (1974).

489. N. Louis and A. Reber, *Anal. Chem. 26,* 936 (1954).

490. K. Nagasawa and O. Ishidaka, *Chem. Pharm. Bull.* (Tokyo) *22,* 375 (1974).

491. E. Ryan, A. E. Pitts, and R. M. Cassidy, *Anal. Chim. Acta, 34,* 491 (1966).

492. Y. Nishikawa, K. Hiraki, K. Morishige, and T. Katagi, *Jap. Anal. 26,* 365 (1977).

493. K. P. Stolyarov and V. V. Firyulina, *Zh. Anal. Khim. 33,* 2102 (1978); *Chem. Abstr. 90,* 47931c.

494. M. A. Matveets and D. P. Shcherbov, *Zh. Anal. Khim. 26,* 823 (1971).

495. K. Watanabe and K. Kawagaki, *Jap. Anal. 23,* 1356 (1974); Chem. Abstr., *82,* 79986a.

496. I. Kasa and J. Korosi, *Period. Polytech. Chem. Eng. 17,* 241 (1973).

497. D. T. Haworth and R. H. Boeckeler, *Microchem. J. 13,* 158 (1968).

498. T. Vesiene and B. Raieinskyte, *Liet. TSR Mokslu Akad. Darb.,* Ser. B, No. 6, 115 (1972); *Chem. Abstr. 79,* 61184r.

499. F. Salinas, C. Genestar, and F. Grases, *Microchem. J. 27,* 32 (1982).

500. A. Navas, F. Sanchez Rojas, and F. Garcia Sanchez, *Mikrochim. Acta I,* 175 (1982).

501. B. K. Pal, F. Toneguzzo, A. Corsini, and D. E. Ryan, *Anal. Chim. Acta 88,* 353 (1977).

502. C. Ti Huu, A. I. Volkova, and T. Getman, *Zh. Anal. Khim. 24,* 688 (1969); *Anal. Abstr. 19,* 3771.

503. H. Taniguchi, T. Yoshida, Y. Adachi, and S. Nakano, *Chem. Pharm. Bull. 28,* 2909 (1980).

504. A. V. Karyakin and G. G. Babicheva, *Zh. Anal. Khim.* *23*, 789 (1968);*Chem. Abstr.* *69*, 38666j.

505. D. Thorburn Burns and P. Hanprasopwattana, *Anal. Chim. Acta 118*, 185 (1980).

506. S. Jaya, T. Prasada Rao, and T. V. Ramakrishna, *Talanta 30*, 363 (1983).

507. G. G. Guilbault and D. Krammer, *Anal. Chem. 37*, 1395 (1965).

508. R. Morgan, G. E. Isom, and J. L. Way, *Proc. West. Pharmacol. Soc. 19*, 392 (1976).

509. J. S. Hanker, R. M. Gamson, and H. Klapper, *Anal. Chem. 29*, 879 (1957).

510. T. Toida, S. Tanabe, and T. Imanari, *Chem. Pharm. Bull. 29*, 3763 (1981).

511. D. E. Ryan and J. Holzbecher, *Int. J. Environ. Anal. Chem. 1*, 159 (1971).

512. F. Vernon and P. Whitham, *Anal. Chim. Acta 59*, 155 (1972).

513. G. Colovos, M. Haro, and H. Freiser, *Talanta 17*, 273 (1970).

514. G. G. Guilbault, P. Brignac, and M. Zimmer, *Anal. Chem. 40*, 190 (1968).

515. G. G. Guilbault, D. N. Kramer, and E. Hackley, *Anal. Biochem. 18*, 241 (1967).

516. J. S. Hanker, A. Gelberg, and B. Whitten, *Anal. Chem. 30*, 93 (1958).

517. G. L. McKinney, H. K. Y. Lau, and P. F. Lott, *Microchem. J. 17*, 375 (1972).

518. M. Bonavita, *Arch. Biochem. Biophys. 88*, 366 (1960).

519. S. Takanashi and Z. Tamura, *Chem. Pharm. Bull. 18*, 1633 (1970); *Chem. Abstr. 73*, 126665w.

520. R. L. Morgan and J. L. Way, *J. Anal. Toxicol. 4*, 78 (1980); *Chem. Abstr. 93*, 62753b.

521. G. G. Guilbault and D. Kramer, *Anal. Chem. 37*, 918 (1965).

522. M. Wronski, *Chem. Anal.* (Warsaw) *15*, 215 (1970).

523. P. M. Zarembaski and A. Hodgkinson, *Biochem. J. 96*, 717 (1965).

524. D. A. Britton and J. C. Guyon, *Anal. Chim. Acta 44*, 397 (1969).

525. J. Possidoni de Albinati, *An. Assoc. Quin. Argen. 55*, 61 (1967).

526. A. N. Razumovich, S. V. Konev, and E. A. Chernitskii, *Biofizika 14*, 597 (1969); *Chem. Abstr. 72*, 226q.

527. S. B. Zamochnick and G. A. Rechnitz, *Z. Anal. Chem. 199*, 424 (1963).

528. D. Thorburn Burns, P. Hanprasopwattana, and B. P. Murphy, *Anal. Chim. Acta 134*, 397 (1982).

529. F. Grases, F. Garcia-Sanchez, and M. Valcarcel, *An. Quim.76B*, 402 (1980).

530. D. Thorburn and P. Hanprasopwattana, *Anal. Chim. Acta 115*, 389 (1980).

531. P. R. Haddad, P. W. Alexander, and L. E. Smythe, *Talanta 23*, 275 (1976).

531a. F. Garcia Sanchez, A. Navas, J. J. Laserna, and M. R. Martinez de la Barrera, *Fresenius' Z. Anal. Chem. 315*, 491 (1983).

532. G. Schenk, K. Dilloway, and J. Coulter, *Anal. Chem. 41*, 510 (1969).

533. E. A. Bozhevol'nov and S. U. Kreingol'd, *Tr. Vses. Nauch. Issled. Inst. Khim. Reakt. Osobo Chist. Khim. Veshchestv.* No. 26, 204 (1964); *Chem. Abstr. 66*, 82087h.

534. I. P. Alimarin, A. M. Medvedeva, and G. P. Tikhonov, *Zh. Anal. Khim. 35*, 1946 (1980); *Anal. Abstr. 40*, 6B179.

535. N. N. Grigor'ev, K. P. Stolyarov, and Z. P. Martynenko, in *Primen. Org. Reagentov Anal. Khim.*, I. A. Tserkovnitskaya (ed.), Leningrad Univ., Leningrad, 1969, p 189; *Chem. Abstr. 74*, 19021g.

536. Z. Holzbecher and J. Ruzicka, *Sb. Vys. Sk. Chem. Tekhnol. Praze. Anal. Chem.* No. H8, 83 (1972); *Chem. Abstr. 77*, 83226s.

537. A. T. Pilipenko, T. L. Shevchenko, and A. I. Volkova, *Zh. Anal. Khim. 32*, 731 (1977); *Chem. Abstr. 87*, 193120.

538. V. Ya Temkina, E. A. Bozhevol'nov, N. M. Dyatlova, S. U. Kreingol'd, G. F. Yaroshenko, V. N. Antonov, and R. P. Lastovskii, *Zh. Anal. Khim. 22*, 1830 (1967).

539. W. E. Hoehne and H. Wessner, *Anal. Chim. Acta 93*, 345 (1977).

540. B. E. Jones and W. W. Brandt, 153rd Meeting, ACS, Miami Florida paper B67, 1963.

541. E. A. Bozhevol'nov, S. U. Kreingol'd, and L. I. Sosenkova, *Tr. Vses. Nauch. Issled. Inst. Khim. Reakt. Osobo Chist. Khim. Veshchestv.* No. 30, 176 (1967); *Chem. Abstr. 69*, 73657d.

542. F. Grases, F. Garcia Sanchez, and M. Valcarcel, *Microchem. J. 25*, 368 (1980).

543. F. Grases, F. Garcia Sanchez, and M. Valcarcel, *Anal. Chim. Acta 119*, 359 (1980).

544. F. Grases, F. Garcia-Sanchez, and M. Valcarcel, *Anal. Chim. Acta 125*, 21 (1981).

545. E. P. Panova, N. P. Checkrii, and E. N. Filatov, *Zavod. Lab. 45*, 805 (1979); *Chem. Abstr. 92*, 134927r.

546. R. Alarcon, M. Silva, and M. Valcarcel, *Anal. Lett. 15*, 891 (1982).

547. L. Zel'tser and Sh. Talipov, *Zh. Anal. Khim. 33*, 1260 (1978); *Chem. Abstr. 89*, 208509t.

548. H. Taniguchi, K. Teshima, K. Tsuge, and S. Nakauo, *Jap. Anal. 24*, 314 (1975); *Chem. Abstr. 83*, 125662u.

549. E. A. Bozhevol'nov, *Sb. Statei, Vses. Nauch. Issled Khim. Reakt. Osobo Chist. Khim. Veshchestv.* No. 24, 24 (1961); *Chem. Abstr. 57*, 27h.

550. L. Zel'tser, Z. Maksimycheva, and S. Talipov, *Akad. Nauk. Uzber SSR* 30 (1969); *Chem. Abstr. 73*, 83532f.

551. E. A. Bozhevolnov and S. U. Kreingol'd, *Metod. Anal. Khim. Reakt. Preparat.* No. 4, 96 (1962); *Chem. Abstr. 61*, 2467g.

552. S. U. Kreingol'd and E. A. Bozhevol'nov, *Zh. Anal. Khim. 18*, 942 (1963); *Chem. Abstr. 60*, 1099f.

553. E. A. Bozhevol'nov and S. U. Kreingol'd, *Metod. Anal. Khim. Reakt. Preparat.* No. 11, 65 (1965).

554. Yu. M. Martynov, E. A. Kreingol'd and B. M. Maevskaya, *Zavod. Lab. 31,* 1447 (1965); *Chem. Abstr. 64,* 1611b.

555. S. U. Kreingol'd and E. A. Bozhevol'nov, *Metod. Anal. Khim. Reakt. Preparat.* No. 11, 68 (1965); *Chem. Abstr. 65,* 1368g.

556. A. V. Konstantinov, L. M. Korobotchkin, and G. V. Anastas'ina, *Tr. Nov. Appart. Metod. 5,* 167 (1967); *Chem. Abstr. 68,* 57314p.

557. E. A. Bozhevol'nov, S. U. Kreingol'd and A. A. Panteieimovova, *Metod. Anal. Khim. Reakt. Preparat.* No. 13, 104 (1966).

557a. K. P. Stolyarov and V. V. Firylina, *Zh. Anal. Khim. 38,* 625 (1983); *Chem. Abstr. 99,* 15742m.

558. B. W. Bailey, R. M. Dagnall, and T. S. West, *Talanta 13,* 1661 (1966).

559. M. Roman, A. Fernandez-Gutierrez, and A. Muñoz de la Peña, *An.Quim. 79B,* 128 (1983).

560. S. Igarashi, T. Yotsuyanagi, and K. Aomura, *Jap. Anal. 28,* 449 (1979); *Chem. Abstr. 91,* 167869f.

561. Y. Yamane, Y. Yamada, and S. Kunihiro, *Jap. Anal. 17,* 973 (1968); *Chem. Abstr. 70,* 43723a.

562. Y. Yamane, M. Miyazaki, and M. Ohtawa, *Jap. Anal. 18,* 750 (1969); *Chem. Abstr. 72,* 18150v.

563. A. S. Saracoglu and H. Erylmaz, *Chim. Acta Turc. 9,* 239 (1981); *Chem. Abstr. 95,* 108003d.

564. K. Ritchie and J. Harris, *Anal. Chem. 41,* 163 (1969).

565. E. L. Melent'eva, M. A. Tishchenko, R. A. Vitkun, L. I. Kononenko, and N. S. Poluektov, *Khim. Prom., Inf. Nauk. Tekh. Zb. 3(23),* 63 (1965).

566. A. I. Kirillov and M. M. Rybakova, *Zavod. Lab. 43,* 1432 (1977); *Chem. Abstr. 89,* 220628p.

567. E. Butter, I. Kolowos, and H. Holzapfel, *Talanta 15,* 901 (1968).

568. M. A. Tishchenko, L. I. Kononenko, R. A. Vitkun, and N. S. Poluektov, *Ukr. Khim. Zh. 32,* 508 (1966); *Chem. Abstr. 65,* 6291h.

569. L. I. Kononenko, M. A. Tishchenko, and N. S. Poluektov, *Zh. Anal. Khim. 19,* 829 (1964); *Chem. Abstr. 61,* 12620f.

570. S. B. Meshkova, T. B. Kravchenko, L. I. Kononenko, and N. S. Poluektov, *Zh. Anal. Khim. 34,* 121 (1979); *Chem. Abstr. 90,* 179517g.

571. E. I. Tselik, N. S. Poluektov, and V. T. Mishchenko, *Zh. Anal. Khim. 34,* 1962 (1979).

572. M. A. Tishchenko, L. A. Alakaeva, and N. S. Poluektov, *Ukr. Khim. Zh. 39,* 482 (1973); *Chem. Abstr. 79,* 132625.

573. N. S. Poluektov, M. A. Tishchenko, and L. A. Alakaeva, *Tr. Khim. Khim. Tekhnol. 5,* 104 (1973); *Chem. Abstr. 81,* 57882r.

574. M. A. Tishchenko, I. I. Zheltvai, and N. S. Poluektov, *Ukr. Khim. Zh.* *41*, 197 (1975).

575. M. A. Tishchenko, G. I. Gerasimenko, and N. S. Poluektov, *Zavod. Lab.* *40*, 935 (1974); *Chem. Abstr. 82*, 38203.

576. S. B. Meshkova, T. B. Kravchenko, L. I. Kononenko, and N. S. Poluektov, *Zh. Anal. Khim. 34*, 121 (1979).

577. E. T. Karaseva and V. E. Karasev, *Zh. Anal. Khim. 37*, 1330 (1982).

578. E. T. Karaseva and V. E. Karasev, *Koord. Khim. 1*, 926 (1975); *Chem. Abstr. 83*, 155006u.

579. E. T. Karaseva, V. E. Karasev, N. L. Sigula, A. P. Golovina, and I. P. Efimov, *Koord. Khim. 5*, 647 (1979); *Chem. Abstr. 91*, 823580p.

580. E. A. Bozhevol'nov, O. A. Fakeeva, V. M. Dziomko, I. A. Krasavin, B. V. Parusnikov, L. N. Shoshmina and V. K. Runov, *Org. Reagenty Anal. Khim., Tezisy. Dokl. Vses. Konf. 4th, 2*, 84 (1976), A. T. Pihipenko (ed.), *Naukova Dumka, Kiev.; Chem. Abstr. 88*, 15443n.

581. A. P. Golovina, S. V. Kachin, and V. K. Runov, *Vestn. Mosk. Univ. Ser. Khim. 18*, 709 (1977); *Chem. Abstr. 89*, 119941k.

582. S. V. Kachin, B. V. Parusnikov, and V. K. Runov, *Depos. Doc. VINITI* 3782, 110 (1979); *Chem. Abstr. 94*, 76154q.

583. V. S. Khomenko, V. V. Kuznetsova, and L. A. Pekarskaya, *Zh. Prikl. Spectr. 6*, 117 (1967); *Chem. Abstr. 67*, 29021b.

584. M. A. Tishchenko, I. I. Zheltvai, N. S. Poluektov, and I. V. Bakshun, *Zavod. Lab. 39*, 671 (1973).

585. R. Belcher, R. Perry, and W. I. Stephen, *Analyst, 94*, 26 (1969).

586. L. I. Kononenko, T. B. Kravchenko, S. V. Bel'tyukova, V. E. Kuz'min, and E. S. Suprinovich, *Ukr. Khim. Zh. 46*, 427 (1980); *Anal. Abstr. 39*, 5B76.

587. R. P. Fisher and J. D. Winefordner, *Anal. Chem. 43*, 454 (1971).

588. D. E. Williams and J. C. Guyon, *Anal. Chem. 43*, 139 (1971).

589. I. M. Lomniczi de Upton and J. F. Possidoni de Albinati, *An. Assoc. Quim. Argent. 68*, 141 (1980).

590. T. Shigematsu, M. Matsui, and T. Suimida, *Bull. Inst. Chem. Res. Kyoto Univ. 46*, 249 (1968).

591. T. Shigematsu, M. Matsui, and R. Wake, *Anal. Chim. Acta 46*, 101 (1969).

592. M. Suzuki, M. Matsui, T. Shigematsu, Y. Matsuo, and T. Nagai, *Bull. Inst. Chem. Res. Kyoto Univ. 58*, 279 (1980).

593. N. S. Poluektov, R. A. Vitkun, and L. I. Kononenko, *Ukr. Khim. Zh. 30*, 629 (1964); *Chem. Abstr. 61*, 10029a.

594. L. I. Kononenko, E. V. Melent'eva, R. A. Vitkun, and N. S. Poluektov, *Prom. Khim. Reaktiv. Osobo. Chist. Veshchestv.* No. 8, 223 (1967); *Chem. Abstr. 69*, 54381.

595. V. T. Mishchenko, E. I. Tselik, and A. P. Koev, *Zh. Anal. Khim. 32*, 71 (1977).

596. L. I. Kononenko, R. S. Lauer, and N. S. Poluektov, *Zh. Anal. Khim. 18*, 1468 (1963).

597. S. V. Bel'tyukova, N. S. Poluektov, I. B. Kravchenko and L. I. Kononenko, *Zh. Anal. Khim. 35*, 1103 (1980).

598. E. V. Melent'eva, N. S. Poluektov and L. I. Kononenko, *Zh. Anal. Khim. 22*, 187 (1967).

599. H. Shi and W. Cui, *Fenxi Huaxue 10*, 561 (1982); *Anal. Abstr. 44*, 6B91.

600. M. A. Tishchenko, N. S. Poluektov, I. I. Zheltvai, and I. V. Bakshun, *Tr. Khim. Khim. Tekhnol.* 112 (1973).

601. M. A. Tishchenko, I. I. Zheltvai, I. V. Bakshun, and N. S. Poluektov, *Zh. Anal. Khim. 28*, 1954 (1973).

602. M. A. Tishchenko, I. I. Zheltvai, G. I. Gerasimenko, and N. S. Poluektov, *Zh. Anal. Khim. 321*, 2147 (1977).

603. Y. Ci, K. Hu, T. Liu, and H. Ma, *Fenzi Huaxue 10*, 232 (1982); *Anal. Abstr. 43*, 6B60.

604. L. I. Kononenko, N. S. Poluektov, and M. P. Nikonova, *Zavod. Lab. 30*, 779 (1964).

605. E. V. Melent'eva, M. A. Tishchenko, R. A. Vitkun, L. I. Kononenko, and N. S. Poluektov, *Khim. Prom. Inf. Nauchn. Tekh. Zb. 3*, 63 (1965).

606. E. V. Melent'eva, M. A. Tishchenko, and N. S. Poluektov, *Sovrem, Metod. Khim. Spektral. Anal. Mater.* 207 (1967).

607. E. V. Melent'eva, L. I. Kononenko and N. S. Poluektov, *Sovrem, Metod. Khim. Spektral. Anal. Mater.* 202 (1967).

608. L. I. Kononenko, M. A. Tischenko, R. A. Vitkun, and E. V. Melent'eva, *Zavod. Lab. 34*, 1432 (1968).

609. H. G. Huang, K. Hiraki, and Y. Nishikawa, *Nippon Kagaku Kaishi* No. 1, 66 (1981); *Chem. Abstr. 94*, 113877a.

610. T. Taketatsu and A. Sato, *Anal. Chim. Acta 108*, 429 (1979).

611. H. G. Huang, K. Hiraki, and Y. Nishikawa, *Jap. Anal. 30*, 452 (1981); *Anal. Abstr., 42*, 2B82.

612. S. V. Bel'tyukova and T. B. Kravchenko, *Zavod. Lab. 48*, 13 (1982).

613. T. Taketatsu, *Talanta 29*, 397 (1982).

614. W. A. Powell and J. H. Saylor, *Anal. Chem. 25*, 960 (1953).

615. G. Hocman, L. Lacko, and L. Hegedus, *Acta Far. Rerum Natur. Univ. Commenianae Chim. 13*, 71 (1968); *Chem. Abstr. 71*, 19424d.

616. H. H. Willard and C. A. Horton, *Anal. Chem. 24*, 862 (1952).

617. R. W. Spence, R. P. Straetz, D. P. Krause, W. M. Byerly, L. W. Safranski, L. E. West, J. I. Waters, and D. E. Wallace, U.S. Atomic Energy Commission, ECD 3089 (1951).

618. J. Bouman, *Chem. Weekbl. 51*, 33 (1955).

619. G. H. Schenk and K. P. Dilloway, *Anal. Lett.* 2, 379 (1969).

620. S. W. Chaikin and T. D. Glassbrook, *Res. Ind.* 5, 2 (1953).

621. T. O. Ivie, L. F. Zieleuski, M. D. Thomas, and C. R. Thompson, *J. Air Pollut. Control Assoc.* 15, 195 (1965).

622. C. R. Thompson, L. F. Zielenski, and T. O. Ivie, *Atmos. Environ.* 1, 253 (1967).

623. D. R. Taves, *Talanta 15,* 1015 (1968).

624. L. H. Tan and T. S. West, *Anal. Chem. 43,* 136 (1971).

625. T. C. Guyon, B. E. Jones and D. A. Britton, *Mikrochim. Acta* 1180 (1968).

626. J. Block and E. Morgan, *Anal. Chem. 34,* 1647 (1962).

627. F. Salinas, F. Garcia-Sanchez, and C. Genestar, *Anal. Lett. 15,* 747 (1982).

628. F. Salinas, C. Genestar, and F. Grases, *Anal. Chim. Acta 130,* 337 (1981).

628a. A. Moreno, M. Silva, D. Perez-Bendito, and M. Valcarcel, *Anal. Chim Acta* (in press).

629. R. Schmidt, W. Weis, V. Klingmuller, and H. Standinger, *Z. Klin. Chem. Klin. Biochem. 5,* 304 (1967).

630. P. Stanchev, *Chem. Anal.* (Warsaw) *16,* 243 (1971); *Chem. Abstr. 75,* 58389y.

630a. D. B. Gladilovich and K. P. Stolyarov, *Zh. Anal. Khim. 38,* 876 (1983); *Chem. Abstr. 99,* 63450n.

631. K. T. Koh and D. E. Ryan, *Anal. Chim. Acta, 54,* 303 (1971).

632. *Tableaux des reactifs pour l'analyse mineral*, Report of the International Commission on New Reaction and Analytical Reagents, Paris, (1948.)

633. E. A. Bozhevol'nov and S. U. Kreingol'd, *Metod. Anal. Khim. Reakt. Preparat.* No. 11, 36 (1965), *Chem. Abstr. 65,* 9726c.

634. S. U. Kreingol'd and E. A. Bozhevol'nov, *Metod. Anal. Khim. Reakt. Preparat.* No. 11, 39 (1965); *Chem. Abstr. 65,* 9726e.

635. M. Laanmaa, M. L. Allsalu, and H. Kokk, *Tartu Riikliku Ultikooli Toim.* No. 219, 199 (1968); *Chem. Abstr. 71,* 77012d.

636. A. A. Obraztsov and V. G. Bocharova, *Tr. Voronezh. Gos. Univ. 82,* 182 (1971); *Chem. Abstr. 77,* 9476g.

637. D. Fink, F. Pivnichny, and W. Ohnesorge, *Anal. Chem. 41,* 833 (1969).

638. L. A. Grigoryan, A. N. Pogosyan, and V. M. Tarayan, *Arm. Khim. Zh. 25,* 931 (1972); *Chem. Abstr. 78,* 131662.

639. A. T. Tashkhodzhaev, L. E. Zelt'ser, Sh. T. Talipov, and Kh. Khikmatov, *Zh. Vses. Khim. Obshch. D. I. Mendeleeva 21,* 114 (1976); *Chem. Abstr. 85,* 116104w.

640. N. Olenovich and L. I. Koval'chuk, *Zh. Anal. Khim. 28,* 2162 (1973).

641. T. T. Laserna, A. Navas, and F. Garcia Sanchez, *Anal. Chim. Acta 121,* 295 (1980).

642. L. Ngog Thu, *Zh. Anal. Khim. 22,* 636 (1967).

643. L. Ngog Thu, R. M. Dranitskaya, and V. A. Nazarenko, *Ukr. Khim. Zh. 34*, 186 (1968).

644. G. Beck, *Mikrochim. Acta* 47 (1939).

645. Y. Nishikawa, *J. Chem. Soc. Jap. 79*, 631 (1958).

646. H. M. Koester, *Ber. Deut, Keram. Ges. 46*, 247 (1969).

647. Nguyen Shi Zuong and F. G. Zharovskii, *Ukr. Khim. Zh. 36*, 1273 (1970).

648. K. Kenyu, H. Kouichi, and I. Nobuhiko, *Jap. Anal. 26*, 246 (1977).

649. Y. Nishikawa, *Jap. Anal. 7*, 549 (1958).

650. K. Kina, K. Shiraishi, and N. Ishibashi, *Jap. Anal. 25*, 501 (1976).

651. M. Deguchi, T. Ogi, and K. Morishige, *Jap. Anal. 30*, 104 (1981); *Anal. Abstr. 41*, 3B74.

652. A. T. Pilipenko and A. I. Volkova, *Ukr. Khim. Zh. 43*, 536 (1977).

653. M. Kato, N. Ban, S. Kawai, and T. Ohno, *Jap. Anal. 20*, 1315 (1971); *Chem. Abstr. 76*, 67853t.

654. Z. Holzbecher, *Z. Sb. Vys. Sk. Chem. Technol. Pr., Anal. Chem. 1*, 63 (1967); *Chem. Abstr. 70*, 73878x.

655. M. Ishihashi, T. Shigematsu, and T. Nishikawa, *J. Chem. Soc. Jap. 78*, 1139 (1957).

656. T. Nishikawa, *J. Chem. Soc. Jap. 79*, 236 (1958).

657. T. Shigematsu, *Jap. Anal. 7*, 787 (1958).

658. E. B. Sandell, *Anal. Chem. 19*, 63 (1947).

659. J. Collat and L. Rogers, *Anal. Chem. 27*, 961 (1955).

660. M. Ishibashi, T. Shigematsu, and Y. Nishikawa, *J. Chem. Soc. Jap. 78*, 1039 (1957).

661. M. Ishibashi, T. Shigematsu, and Y. Nishikawa, *Bull. Inst. Chem. Res., Kyoto Univ. 37*, 191 (1959).

662. M. F. Landi, *Met. Ital. 52*, 366 (1960).

663. Y. Nishikawa, *J. Chem. Soc. Jap. 79*, 236 (1958).

664. S. R. Desai and K. K. Sudhalatha, *Indian J. Appl. Chem. 30*, 116 (1967).

665. I. F. Goryunova, *Sb. Nauch. Tru. Irkut. Nauch. Issledl, Inst. Red. Met., 8*, 33 (1959).

666. S. M. Milaev, *Sb. Tr. Vses. Nauch. Issledl. Gorno Met. Inst. Tsve. Met. 5*, 41 (1959).

667. A. Lukin and E. A. Bozhevol'nov. *Zh. Anal. Khim. 15*, 36 (1960).

668. N. B. Lebed and R. P. Pantaler, *Zh. Anal. Khim. 20*, 59 (1965).

669. Y. Nishikawa, K. Hiraki, K. Morishige, and T. Shigematsu, *Jap. Anal. 16*, 692 (1967).

670. K. Kina and N. Ishibashi, *Microchem. J. 19*, 26 (1974).

671. V. D. Salikov and M. Z. Yampol'skii, *Zh. Anal. Khim. 20*, 1299 (1965).

672. R. P. Pantaler, N. B. Lebed, and L. N. Semenova, *Tr. Kom. Anal. Khim. 16*, 24 (1968).

673. M. A. Matveets and L. P. Shcherbov, *Issled. Razrab. Fotometrich. Metod. Opred. Mikro. Elem. Miner.* 122 (1967); *Chem. Abstr. 71,* 27100k.

674. R. A. Zweidinger, L. Barnett, and C. G. Pitt, *Anal. Chem. 45,* 1563 (1973).

675. T. Imasaka, T. Harada, and N. Ishibashi, *Anal. Chim. Acta 129,* 195 (1981).

676. M. K. Akhmedli, D. A. Efendiev, and F. I. Ruvinova, *Azerb. Khim. Zh.* No. 3, 146 (1972); *Chem. Abstr. 79,* 121575f.

677. E. Herzfeld, *Z. Anal. Chem. 115,* 131 (1939).

678. L. Bradaks, F. Feigl, and F. Hecht, *Mikrochim. Acta* 269 (1951).

679. V. Patrovsky, *Chem. Listy, 47,* 676 (1953); *Chem. Abstr. 48,* 3187g.

680. A. I. Busev and E. P. Shkrobot, *Vest. Mosk. Univ., Ser. Mat., Mekh., Astronl. Fiz. Khim. 14,* 199 (1959).

681. G. Banateanu and P. Costinescu, *Bul. Inst. Pet. Gazl. 2,* 47 (1975); *Chem. Abstr. 87,* 62028v.

682. L. Bark and A. Rixon, *Anal. Chim. Acta 43,* 425 (1969).

682a. E. Requena, J. J. Laserna, A. Navas, and F. Garcia Sanchez, *Analyst 108,* 933 (1983).

683. Sh. T. Talipov, Z. T. Maksimycheva, Z. P. Pakudina, and A. S. Sadykov, *Izv. Vyssh. Ucheb. Zaved., Khim. Khim. Tecknol. 16,* 1154 (1973).

684. K. Watanabe and K. Kawagaki, *Bull. Chem. Soc. Jap. 48,* 1812 (1975); *Chem. Abstr. 83,* 141380z.

685. A. T. Tashkhodzhaev, L. E. Zelt'ser, Sh. T. Talipov, and Kh. Khikmatov, *Zavod. Lab. 41,* 281 (1975); *Chem. Abstr. 83,* 71055w.

686. H. Onishi, *Anal. Chem. 27,* 832 (1955).

687. H. Onishi and E. B. Sandell, *Anal. Chim. Acta 13,* 159 (1955).

688. D. P. Shcherbov and A. I. Ivankova, *Zavod. Lab. 24,* 6 (1958).

689. D. P. Shcherbov and A. I. Ivankova, *Zavod. Lab. 24,* 6 (1958).

690. D. P. Shcherbov and N. V. Kagarlitskaya, *Zavod. Lab. 28,* 30 (1962).

691. G. I. Kuchmistaya, *Zavod. Lab. 27,* 377 (1961).

692. A. I. Chuvileva and I. A. Blyum, *Zavod. Lab. 35,* 1153 (1969).

693. D. P. Shcherbov and I. T. Solov'yan, *Tr. Kaz. Nauch. Issled. Inst. Miner. Syr'ya* 196 (1959); *Chem. Abstr. 55,* 7142d.

694. A. P. Golovina, S. M. Sapezhinskaya, and V. K. Runov, *Zavod. Lab. L17,* 17 (1981); *Chem. Abstr. 94,* 113892b.

695. A. P. Golovina, I. P. Alimarin, N. B. Zorov, and Z. M. Khvathova, *Zh. Anal. Khim. 25,* 2242 (1970); *Chem. Abstr. 71,* 119310g.

696. A. P. Golovina, Z. M. Khvatkova, N. B. Zorov, and I. P. Alimarin, *Vestn. Mosk. Univ. Ser. Khim. 13,* 551 (1972); Chem. Abstr. *78,* 105651.

697. D. P. Shcherbov, *Zavod. Lab. 30,* 1527 (1964).

698. D. P. Shcherbov and A. I. Ivankova, *Issled. Razrab. Fotometrich. Metod. Opred. Mikro. Elem. Miner. Syr'e* 134 (1967).

699. A. P. Golovina, S. M. Sapezhinskaya, and V. K. Runov, *Zavod. Lab. 47*, 17 (1981).

700. B. M. Kondratenok, N. N. Grigor'ev, L. B. Gladilovich, and K. P. Stolyarov, *Izv. Vyssh. Ucheb. Zaved., Khim. Teklnol. 25*, 695 (1982); *Anal. Abstr. 44*, 3B56.

700a. B. M. Kondratenok, N. N. Grigor'ev, L. B. Gladilovich, and K. P. Stolyarov, *Otkr., Izobret., Prom. Obraztsy, Tovarnye Znaki, 23*, 141 (1983); *Chem. Abstr. 99*, 115234r.

701. Sh. T. Talipov, A. T. Tashkodzhaev, L. E. Zel'tser, and Kh. Khikmatov, *Otkr., Izobret., Prom. Obraztsy, Tovarnye Znaki, 15*, 1109 (1972).

702. V. Patrovsky, *Chem. Listy 48*, 537 (1954).

703. R. M. Dagnall, R. Smith, and T. S. West, *Chem. Ind.* (London) *39*, 1499 (1965).

704. M. A. Matveets and D. P. Shcherbov, *Issled. Obl. Khim. Fiz. Metod. Anal. Miner, Syr'ya* 54 (1971); *Chem. Abstr. 77*, 159788z.

705. G. Oshima, *Jap. Anal. 7*, 549 (1958).

706. I. Ladenbauer, J. Korkis, and F. Hecht, *Mikrochim. Acta* 1076 (1955).

707. V. A. Nazarenko and S. Ya. Vinkovetskaya, *Zh. Anal. Khim. 13*, 327 (1958).

708. V. A. Nazarenko and S. Ya. Vinkovetskaya, *Ukr. Khim. Zh. 28*, 726 (1962).

709. M. A. Matveets and D. P. Shcherbov, *Issled. Obl. Khim. Fiz. Metod. Anal. Miner. Syr'ya* 60 (1971); *Chem. Abstr. 77*, 15984k.

710. Z. T. Maksimycheva, Sh. T. Talipov, Z. P. Pakudina, and A. S. Sadykov, *Izv. Vyssh. Ucheb. Zaved. Khim. Tekhnol. 17*, 348 (1974).

711. A. T. Pilipenko, D. I. Bakardzhieva, A. I. Volkova, and T. E. Get'man, *Ukr. Khim. Zh. 37*, 689 (1971); *Chem. Abstr. 76*, 41534.

712. E. C. Stanley, B. I. Kinneberg, and L. P. Varga, *Anal. Chem. 38*, 1362, (1966).

713. G. V. Flyantikova and L. I. Korolenko, *Zh. Anal. Khim. 30*, 1349 (1975).

714. F. Capitan, M. Roman, and A. Fernandez-Gutierrez, *Bol. Soc. Quim. Peru 40*, 65 (1974).

715. G. V. Serebryakova and E. A. Bozhevol'nov, *Metod. Anal. Khim. Reakt. Preparat.* No. 11, 29 (1965).

716. A. M. Lukin, G. V. Serebryakova, E. A. Bozhevol'nov, and G. B. Zavarikhina, *Tr. Vses. Nauch. Issled. Inst. Khim. Reakt. Osobo, Chist. Khim. Veshchestv.* No. 30, 161 (1967).

717. D. P. Shcherbov, R. Plotnikova, and I. N. Astaf'eva, *Zavod. Lab. 36*, 528 (1970); *Chem. Abstr. 73*, 83538n.

718. D. E. Ryan, L. Sommer, M. Katyal, and B. K. Pal, *Mikrochim. Acta* 1181 (1970).

719. T. Ito and A. Murata, *Anal. Chim. Acta 125*, 155 (1981).

720. A. T. Pilipenko, T. U. Kukibaev, A. I. Volkova, and T. E. Get'man, *Zh. Anal. Khim. 27*, 1787 (1972); *Chem. Abstr. 78*, 37542q.

721. A. T. Pilipenko, T. U. Kukibaev, A. I. Volkova, and T. E. Get'man, *Ukr. Khim. Zh. 39*, 813 (1973).

722. A. T. Pilipenko, A. I. Zhebentyaev, and A. I. Volkova, *Zh. Anal. Khim. 29*, 1854 (1974); *Chem. Abstr. 82*, 67836p.

723. Sh. T. Talipov, L. E. Zel'tser, L. A. Morozova, and A. T. Tashkhodzhaev, *Zavod. Lab. 44*, 1052 (1978).

724. A. Brookes and A. Townshed, *Chem. Commun. 24*, 1660 (1968).

725. A. Brookes and A. Townshed, *Analyst 95*, 781 (1970).

726. T. Kouimtzis and A. Townshed, *Analyst 98*, 40 (1973).

727. N. S. Kulikov, Yu. V. Granovskii, and A. P. Golovina, *Izv. Vyssh. Ucheb. Zaved. Khim. Khim. Tekhnol. 16*, 1006 (1973); *Chem. Abstr. 79*, 142604e.

728. A. T. Pilipenko, T. U. Kukibaev, and A. I. Volkova, *Zh. Anal. Khim. 28*, 510 (1973); *Chem. Abstr. 78*, 166379r.

729. L. E. Zel'tser, Sh. T. Talipov, and L. A. Morozova, *Zh. Anal. Khim. 35*, 1747 (1980).

730. A. T. Pilipenko, T. U. Kukibaev, and A. I. Volkova, *Zh. Anal. Khim. 29*, 710 (1974).

731. I. A. Blyum, N. A. Brushtein, and L. I. Oparina, *Zh. Anal. Khim. 26*, 48 (1971).

732. A. Gorgia and D. Monnier, *Anal. Chim. Acta 54*, 505 (1971).

733. T. Naganuma, *Jap. Anal. 27*, 641 (1978).

734. A. I. Ivankova and D. P. Shcherbov, *Tr. Kazo Nauch. Issled. Inst. Miner. Syr'ya* 227 (1962); *Chem. Abstr. 60*, 7450h.

735. A. I. Ivankova, D. N. Perminova, and D. P. Shcherbov, *Prom. Khim. Reakt. Osobo Chist. Veshchestv.* No. 8, 174 (1967); *Chem. Abstr. 69*, 92661s.

736. G. Oshima and K. Nagasawa, *Chem. Pharm. Bull. 18*, 687 (1970).

737. M. Vijayakumar, T. V. Ramakrishna, and G. Aravamudan, *Talanta 27*, 911 (1980).

738. J. Holzhbecher and D. E. Ryan, *Anal. Chim. Acta 64*, 333 (1973).

739. Y. Yamane, M. Miyazaki, T. Kasamatsu, N. Murakami, S. Kito, and H. Komuro, *Jap. Anal. 22*, 192 (1973).

740. Y. Okamoto, S. Inamasu, and T. Kinoshita, *Chem. Pharm. Bull. 28*, 2325 (1980); *Chem. Abstr. 93*, 197116s.

741. A. S. Keston and R. Brandt, *Anal. Biochem. 11*, 1 (1965).

742. M. J. Black and R. Brandt, *Anal. Biochem. 58*, 246 (1974).

743. G. G. Guilbault, P. E. Brignac, Jr., and M. Juneau, *Anal. Chem. 40*, 1256 (1968).

744. K. Zaitsua and Y. Ohkura, *Anal. Biochem. 109*, 109 (1980).

745. W. A. Andreae, *Nature 175*, 859 (1955).

746: H. Perschke and E. Broda, *Nature 190*, 257 (1961).

747. D. A. Bryton and J. C. Guyon, *Microchem. J. 14*, 1 (1969).

748. Z. T. Maksmycheva, V. Ya. Artemova, and Sh. T. Talipov, *Deposited Doc. VINITI*, 105, (1975).

749. M. Ishibashi, T. Shigematsu, and Y. Nishikawa, *J. Chem. Soc. Jap. 77*, 1474 (1956).

750. R. Bock and K. Hochstein, *Z. Anal. Khim. 138*, 337 (1953).

751. L. B. Ginsburg, *Izv. Akad. Nauk. Kaz. SSR, Ser. Khim. 1*, 94 (1957).

752. Y. Nishikawa, K. Hiraka, K. Morishige, K. Takahashi, T. Shigematsu, and T. Nogami, *Jap. Anal. 25*, 459 (1975).

753. Yu. B. Titkov, *Ukr. Khim. Zh. 37*, 502 (1971).

754. A. T. Pilipenko, A. I. Zhebentyaev, and A. I. Volkova, *Ukr. Khim. Zh. 42*, 998 (1976); *Chem. Abstr. 86*, 65000v.

755. V. Patrovsky, *Chem. Listy 47*, 1338 (1953).

756. N. L. Olenovich, L. I. Koval'chuk, and E. P. Lozitskaya, *Zh. Anal. Khim. 29*, 47 (1974).

757. A. Bordea, *Bul. Inst. Politeh. Iasi 13*, 209 (1967); *Chem. Abstr. 69*, 92646r.

758. K. Watanabe, A. Fujiwara, and A. Kawasaki, *Bull. Chem. Soc. Jap. 50*, 1460 (1977).

759. A. T. Pilipenko and A. I. Zhebentyaev, *Ukr. Khim. Zh. 43*, 637 (1977); *Chem. Abstr. 88*, 68554f.

760. Yu. N. Knipovich, V. M. Krasikova, and L. I. Chuenko, *Inf. Sbornik 18*, 11 (1959).

761. A. A. Rozbianskaya, *Tr. Inst. Mineral. Geokhim. Kristallokhim. Redk. Elem. Akad. Nauk. SSR 6*, 138 (1961).

762. A. K. Babko and Z. I. Chalaya, *Zh. Anal. Khim. 18*, 570 (1963).

763. E. P. Mulikovskaya, *Novye Metod. Anal. Khim. Sostava Podzenn. Vod. 78* (1967); *Chem. Abstr. 69*, 5107c.

764. I. A. Blyum and T. K. Dushina, *Bull. All-Union Sci. Res. Inst. Econ. Miner.* No. 3, 1 (1958).

765. Ya. Glovadskii, A. P. Golovina, L. V. Levshin, and Yu. A. Mittsel, *Zh. Anal. Khim. 19*, 693 (1964).

766. I. A. Blyum and T. K. Dushina, *Zavod. Lab. 25*, 137 (1959).

767. K. K. Kabdulkarimova, L. A. Kikharenko, and V. P. Gladyshev, *Reakts. Zhidk. Faze. 52* (1979); *Anal. Abstr. 39*, 6B56.

768. T. B. Vesene, *Zavod. Lab. 35*, 32 (1969).

769. I. M. Ivanova and N. B. Zorov, *Vest. Mosk. Univ. Ser. Khim. 15*, 475 (1974).

770. I. A. Blyum, S. T. Solv'yan, and G. N. Shebalkova, *Zavod. Lab. 27*, 950 (1961).

771. A. I. Pilipenko and A. I. Zhebentyaev, *Ukr. Khim. Zh. 43*, 1314 (1977); *Chem. Abstr. 89*, 16115q.

772. B. S. Garg and R. P. Singh, *Talanta 18*, 761 (1971).

773. M. Deguchi, A. Odachi, K. Morishige, I. Okumura, and K. Yamaguchi, *Jap. Anal. 31*, 212 (1982).

774. A. Fernandez-Gutierrez, A. Muñoz de la Peña, and J. A. Murillo, *Anal. Lett. 16*, 758 (1983).

775. D. Fink and W. Ohnesorge, *Anal. Chem. 41*, 39 (1969).

776. K. Kenyu, K. Shiraishi, and N. Ishibashi, *Jap. Anal. 27*, 291 (1978); *Chem. Abstr. 89*, 99136w.

777. A. Sanz Medel, D. Blanco Gomis, and J. R. Garcia Alvarez, *Talanta 28*, 425 (1981).

777a. S. B. Meshkova, N. V. Rusakova, and N. S. Poluektov, *Zh. Anal. Khim. 37*, 1988 (1982); *Anal. Abstr. 45*, 2B65.

778. A. E. Pitts and D. E. Ryan, *Anal. Chim. Acta 37*, 460 (1967).

779. D. E. Ryan and B. K. Afgham, *Anal. Chim. Acta 44*, 115 (1969).

780. N. S. Poluektov, S. B. Meshkova, and E. V. Melent'eva, *Zh. Anal. Khim. 25*, 1314 (1970); *Chem. Abstr. 73*, 137061x.

781. M. Roman, A. Fernandez-Gutierrez, and C. Martin, *Microchem. J.* (in press).

782. M. Roman, A. Fernandez-Gutierrez, and C. Marin, *Quim. Anal.* (in press).

783. M. Roman, A. Fernandez-Gutierrez, and M. C. Mahedero, *Anal. Lett. 14*, 1579 (1981).

784. C. W. White, M. H. FLetcher, and J. Parks, *Anal. Chem. 23*, 478 (1951).

785. A. I. Markman and S. A. Strel'tsova, *Tr. Tashk. Palitekh. Inst. 42*, 50 (1968); *Anal. Abstr. 17*, 577 (1969).

785a. N. B. Hansen, *Mikrochim. Acta II*, 277 (1983).

786. N. S. Poluektov, R. S. Lauer, and O. F. Gaidarzhi, *Zh. Anal. Khim. 26*, 898 (1971).

787. A. Murata, T. Ito, T. Sakamoto, and H. Kitamura, *Asahi Garasu Kogyo Gijutsu Shoreikai Kenkyu Hokuku 26*, 227 (1975); *Chem. Abstr. 85*, 136736g.

788. E. A. Bozhevol'nov, A. M. Lukin, L. F. Fedorova, G. S. Petrova, and G. V. Serebryakova, *Zh. Anal. Khim. 36*, 1734 (1981).

789. R. M. Dagnall, R. Smith, and T. S. West, *Analyst 92*, 20 (1967).

790. M. G. Brunette and M. E. Crochet, *Anal. Biochem. 65*, 79 (1975).

791. C. E. White and F. Cuttitta, *Anal. Chem. 31*, 2083 (1959).

792. E. A. Bozhevol'nov and G. V. Serebryakova, *Metod. Anal. Khim. Reakt. Preparat.* No. 11, 52 (1965).

793. E. A. Bozhevol'nov and G. V. Serebryakova, *Metod. Anal. Khim. Reakt. Preparat.* No. 11, 55 (1965).

794. E. A. Bozhevol'nov and G. V. Serebryakova, *Metod. Anal. Khim. Reakt. Preparat.* No. 11, 58 (1965).

795. D. P. Shcherbov, R. N. Plotnikova, and T. N. Skvortsova, *Prom. Khim. Reakt. Osobo Chist. Veshchestv.* 8, 166 (1967).

796. E. A. Bozhevol'nov and G. V. Serebryakova, *Metod. Anal. Galogenidov. Shchelochn. Shchelochn. Metal. Vys. Chist.* 2, 20 (1971); *Chem. Abstr.* 79, 87133y.

797. E. Wieteska, *Biul. Wojsk. Akad. Tech.* 29, 91 (1980); *Anal. Abstr.* 42, 1B28.

798. D. Wallach and T. Steck, *Anal. Chem.* 35, 1035 (1963).

799. J. B. Hill, *Clin. Chem.* 11, 122 (1965).

800. S. Zepf, *Clin. Chim. Acta 20*, 473 (1968).

801. M. Roman, A. Fernandez-Gutierrez, and C. Marin, *Anal. Lett.* 15, 1621 (1982).

802. M. Roman, A. Fernandez-Gutierrez, and M. C. Mahedero, *Mikrochim. Acta II*, 85 (1983).

803. H. Diehl, R. Olsen, G. I. Spielhltz, and R. Jensen, *Anal. Chem.* 35, 1144 (1963).

804. R. A. Swanson, D. Hovland, and L. O. Fine, *Soil. Sci.* 102, 244 (1966); *Chem. Abstr.* 65, 20784a.

805. D. G. Oreopoulos, M. Soyannwo, and M. G. McGeown, *Clin. Chim. Acta* 20, 349 (1968).

806. J. J. Laserna, A. Navas, and F. Garcia Sanchez, *Microchem. J.* 27, 312 (1982).

807. F. Capitan, F. Salinas, and L. M. Franquelo, *Quim. Anal.* 31, 275 (1977).

808. T. Hayashi, Sh. Kawai, and T. Ohno, *Chem. Pharm. Bull.* 21, 1147 (1973); *Chem. Abstr.* 79, 102262.

809. J. G. Batsakis, F. Madera-Orsini, D. Stiles, and R. O. Briere, *Tech. Bull. Reg. Med. Tech.* 34, 159 (1964).

810. V. Patrovsky, *Z. Anal. Chem.* 230, 355 (1967); *Chem. Abstr.* 67, 104865k.

811. A. Badrinas, *Talanta 10*, 704 (1963).

812. D. Schachter, *J. Lab. Clin. Med.* 54, 763 (1959); 58, 495 (1961).

813. J. B. Hill, *Ann. N.Y. Acad. Sci.* 102, 1 (1962).

814. M. Oklander and B. Klein, *Clim. Chem.* 12, 243 (1966); 13, 243 (1967).

815. I. Clark and G. How, *Anal. Biochem.* 19, 14 (1967).

816. D. E. Frized, A. G. Malleson, and V. Marks, *Clin. Chim. Acta 16*, 45 (1967).

817. V. Patrovsky, *Coll. Czech. Chem. Comm.* 32, 2656 (1967).

818. D. P. Shcherbov, R. N. Plotnikova, G. P. Gladysheva, and A. I. Ivankova, *Sovrem. Metod. Anal. Miner. Syr'ya* 10 (1979); *Chem. Abstr.* 93, 160607b.

819. E. A. Bozhevol'nov, *Oestrr. Chem. Ztg.* 66, 74 (1965); *Chem. Abstr.* 65, 7889 (1966).

820. P. Baum and R. Czok, *Biochem. Z. 332*, 121 (1959).

821. W. J. Blaedel and G. P. Hicks, *Advances in Analytical Chemistry and Instrumentation*, Vol. 3, Interscience, New York, 1964, p. 105.

822. G. Gusev, *Lab. Delv.* No. 3, 157 (1968); *Chem. Abstr. 69*, 647r.

823. E. A. Bozhevol'nov and G. V. Serebryakova, *Metod. Lyuminests. Anal. Mater. VIII Soveshch. Lyuminests.*, 55 (1959); *Chem. Abstr. 56*, 14918i.

824. G. V. Serebryakova, A. M. Lukin, and E. A. Bozhevol'nov, *Zh. Anal. Khim. 18*, 706 (1963); *Chem. Abstr. 59*, 8099h.

825. A. Korkuc and K. Lesz, *Chem. Anal. 17*, 855 (1972).

826. A. T. Pilipenko, *Ukr. Khim. Zh. 43*, 752 (1977).

827. M. A. Cejas, A. Gomez-Hens, and M. Valcarcel, *Anal. Chim. Acta 130*, 73 (1981).

828. S. Igarashi, T. Yotsuyanagi, and K. Aomura, *Nippon Kagaku Kaishi*, No. 1, 60 (1981).

829. T. Hayashi, S. Kawai, and T. Ohno, *Chem. Pharm. Bull. 18*, 2407 (1970); *Chem. Abstr. 74*, 60585.

830. V. L. Biddle and E. L. Wehry, *Anal. Chem. 50*, 867 (1978).

831. A. Moreno, M. Silva, D. Perez-Bendito and M. Valcarcel, *Talanta 30*, 107 (1983).

832. B. K. Pal and D. Ryan, *Anal. Chim. Acta 47*, 35 (1969).

833. S. U. Kreingol'd, E. A. Bozhevol'nov, and N. S. Petrovskaya, *Metod. Anal. Khim. Reakt. Preparat.* No. 13, 93 (1966); *Chem. Abstr. 67*, 104867n.

834. G. F. Kirkbright, T. S. West, and C. Woodward, *Talanta 13*, 1637 (1966).

835. G. F. Kirkbright, T. S. West, and C. Woodward, *Talanta 13*, 1645 (1966).

836. A. T. Pilipenko, A. I. Zhebentyaev, and A. I. Volkova, *Ukr. Khim. Zh. 41*, 260 (1975); *Chem. Abstr. 83*, 52827h.

837. Y. Titkov, *Ukr. Khim. Zh. 36*, 613 (1970); *Chem. Abstr. 73*, 116040z.

838. P. R. Haddad, P. W. Alexander, and L. E. Smythe, *Talanta 22*, 61 (1975).

839. M. Ikeda and M. Watanabe, *Jap. Anal. 22*, 218 (1973); *Chem. Abstr. 79*, 132119.

840. A. T. Pilipenko, A. I. Volkova, and A. I. Zhebentyaev, *Zh. Anal. Khim. 26*, 2048 (1971).

841. V. V. Khimov and O. S. Didkovskaya, *Zavod. Lab. 29*, 147 (1963).

842. A. I. Volkova and T. E. Get'man, *Ukr. Khim. Zh. 73*, 53 (1971); *Chem. Abstr. 74*, 150774.

843. A. T. Pilipenko, A. I. Zhebentyaev, and A. I. Volkova, *Ukr. Khim. Zh. 38*, 363 (1972); *Chem. Abstr. 77*, 28509.

844. A. T. Pilipenko, A. I. Zhebentyaev, and A. I. Volkova, *Zh. Anal. Khim. 26*, 117 (1971).

845. A. T. Pilipenko, A. I. Zhebentyaev, and A. I. Volkova, *Zh. Anal. Khim. 27*, 84 (1972).

846. S. Belman, *Anal. Chim. Acta 29*, 120 (1965).

847. V. Sardesai and H. Provido, *Mikrochem. J. 14*, 550 (1969).

848. T. Sakano and Y. Amano, *Jap. Anal. 30*, 136 (1981); *Anal. Abstr. 41*, 3B148.

849. M. Rubin and L. Knott, *Clin. Chim. Acta 18*, 409 (1967).

850. B. L. Nazar and A. C. Schoolwerth, *Anal. Biochem. 95*, 507 (1979).

850a. T. Aoki, S. Uemura and M. Munemori, *Anal. Chem. 55*, 1620 (1983).

851. W. S. Gardner, *Limnol. Oceanogr. 23*, 1069 (1978); *Anal. Abstr. 38*, 1H29.

852. N. D. Danielson and C. M. Conroy, *Talanta 29*, 401 (1982).

853. H. D. Axelrod and N. A. Engel, *Anal. Chem. 47*, 922 (1975).

854. G. Oshima and K. Nagasawa, *Chem. Pharm. Bull. 20*, 1492 (1972).

855. L. J. Dombrowski and E. J. Pratt, *Anal. Chem. 44*, 2268 (1972).

856. J. H. Wiersma, *Anal. Lett. 3*, 123 (1970).

857. M. Nakamura, *Anal. Lett. 13*, 771 (1980).

858. N. K. Podberezskaya, E. A. Silenko, and V. A. Sushkova, *Issled. Obl. Khim. Fiz. Metod. Anal. Miner Syr'ya* 90 (1971); *Chem. Abstr. 78*, 79274.

859. C. Sawicki, *Anal. Lett. 4*, 761 (1971).

860. B. K. Afghan and J. F. Ryan, *Anal. Chem. 47*, 2347 (1975).

861. H. D. Axelrod, J. E. Bonelli, and J. P. Lodge, *Anal. Chim. Acta 51*, 21 (1970).

862. M. A. Konstantinova-Shlesinger, *Tr. Fig. Inst. Akad. Nauk. SSSR Fiz. 2*, 7 (1942).

863. M. A. Konstantinova-Shlesinger and V. Krasnova, *Zavod. Lab. 11*, 567 (1945); *Chem. Abstr. 40*, 2414.

864. M. A. Konstantinova-Shlesinger, *Zh. Fiz. Khim. 9*, 6 (1938).

865. H. Watanabe and T. Nakadoi, *J. Air Pollut. Control Assoc. 16*, 614 (1966).

866. D. Amos, *Anal. Chem. 42*, 842 (1970).

867. S. Burchett and C. E. Meloan, *Anal. Lett. 4*, 471 (1971).

868. P. A. Shiryaev and V. I. Rigin, *Zavod. Lab. 41*, 917 (1975).

869. D. P. Shcherbov, A. I. Ivankova, D. N. Lisitsyna, and I. D. Vvdenskaya, *Zh. Anal. Khim. 32*, 1932 (1977); *Chem. Abstr. 88*, 145570n.

870. D. P. Shcherbov, A. I. Ivankova, D. N. Lisitsyna, and I. D. Vvdenskaya, *Org. Reagenty Anal. Khim., Tezisy Dokl. Vses. Konf. 4th, 2*, 129 (1976); *Chem. Abstr. 88*, 15447s.

871. D. B. Land and S. Edmonds, *Mikrochim. Acta* 1013 (1966).

872. J. C. Guyon and W. D. Shults, *J. Am. Water Works Assoc. 61*, 403 (1969).

873. T. Hayashi, S. Ohgaki, C. Yagi, S. Kawai, and T. Ohno, *Chem. Pharm. Bull. 21*, 2141 (1973).

874. G. F. Kirkbright, R. Narayanaswamy, and T. S. West, *Analyst 97*, 174 (1972).

875. G. F. Kirkbright, R. Narayanaswamy, and T. S. West, *Anal. Chem. 43*, 1434 (1971).

876. H. Wachsmuth, *J. Pharm. Belg. 5*, 300 (1950).

877. J. Holzbecher and D. E. Ryan, *Anal. Chim. Acta 64*, 147 (1973).

878. M. G. Brunette, N. Vigneault, and G. Danan, *Anal. Biochem. 86*, 229 (1978).

879. D. B. Schultz, J. V. Passoneau, and O. H. Lowry, *Anal. Biochem. 19*, 300 (1967).

880. F. Grases, J. M. Estela, F. Garcia-Sanchez, and M. Valcarcel, *Anal. Lett. 13*, 181 (1980).

881. L. A. Grigoryan, A. G. Gaibakyan, and V. M. Tarayan, *Dokl. Akad. Nauk. Arm. SSR 54*, 229 (1972); *Chem. Abstr. 78*, 92160.

882. L. A. Grigoryan, A. G. Gaibakyan, and V. M. Tarayan, *Zavod. Lab. 40*, 136 (1974).

883. V. M. Tarayan and A. G. Gaibakyan, *Arm. Khim. Zh. 26*, 812 (1973).

884. A. T. Pilipenko, A. I. Volkova, and T. L. Shvchenko, *Urk. Khim. Zh. 41*, 1190 (1975).

885. I. A. Blyum and N. A. Brushtein, *Zavod. Lab. 36*, 1032 (1970).

886. L. A. Grigoryan, L. G. Mushegyan, and V. M. Tarayan, *Zavod. Lab. 42*, 1038 (1976); *Chem. Abstr. 85*, 201710k.

887. D. P. Shcherbov and A. I. Ivankova, *Zavod. Lab. 29*, 787 (1963).

888. L. A. Grigoryan, S. P. Lebedeva, and V. M. Tarayan, *Arm. Khim. Zh. 28*, 540 (1975); *Chem. Abstr. 84*, 11928u.

889. A. T. Pilipenko, A. I. Volkova, and T. L. Shevchenko, *Zh. Anal. Khim. 28*, 1524 (1973); *Chem. Abstr. 76*, 152570e.

890. A. T. Pilipenko, A. I. Volkova, and T. L. Shevchenko, *Ukr. Khim. Zh. 43*, 653 (1977); *Chem. Abstr. 87*, 210562g.

891. H. Veening and W. Brandt, *Anal. Chem. 32*, 1426 (1960).

892. D. P. Shcherbov, G. P. Gladysheva, and A. I. Ivankova, *Issled. Tsvet. Fluorestsent. Reakts. Opred. Blagorod. Met.* 74 (1969); *Chem. Abstr., 73*, 83538n.

893. D. P. Shcherbov, G. P. Gladysheva, and A. I. Ivankova, *Zavod. Lab. 37*, 1300 (1971); *Chem. Abstr. 76*, 67724b.

894. A. Gruenert and G. Toelg, *Talanta 18*, 881 (1971).

895. A. Gruenert, K. Ballschmiter, and G. Toelg, *Talanta, 15*, 451 (1968).

896. H. D. Axelrod, J. H. Cury, J. E. Bonelly, and J. P. Lodge, *Anal. Chem. 41*, 1856 (1969).

897. B. A. Hardwick, D. K. B. Thistlethwayte, and R. J. Fowler, *Atmos. Environ. 4*, 379 (1970).

898. B. A. Hardwick, D. K. B. Thistlethwayte, and R. J. Fowler, *Atmos. Environ. 5*, 281 (1971).

899. B. A. Hardwick, D. K. B. Thistlethwayte and R. J. Fowler, *Atmos. Environ. 5,* 282 (1971).

900. F. P. Scaringelli, *Atmos. Environ. 5,* 282 (1971).

901. P. K. Zutshi, *Atmos. Environ. 5,* 281 (1971).

902. L. S. Bark and A. Rixon, *Analyst 95,* 786 (1970).

903. V. S. Yanysheva, *Zavod. Lab. 30,* 23 (1964).

904. A. K. Babko, L. V. Markova, and T. S. Tsybina, *Zavod. Lab. 30,* 648 (1964).

905. I. G. Shafran, I. F. Vzorova, M. I. Dorosinskaya, L. K. Fidlon, and V. A. Yur'eva, *Metod. Anal. Khim. Reakt. Preparat.* No. 11, 81 (1965).

906. Ya. V. Samoilov, *Byul. Izobr. Tovarnykh Znakov 24,* 51 (1964).

907. M. Wronski, *Chem. Anal.* (Warsaw) *16,* 439 (1971); *Chem. Abstr. 75,* 5825lx.

908. E. A. Bozhevol'nov, S. U. Kreingol'd, and L. I. Sosenkova, *Metod. Anal. Khim. Reakt. Preparatov,* No. 9, 25 (1971); *Chem. Abstr. 77,* 121767d.

909. P. M. Zaitsev, I. K. Krotova, and L. I. Ustyukova, *Tr. Nauch. Issled. Inst. Udobr. Insektofungits 226,* 226 (1975); *Chem. Abstr. 87,* 15441g.

910. L. S. Tsebrii, V. Z. Belyantseva, and E. I. Vail, *Zavod. Lab. 45,* 499 (1979); *Anal. Abstr. 38,* 1C20.

911. I. Mori, Y. Fujita, M. Goto, S. Furuya, and T. Enoki, *Jap. Anal. 29,* 145 (1980).

912. J. Holzbecher and D. E. Ryan, *Anal. Chim. Acta 68,* 454 (1974).

913. H. D. Axelrod, J. E. Bonelli, and J. P. Lodge, *Anal. Chem. 42,* 512 (1970).

914. T. Nasu, T. Kitagawa, and T. Mori, *Jap. Anal. 19,* 673 (1970).

915. M. H. Fletcher and R. G. Milkey, *Anal. Chem. 28,* 1402 (1956).

916. M. H. Fletcher and R. G. Milkey, *J. Am. Chem. Soc. 79,* 5425 (1957).

917. J. C. Guyon and E. J. Lorah, *Anal. Chem. 38,* 155 (1966).

918. N. A. Vlasov, E. A. Morgen and V. A. Tyutin, *Izv. Nauch. Issled. Inst. Nefte, Uglekhim. Sin. Irkutsk, Univ. 11,* 136 (1969); *Chem. Abstr. 78,* 7657p.

919. N. A. Vlasov, E. A. Morgen and V. A. Tyutin, *Gidrokhim. Mater. 50,* 92 (1969).

920. T. Nasu, *Hokkaido Kyoiku Daigaku Kiyo, Dai-2-Bu, A, 23,* 35 (1972); *Chem. Abstr. 78,* 143429t.

921. D. P. Shcherbov and A. I. Ivankova, *Tezisy Dokladov. Sovesh. Lyuminest. 22,* 38 (1975); *Chem. Abstr. 86,* 100347n.

922. V. A. Nazarenko and M. B. Shustova, *Zavod. Lab. 24,* 1344 (1958); *Chem. Abstr. 54,* 13985b.

923. V. S. Yanysheva and Z. A. Sazonova, *Metod. Anal. Khim. Reakt. Preparatov.* No. 4, 133 (1962); *Chem. Abstr. 61,* 3695a.

924. L. H. Tan and T. S. West, *Analyst 96,* 281 (1971).

925. M.E. Auerbach, H. W. Eckert, and E. Angell, *Ceral. Chem. 26*, 490 (1949).

926. L. A. Grigoryan, Zh. M. Arstamyan, and V. M. Tarayana, *Zh. Anal. Khim. 37*, 629 (1982); *Anal. Abstr. 44*, 2B106.

927. E. A. Bozhevol'nov, *Khim. Reakt. 22*, 60 (1958).

928. A. Murata, T. Omae, and T. Suzuki, *Jap. Anal. 29*, 780 (1980); *Anal. Abstr. 40*, 6B154.

929. D. P. Shcherbov, I. N. Astaf'eva, and R. N. Plotnikova, *Zavod. Lab. 39*, 546 (1973); *Chem. Abstr. 79*, 73291m.

930. I. A. Blyum, F. P. Kalupina, and T. I. Tsenskaya, *Zh. Anal. Khim. 29*, 1572 (1974).

931. I. A. Bochkareva and I. A. Blyum, *Zh. Anal. Khim. 30*, 874 (1975).

932. A. I. Ivankova and D. P. Shcherbov, *Issled. Razrab. Fotometrich. Metod. Opred. Mikrokolichestv. Elem. Miner.* 138 (1967); *Chem. Abstr. 71*, 18492n.

933. M. A. Matveets, D. P. Shcherbov, and S. D. Akhmetova, *Zh. Anal. Khim. 29*, 740 (1974).

934. T. D. Filer, *Anal. Chem. 43*, 725 (1971).

935. Y. Nishikawa, K. Hiraki, and T. Shigematsu, *J. Chem. Soc. Jap. 90*, 483 (1969).

936. Z. Urner, *Coll. Czech. Chem. Comm. 33*, 1078 (1968).

937. K. Morisige, S. Sasaki, K. Hiraki, and Y. Nishikawa, *Jap. Anal. 24*, 321 (1975).

938. M. Nakamura and A. Murata, *Jap. Anal. 22*, 1474 (1973).

939. A. Murata, M. Tominaga, T. Suzuki, and M. Nakamura, *Jap. Anal. 27*, 788 (1978).

940. I. M. Korenman and V. S. Efinychev, *Zh. Anal. Khim. 17*, 425 (1962); *Chem. Abstr., 57*, 15779a.

941. G. F. Kirbright, T. S. West, and C. Woodward, *Analyst 91*, 23 (1966).

942. D. P. Shcherbob and V. P. Nikolaeva, *Prom. Khim. Reakt. Osobo Chist. Veshchestv.* 186 (1967); *Chem. Abstr. 69*, 64380f.

943. D. P. Gladilovich, N. N. Grigor'ev, and K. P. Stolyarov, *Zh. Anal. Khim. 35*, 1283 (1980); *Chem. Abstr. 93*, 197093g.

944. V. Ya Temkina, E. A. Bozhevol'nov, S. U. Kreingol'd, G. F. Yaroshenko, V. N. Antonov, and R. P. Lastovskii, *Zh. Anal. Khim. 25*, 894 (1970); *Chem. Abstr., 73*, 105107a.

945. A. V. Dolgorev, N. N. Pavlova, and V. A. Ershova, *Zavod. Lab. 39*, 658 (1973); *Chem. Abstr. 79*, 142610d.

946. A. V. Dolgorev, N. S. Sivak, T. I. Pla'nikova, and L. M. Gurevich, *Zh. Anal. Khim. 34*, 106 (1979); *Chem. Abstr. 90*, 179516f.

947. V. Nazarenko and V. Antonovich, *Zh. Anal. Khim. 24*, 358 (1969).

948. L. E. Ze''tser, L. A. Morozova, Sh. T. Talipov, and A. T. Tashkhozhaev, *Zh. Anal. Khim. 34*, 896 (1979).

948a. A. P. Golovina and N. Yu. Kuzyakova, *Otkrytiya, Izobret. Prom. Obraztsy, Tovarnye Znaki* No. 18, 90 (1983); *Chem. Abstr. 99*, 63547y.

948b. N. Yu. Kuzyakova and A. P. Golovina, *Zh. Anal. Khim. 38*, 1023 (1983); *Chem. Abstr. 99*, 115132f.

949. M. A. Matveets, S. D. Akhmetova, and D. P. Shcherbov, *Org. Reagenty Anal. Khim. Tez. Dokl. Konf. 4th, 2*, 135 (1976), Chem. Abstr. *88*, 15448t.

950. M. A. Matveets, S. D. Akhmetova, and D. P. Shcherbov, *Zh. Anal. Khim. 32*, 2143 (1977); *Chem. Abstr. 88*, 202473j.

951. S. V. Bel'tyukova and N. S. Poluektov, *Ukr. Khim. Zh. 37*, 1277 (1971); *Chem. Abstr. 76*, 107588.

952. L. I. Kononenko, S. V. Bel'tyukova, V. N. Drobyazko, and N. S. Poluektov, *Zh. Anal. Khim. 30*, 1716 (1975).

953. S. V. Bel'tyukova, V. N. Drobyazko, L. I. Kononenko, and N. S. Poluektov, *Zh. Anal. Khim. 30*, 1321 (1975); Chem. Abstr., *84*, 11846r.

954. E. B. Cousins, *Austral. J. Exp. Biol. Med. Sci. 38*, 11 (1960).

955. J. H. Watkinson, *Anal. Chem. 32*, 981 (1960).

956. C. A. Parker and L. G. Harvey, *Analyst 86*, 54 (1961).

957. M. M. Cost, *Revta Port. Quim. 8*, 136 (1966).

958. V. V. Lushnikov and E. Kondrat'eva, *Novye Metod. Anal. Khim. Sostava Podzemn. Vod.* 84 (1967).

959. F. M. Mamedova, K. I. Nikolaeva, and E. A. Bozhevol'nov, *Stomatologiya 47*, 81 (1968).

960. Y. Hashimoto and Y. Saburo, *Jap. Anal. 17*, 785 (1968); *Chem. Abstr. 69*, 654386n.

961. D. I. Ryabchikov, I. I. Nazarenko, and L. I. Anikina, *Zh. Anal. Khim. 23*, 1242 (1968).

962. H. J. Peters and H. Koehler, *Zentralbl. Pharm. Pharmakoter Laborat., 121*, 127 (1982).

963. C. A. Parker and L. G. Harvey, *Analyst 87*, 558 (1962).

964. P. F. Lott, P. Cukor, G. Moriber, and J. Solga, *Anal. Chem. 35*, 1159 (1963).

965. W. E. Clarke, *Analyst 95*, 65 (1970).

966. J. B. Wilkie and M. Young, *J. Agr. Food Chem. 18*, 946 (1970).

967. W. H. Allaway and E. E. Cary, *Anal. Chem. 36*, 1359 (1964).

968. O. Olson, *J. Assoc. Off. Anal. Chem. 52*, 627 (1969).

969. I. Hoffman, R. J. Westerby, and M. Hidiroglou, *J. Assoc, Off. Anal. Chem. 51*, 1039 (1968).

970. R. J. Hall and P. I. Gupta, *Analyst 94*, 292 (1969).

971. R. C. Ewan, C. A. Baumann, and A. L. Pope, *J. Agr. Food Chem. 16*, 212 (1968).

971a. F. Kirchnawy, G. Kainz, and G. Sontag, *Ernaehrung* (Vienna) *6*, 267 (1982).

972. M. Lamand and C. Astier, *Ann. Fals. Expert. Chim. 62,* 4 (1969).

973. F. W. Kiermeier, *Z. Lebensmit. Forsch. 136,* 158 (1968).

974. W. E. Clarke, *Analyst 95,* 65 (1970).

975. I. I. Nazarenko, A. M. Kislov, and I. V. Koslova, *Zh. Anal. Khim. 25,* 1135 (1970); *Chem. Abstr. 73,* 105194b.

976. I. I. Nazarenko and I. V. Koslova, *Zavod. Lab. 37,* 414 (1971); *Chem. Abstr. 75,* 29658e.

977. G. Patrias and O. E. Olson, *Feedstuffs 41,* 32 (1969).

978. B. D. Luft, I. I. Nazarenko, L. B. Khusid, and V. V. Shemet, *Zh. Anal. Khim. 28,* 536 (1973); *Chem. Abstr. 79,* 13212b.

979. B. D. Luft, I. I. Nazarenko, L. B. Khusid, and V. V. Shemet, *Zavod. Lab. 37,* 1047 (1971); *Chem. Abstr. 76,* 30347e.

979a. D. C. Reamer and C. Veillon, *Anal. Chem. 55,* 1605 (1983).

980. I. I. Nazarenko, A. M. Kislov, and I. V. Kislova, *Metod. Anal. Khim. Reakt. Preparat. 32* (1971); *Chem. Abstr. 77,* 121736t.

981. J. H. Wiersma and G. F. Lee, *Environ. Sci. Technol. 5,* 1203 (1971).

982. B. Plumas and C. Sautier, *Ann. Fals. Expert. Chim. 65,* 322 (1972).

983. P. R. Haddad and L. E. Smythe, *Talanta 21,* 859, (1974).

984. J. M. Rankin, *Environ. Sci. Technol. 7,* 823 (1973).

985. M. W. Brown and J. H. Watkinson, *Anal. Chim. Acta 89,* 29 (1977).

986. C. C. Y. Chan, *Anal. Chim. Acta 82,* 213 (1976).

987. S. Michael and C. L. White, *Anal. Chem. 48,* 1484 (1976).

988. R. Stabel-Taucher, *Finn. Chem. Lett.* 57 (1977).

989. I. I. Nazarenko and I. V. Koslova, *Zh. Anal. Khim. 33,* 1857 (1978); *Anal. Abstr. 36,* 5H42.

990. K. Hiraki, O. Yoshii, H. Hirayama, Y. Nishikawa, and T. Shigematsu, *Jap. Anal. 22,* 712 (1973); *Chem. Abstr. 80,* 63710d.

991. Y. Tamari, K. Hiraki, and Y. Nishikawa, *Jap. Anal. 28,* 164 (1979); *Anal. Abstr. 37,* 6H33.

992. Analytical Methods Committee, *Analyst 104,* 778 (1979).

993. L. Lalonde, Y. Jean, K. D. Roberts, A. Chapdelaine, and G. Bleau, *Clin. Chem. 28,* 172 (1982).

994. T. Moreno-Dominguez, C. Garcia-Moreno, and A. Marine-Font, *Analyst 108,* 505 (1983).

995. W. A. Maher, *Anal. Lett. 16,* 491 (1983).

996. D. P. Shcherbov, A. I. Ivankova, and G. P. Gladysheva, *Issled. Razrab. Fotometrich. Metod. Opred. Mikrokolichestv. Elem. Miner.* 10 (1967); *Chem. Abstr. 71,* 18569t.

997. K. Kasiura, *Chem. Anal. (Warsaw) 14,* 1325 (1969).

998. G. Elliot and J. A. Radley, *Anal. Chem. 33,* 1623 (1961).

999. D. E. Williams and J. C. Guyon, *Mikrochem. Acta II,* 194 (1972).

1000. A. Murata, Y. Sugiyama and T. Suzuki, *Jap. Anal. 21*, 204 (1981).

1001. C. F. Coyle and C. E. White, *Anal. Chem. 29*, 1486 (1957).

1002. M. Nakamura and A. Murata, *Mikrochim. Acta I*, 301 (1980).

1003. I. Mori, Y. Fujita and T. Enoki, *Jap. Anal. 25*, 388 (1976); *Chem. Abstr. 86*, 25496e.

1004. L. B. Ginzburg and E. P. Shkrobot, *Zavod. Lab. 23*, 527 (1957).

1005. B. K. Pal and D. Ryan, *Anal. Chim. Acta 48*, 227 (1969).

1006. A. T. Pilipenko, S. L. Lisichenok, and A. I. Volkova, *Ukr. Khim. Zh. 42*, 976 (1976); *Chem. Abstr. 86*, 64999r.

1007. J. R. A. Anderson and S. L. Lowy, *Anal. Chim. Acta 15*, 246 (1956).

1008. Y. Nishikawa, K. Hiraki, T. Naganuma, and T. Shigematsu, *Jap. Anal. 21*, 390 (1972); *Chem. Abstr. 77*, 109049.

1009. Y. Nishikawa, K. Hiraki, T. Naganuma, and S. Niina, *Jap. Anal. 19*, 1224 (1970); *Chem. Abstr. 74*, 27782.

1010. T. I. Shumova, *Metod. Khim. Anal. Miner. Syr'ya* No. 14, 51 (1975).

1011. T. D. Filer, *Anal. Chem. 43*, 1753 (1971).

1012. I. A. Blyum, T. G. Pronkina, and T. I. Shumova, *Zh. Anal. Khim. 25*, 511 (1970).

1013. A. T. Pilipenko, A. I. Volkova, and A. I. Zhebentyaev, *Ukr. Khim. Zh. 37*, 578 (1971); *Chem. Abstr. 75*, 115501.

1014. A. E. Chistyakova, M. A. Desyatkova, and V. S. Shchvartsev, *Nauch. Tr., Nauch-Issled. Proektn. Inst. Redkomet. Promsti.* No. 42, 161 (1972); *Chem. Abstr. 81*, 145362.

1015. I. P. Alimarin, A. P. Golovina, I. M. Bigalo, and Yu. A. Mittsel, *Zh. Anal. Khim. 20*, 339 (1965).

1016. Y. Tse, M. Nig, and F. Yang, *K'o Hsueh T'ung Pao 26*, 383 (1981); *Anal. Abstr. 42*, 5B59.

1017. Y. Tse, M. Ning, and F. Yang, *Fenxi Huaxue 9*, 647 (1981); *Anal. Abstr. 43*, 2B91.

1018. M. A. Tishchenko, L. A. Alakaeva, and N. S. Poluektov, *Ukr. Khim. Zh. 37*, 591 (1971); *Chem. Abstr. 75*, 115502.

1019. M. A. Tishchenko and N. S. Poluektov, *Ukr. Khim. Zh. 32*, 733 (1966); *Chem. Abstr. 65*, 14425e.

1020. M. A. Tishchenko, L. I. Kononenko, and N. S. Poluektov, *Prom. Khim. Reakt. Osobo Chist. Veshchestv.* 231 (1967).

1021. M. A. Tishchenko, G. I. Gerasimenko, and N. S. Poluektov, *Zh. Anal. Khim. 33*, 77 (1978); *Chem. Abstr. 89*, 35954d.

1022. T. D. Barela and A. D. Sherry, *Anal. Biochem. 71*, 351 (1976).

1023. M. A. Tishchenko, I. I. Zheltvai, and N. S. Poluektov, *Zavod. Lab. 39*, 670 (1973); *Chem. Abstr. 79*, 152550.

1024. V. T. Mishchenko, N. N. Aleksandova, and L. A. Ovchav, *Zh. Anal. Khim. 34*, 902 (1979).

1025. N. S. Poluektov, L. A. Alakaeva, and M. A. Tishchenko, *Zavod. Lab.* *37*, 1077 (1971).

1026. N. S. Poleuktov, L. A. Alakaeva, and M. A. Tishchenko, *Zh. Anal. Khim.* *28*, 1621 (1973).

1027. R. M. Dagnall, R. Smith, and T. S. West, *Analyst 92*, 358 (1967).

1027a. J. L. Burguera, M. Burguera, and M. Gallignani, *Acta Cient. Venez. 33*, 99 (1982).

1028. T. Taketatsu and S. Yoshida, *Bull. Chem. Soc. Jap. 45*, 2921 (1972);*Chem. Abstr. 77*, 159795.

1029. M. A. Tishchenko, N. S. Poluektov, G. F. Yaroshenko, R. P. Lastovskii, G. I. Gerasimenko, I. I. Zheltvai, and L. M. Timakova, *Zh. Anal. Khim.* *33*, 2368 (1978); *Chem. Abstr. 90*, 132218y.

1030. N. S. Poluektov, M. A. Tishchenko, and G. I. Gerasimenko, *Zh. Anal. Khim. 30*, 1325 (1975); *Chem. Abstr. 84*, 11847s.

1031. L. I. Kononenko, S. Mishchenko, and N. S. Poluektov, *Zh. Anal. Khim.* *21*, 1392 (1966).

1032. N. S. Poluektov, L. A. Alakaeva, and M. A. Tishchenko, *Ukr. Khim. Zh.* *38*, 175 (1972); *Chem. Abstr. 76*, 148467x.

1033. N. S. Poluektov, L. A. Alakaeva, and M. A. Tishchenko, *Zh. Anal. Khim.* *25*, 2351 (1970).

1034. V. M. Vladimirova, N. K. Davidovich, G. I. Kuchmistaya, and L. S. Razumova, *Zavod. Lab. 29*, 1419 (1963).

1035. D. P. Shcherbov and A. I. Ivankova, *Zavod. Lab. 24*, 1346 (1958).

1036. Z. M. Khvatkova, A. P. Golovina, N. B. Zorov, N. K. Belova, and I. P. Alimarin, *Vestn. Mosk. Univ. Ser. Khim. 13*, 355 (1972); *Chem. Abstr. 77*, 134673.

1037. D. P. Shcherbov, *Fluorometry in the Chemical Analysis of Mineral Raw Materials*, Nedra, Moscow, 1965.

1038. M. Bovay and M. Marcantonatos, *Anal. Chim. Acta 80*, 180 (1975).

1039. A. P. Golovina and V. K. Runov, *Usp. Anal. Khim. 184* (1974); *Chem. Abstr. 82*, 164398.

1040. I. P. Alimarin, A. P. Golovina, and V. K. Runov, *Izv. AN SSSR, Ser. Khim.* No. 6, 1423 (1974).

1041. Yu. V. Granovskii, V. K. Runov, I. S. Tishchenko, and A. P. Golovina, *Zh. Anal. Khim. 29*, 1959 (1974).

1042. R. S. Bottei and A. D'Alessio, *Anal. Chim. Acta. 37*, 405 (1967).

1043. R. Milkey and M. Fletcher, *J. Am. Chem. Soc. 79*, 5425 (1957).

1044. C. W. Sill, *Anal. Chem. 33*, 1684 (1961).

1045. C. W. Sill and C. P. Willis, *Anal. Chem. 34*, 954 (1962).

1046. C. W. Sill and C. P. Willis, *Anal. Chem. 36*, 622 (1964).

1047. A. K. Babko, T. H. Chan, A. I. Volkova, and T. E. Get'man, *Ukr. Khim. Zh. 35*, 642 (1969); *Chem. Abstr. 71*, 66353a.

1048. A. B. Blank, I. I. Mirenskaya, and L. M. Satonovskii, *Zh. Anal. Khim.* *30*, 1116 (1975).

1049. L. A. Morozova, Sh. T. Talipov, L. E. Zel'tser, A. T. Tashkhodzhaev, and Z. P. Pakudina, *Izv. Vyssh. Uchebn. Zaved. Ser. Khim. Khim. Tekhnol. 21*, 672 (1978); *Chem. Abstr. 89*, 122524g.

1050. A. K. Babko, T. H. Chan, A. I. Volkova, and T. E. Get'man, *Ukr. Khim. Zh. 35*, 292 (1969); *Chem. Abstr. 71*, 9404h.

1051. T. D. Filer, *Anal. Chem. 42*, 1265 (1970).

1052. A. T. Tashkhodzhaev, L. E. Zel'tser, T. Sabirova, and L. A. Morozova, *Dokl. Akad. Nauk. Uzb. SSR 4*, 32 (1976).

1053. T. Ito and A. Murata, *Anal. Chim. Acta 113*, 343 (1980).

1054. Yu. Titkov, *Ukr. Khim. Zh. 35*, 887 (1969); *Chem. Abstr. 72*, 18218y.

1055. M. D. Luque de Castro and M. Valcarcel, *Talanta 27*, 645 (1980).

1056. L. A. Grigoryan, F. V. Mirzoyan, and V. M. Tarayan, *Zh. Anal. Khim. 28*, 1962 (1973); *Chem. Abstr. 80*, 78036.

1057. L. A. Grigoryan, V. Zh. Artsruni, and V. M. Tarayan, *Arm. Khim. Zh. 27*, 188 (1974); *Chem. Abstr. 81*, 85527.

1058. V. M. Tarayan, L. A. Grigoryan, F. V. Mirzoyan, and Zh. V. Sarkisyan, *Arm. Khim. Zh. 26*, 996 (1973); *Chem. Abstr. 81*, 85508.

1059. I. A. Blyum and A. I. Chuvileva, *Zh. Anal. Khim. 25*, 18 (1970).

1060. T. Perez-Ruiz, C. Sanchez-Pedreno, M. Hernandez-Cordoba, and C. Martinez-Lozano, *Anal. Chim. Acta 143*, 177 (1982).

1061. F. Feigl, V. Gentil, and D. Goldstein, *Anal. Chim. Acta 9*, 393 (1953).

1062. H. Onishi, *Bull. Chem. Soc. Jap. 30*, 827 (1957).

1063. E. A. Bozhevol'nov and V. M. Yanishevskaya, *Stsintillyatory i Stsintillyats. Materialy. Vses. Nauch. Issled. Inst. Khim. Reakt.* 252 (1957); *Chem. Abstr. 57*, 11855d.

1064. E. A. Bozhevol'nov, *Metod. Anal. Khim. Reakt. Preparat. No. 4*, 113 (1962).

1065. M. M. Schnepfe, *Anal. Chim. Acta 79*, 101 (1975).

1066. V. M. Vladimirova and N. K. Davidovich, *Metod. Anal. Khim. Reakt. Preparat. No. 4*, 116 (1962).

1067. T. Vesiene, *Tr. Akad. Nauk. Lit. SSR, B, 2*, 101 (1971).

1068. L. I. Zemstova, *Methods for the Chemical Analysis of Raw Minerals*, Nedra, Moscow, No. 11, 43 (1968).

1069. N. R. Anderson and D. M. Hercules, *Anal. Chem. 36*, 2138 (1964).

1070. T. Perez-Ruiz, C. Sanchez-Pedreno, M. Hernandez-Cordoba, and C. Martinez-Lozano, *An. Quim. 77B*, 397 (1981).

1071. H. G. Brittain, *Anal. Lett. 10*, 263 (1977).

1072. Z. T. Maksimycheva, Sh. T. Talipov, and L. E. Zel'tser, *Dokl Akad. Nauk. Uzb. SSR 28*, 39 (1971); *Chem. Abstr. 76*, 54112a.

1073. E. Tomic and F. Hecht, *Mikrochim. Acta* 896 (1955).

1074. E. I. Tselik, S. V. Bel'tyukova, and L. I. Kononenko, *Ukr. Khim. Zh.* 46, 1222 (1980); *Anal. Abstr. 40,* 6B98.

1075. H. H. P. Moeken and W. A. H. van Neste, *Anal. Chim. Acta 37,* 480 (1967).

1076. N. R. Andersen and D. M. Hercules, *Anal. Chem. 36,* 2138 (1964).

1077. Y. S. Kim and H. Zeitlin, *Anal. Chem. 43,* 1390 (1971).

1078. G. Leung, Y. S. Kim, and H. Zeitlin, *Anal. Chim. Acta 60,* 229 (1972).

1079. T. S. Dobrolyubskaya, *Zh. Anal. Khim. 26,* 926 (1971); *Chem. Abstr. 75,* 58320.

1080. F. Salinas, F. Garcia-Sanchez, and C. Genester, *Mikrochem. J. 27,* 25 (1982).

1081. F. Salinas, F. Garcia-Sanchez, F. Grases, and C. Genestar, *Anal. Lett.* 13, 473 (1980).

1082. K. Hiraki, N. Shimizu, Y. Nishikawa, and T. Shigematsu, *Jap. Anal. 30,* 780 (1981).

1083. K. J. Koh and D. E. Ryan, *Anal. Chim. Acta 57,* 295 (1971).

1084. F. Garcia-Sanchez, A. Navas, M. Santiago, and F. Grases, *Talanta 28,* 835 (1981).

1085. A. Navas, M. Santiago, F. Grases, J. J. Laserna, and F. Garcia-Sanchez, *Talanta 29,* 615 (1982).

1086. M. Roman, F. Garcia-Sanchez, and A. Gomez-Hens, *Quim. Anal. 28,* 191 (1974).

1087. A. T. Pilipenko, A. P. Kostyshina, and N. M. Nazarchuk, *Ukr. Khim. Zh. 42,* 633 (1976); *Chem. Abstr. 85,* 136776v.

1088. A. T. Pilipenko, A. I. Volkova, and T. L. Shevchenko, *Otk. Izobret. Prom. Obr. Tovarnye Znaki 55,* 128 (1978); *Chem. Abstr. 90,* 80388d.

1089. T. L. Shevchenko, A. T. Pilipenko, and A. I. Volkova, *Ukr. Khim. Zh. 45,* 456 (1979); *Anal. Abstr. 38,* 2B138.

1090. J. Bognar and O. Jellinek, *Mikrochim. Acta* 1013 (1968).

1091. F. Grases, C. Genestar, and F. Salinas, *Anal. Chim. Acta 147,* 401 (1983).

1092. Kuo-Chen Ch'en and Chu-tzu Chen, *Hsia Men Ta Hsueh Pao, She Hui K'o Hsuch 1,* 121 (1957).

1093. R. S. Bottei and B. A. Trusk, *Anal. Chem. 35,* 1910 (1963).

1094. R. S. Bottei and B. A. Trusk, *Anal. Chim. Acta 37,* 409 (1967).

1095. R. S. Bottei and B. A. Trusk, *Anal. Chim. Acta 41,* 374 (1968).

1096. A. T. Pilipenko, A. I. Zhebentyaev, and A. I. Volkova, *Ukr. Khim. Zh. 41,* 1087 (1975); *Chem. Abstr. 84,* 83689f.

1097. A. Murata and F. Yamaguchi, *J. Chem. Soc. Jap. 77,* 1259 (1956); *Chem. Abstr. 52,* 971b.

1098. S. V. Kachin, *Reakt. Osobo Chist. Veshch. Nauch-Issled Inst. Tekhn. Ekon. Issled.* No. 5, 19 (1978).

1098a. A. P. Golovina, S. V. Kachin, V. K. Runov, and D. A. Fakeeva, *Zh. Anal. Khim. 37*, 1816 (1981); *Anal. Abstr. 45*, 1B79.

1099.   A. Kirillov, R. Lauer, and N. S. Poluektov, *Zh. Anal. Khim. 22*, 1333 (1967); *Chem. Abstr. 68*, 26683d.

1100.   D. E. Ryan, F. Snape, and M. Winpe, *Anal. Chim. Acta 58*, 101 (1972).

1101.   M. Santiago, A. Navas, J. J. Laserna, and F. Garcia-Sanchez, *Mikrochim. Acta II*, 197 (1983).

1102.   F. Garcia-Sanchez, A. Navas, and J. J. Laserna, *Talanta 29*, 511 (1982).

1103.   R. R. Trenholm and D. E. Ryan, *Anal. Chim. Acta 32*, 317 (1965).

1104.   L. Merrit, *Ind. Eng. Chem., Anal. Ed., 16*, 758 (1944).

1105.   G. Smith, R. Jenkins, and J. Gough, *J. Histochem. Cytochem. 17*, 749 (1969).

1106.   K. I. Fridman and O. L. Turchina, *Khim. Prom. Inform. Nauk-Tekn. 2b*, 75 (1975).

1107.   D. Yamamoto, M. Tsukada, and S. S. Lynn, *Meiji Daigaku Kogakubu Kenkyu Hokuku* 25 (1978); *Chem. Abstr. 90*, 114443k.

1108.   Y. Kondoh, M. Kataoka, and T. Kambara, *Jap. Anal. 30*, 109 (1981); *Anal. Abstr. 41*, 3B66.

1109.   A. Moreno, M. Silva, D. Perez-Bendito, and M. Valcarcel, *Analyst 108*, 85 (1983).

1110.   A. Konstantinov, L. M. Korobochkin, and G. V. Anastasina, *Referat. Zh. Biol. Khim.* 1968 Abstr. No. 8F89.

1111.   A. A. Schilt and T. E. Hillison, *Talanta 27*, 1021 (1980).

1112.   E. R. Jensen and R. T. Pfaum, *Anal. Chem. 38*, 1268 (1966).

1113.   M. A. Cejas, A. Gomez-Hens, and M. Valcarcel, *Anal. Lett. 15*, 283 (1982).

1114.   K. Watanabe and K. Kawasaki, *Jap. Anal. 22*, 581 (1973); *Chem. Abstr. 79*, 96757g.

1115.   K. Watanabe and K. Kawasaki, *Bull. Chem. Soc. Jap. 48*, 1945 (1975).

1116.   G. S. Andrushko, A. T. Tashkhodzhaev, P. A. Shakirova, and Sh. T. Talipov, *Deposited Doc. VINITI*, 144 (1975); *Chem. Abstr. 87*, 15473b.

1117.   A. K. Babko and Z. I. Chalaya, *Zh. Anal. Khim. 17*, 286 (1962); *Chem. Abstr. 57*, 9211a.

1118.   A. K. Babko and Z. I. Chalaya, *Zh. Anal. Khim. 16*, 268 (1961); *Chem. Abstr. 56*, 1997g.

1119.   M. M. Alykov, *Zh. Anal. Khim. 35*, 639 (1980); *Chem. Abstr. 93*, 87902n.

1120.   I. Kasa, I. Hornyak, and J. Korosi, *Proc. Conf. Appl. Phys. Chem., 2nd 1*, 45 (1971); *Chem. Abstr. 76*, 148489f.

1120a. A. T. Tashkhodzhaev, L. E. Zel'tser, and Kh. Khikmatov, *Uzb. Khim. Zh. 16*, 22 (1972); *Chem. Abstr. 77*, 96589n.

1121.   N. Shimizu and T. Uno, *Chem. Pharm. Bull. 21*, 762 (1973); *Chem. Abstr. 79*, 26759k.

1122. H. Kato, N. Uemura, S. Kawai, and T. Ohno, *Jap. Anal. 21*, 856 (1972); *Chem. Abstr. 77*, 121849g.

1123. M. R. Martinez de la Barrera, J. J. Laserna, and F. Garcia-Sanchez, Anal. Chim. Acta *147*, 303 (1983).

1124. D. P. Shcherbov and V. V. Kolgomorova, *Zavod. Lab. 28*, 649 (1962); *Anal. Abstr. 10*, 55.

1125. A. V. Zholnin and G. V. Serebryakova, *Tr. Vses. Nauch. Issled. Inst. Khim. Reakt. Osobo Chist. Khim. Veshchestv.* No. 30, 242 (1967).

1126. R. V. Hems, G. F. Kirkbright, and T. S. West, *Anal. Chem. 42*, 748 (1970).

1127. A. P. Golovina, I. P. Alimarin, E. A. Bozhevol'nov, and L. P. Agasyau, *Zh. Anal. Khim. 17*, 591 (1962).

1128. W. C. Alford, L. Shapiro, and C. E. White, *Anal. Chem. 23*, 1149 (1951).

1129. T. Ito and A. Murata, *Jap. Anal. 23*, 274 (1974).

1130. H. O. Schneider and M. E. Roselli, *Analyst 96*, 330 (1971).

1131. A. T. Tashkhodzhaev, Sh. T. Talipov, L. E. Zel'tser, and Kh. Khikmatov, *Dokl. An. Uz. SSR* No. 9, 26 (1973); *Chem. Abstr. 80*, 115663n.

1132. R. Geiger and E. Sandell, *Anal. Chim. Acta 16*, 346 (1957).

1133. R. N. Plotnikova, D. N. Perminova, and D. P. Shcherbov, *Prom. Khim. Reakt. Osobo Chist. Veshchestv.* No. 8, 197 (1967).

1134. A. B. Blank, I. I. Mirenskaya, and L. P. Eksperiendova, *Zh. Anal. Khim. 28*, 1331 (1973).

1135. E. A. Bozhevol'nov, O. A. Fakeeva, and V. S. Ivanova, *Zh. Anal. Khim. 31*, 1916 (1976).

1136. S. R. Desai and K. Sudhalatha, *Anal. Chim. Acta 55*, 395 (1971).

1137. L. Ya Polyak and I. S. Bashkirova, *Zh. Anal. Khim. 19*, 842 (1964).

1138. L. A. Morozova, Sh. T. Talipov, L. E. Zel'tser, A. T. Tashkhodzhaev, and Z. P. Pakudina, *Dokl. Akad. Nauk. Uzb. SSR* No. 1, 43 (1979); *Anal. Abstr. 40*, 2B94.

1139. A. T. Pilipenko, T. U. Kukibaev, and A. I. Volkova, *Zh. Anal. Khim. 29*, 710 (1974).

1140. E. Asmus and W. Klank, *Z. Anal. Chem. 265*, 260 (1973).

1141. E. Asmus and W. Klank, *Z. Anal. Chem. 265*, 267 (1973).

1142. A. T. Tashkhodzhaev, P. A. Shakirova, G. S. Andrushko, and Sh. T. Talipov, Manuscript 1716-74, deposited at Vseoyuznyi Institud Nauchnoi i Tekhnicheskoi Informatisii, Moscow, 1974.

1143. Sh. T. Talipov, A. T. Tashkhodzhaev, L. E. Zel'tser, and Kh. Khikmatov, *Dokl. Akad. Nauk. Uzb. SSR 29*, 34 (1972); *Chem. Abstr. 78*, 92155r.

1144. A. T. Tashkhodzhaev, L. E. Zel'tser, and Kh. Khikmatov, *Uzb. Khim. Zh. 16*, 22 (1972); *Chem. Abstr. 77*, 96589n.

1144a. T. V. Ramakrishna and M. S. Subramanian, *Bull. Chem. Soc. Jap. 56*, 321 (1983); *Anal. Abstr. 45*, 2B107.

1145. T. Filer, *Anal. Chem. 43*, 469 (1971).

1145a. A. Cabrera, J. S. Durand, R. Gallego, and J. Velasco, *Rev. R. Acad. Cienc. Exactas, Fis. Nat. Madrid 77*, 403 (1983); *Chem. Abstr. 99*, 151278s.

1146. M. Roman and A. Fernandez-Gutierrez, *Bol. Soc. Quim. Peru 40*, 190 (1974).

1147. M. Roman, A. Fernandez-Gutierrez, and F. Valdes, *Afinidad 34*, 411 (1977).

1148. M. Roman, A. Fernandez-Gutierrez, and C. Marin, *Afinidad 34*, 481 (1977).

1149. M. Roman, A. Fernandez-Gutierrez, and M. C. Mahedero, *An. Quim. 74*, 439 (1978).

1150. M. Roman, A. Fernandez-Gutierrez, and F. Cardenas, *An. Quim. 76B*, 442 (1980).

1151. M. Roman, A. Fernandez-Gutierrez, and A. Muñoz de la Peña, *An. Quim. 79B*, 269 (1983).

1152. K. P. Stolyarov and A. V. Drobachenko, *Vestn. Leningr. Univ. Fiz. Khim. 2*, 120 (1965).

1153. N. A. Raju and G. G. Rao, *Nature 175*, 167 (1955).

1154. M. Roman, A. Fernandez-Gutierrez, and A. Muñoz de la Peña, *An. Quim. 76B*, 314 (1980).

1155. T. S. West, in *Chemical Spectrophotometry in Trace Characterization, Chemical and Physical*, W. W. Meinke and B. F. Scribner (ed.), National Bureau of Standards Monograph 100, 1967.

1156. G. G. Guilbaul (ed.), *Enzymatic Methods of Analysis*, Pergamon, Oxford, 1970.

1157. M. A. De Luca (ed.), *Methods of Enzymology*, Vol. 57, Academic, New York, 1978.

1158. P. Froehlich, in *Modern Fluorescence Spectroscopy*, E. L. Wehry (ed.), Vol. II, Plenum, New York, 1976, chap. 2.

1159. M. Valcarcel and F. Grases, *Talanta 30*, 139 (1983).

1160. M. Zander, *Phosphorimetry*, Academic Press, New York, 1968.

1161. W. J. McCarthy and J. D. Winefordner, in *Fluorescence: Theory, Instrumentation, and Practice*, G. G. Guilbault (ed.), Dekker, New York, 1967, chap. 10.

1162. J. J. Aaron and J. D. Winefordner, *Talanta 22*, 707 (1975).

1163. J. Vo. Dinh and J. D. Wineforner, *Appl. Spectrosc. Rev. 13*, 261 (1977).

1164. R. T. Parker, R. S. Freelander, and R. B. Dunlap, *Anal. Chim. Acta 119*, 189 (1980).

1165. R. T. Parker, R. S. Freelander, and R. B. Dunlap, *Anal. Chim. Acta, 120*, 1 (1980).

1166. E. A. Solov'ev, E. A. Bozhevol'nov, N. A. Lebedeva, A. F. Mironov, and R. P. Evstigneeva, *Zh. Anal. Khim. 24*, 231 (1969).

1167. E. A. Solov'ev, N. A. Lebedeva, and Z. S. Sidenko, *Zh. Anal. Khim. 29*, 1531 (1974).

1168. E. A. Solov'ev and E. A. Bozhevol'nov, *Metod. Anal. Khim. Reakt. Preparat.* No. 11, 21 (1965).

1169. E. A. Bozhevol'nov and E. A. Solov'ev, *Sovrenen Metod. Anal. Metod. Issled. Khim. Sost. Stroen. Veshch.*, Nauka, Moscow, 1965 p. 75.

1170. E. A. Bozhevol'nov and E. A. Solov'ev, *Priklad. Spectr., Mater.* 16 *Soveshch.* Nauka, Moscow, 1969, p. 166.

1171. M. Marcantonatos, G. Gamba, and D. Monier, *Anal. Chim. Acta 67,* 220 (1973).

1172. M. Marcantonatos, G. Gamba, and D. Monier, *Helv. Chim. Acta 52,* 2183 (1969).

1173. G. Gamba and M. Marcantonatos, *Helv. Chim. Acta 54,* 1509 (1971).

1174. Z. Holzbecher, M. Hejtmanek, and Z. Sobalik, *Coll. Czech. Chem. Comm. 43,* 3325 (1978).

1175. G. F. Kirkbright, J. V. Thomson, and T. S. West, *Anal. Chem. 42,* 782 (1970).

1176. S. B. Meshkova, S. F. Potapova, L. I. Kononenko, and N. S. Poleuktov, *Zh. Anal. Khim. 32,* 1529 (1977).

1177. H. Kautsky and A. Hirsch, *Z. Anorg. Allgem. Chem. 222,* 126 (1935).

1178. J. Franck and P. Pringsheim, *J. Chem. Phys. 11,* 21 (1943).

1179. H. Kautsky and G. Muller, *Z. Naturforsch. 2a,* 167 (1947).

1180. L. J. Tolmach, *Arch. Biochem. Biophys. 33,* 120 (1951).

1181. W. R. Ware, "Transient Luminescence Measurements", in *Creation and Detection of the Excited State,* A. A. Lamola (ed), Dekker, New York, (1971).

1182. H. Kokubun, *Kagaku No. Ryoiki, 24,* 51 (1970).

1183. J. B. Birks and I. H. Munro, in *Progress in Reaction Kinetics,* G. Porter (ed.), Pergamon, Oxford, 1967, chap. 7.

1184. H. P. Haar and M. Hauser, *Rev. Sci. Instrum. 49,* 632 (1978).

1185. J. N. Demas and G. A. Crosby, *Anal. Chem. 42,* 1010 (1972).

1186. L. T. Hundley, E. G. Coburn, and L. Stryer, *Rev. Sci. Instrum. 38,* 488 (1967).

1187. L. J. Cline Love and L. A. Shaver, *Anal. Chem. 48,* 364A (1976).

1188. F. E. Lytle, D. R. Storey, and M. E. Juricich, *Spectrochim. Acta 29A,* 1357 (1973).

1189. Y. Nishikawa, K. Hiraki, K. Morishige, K. Takahasi, T. Shigematsu, and T. Nogami, *Jap. Anal. 25,* 459 (1976).

1190. K. Hiraki, K. Morishige, and Y. Nishikawa, *Anal. Chim. Acta 97,* 121 (1978).

1191. K. Hiraki, N. Watanabe, and Y. Nishikawa, *Kinki Daigaku Ribogakuku; Kenkyu Hokoku 16,* 31 (1981).

1192. T. L. Craven and F. E. Lytle, *Anal. Chim. Acta 107*, 273 (1979).

1193. T. L. Craven and F. E. Lytle, *Spectrosc. Lett. 12*, 559 (1979).

1194. E. R. Hinton, Jr., and L. E. White, *Anal. Lett. 14*, 947 (1981).

1195. R. Kaminiski, F. J. Purcell, and E. Russavage, *Anal. Chem. 53*, 1093 (1981).

1196. J. W. Hass, Jr., *J. Chem. Educ. 44*, 396 (1967).

1197. L. S. Bark and P. R. Wood, *Sel. Ann. Rev. Anal. Sci. 1*, 41 (1971).

1198. W. R. Seitz and D. M. Hercules, *Int. J. Environ. Anal. Chem. 2*, 273 (1973).

1199. U. Isacsson and G. Wettermark, *Anal. Chim. Acta 68*, 339 (1974).

1200. W. R. Seitz and M. F. Neary, *Anal. Chem. 46*, 188A (1974).

1201. J. H. Glover, *Analyst 100*, 449 (1975).

1202. J. W. Hasting and T. Wilson, *Photochem. Photobiol. 23*, 461 (1976).

1203. D. B. Paul, *Talanta 25*, 377 (1978).

1204. A. K. Babko, L. I. Dubovenko, and N. M. Lukovskaya, *Chemiluminescence Analysis*, Tekhnika, Kiev, 1966.

1205. W. R. Seitz, in *Modern Fluorescence Spectroscopy*, Vol. I, E. L. Wehry (ed.), Plenum, New York, 1970, chap. 7.

1205a. J. L. Burguera, M. Burguera, and A. Townshed, *Acta Cient. Venez. 32*, 115 (1981).

1206. M. J. Cormier and D. M. Lee, Jr. (eds.), *Chemiluminescence and Bioluminescence*, Plenum, New York, 1973.

1207. W. R. Seitz and M. P. Neary, in *Methods of Biochemical Analysis,* Vol. 23, D. Glick (ed.), Wiley-Interscience, New York, 1976, p. 161.

1208. A. Fontijn, in *Modern Fluorescence Spectroscopy*, Vol. I, E. L. Wehry (ed.), Plenum, New York, 1976, chap. 6.

1209. E. Schram and P. E. Stanley (eds.), *Proceedings of International Symposium on Analytical Applications of Bioluminescence and Chemiluminescence*, State Printing and Publishing, Westlak Village, Calif., 1979.

1210. M. A. DeLuca and W. D. McElroy, *Bioluminescence* and Chemiluminescence, Academic Press, New York, 1981.

1211. A. K. Babko, A. V. Terletskaya, and L. I. Dubovenko, *Zh. Anal. Khim. 23*, 932 (1968).

1212. N. M. Lukovskaya, A. V. Terletskaya, and T. A. Bogoslovskaya, *Zh. Anal. Khim. 29*, 2268 (1974); *Chem. Abstr. 83*, 132556.

1213. N. N. Lukovskaya and T. A. Bogoslovskaya, *Ukr. Khim. Zh. 41*, 200 (1975).

1214. A. T. Pilipenko, G. V. Angelova, and I. E. Kalinichenko, *Ukr. Khim. Zh. 40*, 1302 (1974); *Chem. Abstr. 82*, 164439.

1215. S. Steig and T. A. Nieman, *Anal. Chem. 49*, 1322 (1977).

1216. S. Steig and T. A. Nieman, *Anal. Chem. 52*, 800 (1980).

1217. N. M. Lukovskaya and V. A. Bilochemko, *Zh. Anal. Khim. 32*, 2177 (1977).

1218. K. Fujiwara, Y. Watanabe, K. Fuwa, and T. D. Winefordner, *Anal. Chem. 54*, 125 (1982).

1219. N. M. Lukovskaya and T. A. Gogoslovskaya, *Ukr. Khim. Zh. 41*, 268 (1975).

1220. N. M. Lukovskaya and T. A. Bogoslovskaya, *Ukr. Khim. Zh. 41*, 529 (1975).

1221. A. T. Pilipenko, N. M. Lukovskaya, and T. A. Bogoslovskaya, USSR Patent 349,341; *Chem. Abstr. 79*, 10243f.

1222. L. I. Dubovenko and E. Ya. Khotinets, *Zh. Anal. Khim. 26*, 784 (1971); *Chem. Abstr. 75*, 14640u.

1223. L. I. Dubovenko and L. M. Korotun, *Vestn. Kiev Univ. Ser. Khim. N15*, 12 (1974).

1224. J. L. Burguera, M. Burguera, and A. Townshend, *Anal. Chim. Acta 127*, 199 (1981).

1225. L. I. Dubovenko, M. M. Tananaiko, and V. G. Drokov, *Ukr. Khim. Zh. 40*, 758 (1974); *Chem. Abstr. 81*, 98896s.

1226. L. I. Dubovenko and Chan Ti Huu, *Ukr. Khim. Zh. 35*, 637 (1969); *Chem. Abstr. 71*, 54723.

1227. F. F. Grigorenko and L. I. Dubovenko, *Ukr. Khim. Zh. 34*, 1294 (1968).

1228. A. K. Babko, A. V. Terletskaya, and L. I. Dubovenko, *Ukr. Khim. Zh. 32*, 728 (1960); *Chem. Abstr. 65*, 13405b.

1229. N. M. Lukovskaya, A. V. Terletskaya, and N. I. Isaenko, *Zavod. Lab. 37*, 897 (1971).

1230. D. F. Marino and J. D. Ingle, Jr., *Anal. Chim. Acta 123*, 247 (1981).

1231. A. V. Terletskaya, N. M. Lukovskaya, and N. L. Anatienko, *Ukr. Khim. Zh. 45*, 1227 (1979); *Anal. Abstr. 39*, 1B130.

1232. A. MacDonald, K. Chan, and T. Nieman, *Anal. Chem. 51*, 2077 (1979).

1233. V. N. Kachibaya, I. L. Siamashvili, and M. V. Mamukashvili, *Zh. Anal. Khim. 26*, 1848 (1971).

1234. S. Musha, M. Ito, Y. Yamamoto, and Y. Inamori, *Nippon Kagaku Zasshi 80*, 1285 (1959).

1235. J. Bognar and L. Sipos, *Mikrochim. Ichnoanal. Acta. 3*, 442 (1963).

1236. L. I. Dubovenko and N. V. Beloshitsky, *Zh. Anal. Khim. 29*, 111 (1974); *Chem. Abstr. 80*, 140814v.

1237. L. A. Montano and J. D. Ingle, *Anal. Chem. 51*, 919 (1979).

1238. L. A. Montano and J. D. Ingle, *Anal. Chem. 51*, 926 (1979).

1239. A. K. Babko and N. M. Lukovskaya, *Ukr. Khim. Zh. 30*, 388 (1964); *Chem. Abstr. 61*, 3677e.

1240. A. K. Babko and N. M. Lukovskaya, *Zavod. Lab. 29*, 404 (1963); *Chem. Abstr. 59*, 3301d.

1241. S. Pantel and H. Weisz, *Anal. Chim. Acta 74*, 275 (1975).

1242. V. Nall and T. A. Nieman, *Anal. Chem. 51*, 424 (1979).

1243. V. I. Rigin and A. I. Blokhin, *Zh. Anal. Khim. 32*, 312 (1977).

1244. N. M. Lukovskaya, A. V. Terletskaya and T. A. Bogoslovskaya, *Ukr. Khim. Zh. 45*, 1239 (1979); *Anal. Abstr. 39*, 1B29.

1245. J. Lasovsky and F. Grambal, *Acta Univ. Palacki Olumuc., Fac. Rerum Nat. 61/65*, 57 (1979–1980).

1246. Ya. P. Skorobogatyi, *Ukr. Khim. Zh. 48*, 309 (1982).

1247. G. B. Angelova, A. K. Panova, and D. I. Bakurdzhieva, *Dokl. Bolg. Akad. Nauk. 31*, 323 (1978).

1248. Sh. Nakahara, M. Yamada, and Sh. Suzuki, *Anal. Chim. Acta 141*, 255 (1982).

1249. A. T. Pilipenko, V. A. Barovskii, and I. E. Kalinichenko, *Zh. Anal. Khim. 33*, 1880 (1978); *Anal. Abstr. 36*, 5B156.

1250. D. F. Marino, F. Wolff, and J. D. Ingle, Jr., *Anal. Chem. 51*, 2051 (1979).

1251. A. V. Terletskaya, N. M. Lukovskaya, and N. L. Anatienko, *Ukr. Khim. Zh.* 1111 (1979); *Anal. Abstr. 38*, 6B6.

1252. R. J. Miller and J. D. Ingle, Jr., *Talanta 29*, 303 (1982).

1253. A. D. Snyder and G. W. Wooten, Final Report, Environmental Protection Agency, Research Triangle Park, N.C., Contract No. CPA-22-69-8 NTIS PB-1880-103 (1969).

1253a. I. Rubinstein, Ch. R. Martin, and A. J. Bard, *Anal. Chem. 55*, 1580 (1983).

1254. L. I. Dubovenko, A. M. Guta, and D. M. Kryzhanovskaya, *Visn. Lviv. Univ. Ser. Khim.* No. 13, 90, 30 (1972); *Chem. Abstr. 79*, 61217d.

1255. J. L. Bowling, J. A. Dean, G. Goldstein, and T. M. Dale, *Anal. Chim. Acta 76*, 47 (1975).

1256. S. D. Hoyt and J. D. Ingle, *Anal. Chim. Acta 87*, 163 (1976).

1257. W. R. Seitz, W. W. Suydam, and D. M. Hercules, *Anal. Chem. 44*, 957 (1972).

1258. V. I. Rigin and A. S. Bakhmurov, *Zh. Anal. Khim.* 31, 93 (1976); *Chem. Abstr. 82*, 80014.

1259. D. E. Bause and H. H. Patterson, *Anal. Chem. 51*, 2288 (1979).

1260. L. I. Dubovenko and A. M. Guta, *Izv. Vyssh. Ucheb. Zaved., Khim. Tekhnol. 18*, 1211 (1975); *Chem. Abstr. 84*, 38257m.

1261. A. M. Guta, *Verstn. L'vov. Univ. Ser. Khim.* No. 15, 40 (1974).

1262. A. V. Terletskaya, N. M. Lukovskaya, and N. M. Anatienko, *Zh. Anal. Khim. 34*, 1460 (1979); *Anal. Abstr. 38*, 3B144.

1263. Z. Holzbecher, L. Kabrt, and J. Jansta, *Coll. Czech. Chem. Comm. 47*, 1606 (1982).

1264. E. H. Huntress, L. N. Stanley, and A. S. Parker, *J. Chem. Educ. 11*, 142 (1934).

1265. A. A. Ponomarenko, N. A. Markar'yan, and A. I. Komlev, *Dokl. Akad. Nauk. SSSR 86*, 115 (1952).

1266. A. F. Samsonyuk, A. A. Ponomarenko, and R. E. Shindel', *L'vovski Torgovo-ekonomicheskii Institut. Zapiski Nauchnogo Studencheskogo Obshchestra*, No. 1, L'vov (1957).

1267. A. K. Babko and N. M. Lukovskaya, *Zh. Anal. Khim. 17*, 50 (1962).; *Chem. Abstr. 57*, 5293b.

1268. A. K. Babko and L. I. Dubovenko, *Zavod. Lab. 30*, 1325 (1964); *Chem. Abstr. 62*, 5881c.

1269. L. I. Dubovenko and L. A. Pilipenko, *Visnik Kiivs'k Univ. Ser. Fiz. Khim. 11*, 75 (1970); *Chem. Abstr. 76*, 67823h.

1270. A. K. Babko and L. I. Dubovenko, *Zh. Anal. Chem. 200*, 428 (1964).

1271. N. A. Truba and B. I. Nabivanets, *Gidrobiol. Zh. 11, N2*, 125 (1975).

1272. A. T. Pilipenko, G. V. Angelova, and I. E. Kalinichenko, *Dokl. Bolg. Akad. Nauk. 31*, 315 (1978); *Anal. Abstr. 36*, 1B41.

1273. A. P. Tovmasyan, *Mater. Konf. Molodykh. Uch. Spets., Akad. Nauk. Arm. SSR*, 188 (1972); *Chem. Abstr. 83*, 187824a.

1274. A. K. Babko, L. I. Dubovenko, and A. V. Terletskaya, *Ukr. Khim. Zh. 32*, 1326 (1966); *Chem. Abstr. 66*, 80637.

1275. V. G. Drokov and L. I. Dubovenko, *Ukr. Khim. Zh. 40*, 549 (1974).

1276. A. P. Tovmasyan and L. I. Dubovenko, *Metod. Nauch. Ratobn. Ser. Estestn. Nauk.* No. 2, 100 (1973).

1277. M. Yamada and S. Suzuki, *Chem. Lett.* No. 11, 1747 (1982).

1277a. L. I. Dubovenko and T. I. Rymar, *Ukr. Khim. Zh. 49*, 395 (1983); *Chem. Abstr. 99*, 15712b.

1278. L. I. Dubovenko and A. Yu. Nazarenko, *Ukr. Khim. Zh. 41*, 1205 (1975); *Chem. Abstr. 85*, 13306.

1279. A. K. Babko and I. E. Kalinichenko, *Ukr. Khim. Zh. 31*, 1316 (1965); Chem. Abstr. *64*, 13374b.

1280. W. R. Seitz and D. M. Hercules, *Anal. Chem. 44*, 2143 (1972).

1281. I. E. Kalinichenko and O. M. Grischchenko, *Ukr. Khim. Zh. 36*, 610 (1970); *Chem. Abstr. 73*, 105171.

1282. A. K. Babko and I. E. Kalinichenko, *Ukr. Khim. Zh. 29*, 527 (1963), *Chem. Abstr. 59*, 12178d.

1283. K. Weber and J. Matkovic, *Arch. Hig. Rada Toksikol. 15*, 141 (1964).

1284. N. M. Lukovskaya and V. A. Bilochenko, *Zh. Anal. Khim. 34*, 477 (1979); *Anal. Abstr. 37*, 4B89.

1285. L. I. Dubovenko and V. A. Bilochenko, *Ukr. Khim. Zh. 40*, 423 (1974); *Chem. Abstr. 81*, 32896k.

1286. L. I. Dubovenko and T. A. Bogoslovskaya, *Ukr. Khim. Zh. 37*, 1057 (1971); *Chem. Abstr. 76*, 67862v.

1287. G. L. Kok, T. P. Holler, M. B. Lopez, M. A. Nachtrieb, and M. Yuan, *Environ. Sci. Technol. 12,* 1072 (1978).

1288. Kh. I. Agranov and L. V. Reiman, *Zh. Anal. Khim. 34,* 1533 (1979).

1289. T. N. Das, P. N. Moorthy, and K. N. Rao, *J. Indian. Soc. 59,* 85 (1982).

1290. V. K. Zinchuk and A. I. Komlev, *Zh. Anal. Khim. 28,* 616 (1973); *Chem. Abstr. 69,* 13219j.

1291. F. Shaw, *Analyst, 105,* 11 (1980).

1292. W. Langenbeck and U. Ruge, *Ber. 70,* 367 (1937).

1293. L. I. Dubovenko, N. S. Rigun, and V. O. Bilochenko, *Visn. Kiiv. Univ. Ser. Khim.* No. 13, 25, 68 (1972); *Chem. Abstr. 79,* 4888d.

1294. J. Kubal, *Chem. Listy, 62,* 1478 (1968); *Chem. Abstr. 70,* 74021z.

1295. A. K. Babko and N. M. Lukovskaya, *Zh. Anal. Khim. 20,* 1100 (1965); *Chem. Abstr. 64,* 5751h.

1296. A. P. Tovmasyan, G. G. Galstyan, and S. M. Uloyan, *Gig. Tr. Prof. Zabol.* (*3*), 58 (1979).

1297. A. G. Mohan and N. J. Turro, *J. Chem. Ed. 51,* 528 (1974).

1298. A. K. Babko, L. V. Markova, and N. M. Lukovskaya, *Zh. Anal. Khim. 23,* 401 (1968).

1298a. J. L. Burguera and M. Burguera, *An. Quim.* 78B, 307 (1982).

1299. N. M. Lukovskaya, L. V. Markova, and N. F. Evtushenko, *An. Quim. 29,* 767 (1974); *Chem. Abstr. 81,* 44896.

1300. N. M. Lukovskaya, N. F. Kushchevskaya, *Ukr. Khim. Zh. 42,* 87 (1976); *Anal. Abstr. 31,* 136.

1300a. A. T. Pilipenko, N. M. Lukovskaya, A. V. Terletskaya, N. L. Anatienko, T. A. Bogoslovskaya, and N. F. Kushchevskaya, *Zh. Anal. Khim. 38,* 1071 (1983); *Chem. Abstr. 99,* 115135j.

1301. N. M. Lukovskaya, T. A. Bogoslovaskaya, and N. F. Kushchevskaya, *Zh. Anal. Khim. 48,* 842 (1982).

1302. N. M. Lukovskaya, A. V. Terletskaya, and N. F. Kushchevskaya, *Zh. Anal. Khim. 33,* 750 (1978).

1303. J. L. Burguera and A. Townshend, *Talanta 28,* 731 (1981).

1304. B. I. Nabivanets and N. A. Truba, *Gidrobiol. Zh. 9, N5,* 90 (1973).

1305. I. E. Kalinichenko, *Ukr. Khim. Zh. 35,* 755 (1969).

1306. L. I. Dubovenko and A. P. Tovmasyan, *Zh. Anal. Khim. 25,* 940 (1970).

1307. L. I. Dubovenko and A. P. Tovmasyan, *Ukr. Khim. Zh. 37,* 943 (1971).

1308. A. K. Babko, L. I. Dubovenko, and L. S. Mikhailova, *Sov. Prog. Chem. 32,* 471 (1966); *Chem. Abstr. 65,* 14419d.

1309. A. Hasegawa, T. Somiya, and E. Niki, *Kogakuin Daigaku Kenkyu Hokoku 37,* 119 (1974).

1310. L. I. Dubovenko and A. Yu. Nazarenko, *Zh. Anal. Khim. 32,* 1345 (1977).

1311. W. R. Seitz, in *Chemiluminescence and Bioluminescence,* M. J. Cormier, D. M. Hercules, and J. Lee (eds.), Plenum, New York, 1973, p. 427.

1312. A. T. Pilipenko, L. V. Markova, N. M. Lukovskaya, and N. F. Evtushenko, *Ukr. Khim. Zh. 40*, 1205 (1974); *Chem. Abstr. 83*, 80014p.

1313. A. T. Pilipenko, G. Angelova-Todorova, and I. E. Kalinichenko, *Dokl. Bolg. Akad. Nauk. 31*, 209 (1978); *Chem. Abstr. 89*, 156872j.

1314. Ya. P. Skorobogatyj and V. K. Zinchuk, *Zh. Anal. Khim. 30*, 819 (1975).

1315. V. K. Zinchuk and Ya. P. Skorobogatyj, *Tr. po. Khim. i Khim. Tekhnol. Polucheme i Analiz. Chist. Veshch.*, N4(35), Gor'kij, 1973, pp. 117.

1316. L. I. Dubovenko and N. F. Evtushenko, *Vestn. Kiev. Univ. Ser. Khim. N14*, 12 (1973).

1317. L. I. Dubovenko, O. K. Notsyk, and L. V. Kononenko, *Ukr. Khim. Zh. 46*, 854 (1980); *Anal. Abstr. 40*, 5B199.

1318. R. E. Baumgardner, Jr., W. A. McClenny, and R. K. Stevens, U.S. Environ. Prot. Agency, Off. Res. Dev., [Rep.], EPA/600-2-79/028 (1979), pp. 48; *Anal. Abstr., 40*, 3H15.

1319. R. K. Stevens and J. A. Hodgeson, *Anal. Chem. 45*, 443A, (1973).

1320. A. Fontijn, A. J. Sabadell, and R. J. Ronco, *Anal. Chem. 42*, 575 (1970).

1321. N. V. Artishcheva, S. A. Krapivina, and V. A. Ershov, *Zh. Vses. Khim. O-va. 25*, 116 (1980).

1322. K. J. Krost, J. A. Hodgeson, and R. K. Stevens, Environmental Protection Agency, Research Triangle Park, N.C., 1972, publication preprint.

1323. A. Sh. Agaverdiev, *Uch. Zap. Perm. Gos. Univ. 222*, 165 (1969).

1324. Y. Maeda, K. Aoki, and M. Munemori, *Anal. Chem. 52*, 307 (1980).

1325. R. D. Cox, *Anal. Chem. 52*, 332 (1980).

1326. C. Garcide, *Mar. Chem. 11*, 159 (1982).

1327. A. Burr and D. Mayzerall, *Biochim. Biophys. Acta, 153*, 614 (1968).

1328. B. F. MacDonald and W. R. Seitz, *Anal. Lett. 15*, 75 (1982).

1329. G. W. Nederbragt, A. Van der Horst, and J. Van Duijn, *Nature 206*, 87 (1965).

1330. J. A. Hodgeson, B. E. Martin, and R. E. Baumgardner, Eastern Analytical Symposium, New York, N.Y., paper no. 77 (1970).

1331. J. N. Pitts Jr., W. A. Kummer, R. P. Steer, and B. J. Finlayson, *Am. Chem. Soc. Adv. Chem. Ser. 113*, 246 (1972).

1332. V. H. Regener, *J. Geophys. Res. 69*, 3795 (1964).

1333. J. A. Hodgeson, K. J. Krost, A. E. O'Keeffe, and R. K. Stevens, *Anal. Chem. 42*, 1795 (1970).

1334. L. G. Bol'shakova, *Probl. Fiz. Atmos. No. 7*, 113 (1969); *Chem. Abstr. 73*, 62229x.

1335. R. Guicherit, *Z. Anal. Chem. 256*, 177 (1971).

1336. D. Bersis and E. Vassiliou, *Analyst 91*, 499 (1966).

1337. V. H. Regener, *J. Geophys. Res. 65*, 3975 (1960).

1338. A. G. Stepanova and E. A. Bozhevol'nov, USSR Patent 329, 451; *Chem. Abstr. 77*, 42859g.

1339. E. G. Janzen, I. G. Lopp, and J. W. Happ, *J. Chem. Soc. D*, 1140 (1970).

1340. N. M. Lukovskaya and A. V. Terletskaya, *Ukr. Khim. Zh. 40*, 1311 (1974).

1341. N. M. Lukovskaya and A. V. Terletskaya, *Ukr. Khim. Zh. 40*, 1311 (1974).

1342. A. T. Pilipenko and A. V. Terletskaya, *Zh. Anal. Khim. 27*, 1570 (1972), *Chem. Abstr. 77*, 172304h.

1343. A. Ya. Brin, S. E. Strel'nikova, and V. I. Petrovicheva, *Anal. Tekhnol. Blagorod. Met.* 86, (1971); *Chem. Abstr. 78*, 66575h.

1344. A. Ya. Brin, S. E. Kashlinskaya, N. P. Strel'nikova, and V. I. Petrovicheva, *Nauch. Tr. Sib. Nauch.-Issled. Proekt. Inst. Tsvet. Met.* No. 4, 29 (1971); *Chem. Abstr. 76*, 41657g.

1345. N. M. Lukovskaya and V. A. Bilochenko, *Zavod. Lab. 40*, 936 (1974).

1346. N. M. Lukovskaya and V. A. Bilochenko, *Ukr. Khim. Zh. 43*, 756 (1977).

1347. L. I. Dubovenko and L. D. Guz, *Ukr. Khim. Zh. 36*, 1264 (1970); *Chem. Abstr. 74*, 134595.

1348. L. I. Dubovenko, N. V. Shvydak, and V. T. Lenets, *Ukr. Khim. Zh. 48*, 1294 (1982); *Anal. Abstr. 44*, 6H18.

1349. A. V. Terletskaya, *35*, 1065 (1969); *Chem. Abstr. 72*, 38565n.

1350. V. I. Rigin, A. S. Bakhmurov, and A. I. Blokhin, *Zh. Anal. Khim. 30*, 2413 (1975); *Chem. Abstr. 85*, 71630w.

1351. N. M. Lukovskaya and N. F. Kushchevskaya, *Ukr. Khim. Zh. 41*, 643 (1975).

1352. N. M. Lukovskaya and A. V. Terletskaya, *Zh. Anal. Khim. 31*, 751 (1976).

1353. G. S. Lisetskaya and N. S. Bilenko, *Metody Anal. Kontrolya Proizvod. Khim. Prom-sti 9*, 49 (1977); *Anal. Abstr. 36*, 1B144.

1354. N. M. Lukovskaya, A. V. Terletskaya, and N. S. Bilenko, *Ukr. Khim. Zh. 45*, 886 (1979); *Anal. Abstr. 38*, 4B181.

1355. I. Ya. Koshcheeva, G. M. Varshal, and V. I. Ejger, in *Novye Metodi Vydel i Opred. Blagorodn. Elementov*, Geokhi in V. I. Vernadskogo, Moscow (1975), p. 89.

1356. N. M. Lukovskaya and L. V. Markova, *Zh. Anal. Khim. 24*, 1862 (1969).

1357. J. L. Burguera and A. Townshend, *Talanta 27*, 309 (1980).

1358. D. Klochkow and J. Teckentrup, *Talanta 23*, 889 (1976).

1359. O. I. Komlev and V. K. Zinchuk, *Visn.l'vivsk'k. Univ. Ser. Khim. 9*, 50 (1967); *Anal. Abstr. 15*, 6604.

1360. V. K. Zinchuk and L. M. Rekhlitskaya, *Vestn. L'vov. Univ. Ser. Khim. N15*, 35 (1974).

1360a. M. Yamada, T. Nakada, and Sh. Suzuki, *Anal. Chim. Acta 147*, 401 (1983).

1361. M. F. R. Mulcahy and D. J. Williams, *Chem. Phys. Lett. 7*, 455 (1970).

1362. L. I. Dubovenko and Chan Ti Huu, *Ukr. Khim. Zh. 35*, 957 (1969); *Chem. Abstr. 72*, 36224.

1363. L. I. Dubovenko and A. P. Tovmasyan, *Ukr. Khim. Zh. 37*, 845 (1971); *Chem. Abstr. 75*, 126070g.

1364. U. Fritsche, W. Balzer, and H. Steinhanses, *Fresenius' Z. Anal. Chem. 311*, 234 (1982).

1365. V. P. Kazakov, G. S. Parshin, R. G. Bulgakov, L. A. Khamidullina, and D. D. Afonichev, *Tezisy Dokl. Konf. Anal. Khim. Radioakt. Elem.* 45 (1977); *Chem. Abstr. 91*, 48872s.

1366. A. T. Pilipenko, E. V. Mitropolitska, and N. M. Lukovskaya, *Ukr. Khim. Zh. 41*, 525 (1975).

1367. A. T. Pilipenko, E. V. Mitropolitska, and N. M. Lukovskaya, *Ukr. Khim. Zh. 41*, 1196 (1975).

1368. A. T. Pilipenko, E. V. Mitropolitskaya, and N. M. Lukoskaya, *Ukr. Khim. Zh. 39*, 73 (1973).

1369. F. F. Grigorenko and L. I. Dubovenko, *Ukr. Khim. Zh. 38*, 841 (1972).

1370. L. Erdley and L. Buzas, "Luminescence Redox Titrations," AD 627166, avail. CFSTI (1965).

1371. E. Bishop (ed.), "Indicators," *Int. Ser. Monogr. Anal. Chem.*, Vol. 51, Pergamon Press, Oxford, 1972.

1372. J. A. Bishop, *Talanta 22*, 617 (1975).

1373. J. A. Bishop, *Talanta 22*, 619 (1975).

1374. L. S. A. Difshitulu and G. G. Rao, *Talanta 9*, 289 (1962).

1375. R. Geyer and H. Steinmtzer, *Angew. Chem. 72*, 634 (1960); *Chem. Abstr. 55*, 15407f.

1376. B. Kratochvil and D. A. Zatko, *Anal. Chem. 36*, 527 (1964).

1377. B. Kratochvil and M. C. White, *Anal. Chim. Acta 31*, 528 (1964).

1378. A. B. Crawford and E. J. Bishop, *Roy. Tech. Coll. Glasg. 5*, 52 (1950); *Chem. Abstr. 44*, 6761.

1379. I. U. Buric and B. Draskovic, *Glas. Hem. Drus. Beograd. 27*, 271 (1962); *Chem. Abstr. 569*, 13325.

1380. M. Roman and F. Ales-Barrero, *Quim. Anal. 29*, 323 (1975).

1381. F. Grases, F. Garcia-Sanchez, and M. Valcarcel, *An. Quim. 76B*, 124 (1980).

1382. F. Salinas, F. Garcia-Sanchez, and C. Genestar, *Afinidad 39*, 261 (1982).

1383. D. H. Wilkins, *Anal. Chim. Acta 23*, 309 (1960).

1384. L. A. Solovjeva, K. P. Stolyarov, and N. N. Grigorjev, *Instrum. Khim. Metody Anal.* 98 (1973); *Chem. Abstr. 82*, 38241k.

1385. D. H. Wilkins, *Talanta 4*, 182 (1960).

1386. V. Ya. Temkina, N. M. Dyatlova, G. F. Yaroshenko, O. Yu. Lavrova, and R. P. Lastovskii, *Zh. Anal. Khim. 24*, 240 (1969); *Chem. Abstr. 70*, 111297s.

1387. V. Ya. Temkina, G. F. Yaroshenko, Yu. F. Belugin, and R. P. Lastovskii, *Tr. Vses. Nauch-Issled., Inst. Khim. Reakt. Osobo Chist. Khim. Veshchestv. 32*, 60 (1970); *Chem. Abstr. 77*, 69660x.

1388. R. L. Clements, J. I. Read, and G. A. Sergeant, *Analyst 96*, 656 (1971).

1389. Ch. Nagasaki and K. Toei, *Jap. Anal. 21*, 87 (1972); *Chem. Abstr. 76*, 107582n.

1390. A. B. Borle and F. N. Briggs, *Anal. Chem. 40*, 339 (1968).

1391. G. F. Kirkbright and W. I. Stephen, *Anal. Chim. Acta 27*, 294 (1962).

1392. A. M. Escarrilla, *Talanta 13*, 363 (1966).

1393. J. H. Eggers, *Talanta 4*, 38 (1960).

1394. G. K. Singhal and K. N. Tandon, *Talanta 14*, 1351 (1967).

1395. J. L. Beck, J. M. Fitzgerald, and J. A. Bishop, *Anal. Chim. Acta 51*, 191 (1970).

1396. F. Bermejo-Martinez. A. Margalet, A. Badrinas, and A. Prieto, *Rev. Univ. Ind. Santander 6*, 92 (1964); *Chem. Abstr. 652*, 15397g.

1397. F. Bermejo-Martinez, *Chemist-Analyst 53*, 45 (1964); *Chem. Abstr. 61*, 12620c.

1398. F. Bermejo-Martinez, A. Badrinas, and A. Prieto, *Inf. Quim. Anal. 14*, 151 (1960); *Chem. Abstr. 57*, 2828d.

1399. D. H. Wilkins, *Talanta 2*, 355 (1959).

1400. V. Ya. Temkina, E. A. Bozhevol'nov, N. M. Dyatlova, G. F. Yaroshenko, O. Yu. Lavrova, and R. P. Lastovskii, *Zh. Anal. Khim. 24*, 240 (1960).

1401. F. Bermejo-Martinez and M. Gras G. de Lopidana, *Anal. Chim. Acta 47*, 139 (1969).

1402. M. Gras. G. de Lopidana, *Acta Cient. Compostelana 3*, 173 (1966); *Chem. Abstr. 69*, 8009e.

1403. F. Bermejo-Martinez and A. Margalet, *Inform. Quim. Anal. 19*, 39 (1965); *Chem. Abstr. 63*, 12311e.

1404. D. H. Wilkins and L. E. Hibbs, *Talanta 2*, 201 (1959).

1405. M. A. Salam Khan and W. I. Stephen, *Anal. Chim. Acta 49*, 255 (1970).

1406. M. M. Vinnik, *Novje Metod. Anal. Issled. Osnovn. Moscow, Sb.* 80 (1962); *Chem. Abstr. 61*, 8887h.

1407. H. N. Elsheiner, *Talanta 14*, 97 (1967).

1408. E. A. Bozhevol'nov, A. M. Lukin, and M. N. Gradinarskaya, *Avtomat. Svarka SSSR 116*, 838 (1959); *Anal. Abstr. 7*, 3164.

1409. V. Ya. Temkina, M. N. Rusina, G. F. Yaroshenko, M. Z. Branzburg, L. M. Timakova, and N. M. Dyatlova, *Zh. Obshch. Khim. 45*, 1564 (1975).

1410. V. Ya. Temkina, G. F. Yaroshenko, L. M. Timakova, N. E. Khavchenko, M. N. Rusina, and R. P. Lastovskii, *Otkrytiya, Izobret. Prom. Obraztsy, Tovarnye Znaki 52*, (48), 121 (1975); *Chem. Abstr. 85*, 28285a.

1411. F. Bermejo-Martinez, A. Badrinas, and A. Prieto, *Inform. Quim. Anal. 14*, 151 (1960).

1412. F. Bermejo and A. Margalet, *Inform. Quim. Anal. 19*, 171 (1965).

1413. K. P. Stolyarov, V. V. Firyulina, and I. Z. Gabajdulin, *Zavod. Lab. 38*, 664 (1972); *Chem. Abstr. 77*, 134789n.

1414. R. Van Slageren, G. den Boef, and W. E. Van der Linden, *Talanta 20*, 739 (1973).

1415. W. E. Van der Linden, G. den Boef, and W. Ozinga, *Mikrochim. Acta I*, 83 (1975).

1416. L. Erdey, *Ind. Chem. 33*, 459, 523, 575 (1957).

1417. L. Erdey, *Acta Chim. Acad. Sci. Hung. 3*, 81 (1953); *Chem. Abstr. 47*, 10399e.

1418. F. Kenny and R. B. Kurtz, *Anal. Chem. 23*, 339 (1951).

1419. F. Kenny and R. B. Kurtz, *Anal. Chem. 24*, 1218 (1952).

1420. L. Erdey and I. Buzas, *Anal. Chim. Acta 15*, 322 (1956).

1421. J. Slawinski, *Rocz. Glebozn. 18*, 191 (1967); *Chem. Abstr. 68*, 119065u.

1422. D. Slawinska and S. Slawinski, *Zeszyty Nauk. Wyz. Szk. Rol. Szc.* No. 10, 163 (1963); *Chem. Abstr. 61*, 6362e.

1423. L. Erdey, J. Takacs, and I. Buzas, *Acta Chim. Acad. Sci. Hung. 39*, 295 (1963).

1424. L. Erdey, W. F. Pickering, and C. L. Wilson, *Talanta 9*, 371 (162).

1425. L. Erdey, *Magyar Kem. Lapja 13*, 7 (1958); *Chem. Abstr., 52*, 18071g.

1426. L. Erdey and I. Buzas, *Anal. Chim. Acta 22*, 524 (1960).

1427. I. Sarudi, *Z. Anal. Chem. 260*, 114 (1972).

1428. A. K. Babko and N. M. Lukovskaya, *Ukr. Khim. Zh. 35*, 1060 (1969).

1429. F. Kenny and R. B. Kurtz, *Anal. Chem. 25*, 1550 (1953).

1430. F. Kenny, R. B. Kurtz, I. Beck, and I. Lukosevicius, *Anal. Chem. 29*, 543 (1957).

1431. F. Kenny, R. B. Kurtz, A. C. Vandenoever, C. J. Sanders, C. A. Novarro, L. E. Menzel, R. Kukla, and K. M. McKenny, *Anal. Chem. 36*, 529 (1964).

1432. L. Erdey and I. Buzas, *Acta Chim. Acad. Sci. Hung. 6*, 77, 93, 115, 123 (1955); *Chem. Abstr. 49*, 13012a.

1433. L. Edey and L. Buzas, *Magyar Tudomanyos Akad. Kem. Tudomanyok Osztalaynak Kozlemenyei 5*, 279, 293, 313, 321, 325 (1954); *Chem. Abstr. 49*, 9431d.

1434. U. Fritsche, U. Mihm, and A. Koenig, *Mikrochim. Acta, II*, 85 (1980).

1435. W. M. Hardy, W. R. Seitz, and D. M. Hercules, *Talanta 24*, 297 (1977).

1436. N. M. Lukovskaya and L. V. Markova, *Zh. Anal. Khim. 34*, 1893 (1969).

1437. L. Erdey, I. Buzas, and L. Polos, *Z. Anal. Chem. 169*, 187 (1959).

1438. F. Kenny and R. B. Kurtz, *Anal. Chem. 22*, 693 (1950).

1439. L. Erdey, I. Buzas, and L. Polos, *Z. Anal. Chem. 169*, 263 (1959).

1440. I. Buzas and L. Erdey, *Talanta, 10*, 467 (1963).

1441. F. F. Grigorenko, L. I. Dubovenko, and G. I. Kovalenko, *Zavod. Lab.* *39*, 133 (1973); *Chem. Abstr. 78*, 154553e.
1442. P. Szarvas, I. Korondan, and I. Raisz, *Magy. Kem. Foly 72*, 441 (1967); *Anal. Abstr. 15*, 697.
1443. L. Erdey and I. Buzas, *Anal. Chim. Acta 22*, 524 (1960).
1444. L. Erdey, O. Weber, and I. Buzas, *Talanta 17*, 1221 (1970).

CHAPTER

5

BIOINORGANIC LUMINESCENCE SPECTROSCOPY

HARRY G. BRITTAIN

Department of Chemistry
Seton Hall University
South Orange, NJ 07079

5.1. INTRODUCTION

Studies of the intrinsic fluorescence associated with biomolecular systems have provided a great deal of information regarding the rotational motion, solvent reorientation, complex formation, and atom transfer processes associated with the biochemical systems of interest (1). Both steady-state and time-resolved luminescence methods are of great utility in the study of the immediate environment of the fluorophore, and extrapolation of the information obtained on the emitting chromophore can provide important details regarding the structure and function of a wide variety of biopolymers. The intrinsic fluorescence of proteins and enzymes originates from the aromatic residues of tyrosine and tryptophan, but the difficulty in specifying the exact position and location of these residues within a folded biopolymer can create problems in providing general theories of intrinsic protein fluorescence. Furthermore, it has been noted that fluorophores are often not located in positions where one would like them to be (2).

As a result of these difficulties, a number of workers have labeled biopolymers with a variety of extrinsic luminescent molecules of known emissive properties in an attempt to overcome the problems associated with the intrinsic fluorescence. Edelman and McClure have defined these fluorescence probes as "small molecules which undergo changes in one or more of their fluorescence properties as a result of noncovalent interaction with a protein or other macromolecule" (3) and have noted that one may then study sites of particular interest if these can be labeled with a suitable probe molecule. Excellent reviews exist on the use of organic reagents as fluorescence probes for systems of biomolecular interest (4,5).

The use of luminescent metal ions as luminescence probes has received far less attention than has the use of organic fluorescence probes, with this lack of activity being due to the unfortunate situation that most metal ions simply do not emit strongly in fluid solution at room temperature. The most notable exceptions to this general trend are the lanthanide ions and the uranyl ion. However, at cryogenic temperatures quite a few metal ions are found to emit with strong intensities, and the study of biomolecular luminescence at low temperatures could prove to be an area of future interest.

In spite of these limitations, a large number of studies have been carried out in which the probe luminescence originates from the excited state of a metal ion. Essentially all of these works have involved the substitution of a lanthanide ion into the biomolecular system; as a result, the bulk of the review is concerned with this particular area of chemistry. However, it will prove useful to outline general features of transition metal luminescence in the hope of defining and suggesting work that can and should be performed.

5.2. LIMITATIONS AND ADVANTAGES TO THE USE OF LUMINESCENT METAL IONS AS PROBES OF BIOMOLECULAR SYSTEMS

It has become well known that often the tertiary structure of a biopolymer can be partially determined by the presence of metal ions bound at certain well-defined sites along the polymer chain. Many times these can be removed without causing collapse of the biopolymer structure, and in these situations the metal ion usually can be reintroduced without altering the function of the polymer. In other systems, removal of the metal irreversibly destroys the tertiary structure of the material and leads to a total loss of its function. The number of metalloproteins currently known numbers into the hundreds, with new examples being added continually.

Naturally, the structure of these metal sites is of great interest to the bioinorganic chemist, and great efforts are being expended to learn more about the role of essential metal elements in biochemistry. Several of these (Fe, Mn, and Cu, for example) possess chromophores that are suitable for spectral investigations and may thus be studied directly in a variety of biomolecular and model systems. Other metal ions (K, Na, Mg, Ca, and Zn) do not possess suitable chromophores that permit direct study of the biochemistry associated with these ions. Obtaining data on these spectroscopically inert metal ions is of great importance (consider, for example, the regulatory functions of Ca), and progress has been made in understanding the functions of these metals (and of others as well) by replacing them with other metal ions that are capable of binding at the same sites without causing large perturbations on the existing biopolymer structure.

The introduction of a metal ion containing unpaired electrons into a biopolymer site permits one to carry out a variety of spectroscopic studies in which data are obtained pertaining to the immediate environment of the metal ion. Most of the work carried out to date has been concerned with absorption in the visible region (including studies of circular dichroism) or with magnetic resonance studies (both NMR and EPR). With the exception of the lanthanide ions, little work has been concerned with the use of luminescence spectroscopy in the study of metal sites. This situation has arisen as a result of the difficulties already mentioned, and the lack of suitable luminescence probes for metal ion sites in biomolecular systems must be reviewed as the most serious limitation to the application of emission spectroscopy to bioinorganic chemistry.

In spite of the aforementioned limitation, it still is highly desirable to use the existing luminescent metal ions as probes for metal sites since the information available from such studies can be very site-specific. In addition, one is also able to use the emission analog of circular dichroism, circularly polarized luminescence (CPL) spectroscopy, to obtain chiroptical information within the emission spectra. With the lanthanide ions, one can sensitize the emission by irradiation of nearby aromatic residues in the biopolymer chain and thus differentiate between bound and free metal ions. Another useful technique involves the measurement of distances between bound metal ions by measuring the efficiencies of energy transfer between the metal ions, thus effectively making use of a "spectroscopic ruler."

Other advantages to the use of luminescent metal ion probes could be listed, but these will become apparent in the work that is described in the following sections. No effort is made to cover absolutely every study that has been made; instead selected investigations that feature a particular technique or area of bioinorganic chemistry are discussed. As a result, this review is intended to be representative rather than exhaustive in its scope. Whenever possible (especially in the case of luminescent transition probes) future directions for work are suggested.

5.3.   AVAILABLE EXPERIMENTAL METHODS

The techniques available to the bioinorganic chemist who wishes to investigate the photoluminescence of metal ions bound to biopolymer systems must naturally fall into either the steady-state or time-resolved category. It is far beyond the scope of the present review to detail the full range and capabilities of these areas of spectroscopy, but it is useful to touch on several techniques and methods suited for the study of inorganic photoluminescence in biomolecular systems.

Undoubtedly, the two most useful measurements that can be made via steady-state luminescence are determinations of emission spectral characteristics and emission intensities. For metal ions, the luminescence is usually associated with a spin-forbidden transition (when, indeed, the spin quantum number can be well defined in the absence of strong spin-orbit coupling), with the observed lifetimes being in the range that defines phosphorescence. This situation can be extremely fortunate, as ligand fluorescence (from aromatic residues in the biopolymer) generally is located at much higher energy than is the metal ion emission. As a result, one can usually differentiate the two sources of emission merely on the basis of wavelength, and there is normally no necessity for using a phosphorimeter to separate the differing emissions. In addition, emission from metal ions can be quite narrow in bandshape [e.g., Cr(III) and the lanthanide elements], and this distinction relative to the broad-band fluorescence of the aromatic residues within the biopolymer serves as another means to identify the differing emission bands.

In oriented samples, single crystals, or solutes dissolved in viscous media, polarization within the luminescence spectra can contain extremely valuable information regarding transition moments of the emitting species. The technique of fluorescence depolarization is perhaps the most direct method available for the study of structural mobility in proteins [(6); also see Ref. 7 for an excellent review.] Chiral luminescent molecules will spontaneously emit either left- or right-circularly polarized light to a greater extent on irradiation with unpolarized light, and this phenomenon of circularly polarized luminescence (CPL, which is the emission analog of circular dichroism) has recently received a great deal of attention (8,9). This spectroscopic technique is reviewed in Chapter 6 of the present text. Another chiroptical technique available for the study of luminescent molecules is that of fluorescence detected circular dichroism (FDCD), and in this method the sample is irradiated with circularly polarized light and variations in luminescence intensities are used to obtain an excitation spectrum for the differential absorbance (10). While no measurements of FDCD have yet been performed on luminescent metal ions in biomolecular systems, a substantial amount of work has already appeared regarding the CPL spectra of Tb(III) in proteins and enzymes (9,11).

Generally, the emission spectrum of a luminescent molecule is expected to be a time-dependent property if any one of the following pathways can be followed after excitation:

1. Photochemical reaction within the excited molecular system
2. Nonradiative energy transfer to suitable quencher molecules
3. Formation of excited dimers (eximers or exiplexes).
4. Transformation of molecular electronic states.
5. Relaxation of the solvent shell surrounding the luminescent chromophore.

Study of these processes falls under the general category of time-resolved techniques, and an excellent review of the field as applied to the study of intrinsic and extrinsic polymer luminescence has appeared (12). These authors have subdivided the experimental techniques as pulsed sampling methods, phase modulation methods, time-correlated single-photon counting methods, time-resolved fluorescence spectra, and time-resolved polarization methods.

Extensive use of luminescence lifetimes (obtained by either pulsed excitation or phase modulation techniques) has been a feature of inorganic photoluminescence studies, but comparatively little work has appeared in the study of inorganic biopolymers. Nevertheless, this situation can be expected to change as workers in the field become more interested in using the bound metal ion luminescence probes as means to study polymer structure and mobility in the same manner as intrinsic and extrinsic fluorescence of aromatic fluorophores has been employed in the past.

5.4. RESULTS OBTAINED ON d-BLOCK COMPOUNDS

The binding of heavy metals by proteins and enzymes is often only considered in context of their toxic physiological effects (13), but it is recognized that several metal ions play important roles in biological processes. It is also known that the electronic structures of these same ions are such that rapid radiationless loss of absorbed excitation energy takes place and no photoluminescence is usually observed (14). In addition, the excited states of transition ions are often observed to be quite photoreactive (15), and the implications for bioinorganic systems of this pathway out of the excited state have been discussed (16). Photoluminescence can be obtained for certain metal ions, although cryogenic temperatures are often required and the emission is found to occur in the red or near-infrared region of the spectrum. To our knowledge, no luminescence has been measured that originates from a transition metal ion bound to a biopolymer. However, in the following sections we briefly discuss possible candidates for such studies and a few investigations where suitable metal ions have been employed (although no luminescence results were obtained).

5.4.1. Luminescent Transition Metal Ions

The types of emission associated with transition metal ions can best be described as to the nature of the excited state yielding and observed emission. Such a classification has been provided by Crosby (14), and it is worthwhile to review the characteristics of these excited states at this time. We shall not attempt to include any spin labels in the discussion, as spin-orbit coupling renders this particular quantum number irrelevant for heavy atoms.

If the average ligand-field strength is small and the oxidation potential of the metal is sufficiently high, then the lowest excited state will be predominantly dd in character. The emission from such a level is generally broad and featureless and only rarely is observed at room temperature. However, if one examines metal ions having 4d or 5d configurations that can be oxidized easily, then one finds that the lowest excited state is essentially $d\pi^*$ in character. The best example exhibiting this behavior is tris(bipyridyl) ruthenium(II). With this type of excited state, emission can often be quite intense, even in fluid solution at room temperature. On cooling to cyrogenic temperatures, vibrational progressions usually can be observed in the emission bands. Finally, when one considers metal ions possessing full electron shells, the emitting state is essentially π,π^* in nature. Here the emission tends to resemble greatly the spectra obtained for the uncoordinated ligands. However, should the crystal field stabilization energy be very large and the metal ion difficult to oxidize, the π,π^* state of a paramagnetic ion can become the lowest-energy excited state [as is the case for Rh(III) complexes of bipyridine].

One can see from the preceding classification that given the varying electronic properties of transition metal ions, a wide variety of emissive processes may exist and these spectroscopic transitions would be of use in the study of bioinorganic chemistry. Fleischauer (17) has provided an excellent review of the photoluminescence of transition metal coordination compounds and has arranged the known luminescent species by electron configuration. Luminescence has been noted out of the d^0, d^2, d^3, d^5, d^6, and d^8 electron configurations, but a quick examination of the literature reveals that considerable work has been carried out on metal ions having the d^3 and d^6 configuration. Less work has been carried out with Pt(II) (which possesses the d^8 configuration), but the excited states of several Pt(II) complexes have been well detailed.

In the following section, we describe a few of the studies that have used metal ions that would be suitable for luminescence studies. In essentially every case, the luminescence work has not been performed, but it is our intent to point out how inorganic photoluminescence (even if it requires cyrogenic temperatures) could aid in the investigation of the bioinorganic problem.

5.4.2. Bioinorganic Studies

Following the example of Fleischauer (17), we propose to group the transition metal ion discussions as to the electron configuration of the metal ion.

Closed-Shell Metal Ions

It has been noted that luminescence can be obtained from the solutions of Tl(I), Pb(II), Sn(II), Sb(III), and Cu(I) at room temperature (18), with the emission of Tl(I) receiving the greatest degree of attention (19–21).

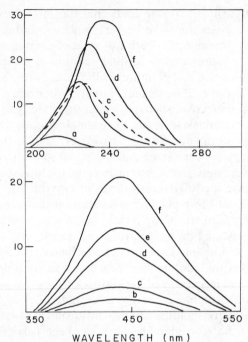

Fig. 5.1. Excitation spectra (upper) and emission spectra (lower) for TlCl solutions (concentration = 0.133 mM) containing various amounts of KCl. The excitation spectra were obtained upon observation of the emission at 430 nm, and the emission spectra employed an excitation wavelength of 250 nm; in both situations, the spectra are shown in arbitrary units. The KCl concentrations were: $a = 0.0\ M,\ b = 0.5\ M,\ C = 1.0\ M,\ d = 2.0\ M,\ e = 3.0\ M$, and $f = 4.0\ M$. (Spectra from Fig. 1 of Ref. 20, with permission of the copyright owner.)

The ionic radius of Tl(I) is found to be only slightly larger than that of K(I), and one may conclude that partial replacement of the spectroscopically inert K(I) by Tl(I) could permit studies of K(I) transport. As illustrated in Figure 5.1, the excitation and emission maxima of Tl(I) complexes are found to depend strongly on the nature of the species existing in solution (20), but in general Tl(I) solutions may be excited efficiently at 250 nm and the luminescence observed at 450 nm. The emission lifetime is also found to vary with the nature of the counter-ion associated with the Tl(I) ion and can range from 0.4 μsec in pure H$_2$O to 0.6 μsec in 2 M NaCl (21). One must, of course, remember the extremely toxic nature of this metal and this effect on any system under study.

Ions Having the d^3 Electron Configuration

While quite a few transition metal ions having a d^3 electron configuration have been studied by luminescence techniques, it is certainly true that coordination compounds containing Cr(III) have received the greatest degree of attention. In addition, it has been noted that more photochemical studies have been carried out with Cr(III) than with any other metal (22). In the solid state, emission has been observed from the ^2E and the $^4T_2$ excited states, and laser emission has been obtained within the ruby system. With Cr(III) complexes measurements have been obtained in the solid state and in solution phase, with all possible temperature ranges being employed to obtain the desired data. The circularly polarized luminescence spectrum of resolved tris(ethylenediamine)Cr(III) has been obtained at room temperature in water–ethylene glycol solvent (23).

The Cr(III) ion is also known to be kinetically inert, and this property has been used to advantage by several groups in the study of metal–ATP complexes. The ability of substitution-inert complexes of ATP to permit a study of the sites on ATP and related enzymes has shed valuable light on the role of the activating metal ions involved in various biochemical processes. Cleland and co-workers have reported the preparation and properties of a variety of Cr(III)–ATP and Cr(III)–ADP complexes (24) and have also examined the interaction of these complexes with several kinase enzymes (25,26). More recently, Grisham and co-workers have detailed the interaction between $Cr(H_2O)_4ATP$ and several ATPase enzymes (27). However, the luminescence properties of the Cr(III) ion have not been exploited except for the work of Dooley (28). While no Cr(III)–ATP complex was found to emit at room temperature, at 77K Cr(NH$_3$)$_3$–ATP and CR(NH$_3$)$_4$–ATP exhibit luminescence from the ^2E level and $Cr(H_2O)_4$–ATP emits from the $^4T_2$ level (28). Clearly, further work should be carried out to develop the Cr(III)–nucleotide complexes as probes of the active sites in enzymes.

When doped into transparent metal oxide host systems, V(II) and Mn(IV) can exhibit luminescence under suitable conditions (17). V(II) is of no interest to the bioinorganic chemist, and the only significant Mn(IV) compound prepared has been that of Mn(IV)–hematoprophyrin (29). Mo(III) has been found to be luminescent in a few complexes (31), and this property of Mo(III) may prove eventually to be of use in the study of the various oxidation states associated with the redox processes associated with Mo enzymes and related proteins. The most stable oxidation states of this element are known to be Mo(V) and Mo(VI), but the Mo(III) cónfiguration can be reached under extreme circumstances.

Ions Having the d^6 Electron Configuration

While early photoluminescence work concentrated on Cr(III), subsequent studies have revealed that more luminescent transition metal ions have the d^6 configuration than any other. Luminescence has been detected in complexes containing Fe(II), Ru(II), Os(II), Co(III), Rh(III), Ir(III), and Pt(IV), and in essentially every case the metal ion experiences strong crystal fields leading to a diamagnetic complex (17). Current interest in photovoltaic harnessing of sunlight has led to extensive efforts employing Ru(II) complexes as donors in excited-state redox processes. However, the vast majority of these luminescent complexes requires the presence of nitrogen-coordinated heterocyclic ligands in order for luminescence to occur.

While the luminescence requirments for d^6 metal ions may preclude the use of these ions as bioinorganic probes, one should remember that low-spin d^6 transition metal ions are also kinetically inert, and these may be used with success as diamagnetic substitution inert complexes. The preparation of and reactions associated with Co(III)–ATP complexes have been reported along with the Cr(III) work (24,25), and the interaction of $Co(NH_3)_4$–ATP with several ATPase enzymes has also been investigated (27). Clearly, the use of other d^6 metal ions as metal ion probes is called for, and workers should attempt emission studies to learn under what conditions luminescence might occur and how it could be used to obtain more information about the role of the activating metal ions.

Ions Having the d^8 Electron Configuration

Ni(II) ions doped into MgO are known to emit both in the visible (31) and in the infrared (32) regions of the electromagnetic spectrum, but no solution phase luminescence has yet been obtained with this ion. Essentially all the photoluminescence work associated with the d^8 configuration has been carried out on Pt(II), and a considerable amount of work has been centered on $Pt(CN)_4^{2-}$ (17). This particular complex has also been observed to emit in aqueous solution, either in the liquid or frozen phase. Luminescence has also been obtained in Pt(II) complexes containing thiourea, dipyridine, phenanthroline, and glycine (17).

Of perhaps greatest interest to the bioinorganic chemist is the observation that the $PtCl_4^{2-}$ ion will emit under suitable conditions (33), as shown in Fig. 5.2. This information is of great importance when one considers that cis-$PtCl_2(NH_3)_2$ has antitumor properties (34), and this material is currently being used to treat a variety of forms of cancer. The use of Pt(II) complexes as probes of polynucleotide structure and antitumor

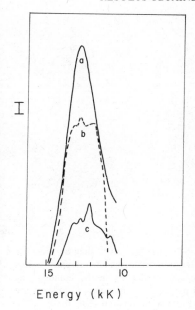

Fig. 5.2. Luminescence spectra of $PtCl_4^{2-}$ at 77°K. (a) Emission from powdered material, and essentially that of emission obtained from a single crystal in the $\perp$ polarization, (b) spectrophotographic spectrum of the powdered sample, and (c) single-crystal emission, but polarized in the $\parallel$ direction. (Data from Fig. 1 of Ref. 33, with permission of the copyright owner.)

Energy (kK)

drugs has been reviewed by Lippard (35). While no reports have yet appeared relating to the intrinsic luminescence of the Pt(II) ion in these complexes, quenching of ethidium bromide fluorescence (after being bound to calf thymus DNA) by Pt(II) reagents has been used to study the intercalation of the dye molecules (36). It is clear from the published work that the Pt(II) complexes interact with DNA by intercalative binding, and one would anticipate that studies of intrinsic Pt(II) luminescence could aid in these studies. Chiroptical studies (e.g., CPL spectroscopy) should prove even more useful to examine how the Pt(II) complexes affect the helical backbone of the DNA polymer after intercalation occurs.

5.5.   RESULTS OBTAINED ON F-BLOCK COMPOUNDS

The bioinorganic chemistry of lanthanide ions has assumed a great deal of importance since the report by Darnall and Birnbaum (37) that these metal ions could be used as probes of electrostatic binding sites in proteins. With most of the lanthanide ions being paramagnetic because of the presence of unfilled shells of f electrons, it is not surprising that Pr(III), Eu(III), and Gd(III) have been used as paramagnetic shift reagents for nmr spectroscopic studies of biopolymers (38,39). Perhaps the most useful application of lanthanide ions has been in the study of calcium binding

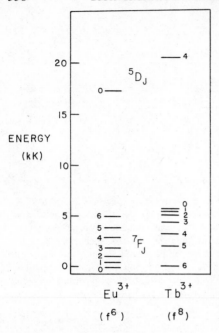

Fig. 5.3. Energy level diagram for the Eu(III) and Tb(III) free ions. Because of low degrees of covalency in the bonding, the band positions in analogous complexes are found at values very close to those of the free ions.

proteins (11), since this spectroscopically inert ion is exceedingly important in biochemistry and quite difficult to study directly (40).

Lanthanide complexes of biopolymers and model systems have been studied by optical absorption and emission methods, NMR and EPR spectroscopies, Mossbauer spectroscopy, and x-ray crystallography (41,42). Given the interest that has existed in all aspects of lanthanide ion emission, much activity has centered about the use of luminescence techniques as a means of gathering information about the role of these ions in the biopolymers (11,41–43). Quite a bit of work has also been carried out on lanthanide complexes that serve as model compounds for the more complicated biopolymer systems, and we shall focus on these studies before discussing the work carried out on biomolecular assemblies. To aid the discussions of the following sections, energy level diagrams for the Tb(III) and Eu(III) ions have been provided in Fig. 5.3.

5.5.1. Lanthanide Complexes of Amino Acids

Lanthanide ions generally display only a small affinity for α-amino acids, since these ligands bind solely at the carboxyl portion of the ligand. However, when the ligand contains two carboxyl groups capable of forming a chelate (as in the case of L-aspartic acid), the formation constants can be substantial (44,45). In spite of the weakness of the complexes, the amino acid complexes of lanthanide ions have received some degree of attention since the paramagnetic members of the series may be used as aqueous shift reagents for the study of proteins. Central to an understanding of the lanthanide–protein complexes is a comprehension of the simpler complexes, and the approaches have been made along several lines.

Potentiometric titrations have proved extremely useful in the determination of stability constants, and these have been tabulated (46,47). Of course, these measurements do not provide stereochemical information but are useful in evaluations of thermodynamic properties. Martin and co-workers (48) have observed that even in the presence of a 10-fold excess of ligand, lanthanide ions titrate 2.4–2.8 equivalents of base per mole of Ln(III) associated with hydrolysis of the metal ion. These workers suggested that this nonintegral portion of the titration curves was due to the formation of polymeric species, with the bridging being effected by hydroxide ions. This polymeric association was studied by Brittain, who used intermolecular energy transfer between Tb(III) donor complexes and Eu(III) acceptor complexes to study the influence of aspartic acid (49), histidine (50), and hydroxyamino acids (51) on the polynuclear association. A very interesting result found during the course of the quenching studies was the existence of stereoselectivity in the energy transfer: complexes prepared from D- or L-ligands exhibited different transfer efficiencies when compared to complexes prepared from D, L-ligands in the high pH region.

In conjunction with efforts to develop the use of lanthanide ions as reagents to aid in the study of protein and polypeptide conformations in solution, a wide variety of nmr studies has been carried out on simple amino acid complexes of lanthanide ions. The applicability of nitrotyrosine residues as potential specific lanthanide ion binding sites was evaluated from studies on the Eu(III), Pr(III), Gd(III), and La(III) complexes with N-acetyl-L-3-nitrotyrosine (52). During the course of this work, several limitations regarding lanthanide-induced shifts were uncovered, but the probe technique was shown to be a valuable conformational aid. General methods for the determination of the conformation of small molecules in solution by means of paramagnetic shifts and relaxation perturbations have been described (53), and specific examples of amino acids and peptides were quoted.

Complexes of Nd(III) with alanine, histidine, threonine, and serine have been studied by NMR methods (54), and during the course of this investigation it was learned that formation constants of the amino acid complexes calculated from both NMR and potentiometric methods agreed quite well. Sherry and Pascual (55) examined the proton- and carbon-lanthanide-induced shifts obtained when L-alanine was complexed to each of the members of the lanthanide series and concluded that structural changes took place across the series. A monodentate coordination geometry was proposed for Pr(III) to Tb(III), while bidentate carboxyl coordination appeared to dominate for Dy(III) to Yb(III). This behavior was thought to parallel known hydration sphere alterations that are known to take place along the lanthanide series. However, such changes were not found in a subsequent work involving sarcosine (56), and a number of explanations for the discrepancy were advanced.

Other spectroscopic techniques have been used with great success to study the solution chemistry of the lanthanide–amino acid complexes. Birnhaum and Darnall have used difference absorption spectroscopy to uncover details in the Nd(III) binding of alanine, histidine, and other ligands (57). Legendziewicz and co-workers have used a combination of NMR, absorption, and luminescence spectroscopies to study a wide variety of lanthanide complexes with L-aspartic acid, L-glutamic acid, L-asparagine, L-alanine, and L-glutamine (58,60).

As would be expected, chiroptical techniques can provide extremely detailed stereochemical information, if the structure of the chiral complex is known with some degree of certainty. Katzin and Gulyas studied the pH dependence of the CD spectra associated with the Pr(III) complexes of alanine, valine, leucine, serine, asparagine, ornithine, lysine, aspartic acid, and glutamic acid (61). However, in this work the presence of 1:1 complexes was assumed even though large excesses of ligand were used, and the possible polynuclear association of the complexes was not even considered. As a result, many incorrect conclusions were reached, such as bidentate coordination by glutamic acid via its two carboxyl groups. The CD of these ligands with a variety of metals was reported by Martin and co-workers (48), who carefully considered all the parameters necessary to comprehend the data.

One approach to overcoming the problem of polynuclear complex association is to use a lower concentration of lanthanide complex for the chiroptical determinations. Because of the low absorptivity of the f–f transitions, CD studies could not be carried out at lower concentrations, but CPL studies of complex chirality were quite possible. Luk and Richardson carried out the first studies involving chiral lanthanide amino acid complexes in aqueous media (62,63). These survey studies demonstrated that CPL spectroscopy was an excellent probe of the solution-phase stereochemistry of Eu(III) and Tb(III) complexes. Most of the work involved studies of the Tb(III) complexes because of the greater quantum

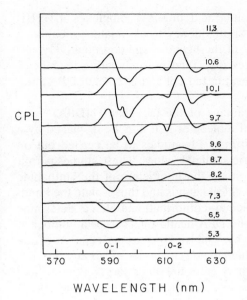

CPL

WAVELENGTH (nm)

Fig. 5.4. CPL spectra for Eu(III)/L-threonine (1:5 mole ratio) solutions in D_2O at various pH values. (Spectra from Fig. 4 of Ref. 64, with permission of the copyright owner.)

yield of emission for this metal ion, although the chemical similarity of lanthanide ions insures that the results would be translatable to most of the members of the series. A 5:1 ratio of ligand–metal was used for much of the work, with the chiral ligands under study being L-malic, L-aspartic, L-glutamic, and L-lactic acids, as well as L-alanine and L-serine.

Subsequently, CPL spectroscopy was used to detail the interaction between Tb(III) and Eu(III) ions and a series of potentially terdentate amino acids (L-aspartic acid, L-serine, L-threonine, and L-histidine) (64). In both of these works, the 5:1 ratio of ligand to metal was used, and it was found that while direct correlations between observed spectra and plausible solution-phase structures were elusive, nonetheless the CPL spectra were extraordinarily sensitive to details of pH. Generally, it was noted that the most intense CPL was found above neutral pH values, as shown in Fig. 5.4.

However, with the observations of Martin et al. (48) indicating the presence of polynuclear species at elevated pH and the energy transfer results of Brittain (73–75) actually proving the presence of these compounds, the observation that most CPL was obtained at high pH values compromised some of the results. For the results to have the greatest utility, one had to obtain information on complexes known to be monomeric in the pH ranges of interest. One ligand system was found in which the complexes were determined to be monomeric at all pH values. Pyridine-2,6-dicarboxylic acid (dipicolinic acid, or DPA) complexes of Tb(III) and Eu(III) apparently displayed no tendency to associate, probably as a result of the extreme bulkiness of the ligands. The coordinatively saturated 1:3 metal–DPA complexes displayed no associative tendencies (65), nor did the coordinatively unsaturated 1:1 and 1:2 metal–DPA complexes (66).

Brittain was able to take advantage of the fact that Tb(DPA)$^+$ and Tb(DPA)$_2^-$ contain bound solvent molecules that may be displaced by a second chiral ligand, thus permitting CPL studies to be carried out on complexes known to be monomeric at all pH values. These studies permit an easy method where one may study the interaction of the lanthanide ion and only one (or at most two) chiral ligands and thus isolate the bonding situations where vicinal or vicinal–conformational effects determine the chirality of the Tb(III) ion. The DPA ligand thus serves a number of purposes: (1) it acts as a means by which polymeric association of the complexes is prevented, even at high pH; (2) it acts as a filler and restricts the number of chiral ligands that may enter the inner coordination sphere of the Tb(III) ion; and (3) it serves as an efficient sensitizer of the Tb(III) emission (the DPA ligand is irradiated, and the absorbed energy is transferred nonradiatively to the metal ion). The very large binding constants of DPA for lanthanide ions (67) ensure that it will not be displaced from the metal coordination sphere.

Brittain studied the CPL spectra obtained in mixed-ligand Tb–DPA complexes containing L-alanine, L-valine, L-leucine, L-isoleucine, L-serine, L-threonine, and L-aspartic acid and in this work was able to see that the CPL spectra associated with chirality due solely to vicinal effects was quite different in sign and magnitude when compared to chirality due to a combination of vicinal and conformational effects (68).

In these works, the CPL spectra obtained within the $^5D_4 \rightarrow {}^7F_5$ Tb(III) transition were contrasted for Tb(DPA)$_2$(L-alanine) and Tb(DPA)$_2$(L-aspartic acid), and one could easily note the difference between the situations where the chiral ligand can only bind in a monodentate fashion (leading to the vicinal effect determining the chirality) as opposed to a bidentate fashion (permitting the conformational effect to contribute). It was also demonstrated that both serine and threonine could bind to the Tb(III) ion in a bidentate manner (68), as the spectra in Fig. 5.5 demonstrate. The distinction between CPL spectra associated with pure vicinal effects versus spectra obtained when vicinal–conformational effects could contribute was subsequently addressed, and a general trend was found to exist for the Tb(DPA) complexes: if the CPL within the $^5D_4 \rightarrow {}^7F_5$ transition is single-signed only, then the vicinal effect is the only contributor to the overall optical activity experienced by the metal ion

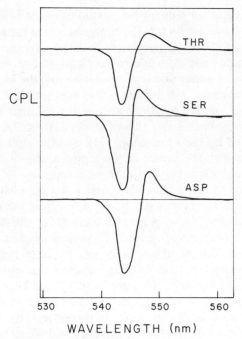

Fig. 5.5.   CPL spectra for Tb(DPA)$_2^-$ complexes with L-threonine, L-serine, and L-aspartic acid. (Data from Figs. 2 and 3 of Ref. 68, with permission of the copyright owner.)

(69). The sign of the CPL associated with this vicinal effect is directly relatable to the absolute configuration of the asymmetric atom in the chiral ligand.

5.5.2.   Lanthanide Complexes of Nucleosides, Nucleotides, and Nucleic Acids

The lanthanide complexes of nucleotides have received extensive attention, since the study of lanthanide-induced shifts and line broadening within nmr spectra can be used to deduce conformations in solution (70). The lanthanide ions are known to bind at the phosphate group (which functions in a bidentate manner) in the case of nucleotide monophosphates (70), and in adenoside triphosphate the binding appears to be with the terminal phosphate groups (71). With nucleoside complexes the bonding undoubtedly exists on the sugar group as the lanthanide ions have little tendency for bonding with the nitrogen donor groups of the base (72).

The utility of the shift reagent technique was demonstrated for cyclic adenosine-3',5'-phosphate (73). Formation of the Pr(III) and Ho(III) complexes at pH 5.3 and subsequent NMR studies of the observed chemical shifts permitted a conclusion that the conformation of the ribose and phosphate groups in solution was consistent with the structure known for the solid state. Williams and co-workers have carried out very detailed and extensive studies of nucleotide conformation using the lanthanide ions as chemical shift regeants. For instance, the conformation of cytidine 5'-monophospate in solution has been shown to be similar to that of adenosine 5'-monophosphate under the same conditions (73). Dinucleotides have been studied by the same group (74), and the shift reagent method was used to establish the nature of the temperature- and pH-dependent conformational equilibria existing in these complexes.

Cleland and co-workers have also carried out detailed studies of lanthanide nucleotide complexes but have approached their investigations from a kinetic point of view. A general method for the determination of the dissociation constants of metal–ATP complexes has been advanced (75), and this kinetic method works for any metal complex that can act as an inhibitory substrate analog for any enzyme that utilizes $MgATP^{2-}$. Since dissociation constant values are available for MgATP under a variety of conditions, the corresponding constants for lanthanide–ATP complexes are easily calculated. The utility of this method has been illustrated in the interaction of Ln(III)–ATP complexes with yeast hexokinase, and dissociation constants were obtained for nine members of the lanthanide series (76).

The utility of lanthanide complexes of ethylenediaminetetraacetic acid as aqueous shift reagents has been demonstrated (77), and the ternary complexes that can be formed with Ln(EDTA) are useful as conformational probes of nucleotide structure in solution. For instance, Williams and co-workers have used the La(III), Pr(III), Eu(III), and Gd(III) complexes of EDTA to study the solution phase conformation of adenosine and cytidine 5'-monophosphates (78) and have shown that the same information is obtained as when the aquo ions are used as the NMR shift reagents. Evidence has been presented that indicates that effective axial symmetry is present in the bidentate chelates formed between lanthanide–EDTA complexes and cytidine 5'-monophosphate (79).

Marzilli and co-workers have examined the lanthanide complexes of nucleosides in detail, using ^{13}C NMR and Raman spectroscopies to study the details of the binding in aqueous and nonaqueous solution (80–82). The interactions of cations and anions with the four common nucleosides (uridine, adenosine, guanosine, and cytidine) were summarized with respect to the favorable and unfavorable interactions, and these workers proposed that all NMR observations on metal–nucleoside interactions could be explained using the binding criteria specified (81). It was also

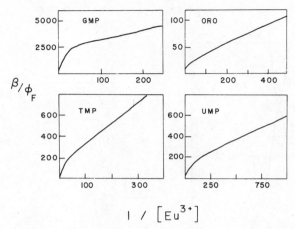

Fig. 5.6. Sensitization of Eu(III) luminescence in D_2O at room temperature by guanine monophosphate (GMP), thymine monophosphate (TMP), uridine monophosphate (UMP), and orotic acid (ORO). β is the probability that the Eu(III) ion will be de-excited radiatively, and ϕ_F is the quantum yield for the luminescence process. (Data from Fig. 1 of Ref. 83, with permission of the copyright owner.

determined that the nature of the metal–nucleoside interaction can change substantially when the solution medium was made more basic (82).

The excited states of these compounds have received much less attention, since neither DNA nor any of its constituent nucleotides exhibits easily observable emission in fluid solution at room temperature. However, as illustrated in Fig. 5.6, the Eu(III) ion luminescence may be sensitized after being bound to nucleotides (83), and this property has been used to predict excited-state properties and energy transfer characteristics of polynucleotides in aqueous solution. The emission of Tb(III) may also be sensitized, and a fluorimetric method was used to determine the binding constant of Tb(III) with nucleotide monophosphates (84).

No CD investigations of the lanthanide complexes of nucleotides or nucleosides have yet been reported, but Richardson and co-workers have reported CPL studies of Tb(III) complexes with nucleosides in aqueous solution (72). Not all nucleosides were found to induce CPL in the Tb(III) emission, but three systems were found that yielded strong CPL: uridine, cytidine, and inosine. Common to all three systems was the presence of two donor hydroxyl groups on the ribosyl group and a carbonyl group on the base with suitable conformations that permit terdentate bonding with the lanthanide ion. Further CPL studies have been carried out in both aqueous and nonaqueous solution by the same group, with the Tb(III)

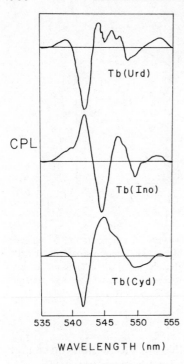

Fig. 5.7. CPL spectra associated with the $^5D_4 \rightarrow$ $^7F_5$ emission band system of 1 : 2 Tb(III)–nucleoside complexes at pH 6.8. Results are shown for uridine (Urd), inosine (Ino), and cytidine (Cyd). (Data from Fig. 1 of Ref. 85, with permission of the copyright owner.)

and Eu(III) complexes of nucleosides (85) and nucleotides (86) being studied. Examples of the CPL spectra are found in Fig. 5.7.

The luminescence of Tb(III) and Eu(III) has been shown to be dramatically sensitized when bound to nucleic acids. Mushrush and co-workers have examined the use of Tb(III) as a luminescent probe for DNA and chromatin (87) and were able to determine that while one equivalent of Tb(III) could be bound per phosphate group in calf thymus DNA, only 0.48 equivalents of Tb(III) could be bound by chromatin. From these data it was concluded that 52% of the phosphate groups in chromatin were unavailable for lanthanide ion bonding. These workers have also developed several other methods that use Tb(III) binding as a means to study the properties of DNA and RNA (88).

Enhancement of Tb(III) and Eu(III) luminescence has been used to study the transfer RNA isolated from *Escherichia coli*. This particular nuclei acid is unusual in that it contains a quantity of 4-thiouridine, and the excitation spectrum of Tb(III)–tRNA is identical with that of this uncommon base (89). This emission enhancement is observed with tRNA[Phe], tRNA[Glu], and tRNA[fMet] isolated from *E. coli*, but not with yeast tRNA^{Phe} (which does not contain 4-thiouridine). Wolfson and Kearns (90) have carried out further investigations on tRNA isolated from *E. coli* and have found that binding of the first three to four Eu(III) ions is independent and sequential, and the binding of these is approximately 600 times stronger than the binding of Mg(II). In addition, the binding sites are believed to be located near the 4-thiouridine residue found at position 8 in a number of *E. coli* tRNA molecules.

The binding of lanthanide ions by yeast tRNA has also been examined by nmr techniques. It was found that the binding of Eu(III) by yeast tRNA^{Phe} shifts several resonances that have been assigned as being due to protons on the ring nitrogens of the base units (91). The charges in the nmr spectra as the tRNA is titrated with Eu(III) indicate that four or five Eu(III) ions are tightly bound, that the metal binding is in the fast exchange limit on the nmr time scale, and that the binding to different sites is sequential rather than cooperative.

Tb(III) has been shown to bind to ribosomes and ribosomal RNA (92), and evidence was presented that indicated that the Tb(III) bound to the ribosomes primarily through tRNA interactions. Subsequent studies on DNA and RNA employing Tb(III) as a luminescent probe revealed that only guanine-containing nucleotides (Guo-5'-P and dGuo-5'-P) enhanced Tb(III) emission significantly (93). The guanine moiety in the polymers was found to be much more efficient in its ability to enhance Tb(III) luminescence relative to the guanine moiety as a free nucleotide.

The binding of Tb(III) in nucleic acid complexes appears to be a specific probe that can provide information regarding the structure of these materials. Unpaired residues in nucleic acids produced from a break in the hydrogen bonding appear to be particularly able to sensitize Tb(III) luminscence (94). Base-paired residues of nucleic acids induce no emission enhancement. This feature of lanthanide ion binding has led to the development of a method whereby Tb(III) emission enhancement may be used to assess the single-strand content of DNA (95,96).

5.5.3. Lanthanide Complexes of Proteins and Enzymes

Generally, the lanthanide ions bind more strongly to proteins and enzymes than does Ca(II), an observation that is undoubtedly due to the higher positive charge of the Ln(III) ions (since the ionic radii are quite comparable). Martin and Richardson (11) have provided a convenient classification for lanthanide–protein interactions: (1) proteins inhibited by Ln(III) substitution, (2) proteins whose action is largely unaffected by substitution of Ln(III) for Ca(II), (3) specific interactions of Ln(III) substitution for ions other than Ca(II) or Mg(II). While this method of classification is extremely useful, we shall adopt a slightly different mode and will classify the luminescence work as to protein type instead. Consistent with the purpose of this review, we shall not be all-inclusive in the scope of work covered, but will instead focus on several interesting classes of Ln(III)-substituted proteins and comments on how luminescence techniques have been able to provide useful information on the metal ion binding sites.

Of all the lanthanide ions, Tb(III) has received the greatest use as a fluorimetric probe of Ca(II) binding sites in proteins. In a comprehensive study, Brittain et al. studied the addition of Tb(III) to 40 proteins and observed characteristic luminescence in 36 of the systems (97). The emission in these proteins is obtained by irradiating the aromatic residues present in the protein chain (tyrosine, tryptophan, or phenylalanine) and then taking advantage of the fact that this absorbed electronic energy can be transferred effectively to the Tb(III) ion. With Tb(III), f–f states alone are involved in the energy reception, but with Eu(III) a charge transfer transition interferes with the energy transfer (98) and as a result the Eu(III) ion cannot exhibit sensitized emission. This sensitization of Tb(III) luminescence is a very useful property of these systems, as unbound Tb(III) exhibits a much lower quantum yield for emission. At micromolar Tb(III) concentrations, only the metal ions actually bound to the protein and receiving the nonradiative energy transfer can be observed in the emission spectrum.

Martin and co-workers have examined the energy transfer process by recording the excitation spectrum of Tb(III) bound in a wide variety of biomolecular systems and have attempted to determine what ligand transitions are most suitable for sensitization of the Tb(III) emission (99). In particular, it is highly desirable to determine whether the transferred energy originated with a phenylalanine, tyrosine (nonionized), tyrosine (ionized), or tryptophan residue, as this information will indicate which residue is closest to the Tb(III) ion and thus indicate its approximate position. When present, tryptophan appears to be most efficient at sensitizing Tb(III) emission as a result of favorable overlapping between Tb(III) absorption bands and the region of tryptophan fluorescence.

Horrocks and co-workers have introduced a series of methods for the study of Ln(III)-substituted biopolymers that involve direct excitation of metal ion levels. Since the Ln(III) ions are excited directly and not sensitized through an energy transfer process, the emission of Eu(III) within biomolecular systems can also be examined. Using pulsed dye laser sources as excitation sources and variation of the solvent composition (from H_2O to D_2O), a method has been developed whereby one is able to determine the number of water molecules bound to the Ln(III) in a complex (100,101). CW excitation (via an Ar^+ laser) of Tb(III)-substituted protein complexes has also been used to obtain details of the interactions, and in this method one monitors the fine structure within the Tb(III) emission bands (102). Finally, energy transfer between lanthanide ions has been used to determine distances between these metal ions in a biopolymer capable of multiple metal ion binding (102).

Ln(III) Substitution in Muscle Proteins

The white muscles of fishes and amphibians contain significant amounts of a class of light proteins (molecule weights range between 9000 and 13,000) that have been labeled as parvalbumins (103,104). These proteins contain two Ca(II) ions, both of which may be replaced by Tb(III), and the Tb(III) emission may be sensitized via the phenylalanine residues (105). The Ca(II) removal and the Tb(III) substitution have been shown to be noncooperative in nature, and circular dichroism studies of native and Tb(III)-substituted carp parvalbumins have shown that no large conformational changes take place on lanthanide substitution (106). A number of well-resolved resonance lines can be observed in ^{13}C NMR spectra of parvalbumin proteins, and substitution of lanthanide ions for Ca(II) has been used to determine the origin of these lines (107). Such studies are of great value, since it appears that the parvalbumin proteins (because of their ease of isolation and purification and their considerable natural abundance) constitute a convenient model system for the exploration of structure–function interrelationships in Ca(II)-binding proteins.

The crystal structure of carp parvalbumin B has been determined (108) and is found to contain six helical regions of similar length (labeled A–F). Loops occur between the helical regions, and the Ca(II) ions are bound by aspartate and glutamate residues within the CD and EF helical regions. The Ca(II) coordinated in the EF site can be displaced at low Tb(III) concentrations, while the Ca(II) at the CD site is replaced only at higher Tb(III) concentrations (109). The suggestion of a third site (110) has been made from analysis of Tb(III) self-quenching data.

Lee and Skykes (III) have evaluated metal–proton distances in the EF site of carp parvalbumin using the susceptibility contribution to the line broadening of lanthanide ion-shifted resonances and have attempted to describe the structure of the EF metal ion binding site in solution. Horrocks and Collier have determined distances from tryptophan energy donors to lanthanide ion energy acceptors from quenching studies and have obtained results in good agreement with the x-ray structural studies (98). Spectroscopic evidence for the third Tb(III) site has been presented (112), and the distance between metal ion sites has been evaluated via the in-

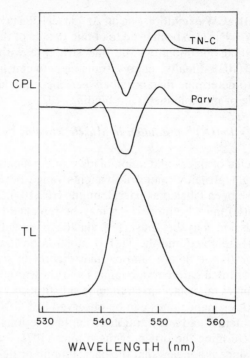

Fig. 5.8. CPL spectra obtained for Tb(III) substituted carp parvalbumin (Parv) and bovine cardiac TN-C protein systems. The spectra illustrated were obtained within the $^5D_4 \rightarrow {}^7F_5$ transitions.

ternal energy transfer technique. The circularly polarized luminescence of Tb(III)-substituted carp parvalbumin has been reported (105,113) and will be discussed in conjunction with other muscle protein studies.

In mammalian skeletal muscle the analog of parvalbumin is troponin. The Ca(II)-binding subunit is troponin C (TN-C), and this protein contains four Ca(II) binding sites (114). Titrations indicate that Tb(III), like Ca(II), binds to the two high-affinity sites at low concentrations of lanthanide ion (115), and that the Tb(III) emission may be sensitized by the tyrosine residue of TN-C. From lifetimes of laser-induced luminescence in H_2O and D_2O, the number of water molecules coordinated to bound lanthanide ions was two at the high-affinity sites and three at the lower-affinity sites (116).

Quite interesting results were obtained by examining the circularly polarized luminescence spectrum of Tb(III)-substituted TN-C (117). As illustrated in Fig. 5.8, the CPL spectra of Tb(III)-substituted carp parvalbumin and bovine cardiac TN-C were essentially identical in lineshape and magnitude, indicating an extremely close correspondence in stereochemistry about the emitting Tb(III) ion. Examination of the known amino acid sequences revealed a consistent pattern of homologies associated with the most easily substituted Ca(II) site, and in each case an aromatic residue appears to overlie the metal ion site. Conformational changes became evident on addition of other troponin subunits, as binding of TN-I and TN-T weakens the total emission and completely quenches the circularly polarized luminescence.

Calmodulin is a multifunctional regulator, known to bind four Ca(II) ions, whose mechanisms are not yet totally understood (118,119). The Ca(II) ions in calmodulin can be replaced by Tb(III) ions, and it has been found that Tb(III) is capable of activating the protein (120). The relative binding strength of the four binding sites has been established for calmodulin proteins isolated from various sources as a result of Tb(III) emission studies, and it is now concluded that sites I and II exhibit the highest affinity for Tb(III) and that sites III and IV are of lower affinity (121–123). This behavior may be contrasted with that of troponin-C, in which the high-affinity sites are III and IV. The role of calmodulin's tyrosine residue in sensitizing Tb(III) luminescence has also been discussed as part of a general study of the Ca(II) binding domains of calmodulin (124).

A very real danger associated with the use of lanthanide ions as probes for Ca(II) binding sites has been pointed out by Kay (125). The effects of La(III) and Tb(III) on the ATPase activity of the myosin S-1 enzyme system alone and in combination with F-actin have been evaluated, and it was found that these metal ions had a pronounced inhibitory effect on the enzymatic reaction. The authors suggested that any study of Tb(III) substitution for Ca(II) in a biomolecular system be accompanied by some sort of bioassay to ascertain the degree of preservation of the intact, native, active structure. Any modification of the native structure should be considered in application of the results to wider generalizations or to explanations for Ca(II) ion-mediated behavior. Such advice should be heeded by all who are working in the field.

Ln(III)-Substituted Proteolytic Enzymes

Bovine trypsin has been shown to bind one Ca(II) ion, but the presence of this ion in the biopolymer apparently does not alter its catalytic activity toward hydrolysis of esters and peptides containing lysine or arginine residues (126). However, the binding of the Ca(II) ion does serve to retard denaturation of the protein and also inhibits autocatalytic degradation of the enzyme (127). Commercially available trypsin contains a mixture of

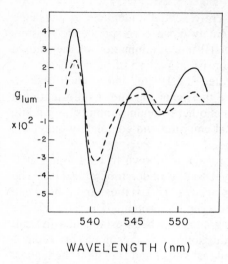

Fig. 5.9. CPL spectra obtained for Tb(III) substituted trypsin of porcine (– – –) and bovine (——) origin. The solutions were buffered at pH 6.3 in 0.2 *M* Pipes, and the dissymmetry factors were computed point by point from separate measurements of the TL and CPL. (Data from Fig. 9 of Ref. 131, with permission of the copyright owner.)

proteins, designated as α-trypsin, β-trypsin, and "inert proteins," and the substitution of lanthanide ions for Ca(II) has enabled the location of the metal ion binding sites in both α-trypsin and β-trypsin (128). Ca(II) ions are also known to accelerate the conversion of trypsinogen to trypsin, and this process may also be catalyzed by lanthanide ions (129).

Measurements of Tb(III) excitation spectra have shown that the sensitized Tb(III) luminescence arises after energy transfer from one or more tryptophan residues located near the bound metal ion (130). Commercially obtained trypsin was found to bind Tb(III), but no circularly polarized emission could be observed within the luminescence spectrum (97). However, separation of trypsin into its component fractions and formation of the Tb(III) complexes of these did enable the performance of chiroptical studies on the metal ion sites (131) (see Fig. 5.9). Martin and co-workers have determined the binding strengths for the addition of Tb(III) to porcine and bovine trypsins and have used luminescence and excitation spectra to study the Ca(II) sites in these proteins (132). In addition, these workers have also obtained the binding constants of Ca(II) to α-chymotrypsin, bovine and porcine trypsin, and elastase as 1 : 12 : 24 : 23 (132).

Pancreatic elastase has received attention in that it had been reported that, unlike trypsin, elastase had a very low affinity for Ca(II) (133). However, Brittain et al. (97) showed that strong Tb(III) luminescence and binding could be obtained after addition of the lanthanide ion to the protein, and Dimicoli and Bieth (134) confirmed that stable Ca(II) and Gd(III) complexes could be formed with elastase. Subsequently, variations in

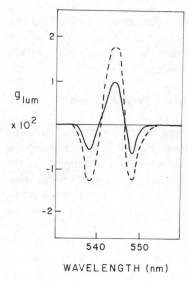

Fig. 5.10. CPL spectra of Tb(III) substituted elastase (– – –) and Tb(III)-elastase complexes with α_1-proteinase inhibitor (——). (Spectra taken in part from Fig. 2 of Ref. 136, with permission of the copyright owner.)

Tb(III) luminescence (135) and circularly polarized luminescence (136) were used to study the binding of inhibitors on Tb(III)-substituted elastase (see Fig. 5.10). Measurements of the binding constants for Mg(II) and Zn(II) to elastase demonstrated that under physiological conditions, the protein exists as the Ca(II) complex (135).

Thermolysin is a heat-stable proteolytic enzyme of known structure (137) that has been shown to bind lanthanide ions (138). Four Ca(II) sites are known for this protein (one being a double site), but only three lanthanide ions can be bound (only one ion can be substituted at the double site). The crystal structure of Ln(III)-substituted thermolysin has also been determined (138). Strong Tb(III) luminescence may be obtained after formation of Tb(III)-substituted thermolysin (sensitized by tryptophan residues), and weak circularly polarized luminescence has been obtained within the Tb(III) emission (97).

Given the degree of structural information available on Ln(III)-substituted thermolysin systems, it is not surprising that the use of luminenscence spectroscopy as a "spectroscopic ruler" for the determination of distances between metal centers originated with the thermolysin system. At essentially the same time Darnall and Birnbaum (139) and Horrocks et al. (140) used the quenching of Tb(III) emission by Co(II) to ascertain the distance between the two metal ion sites. Both groups obtained excellent agreement with the known crystal structural value of 13.7 Å (138). More recently, Horrocks and co-workers have used quenching measurements to obtain the distance between fluorescent amino acid residues and metal ion binding sites in thermolysin (141) and have also employed site selection spectroscopy to measure the distances between the other metal ion sites in the same protein system (142).

Ln(III) Substitution in Plasma–Serum Protein Systems

Human serum transferrin is an iron-binding protein whose major function is that of an iron carrier. Transferrin possesses two iron binding sites, and many other cations may be bound at these sites after removal of the iron ions. However, the binding of an anion (such as bicarbonate) must accompany the binding of the cations (143). The two metal ion binding sites are essentially identical and contain two histidine and three tyrosine residues for cation binding (144).

Luk replaced the iron cations of transferrin with lanthanide ions and demonstrated that considerable enhancement of Tb(III) emission was achieved after binding (145). This work was the first to employ Tb(III) as a luminescent biomolecular probe and attempted to use Fe(II) quenching of the Tb(III) emission in mixed systems to determine the separation between the metal sites. Gafni and Steinberg studied the circularly polarized luminescence within the Tb(III) emission after forming the transferrin complex and also examined the CPL spectra associated with ternary Ho(III)–Tb(III)–transferrin and Fe(III)–Tb(III)–transferrin complexes (146). In these latter studies, it was found that the two metal ion binding sites of the transferrin molecule were equivalent in structure and conformation. This last result has been confirmed in a series of CPL titrations in which Tb(III) was added incrementally to apo-transferrin: no evidence of cooperative behavior (in either the total or circularly polarized emission) could be noted at any point along the titration curves (147). (Examples of the CPL spectra within various Tb(III) emission bands are found in Fig. 5.11.)

More detailed understanding of the processes involved when metal ions add to transferrin indicated that the preparative method used by Luk could lead to low yields and complicated side reactions. As a result, Meares and Ledbetter reinvestigated the problem and determined that the separation between sites was 25 Å (148), rather than the 43 Å measured by Luk (145). But it was pointed out subsequently that even this new determination also suffered from systematic experimental errors [chiefly in the accurate measurement of the quantum yield for emission of protein-bound Tb(III) ions], yet one more study of energy transfer was reported. In this latest work, energy transfer from bound Tb(III) to Fe(III) or Mn(III) was used to evaluate the site separation, and a value of 35.5 Å

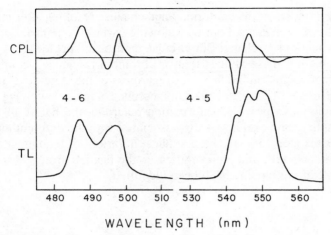

WAVELENGTH (nm)

Fig. 5.11. TL and CPL spectra within the $^5D_4 \rightarrow {}^7F_6$ and $^5D_4 \rightarrow {}^7F_5$ luminescence transitions of Tb(III) substituted transferrin.

was obtained this time (149). The detailed data analysis of this last study appears to contain enough understanding of the photophysics to yield a value that should be correct.

In a series of novel experiments the distance between the protein surface to the cation binding sites was measured by diffusion-enhanced energy transfer (150). In this technique the energy transfer efficiency between freely diffusing Tb(III) chelates and bound metal ions was followed, and it could be determined that the iron chromophores are not located on the surface of the protein but are more than 17 Å below the surface. Substitution of oxalate for bicarbonate as the required anion reduced the accessibility of the Tb(III) chelates for the cation sites.

Prothrombin is a circulating plasma protein that is involved in the final stages of blood coagulation (151) and that requires the presence of Ca(II) ions to become activated to thrombin. Thrombin converts fibrinogen to fibrin, resulting in the formation of a fibrin clot. Lanthanide ions have been shown to substitute for Ca(II) in the Ca(II)-binding sites, and it has been shown that Ce(III) may be substituted for Ca(II) in the generation of thrombin from prothrombin (152). Prothrombin is known to bind about 11 Ca(II) ions, and the first three of these are bound in a cooperative manner (153). Both prothrombin and prothrombin fragment I have been shown to bind Tb(III), and the luminescence titrations associated with these protein systems reflects the Ca(II) binding behavior very closely (97). Quantitative comparison of the luminescence titration curves reveals that fragment I appears to contain the strongly binding metal ion sites.

Further studies of the conformational changes resulting after metal ion binding have been carried out by following variations in intrinsic protein fluorescence and in changes in circular dichroism spectra (154). It was concluded from these studies that prothrombin fragment I contains six well-defined metal ion binding sites regardless of the state of aggregation of the protein. Time-resolved luminescence studies involving bound Eu(III) ions have provided information regarding the nature of the hydration sphere and exchange rates associated with the different metal ion binding sites (155,156). In these studies, a variety of binding sites have been characterized, and in several cases the ligating groups responsible for the metal ion binding have been identified.

5.6. SUMMARY

From the applications that have been presented in the preceding sections, it should be quite clear that luminescence spectroscopy can be quite useful to the bioinorganic chemist. One of the distinct disappointments that nature has provided is the unwillingness of most transition metal ions to emit in fluid solution at room temperature. However, a considerable body of work now exists employing lanthanide ions [and Tb(III) in particular] as luminescent metal ion probes of biomolecular systems, and these studies indicate possible pathways that could be followed if one works with transition metal ions that do display luminescence under conditions that could be considered physiological. Certainly, the use of internal energy transfer as a means to compute distances between metal ion sites is an important use of luminescence spectroscopy, and the utility of luminescence titrations to indicate the number of metal ions that can be bound by a particular biopolymer is evident (97).

The application of cryogenic techniques to bioinorganic chemistry should greatly extend the application of luminescence spectroscopy in this area. While low-temperature studies do not necessarily provide data that can be extended to interpret data obtained under physiological conditions, it is a given fact that such work could provide an opening through which additional insight could be obtained. One may argue that the species existing at low temperatures do not reflect the composition of solutions at 37°C, but an equally persuasive argument can be made regarding the interrelation of solution phase studies and results of x-ray crystallographic determinations.

Some of the most exciting results should come about as new luminescence techniques are applied to bioinorganic chemistry. The possible uses of luminescence and excitation spectra, intramolecular energy transfer, circularly polarized luminescence, fluorescence detected circular dichroism, and other techniques have only been hinted at in the studies featured in the present work, and one would anticipate that a great deal of useful information will be forthcoming in the near future.

5.7. ACKNOWLEDGMENTS

Support by the Camille and Henry Dreyfus Foundation through a Teacher–Scholar award is gratefully acknowledged. Special thanks are due to Professors F. S. Richardson and R. B. Martin for first introducing the author to the use of lanthanide ions as bioinorganic probes and for their support of subsequent work.

REFERENCES

1. S. Georghiou, "Fluorometric Studies of Biologically Important Molecular Complexes", in *Modern Fluorescence Spectroscopy*, vol. 3, E. L. Wehry (ed.), Plenum Press, New York, 1981.
2. L. Stryer, *Science 162,* 526 (1968).
3. G. M. Edelman and W. O. McClure, *Acc. Chem. Res. 1,* 65 (1968).
4. L. Brand and J. R. Gohlke, *Ann. Rev. Biochem. 41,* 843 (1972).
5. Y. Kanaoka, *Angew. Chem. Int. Ed. 16,* 137 (1977).
6. G. Weber, *Biochem. J. 51,* 145 (1962).
7. G. G. Guilbault, *Practical Fluorescence,* Dekker, New York, 1973, pp. 468–497.
8. F. S. Richardson and J. P. Riehl, *Chem. Rev. 77,* 773 (1977).
9. H. G. Brittain, *Coord. Chem. Rev. 48,* 243 (1983).
10. I. Tinoco, Jr., and D. H. Turner, *J. Am. Chem. Soc. 98,* 6453 (1976).
11. R. B. Martin and F. S. Richardson, *Quart. Rev. Biophys. 12,* 181 (1979).
12. K. P. Ghiggino, A. J. Roberts, and D. Phillips, *Adv. Polym. Sci. 40,* 69 (1981).
13. T. L. Blundell and J. A. Jenkins, *Chem. Soc. Rev. 6,* 139 (1977).
14. Glenn A. Crosby, *Acc. Chem. Res. 8,* 231 (1975).
15. A. W. Adamson and P. D. Fleischauer (eds.) *Concepts of Inorganic Photochemistry,* Wiley-Interscience, New York, 1975.
16. E. A. Koerner, T. I. von Gustorf, L. H. G. Leenders, I. Fischler, and R. N. Peruto, *Adv. Inorg. Chem. Radiochem. 19,* 65 (1976).
17. P. D. Fleischauer and P. Fleischauer, *Chem. Rev. 70,* 199 (1970).
18. C. W. Sill and H. E. Peterson, *Anal. Chem. 21,* 1266 (1949).

19. G. F. Kirkbright, T. S. West, and C. Woodward, *Talanta 12*, 517 (1965).

20. R. E. Curtice and A. B. Scott, *Inorg. Chem. 3*, 1383 (1964).

21. P. J. Mayne and G. F. Kirkbright, *J. Inorg. Nucl. Chem. 37*, 1527 (1975).

22. A. W. Adamson, W. L. Waltz, E. Zinato, D. W. Watts, P. D. Fleishauer, and R. D. Lindholm, *Chem. Rev. 68*, 541 (1968).

23. G. L. Hilmes, H. G. Brittain, and F. S. Richardson, *Inorg. Chem. 16*, 528 (1977).

24. D. Dunaway-Mariano and W. W. Cleland, *Biochemistry 19*, 1496 (1980).

25. D. Dunaway-Mariano and W. W. Cleland, *Biochemistry 19*, 1506 (1980).

26. R. E. Viola, J. F. Morrison, and W. W. Cleland, *Biochemistry 19*, 3131 (1980).

27. M. L. Gantzer, C. Klevickis and C. M. Grisham, *Biochemistry 21*, 4083 (1982).

28. D. M. Dooley, personal communication.

29. P. A. Loach and M. Calvin, *Biochemistry 2*, 361 (1963).

30. P. M. Plaskin, R. C. Stouffes, M. Matthew, and G. J. Palenik, *J. Am. Chem. Soc. 94*, 2121 (1972).

31. F. A. Kroger, H. J. Vink, and J. v.d. Boomgaard, *Physica 18*, 77 (1952).

32. J. E. Ralph and M. G. Townsend, *J. Chem. Phys. 48*, 149 (1968).

33. D. L. Webb and L. A. Rossiello, *Inorg. Chem. 9*, 2622 (1970).

34. J. M. Hill, E. Loeb, A. Maclellan, N. O. Hill, A. Khan, and J. J. King, *Cancer Chemother. Rep. 59*, 647 (1975).

35. S. J. Lippard, *Acc. Chem. Res. 11*, 211 (1978).

36. K. W. Jennette, S. J. Lippard, G. A. Vassiliades, and W. R. Bauer, *Proc. Natl. Acad. Sci. USA 71*, 3839 (1974).

37. E. R. Birnbaum, J. E. Gomez, and D. W. Darnall, *J. Am. Chem. Soc., 92*, 5279 (1970).

38. J. A. Glasel, *Prog. Inorg. Chem. 18*, 383 (1973).

39. R. J. P. Williams, *Struct. Bonding 50*, 79 (1982).

40. R. H. Kretsinger and D. J. Nelson, *Coord. Chem. Rev. 18*, 29 (1976).

41. J. Reuben, "Bioinorganic Chemistry: Lanthanides as Probes in Systems of Biological Interest", in *Handbook on the Physics and Chemistry of Rare Earths*, K. A. Gschneider and L. Eyring (ed.), North-Holland, Amsterdam, 1979.

42. E. Nieboer, *Struct. and Bonding 22*, 1 (1975).

43. W. De W. Horrocks and D. R. Sudnick, *Acc. Chem. Res. 14*, 384 (1981).

44. M. Cefola, A. S. Tompa, A. V. Celiano, and P. S. Gentile, *Inorg. Chem. 1*, 290 (1962).

45. R. Dreyer, J. Redlich, and R. Syhre, *Z. Phys. Chem. (Leipzig) 238*, 417 (1968).

46. A. E. Martell and R. M. Smith, *Critical Stability Constants*, Plenum, Vol. I, New York, 1974.

47. T. Moeller, D. F. Martin, L. C. Thompson, R. Ferrus, G. R. Feistel, and W. J. Randall, *Chem. Rev. 65*, 1 (1965).

48. R. Prados, L. G. Stadtherr, H. Donato, and R. B. Martin, *J. Inorg. Nucl. Chem. 36*, 689 (1974).

49. H. G. Brittain, *Inorg. Chem. 18*, 1740 (1979).

50. H. G. Brittain, *J. Lumin. 21*, 43 (1979).

51. H. G. Brittain, *J. Inorg. Nucl. Chem. 41*, 1775 (1979).

52. T. D. Marinetti, G. H. Snyder, and B. D. Sykes, *J. Am. Chem Soc. 97*, 6562 (1975).

53. B. A. Levine and R. J. P. Williams, *Proc. Roy. Soc. Lond. A345*, 5 (1975).

54. A. D. Sherry, C. Yoshida, E. R. Birnbaum, and D. W. Darnall, *J. Am. Chem. Soc. 95*, 3011 (1973).

55. A. D. Sherry and E. Pascual, *J. Am. Chem. Soc. 99*, 5871 (1977).

56. G. A. Elgavish and J. Reuben, *J. Am. Chem. Soc. 100*, 3617 (1978).

57. E. R. Birnbaum and D. W. Darnall, *Bioinorg. Chem. 3*, 15 (1973).

58. J. Legendziewicz, H. Kozlowski, and E. Burzala, *Inorg. Nucl. Chem. Lett. 14*, 409 (1978).

59. J. Legendwiewicz, H. Kozlowski, B. Jezowska-Trzebiatowska, and E. Huskowska, *Inorg. Nucl. Chem. Lett. 15*, 349 (1979); *17*, 57 (1981).

60. J. Legendziewicz, E. Huskowska, W. Strek, and B. Jezowska-Trzebiatowska, *J. Lumin. 24/25*, 819 (1981).

61. L. I. Katzin and E. Gulyas, *Inorg. Chem. 7*, 2442 (1968).

62. C. K. Luk and F. S. Richardson, *Chem. Phys. Lett. 25*, 215 (1974).

63. C. K. Luk and F. S. Richardson, *J. Am. Chem. Soc. 97*, 6666 (1975).

64. H. G. Brittain and F. S. Richardson, *Bioinorg. Chem. 7*, 233 (1977).

65. H. G. Brittain, *Inorg. Chem. 17*, 2762 (1978).

66. R. A. Copeland and H. G. Brittain, *J. Inorg. Nucl. Chem. 43*, 2499 (1982).

67. I. Grenthe, *J. Am. Chem. Soc. 83*, 360 (1961).

68. H. G. Brittain, *J. Am. Chem. Soc. 102*, 3693 (1980).

69. H. G. Brittain, *Inorg. Chim. Acta 53*, L7 (1981).

70. C. D. Barry, A. C. T. North, J. A. Glasel, R. J. P. Williams, and A. V. Xavier, *Nature 232*, 236 (1971).

71. P. Tanswell, J. M. Thornton, A. V. Korda, and R. J. P. Williams, *Eur. J. Biochem. 57*, 135 (1975).

72. S. A. Davis and F. S. Richardson, *J. Inorg. Nucl. Chem. 42*, 1793 (1980).

73. C. D. Barry, C. M. Dobson, R. J. P. Williams, and A. V. Xavier, *J. Chem. Soc. Dalton Trans.* 1765 (1974).

74. C. F. G. C. Geraldes and R. J. P. Williams, *Eur. J. Biochem. 97*, 93 (1979).

75. J. F. Morrison and W. W. Cleland, *Biochemistry 19*, 3127 (1980).

76. R. E. Viola, J. F. Morrison, and W. W. Cleland, *Biochemistry 19*, 3131 (1980).

77. G. A. Elgavish and J. Reuben, *J. Am. Chem. Soc. 98*, 4755 (1976); *99*, 1762 (1977).

78. C. M. Dobson, R. J. P. Williams, and A. V. Xavier, *J. Chem. Soc. Dalton Trans.* 1762 (1974).

79. A. D. Sherry, C. A. Stark, J. R. Ascenso, and C. F. G. C. Geraldes, *J. Chem. Soc. Dalton Trans.* 2078 (1981).

80. L. G. Marzilli, R. C. Stewart, C. P. van Vuuren, B. de Castro, and J. P. Caradonna, *J. Am. Chem. Soc. 100*, 3967 (1978).

81. L. G. Marzilli, D. de Castro, J. P. Caradonna, R. C. Stewart, and C. P. van Vuuren, *J. Am. Chem. Soc. 102*, 916 (1980).

82. L. G. Marzilli, B. de Castro, and C. Solorzano, *J. Am Chem. Soc. 104*, 461 (1982).

83. A. A. Lamola and J. Eisinger, *Biochem. Biophys. Acta 240*, 313 (1971).

84. C. Formoso, *Biochem. Biophys. Res. Comm. 53*, 1084 (1973).

85. S. Davis, D. Foster, and F. S. Richardson, unpublished results.

86. D. Foster and F. S. Richardson, unpublished results.

87. G. Yonuschot and G. W. Mushrush, *Biochemistry 14*, 1677 (1975).

88. G. Yonuschot, G. Robey, G. W. Mushrush, D. Helman, and G. Van de Woude, *Bioinorg. Chem. 8*, 397, 405 (1978).

89. M. S. Kayne and M. Cohn, *Biochemistry 13*, 4159 (1974).

90. J. M. Wolfson and D. R. Kearns, *J. Am. Chem. Soc. 96*, 3653 (1974); *Biochemistry 14*, 1436 (1975).

91. C. R. Jones and D. R. Kearns, *J. Am. Chem. Soc. 96*, 3651 (1974); *Proc. Natl. Acad. Sci. US, 71*, 4237 (1974).

92. T. D. Barela, S. Burchett, and D. E. Kizer, *Biochemistry 14*, 4887 (1975).

93. D. P. Ringer, S. Burchett, and D. E. Kizer, *Biochemistry 17*, 4818 (1978).

94. M. D. Topal and J. R. Fresco, *Biochemistry 19*, 5531 (1980).

95. D. P. Ringer, B. A. Howell, and D. E. Kizer, *Anal. Biochem. 103*, 337 (1980).

96. T. Haertle, J. Augustyniak, and W. Guschlbauer, *Nucl. Acids Res. 9*, 6191 (1981).

97. H. G. Brittain, F. S. Richardson, and R. B. Martin, *J. Am. Chem. Soc. 98*, 8255 (1976).

98. W. De W. Horrocks, Jr., and W. E. Collier, *J. Am. Chem. Soc. 103*, 2856 (1981).

99. John de Jersey, P. J. Morley, and R. B. Martin, *Biophys. Chem. 13*, 233 (1981).

100. W. De W. Horrocks, Jr., G. F. Schmidt, D. R. Sudnick, C. Kittrell, and R. A. Bernheim, *J. Am. Chem. Soc. 99*, 2378 (1977).

101. W. De W. Horrocks, Jr., and D. R. Sudnick, *J. Am. Chem. Soc. 101*, 334 (1979).

102. W. De W. Horrocks, Jr., M.-J. Rhee, A. P. Snyder, and D. R. Sudnick, *J. Am. Chem. Soc. 102*, 3650 (1980).

103. G. Hamoir and S. Konosu, *Biochem. J. 96*, 85 (1965).

104. S. Konosu, G. Hamoir, and J. F. Pechere, *Biochem. J. 96*, 98 (1965).

105. H. Donato, Jr., and R. B. Martin, *Biochemistry 13*, 4575 (1974).

106. D. J. Nelson, T. L. Miller, and R. B. Martin, *Bioinorg. Chem. 7*, 325 (1977).

107. D. J. Nelson, A. D. Theoharides, A. C. Nieburgs, R. K. Murray, F. Gonzalez-Ferandez, and D. S. Brenner, *Int. J. Quantum Chem. 16*, 159 (1979).

108. R. H. Kretsinger and C. E. Nockolds, *J. Biol. Chem. 248*, 3313 (1973).

109. J. Sowadski, G. Cornick, and R. H. Kretsinger, *J. Mol. Biol. 124*, 123 (1978).

110. T. L. Miller, R. M. Cook, D. J. Nelson, and A. D. Theohardies, *J. Mol. Biol. 141*, 223 (1980).

111. L. Lee and B. D. Sykes, *Biochemistry 19*, 3208 (1980).

112. M.-J. Rhee, D. R. Sudnick, V. K. Arkle, and W. De W. Horrocks, Jr., *Biochemistry 20*, 3328 (1981).

113. T. L. Miller, D. J. Nelson, H. G. Brittain, F. S. Richardson, and R. B. Martin, *FEBS Lett. 58*, 268 (1975).

114. J. H. Collins, J. D. Potter, M. J. Horn, G. Wiltshire, and N. Jackman, *FEBS Lett. 36*, 268 (1973).

115. P. C. Leavis, B. Nagy, S. S. Lehrer, H. Bialkowska, and J. Gergely, *Arch. Biochem. Biophys. 200*, 17 (1980).

116. C.-L. A. Wang, P. C. Leavis, W. De W. Horrocks, Jr., and J. Gergely, *Biochemistry 20*, 2439 (1981).

117. H. G. Brittain, F. S. Richardson, R. B. Martin, L. D. Burtnick, and C. M. Kay, *Biochem. Biophys. Res. Comm. 68*, 1013 (1976).

118. W. Y. Cheung, *Science 207*, 19 (1980).

119. A. R. Means and J. R. Dedman, *Nature 285*, 73 (1980).

120. E. A. Tallant, R. W. Wallace, M. E. Dockter, and W. Y. Cheung, *Ann. NY Acad. Sci. 356*, 436 (1980).

121. M.-C. Kilhoffer, D. Gerard, and J. G. Demaille, *FEBS Lett. 120*, 99 (1980).

122. C. L. A. Wang and R. Aquaron, *Dev. Biochem. 14*, 251 (1980).

123. M.-C. Kilhoffer, J. G. Demaille, and D. Gerard, *FEBS Lett. 116*, 269 (1980).

124. R. W. Wallace, E. A. Tallant, M. E. Dockter, and W. Y. Cheung, *J. Biol. Chem. 257*, 1845 (1982).

125. K. Oikawa, W. D. McCubbin, and C. M. Kay, *FEBS Lett. 118*, 137 (1980).

126. M. Delaage and M. Lazdunski, *Biochem. Biophys. Res. Comm. 28*, 390 (1967).

127. D. Gabel and V. Kasche, *Acta Chem. Scand. 27*, 1971 (1973).

128. F. Abbott, J. E. Gomez, E. R. Birnbaum, and D. W. Darnall, *Biochemistry 14*, 4935 (1975).

129. J. E .Gomez, E. R. Birnbaum, and D. W. Darnall, *Biochemistry 13*, 3745 (1974).
130. M. Epstein, A. Levitzki, and J. Reuben, *Biochemistry 13*, 1777 (1974).
131. M. Epstein, J. Reuben, and A. Levitzki, *Biochemistry 16*, 2449 (1977).
132. J. de Jersey, R. S. Lahue, and R. B. Martin, *Arch. Biochem. Biophys. 205*, 536 (1980).
133. D. W. Darnall, F. Abbott, J. E. Gomez, and E. Birnbaum, *Biochemistry 15*, 5017 (1976).
134. J. L. Dimicoli and J. Bieth, *Biochemistry 16*, 5532 (1977).
135. J. de Jersey and R. B. Martin, *Biochemistry 19*, 1127 (1980).
136. G. Duportail, J.-F. LeFeure, P. Lestienne, J.-L. Dimicoli, and J. G. Bieth, *Biochemistry 19*, 1377 (1980).
137. B. W. Matthews, P. M. Colman, J. N. Jansonius, K. Titani, K. A. Walsh, and H. Neurath, *Nature 238*, 37, 41 (1972).
138. B. W. Matthews and L. H. Weaver, *Biochemistry 13*, 1719 (1974).
139. V. G. Berner, D. W. Darnall, and E. R. Birnbaum, *Biochem. Biophys. Res. Comm. 66*, 763 (1975).
140. W. De W. Horrocks, Jr., B. Holmquist, and B. L. Vallee, *Proc. Natl. Acad. Sci. USA 72*, 4764 (1975).
141. W. De W. Horrocks, Jr., and A. P. Snyder, *Biochem. Biophys. Res. Comm. 100*, 111 (1981).
142. A. P. Snyder, D. R. Sudnick, V. K. Arkle, and W. De W. Horrocks, Jr., *Biochemistry 20*, 3334 (1981).
143. P. Aisen and E. B. Brown, *Semin. Hematol. 14*, 31 (1977).
144. A. Bezkorovainy and D. Grohlich, *Biochem. J. 123*, 125 (1971).
145. C. K. Kuk, *Biochemistry 10*, 2838 (1971).
146. A. Gafni and I. Z. Steinberg, *Biochemistry 13*, 800 (1974).
147. H. G. Brittain, unpublished results.
148. C. F. Meares and J. E. Ledbetter, *Biochemistry 16*, 5178 (1977).
149. P. O'Hara, S. M. Yeh, C. F. Meares, and R. Bersohn, *Biochemistry 20*, 4704 (1981).
150. S. M. Yeh and C. F. Meares, *Biochemistry 19*, 5057 (1980).
151. C. M. Heldebrant, R. J. Butkowski, S. P. Bajaj, and K. G. Mann, *J. Biol. Chem. 248*, 7149 (1973).
152. B. C. Furie, K. G. Mann, and B. Furie, *J. Biol. Chem. 251*, 3235 (1976).
153. S. P. Bajaj, R. J. Butkowski, and K. G. Mann, *J. Biol. Chem. 250*, 2150 (1975).
154. G. L. Nelsesteun, R. M. Resnick, G. J. Wei, C. H. Plecher, and V. A. Bloomfield, *Biochemistry 20*, 351 (1981).
155. M. E. Scott, M. M. Sarasua, H. C. Marsh, D. L. Harris, R. G. Hiskey, and K. A. Koehler, *J. Am. Chem. Soc. 102*, 3413 (1980).
156. H. C. Marsh, M. M. Sarasua, D. A. Madar, R. G. Hiskey, and K. A. Koehler, *J. Biol. Chem. 256*, 7863 (1981).

CHAPTER

6

EXCITED-STATE OPTICAL ACTIVITY

HARRY G. BRITTAIN

Department of Chemistry
Seton Hall University
South Orange, NJ 07079

6.1. INTRODUCTION

The phenomenon of optical activity has been known since 1810, and since that time a great deal of research effort has been expended to understand the nature of the effect and its molecular origins. An excellent account of the history of these investigations and of the early theories associated with optical activity may be found in the monograph by Charney (1). In

583

spite of the availability of texts and review articles dealing with molecular chirality, many workers incorrectly assign the basis of optical activity as being due to asymmetry. The actual origin of the effect was correctly assigned by Pasteur, who properly recognized that optical activity was a manifestation of molecular dissymmetry.

During the course of these studies, many physical techniques for the detection and measurement of optical activity were designed. Essentially all early work involved a measurement of the angle by which a beam of plane-polarized light could be rotated by a particular sample. This effect was termed *optical rotation* and is actually due to circular birefringence (circular double refraction). It was learned that the optical rotation of a sample varied with the wavelength of the analyzing light beam used in the measurement, and thus by sweeping the wavelength one could obtain the optical rotatory dispersion (ORD) of the sample as well. In the region of absorption transitions, the ORD curves were observed to change sign (the Cotton effect) and this observation led to yet another method for the study of optical activity. By measuring the differential absorption of left- and right-circularly polarized light by a sample, one could obtain the circular dichroism (CD) spectrum of that material. CD possessed a great advantage over ORD in that it vanished outside of an absorption band, and it has essentially supplanted ORD as the preferred method for the study of molecular optical activity.

While CD spectroscopy in the ultraviolet and visible regions of the electromagnetic spectrum has received the greatest degree of attention, an entire series of alternative chiroptical techniques is in the process of development at the present time. The optical activity associated with molecular vibrations has been measured via infrared absorption (2) and Raman scattering (3) methods. CD has been measured by exciting with circularly polarized light and measuring the resulting fluorescence as a function of the excitation wavelength. This latter technique is termed fluorescence detected circular dichroism (FDCD), and has the advantage that low concentrations of analyte may be examined. Yet another chiroptical method involves a measurement of the differential emission of left- and right-circularly polarized light by a chiral luminescent compound, and is termed circularly polarized luminescence (CPL) (4).

It is the purpose of this review to detail the current state of affairs regarding chiroptical methods that involve the excited state of the dissymmetric molecule under study. We focus first on CPL spectroscopy and the inherent optical activity of the lowest excited state associated with the luminescent molecule. In spite of the recent introduction of this technique, considerable work has been done with chiral organic molecules, metal complexes, and biochemical systems. Next, the FDCD experiment is analyzed, and we see how the ground-state chirality of a molecule may be studied through fluorescence processes.

It is hoped that the current review will provide a useful outline of how each technique has developed, what each has accomplished, and what future trends might be. Clearly, these new chiroptical methods will prove to be fruitful areas of research, and considerably more work remains to be done with each technique before each reaches the status of being accepted in the manner that CD and ORD now are.

6.2. CIRCULARLY POLARIZED LUMINESCENCE SPECTROSCOPY

The CPL experiment involves a measurement of the spontaneous differential emission of left- and right-circularly polarized light by a naturally optically active molecule. The technique therefore yields the molecular chirality of the luminescent excited state in exactly the same manner as CD spectroscopy is related to the molecular chirality of the ground state. If the geometry of the molecule remains invariant during the excitation process, then the two chiroptical methods will yield the same results regarding the optical activity of the molecule under study. However, should a geometrical change accompany the excitation, then the CD and CPL will be quite different. As will be seen in the following sections, such possibilities have been realized.

The one restriction associated with CPL spectroscopy is that it is confined to luminescent molecules only. However, if properly exploited, this limitation may actually turn out to be an advantage. The CPL technique is highly selective, in that it will originate only in the emitting chromophore of the molecule in question. Should the absorptivity of a molecule be too weak to permit CD measurements at reasonable concentrations (such as in the f–f transitions of lanthanide complexes, or singlet–triplet transitions of organic molecules), use of the CPL technique is possible. One may easily see that CPL spectroscopy combines the structural selectivity of CD spectroscopy with the instrumental sensitivity of luminescence spectroscopy. The strengths of this technique will become apparent in the sections that follow.

The theory associated with the CPL experiment has been presented in a general fashion by Richardson and Riehl (4,5,6,7), and interested readers should consult these sources for additional information. The present review is concerned mainly with the experimental aspects of the investigations that have been carried out.

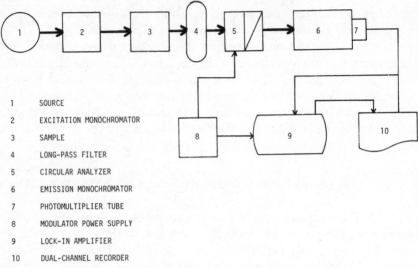

1	SOURCE
2	EXCITATION MONOCHROMATOR
3	SAMPLE
4	LONG-PASS FILTER
5	CIRCULAR ANALYZER
6	EMISSION MONOCHROMATOR
7	PHOTOMULTIPLIER TUBE
8	MODULATOR POWER SUPPLY
9	LOCK-IN AMPLIFIER
10	DUAL-CHANNEL RECORDER

Fig. 6.1. Block diagram of an emission spectrometer capable of measuring circularly polarized luminescence simultaneously with the total emission.

6.2.1. Experimental Methods

While CD spectrometers are commercially available at a high level of sophistication (including computer control), CPL instrumentation has not yet been developed by any commercial instrument company. As a result, all measurements to date have been carried out on laboratory-constructed devices that are generally adapted for the particular area of interest to the research group. Two publications have appeared that contain detailed descriptions of CPL instrumentation developed in the laboratories of Steinberg (8) and McCaffery (9), and Hipps has described further aspects of instrumental design in his Ph.D. dissertation (10). Richardson has provided details of the setup used at the University of Virginia (4), and in the following discussion we describe the instrumentation set up in our own laboratory. The version we use appears to be the simplest currently in use, and one that illustrates the necessary features associated with the CPL technique.

A block diagram of the basic CPL spectrometer is shown in Figure 6.1. The source of excitation energy can be either a laser or an arc lamp, but the best operation is obtained when the irradiation beam is both continuous and depolarized. Ordinarily, the excitation energy is passed through a monochromator–filter system to isolate the desired wavelength from the background or laser plasma lines before being used to excite the sample. The sample itself can be a fluid solution in a cuvette, or a single crystal mounted on a silica plate or held in a plastic film. For chiral compounds, the sample will spontaneously emit left- or right-circularly polarized light, but optical activity can also be induced in an achiral sample if the sample is placed in a longitudinal magnetic field.

While most luminescence spectrometers collect the emitted light at 90° to the excitation beam, it is advantageous in the CPL experiment to work at 0° to the excitation beam. This "head-on" method of detection has the advantage that it will eliminate any problems arising from linear polarization in the emission. The problems associated with linear polarization and photoselection have been discussed by Steinberg (11,12). It is trivial to remove any unabsorbed exciting light from the emission by placing a long-pass filter immediately after the sample.

The emission is then passed through a circular analyzer, which consists of a dynamically operated quarter-wave retardation element followed by a linear polarization element. The quarter-wave plate can be either a Pockels cell or photoelastic modulator, but the latter is preferable since it can function with a much wider acceptance angle. The photoelastic modulator contains an isotropic material (silica) that is rendered optically anisotropic by application of a periodic external stress. The effect of this element is to transform the small quantity of circular polarization in the emission beam into periodically interconverting orthogonal planes of linearly polarized light by introducing phase shifts of ±90°. The linear polarizer following the modulator extinguishes one of the linear components and thereby produces an ac ripple (proportional to the CPL intensity) on top of the large dc signal (proportional to the total emission, or TL). If at all possible, the number of optical elements preceding the circular analyzer should be reduced to the absolute minimum to avoid any possible depolarization of the emission by these optical elements.

The emission is then analyzed in the usual manner by the emission monochromator, at whatever degree of resolution is required by the spectroscopy of the system under question. The light intensity is typically detected by a photomultiplier tube, whose current is converted to voltage and then divided. One single can be fed directly to one input of a dual-channel recorder and displayed as the TL spectrum. The other signal is fed into a lock-in amplifier set to amplify only signals having the same frequency as that of the photoelastic modulator, with the output of the lock-in being fed into the other channel of the recorder. Alternately, one can take the analog ratio of the CPL and TL signals and display this quantity if desired.

The experiment thus yields two observables, which are measured in totally arbitrary units. One of these is the total luminescence emission intensity, or TL, defined as

$$I = I_L + I_R \tag{6.1}$$

and the other is the CPL intensity, defined as

$$\Delta I = I_L - I_R \tag{6.2}$$

In these relations, I_R and I_L represents the emitted intensities of left- and right-circularly polarized light, respectively. The ratio of these quantities is thus seen to be dimensionless and the taking of this ratio removes any unit dependence. In analogy to the Kuhn anisotropy factor used in CD spectroscopy (1), one defines the luminescence dissymmetry factor for CPL spectroscopy (4):

$$g_{\text{lum}} = \Delta I / \frac{1}{2} (I) \tag{6.3}$$

The dissymmetry factor has theoretical significance in that it may be related to the rotational strength of the transition:

$$g_{\text{lum}} = 4R_{ab}/D_{ab} \tag{6.4}$$

where R_{ab} is the rotatory strength:

$$R_{ab} = \text{Im} \langle \Psi_a/\mu/\Psi_b \rangle \langle \Psi_b/m/\Psi_a \rangle \tag{6.5}$$

and D_{ab} is the dipole strength:

$$D_{ab} = \langle \Psi_a/\mu/\Psi_b \rangle^2 \tag{6.6}$$

Equations (6.4–6.6) are valid for randomly oriented emitting systems in which the luminescent excited state is thermally equilibrated prior to emission.

The calibration of CPL spectra is an important question in order that results from different laboratories may be compared. Steinberg and Gafni have discussed a simple and accurate method that can be used to obtain the necessary calibration factor required to transform empirically measured dissymmetry factors into absolute quantities (8). Alternatively, one can measure the dissymmetry factor of a compound whose exact values are known and then obtain the calibration factor by comparison of the results. No generally accepted standard currently exists, but the optical activity associated with the $^5D_0 \rightarrow {}^7F_1$ transition at 595 nm of tris(trifluoroacetyl-d-camphor)Eu(III) dissolved in dimethyl sulfoxide may well be an acceptable candidate. The compound is commercially available, and its CPL spectra consist of only a few well-defined bands of high optical activity (13).

6.2.2. Results Obtained on Chiral Organic Molecules

Aside from a few isolated studies on the CPL of sodium uranyl acetate (which happens to crystallize in an optically active space group) (14,15), the initial measurements of CPL involved studies on fluorescent organic molecules. The first reported study detailed the CPL associated with the emission from sodium 1,3,5-triphenylpyrazoline sulfonate (16). However, it was not until Oosterhoff (University of Leiden) and his students undertook systematic studies of the CPL effect that the method began to attract much attention. The work began with a demonstration that optical activity could be observed in racemic mixtures if one excited the sample with circularly polarized light (17). Racemic *trans*-β-hydrindanone was used in this particular work; subsequently, CPL was obtained for resolved *trans*-β-hydrindanone and *trans*-β-thiohydrindanone after irradiation with unpolarized light (18,19).

Shortly thereafter, the research groups led by Steinberg (Weizmann Institute) and Richardson (University of Virginia) began to exploit the CPL technique for its chemical information. Steinberg's group reported that CPL emitted by resolved 1,1'-bianthracene-2,2'-dicarboxylic acid (20) and a series of diketopiperazines containing aromatic side chains (21). A very interesting study was carried out by Luk and Richardson, in which CD and CPL were obtained on D-camphorquinone: a careful comparison of two sets of spectra coupled with suitable theoretical analysis yielded the geometry change associated with the excitation process (22). While the CD and CPL spectra almost appear to be mirror images of each other (as is shown in Fig. 6.2), analysis of the data revealed that the dihedral twist angle of the dicarbonyl group was smaller in the fluorescent state relative to that of the ground state.

Steinberg and co-workers subsequently examined the optical activiy of chlorophyll dimers in solution, subchloroplast particles, and chloroplasts by both CD and CPL spectroscopies (23). While the chlorophyll dimers show strong CD, the optical activity is lost in the electronically excited state (no CPL can be detected), reflecting a marked change in the

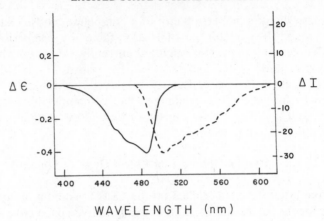

Fig. 6.2. Room-temperature CD (———) and CPL (– – –) spectra of *d*-camphorquinone dissolved in cyclohexane. (Data adapted from Fig. 1 of Ref. 22, with permission of the copyright owner.)

structure of the dimers on excitation. However, in solutions of high viscosity, the excited-state optical activity could be observed, and it was also determined that the chloroplasts and subparticles could not be used as models for associated chlorophyll in electronically excited native states.

Brittain and Richardson (24) showed that achiral molecules could be rendered chiral by dissolution into an optically active solvent. Fluorescein dissolved in resolved phenylethylamine exhibited relatively large values for the luminescence dissymmetry factors, and it was also learned that the g_{lum} values were not constant across the emission band. This result suggested that different emitting species were probably present in the solution. In a similar vein, Stegemeyer and co-workers have found that optical activity may be induced in achiral molecules when these are dissolved in cholesteric liquid crystals (25). Both of these approaches involving induced optical activity indicate methods whereby one may perform chiroptical studies (and thus extract the additional information available from such works) on molecules that do not contain the necessary natural optical activity.

Dekkers and Closs (26) have carried out a detailed series of CD–CPL investigations on chiral ketones, and thus have been able to learn a great deal about the $n \rightarrow \pi^*$ transition of the ketone group. For a large number of chiral ketones, the CD and CPL spectra are markedly different, which in turn indicates that the ground- and excited-state optical activities are quite different. Dekkers and Closs attributed this difference to a distortion

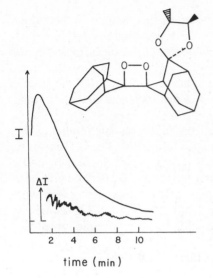

I

ΔI

2 4 6 8 10

time (min)

Fig. 6.3. Chemiluminescence and circular anisotropy in the chemiluminescence associated with an optically active dioxetane, whose structure is illustrated. (Data from Fig. 2 of Ref. 28, with permission of the copyright owner.)

of the $^1n\pi^*$ excited state but pointed out that the relatively simple theory of Moffitt and Moscowitz (27) could still be used to understand the optical activity of chiral ketones if allowance was made for the role of the excited state.

Optical activity detected in chemiluminescence has been reported (28) and represents another front where CPL spectroscopy can develop. The chemiluminescence that accompanies the thermal decomposition of resolved 1,2-dioxetane (whose structure is shown in Fig. 6.3) contains a small degree of CPL, which is observed to decay with the chemiluminescence.

The optical activity associated with chiral excimer complexes has been reported by Brittain and Fendler (29). At low concentrations where excimer formation does not take place, no CPL can be detected for either (+)- or (−)-1-(1-hydroxyhexyl)-pyrene. This result is easily interpreted as indicating that the planar emitting pyrene ring does not sense the chirality of the solitary asymmetric atom bound at its periphery. However, when the concentration is increased beyond the value where excimer formation becomes important, new emission appears to the red of the monomer emission and this new emission shows strong CPL. It is clear that the excimers have a definite preferred orientation of the pyrene rings, and it is this configurational preference that leads to the observed optical activity.

Another manner in which achiral molecules can be rendered chiral is by insertion of these into the hydrophobic interior of a cyclodextrin molecule. CPL has been induced in fluorescein in this fashion by including the carboxyphenyl portion of the molecule in β-cyclodextrin (30), while no CPL could be obtained with the inclusion of acridine. These results suggest that the formation of the fluorescein inclusion complex stabilizes a particular conformation of the molecule that is chiral and that the simple presence of an asymmetric environment is insufficient to induce optical activity in a planar fluorophore.

The results obtained so far on small organic molecules indicate the potential of the method and point the way for future directions of research. Many chiral luminescent molecules display only small degrees of CPL, reflecting the difficulty that the fluorophores are often planar and aromatic and are only slightly perturbed by their surroundings. However, with efficient data processing it should be possible to use the CPL technique to study the excited-state chirality of any luminescent organic molecule.

6.2.3. Results Obtained on Chiral Metal Complexes

Probably the largest number of CPL studies that have been carried out to date have dealt with chiral metal complexes, and essentially all of these studies have been concerned with the f–f optical activity of lanthanide complexes. The first measurement of CPL was made by Samoilov (14), who observed circular polarization in the luminescence spectrum of sodium uranyl acetate at cryogenic temperatures. The sharpness of the emission bands observed with lanthanide complexes and the paramagnetic nature of the metal ions combine to make CPL spectroscopy the preferred technique for the study of f–f optical activity. It is well known that f–f absorption bands are exceedingly weak, and thus CD measurements must be made at such high concentrations that polynuclear association of the complexes cannot be ignored. Consequently, one is able to obtain more reliable data using the sensitivity of luminescence spectroscopy in the CPL technique. Of all the lanthanide ions, Tb(III) and Eu(III) are found to emit with the highest intensity, and these are observed to luminesce in a series of well-defined bands even in fluid solution at room temperature.

The optical activity of a metal complex may arise from a variety of mechanisms:

1.   Dissymmetric placement of monodentate ligands about a metal ion
2.   Dissymmetry created by the simple presence of an asymmetric atom in the ligand bound to the metal (the vicinal effect)
3.   Dissymmetry created when an asymmetric atom of a ligand is also contained in a chelate ring bound by the metal (the conformational effect)

4. Chirality arising from a dissymmetric placement of chelate rings about the metal ion (the configurational effect).

CPL spectra have been obtained for systems in which the latter three mechanisms are found to be operative.

The theory associated with f–f optical activity can still be regarded as being in a state of flux, but the situation should improve as more high-quality experimental data become available. Richardson (31) has provided a series of selection rules obtained from consideration of the S, L, and J angular momentum quantum numbers of lanthanide 4f states perturbed by spin-orbit coupling and crystal field interactions, and these rules have proved quite effective in the prediction of relative magnitudes associated with the chiroptical transitions of lanthanide complexes. More recently, detailed theory and calculations have been developed for trigonal dihedral lanthanide complexes of D_3 symmetry (32,33) in which the configurational effect provides all the molecular chirality.

Transition Metal Complexes

In spite of the vast number of transition metal complexes that conceivably could be studied by means of CPL spectroscopy, the number of studies that actually have been performed is strikingly small. Emeis and Oosterhoff measured the CPL associated with the $^2E \rightarrow {}^4A_2$ phosphorescence of resolved tris(ethylenediamine)Cr(III) in a mixture of ethylene glycol–water at room temperature (18,34), but because of the weakness of the CPL these workers were unable to obtain a conventional spectrum. However, improvements in instrumentation and the use of laser excitation enabled Hilmes, Brittain, and Richardson to obtain the CPL spectrum of the same compound in the same solvent (35). In this study the CD of the identical transition was obtained and it was found that the two sets of chiroptical spectra were of similar magnitude and identical sign, as may be seen in Fig. 6.4. This result indicates that the structure of the $Cr(en)_3^{3+}$ species is essentially the same in the ground ($^4A_2$) and excited (^2E) states.

The only other CPL study involving a transition metal complex was performed by Gafni and Steinberg (36), who resolved tris(2,2′-bipyridine)Ru(II) and then studied the CD and CPL spectra of this compound. Again, it was found that the dissymmetry factors associated with the CD and CPL were identical within experimental error, indicating that the chirality of the emitting and ground states is similar.

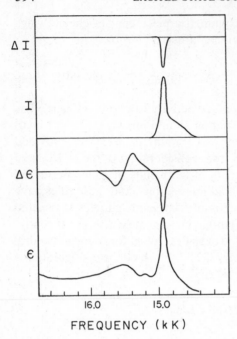

Fig. 6.4. Absorption (ϵ), circular dichroism ($\Delta\epsilon$), total emission (I), and circularly polarized emission (ΔI) spectra for $(-)-[Cr(en)_3]Cl_3$ in 2:1 ethylene glycol–water solution at room temperature. (Data from Fig. 2 of Ref. 35, with permission of the copyright owner.)

Lanthanide Complexes Containing Chiral Amino and Carboxylic Acid Ligands

The CPL technique has been used with great success to study the solution-phase stereochemistry of lanthanide complexes containing chiral carboxylate ligands. It had been observed earlier that the CD intensities associated with a variety of lanthanide compounds were quite small (37), but the strong emission intensities of the corresponding Eu(III) and Tb(III) complexes permitted studies to be carried out at quite good signal–noise ratios over reasonable concentration ranges.

Luk and Richardson carried out the first studies involving chiral lanthanide carboxylate complexes in aqueous media (38,39). These survey studies demonstrated that CPL spectroscopy was an excellent probe of the solution-phase stereochemistry of Eu(III) and Tb(III) complexes. Most of the work involved studies of the Tb(III) complexes because of the greater quantum yield of emission for this metal ion, although the chemical similarity of lanthanide ions ensures that the results would be translatable to most of the members of the series. A 5:1 ratio of ligand–metal was used for much of the work, with the chiral ligands under study being L-malic, L-aspartic, L-glutamic, and L-lactic acids, as well as L-alanine and L-serine.

These studies indicated further in-depth studies were required. Brittain and Richardson carried out a detailed study of the pH dependence of the Tb(III) and Eu(III) complexes of L-malic acid in H_2O and D_2O solvent (40). Subsequently, CPL spectroscopy was used to detail the interactions between the same two lanthanide ions and a series of potentially terdentate amino acids (L-aspartic acid, L-serine, L-threonine, and L-histidine) (41). In both of these works, the 5:1 ratio of ligand–metal was used and it was found that while direct correlation between observed spectra and plausible solution-phase structures were elusive, nonetheless the CPL spectra were extraordinarily sensitive to details of pH. Generally, it was noted that the most intense CPL was found above neutral pH values.

It has been suggested by Martin and co-workers (42) from examinations of potentiometric titration curves that many lanthanide complexes were associated into polynuclear species above neutral pH. Presumably, the complexes associated through hydroxide bridges. Brittain studied the energy transfer from Tb(III) to Eu(III) complexes of many of the same ligands as used in the CPL studies at similar ligand–metal ratios and learned that under the conditions where strongest CPL was observed, the complexes were extensively associated into polymeric species (43–45). This result indicated that many of the CPL changes observed in the earlier works could have been due to changes in the nature of the emitting species and might have nothing to do with the nature of the metal–ligand bonding.

However, one ligand system was found in which the complexes were found to be monomeric at all pH values. Pyridine-2,6-dicarboxylic acid (dipicolinic acid, or DPA) complexes of Tb(III) and Eu(III) apparently displayed no tendency to associate, probably as a result of the extreme bulkiness of the ligands. The coordinatively saturated 1:3 metal–DPA complexes displayed no associative tendencies (46), nor did the coordinatively unsaturated 1:1 and 1:2 metal–DPA complexes (47).

Brittain was able to take advantage of the fact that $Tb(DPA)^+$ and $Tb(DPA)_2^-$ contain bound solvent molecules that may be displaced by a second chiral ligand, thus permitting CPL studies to be carried out on complexes known to be monomeric at all pH values. These studies permit an easy method where one may study the interaction of the lanthanide ion and only one (or at most two) chiral ligands, and thus isolate the bonding situations where vicinal or vicinal–conformational effects determine the chirality of the Tb(III) ion. The DPA ligand thus serves a number of purposes: (1) it acts as a means by which polymeric association of the complexes is prevented, even at high pH; (2) it acts as a filler and restricts the number of chiral ligands that may enter the inner coordination sphere

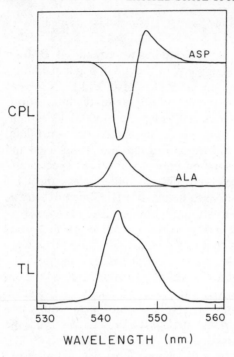

Fig. 6.5. CPL spectra obtained within the $^5D_4 \rightarrow {}^7F_5$ emission band system of the L-aspartic acid and L-alanine complexes with $Tb(DPA)_2^-$. The TL spectrum was essentially the same for each complex and is shown as the lowest trace. (Spectra adapted from Figs. 1 and 2 of Ref. 49, with permission of the copyright owner.)

of the Tb(III) ion, and (3) it serves as an efficient sensitizer of the Tb(III) emission (the DPA ligand irradiated, and the absorbed energy is transferred nonradiatively to the metal ion). The very large binding constants of DPA for lanthanide ions (48) ensure that it will not be displaced from the metal coordination sphere.

Brittain studied the CPL spectra obtained in mixed-ligand Tb–DPA complexes containing L-alanine, L-valine, L-leucine, L-isoleucine, L-serine, L-threonine, and L-aspartic acid, and in this work was able to see that the CPL spectra associated with chirality due solely to vicinal effects was quite different in sign and magnitude when compared to chirality due to a combination of vicinal and conformational effects (49). In Fig. 6.5 the CPL spectra obtained within the $^5D_4 \rightarrow {}^7F_5$ Tb(III) transition is shown for Tb(DPA)$_2$(L-alanine) and Tb(DPA$_2$)L-aspartic acid), and one may easily note the difference between the situations where the chiral ligand can only bind in a monodentate fashion (leading to the vicinal effect determining the chirality) as opposed to a bidentate fashion (permitting the conformational effect to contribute). It was also demonstrated that both serine and threonine could bind to the Tb(III) ion in a bidentate manner (49). The distinction between CPL spectra associated with pure vicinal effects versus spectra obtained when vicinal–conformational effects can contribute was subsequently addressed and found to be a general effect for the Tb(DPA) complexes: If the CPL of the $^5D_4 \rightarrow {}^7F_5$ transition is single-signed only, then only the vicinal effect is operative (50). The sign of the vicinal effect CPL is directly relatable to the absolute configuration of the asymmetric atom in the chiral ligand.

Brittain has also carried out a series of studies in which the mixed-ligand complexes formed between Tb–DPA and chiral α-hydroxycarboxylic acids have been examined by CPL spectroscopy. One sequence of studies examined the optical activity of the Tb(DPA) and Tb(DPA)$_2$ complexes containing L-lactic acid (51,53), L-mandelic acid (51,53), L-malic acid (52,53), L-hydroxyisocaproic acid (53), L-arginnic acid (53), L-phen-yllactic acid (51,53), L-hydroxyglutaric acid (53), and D-isocitric acid (53). In these studies examination of the CPL spectra demonstrated bidentate attachment of the chiral ligands, and with malic acid a transition from the bidentate to terdentate mode was observed (52). This bonding change was indicated by a total inversion of the CPL spectra, as may be seen in Fig. 6.6. Das Gupta and Richardson also studied mixed-ligand complexes containing DPA and also employed N,N'-ethylenebis-α-(o-hydroxyphenyl)-glycine) as the achiral ligand (54). In this latter work, the sequence of hydroxycarboxylic acids was extended to include tartaric acid.

The effect of the achiral ligand on the observed CPL has been investigated further, and it has been found that replacing the DPA ligand by either sulfosalicylic (55) or phthalic acids (56) can not only alter the optical activity for a given mode of coordination, but in fact prevent malic acid from binding in the terdentate mode. Additional work using aminopoly-carboxylate ligands as the achiral portion of the coordination sphere confirms these observations (57). It is presumed that the conformation of the chiral ligand can be perturbed by the other ligands and that these conformational changes lead to differing f–f chirality as experienced by the metal ion.

The excited-state optical activity of lanthanide complexes with inherently chiral aminopolycarboxylate ligands has only recently received study, although CD studies on such complexes were the first chiroptical studies carried out on lanthanide complexes (58). Murata and Morita have recently reported studies involving the Tb(III) complex of L-ammoni-aisopropionicdiacetic acid (59) in communication form. Brittain and Pearson (60) have carried out a detailed investigation of the Tb(III) and Eu(III) complexes of R-propylenediaminetetraacetic acid and R,R-$trans$-1,2-cyclohexanediaminetetraacetic acid. The utility of CPL spectroscopy in the protonation equilibria associated with these ligands was demonstrated in that extensive conformational changes within the ligand system could

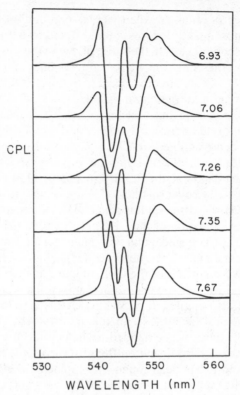

Fig. 6.6. CPL spectra accompanying the bonding transition from bidentate (low pH) to terdentate (high pH) associated with the L-malic acid complex of Tb(DPA)$_2^-$. (Data from Fig. 4 of Ref. 52, with permission of the copyright owner.

be detected between pH 10 and 11.5, where the ligand becomes totally deprotonated and new mixed-ligand hydroxy complexes form.

Richardson and co-workers have carried out a series of studies in which the optical activity of lanthanide ions in a variety of environments has been studied. The CPL spectra of Tb(III) and Eu(III) complexes with nucleosides (61,62), nucleotides (63), carbohydrates (64), and chiral crown ethers (65) have been obtained, and the results have been used to obtain information regarding the stereochemistry of these complexes in aqueous and nonaqueous media.

Lanthanide Complexes Containing β-Diketone Ligands

Since the first demonstration by Hinckley (66) that tris(β-diketonate) lanthanide complexes could be used as shifting reagents for the simplification of overlapping nuclear magnetic resonance lines, the utility of these materials has been demonstrated many times. The ability of lanthanide ions to expand their coordination numbers beyond six makes it possible to form ternary complexes:

$$Ln(DK)_3 + S \rightleftharpoons Ln(DK)_3(S) \tag{6.7}$$

and to thus bring the nuclei of the substrate into the magnetic environment of the paramagnetic lanthanide ion.

Many of the studies that have been carried out have involved β-diketone ligands that are inherently achiral and in which the substrate molecule is chiral (one example was that Hinckley's original study, in which a ternary complex was formed with cholesterol), thus rendering the adduct chiral. Other workers have synthesized lanthanide complexes containing chiral β-diketone ligands (67,68) for the nmr determination of enantiomeric purity, with the chiral ligands usually being derived from d-camphor. Structures of the two classes of β-diketone ligands may be found in Fig. 6.7.

The first CPL studies carried out on lanthanide β-diketonate complexes concerned the Eu(III) derivative of 3-trifluoroacetyl-d-camphor (TFAC, or where $R_3 = CF_3$ in structure II of Fig. 6.7). The optical activity of Eu(TFAC)$_3$ was studied in 28 neat solvents, but CPL was only observed in eight of these (69). When CPL could be observed, it was found that the lineshapes were solvent independent and differed only in magnitude as the solvent was changed. With dimethyl sulfoxide, one of the components of $^5D_0 \rightarrow {}^7F_1$ transition was found to be totally circularly polarized.

The origin of this optical activity, and an explanation for the lack of CPL in many solvents, requires that an understanding of the processes accompanying the solvation of these complexes be understood. Brittain and Richardson also examined the behavior of Eu(HFBC)$_3$(HFBC = 3-heptafluorobutyryl-d-camphor, $R_3 = C_3F_7$ in Fig. 6.7) in a variety of solvents (70) and found that in diemthyl sulfoxide solvent the sign of the optical activity was opposite for Eu(HFBC)$_3$ relative to Eu(TFAC)$_3$ even though the β-diketone ligands were both prepared from d-camphor. These results suggested that the metal chirality was due primarily to configurational effects, as the sign of the vicinal effect was required to be the same in each case. (No contribution could be expected from the conformational effect since the β-diketone ring is planar.)

The configurational nature of this solvent-induced optical activity was conclusively demonstrated by Chan and Brittain, who prepared and characterized a series of mixed-ligand Eu(III) compounds containing chiral

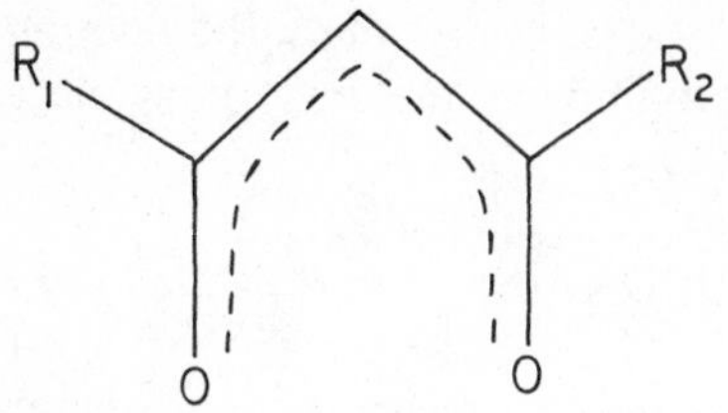

DPM	$R_1 = C(CH_3)_3$	$R_2 = C(CH_3)_3$
FOD	$R_1 = C(CH_3)_3$	$R_2 = C_3F_7$
TTFA	$R_1 = CF_3$	$R_2 = C_4H_3S$
BZAC	$R_1 = CH_3$	$R_2 = C_6H_5$
DBM	$R_1 = C_6H_5$	$R_2 = C_6H_5$

ATC	$R_3 = CH_3$
FACAM (TFAC)	$R_3 = CF_3$
HFBC	$R_3 = C_3F_7$

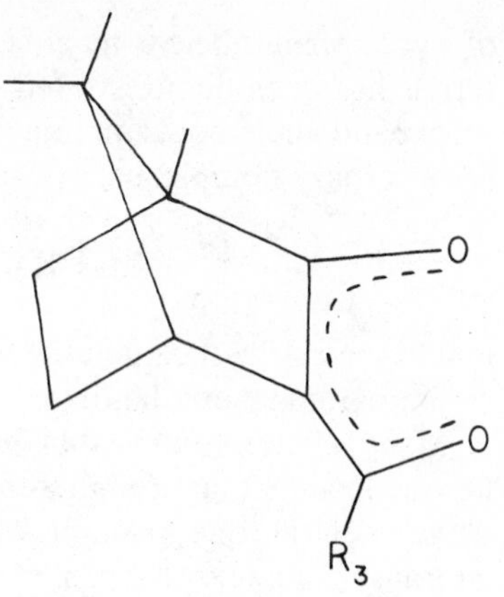

Fig. 6.7. Structures of the achiral and chiral β-diketone ligands discussed in the present review.

and achiral β-diketones (71). The CPL spectra of Eu(DK)$_2$(TFAC) or Eu(DK)$_2$(HFBC) could be obtained in noncoordinating solvents, the signs of the observed CPL bands were solvent independent, and the magnitude of the observed optical activity was only 10% that seen in the tris TFAC or HFBC chelates in the coordinating solvents. In the mixed-ligand chelate systems, only vicinal contributions are possible and the weak CPL associated with these systems identifies the source of the optical activity in the pure complexes. It was hypothesized that formation of the solvent adduct causes crowding of the bulky camphorato ligands, and these adopt the configuration of lowest energy. However, since all three ligands in the pure compounds are inherently chiral, the preference for any isomer is in fact a partial resolution of the complex.

With this assumption it is clear that the solvents which do not cause extensive crowding of the ligands (and therefore have small steric requirements themselves) will not lead to measurable CPL. The substrates that lead to the largest CPL actually are observed to form 1:2 chelate–substrate adducts, and these have been studied recently in great detail. The effect associated with the steric nature of the substrate was investigated by measuring the CPL of Eu(TFAC)$_3$ and Eu(HFBC)$_3$ with a series of sulfoxides, sulfones, and phosphate esters (72) and with a series of formamide and acetamide derivatives (73). The stereoselective interaction of these chelates containing chiral β-diketones has also been studied with Tb(ATC)$_3$ (ATC = 3-acetyl-d-camphor, with R$_3$ = CH$_3$ in Fig. 6.7), and the different acidity of the nonfluorinated ligand permitted quite different chemistry (74). Depending on the steric nature of the substrate, different diastereomers of the Tb(III) complex could be produced in excess.

An entirely separate sequence of studies involving lanthanide β-diketone complexes involves the induction of dissymmetry in achiral chelates by the coordination of a chiral substrate. Such a possibility was first studied via CPL spectroscopy by Brittain and Richardson (70), who examined the chirality obtained when Eu(DPM)$_3$ (DPM = dipivaloylmethane, and where R$_1$ = R$_2$ = C(CH$_3$)$_3$ in Fig. 6.7) and Eu(FOD)$_3$ (FOD = 6,6,7,7,8,8,8-heptafluoro-2,2-dimethyl-3,5-octane, and where R$_1$ = C(CH$_3$)$_3$ and R$_2$ = C$_3$F$_7$ in Fig. 6.7) were dissolved in resolved α-phenethylamine. In a later study it was shown that the sign of the CPL within the $^5D_4 \rightarrow {}^7F_5$ transition of Tb(DPM)$_3$ could be used to determine the absolute configuration of the chiral substrate inducing the optical activity (75).

By studying the CPL spectra of complexes in which the chiral substrate is able to attach to the Eu(III) β-diketone complexes in a bidentate fashion, it has been possible to examine the conformational effect in these chelate systems. Adducts of Eu(TTFA)$_3$ (TTFA = theonyltrifluoroacetone, and has R$_1$ = CF$_3$ and R$_2$ = C$_4$H$_3$S) with cinchona alkaloids revealed that the sign of the observed CPL could be correlated with the absolute configuration of one of the substrate asymmetric atoms and that the different alkaloids were able to induce varying degrees of chirality at the metal ion (76). CPL studies were also carried out using aliphatic amino alcohols to induce optical activity in Eu(TTFA)$_3$ (77) and Eu(FOD)$_3$ (78), and in these works the emission intensities were also used to calculate the formation constants of the adduct complexes. Finally, in a comprehensive study involving 10 chiral phenylalkylamines and phenylalkylamino alcohols and their adducts with Eu(FOD)$_3$ and Eu(TTFA)$_3$, it was shown that the optical activity associated with monodentate substrate binding was significantly different from that observed on bidentate attachment and different yet from chirality arising from terdentate bonding by a substrate (79).

It is always assumed in NMR spectroscopy that formation of the adduct complex does not alter either the conformation or structure of the substrate. This is not always the case, as was demonstrated during the course of CPL investigations in which chiral amino alcohols were bound to Eu(BZAC)$_3$ (BZAC = benzoylacetone, and has R$_1$ = CH$_3$ and R$_2$ = C$_6$H$_5$ in Fig. 6.7) and Eu(DBM)$_3$ (DBM = dibenzoylmethane, with R$_1$ =

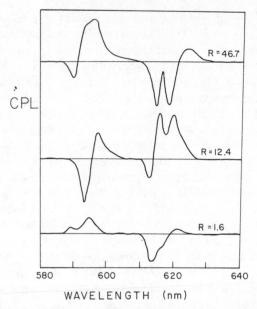

CPL

R = 46.7

R = 12.4

R = 1.6

580   600   620   640

WAVELENGTH (nm)

Fig. 6.8. CPL spectra obtained during the titration of Eu(BZAC)$_3$ with D-phenylglycinol. Spectra are shown for the $^5D_4 \rightarrow {}^7F_5$ emission bands, and the mole ratio of phenylglycinol to Eu(III) chelate accompanies the corresponding spectrum. (Data from Fig. 4 of Ref. 80, with permission of the copyright owner.)

$R_2 = C_6H_5$) (80). In the normal situation (75–79), addition of the chiral substrate results in the development of a particular CPL lineshape that intensifies as the adduct becomes more fully formed. Full complexation of the chelate results in no further CPL changes. However, with Eu(BZAC)$_3$ and Eu(DBM)$_3$, the CPL spectra underwent drastic changes as a function of substrate concentration, as has been illustrated in Fig. 6.8. These CPL changes were finally determined to arise as a result of Schiff-base formation between the amino alcohol substrates and the co-ordinated β-diketone ligand. Such behavior was not observed with the fluorinated β-diketone ligands, as the Schiff-base complexes of these ligands are prone to hydrolysis.

Very recently, Calienni and Brittain have begun to synthesize and characterize a series of Schiff-base complexes, in which salicylaldehydes and β-diketones have been condensed with chiral amines and amino alcohols. The Eu(III) derivatives of these are often highly luminescent and display very strong optical activity characteristic of a merged sum of vicinal, conformational, and configurational effects (81).

CPL Studies of the Pfeiffer Effect

The observation of optical activity in a racemic mixture of a labile metal complex on addition of a chiral material was first observed by Pfeiffer (82), and the effect has been named by Brasted and Dwyer after its discoverer. In most instances, the effect is produced when outer-sphere complexation takes place between the metal complex and the chiral environment substance, and this interaction perturbs the enantiomeric equilibrium. The net effect of this perturbation is the production of measurable optical activity, which often is identical in lineshape to that obtained when the metal complex is resolved by chemical means (83). While CD studies of a wide variety of Pfeiffer-active transition metal complex systems have been reported, no CPL investigations have yet been made.

On the other hand, while no CD studies of Pfeiffer-active lanthanide systems have been reported, CPL studies on trigonal complexes have appeared in the literature. It is known that the symmetry of the $Ln(DPA)_3^{3-}$ series of complexes is approximately trigonal in solution, and therefore the complex consists of a pair of rapidly interconverting enantiomers (84). That the enantiomeric equilibrium can be upset was first shown by the addition of resolved tris(ethylenediamine)Cr(III) to a solution of $Tb(DPA)_3^{3-}$ (85). Formation of the expected ion pair yielded strong CPL, and it was shown that decreasing the dielectric constant of the solvent resulted in an intensification of the CPL intensity (as would be expected for an electrostatic attraction).

However, a second mode of bonding between $Tb(DPA)_3^{3-}$ and a potential environment substance was shown when CPL was observed on addition of L-ascorbic acid to the Tb(III) complex (86). Here the Pfeiffer effect could be observed only as long as the lactone ring remained intact; hydrolysis of this functionality to yield L-diketogluconic acid led to complete loss of CPL. This effect suggested that a hydrophobic site might exist between the DPA ring bound to the metal ion. Positively charged substrates interact favorably with the $Tb(DPA)_3^{3-}$ complex, and the Pfeiffer effect induced by amino acids containing nitrogen atoms bound in a ring system (histidine and proline derivatives) has been reported (87).

The nature of the hydrophobic site has been examined in great detail using a wide variety of environment substances to determine the conditions under which the outer-sphere bonding might take place. The $Tb(DPA)_3^{3-}$ and $Eu(DPA)_3^{3-}$ complexes were all found to exhibit the Pfeiffer effect with monoamino- and diaminocarboxylic acids (88), derivatives of tartaric acid (89), and phenylalkylamines, phenylalkylamino alcohols, and phenylalkylamino acids (90). In these studies, CPL spectroscopy has been used in conjunction with NMR and absorption spectroscopies to verify the conditions under which binding takes place, what the probable mechanism is, and where the bonding is outer sphere in nature. A very interesting effect was noted with long-chain diamino carboxylic acids; here the CPL spectra were found to invert sign at elevated pH values, indicating a change in the mechanism for the association not observed for the short-chain substances.

Solid-State CPL Studies

While the first CPL studies were made on crystalline samples of sodium uranyl acetate (14,15), comparatively little work has been carried out using CPL spectroscopy in the solid state. A partial CPL spectrum of the uranyl ion in the sodium uranyl acetate has been presented recently (91), however, and several other investigations have recently been reported.

Peacock and Stewart have studied the CPL spectrum of $Cr(en)_3^{3+}$ in the uniaxial host crystal 2 $Rh(en)_3Cl_3 \cdot \frac{1}{4} \cdot NaCl \cdot 6H_2O$ (92) and have thus been able to improve on the solution phase studies mentioned earlier (35). Two distinct emitting species were identified during the course of this study. Like sodium uranyl acetate, benzil happens to crystallize in a chiral space group, and the CPL spectrum of this crystalline material has been obtained (93).

Perhaps the most detailed studies have involved the lanthanide complexes of oxydiacetic (diglycolic) acid. The CD spectra of the Eu(III) and Pr(III) ions have been obtained in the crystalline material that crystallizes in the R32 point group (94), and the sign of the CD associated with the Eu(III) complex has been correlated with the absolute configuration of the metal ion (95). Richardson and co-workers have carried out extremely detailed chiroptical investigations on the Eu(III) (96) and Tb(III) (97) oxydiacetate complex systems and have attempted to correlate the results of their investigations with predictions of a theoretical model. While good agreement is obtained in a number of situations, the authors indicate that improvement of the calculated results is desirable.

6.2.4. Results Obtained on Biomolecular Systems

CPL spectroscopy forms a useful complement for CD spectroscopy in the study of biomolecular systems in that the chiroptical information obtained is invariably associated with the immediate environment of the fluorophore. Thus, while the CD signal could contain contributions from many sources and therefore cancellation of CD due to opposing rotational strengths might be possible, this situation is much less likely within the emission spectrum. It is entirely possible that the magnitude of the CPL might actually be much larger than the CD of the same system. For these reasons CPL studies have been carried out on a variety of biomolecular systems where the luminescence either is intrinsic to the system or arises from the binding of a probe.

CPL Studies of Intrinsic Fluorescence

Intrinsic protein fluorescence is usually associated with the tyrosine (phenol fluorophore) or tryptophan (indole fluorophore) residues, and if both aromatic types are present then the fluorescence is often dominated by emission from the tryptophan moiety. The group at the Weizman Institute (led by Steinberg) has carried out extensive and detailed CPL studies and has found that in every case the CPL originating from fully reduced or denatured proteins was equal to zero within experimental error (98). On the other hand, the CPL observed in native proteins can be sizable and thus reflects the effects experienced by the fluorophore as a result of the secondary and tertiary structure of the protein.

Homogeneity and variability in the structure of azurin molecules was studied by CPL spectroscopy through fluorescence originating out of the tryptophan residue (99). It was determined from additional CD measurements that the excited-state geometry is essentially the same as the ground-state geometry and that restricted portions of the molecule may not be homogeneous across the entire population. The fluorescence associated with the Novo and Carlsberg type subtilisins is found to contain contributions from both the tyrosine and tryptophan residues, and the CPL spectra of these protein systems demonstrated that the overlapping emission could be readily resolved (100). In this situation CPL spectroscopy was able to probe a specific fluorophore and thus determine stereochemical information regarding the local environment of this fluorophore.

CPL studies of antibody and antibody–ligand systems have been carried out and the results have been used to understand the processes associated with these systems. The CPL of protein 315 and its fragments (Fab' and Fv) has been obtained (101) and used to study the subunit interactions. It had been known that the interaction of heavier chains in the fragments led to greater resistance toward denaturation by urea, and the CPL spectra of the tryptophan residues were markedly different with the different chain lengths. Conformational changes induced in antibody molecules and in their Fab fragments on binding of antigens (102) and oligosaccharides (103) have also been examined using CPL spectroscopy.

The CD and CPL spectra of nicotinamide adenine dinucleotide (NADH) in aqueous solution were found to be markedly different, and apparently were not affected by cleavage of the coenzyme by phospho-diesterase (104). It was thus concluded that the spectral differences could not be ascribed to extended versus folded conformations of the NADH and thus had to be due to excited-state conformational changes. Binding of the NADH to horse liver alcohol dehydrogenase and beef heart lactate dehydrogenase revealed structural differences between the binding sites. The utility of the two chiroptical techniques was indicated in this latter study, and workers contemplating future photophysical studies of protein stereochemistry would be well advised to examine their systems by both means.

Studies Involving Fluorescent Dye Probes

The utility of organic fluorescence probes in studies of the structure of proteins and enzymes is well known (105,106), with derivatives of naph-thalene sulfonic acid receiving the widest use. When bound to inherently chiral polymolecular structures, these fluorescent dyes can acquire optical activity. Schlessinger and Steinberg studied the CPL spectra of 2-p-to-luidinylnaphthalene-6-sulfonate (which does not bind at the active site) bound to chymotrypsin and of anthraniloyl chymotrypsin (with the flu-orescent probe bound at the active site); they found that in both fluoro-phores a conformational change accompanied the excitation process (107). However, the effect was found to be particularly large for the TNS chrom-ophore, indicating that the orientation of the dye, its freedom of rotation, its strength of binding, or even the binding site itself could have changed on electronic excitation. Comparison of the dissymmetry factors asso-ciated with the CPL and CD spectra may be facilitated by examination of Fig. 6.9.

The interaction of acridine dyes (acridine orange and 9-amino-acridine) with polyadenylic acid and calf thymus DNA also has been studied by both CD and CPL spectroscopies (108). It was found that the optical activity of acridine orange bound to DNA was essentially the same in the ground state as in the excited state, but with the poly-A complexes the state of aggregation of the dye yielded varying behavior in the CPL spec-tra. A very interesting result was obtained when proflavine was com-plexed with DNA; here appreciable ground-state optical activity was noted while the complex was optically inactive in the excited state (108).

The binding of several dansyl derivatives (dansylamide, dansyl-L-ly-sine, dansyl-L-alanine, and dansyl-L-alanine amide) by antidansyl anti-bodies was studied by means of fluorescence and CPL spectroscopy (109). These studies were used to examine the nature of the interaction between the antibody and the hapten. Linear polarization within the emission

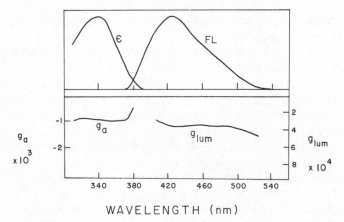

Fig. 6.9. Spectroscopic data obtained for anthraniloyl-chymotrypsin. Upper: absorption (ϵ) and fluorescence (FL) spectra. Lower: absorption anisotropy factor [$g_a = \Delta\epsilon/\epsilon$)] and luminescence anisotropy factor [$g_{lum} = (2 \Delta I/I)$]. All data were obtained at a protein concentration of 40 μM. (Data from Fig. 3 of Ref. 107, with permission of the copyright owner.)

bands was also used to study the complexes and indicated that the transition dipole moments related to the various vibronic states do not have the same spatial direction.

The use of N^6-etheno-adenine dinucleotide as a fluorescent probe in the study of various dehydrogenase enzymes has been investigated, and the possible structural information extracted with CPL spectroscopy. This probe is the fluorescent analog of nicotinamide adenine dinucleotide and was interacted with rabbit muscle glyceraldehyde-3-phosphate dehydrogenase (110), horse liver alcohol dehydrogenase, beef liver glutamate dehydrogenase, and pig heart malate dehydrogenase (111). It was found in these studies that binding of the first mole of coenzyme produces the greatest structural change and that the structure of the adenine subsite in the rabbit dehydrogenase is substantially different from that of the other three enzymes.

Studies Using Tb(III) as a Luminescent Probe

The biochemical requirements for Ca(II) cannot be overstated as this ion plays a significant role in a wide variety of regulatory processes. But since the electronic transitions of Ca(II) do not fall into readily accessible regions of the spectrum and since it does not possess the unpaired electrons or suitable magnetic nucleus required for magnetic resonance spectroscopy, studies involving Ca(II) have been limited to radiotracer experiments. However, the lanthanide ions possess chemical and physical properties that allow these metal ions to function as excellent probes for Ca(II) on replacement (112). Martin and Richardson have classified lanthanide ion interactions as belonging to four main categories (113): Ca(II) proteins inhibited by substitution of Ln(III); Ca(II) proteins, which function similarly on Ln(III) substitution; specific Ln(III) interactions with proteins not normally considered to contain Ca(II); and Ln(III) substitution for metal ions other than Ca(II).

Of all the lanthanide ions, only Tb(III) is found to exhibit luminescence on UV excitation. The luminescence process actually can be sensitized by irradiation between 290 and 300 nm of tryosine or tryptophan aromatic residues, with the absorbed energy being readily transferred in a nonradiative fashion to the Tb(III) ion (113). Eu(III) is not found to experience this emission enhancement [which can be as large as 10^6 for Tb(III) in some situations] because of an unfortunate charge transfer absorption (114). However, both Tb(III) and Eu(III) can be directly excited with a suitable laser line to obtain a considerable amount of information regarding the structure of the metal site (115).

The two Fe(III) atoms of transferrin and conalbumin were replaced by Tb(III), and CPL spectra were then obtained within the Tb(III) emission (116). In this study (and all subsequent works), only data obtained within the $^5D_4 \rightarrow \, ^7F_5$ were obtained. Examination of several mixed-metal transferrin preparations provided evidence that the two metal sites were equivalent in structure and conformation. It was also found that the metal binding sites of conalbumin were essentially identical in their stereochemical properties, in spite of the disparity in the origin of these proteins.

Donato and Martin (117) then described the conformation of the metal binding site of carp muscle parvalbumin B by replacing the Ca(II) by Tb(III) and using CPL spectroscopy as the chiroptical probe. In this protein the excitation energy was channeled into the Tb(III) ion by means of a phenylalanine side chain as there are no tyrosines or tyrptophans in this protein. Examination of the CD spectra in the UV region provided evidence that no structural change in the protein secondary and tertiary structure accompanied the binding of the Tb(III) ion, even though the lanthanide ion was found to bind much more tightly than did the Ca(II) ion. It should be pointed out that the CPL spectrum shown in Fig. 2 of Ref. 117 is inverted, so that the major CPL peak is actually negative and not positive as shown.

The CPL spectrum of rabbit troponin was obtained after Tb(III) replacement, and the resulting optical activity was compared to that of parvalbumin (118). These two muscle proteins share a number of homologies, and it was found that the CPL of the TN-C component of troponin

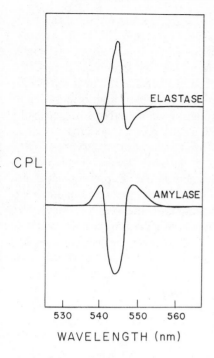

CPL

530   540   550   560

WAVELENGTH (nm)

Fig. 6.10. CPL spectra associated with the $^5D_4 \rightarrow {}^7F_5$ emission band systems of the Tb(III) complexes with porcine elastase and bacterial α-amylase at pH 6.5 in piperazine buffer. (Data from Fig. 2 and 5 of Ref. 120, with permission of the copyright owner.)

was essentially identical to that of the parvalbumin. In a subsequent study, it was found that the CPL of Tb(III)-substituted bovine cardiac troponin-C was also identical with that of the carp parvalbumin and rabbit troponin-C (119). An overlay of the amino acid sequences of the three protein systems revealed that in each system, the most easily substituted metal site was near to one of the aromatic residues. Addition of the other troponin subunits, TC-I and TN-T, was found to weaken the total emission and completely quench the CPL.

The most detailed examination of CPL in Tb(III)-substituted proteins was reported by Brittain, Richardson, and Martin (120). In this study, 40 different protein systems were examined [not all of which were known to bind Tb(III) or Ca(II)], and 36 of these were found to exhibit enhanced Tb(III) emission characteristic of binding at a definite site. However, only nine of these protein systems were found to exhibit CPL: carp parvalbumin, rabbit troponin-C, bovine cardiac troponin-C, pronase, porcine elastase, collagenase, bacterial α-amylase, porcine α-amylase, and thermolysin. Interestingly, the CPL lineshape invariably took one particular form and occurred as either mirror image of this form (as is shown in Fig. 6.10). Comparison of the CPL lineshapes and magnitudes with situations of configurational optical activity (85–90) indicates the origin of the molecular chirality, and a possible explanation as to why CPL is not always observed. If the Tb(III) ion is bound at a site where it is able to experience a chiral environment created by the protein tertiary structure, then CPL will be observed. Binding of the Tb(III) ion at a surface site or in a pleated sheet (and not in a helical environment) does not lead to configurational effect, and apparently pure conformational or vicinal effects are not sufficient to permit the observation of CPL in the protein systems.

However, other reasons may exist for the lack of observable CPL. Brittain et al. (120) did not observe CPL in Tb(III)-substituted trypsin when the protein was obtained from a commercial source. Epstein and co-workers (121) carefully purified trypsin and separated components that did exhibit CPL on Tb(III) binding. Presumably, the negative result obtained in the earlier study (120) arose from cancellations within overlapping CPL spectra, which in turn were derived from the presence of different types of binding sites in the different protein types present.

The CPL of Tb(III)-substituted elastase were reexamined by Bieth and co-workers, and the conformational changes induced by the binding of inhibitors were also considered (122). Turkey ovomucoid, human plasma α_1-proteinase inhibitor, and α_2-macroglobulin did not prevent the binding of Tb(III), although the latter two proteins induced considerable conformational changes in elastic. These conformational changes could be studied from detailed studies of the CPL spectra of the free and inhibited enzymes.

The CPL spectrum of lasolocid A (X537A) has been used to study the conformation of this antibiotic, and the conformational changes that accompany the binding of K(I), Na(I), Rb(I), Cs(I), Mg(II), Ca(II), Sr(II), and Ba(II) (123). The CPL spectra were found to be solvent dependent, as well as being metal ion dependent. Richardson (124) used CPL spectroscopy in conjuction with a variety of photophysical measurements to examine the binding of lanthanide ions by the same antibiotic system. During the course of these latter studies it was established that the probable metal ion binding site was at the salicylate functionality.

6.3. OPTICAL ACTIVITY OF THE TRIPLET STATE

The optical activity of singlet–triplet transitions in organic molecules has not received much attention, as the absorption intensity of these bands is quite weak and resulting CD magnitudes are feeble. Nevertheless, such studies are of obvious importance in that stereochemical information is, in principle, available from such studies. Given that many photochemical reactions take place in the excited triplet state, acquisition of any geometrical information regarding this state would be useful to a wide range of chemists.

At one point it was believed that no technique could be developed that would place a sufficient number of chiral molecules in their excited states to permit a measurement of optical activity (125). However, the study of organic molecules by flash photolysis demonstrated that it was indeed possible to obtain a sufficient population of excited molecules, and as a result several approaches have been made in the study of the optical activity of the triplet state.

The first attempt to obtain chiroptical data on excited systems involved a measurement of optical rotatory dispersion (ORD) during the flash photolysis of benzoin (126). The optical activity of triplet benzoin was obtained in a commercial spectropolarimeter after excitation by a microsecond-lifetime pulsed Xe flashlamp. The intense excitation was found to induce photoracemization of the sample, but it could be determined that the ground- and excited-state optical activities were somewhat similar. It was also found that the excited-state ORD curves obtained with the two enantiomers of benzoin were roughly mirror images of each other, as would be expected.

Freezing the chiral molecules in a glass at cryogenic temperatures (and purging of molecular oxygen) extends the lifetime of the triplet state to at least several seconds, and thus with continuous optical pumping of the sample time resolution ceases to be a problem. Tetreau and Lavalette (127) have taken this approach to obtain CD spectra of chiral molecules in the excited triplet state and have also provided a discussion of the pitfalls that accompany work at low temperatures in glasses. One must be extremely careful to avoid artifacts when taking this approach; one must also be prepared for the possibility that the observed optical activity only exists at the cryogenic temperature.

Another method by which the optical activity of the triplet state can be obtained is of CPL spectroscopy. Circular polarization within the phosphorescence spectrum contains the same chiroptical information as a CD measurement on the triplet state but is somewhat simpler to perform. The spectra still need to be obtained at cryogenic temperatures, however, and as a result the data interpretation must be handled carefully.

Dekkers (128) has obtained the CPL spectrum of thiocamphor at 77 K in a methylcyclohexane–isopentane glass, as well as the CD spectrum under the same conditions. Combination of these studies provided evidence that the weak absorption band observed at the onset of the $^1(n \rightarrow \pi^*)$ transition was that of a singlet–triplet transition. The dissymmetry factors associated with the CD and CPL spectra were very similar in lineshape and magnitude.

Further investigations of the CPL spectra associated with the triplet states of organic molecules have been carried out by Steinberg and co-workers (129). In the case of d-camphorquinone, the optical activities of the lowest excited singlet state (measured by circular polarization of fluorescence) and the lowest triplet state (measured by circular polarization of phosphorescence) were found to be quite different at cryogenic temperatures. The chirality of the triplet state was found to be affected by the nature of the solvent making up the glass, and on the concentration of solute material. Optical activity was induced in benzophenone and acetophenone by formation of 1:1 complex with L-menthol, and the CPL of the triplet states of these two complexes was also obtained. The effect was much more pronounced with benzophenone compared to acetophenone, and this difference was ascribed to differences between the two ketones in the various solvents. It was generally found that inherent differences existed between the optical activity of triplet–singlet transitions relative to that of singlet–singlet transitions.

6.4. FLUORESCENCE DETECTED CIRCULAR DICHROISM

Tinoco and co-workers at the University of California (Berkeley) have developed a new chiroptical technique that uses molecular fluorescence in the determination of the ground-state CD spectrum of a chiral molecule. This method has been termed fluorescence detected circular dichroism (FDCD) and does not provide information regarding excited-state chirality as does CPL spectroscopy. However, since the pathway by which one obtains the CD spectrum is through a fluorescent excited state, it seemed appropriate to include a brief discussion of the experimental method in the present review.

In FDCD the difference in absorption for left- and right-circularly polarized light is obtained by measuring the difference in total luminescence intensity after excitation of the sample with left- and right-circularly polarized light. Crucial to the data interpretation is the assumption (which is usually true) that the excitation spectrum of the molecule matches the absorption spectrum. The CD spectrum is obtained by taking the ratio of FDCD signal to the total luminescence signal:

$$FDCD = \frac{I_R - I_L}{I_R + I_L} \qquad (8)$$

Further experimental details will be presented in the following section.

A general theory relating FDCD and the standard method of measuring CD has been presented (131), with the effects of concentration on the observed FDCD signals and signal-to-noise ratios being considered. As would be expected, only transitions that ultimately lead to fluorescence after absorption of light can contribute to the observed FDCD. The interpretation of FDCD data has caused some concern and has been addressed in a different communication (132). In this latter work three situations that could lead to FDCD were considered, and the experimental observables evaluated in terms of theoretical parameters. Considered were cases where the optically active material was nonfluorescent but coupled with a nonoptically active fluorescent emitter, where a nonfluorescent nonoptically active material was coupled to a fluorescent optically active material and where nonfluorescent optically active material was coupled to optically active material. These situations are of importance in that essentially all fluorescent centers are planar aromatic systems who sense the chirality of their environment. Thus, for the situation of a tyrosine residue in a protein, the FDCD would be described by the first case.

The effect of photoselection on the FDCD spectrum has been pointed out (133,134). The success of the FDCD experiment rests with the assumption (usually valid) that the excitation spectrum of a fluorescent molecule will parallel the absorption spectrum of the same molecule. However, when the molecule is placed in viscous (where rotary Brownian motion is restricted) or rigid media, the exciting light will not be absorbed equally by molecules of all orientations. As a result, the fluorescence will depend not only on the total number of molecules, but also on the distribution in space of these. If this is the case, then the excitation spectrum will not be similar to the absorption spectrum, and the FDCD spectrum will not match the CD spectrum.

6.4.1 Experimental Methods

A block diagram of a typical FDCD spectrometer is shown in Fig. 6.11. The best source of excitation light would be a high-power Xe lamp (to permit wavelength tuning across the UV and near the UV spectrum), although for determinations in the visible region a tunable laser would function as well. The irradiation wavelength is selected by a scanning monochromator, whose output is then polarized via a linear polarizer. The modulated circularly polarized light is obtained by passing this excitation beam through a driven quarter-wave plate, which may be either a pockels cell or a photoelastic modulator. The light is then allowed to enter the sample cell and produce the fluorescence to be analyzed.

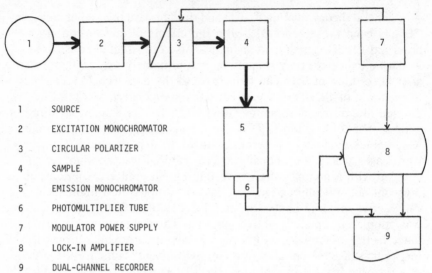

1   SOURCE

2   EXCITATION MONOCHROMATOR

3   CIRCULAR POLARIZER

4   SAMPLE

5   EMISSION MONOCHROMATOR

6   PHOTOMULTIPLIER TUBE

7   MODULATOR POWER SUPPLY

8   LOCK-IN AMPLIFIER

9   DUAL-CHANNEL RECORDER

Fig. 6.11. Block diagram of an emission spectrometer capable of obtaining measurements of fluorescence-detected circular dichroism.

The emitted light is detected at 90° to the excitation beam by a photomultiplier tube. The output of this tube (after the current is converted to voltage) will contain an ac signal proportional to the $I_L - I_R$ fluorescence, while a measurement of the average DC output is proportional to $I_L + I_R$. Taking the analog ratio of these signals yields the FDCD signal, which is displayed as the recorder output. As in the CPL experiment, accurate measurements can be obtained only if the two measurements are made at the same time.

6.4.2   Applications of FDCD

The first report of FDCD presented spectra obtained for morphine, d-10-camphorsulfonic acid, and L-tryptophan (130), and it was found that after suitable data reduction the FDCD and CD spectra were essentially identical. These results indicate that with these systems, the excitation spectrum does indeed match the true absorption spectrum. Addition of L-cystine (which is nonfluorescent) was found to have a profound influence on the observed CD spectrum (which represents the sum of the CD spectra due to the two components), while the FDCD spectrum remained unaffected by the addition of the L-cystine. Since the two components do not form any sort of complex, the experiment clearly demonstrates how the FDCD technique can focus on only the chirality of fluorescent chromophores in a mixture.

FDCD studies on the conformations of the anticodon loop of yeast tRNA have been carried out (135), and several advantages of the fluorescence method have been outlined. The base that is immediately adjacent to the 3' end of the anticodon (termed the Y base) is moderately fluorescent, and thus the FDCD obtained within this fluorescence is specific for conformational changes near the anticodon loop. Conformational changes were studied as a function of temperature to examine the melting of the tRNA, and it was also found that removal of Mg(II) from the tRNA resulted in a decrease in the conformational rigidity of the anticodon loop.

The FDCD spectrum of tryptophan in a series of proteins has been obtained, and the results have been used to study the environment of the fluorophore (137). The protein systems examined were adrenocorticotropic hormone, glucagon, human serum albumin, monellin, and ribonuclease T1, each of which is distinguished in that it contains only a single fluorescent tryptophan. The results indicated that combination of CD and FDCD techniques can yield extremely detailed information regarding macromolecular conformation. In addition, the authors have described a dual photomultiplier detection geometry that is able to eliminate artifacts due to imperfectly circularly polarized light.

The question of artifacts and polarization effects had been addressed earlier (133), and it had been pointed out that photoselection effects could produce large artifacts in the FDCD measurement. These effects were verified by Turner and co-workers, who were able to obtain photoselected FDCD spectra with d-10-camphorsulfonic acid and morphine (137). Genuine FDCD measurements were then obtained on morphine and L-tryptophan by the same group (138) using the dual photomultiplier detection system.

FDCD spectroscopy can be viewed as a highly selective chiroptical technique whose potential has yet to be fully utilized. The main advantage of this method is that, like CPL spectroscopy, it is restricted to the study of luminescent chromophores. While this may actually sound like a limitation, the situation actually represents a great advantage: CD spectra represent the chirality of all absorbing species, but FDCD and CPL only yield data on the emitting species. If information on a single component in a complex mixture is desired, these data may be obtained if that component is luminescent or can be rendered luminescent.

6.5. SUMMARY

The instrumental sensitivity of luminescence spectroscopy has been combined with the stereochemical selectivity of chiroptical spectroscopy in the two techniques that have been discussed in this review. The studies described have all involved natural optical activity, but both CPL (4) and FDCD (139) spectra can be obtained an achiral fluorescent molecules in the presence of a magnetic field. Hampering future efforts in the areas is a lack of commercial instrumentation for measurement of these effects, but this situation could change as the fields continue to grow. Further developments will focus on improved methods of data collection and processing, improvements of the theories and models needed to interpret the observed results, and extension of the methods to include new luminescent systems.

REFERENCES

1. E. Charney, *The Molecular Basis of Optical Activity*, Wiley, New York, 1979.

2. L. A. Nafie and M. Diem, *Acc. Chem. Res. 12*, 296 (1979).

3. L. D. Barron, *Acc. Chem. Res. 13*, 90 (1980).

4. F. S. Richardson and J. P. Riehl, *Chem. Rev. 77*, 773 (1977).

5. J. P. Riehl and F. S. Richardson, *J. Chem. Phys. 65*, 1011 (1976).

6. F. S. Richardson, "Circular Polarization Differentials in the Luminescence of of Chiral Systems," in *Optical Activity and Chiral Discrimination*, S. F. Mason (ed.), D. Reidel, Dordrecht, 1978, p. 189.

7. F. S. Richardson, "Emission Circular Intensity Differentials," in *ORD and CD: Theory, Chemical Practice, and Biochemical Applications*, F. Ciardelli, S. F. Mason, P. Salvadori, and G. Snatzke (eds.), Wiley, New York, 1982.

8. I. Z. Steinberg and A. Gafni, *Rev. Sci. Instr. 43*, 409 (1972).

9. R. A. Shatwell and A. J. McCaffery, *J. Phys. E. 7*, 297 (1974).

10. K. W. Hipps, Ph.D. Dissertation, Washington State University, Pullman, Washington, 1976.

11. I. Z. Steinberg and B. Ehrenberg, *J. Chem. Phys. 61*, 3382 (1974).

12. I. Z. Steinberg, Fluorescence Polarization: Some Trends and Problems, in *Concepts in Biochemical Fluorescence*, R. F. Chen and H. Edelbroth (eds.), Dekker, New York, 1975, p. 79.

13. H. G. Brittain and F. S. Richardson, *J. Am. Chem. Soc. 98*, 5858 (1976).

14. B. N. Samoilov, *J. Exp. Theor. Phys. 18*, 1030 (1948).

15. M. S. Brodin and V. Reznichenko, *Ukr. Phys. J. 10*, 178 (1965).

16. O. Neunhoeffer and H. Ulrich, *Z. Electrochem. 59*, 122 (1955).

17. H. P. J. M. Dekkers, C. A. Emeis, and L. J. Oosterhoff, *J. Am. Chem. Soc. 91*, (41).

18. C. A. Emeis and L. J. Oosterhoff, *Chem. Phys. Lett. 1*, 129, 268 (1967). H. G. Brittain and F. S. Richardson, *Bioinorg. Chem. 7*, 233 (1977).

19. C. A. Emeis and L. J. Oosterhoff, *J. Chem. Phys. 54*, 4809 (1971).

20. A. Gafni and I. Z. Steinberg, *Photochem. Photobiol. 15*, 93 (1972).

21. J. Schlessinger, A. Gafni, and I. Z. Steinberg, *J. Am. Chem. Soc. 96*, 7396 (1974).

22. C. K. Luk and F. S. Richardson, *J. Am. Chem. Soc. 96*, 2006 (1974).

23. A. Gafni, H. Hardt, J. Schlessinger, and I. Z. Steinberg, *Biochim. Biophys. Acta, 387*, 256 (1975).

24. H. G. Brittain and F. S. Richardson, *J. Phys. Chem. 80*, 2590 (1976).

25. P. Pollmann, K.-J. Mainusch, and H. Stegemeyer, *Z. Phys. Chem. 103*, 295 (1976).

26. H. P. J. M. Dekkers and L. E. Closs, *J. Am. Chem. Soc. 98*, 2210 (1976).

27. W. Moffitt, R. B. Woodward, A. Moscowitz, W. Klyne, and C. Djerassi, *J. Am. Chem. Soc. 83*, 4013 (1961).

28. H. Wynberg, H. Numan, and H. P. J. M. Dekkers, *J. Am. Chem. Soc. 99*, 3870, (1977).

29. H. G. Brittain, D. L. Ambrozich, M. Saburi, and J. H. Fendler, *J. Am. Chem. Soc. 102*, 6372 (1980).

30. H. G. Britain, *Chem. Phys. Lett. 83*, 161 (1981).

31. F. S. Richardson, *Inorg. Chem. 19*, 2806 (1980).

32. F. S. Richardson and T. R. Faulkner, *J. Chem. Phys. 76*, 1595 (1982).

33. J. D. Saxe, T. R. Faulkner, and F. S. Richardson, *J. Chem. Phys. 76*, 1607 (1982).

34. C. A. Emeis, Ph.D. thesis, The University of Leiden, The Netherlands, 1968.

35. G. L. Hilmes, H. G. Brittain, and F. S. Richardson, *Inorg. Chem. 16*, 528 (1977).

36. A. Gafni and I. Z. Steinberg, *Isr. J. Chem. 15*, 102 (1977).

37. L. I. Katzin, *Coord. Chem. Rev. 5*, 279 (1970).

38. C. K. Luk and F. S. Richardson, *Chem. Phys. Lett. 25*, 215 (1974).

39. C. K. Luk and F. S. Richardson, *J. Am. Chem. Soc. 97*, 6666 (1975).

40. H. G. Brittain and F. S. Richardson, *Inorg. Chem. 15*, 1507 (1976).

41. H. G. Brittain and F. S. Richardson, *Bioinorg. Chem. 7*, 233 (1977).

42. R. Prados, L. G. Stadtherr, H. Donato, and R. B. Martin, *J. Inorg. Nucl. Chem. 36*, 689 (1974).

43. H. G. Brittain, *Inorg. Chem. 18*, 1740 (1979).

44. H. G. Brittain, *J. Inorg. Nucl. Chem. 41*, 561, 567, 721, 1775 (1979).

45. H. G. Brittain, *J. Lumin. 21*, 43 (1979).

46. H. G. Brittain, *Inorg. Chem. 17*, 2762 (1978).

47. R. A. Copeland and H. G. Brittain, *J. Inorg. Nucl. Chem. 43*, 2499 (1982).

48. I. Grenthe, *J. Am. Chem. Soc. 83*, 360 (1961).

49. H. G. Brittain, *J. Am. Chem. Soc. 102*, 3693 (1980).

50. H. G. Brittain, *Inorg. Chim. Acta 53*, L7 (1981).

51. H. G. Brittain, *Inorg. Chem. 20*, 959 (1981).

52. H. G. Brittain, *Inorg. Chem. 19*, 2136 (1980).

53. H. G. Brittain, *Inorg. Chem. 20*, 4267 (1981).

54. A. DasGupta and F. S. Richardson, *Inorg. Chem. 20*, 2616 (1981).

55. R. A. Copeland and H. G. Brittain, *Polyhedron 1*, 693 (1982).

56. R. A. Copeland and H. G. Brittain, *J. Lumin. 27*, 307 (1982).

57. M. Ransom and H. G. Brittain, *Inorg. Chem. 22*, 2494 (1983).

58. S. Misumi, S. Kida, and T. Isobe, *Spectrochim. Acta 24A*, 271 (1968).

59. K. Murata and M. Morita, *J. Lumin. 26*, 207 (1981).

60. H. G. Brittain and K. H. Pearson, *Inorg. Chem. 22*, 78 (1983).

61. S. A. Davis and F. S. Richardson, *J. Inorg. Nucl. Chem. 43*, 1793 (1980).

62. S. A. Davis, D. Foster, and F. S. Richardson, unpublished results.

63. D. Foster and F. S. Richardson, unpublished results.

64. J. Morley, D. Foster, and F. S. Richardson, unpublished results.

65. D. Foster, R. A. Palmer, and F. S. Richardson, unpublished results.

66. C. C. Hinckley, *J. Am. Chem. Soc. 91*, 5160 (1969).

67. M. D. McCreary, D. W. Lewis, D. L. Wernick, and G. M. Whitesides, *J. Am. Chem. Soc. 96*, 1038 (1974).

68. H. L. Goering, J. N. Eikenberry, G. S. Koermer, and C. J. Lattimer, *J. Am. Chem. Soc. 96*, 1493 (1974).

69. H. G. Brittain and F. S. Richardson, *J. Am. Chem. Soc. 98*, 5858 (1976).

70. H. G. Brittain and F. S. Richardson, *J. Am. Chem. Soc. 99*, 65 (1977).

71. C. K. Chan and H. G. Brittain, *J. Inorg. Nucl. Chem. 43*, 2399 (1981).

72. H. G. Brittain, *Polyhedron, 2*, 261 (1983).

73. H. G. Brittain, *J. Chem. Soc. Dalton Trans.* 2059 (1982).

74. H. G. Brittain, *Inorg. Chem. 19*, 2233 (1980).

75. H. G. Brittain, *J. Am. Chem. Soc. 102*, 1207 (1980).

76. H. G. Brittain, *J. Chem. Soc. Dalton Trans.* 2369 (1980).

77. H. G. Brittain, *Inorg. Chem. 19*, 3473 (1980).

78. X. Yang and H. G. Brittain, *Inorg. Chim. Acta 57*, 165 (1982).

79. X. Yang and H. G. Brittain, *Inorg. Chem. 20*, 4273 (1981).

80. X. Yang and H. G. Brittain, *Inorg. Chim. Acta 59*, 261 (1982).

81. J. Calienni and H. G. Brittain, unpublished results.

82. P. Pfeiffer and K. Quehl, *Ber. 64*, 2667 (1931); *65*, 560 (1932).

83. S. Kirschner, N. Ahmad, C. Munir, and R. J. Pollock, *Pure Appl. Chem. 51*, 913 (1979).

84. H. Donato, Jr., and R. B. Martin, *J. Am. Chem. Soc. 94*, 4129 (1972).

85. J. S. Madaras and H. G. Brittain, *Inorg. Chem. 19*, 3841 (1980).

86. J. S. Madaras and H. G. Brittain, *Inorg. Chim. Acta 42*, 109 (1980).

87. H. G. Brittain, *Inorg. Chem. 20*, 3007 (1981).

88. F. Yan, R. A. Copeland, and H. G. Brittain, *Inorg. Chem. 21*, 1180 (1982).

89. F. Yan and H. G. Brittain, *Polyhedron* (in press).

90. H. G. Brittain, *Inorg. Chem. 1*, 195 (1982).

91. K. Murata, Y. Yamazaki, and M. Morita, *J. Lumin. 18/19*, 407 (1979).

92. R. D. Peacock and B. Stewart, *J. Chem. Soc. Chem. Comm. 295* (1982).

93. H. G. Brittain and F. S. Richardson (unpublished results).

94. A. C. Sen, M. Chowdhury, and R. W. Schwartz, *J. Chem. Soc. Far. Trans. II 76*, 620 (1980); *77*, 1293 (1981).

95. R. R. Fronczek, A. K. Banerjee, S. F. Watkins, and R. W. Schwartz, *Inorg. Chem. 20*, 2745 (1981).

96. J. P. Morley, J. D. Saxe, and F. S. Richardson, *Mol. Phys. 47*, 379 (1982).

97. J. D. Saxe, J. P. Morley, and F. S. Richardson, *Mol. Phys. 47*, 407 (1982).

98. I. Z. Steinberg, J. Schlessinger, and A. Gafni, in *Peptides, Polypeptides, and Proteins*, E. R. Blout, F. A. Vobey, M. Goodman, and N. Lotan (eds), Wiley, New York, 1974, p. 351.

99. A. Grinvald, J. Schlessinger, I. Pecht, and I. Z. Steinberg, *Biochemistry 14*, 1921 (1975).

100. J. Schlessinger, R. S. Roche, and I. Z. Steinberg, *Biochemistry 14*, 255 (1975).

101. J. Schlessinger, I. Z. Steinberg D. Givol, and J. Hochman, *FEBS Lett. 52*, 231 (1975).

102. J. Schlessinger, I. Z. Steinberg, D. Givol, J. Hochman, and I Pecht, *Proc. Natl. Acad. Sci. USA 72*, 2775 (1975).

103. J. C. Jaton, H. Huser, D. G. Braun, D. Givol, I. Pecht, and J. Schlessinger, *Biochemistry 14*, 5312 (1975).

104. A. Gafni, *Biochemistry 17*, 1501 (1978).

105. L. Brand and J. R. Gohlke, *Ann. Rev. Biochem. 41*, 843 (1972).

106. Y. Kanaoka, *Angew. Chem. Int. Ed. 16*, 137 (1977).

107. J. Schlessinger and I. Z. Steinberg, *Proc. Natl. Acad. Sci. USA 69*, 769 (1972).

108. A. Gafni, J. Schlessinger, and I. Z. Steinberg, *Isr. J. Chem. 11*, 423 (1973).

109. J. Schlessinger, I. Z. Steinberg, and I. Pecht, *J. Mol. Biol. 87*, 725 (1978).

110. J. Schlessinger and A. Levitzki, *J. Mol. Biol. 82*, 547 (1974).

111. J. Schlessinger, I. Z. Steinberg, and A. Levitzki, *J. Mol. Biol. 91*, 523 (1975).

112. E. Nieboer, *Struct. Bonding 22*, 1 (1975).

113. R. B. Martin and F. S. Richardson, *Quart. Rev. Biophys. 12*, 181 (1979).

114. W. DeW. Horrocks, Jr., and W. E. Collier, *J. Am. Chem. Soc. 103*, 2856 (1981).

115. W. DeW. Horrocks, Jr., and D. R. Sudnick, *Acc. Chem. Soc. 14*, 384 (1981).

116. A. Gafni and I. Z. Steinberg, *Biochemistry 13*, 800 (1974).

117. H. Donato, Jr., and R. B. Martin, *Biochemistry 13*, 4575 (1974).

118. T. L. Miller, D. J. Nelson, H. G. Brittain, F. S. Richardson, R. B. Martin, and C. M. Kay, *FEBS Lett. 58*, 262 (1975).

119. H. G. Brittain, F. S. Richardson, R. B. Martin, L. D. Burtnick, and C.M. Kay, *Biochem. Biophys. Res. Comm. 68*, 1013 (1976).

120. H. G. Brittain, F. S. Richardson, and R. B. Martin, *J. Am. Chem. Soc. 98*, 8255 (1976).

121. M. Epstein, J. Reuben, and A. Levitzki, *Biochemistry 16*, 2449 (1977).

122. G. Duportail, J.-F. LeFevre, P. Lestienne, J.-L. Dimicoli, and J. G. Bieth, *Biochemistry 19*, 1377 (1980).

123. B. Ehrenberg, I. Z. Steinberg, R. Panigel, and G. Navon, *Biophys. Chem. 7*, 217 (1977).

124. F. S. Richardson and A. Das Gupta, *J. Am. Chem. Soc. 103*, 5716 (1981).

125. E. V. Condon, W. Altar, and H. Eyring, *J. Chem. Phys. 5*, 753 (1937).

126. P. A. Carapellucci, H. H Richtol, and R. L. Strong, *J. Am. Chem. Soc. 89*, 1742 (1967).

127. C. Tetreau and D. Lavalette, *Novveau J. Chem. 4*, 423 (1980).

128. H. P. J. M. Dekkers, Ph.D. thesis, The University of Leiden, The Netherlands, 1975.

129. N. Steinberg, A. Gafni, and I. Z. Steinberg, *J. Am. Chem. Soc. 103*, 1636 (1981).

130. D. H. Turner, I. Tinoco, Jr., and M. Maestre, *J. Am. Chem. Soc. 96*, 4340 (1974).

131. I. Tinoco, Jr., and D. H. Turner, *J. Am. Chem. Soc. 98*, 6453 (1976).

132. T. G. White, Y.-H. Pao, and M. M. Tang, *J. Am. Chem. Soc. 97*, 4751 (1975).

133. B. Ehrenberg and I. Z. Steinberg, *J. Am. Chem. Soc. 98*, 1293 (1976).

134. I. Tinoco, Jr., B. Ehrenberg, and I. Z. Steinberg, *J. Chem. Phys. 66*, 916 (1977).

135. D. H. Turner, I. Tinoco, Jr., and M. F. Maestre, *Biochemistry 14*, 3794 (1975).

136. E. W. Lobenstine, W. C. Schafer, and D. H. Turner, *J. Am. Chem. Soc. 103*, 4936 (1981).

137. E. W. Lobenstine and D. H. Turner, *J. Am. Chem. Soc. 101*, 2205 (1979).

138. E. W. Lobenstine and D. H. Turner, *J. Am. Chem. Soc. 102*, 7786 (1980).

139. J. C. Sutherland and H. Low, *Proc. Nat. Acad. Sci. USA 73*, 276 (1976).

CHAPTER

7

FLUORESCENCE DETECTION IN CHROMATOGRAPHY

A. HULSHOFF and H. LINGEMAN

Pharmaceutical Laboratory
Department of Analytical Pharmacy
State University of Utrecht
Catharijnesingel 60
3511 GH Utrecht
THE NETHERLANDS

7.1. INTRODUCTION

The various modes of chromatography are among the most frequently applied separation methods in many laboratories. Chromatographic systems are characterized by a mobile phase moving through an open bed or a column with a stationary phase. The components of mixtures brought into the chromatographic system are separated because of their different distribution behavior between the two phases. Solute compounds to be separated with a low distribution coefficient K (K = concentration in stationary phase/concentration in mobile phase) migrate with greater velocity through the system (column or bed) than compounds with a high K value. The migration rates depend on the physicochemical properties of the solute molecules, the composition of the stationary and mobile phases, and other experimental variables, such as the temperature.

Chromatographic systems can be classified by the aggregation state of the mobile phase, which is either a gas or a liquid, hence the distinction between gas chromatography (GC) and liquid chromatography (LC). The stationary phase can be a liquid or a solid. Frequently applied systems are gas–liquid chromatography (GLC), liquid–liquid chromatography (LLC), and liquid–solid chromatography (LSC). In paper chromatography (PC) paper is used as support for the (liquid) stationary phase. In thin-layer chromatography (TLC) an adsorbent layer, such as silica gel, is fixed to a suitable, inert plate (e.g., of glass); the adsorbent can either function as the stationary phase or as support for a liquid stationary phase. Contrary to these two flat bed systems, the system in column chromatography is a closed one; the stationary phase is contained in a glass or metal column, through which the mobile phase moves. For obvious reasons, gas chromatography is restricted to the closed column system.

The enormous success of chromatography has been triggered by the development of chromatographic systems with a very high degree of separation power, together with the introduction of extremely sensitive detectors suitable for the in situ (PC, TLC) or on-line column chromatography detection and quantitation of solutes. Chromatographic methods are therefore particularly well suited for the analysis of microquantities of material in complex matrices.

Among the various detection principles used in chromatography, fluorescence detection has gained a prominent position because of its inherent selectivity and sensitivity. Although fluorescence detection can be applied in every chromatographic mode, it has been used mainly in TLC and column LC systems. The number of reports in PC methods is steadily declining, mainly because the modern TLC systems are superior in separation efficiency and the (much shorter) analysis time. Fluorescence detection has found only very limited application in GC. The reasons for this are probably the availability of a number of very sensitive and selective detectors in GC (e.g., the NP-sensitive thermionic detector and the electron capture detector) and, until recently, the inadequate fluorescence detector design, which has prevented the realization of its full potential. Recent studies, however, may induce a renewal of interest in on-line GC fluorescence detectors (1–4). A survey of GC fluorescence detection has been given by Froehlich and Wehry (see Bibliography). Fundamental aspects of liquid chromatographic separation systems, particularly of high-performance liquid chromatography (HPLC) and high-performance) thin-layer chromatography ((HP)TLC), are described in Section 7.2. In Sections 7.3–7.5 the principles and applications of fluorescence detection in TLC and column LC systems will be discussed. For applications the emphasis will be on drug analysis in biological fluids (serum, urine, etc.).

7.2. INTRODUCTION TO LIQUID CHROMATOGRAPHY

A short introduction to the fundamentals and practice of liquid chromatography is presented in the next paragraphs. For a more comprehensive discussion the reader is referred to the extensive literature on chromatography. A selection of books on TLC and column LC appears in the Bibliography.

7.2.1. Chromatographic Parameters in TLC and Column LC

TLC Parameters

A thin layer (about 0.2–0.3 mm) of adsorbent is attached onto a glass plate or other suitable material. The usual procedure is as follows. A solution of the sample is spotted onto the plate, which is then treated with an eluent by placing the TLC plate in a closed tank with a layer of the eluent on the bottom. The mobile phase—through capillary action—passes the sample spot and the components of the sample migrate in the same direction as the ascending mobile phase. When the eluent has covered a certain distance, the chromatographic process is stopped. All components have spent the same time period in the chromatographic system, but they have traveled different distances (Fig. 7.1).

The distance of migration is expressed by the R_f value:

$$R_f = \frac{\text{migration distance of solute}}{\text{migration distance of eluent}} = \frac{\chi}{\chi_{el}} \tag{7.1}$$

The R_f value is related to the distribution coefficient K:

$$R_f = \frac{1}{1 + K\varphi} \tag{7.2a}$$

or

$$R_f = \frac{1}{1 + k'} \tag{7.2b}$$

where φ is the volume of the stationary phase per volume of the mobile phase (in case of a liquid stationary phase). When the stationary phase is a solid, a similar expression can be derived with different dimensions for K. The product $K\varphi$ is often replaced by the symbol k' and is called the capacity factor. R_f and k' are parameters that describe the chromatographic behavior of a compound in a TLC system.

An important conception in chromatography is resolution. In a TLC system the resolution R_s can be defined as

$$R_s = \frac{2\Delta x}{d_1 + d_2} \tag{7.3}$$

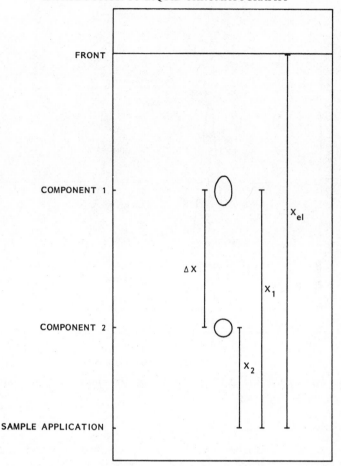

Fig. 7.1. Measurement of R_F in TLC. The R_F values of components 1 and 2 are calculated as x_1/x_{el} and x_2/x_{el}, respectively.

in which Δx is the center-to-center separation of two compounds and d_1 and d_2 are the average diameters of the two spots (Fig. 7.1). If $R_s = 1$, the spots are just separated. The resolution is a measure of the separation efficiency and the selectivity of the chromatographic system.

Column LC Parameters

Contrary to the open packed-bed systems characteristic of TLC, the particles providing or supporting the stationary phase in column chromatog-

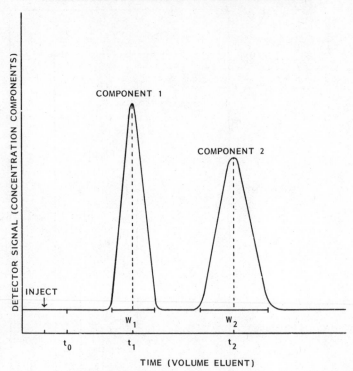

Fig. 7.2. Measurement of k' in HPLC. The k' values of components 1 and 2 are calculated as $(t_1 - t_0)/t_0$ and $(t_2 - t_0)/t_0$, respectively.

raphy are packed in a glass or metal column through which the mobile phase travels, either by gravitational force or under pressure. In modern column LC high pressures are applied to force the eluent through columns filled with microparticle packing materials, resulting in highly efficient separation systems, hence the terms high-pressure liquid chromatography or high-performance liquid chromatography (HPLC). In this chapter only HPLC systems will be considered.

After a solution of the sample is brought into the system the components of the sample travel—each with its own velocity—through the column and are detected after leaving the column. All sample components migrate the same distance, but spend different times in the chromatographic system.

The on-line registration of solute concentrations in the column effluent yields chromatograms of which an example is shown in Fig. 7.2. In the case of much used optical detectors such as the UV absorbance and the fluorescence detector, the detector signal is within a certain dynamic range proportional to the concentration of the solute.

In constant-flow systems the retention of solutes can be expressed by the capacity factor k' as

$$k' = \frac{t_r - t_0}{t_0} \tag{7.4}$$

in which t_r is the retention time of the solute (t_1 and t_2 for compounds I and II, respectively, in Fig. 7.2) and t_0 is the hold-up time, which is the time needed for a nonretained compound to pass through the column.

In liquid–liquid partition chromatography the relationship between retention time and distribution coefficient K is as follows:

$$t_r = t_0(1 + K \cdot \varphi) \tag{7.5}$$

Substitution of $k' = K'\varphi$ into Eq. (7.5) yields Eq. (7.4). The same equations, (7.4) and (7.5), hold for liquid-solid chromatographic systems having different dimensions for K.

The resolution R_s of the chromatographic peaks in Fig. 7.2 is defined

$$R_s = \frac{2(t_2 - t_1)}{W_1 + W_2} \tag{7.6}$$

W_1 and W_2 are the peak–base widths of peaks I and II, respectively. The values for W_1 and W_2 can be obtained by extrapolation of the tangents at the points of inflection to the base line.

Band Broadening and Efficiency in Liquid Chromatography

The bands or zones containing the solute molecules will expand with increasing migration distances of the solutes through the chromatographic system. The concentration profiles in the bands take on a Gaussian curve shape during the separation process (providing the band concentration is not in excess of the linear range of the distribution isotherm). The narrower the bands remain during chromatography, the higher is the separation efficiency of the system.

In column LC the efficiency of the system is often expressed by the number of theoretical plates N; N can be derived from

$$N = 16 \left(\frac{t_r}{W} \right)^2 \tag{7.7}$$

The main causes of band broadening in HPLC are the resistance to mass transfer and inequalities in the flow pattern of the mobile phase.

Resistance to Mass Transfer.   During the chromatographic process the solute molecules stay alternately in the mobile phase and the stationary phase. The time periods spent by the molecules in these phases is not infinitely small. When staying in the stationary phase the molecules will fall behind, while during their stay in the mobile phase they will move ahead. The contribution to band broadening by these processes is proportional to the mean residence time spent each time by the solute molecules in the stationary phase and to the velocity of the mobile phase.

Inequalities in the Flow Pattern of the Mobile Phase.   The flow of a fluid through a packed bed is a very complicated process. The molecules of the fluid will move along tortuous flow paths taking the line of least resistance. The flow velocity is higher in wider paths than in narrow ones. These flow inequalities contribute to band broadening. This contribution increases with longer residence times of molecules in the same flow path.

In the case of porous packing materials part of the mobile phase is stagnant in the pores of the particles. The solute molecules can only reach the moving part of the mobile phase by diffusion. The stagnant mobile phase thus contributes to the overall band broadening. The molecules of the mobile phase may be moved from one flow path into another by lateral diffusion and by convection, which results in a relative decrease in the amount of band broadening.

In general, the contributions due to flow pattern inequalities in the mobile phase increase with the diameter of the packed-bed particles and with flow velocity, and decrease with solute diffusivity.

Band broadening is not confined to the separation process in the packed bed. The introduction of the sample also contributes to band broadening, because it is impossible to introduce the sample as an infinitely narrow band. In HPLC extracolumn band broadening also stems from the connections between the column and other parts of the system and from the detector cell.

In TLC widening of the spot area is a two-dimensional process. The velocity of the mobile phase decreases during the chromatographic process. As the mobile-phase velocity is comparatively low, resistance to mass transfer is usually not a very important factor in TLC band broadening. The elongation of the spots in conventional TLC is the result of resistance to mass transfer. The main contribution to band broadening is caused by diffusion of the solute molecules in the mobile phase. This is the overriding factor in HPTLC, resulting in circular spots of more or less similar diameters (5,6).

High efficiency LC systems are necessary when dealing with samples containing many compounds of comparable physicochemical properties. Furthermore, the sensitivity of detection is better when the solutes are contained in narrow bands, because the peak concentration is then higher.

Increasing the efficiency is therefore an important goal in the development of liquid chromatographic systems. This can be achieved in a number of ways, but for each of these one has to pay a price. Important possibilities are:

The separation length of the system can be increased. In column LC, N, the number of theoretical plates, increases proportionally with the column length. However, the pressure needed to maintain suitable flow rates through the column and the analysis time will also increase. In TLC there is a maximum to the number of theoretical plates to be achieved by increasing the migration distance, because during development of the plates the plate number becomes constant. The chromatogram then merely expands without further gain in resolution (5,6).

The flow rate of the eluent can be decreased in column LC, which can give significant improvements in the efficiency of the system, again at the cost of increased analysis time.

Using well-packed beds of particles with uniform size results in less pronounced mobile-phase flow inequalities and therefore in narrower bands.

In LC the efficiency tends to increase with decreasing particle size. In column LC the pressure drop increases strongly with smaller sized particles, and column packing also becomes more difficult for very small particles.

Resolution in HPLC and TLC

The extent to which bands are separated in TLC and HPLC can be expressed by the resolution as defined in Eqs. (7.3) and (7.6), respectively.

In column LC the following relation can be derived between resolution (R_s), capacity factors $(k'_1$ and $k'_2)$ of the two peaks which are least well separated and the theoretical plate number N:

$$R_s = \frac{\sqrt{N}}{4} \left(1 - \frac{k'_1}{k'_2} \right) \left(\frac{k'_2}{1 + k'_2} \right) \qquad (7.8)$$

Three factors contribute to the resolution: the efficiency (N), the relative retention (k_1'/k_2'), and the capacity factor (k_2').

The resolution can be increased by using a column with high N value (see Section 7.2.1, under "Band Broadening and Efficiency in Liquid Chromatography"). As the resolution is proportional to the square root of N, a fourfold increase in N will only increase R_s by a factor 2. The resolution increases with retention time. R_s becomes maximal for infinite values of k_2'; then the ratio $k_2'/(1 + k_2')$ approaches 1. Very long analysis times are, of course, unattractive, and also result in comparatively broad solute bands and therefore in decreased sensitivity. The conditions are usually chosen in such a way that k' values are between 1 and 10.

From Eqs. (7.4) and (7.5) it follows that $1 - k_1'/k_2' = 1 - K_1/K_2 = \Delta K/K$. The factor $1 - k_1'/k_2'$ is therefore equal to the relative difference in distribution coefficients and thus reflects the selectivity of the system. The selectivity can often be improved by changing the mobile phase and/ or stationary phases.

For thin-layer chromatography a similar expression can be derived for the resolution:

$$R_s = \frac{\sqrt{N}}{4} \left(1 - \frac{k_1'}{k_2'}\right) \left(\frac{k_2'}{1 + k_2'}\right) \sqrt{\frac{1}{(1 + k_2')}} \qquad (7.9)$$

From Eq. (7.2b) it follows that

$$\left(\frac{k_2'}{1 + k_2'}\right) \sqrt{\frac{1}{1 + k_2'}} = (1 - R_F) \sqrt{R_f} \qquad (7.10)$$

Upon differentiation it can be derived that R_s is maximal for $R_f = 1/3$. However, this result is obtained with—among others—the incorrect assumptions that N is independent of the R_f value and that the sample at the start is infinitely small. In practice the maximum resolution is usually observed at somewhat higher R_f values.

In the past years the efficiency of TLC systems has been considerably improved by the development of thin-layer plates covered with well-packed beds of small (~ 5 μm) particles with narrow size distribution. This led to the introduction of the term high-performance thin-layer chromatography (HPTLC).

7.2.2.   Principles of Liquid Chromatographic Processes

The basis for separation of solutes in LC systems can stem from differences in one or more of the following properties: ease of adsorption onto solid surfaces, partition coefficients in liquid–liquid systems, strength of interaction (in the case of ionic solutes) with surface ion exchange sites, molecular size and shape. In the following paragraphs the principles of the four corresponding modes of LC systems, adsorption LC, partition LC, ion exchange chromatography (IEC), and gel permeation chromatography (GPC), will be discussed. In addition, an introduction to the newly developed paired-ion chromatographic systems will be given.

Liquid–Liquid Partition Chromatography

The separation between solutes is based on differences in partition coefficients in chromatographic systems consisting of (almost) immiscible mobile and stationary liquid phases. The liquid stationary phase is dispersed onto a finely divided support.

The partition coefficient is determined, on the one hand, by the interactions between the solute molecules and the molecules of the stationary and mobile phases and, on the other hand, by the interactions between the molecules of these phases themselves. As the solvents of the phases are immiscible, their polarity is quite different. Generally speaking the solutes will partition into the liquid phase of which the polarity agrees best with the solute polarity. Reasonable R_f values and retention times are obtained if the polarity of the solute is roughly intermediate between the polarities of the mobile and stationary phases (7).

In normal-phase LLC the stationary phase is a polar liquid (e.g., water or formamide), while the mobile phase is comparatively nonpolar. Reversed-phase systems consist of a nonpolar stationary phase, for example, hydrocarbon, and a polar mobile phase such as a mixture of methanol and water.

The liquid phases of the LLC systems are almost always mutually miscible to such an extent that the stationary phase is gradually stripped from the support material if no precautions are taken. Loss of the stationary phase can be prevented by presaturation of the eluent with the stationary phase. In column LC this is usually done by pumping the eluent through a precolumn containing packing material that is heavily loaded with the stationary phase. In TLC presaturation is often achieved automatically when the eluent passes the first part of the TLC plate below the line of the spotted samples.

To overcome the necessity of presaturation of the eluent, column packing materials with chemically bonded phases have been developed (see the following subsection). In column LC the use of these materials has largely replaced liquid–liquid partition chromatography.

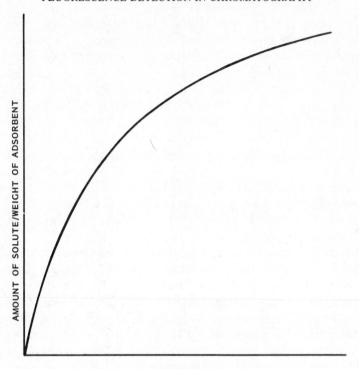

CONCENTRATION OF SOLUTE

Fig. 7.3.   Adsorption isotherm.

Adsorption Liquid Chromatography

In TLC and in HPLC a porous bed of a finely divided adsorbent with a large specific surface area can be used as the stationary phase. When a compound in solution (mobile phase) is brought into contact with an adsorbent, solute molecules will be adsorbed onto the solid phase until a state of equilibrium is reached, where the rates of desorption from the solid and adsorption onto the solid phase are equal. At low solute concentrations the amount of solute bound per unit weight of adsorbent is proportional to the concentration of solute in solution. At higher solute concentrations the increase of the amount adsorbed per unit weight is ordinarily less than proportional with concentration (Fig. 7.3). In adsorption LC the peak form of a solute in the chromatogram depends on the concentration of the solute in the zone moving through the system. If this solute concentration remains in the linear part of the adsorption

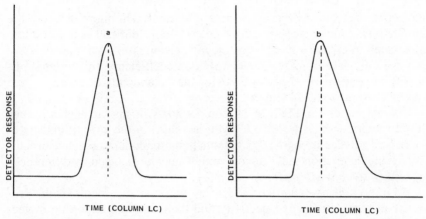

Fig. 7.4. Symmetrical peak (*a*) and asymmetrical peak (*b*), due to solute concentrations in the linear and nonlinear parts of the adsorption isotherm (Fig. 3), respectively.

isotherm (at low concentrations of solute), the peak in the chromatogram will be practically symmetrical (Fig. 7.4*a*). However, if the solute concentration is in the nonlinear part of the isotherm, the chromatographic peak is asymmetrical. The resulting peak shape in the case of an adsorption isotherm as depicted in Fig. 7.3 is shown in Fig. 7.4*b*. The adsorption equilibrium is the result of the following interactions:

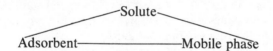

The commonly used adsorbents can be divided into two groups: the polar adsorbents such as silica gel and alumina, and the nonpolar adsorbents, charcoal and synthetic polymer beads.

A special case of interest is chemically modified silica gel.

Polar Adsorbents.   Solute and mobile phase molecules interact with the polar adsorbents mainly by forces resulting from molecular polarity and hydrogen bonding; the van der Waals and London forces are less important. Consequently, chromatographic retention of solute molecules increases with polarity. The most important interactions are those between the solute and the adsorbent and those between the mobile-phase molecules and the adsorbent. In other words, the solute molecules and the mobile-phase molecules compete for the available adsorption sites at the interface of the eluent and the adsorbent; interactions between solute molecules and mobile phase are less important. The more strongly the mobile-phase molecules are adsorbed, the more difficult it is for solute molecules to displace them from the adsorption sites. When solutes are arranged in order of increasing distribution coefficients, this order is primarily dependent on the properties of the compounds themselves, and not on the nature of the eluent.

The R_f values (in TLC) of all compounds will shift to higher values [their retention times (in HPLC) will decrease] in the case of eluents of increased solvent strength (i.e., eluents that are more strongly adsorbed). However, the elution order of compounds may be changed by differences in solute–solvent interactions.

An important characteristic of LSC (with polar adsorbents) is the pronounced compound type selectivity and the much less important molecular-weight selectivity. Separation into different classes of compounds is therefore comparatively easy (e.g., the separation of barbiturates into N-unsubstituted barbiturates, N-monoalkyl barbiturates, and N,N-dialkyl barbiturates), but the separation of members of a homologous series is much more difficult.

Polar adsorbents are partially deactivated by the adsorption of water from the atmosphere or the eluent. Careful control of the conditions are therefore necessary to ensure reproducible retention characteristics.

Nonpolar Adsorbents.   For nonpolar adsorbents the dispersion forces are the dominant contributors to the adsorption energy. Therefore higher molecular-weight compounds and the more polarizable compounds (aromatic compounds) are the most strongly retained.

Recently, macroporous poly(styrene divinylbenzene) adsorbents have been introduced that are suitable for reversed-phase HPLC systems. Their main advantages over the silica-based materials is the improved chemical stability allowing the use of mobile phases over a wide pH range, including very alkaline conditions (8–11).

Eluotropic Series.   For a given adsorbent the eluents can be arranged in the order of increasing interaction (adsorption) with the adsorbent, that is, in the order of increasing solvent strength. The eluotropic series is not the same for all adsorbents, but is more or less similar for polar adsorbents (silica gel, alumina).

Adsorption chromatography with inorganic polar adsorbents (e.g., silica gel) is particularly useful in TLC because of the relative ease of preparation of the thin layers and the possibility of using aggressive reagents and rather extreme conditions for the detection of the spots.

Chemically Modified Silica Gel. The nature of the adsorption sites of silica gel can be drastically changed by chemical modification of its surface. Column packing materials with so-called bonded phases are now the chromatographer's first choice. The typical LLC problems related to the partial miscibility of liquid stationary phases with the eluent (see Section 7.2.2 under "Liquid–Liquid Partition Chromatography") are thus circumvented.

Reversed-phase systems consisting of apolar modified silica gel and a mixture of water or buffer solution and an organic solvent—usually methanol or acetonitrile—as the mobile phase are most frequently applied (12). The retention time of solutes depends mainly on the type of bonded phase packing, the type and concentration of organic co-solvent in the eluent, the temperature, and, in the case of weak acids and bases, the pH of the mobile phase. The success of these column packing materials has also led to the development of TLC plates covered with apolar modified silica gel.

The surface silanol groups of silica gel can react with various agents in different ways. A typical reaction, resulting in an apolar surface, is

$$\equiv Si-OH + Cl-\underset{\underset{X}{|}}{\overset{\overset{X}{|}}{Si}}-C_nH_{2n+1} \rightarrow \equiv Si-O-\underset{\underset{X}{|}}{\overset{\overset{X}{|}}{Si}}-C_nH_{2n+1} + HCl$$

$$(7.11)$$

X can be Cl, OCH_3, OC_2H_5, or CH_3. Especially C_{18} and C_8 alkyl chains have become popular.

One should realize that not all surface silanol groups of the silica gel react; conversion of about 50–60% is usual. The remaining free silanol groups can contribute significantly to the retention behavior of solutes.

The polarity of the modified silica gel surface can be varied by the reaction with different agents. Apart from the hydrophobic (apolar) reversed-phase packings mentioned above, polar materials are also available. The surface is then modified by the introduction of groups with polar substituents, such as —OH, —NH₂ and —CN. These packings are frequently used in the normal-phase mode. Contrary to unmodified silica gel the retention behavior of solutes on these bonded phases is hardly influenced by small amounts of water in the eluent. A third group of bonded phase materials has been introduced, in which the surface modifying reagents carry an ionizable group (e.g., —COOH); these packings are used as ion-exchange materials (see the following subsection).

Ion-Exchange Chromatography

The usual ion-exchange materials consist of macromolecular organic material carrying ionogenic functional groups. Cation exchangers carry groups such as $-SO_3^-$ or $-COO^-$; anion exchangers contain amine groups, for example, $-NH_3^+$ or quaternary ammonium groups, $-\overset{|}{\underset{|}{N}}-^+$.

The eluent is usually an aqueous solution of an electrolyte. In water the ion-exchange matrix gets swollen by the hydration of the exchange sites and the counter-ions.

In anion exchange the positively charged exchange sites, R^+, on the polymer matrix attract and hold negative counterions, Y^-, of the mobile phase. Solute anions (X^-) can be exchanged with Y^-

$$R^+Y^- + X^- \rightleftarrows R^+X^- + Y^- \qquad (7.12)$$

In cation exchange the solute cations X^+ can be exchanged with mobile-phase cations Y^+ attracted by the negatively charged exchange sites R^-

$$R^-Y^+ + X^+ \rightleftarrows R^-X^+ + Y^+ \qquad (7.13)$$

The equilibrium constant K_e of these ion-exchange processes is proportional to the distribution coefficient and thus determines the migration rate of the solute ions through the chromatographic system.

The distribution coefficient K for a positively charged solute ion X^+ in an anion exchange system R^-Y^+ is related to the equilibrium constant K_e as follows:

$$K = K_e \frac{[R^-Y^+]}{[Y^+]} \qquad (7.14)$$

The distribution coefficient is inversely proportional to the counter-ion concentration of the mobile phase; higher counter-ion concentrations result in shorter retention times. The charge of the solute ions and counterions is also of importance. Solute ions with higher charge have larger distribution coefficients in IEC; bivalent counter-ions compete more effectively with solute ions for the exchange sites than monovalent counterions.

The pH of the mobile phase often has a profound effect on the retention behavior of solutes, particularly in the case of the so-called weak cation and anion exchangers. These materials carry weakly acidic ($-COOH$)

and basic (amine) groups, respectively. If for instance, in the case of a weak cation exchanger with —COO$^{(-)}$ exchange sites, the pH of the mobile phase is gradually lowered and the net charge of the stationary phase decreases because of protonation of the carboxylate ions. Consequently, the ability of the stationary phase to retain solute ions is diminished and the retention time is decreased.

If the solute is a weak acid or base, the pH of the mobile phase determines the fraction of the compound in its ionized form. When the pH of the mobile phase is raised, protonated amines will become deprotonated with a concomitant decrease in retention.

The ion-exchange equilibrium is usually the most important one in IEC, but it is not the only process contributing to the distribution of the solute ions. The swollen matrix forms a porous medium, into which ions can penetrate if their dimensions allow that. Size exclusion effects (see the following subsection) will therefore frequently play a role in IEC. The degree of swelling, and therefore the pore diameter, depends on the ionic strength and the pH of the mobile phase, and on the presence of organic cosolvents.

Another secondary distribution mechanism is hydrophobic interaction between the hydrophobic area of the ion-exchange matrix and lipophilic groups in organic solute ions. Apolar ions will be retained more strongly than less apolar ones of similar charge.

As the degree of swelling of the exchange material depends on the composition of the mobile phase, a significant increase or decrease of the pore size can occur during the ion-exchange process, particularly in the case of gradient elution, when the composition of the eluent is continually changed.

Swelling of the stationary phase can diminish IEC efficiency resulting in poor separations. Shrinking of the stationary phase can cause the formation of channels in the packed bed, resulting in peak broadening. These problems can be overcome by the use of ion-exchange groups chemically bonded on silica gel. However, the exchange capacity of these materials is comparatively small and, as they are silica gel based, they can be used over only a limited pH range (pH 2–8).

IEC is the predominant chromatographic separation mode for the analysis of inorganic cations. It is also widely applied in separations of amino acids, nucleic acid constituents, proteins, vitamins, drugs, and endogenous compounds in body fluids.

Gel Permeation Chromatography

In gel permeation chromatography (GPC) (also known as size exclusion chromatography) solutes are separated mainly according to their molecular size and shape. The stationary phase is a porous, three-dimensional macromolecular material; the pores are filled with eluent. The eluent in the pores of the matrix is considered to be the stationary phase; the eluent outside the pores is defined as the mobile phase.

Large molecules, which can not enter the pores, are completely excluded from the stationary phase and move through the system with the same velocity as the mobile phase. These large molecules are said to have a distribution coefficient $K = 0$. Molecules of sufficiently small size can enter the pores, at least to some extent. These molecules are therefore distributed between the mobile and stationary phases.

If the molecules are so small that they can penetrate completely into the pores, their concentration inside and outside the pores is the same; consequently, $K = 1$. Intermediate-sized molecules have K values between 0 and 1.

Solute molecules are therefore eluted in order of decreasing molecular size. The solvent molecules of the eluent are usually much smaller than the solute molecules and are therefore eluted last.

The stationary phases can be divided into two main types. The first group is made up of the organic gels; these are polymeric materials that form a three-dimensional structure through crosslinking between the polymer chains. The degree of crosslinking determines the mechanical strength of the material. "Soft" gels with limited crosslinking are not resistant at all to pressures exceeding atmospheric pressure. "Hard" gels are more resistant, but their use in HPLC is still limited to comparatively low pressure systems.

Rigid silica gel based materials and porous glassy materials have been developed that can withstand even high pressures. Deactivation of the surface can be necessary to prevent adsorption of the solute molecules. The mean pore size of the packing determines the molecular-weight range of the compounds with K values between 0 and 1. The molecular-weight range from 500 to 600,000 is covered by materials with varying pore sizes.

GPC is mainly used for the analysis of high-molecular-weight compounds such as proteins and nucleic acids and for the determination of the molecular-weight distribution of high polymers.

Paired-Ion Chromatography

In paired-ion chromatography (PIC) solute ions are distributed between an aqueous or semiaqueous phase and an organic liquid phase or an apolar surface. Organic or inorganic counter-ions of opposite charge sign are added in excess to the aqueous phase.

PIC can be executed in various ways; reversed-phase adsorption and liquid–liquid partition systems, in the normal phase as well as the reversed-phase modes, have been described. In LLC systems the solute ions partition into the organic phase accompanied by an equivalent amount of counter-ions.

Reversed-phase adsorption LC on apolar modified silica gel has become the preferred PIC system. The eluent consists of a water-organic solvent mixture containing a buffer and organic counter-ions. Solute anions can be chromatographed in the presence of protonated amines or quaternary ammonium ions (e.g., cetrimide, tetrabutylammonium); solute cations are often eluted with the negatively charged alkyl sulfates or alkyl sulfonates added to the mobile phase.

The retention mechanism of solute ions in these systems has been described by different models. One view is that the solute ions are adsorbed as ion pairs with the counter-ions of the eluent onto the apolar surface of the stationary phase. On the other hand, it has been stated, that the organic counter-ions are adsorbed onto the stationary phase with the ionic groups oriented toward the mobile phase, forming a double layer with the ions in the mobile phase. The column can then act as an ion exchanger.

Although this second hypothesis is probably more in accordance with reality, not all phenomena observed with these PIC systems can thus be satisfactorily explained. Potential complicating factors are the surface charge of the silica gel matrix due to dissociated free silanol groups (13) and micelle formation of the organic counter-ions in the eluent (14). In the reversed-phase PIC systems the retention behavior of solute ions can be regulated through the adjustment of a number of parameters, the type and concentration of the counter-ions, the pH, the type and concentration of the organic cosolvent of the eluent, the ionic strength of the eluent, and the temperature. The influence of these parameters on the retention behavior of organic ions and weak organic acids, bases, and neutral compounds has been discussed in recent reviews (15,16). Polystyrene divinylbenzene reversed-phase adsorbents have also been found suitable for PIC (17,18).

PIC can be applied to the analysis of virtually all types of ionogenic organic compounds. In certain areas, for instance in drug analysis, PIC has largely supplanted IEC.

7.3. THIN-LAYER CHROMATOGRAPHY AND FLUORESCENCE DETECTION

TLC is a very versatile separation technique, which is well suited to the separation with subsequent characterization, identification, or quantitation of trace amounts of organic compounds.

The spots of the separated compounds can be detected and quantified by a number of methods; among these, fluorescence-based techniques play an important role.

For an understanding of the various ways in which fluorescence detection can be applied in TLC and HPTLC, some insight into the practical aspects of these chromatographic techniques is indispensable. TLC sorbents and plates, the spotting of the sample solutions onto the plates, and the development of the plates, are therefore discussed in this section before the various modes of fluorescence detection are described.

Finally derivatization methods aimed at the conversion of nonfluorescent compounds into fluorescent products are dealt with.

7.3.1. Practical Aspects of TLC

TLC and HPTLC Plates

TLC plates are usually prepared by the application of a slurry of the adsorbent to an inert plate made of glass, aluminum, or polyethylene. Silica gel, alumina, and polyamide are the most frequently used sorbents for adsorption chromatography, while kieselguhr and cellulose serve as the support materials for stationary liquid phases in LLC systems. The adsorbents often contain a binder, for example, starch or calcium sulfate. For analytical purposes the layer thickness is about 0.25 mm. The size of the plates depends on the number of samples and the required elution distance. Square plates and rectangular plates with sides of 5–20 cm are customary, but other shapes, for example, triangular plates (see Section 7.3.1, under "Development of the Chromatograms") have been recommended too.

After the application of the slurry, the plates are dried. If high activity of the adsorption layers is required, the thin layers are activated by heating at 110–130°C for at least 2 hours. Only strongly bound water then remains. As these activated layers adsorb rapidly water and other vapor compounds, rigorous standardization of the entire chromatographic procedure is essential to obtain reproducible R_f values. In LLC the adsorbent layers are covered with a stationary liquid phase. Cellulose plates can adsorb water as the stationary phase by bringing the plates into contact with the vapor of the water containing eluent in the closed TLC chamber.

For reversed-phase conditions the adsorbent is impregnated with a nonpolar solvent, such as paraffin. This can be achieved by placing the plate in a chromatography chamber with a solution of the stationary phase in a volatile solvent and permitting the liquid to ascend the plate. After evaporation of the solvent the plate is ready for use.

Commercially available TLC plates frequently possess a higher degree of uniformity with respect to the layer thickness and density than manually prepared home-made plates, resulting in better plate-to-plate reproducibility of separations and quantitation of the spots. From a practical point of view, the elimination of time and effort spent in preparing TLC plates, and afterward the cleaning of the glass plates (and the lab as well), may be an even greater advantage.

Recently, reversed-phase-type TLC plates precoated with modified silica gel have become available, which can replace many of the home-made impregnated TLC plates.

In HPTLC (19) the chromatographic process proceeds considerably faster—due to the reduced migration distance—and can be standardized much better than in conventional TLC. Moreover, the separation efficiency is much improved as compared to TLC, and is of the same order as obtained with (conventional) HPLC columns. An additional advantage of HPTLC over TLC is the increased number of sample spots that can be applied onto the plate. These improvements have been realized by the use of very small particles of narrow particle size distribution, together with coating procedures that guarantee tightly packed beds.

The optimal resolution is achieved after migration of the eluent over a distance of only a few centimeters, resulting in sharply decreased analysis times. Provided the diameter of the applied spots is small (less than about 1 mm), the diameter of the separated spots will also remain small and rarely exceeds 4–5 mm.

Application of Samples

Samples are applied to the plates by spotting or by streaking as a narrow band with the help of a sample applicator. For optimal resolution the diameter of the spots, or the width of the bands, should be kept as small as possible. The solvent strength of the solvent in which the sample is dissolved should therefore be as low as possible. That means that for instance in adsorption chromatography with polar adsorbents (silica gel) an apolar sample solvent is best used (e.g., toluene or hexane).

The volume that can be spotted onto the plate is restricted (1–5 μl in conventional TLC; 0.1–0.2 μl in HPTLC). The amount of material that can be brought onto the plates is also limited; overloading of the TLC system by excessive amounts of sample results in tailing of the spots and, frequently, in poor separations.

The thin layers are easily damaged, especially in the case of manually prepared plates. If this happens during sample spotting, errors in the amount of sample introduced in the thin layer can be the result. Therefore, the tip of the spotting device should preferably not touch the surface of the layer. On the other hand, the tip should be kept as close to the surface as possible to avoid the formation of large drops.

Sample solutions can be applied manually or with automatic devices. For quantitative purposes the volume to be delivered onto the plate must be accurate and precise. Several types of calibrated disposable and non-disposable syringes are available. Excellent results can be obtained with simple disposable glass capillary pipettes of 1 µl or more. These pipettes are readily filled, and, when contacted with the layer at the correct angle, are emptied completely by the capilary action of the adsorbent.

In HPTLC very small sample volumes are essential for good resolution. Microsyringes and capillary applicators are available, which can deliver volumes of 5–100 nl. Platinum–iridium capillaries have been recommended instead of glass capillaries. The requirement of very small sample zones can be overcome to some degree by the use of silica gel TLC plates with preadsorbent spotting areas (20). An apparatus and method to increase the application volume to 5–50 ml per spot, allowing the detection and semiquantitative determination of trace substances, has been described (21). A new approach to sample application has been reported (22), called contact spotting, which enables rapid concentration of sample solutions of up to 50 µl with direct transfer of the residue to the plate. By this procedure many samples can be applied simultaneously and the spot diameter can be kept small enough for HPTLC.

The solvent of the sample solution must be removed completely from the thin layer before elution. Evaporation of the solvent can be promoted by heating the plate and by permitting an inert gas to flow over the plate surface, for instance with a hot air dryer.

Automatic devices for sample application, either as single spots or in narrow bands, can be very useful for the handling of large numbers of samples, provided the sample volumes to be applied onto the plate are not too small. Automatic spotters are usually equipped with a heater and/ or gas flow system for the ready evaporation of sample solvents.

Development of the Chromatograms

The most frequently applied TLC mode is ascending TLC in closed glass chambers containing the eluent, which, through the capillary action of the adsorbent layer, ascends the plate.

Elution is usually carried out under saturated conditions, in order to achieve reproducible results. The TLC plate in the chamber is brought into contact with the eluent vapor phase. To obtain rapid equilibrium between adsorbed and vapor phase molecules, the walls of the chamber are lined with filter paper that is saturated and in contact with the eluent. After chamber saturation the plate is placed in the eluent. Direct elution without chamber saturation is sometimes useful. Saturation of the chamber then occurs during the separation process; the composition of the vapor phase changes continually, and with it the condition of the adsorbent layer.

The migration distance of the eluent in conventional TLC is 10–15 cm. In HPTLC the migration distance of the solvent is often not more than a few centimeters. The position of the solvent front z is related to analysis time t by

$$z^2 = kt \qquad (7.15)$$

in which k is a constant. Therefore, the analysis time is sharply reduced in the case of smaller migration distances. Separation times in HPTLC systems are thus considerably shorter than in conventional TLC, despite the decreased velocity of the eluent through the HPTLC bed.

A number of special techniques have been developed to improve the efficiency, speed of development, or sensitivity of the TLC process.

A development technique that yields zones with sharply reduced band widths is called programmed multiple development. The plate is repeatedly developed with the same solvent in the same direction. With each successive development the eluent is allowed to migrate over a longer distance. Between developments the solvent is evaporated in a controlled way. Because of the repetitive traversing of the sample spots by the solvent, zone reconcentration takes place, resulting in narrow bands.

Alternatives to the linear TLC systems (in which the flow rate at any given moment is the same across the entire plate) are such techniques as circular TLC, anticircular TLC, and TLC with triangular plates (23–25).

A constant high flow rate of the eluent through the packed bed can be achieved in so-called short bed continuous development chambers. After migration over a comparatively short distance (1–5 cm), the solvent is evaporated. This causes a constant flow through the system. The solvent velocities thus achieved are much higher than in conventional linear TLC; therefore, solvents of low solvent strength can be used without the drawback of very low R_f values. With these solvents the selectivity of the TLC system is often considerably improved. Another way to obtain substantially shorter analysis times is possible through the use of a pressurized system, in which the vapor phase is practically eliminated by covering the sorbent layer with a membrane under external pressure; the solvent is pumped into the system under pressure (26,27).

A marked advantage of the flat bed systems used in TLC is the possibility of the successive development of the plates in two directions—perpendicular to each other—with different eluents. Very complex samples containing hundreds of compounds can thus be analyzed with each of the components separated from the others.

7.3.2.   Fluorescence Detection in TLC

Fluorescence-based techniques after TLC separation are frequently applied to the detection and quantitation of sample components. The qualitative and quantitative aspects of fluorescence detection in TLC are discussed in this section.

Qualitative Aspects

The comparatively few compounds showing native fluorescence upon UV irradiation can be sensitively and selectively detected after TLC separation.

Many organic compounds lack native fluorescence but do absorb UV radiation. Spots of these compounds can be visualized upon UV irradiation by the addition of a fluorescent indicator (e.g., zinc silicate) to the adsorbent layer. If excited with UV radiation of 254 nm (the most intensive line of the mercury lamps), the indicator fluoresces green. UV-absorbing compounds act as a filter, absorbing part of the fluorescence excitation radiation; this results in a diminution of the intensity of the emitted radiation. Spots of UV-absorbing compounds are thus apparent as dark areas against the brightly fluorescing background.

Although many compounds can be detected by this fluorescence diminution technique, it is not a universal method. Compounds that lack UV absorption at 254 nm fail to show up in the chromatograms.

Undetected compounds can be visualized after chemical derivatization by spraying the plate with a reagent solution and/or heating. Usually the result of this treatment is the development of colored products. However, a number of techniques have been developed to convert the compounds into fluorescent products. These procedures are discussed in Section 7.3.3.

Occasionally compounds with native fluorescence can not be visualized by irradiation of the spots with the usual low-energy sources of ultraviolet energy because of the small quantum yield of fluorescence of these compounds. The use of a laser as the energy source can help to increase the intensity of the emitted radiation and therefore make the fluorescence detection of these compounds possible. This has been demonstrated, for example, by the detection of dyes by means of the argon ion laser (28).

TLC scanning devices are available that enable the chromatographer to measure in situ the fluorescence of spots on the thin-layer chromatogram (see the following subsection). Some of these scanning devices are also capable of registering in situ the excitation and emission spectra, thus providing valuable information regarding the identity of the compounds (29).

A recent development is the use of an imaging photometer as a two-dimensional fluorescence detector (30). With this device emission spectra at each spot along the migration axis of a TLC chromatogram can be registered without mechanical scanning (see the following discussion).

Quantitative Fluorescence Measurement in TLC

Quantitation of separated compounds can be effected in two ways. The older method is to scrape the adsorbent layer containing the fluorescent spot from the plate, then elute the compound from the adsorbent material with a suitable solvent and measure the fluorescence of the resulting solution with a standard type of fluorimeter. With the development of modern scanning devices, in situ measurement of the spots has become possible. The direct in situ measurement method has largely supplanted the elution method, which suffers from a number of drawbacks: it is time-consuming and errors can easily be introduced due to incomplete or irreproducible recovery from the adsorbent. The worst disadvantage of the elution method is its limited sensitivity, because at least a few milliliters of solvent are needed for elution of a TLC spot. These problems can be partly overcome by the use of automatic elution devices, which enable the chromatographer to elute the compounds directly from the plates, without the removal of the adsorbent layer, using small volumes of eluting solvent (less than 1 ml per spot).

However, in situ measurement of the fluorescing spots with a scanning densitometer has become the method of choice, as it offers a fast and simple means of quantitating directly the chromatographic spots with greater sensitivity and without the risk of incomplete recovery. Quantitation by in situ measurement of the fluorescence can be based either on the native fluorescence of the separated compounds (or of fluorescing derivatives of these) or on the fluorescence diminution of an indicator by UV-absorbing compounds. The measurement of the fluorescence diminution is less selective and often less sensitive than the direct fluorescence measurement; it also suffers markedly from background fluctuation (31).

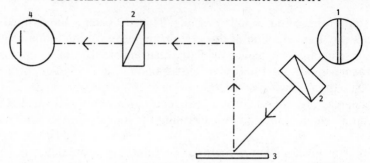

Fig. 7.5. Optical arrangement for quantitative TLC fluorescence measurement. 1, UV radiation source; 2, filter or monochromator; 3, TLC plate; 4, photomultiplier.

For these reasons the direct measuring of fluorescent compounds (or derivatives) is to be preferred.

Several scanning devices are available for in situ fluorescence measurement on TLC plates. Usually these combine the possibilities of fluorimetric and UV absorption measurements. An optical arrangement for measuring the fluorescence in situ in the reflectance mode is shown in Fig. 7.5. The plate is irradiated, under an angle of, for example, 45°. The main radiation sources are the mercury lamp and the xenon arc lamp. Filters or monochromators are used for obtaining spectral selectivity of the excitation radiation and the emitted radiation.

With a photomultiplier tube the emitted radiation is measured. A single-beam mode of operation is usually satisfactory (contrary to absorbance densitometry).

The TLC plate is placed on a positioning table, which is moved—usually motor driven—under the beam of the incident radiation. The scan area is preferably somewhat larger than the chromatographic spots to be measured. The emitted fluorescence radiation is usually measured at the illuminated side of the plate, that is, in the remission (reflectance) mode of scanning. Measuring the fluorescence at the other side of the plate in the transmission mode is also possible, but less desirable (32). The use of on-line coupled programmable electronic desk calculators and computers for the quantitative evaluation of the chromatograms has been described (33–35). Measuring in situ the fluorescence of separated spots has considerable advantages over the also frequently applied in situ UV absorbance measurements (36). The selectivity is enhanced because non-fluorescent UV-absorbing compounds are not measured. The spectral selectivity can be further optimized through the careful selection of the excitation and emission wavelengths. The sensitivity is often a factor 10 to 1000 better than with UV absorption. Analogous to the fluorescence measurement in solutions the fluorescence intensity is—for low amounts—directly proportional to the amount of fluorescing compound. Consequently, the distribution of the analyte molecules over the spot surface, and therefore the spot shape, is of no consequence, provided the total spot area is scanned and the effects of a concentration gradient, as the result of secondary chromatography, in different sublayers of the adsorbent are avoided (32). Finally, less optical background noise is obtained than in the case of in situ UV absorption densitometry.

The application of the samples in narrow bands instead of in spots has been advocated for the automatic scanning of TLC chromatograms, where the fluorescence is monitored as a function of the position of the scan area on the chromatogram. Difficulties related to the proper adjustment of the plate under the scanning beam are thus avoided.

If the table holding the plate is driven with constant velocity, concentration–position curves are obtained that are very similar in appearance to the concentration–time curves in column LC.

A number of factors affect the accuracy and the precision of quantitative TLC with in situ fluorescence measurement. The sample application is a critical step in the procedure. For optimal resolution and sensitivity, the sample spots should be as small as possible. The amounts of sample component(s) under investigation should be sufficiently small to avoid overloading with subsequent tailing upon elution—and nonlinearity of the fluorescence intensity–amount of substance relationship. The surface of the adsorbent layer must not be damaged during sample application. Serious errors can be introduced by the curling back around the tip of the applicator of part of the drop to be delivered onto the plate, and also by the withdrawal of more sample fluid than intended through capillary action upon touching of the plate by the applicator. Addition of an internal standard to the sample solution helps to avoid errors due to inconsistent volume deliveries in the same way as in GLC and column LC.

Plate-to-plate variations in fluorescence yield per spot, with the same amounts of sample applied, can be considerable. A sufficient number of calibration samples should therefore be applied to allow the construction of a calibration line for each plate.

The plates should be free of eluting solvent before measuring the fluorescence. Nevertheless, the type of adsorbent and the amount of adsorbed water (or other vapors) still influence the fluorescence yield. The moisture content can increase between evaporation of the solvent and the moment of measurement. Therefore, a strict time schedule for the entire procedure must be maintained; the effect due to the adsorption of water can also be suppressed by spraying the plate with triethanolamine-iso-propanol (37).

In situ multichannel image detection of fluorescing spots in TLC has been described recently (30), obviating the need for mechanical scanning. A digital imaging photometer based on a SIT vidicon was used as a two-dimensional fluorescence detector, connected with a computer system for data analysis. Two optical arrangements were tested. One system was designed to produce an emission spectrum from every point along the development axis of a one-dimensional chromatogram. The second optical arrangement allows the quantitative analysis—again without scanning—of the entire TLC plate; that is, the emitted fluorescence radiation for a fixed excitation–emission wavelength pair can be measured simultaneously at each position of the TLC plate.

Despite some shortcomings—the worst problem is the high blank value of the plate—the in situ quantitation of fluorescent compounds at the picogram level is quite feasible with this system. Combined with two-dimensional TLC a fast and highly selective analytical method is obtained as it counts no less than four variables to provide distinction between sample components, i.e., the wavelengths of excitation and emission, and the migration along two axes.

7.3.3. Fluorescence Enhancing and Inducing Methods; Fluorescence Labeling

The fluorescence intensity of compounds separated by TLC can be en-hanced in various ways. Nonfluorescent or weakly fluorescent com-pounds can be converted into strongly fluorescent derivatives, either by a comparatively nonselective treatment (e.g., acid treatment) or by the introduction of a fluorophore (fluorescence labeling).

Fluorescence Enhancing Methods

Changing the environment of the fluorescing compounds on the TLC plates strongly affects the quantum yield of fluorescence. The stationary phase exerts a definite influence (38).

The quenching effect by atmospheric oxygen can be diminished by flushing nitrogen over the TLC plate. In this way an increase of the flu-orescence intensity of some butyrophenones of up to a factor 4 could be obtained (39).

Enhanced fluorescence intensities can also be obtained by uniformly spraying the thin-layer chromatograms with a nonvolatile and viscous organic solvent after development (40,41). The polarity and viscosity of the spray solvents play an important role, as well as the extent of trans-migration of the fluorescing compounds from the adsorbent into the spray reagent medium. The fluorescence intensities of spots of some dansylated amines could be enhanced more than a hundredfold by spraying with a mixture of Triton X-100 and chloroform (41).

The effect of polarity changes of the adsorbent layer caused by the presence of polar organic compounds has also been exploited (42). Some 3-hydroxy flavines (e.g., fisetin) are almost nonfluorescent in apolar sur-roundings but exhibit strong fluorescence in polar media. After spraying TLC plates with a fisetin solution, polar compounds show up as fluores-cent spots against a weakly fluorescing background. Another example is the use of pH-sensitive fluorogenic spray reagents for the detection of, for example, sulfur-containing pesticides (34). After development the TLC plate is placed in bromine vapor. The pesticides are oxidized with si-multaneous production of acid. The presence of protons in the pesticide spot area is then detected by spraying the plate with a solution of 8-hydroxyquinaldine or another suitable reagent. This reagent is relatively nonfluorescent, creating a low fluorescence background, but shows strong fluorescence upon protonation.

Fluorescence Induction

Many organic compounds such as steroids, lipids, carbohydrates, amino acids, and pesticides, which themselves lack native fluorescence, can be converted into fluorescing products, after separation on inorganic adsor-bent layers, by a number of more or less aggressive treatments. One such method is the exposure of the thin-layer chromatogram for several hours to the vapors of ammonium hydrogen carbonate at temperatures of 130–150°C (44). The long duration of this method and its limited sensitivity are serious drawbacks and incited the investigation of the usefulness of other inorganic fluorogenic compounds. Some metal salts were found to be very well suited for this purpose (45); zirconium derivatives show many advantages over other reagents. The plates are sprayed with or dipped in an aqueous solution of zirconyl chloride or zirconyl sulfate and placed in an oven at 150–200°C for a certain period, usually not more than a few minutes. A number of compounds of widely different structure could be detected with considerable sensitivity by this method (45). Simply ex-posing the chromatograms to vapors of strong inorganic acids, such as hydrochloric acid or nitric acid, at room temperature, followed by heating for 10 min at 160°C has been found to induce intense fluorescence in many compounds (46). Nitric acid vapor has the additional advantage of a sig-nificantly reduced background fluorescence as compared with hydrochloric acid and hydrobromic acid treatment. Comparable results were obtained in an entirely different way, that is, by the use of a gaseous electrical discharge (47). The TLC plate is placed in a discharge chamber, which is partially evacuated to about 0.2 torr, and exposed to the plasma discharge in the presence of nitrogen or another gas for a period of 5 sec to 5 min. After this treatment submicrogram quantities of several types of compounds can be detected with the unaided eye; subsequent heating of the exposed chromatogram for a few minutes at 130°C greatly increases the fluorescence intensity. This method has been applied to the detection of pesticides on silica gel TLC plates (48).

The exact nature of the fluorescent products induced by the methods described above is not known. Molecular rearrangement and/or decomposition probably occurs. In view of the similarity between resultant fluorescing materials a parent structure might be responsible. It is interesting to note that the various treatments produce fluorescent products on a silica gel adsorbent layer but not on a ground glass surface. Possibly the silica gel surface induces a common type of structure and confers rigidity to this structure by its adsorption on the solid substrate, resulting in enhanced fluorescence (46,49).

Fluorescence Labeling

Fluorescence labeling involves the introduction of a substituent (fluorophore) into nonfluorescent compounds through a chemical derivatization reaction. The reaction mechanism and the structure of the resulting fluorescent products are usually known.

Fluorescence labeling may be done either prior to or following TLC separation. Usually fluorophores are introduced into the molecules of the compound(s) under investigation after TLC separation, with the sole purpose of enhancing the detectability.

The following points should be considered in postchromatographic fluorescence labeling. The reagent must be practically nonfluorescent and should not be decomposed into fluorescent products upon further treatment of the plates before measurement. For quantitative purposes it is also important that the resulting fluorescent intensity be reproducible. Although not absolutely necessary this is best guaranteed by a reaction in which the parent compound is completely converted into one and only one derivative. The usual method of bringing the reagent into contact with the spots is by spraying the plates. Uniformly spraying is essential for reproducible results. The reproducibility can be improved by mechanized spraying (50).

Alternatively, the TLC plates can be dipped in a reagent solution. Although this is a simple procedure it suffers from some limitations. The reagent solution can become contaminated with impurities from the thin layer, resulting in diminished fluorescence of the spots. The reagent solution might elute the compounds to be measured. Dipping the plates can easily lead to damaging the layer.

The conditions of drying the plates prior to derivatization, the choice of the solvent for the reagent, and the length of the period between spraying or dipping and fluorescence measurement affect the fluorescence intensity of the spots and should be rigorously standardized (51).

Postchromatographic fluorescence detection may be hindered by high background fluorescence radiation from the TLC plates after treatment with the reagent. In these cases prechromatographic derivatization is indicated. There can be other motives as well why one would wish to label the compounds before TLC separation.

Prechromatographic derivatization is useful in the case of unstable or volatile compounds; it may improve the separation between certain compounds, reduce the reactivity toward the stationary phase, or diminish the polarity of the compounds (resulting in better migration). Prechromatographic labeling can be carried out in the sample solution before spotting, or after spotting onto the plate. Derivatization before spotting has the advantage of exactly standardized conditions and comparative freedom of choice of the reaction conditions. If the reagent is fluorescent, excess reagent must be separated from the derivatization products either by the TLC process or by a prechromatographic clean-up procedure. In situ derivatization prior to elution can be performed by spotting the reagent solution on top of the sample spots. This solution should, of course, not elute the sample compounds (52). Furthermore, the same precautions against damaging the thin layer should be taken as in postchromatographic derivatization.

Most derivatization procedures are one-step reactions. However, procedures have also been reported that include two or more steps. Young (53) described a method for the detection of aldehydes at the picomole level on silica gel layers. The aldehyde sample spots were converted on the plate into imines by treatment with an aniline solution. After TLC separation of the derivatives, the plates were UV-irradiated causing the imine spots to regenerate aniline, which was subsequently detected by reacting the spots with fluorescamine.

The selectivity of fluorescence labeling is determined by the presence or absence of certain functional groups in the molecules of the sample constituents. For instance, fluorescamine reacts with all primary amines. Increased selectivity can sometimes be achieved, as shown by Nakamura

Table 7.1. Fluorescence-Introducing Derivatization Reagents and Functional Groups Allowing Fluorescence Labeling with These Reagents

Reagent	Abbreviation	Functional group(s)
(D)-(L)-1-Aminoethyl-4-dimethylaminonaphthalene	DANE	Carboxyl
Anthracene-isocyanate	AIC	(Amine), hydroxyl
9-Anthryldiazomethane	ADAM	Carboxyl (and other acidic groups)
4-Bromo-methyl-7-acetoxycoumarin	Br-Mac	See Br-Mmc
4-Bromo-methyl-7-methoxycoumarin	Br-Mmc	Carboxyl, imide, phenol, thiol
N-Chloro-5-dimethylaminonaphthalene-1-sulfonamide	NCDA	Amine (prim, sec), thiol
9-(Chloromethyl)anthracene	9-CIMA	See Br-Mmc
4-Chloro-7-nitrobenzo(c)-1,2,5-oxadiazole	NBD-Cl	Amine (prim, sec), phenol
2-p-Chlorosulfophenyl-3-phenylindone	DIS-Cl	See Dns-Cl
9,10-Diaminophenanthrene	DAP	Carboxyl
4-Diazomethyl-7-methoxycoumarin	DMC	See ADAM
5-Di-n-butylaminonaphthalene-1-sulfonylchloride	Bns-Cl	See Dns-Cl
Dicyclohexylcarbodiimide	DCC	Carboxyl
N,N'-Dicyclohexyl-O-(7-methoxycoumarin-4-yl)methylisourea	DCCI	Carboxyl
N,N'-Diisopropyl-O-(7-methoxycoumarin-4-yl)methylisourea	DICI	Carboxyl
4-Dimethylaminoazobenzene-4'-sulfonylchloride	Dbs-Cl	See Dns-Cl
N-(7-Dimethyl)amino-4-methyl-3-coumarinylmaleimide	DACM	Thiol
5-Dimethylaminonaphthalene-1-sulfonylaziridine	Dns-aziridine	Thiol
5-Dimethylaminonaphthalene-1-sulfonylchloride	Dns-Cl	Amine (prim, sec, (tert)), (hydroxyl), imidazole, phenol, thiol
5-Dimethylaminonaphthalene-1-sulfonylhydrazine	Dns-hydrazine	Carbonyl
4-Dimethylamino-1-naphthoylnitrile	DMA-NN	Hydroxyl
9,10-Dimethoxyanthracene-2-sulfonate	DAS	Amine (sec, tert)
1-Ethoxy-4-(dichloro-s-triazinyl)naphthalene	EDTN	Amine, hydroxyl (prim)

652

Table 7.1. (*continued*)

Reagent	Abbreviation	Functional group(s)
9-Fluorenyl-methylchloroformate	FMOCCl	Amine (prim, sec)
7-Fluoro-4-nitrobenzo-2-oxa-1,3-diazole	NBD-F	Amine (prim, sec), phenol, thiol
4-Hydroxymethyl-7-methoxycoumarin	Hy-Mmc	Carboxyl
4-(6-Methylbenzothiazol-2-yl)-phenyl-isocyanate	Mbp	Amine (prim, sec), hydroxyl
1,2-Naphthoylenebenzimidazole-6-sulfonylchloride	NBI-SO$_2$Cl	See Dns-Cl
2-Naphthylchloroformate	NCF	Amine (tert)
Naphthyl-isocyanate	NIC	(Amine), hydroxyl
Ninhydrin	—	Amine (prim)
4-Phenylspiro(furan-2(3H), 1'-phthalan)-3,3' dione (fluorescamine)	Flur	Amine (prim, sec), hydroxyl, (thiol)
o-Phthaldialdehyde(o-phthalaldehyde)	OPA	Amine (prim, sec), thiol
N-(1-Pyrene)maleimide	PM	Thiol

(54), who used the fluorescamine reagent for the labeling of histidine, histamine, and histidyl peptides at the picomole level. These 2-(4-imidazolyl)ethyl amines were first derivatized with fluorescamine and then converted into new fluorophores by heating in an acidic medium. Other primary amines, amino acids, and peptides lacking histidine residues did not fluoresce.

Although fluorescence labeling reagents are much less abundant than chromophoric reagents, a fair number of fluorophoric reagents have been described in the literature. The majority of the derivatization reactions so far have been aimed at the sensitive detection of primary and secondary amines and amino acids. Popular reagents are: dansyl chloride, fluorescamine, o-phthalaldehyde, and NBD chloride (Table 7.1). Reagents have been developed for other functional groups as well, such as dansyl hydrazine for compounds with a carbonyl function (55) and 4-bromomethyl-7-methoxycoumarin for acidic compounds (56). Acid chlorides (e.g., dansyl chloride) react with several groups carrying active hydrogen, for instance, phenols. An overview of reagents is given in Table 7.1.

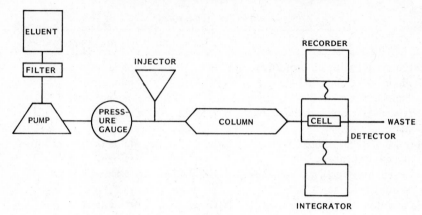

Fig. 7.6. Block diagram of a HPLC system.

7.4 FLUORESCENCE DETECTION IN HPLC

As compared with TLC and GLC, HPLC is a young branch of chromatography. It is a rapidly growing technique, not least because of continuing advances in instrumentation and column technology. The combination of high-efficiency columns with on-line fluorescence detection is one of the most powerful analytical tools today. In the following paragraphs HPLC instrumentation, fluorescence detection, fluorescence enhancement, and derivatization techniques are discussed.

7.4.1. HPLC Instrumentation and Columns

Figure 7.6 shows a block diagram of the components of a simple HPLC system. The eluent is pumped from a reservoir, passing a filter, through the column, where separation of the mixture—introduced into the fluid stream by an injector—takes place. The separated compounds are successively detected as they pass the detector cell. The detector response is registered by a recorder. Frequently, a data handling and computing system is coupled to the detector to record retention times and integrate peak areas for quantitative measurements.

A more complicated setup is often needed, for example, for solvent programming, recycling, sample cleanup with precolumns, reactors for postcolumn derivatization, multidetection systems, and automated sample injection.

The Pump

HPLC pumps can be divided into two groups, the constant-pressure pumps and the constant-flow pumps. In constant-pressure pumps the flow rate of the eluent varies, depending on the resistance of the system; thus, the viscosity of the eluent, the temperature, and the resistance in the filters, detector cell and, above all, the column determine the flow rate at a given pressure. For practical purposes it is more important to sustain a constant flow through the system in order to obtain reproducible retention times and a stable base line. Constant-flow pumps have therefore become far more popular than constant-pressure pumps. The motor-driven constant-flow pumps with small reciprocating pistons are the most frequently applied type. A piston moves forward and backward in a cylindrical space equipped with an inlet and an outlet valve. Eluent is alternatively sucked into the pump and pushed out. The main advantages of reciprocating pumps are ease of operation, small internal volume (therefore, rapid solvent change and compatibility with gradient elution), and continuous delivery. The main problem is to obtain a pulsefree flow of solvent. Many pumps are therefore of the dual-piston type, designed to minimize flow pulsations. Flow controllers and pulse damping systems are frequently introduced to ensure a constant, smooth, and pulsefree eluent flow. Modern pumps are able to deliver a flow of several milliliters per minute against a pressure of up to 500 atm.

Chromatographers are frequently confronted with the problem of designing an HPLC system for the simultaneous analysis of compounds with widely different physicochemical properties. Isocratic conditions, that is, use of the same mobile phase throughout the analysis, may then lead to excessive analysis times and peak broadening. In these cases solvent programming (gradient elution) is very helpful. The composition of the eluent is changed during the chromatographic run in a preset, programmable way. This can be achieved most effectively at the high-pressure side of the system by the use of two or more HPLC pumps, one for each eluent component. The eluent components are pumped with continuously changing flow rates into the mixing chamber connected to the inlet side of the column. Another possibility is the use of a single HPLC pump, which is connected with two (or more) solvent reservoirs through an electronically controlled proportioning valve, of which the position is changed with high frequency. The position of the valve is controlled by a programming unit.

Sample Introduction Systems

The injection of a sample against the high pressure of the mobile phase poses special problems. Furthermore, the sample must be introduced into the solvent stream as a narrow band to limit extracolumn band broadening. One way of overcoming the difficulty of injecting samples into a high-pressure system is by stopping the flow, injecting at atmospheric pressure, and reinitiating the flow. Usually, however, sample introduction occurs in such a way that the flow need not be stopped. This can be achieved by the use of loop injection valves. The solvent flow is guided directly into the column bypassing the sample loop and the sample solution is injected into the sample loop, which is filled with eluent under atmospheric pressure. By switching the valve the eluent passes through the sample loop and the sample is swept onto the column. Depending on the size of the sample loop, volumes of up to a few milliliters can be injected in this way. If the valve is equipped with an actuator and the sample port is connected to an automated sample injection system (autosampler), large numbers of samples can be analysed consecutively.

Loop sample injection with calibrated syringes allows the introduction of precisely controlled injections with high reproducibility and is thus well suited for quantitative analyses.

Columns

The column is the heart of the HPLC system. The inherent separation power of columns packed with 5 or 10-μm microparticles is impressive, but can only be made fully effective if the entire system is optimized. Modern analytical columns usually have inner diameters of 2–5 mm; larger column diameters are used in preparative HPLC.

The column length ranges from 5 to 30 cm for packed columns. Typically a 25-cm column packed with 5-μm particles has about 16,000 or more theoretical plates. This number can be further increased by coupling columns in series. Much work has been done in recent years to further improve the efficiency of HPLC columns. This can be achieved by the use of even smaller particles. Columns packed with 3 μm particles have been shown to result in higher efficiencies than columns with 5 μm particles. However, for a column of a certain length the pressure drop increases with decreasing particle size.

Eventually the thermal effects due to high frictional flow resistance or the finite kinetics of solute transfer will prohibit further gain in efficiency by using particles of more and more reduced size. Particles of about 1 μm are probably the limit.

Columns in HPLC are usually made of stainless steel tubing, although glass columns are also available. A very smooth inner surface of the column wall is essential; wall irregularities can cause channel formation in the packing near the wall-packing interface, resulting in decreased efficiencies.

In recent years much progress has been made in the development of microbore columns (57,58). Except for the smaller inner diameter of 0.05–1 mm, microbore packed columns are not essentially different from the usual 2–5-mm I.D. packed columns in terms of geometrical character-istics. Apart from the possibility of allowing fast separations the main advantages of microbore columns are as follows: They are economical to use with respect to the amount of column packing material and eluent consumption; they can be joined together in series without appreciable loss of efficiency per unit of length, yielding efficiencies of several hundreds of thousands of theoretical plates per meter; columns of high efficiency and small cross-sectional area produce very narrow packs as the solute may be contained in a volume of only a few microliters. The small intracolumn volume of microbore columns, however, puts strains on the engineering of small dead-volume injectors and, particularly, de-tector cells, to prevent the loss of much of the increased column efficiency through extracolumn band broadening (57).

These extracolumn band broadening effects are even more liable to occur in the case of open tubular microbore columns with inner diameters of 50 μm or less. Because of the much smaller pressure drop, the length of these columns can be considerably increased with accompanying in-creasing theoretical plate numbers. The main limitations to increasing the column length are long separation times. Again there is the necessity of very low extracolumn dead volumes. Ideally, the detector cell volume should be of the order of 1 nl (59). This not only poses technological difficulties; there is also the problem that in these very reduced volume cells only a limited number of molecules at a certain eluate concentration can be present, thus making sensitive detection more difficult. One should also realize that the capacity of these microcolumns is very much reduced as compared with the usual packed columns. Consequently, column ov-erloading can cause severe problems; particularly in the analysis of con-centrated extracts of blood plasma or other complex mixtures the danger of column overloading is imminent.

To protect the analytical column the eluent can be passed through a guard column, packed with larger particles and situated between the eluent container and the injection system. The guard column filters small particles from the eluent, thus preventing the analytical column from be-coming clogged up. Alternatively a short precolumn of comparable effi-ciency as the analytical column can be fixed between the injection system and the analytical column, again with the intention of protecting the ana-lytical column. In combination with a multichannel valve, precolumns can also be used for the cleanup of complex mixtures (e.g., 60,61).

7.4.2. Fluorescence Detectors

Fluorescence detection of column effluents was originally operated by incorporation of simple, rather large volume flow cells into existing fluorimeters. In the past decade, however, a number of instruments were designed specifically for HPLC analysis (e.g., 62).

Conventional HPLC fluorescence detectors today are not basically different from fluorimeters used for batch measurements. The design and use of these detectors have been described by several authors (e.g., 62–64). For the selection of excitation and emission wavelengths the detector can be equipped with filters or monochromators. It can be argued that because of the selectivity provided by the HPLC column, wavelength selection with filters is adequate; it is certainly more economical. Another advantage of filter instruments is the higher excitation and emission radiation intensities and therefore better sensitivity (65). On the other hand, the monochromator-equipped detector offers the possibility of further enhancing the selectivity through the proper adjustment of the excitation and emission wavelengths. This can be very helpful and frequently even essential when dealing with the analysis of compounds in complicated matrices such as serum and urine.

A compromise is found in instruments with monochromatic excitation radiation and broad spectral band pass of the emitted radiation by blocking or cutoff filters.

In the HPLC detector a low-volume flow cell—less than 20 µl—replaces the cuvette. These flow cells are usually made from quartz tubing. Apart from having a low dead volume the flow cell must be resistant to pressure of up to 35 atm (63). Other devices have also been reported (section 7.4.2, under "Flow Cells"). The excitation radiation source is practically always either the mercury lamp with a few narrow bands of high intensity radiation or a continuous source lamp: the deuterium lamp or the xenon lamp. The use of the mercury lamp is restricted to the detection of compounds excited by radiation of one of the mercury lines. The xenon lamp provides a continuous spectrum of fairly high intensity from 260 to 650 nm and is at present the most frequently applied radiation source. The deuterium lamp also delivers a continuum, with the useful wavelength range going down to 190 nm. Short excitation wavelength fluorimetric detection with the deuterium source has been found useful for a number of compounds, which show weak fluorescence at excitation wavelengths above 260 nm (66). On the other hand, the selectivity of detection is more limited in the short wavelength range. Other excitation sources—lasers, β radiation—have been developed (Section 7.4.2, under "Laser Fluorimetric Detection").

Increased spectrophotometric selectivity can be obtained by the use of the second derivative (instead of zero order) spectra for monitoring the column effluent fluorescence, yielding better resolved peaks in the chromatogram (67).

Improved spectral selectivity can also be obtained by the application of time resolved fluorescence detection, which has been introduced recently (68,69).

Flow Cells

As has already been stressed, the volume of the flow cell must be kept small to minimize postcolumn band broadening. The basic need to supply the cell with sufficient excitation energy and to collect the (nonfocused) emitted radiation requires transparency on more than one axis of the cell. Focusing the excitation light beam and collecting the emitted photons is more difficult in the case of small volume cells. Furthermore, light scattering by reflection and refraction from the walls of the flow cells is a major potential source of interference, particularly with laser-based detectors.

Flow cells are usually constructed of narrow-bore quartz tubing. Stray light interference can be fairly effectively eliminated by detecting fluorescence with right-angle geometry from flow cells with a linear bore of square cross section (64,70).

In view of the advantages of microcolumns over conventional 2–5 mm I.D. columns the development of flow cells with volumes less than 1 μl is of current interest. In a recent paper a flow-through microcell with a volume of 300 nl and an excitation path length of 1 cm was described (71); the excitation energy generated by a mercury lamp is guided to the cell by an optical fiber.

Some of the potential sources of error connected with quartz flow cells can be eliminated by using the free-falling drop detector as designed by Martin et al. (72). The column effluent is fed to a Teflon needle and forms drops of uniform size, which upon falling from the needle pass through the excitation light beam. The pulse of scattered light thus produced is detected by a photomultiplier. The output of this multiplier is used as a timing signal to process the fluorescence signal generated with a second photomultiplier. This detector has the following advantages: a much more intense excitation energy can be used, resulting in improved sensitivity; because of the incorporation of an optical integration sphere the output of the detector is proportional to the total light fluoresced by each drop (flow-through fluorimeters normally detect only a fraction of the total fluoresced light); the detector is immune to the detrimental effects produced by changes in the primary absorbance, which are frequently observed in quartz flow-through systems; the internal reflections, refractions, and scattered-light phenomena common to flow-through cells are absent.

Instead of allowing the eluent to form falling drops, the eluent can also be fed from a capillary tube to a rod positioned a few millimeters lower. A small-volume (4 μl) windowless cell is thus formed (73,74). In a recent study the effluent from conventional HPLC columns was arranged as a free-falling thin jet (75).

Another possibility is the application of the sheath flow principle in order to confine the column effluent in the center of an ensheathing solvent stream under laminar conditions (76). The effluent is isolated from the cell walls, and stray light from the walls is therefore reduced to a minimum. These windowless cells are especially useful in combination with laser excitation sources.

It has been shown that transference of solutes from solution to an adsorbed state results in an extensive modification of the fluorescence emission characteristics (77). The fluorescence of organic compounds may be intensified by adsorption onto silica gel or another adsorbent. This phenomenon has been exploited in HPLC detector flow cells packed with an adsorbent, resulting in detection with increased sensitivity of polynuclear aromatic hydrocarbons (78) and aflatoxins (79).

Laser Fluorimetric Detection

Laser-induced fluorescence detection offers some definite potential advantages over the conventional (incoherent) light sources (80,81). Lasers produce a very high photon flux. Because of the direct proportionality between fluorescence and excitation energy, this can be lead to remarkably increased detector signals. The signal-to-noise ratios are also improved, but these improvements are limited by the Rayleigh scattering from eluents and optical components, by Raman scattering from the solvents, and by fluorescence from contaminants in the eluent and from the optics. Although much of the disturbing scattered light can be filtered off, solvent Raman bands can frequently cause problems because these may overlap the emission bands of the eluted compound.

Laser beams can be accurately positioned and focused, which makes them ideal radiation sources in combination with very low volume flow cells. To circumvent problems due to light scattering, various types of windowless cells (Section 7.4.2, under "Flow Cells") have been advocated. An alternative is the use of a capillary tube cell coupled to an optical fiber (82,83). Short-duration pulses, with pulse widths ranging from microseconds to picoseconds, can be generated by lasers because of the temporal coherence of the laser emission beam. This makes the resolution of components with different fluorescence life times possible, thus adding a new dimension to fluorescence detection (68,69). A distinct advantage of time-resolved fluorescence is the possibility of discriminating analyte fluorescence against Raman scattering and short-lived fluorescence from the eluting solvent, resulting in the efficient reduction of background emission.

Because of the high radiation energy emitted by lasers, the simultaneous absorption of two photons by atoms or molecules, resulting in two-photon-excited fluorescence, can be exploited in HPLC detection in spite of the inefficiency of the process (84).

Because of the absence of background signal the emitted fluorescence radiation is sufficiently intense for the detection of species at concentration levels that are typical for HPLC analysis, but offering increasing selectivity. This was demonstrated in the separation and optical resolution of three oxadiazoles from interfering compounds (84). Another recent laser-based development is the use of sequentially excited fluorescence detection (85). Fluorescence from a highly excited molecular state following sequential resonant excitation is monitored. Again, the inherent inefficiency of the process is overcome by the use of a high-powered tunable laser. Contrary to "conventional" laser fluorimetric detection, increasing the excitation radiation intensity does not lead to an appreciable increase in background emission levels, due to a large blue shift between excitation and emission wavelengths. Sequential excited fluorescence detection is therefore a potentially sensitive (and selective) method.

Laser fluorimetric detection in HPLC has been applied among others to the determination of adriamycin and daunorubicin in urine (83) and to the determination of aflatoxins (86) and zearalenon in corn (87).

Beta-Induced Fluorescence Detection

Malcolme-Lawes and Warwick (88–91) introduced a new excitation radiation energy source. β^- particles, produced by the decay of a radioisotope (e.g., ^{63}Ni or ^{147}Pm) are allowed to enter a small-volume flow cell through which the eluent from an HPLC column is passing. The emitted fluorescence radiation from excited constituents of the eluent is detected in the usual way by a photomultiplier tube after passing a filter or monochromator.

The main advantages are that this type of detector can be operated for long periods without deterioration and without short-term variations in source intensity. The detector is still in the experimental stage. With the realization of some improvements indicated by the inventors (90), it has the potential to compete with respect to sensitivity with modern conventional fluorescence detectors. Its main limitation from the chromatographers point of view is that optimum sensitivity is obtained with eluents consisting of aromatic hydrocarbons (benzene, toluene), either in pure form or diluted with hexane; these solvents are best suited to transfer efficiently the excitation energy to the fluorescent compounds. The detector can also be used for the detection of nonfluorescent compounds that quench the fluorescence emitted by molecules in the eluent (91).

Monitoring of Emission–Excitation Spectra in the HPLC Effluent

In addition to instruments employing single excitation and emission wavelengths, detectors have been designed to scan the excitation or emission spectra of solutes eluted from an HPLC column. Scanning can be performed after stopping the flow (92). In principle, this is the easiest way of obtaining the emission and/or excitation spectra of fluorescing compounds in the detector zone. Well-resolved spectra can be produced because the scan speed can be kept sufficiently small. A disadvantage is the possible diffusion of the solute from the detection zone during scanning. This causes irreproducible distortions of the spectra and loss or resolution between adjacent solute bands. Practical solutions to diminish this and other problems connected with stopped-flow scanning have been proposed.

However, the scanning of fluorescence spectra "on-the-fly" in the HPLC effluent, without altering the flow, would seem to be more desirable. The possibility of rapid scanning (about 10 nm/sec) has been discussed (93). Although some spectroscopic resolution is lost, the spectra obtained in this scanning mode contain usually sufficient information for identification of the solute. On-the-fly rapid scanning can also serve to determine whether a chromatographic peak results from one or more eluted compounds. Emission spectra can be scanned a number of times during passage of the zone through the detector cell. The presence of more than one compound in the chromatographic peak will become evident by appreciable spectral changes, provided the compounds possess different fluorescence properties.

On-the-fly mechanical scanning of spectra will always involve the loss of spectroscopic resolution if chromatographic resolution is to be kept unimpaired. This problem can be overcome by the use of electronic array detectors (94). With this type of detector entire spectra—or a large portion of these—can be registered simultaneously. Jadamec et al. (95) described an intensified vidicon multichannel analyzer system for the characterization of petroleum fractions.

A similar type of instrument, the rapid scanning video fluorimeter, was described by others and tested as a potential HPLC detector (96–98). These detectors provide multiple excitation and emission spectra in a single (nonmechanical) scan without any need for stopping the eluent flow. Thus, three-dimensional fluorescence intensity plots can be constructed at any time during elution. Combined with algorithms for the qualitative and quantitative evaluation of the emission–excitation matrix these detectors can identify and quantify many fluorescence compounds in very complex samples, such as crude oil and ash residue samples (97). Strategies for analyzing data from video fluorimetric monitoring of HPLC effluents have been proposed (99).

Because of the high costs of acquiring these devices and their limited sensitivity as compared with photomultipliers, these detectors until today have found only limited application.

7.4.3. Fluorescence Detection of Nonfluorescent Compounds and Fluorescence Enhancement by Eluent Manipulation

Relatively few compounds show a sufficiently high quantum yield of fluorescence to allow detection at levels comparable to or lower than those obtained with UV absorption detection. Weakly fluorescing or nonfluorescent compounds can be converted into fluorescing derivatives either before or after HPLC separation. These methods will be discussed in the next sections. Another possibility is to make use of the dependence of the emitted radiation intensity on the environment of the fluorescing compounds. One way of accomplishing this is the addition of fluorescing compounds to the eluent. Asmus et al. (100) developed an HPLC detection method for lipids based on this principle. A continuous stream of an aqueous solution of the fluorescent dye 1-anilinonaphthalene-8-sulfonic acid (ANS) was mixed with the column effluent, giving rise to a certain level of base-line fluorescence. The fluorescence intensity of ANS in aqueous solutions is increased in the presence of lipids. Micelles are probably formed by the lipids, into which the ANS molecules partition. In these hydrophobic surroundings ANS exhibits enhanced fluorescence, allowing the lipids to be detected.

The addition of small amounts of aniline to the methanol–water and acetonitrile–water mixtures used as eluents in reversed-phase HPLC can make the fluorescence detector respond to nonfluorescent compounds (101). The aniline in the eluent gives rise to a base-line fluorescence signal. The passage through the detector of sample components eluting from the column results in a positive peak for a more highly fluorescent species than aniline and in a negative peak for a less-fluorescent compound. The composition of the eluent—solvents; added salts and buffers, pH and ionic strength—and its oxygen gas content and temperature are of paramount importance for the fluorescence detection. The fluorescence of dansylated phenols decreases in solvents with increasing dielectric constant, with a concomitant red shift of the excitation–emission spectra (102). The pH of an aqueous solution of dansyl derivatives also exerts a considerable effect; at low pH values the fluorescence is markedly decreased (103).

Froehlich and Yeats (104) studied the effect of mixed aqueous solvents on the fluorescence of indoles and aromatic amino acids and concluded that the fluorescence of these compounds is enhanced in ethanol–water and DMSO–water mixtures as compared with pure water. This fluorescence enhancement is probably due to a decrease in the formation of exciplexes between the excited state of the fluorophore and water; hence the quantum yield of fluorescence is increased upon the addition of organic solvents such as DMSO or ethanol.

When choosing an HPLC system with fluorescence detection an investigation of the type of solvent mixture (eluent) required for optimal quantum yield of fluorescence should come first, especially if high sensitivity is requested. Chromatographers, however, tend to give top priority to the search for optimal separation conditions and very often have a strong preference for reversed-phase systems. The eluents commonly used in these system—mixtures of an aqueous solution with a polar organic solvent—are frequently unsuited for sensitive fluorescence detection, because of the relatively low quantum yield of fluorescence of many organic compounds under these conditions. Changes in the composition of the eluent or the effluent have been shown to be of advantage in some cases. The addition of a solution of β-cyclodextrin to the column effluent resulted in a significant increase of the fluorescence of some derivatized thiols (105). The fluorescence of polynuclear aromatic hydrocarbons could be enhanced up to 10 times when these compounds were eluted with a mobile phase containing micelles of sodium dodecyl sulfate (106).

As the fluorescence of many compounds in aqueous solution strongly depends on the pH, post column acid–base manipulation can help increase the sensitivity and/or selectivity of fluorescence detection (107).

7.4.4. Conversion of Nonfluorescent Compounds into Fluorescent Products

Many nonfluorescent compounds can be converted into fluorescent derivatives either before or after HPLC separation. Reasons for derivatization other than gaining increased sensitivity of detection may exist; these are in principal the same as those outlined in Section 7.3.3.

The compounds of interest can be converted into fluorescent products by a number of methods. Fluorescence labeling, that is, the attachment of one or more fluorophoric groups to the molecule through covalent binding, is discussed in Section 7.4.4, under "Fluorescence Labeling."

Many other types of derivatization reactions have been described as well. Some examples serve to demonstrate the braod range of possible reactions. Phenothiazines can be detected after oxidation with permanganate added to the column effluent (108).

Indomethacin and its main metabolite, desmethyl indomethacin, yield fluorescent products by deacylation prior to HPLC analysis (109). 8-Hydroxyquinoline and related compounds form strongly fluorescing metal chelates with Mg(II) (110). Upon heating with an ethanol–sulfuric acid mixture cortisol is converted into fluorescing product(s) (111). Digitalis glycosides form fluorescing derivatives when heated with concentrated hydrochloric acid (112). Dimethoxyanthracene sulfonate (DAS) can serve as a fluorescent counter-ion for the postcolumn ion pair extraction of some basic drugs and pesticides (113).

Postcolumn photochemical reactions by UV irradiation are a useful means of converting certain compounds into strongly fluorescent products (114). Another approach is using nonfluorescent reagents that are converted upon reaction with the solutes in the column effluent into fluorescent products. Many compounds, such as the carbohydrates, can be oxidized by cerium (IV), which is reduced to the strongly fluorescent cerium (III) (115).

Another example is the ligand exchange reaction of thiols and other organosulfur compounds with the nonfluorescent palladium(II)–calcein complex. The reactive sulfur group competes with calcein for the palladium(II) ions, resulting in a decomposition of the palladium(II)–calcein complex and the release of the fluorescent calcein (116).

In the next sections several aspects of fluorescence labeling and precolumn and postcolumn derivatization are discussed.

Fluorescence Labeling

The recent upsurge of interest in fluorescence-detection-oriented techniques has resulted in a score of new labeling reagents. Together with the older reagents they offer the chromatographer ample choice. Many reagents have been developed for the labeling of primary and secondary amines and amino acids. Fluorescence labels are now also available for several functional groups. A survey of some well-known reagents and newer ones is given in Table 7.1.

As with any other method in which a reagent reacts with a certain functional group, fluorescence labeling limits the selectivity of the detection. The relative loss in selectivity in favor of enhanced sensitivity can be minimized by the use of fluorescence reagents that react with one type of functional group of structural moiety only. On the other hand, the separation power of modern HPLC columns is such that selectivity of the labeling reagents is not always needed. Furthermore, for the chromatographer there are defined advantages in using only a few reagents, which cover a wide range of compounds with different functional groups and for which the optimum reaction conditions have been well established. Acylating reagents such as dansyl chloride and alkylating reagents such as 4-bromo-methyl-7-methoxycoumarin react with several functional groups possessing an active hydrogen. Much knowledge and experience has been accumulated in the literature on GLC and HPLC with respect to the reaction modes and conditions with those types of reagents (117,118). Ideally the following conditions are fulfilled in fluorescence labeling procedures:

The absorption transition of lowest energy of the introduced fluorophore is intensive (large ϵ_{max}).

The fluorophore does not contain structural moieties of functional groups, which increase the rates of radiationless transitions (large quantum yield of fluorescence).

The reagent is nontoxic.

The reagent solution used for the derivatization reaction is stable for prolonged periods.

The fluorogenic reagent and its degradation products formed during the reaction are nonfluorescent or else they are well separable from the derivatized compounds of interest.

The compound(s) of interest is rapidly and quantitatively derivatized under mild conditions, yielding a single product.

The derivatives are stable and possess favorable chromatographic properties.

When developing a derivatization procedure for a certain compound the influence of all reaction conditions—choice of solvent, concentrations of reactants and catalyst(s), temperature and reaction time—must be thoroughly investigated to obtain optimal reaction conditions. The reaction product(s) must be identified and the fluorescence properties of the derivative—including its quantum yield of fluorescence—in possible HPLC eluents must be determined.

The reproducibility of the labeling procedure should be thoroughly tested, especially if the yield of the reaction is less than 100%.

For quantitative analyses the use of an internal standard helps to limit the number of manipulations that have to be carried out with exact reproducibility. In fluorescence labeling the choice of an internal standard that is structurally related to the compound of interest and therefore derivatized in the same way is common practice.

Precolumn Derivatization

A major advantage of derivatizations prior to separation is that no restrictions are imposed on the reaction kinetics, provided that the reaction goes to completion within a reasonable time, yielding one derivatization product for the compound in question without the formation of side products. One is free to vary the conditions in order to optimize reaction time and yield.

Contrary to on-line postcolumn derivatization techniques, the solvent in which the precolumn reaction takes place need not be compatible with the mobile phase of the HPLC system. Furthermore, side products formed during the derivatization that might quench the fluorescence of the derivation can be separated either on the HPLC column or by a prechromatographic clean-up step.

The derivatization reactions are carried out in small volumes with only minor amounts of material; the equipment for the reactions must therefore be on microscale too. Several types of reaction vials in varying sizes are available, as well as plastic-capped sample bottles and glass and plastic low-volume centrifuge tubes. As heating is often necessary, the sample vials should be resistant to the solvents and reagents even at elevated temperatures. The number of manipulations in the entire derivatization procedure should be kept as small as possible, not only to avoid lengthy procedures but also to prevent sample loss.

One of the most frequently encountered problems in derivatization procedures is the occurrence of disturbing peaks in the chromatograms, due to degradation products or impurities of reagents and/or solvents. Impurities can be present in freshly delivered materials or they may have been formed upon storage. Great care should therefore be exercised in checking the purity of the reagents and solvents—the analysis of control samples (reagent blanks) is essential—and in storing them under suitable conditions in appropriate containers. Impurities in solvents causing interference can be hard to identify and to remove. Despite the high cost, it may be worthwhile to use high-purity solvents prepared in all-glass systems, only.

Ideally, the derivatization mixture components do not cause changes in the chromatographic system through chemical reactions, precipitation,

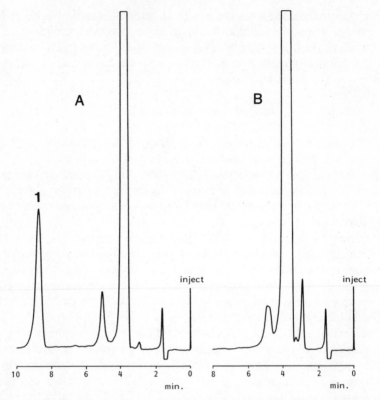

Fig. 7.7. Chromatograms obtained after derivatization of benzoic acid (*a*) with 9-hydroxymethylanthracene and (*b*) from the reagent blank. Peak 1, benzoic acid derivative. Column, Lichrosorb RP 18(10 μm); eluent, methanol–water (9:1); excitation wavelength, 365 nm; emission wavelength, 415 nm; flow, 1 ml/min; room temperature. (H. Lingeman and A. Hulshoff, unpublished results.)

or demixing of the eluent upon injection or otherwise, and do not interfere with the detection of the derivatized compounds. If these conditions are met, aliquots of the derivatization mixture can be directly injected into the chromatographic system. An example is shown in Fig. 7.7. However, it is frequently necessary to remove the excess reagent or the solvent or other components of the reaction mixture before HPLC analysis.

Evaporation of the reagent mixture under a stream of dry nitrogen, either at room temperature or under heating, is a simple and convenient way of removing the solvents and, if volatile, the reagent. This procedure is also a concentration step, because the residue after evaporation can be reconstituted in a small volume of the mobile phase with subsequent injection of a major part of the final solution.

Precolumn cleanup of the derivatization can also be achieved by liquid–liquid extraction or other precolumn separation steps. Excess reagent is sometimes removed by allowing it to react with an excess of another compound; neither that compound nor its derivative must interfere with the analysis. For instance, excess of dansyl chloride can be eliminated by letting it react with an excess of proline.

Postcolumn Derivatization; Reaction Detectors

Considerable efforts by several research groups have been put into the development of on-line postcolumn derivatization methods and a number of reviews on this topic have been published recently (119–124). Postcolumn reactions take place in a part of the HPLC system, called the reactor, between the column and the detector. Postcolumn derivatization techniques have some important advantages as compared with the precolumn methods. The formation of side products (artifacts) is not a problem. The reaction need not go to completion, nor should the reaction product be well defined, as long as the procedure is reproducible. The underivatized compounds are separated and eluted from the column and can therefore be detected with other, nondestructive, detection methods before being derivatized.

Obvious disadvantages are the restricted freedom in choosing the reaction conditions, band broadening in the reactor with a resulting loss of chromatographic resolution, and possible interference by excess reagent or reagent degradation products with the detection of the derivatized compounds. Another disadvantage is the need for instrument modifications to suit the postcolumn reaction.

Reactors consist of one or more units (made of an inert material), depending on the chemical operations to be carried out. Low-pressure peristaltic pumps and low dead volume mixing tees are used to add reagent solutions, solvents or, air to the column effluent. Three different types of postcolumn reactors have been described. These are the tubular or capillary reactors with a nonsegmented reaction medium, bed reactors, and segmented-stream reactors. The tubular or capillary reactor is the simplest reactor to handle and is well suited for fast reactions with reaction times up to 30 sec. This reactor is to be preferred over the other types only for very fast reactions that take place in a few seconds.

Coiling of the capillary helps to minimize band broadening. Coils of tubes of 0.2–0.3 mm I.D. are commonly used, but band broadening could be diminished even further by using capillaries with smaller internal diameters. Fast reactions that can be performed with this type of reactor are the reactions of primary amines with o-phthalaldehyde and with fluorescamine.

When the reaction times exceed 20–30 sec band broadening can become unacceptably large (>20%).

Bed reactors are columns filled with small glass beads to provide an extra surface area for the reaction to proceed. A bed reactor does not contribute to the retention of the solutes. The particle size of the glass beads is important and should be small to avoid excessive band broadening. The bed reactor is a suitable device for reaction times up to 4 min.

In a recent study (125) it was concluded that with nonsegmented flow systems the capillary reactor of about 0.3 mm I.D. is best used for very fast reactions only (less than 5 sec) and the packed bed reactor for reactions of 5–20 sec. Solvent or gas segmented-flow systems perform better in the case of reactions that take more than about 20 sec to be completed, provided that all parts of the segmented-flow system, in particular the phase separators, are properly constructed.

In flow-segmented systems air bubbles or solvent plugs that are immiscible with the mobile phase, are introduced into the column effluent. The segmentation of the flow serves to reduce band broadening in case of comparatively long residence times of the solutes in the reactor system. Slow on-line performed postcolumn reactions are then still feasible. Segmentation is very efficient in preventing band broadening if no significant transfer from one segment to the next takes place through wall wetting or partial solubility in the solvent plugs. A critical part of the system is the phase separator or debubbler, which separates the air or the immiscible solvent from the eluent stream. Proper construction of the phase separators is important in order to limit band broadening (125). The choice of reactor type does not only depend on the reaction time. Extreme reaction conditions, for example, high temperatures or the use of aggressive reagents, limits the freedom of reactor design. Solvent-segmented systems must be used in case of extraction detection systems (see below). Packed-bed reactors are particularly well suited for reactions taking place at solid–liquid interfaces as illustrated by the immobilized enzyme bed reactor (126). The beads of this packed-bed reactor are coated with an enzyme, which interacts with the solute(s) in the column effluent.

Excess reagent frequently interferes with the fluorescence signal of the derivatized compound. If the difference between the partition coefficients of the derivative and the reagent is sufficiently large, separation by liquid–liquid extraction prior to detection is possible. In such an extraction detector the eluent flow is segmented with nonmiscible solvent plugs into

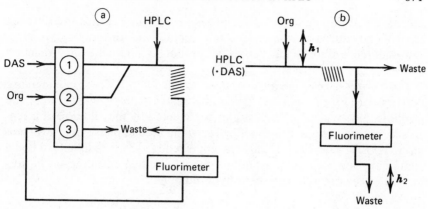

Fig. 7.8. Postcolumn ion-pair extraction systems. (*a*) Flow rates (ml/min): 10^{-4} *M* DAS, 0.6; organic solvent, 0.6; through the flow cell, 0.3; HPLC, 1.0. (*b*) Flow rates (ml/min): organic solvent, 0.2; through the flow cell, 0.14; HPLC, 1.0 $h_1 = 60$ cm; $h_2 = 80$ cm (Ref. 390).

which derivative or reagent partition; the plugs also serve to diminish band broadening.

After phase separation the liquid phase containing the derivative flows through the detector cell. Several organic bases can be analyzed by HPLC with fluorescence detection via postcolumn ion pairing with a fluorescent counter-ion such as dimethoxyanthracene sulfonate (DAS) (127). Figure 7.8*a* shows an extraction detection system suitable for the analysis of amines (e.g., secoverine). The reagent (DAS) solution and the organic extractant solvent are pumped into the column effluent. Reaction and extraction take place in a coil. After phase separation the organic solvent containing the organic base–DAS ion pair passes the detector; the excess reagent is partitioned into the aqueous phase going to waste. This design can be simplified by chromatography of the organic base in a reversed-phase ion pair system with DAS as the counter-ion reagent in the eluent (Fig. 7.8*b*).

Extraction detectors have also been used in normal-phase HPLC systems (128). The effluent consisting of an organic solvent mixture is then segmented by an aqueous solution of the reagent.

Several investigators have applied postcolumn photochemical reactors to improve the fluorescence detection properties of selected groups of compounds (e.g., 129,130). The simplest way of achieving this is UV irradiation of the HPLC effluent flowing through a reactor coil with good UV transparancy, such as Teflon. The tubular reactor system—segmented or nonsegmented—can be coiled around the UV source. The selectivity of this detection principle is very high, because interference can only be caused by compounds that coelute with the solute of interest and undergo a fluorescent derivative producing photochemical reaction as well.

Furthermore, photochemical reagents added to the eluent can increase the applicability of this type of reaction detector. For instance, aliphatic alcohols, aldehydes, and ethers can be detected by the addition of a sensitizer, anthraquinone-2,6-disulfonate, to the mobile phase of a reversed-phase HPLC system (131). This reagent is reduced by the aliphatic oxygen containing compounds in a photochemical reaction resulting in the formation of the highly fluorescent hydroquinone.

7.5. APPLICATIONS

An up-to-date survey of the many applications of fluorescence detection in liquid chromatography is presented in Table 7.2, with the emphasis on the quantitative determination of drugs and drug metabolites in biological fluids. The drugs mentioned in Table 7.2 are divided into five main groups: four of these groups are: acidic, amphoteric, basic, and neutral compounds. Cytostatic agents, steroids, and vitamins are dealt with in a separate group (V). Each group is subdivided into one or more structurally related subgroups and/or a number of miscellaneous compounds.

In the first column of Table 7.2 the name of the drug or group of drugs is mentioned, for which the procedure referred to was originally designed. In some cases the internal standard and/or other drugs mentioned in the original publication are also reported.

"Sample" (second column of Table 7.2) refers to the type of biological fluid (blood, urine, etc.) or tissue to which the method is applicable and the volume required per analysis.

In the third column the clean-up procedure used in the analysis, prior to chromatographic separation, is denoted. The various clean-up procedures are classified according to type and number of clean-up steps in the prechromatographic sample treatment. Simple "washings" of samples or extracts, in which the analyte is not transferred to another phase, is not considered a clean-up step here. The meaning of the symbols is as follows:

No clean-up step is used in the prechromatographic sample treatment

a   The only clean-up step is some form of protein removal (ultrafiltration or precipitation with ammonium acetate, acetonitrile, ethanol, etc.)

Table 7.2. Bioanalytical Applications of Liquid Chromatography with Fluorescence Detection

Compound(s)	Sample (Volume Required)	Clean-up Chromatography Procedure[a]		Fluorescence Method	Ref.
I. Acidic Compounds					
A. Carboxylic acids					
Bile-(acids)	Bile (0.01 ml)	d HPLC	RP	Pre-Bromoacetylpyrene	132
	Serum (0.1 ml)	d HPLC	RP	Pre-Bromoacetylpyrene	132
	Serum (0.1 ml)	d HPLC	RP	Pre-Dns-hydrazine	133
Bromo-lasalocid	Plasma (1 ml)	b HPLC	RP	Native	134
Caprylic-	Plasma (0.5 ml)	d HPLC	RP	Pre-Br-Mmc	135
Carboxylic-		HPLC	NP	Pre-DANE	136
		HPLC	RP	Pre-Br-Mac	137
		HPLC	RP	Pre-Br-Mmc	138
		HPLC	RP	Pre-DMC	139
		TLC	NP	Pre-Br-Mmc	140
		TLC	NP	Pre-DCCI	141
		TLC	NP	Pre-Hy-Mmc	142
Celastrol		TLC	NP	Post-Oxidation spray	143
Dicarboxylic-		HPLC	RP	Pre-Br-Mmc	138, 144
Di-, tricarboxylic-		TLC	NP	Post-DCC spray	145
Fatty-		HPLC	RP	Pre-ADAM	146, 147
	Serum (1 ml)	b HPLC	RP	Pre-Br-Mmc	148
		HPLC	RP	Pre-Naphthacylbromide	149
		HPLC	RP	Pre-9-CIMA	150
		HPLC	RP	Pre-DAP	151
		TLC	NP	Pre-Br-Mmc	56

Table 7.2. (continued)

Compound(s)	Sample (Volume Required)	Clean-up Chromatography Procedure[a]		Fluorescence Method	Ref.
Gentisic-	Urine	—	HPLC	RP-PI	Native	152
Gibberellins			HPLC	NP/RP	Pre-Br-Mmc	153
			HPLC	RP	Native	154
Hippuric-	Urine (1 ml)	a	HPLC	NP-PI	Native	155
p-Hydroxybenzoic-	Urine (1 ml)	d	HPLC	NP	Pre-Dns-Cl	156
Hydroxy-substituted car-boxylic-	Urine		HPLC	IE	Post-Oxidation	157
Ibuprofen	Plasma (1 ml)	a	HPLC	RP	Native	158
3-Indolacetic-(5-hydroxy-)	Urine		HPLC	IE	Native	159
	Urine (0.05 ml)	b	HPLC	RP	Native	160
	Urine (0.2 ml)	—	HPLC	RP-PI	Native	161
Indomethacin	Plasma (1 ml)	a	HPLC	RP	Post-Hydrolysis	162
	Urine (0.3 ml)	a	HPLC	RP	Post-Hydrolysis	162
	Plasma (0.1 ml)	b	HPLC	RP	Pre-Deacylation	109
	Urine (0.5 ml)	b	HPLC	RP	Pre-Deacylation	109
Isoxepac	Plasma (1 ml)	b	HPLC	RP	Native	163
α-Ketocarboxylic-			HPLC	RP	Pre-Phenylenediamine	164
			TLC	NP	Pre-DCCI/DICI	165
Lonazolac	Serum (0.2 ml)	c	HPLC	RP	Native	166
Naproxen	Serum (0.02 ml)	d	HPLC	NP	Pre-DANE	167
			HPLC	RP	Native	168
	Plasma (0.01 ml)	—	HPLC	RP	Native	169
Phenylpyruvic-	Serum (0.2 ml)	d	HPLC	RP	Pre-Hydrazinostilbazole	170
	Urine (0.2 ml)	d	HPLC	RP	Pre-Hydrazinostilbazole	170

674

Compound	Sample	Method	Mode	Derivatization	Ref.
Prostaglandins		HPLC	NP	Pre-Br-Mmc	171
		HPLC	RP	Pre-ADAM	172
		HPLC	RP	Pre-Panacylbromide	173
	Seminal fluid	b HPLC	RP	Pre-Br-Mac	174
		TLC	NP	Pre-Br-Mmc	171
Salicylic-	Plasma (0.01 ml)	— HPLC	RP	Native	169
	Urine	— HPLC	RP-PI	Native	152
	Saliva (2 g)	b HPTLC	NP	Native	175
Valproic-	Plasma (0.5 ml)	d HPLC	RP	Pre-Br-Mmc	135
	Plasma (0.5 ml)	a HPLC	RP	Pre-Naphthacylbromide	176
Vanillylmandelic-	Urine (1 ml)	d HPLC	NP	Pre-Dns-Cl	156
	Urine (5 ml)	d HPLC	RP-PI	Native	177
B. Miscellaneous					
Barbiturates	Serum (1 ml)	HPLC	RP	Native	178
	Blood (0.02 ml)	b HPLC	RP	Pre-Br-Mmc	179
Captopril	Plasma (1 ml)	b HPLC	RP	Pre-Dns-Cl	180
Cysteine derivatives		b HPLC	RP	Pre-PM	181
		HPLC	RP	Pre-OPA	182
	Plasma (0.2 ml)	a HPLC	RP-PI	Pre-OPA	183
	Urine (5 ml)	d HPLC	RP-PI	Pre-DACM/PM	184
Glutathione		HPLC	RP-PI	Post-NCDA	185
Penicillin G		HPLC	RP	Pre-Flur	186
Porphyrins	Biological fluids	HPLC	RP	Native	187
	Urine (0.025 ml)	— HPLC	RP	Native	188, 189
	Blood (0.1 ml)	a HPLC	RP-PI	Native	190
		HPTLC	RP	Native	30
Thiol derivatives		HPLC	IE	Post-Ligand exchange	116
		HPLC	IE	Post-OPA	191
		HPLC	IE/RP	Pre-Monobrobimane	192, 193
		HPLC	RP	Pre-Dns-aziridine	194
		HPLC	RP	Pre-OPA	195

675

Table 7.2. (continued)

Compound(s)	Sample (Volume Required)	Clean-up	Chromatography Procedure[a]	Fluorescence Method	Ref.
C. Phenolic compounds					
Acetaminophen		TLC	NP	Pre-Dns-Cl	196
Alkylphenols		HPLC	NP/RP	Native	197
Cannabinoid derivatives	Urine (10 ml)	b HPLC	NP	Post-Irradiation	198
		HPLC	RP	Pre-Dns-Cl	199
	Plasma (5 ml)	b TLC	NP	Pre-DIS-Cl	200
Capsaicin		a HPLC	RP	Native	201
o-Cresol	Blood–tissue	d HPLC	NP-PI	Native	155
8-Hydroxyquinoline	Urine (10 ml)	HPLC	IE	Post-Metal chelates	110
Meptazinol	Plasma (1 ml)	b HPLC	RP	Native	202
1-Naphthol	Blood (0.25 ml)	b HPLC	RP	Native	203
Nitrofur-ans(nitrofurantoin)		TLC	NP	Post-Pyridine spray	204
Phenols		HPLC	RP	Native	205, 206
		HPLC	RP	Post-Oxidation	207
Phenprocoumon	Urine (5 ml)	b HPLC	NP	Pre-Dns-Cl	208
Salicylamide	Plasma (0.2 ml)	b TLC	NP	Native	209
	Plasma (1 ml)	— HPLC	RP	Native	210
Warfarine	Plasma, urine	a HPLC	NP	Post-Acid–base manipulation	107
		TLC	NP	Post-Heating	211

II. Amphoteric Compounds

A. Amino acids (peptides, proteins)

Sample	Technique	Mode	Detection	Ref.
	HPLC	IE	Post-Flur	212
	HPLC	IE	Post-NBD-Cl	213
	HPLC	IE	Post-Ninhydrin	214
	HPLC	IE	Post-OPA	215, 216
	HPLC	IE	Pre-OPA	217, 218
	HPLC	IE	Pre-Pyridoxal	219
	HPLC	NP	Post-Flur	220
	HPLC	NP	Pre-Dns-Cl	221
	HPLC	RP	Post-Flur	222, 223
	HPLC	RP	Post-OPA	224, 225
	HPLC	RP	Pre-Dbs-Cl	226
	HPLC	RP	Pre-Dns-Cl	227, 228
Serum (1 ml)	a HPLC	RP	Pre-Dns-Cl	229
	HPLC	RP	Pre-Flur	230
	HPLC	RP	Pre-NBD-F	231
	HPLC	RP	Pre-NCDA	232
	HPLC	RP	Pre-OPA	233, 234
Plasma (0.01 ml)	a HPLC	RP	Pre-OPA	235, 236
Urine (0.02 ml)	a HPLC	RP	Pre-OPA	237, 238
Urine	d HPLC	RP-PI	Post-Flur	239
	HPLC	RP-PI	Post-OPA	240
	HPLC	RP-PI	Pre-Flur	241
	HPLC	RP-PI	Pre-OPA	242
	HPTLC	NP	Post-NBD-Cl spray	243
	TLC	NP	Various	244
	TLC	NP	Post-Flur spray	51, 245
	TLC	NP	Post-NCDA spray	246

677

Table 7.2. (*continued*)

Compound(s)	Sample (Volume Required)	Clean-up Chromatography Procedure[a]	Fluorescence Method	Ref.
		TLC	NP	Post-NaClO + thiamine spray	247
		TLC	NP	Pre-Flur	54
		TLC	NP	Pre-DNS-Cl	248, 249
Histidine	Urine (1 ml)	d TLC	NP	Pre-NBD-Cl	250
Hydroxyphenylalanine		HPLC	RP	Native	251
Hydroxyproline		HPLC	IE	Post-NBD-Cl	213
	Serum (0.1 ml)	a HPLC	IE	Post-OPA	252
		HPLC	RP	Pre-NBD-Cl	253
	Urine (1 ml)	b TLC	NP	Pre-NBD-Cl	254
5-Hydroxytryptophan	Plasma (0.75 ml)	d HPLC	IE	Post-OPA	255
Norleucine	Serum (0.01 ml)	a HPLC	RP	Pre-Dns-Cl	256
Phenylalanine	Plasma (2 ml)	a HPLC	RP	Native	257
	Serum (0.01 ml)	a HPLC	RP	Pre-Dns-Cl	258
Proline		HPLC	IE	Post-NBD-Cl	213
Taurine		HPLC	RP	Pre-Dns-Cl	259
		HPLC	RP	Pre-OPA	260
Tryptophan	Brain tissue	a HPLC	IE	Native	261, 262
		HPLC	RP	Native	263, 264
	Plasma (0.02 ml)	a HPLC	RP	Native	265
	Urine (0.02 ml)	a HPLC	RP	Native	265
		TLC	NP	Post-Flur dip	266
		TLC	NP	Post-Formaldehyde gas	267
m-Tyrosine	Plasma (2 ml)	a HPLC	RP	Native	257

678

B. Miscellaneous amphoteric compounds

Compound	Sample (volume)		Method	Mode	Derivatization	Page(s)
Aminobenzoic acid	Plasma (0.1 ml)	a	HPLC	RP-PI	Native	268
4-Aminobutyric acid	csf	a	HPLC	IE	Post-OPA	269
			HPLC	RP	Pre-Dns-Cl	270
			HPLC	RP-PI	Pre-OPA	242
Aminocaproic acid	Serum (0.01 ml)	a	HPLC	RP	Pre-Flur	271
6-Aminopenicillanic acid			HPLC	RP	Pre-Flur	186
Aminosalicylic acid	Plasma (1 ml)	a	HPLC	NP-PI	Native	272
	Urine (1 ml)	a	HPLC	NP-PI	Native	272
	Plasma (0.5 ml)	b	HPLC	RP	Pre-Acetylation	273
	Urine (0.5 ml)	b	HPLC	RP	Pre-Acetylation	273
	Plasma (0.1 ml)	a	HPLC	RP-PI	Native	268, 274
Bumetanide	Plasma (0.2 ml)	a/b	HPLC	RP	Native	275, 276
	Urine (0.2 ml)	a/b	HPLC	RP	Native	275, 277
Cefotaxine			TLC	NP	Native	278
Cephradine }	Serum (0.2 ml)	a	HPLC	RP	Post-Flur	279
Cephatrizine }	Urine (0.2 ml)	a	HPLC	RP	Post-Flur	279
Furosemide	Plasma (0.2 ml)	a/b	HPLC	RP	Native	280, 281
	Urine (0.5 ml)	a/b	HPLC	RP	Native	280, 281
	Plasma (1 ml)	b	TLC	NP	Post-Propyleneglycol dip	282
	Urine (0.05 ml)	b	TLC	NP	Post-Propyleneglycol dip	282
Glyburide	Serum (1 ml)	b	HPLC	RP	Native	283
Metolazone	Plasma (2 ml)	d	HPLC	RP	Native	284
	Urine (0.5 ml)	b	HPLC	RP	Native	284
D-Penicillamine	Serum (0.5 ml)	a	HPLC	RP	Pre-Dns-aziridine	285
Sulfapyridine	Plasma	a	HPLC	RP-PI	Native	274
Sulfasalazine	Plasma	a	HPLC	RP-PI	Native	274
Sulpiride	Serum (4 ml)	b	HPLC	RP	Native	286
	Urine (1 ml)	b	HPLC	RP	Native	286
YM-09538	Plasma (1 ml)	b	HPLC	NP	Pre-Bns-Cl	287

Table 7.2. *(continued)*

III. Basic Compounds

Compound(s)	Sample (Volume Required)	Clean-up Chromatography Procedure[a]		Fluorescence Method	Ref.
A. Amines					
Aliphatic amines		TLC	NP/RP	Pre-Dns-Cl	288
		HPLC	RP	Pre-NBI-SO$_2$Cl	289
		TLC	NP	Pre-Dns-Cl	288
		TLC	NP	Pre-NBI-SO$_2$Cl	289
Alizapride	Plasma (1 ml)	a HPLC	RP	Native	290
Amilorid	Plasma (1 ml)	b HPTLC	NP	Native	291
Amines		HPLC	NP	Native	178, 292
		HPLC	RP	Post-DAS	127
		HPLC	RP	Pre-Mbp	293
	Blood	HPLC	RP	Pre-NBD-F	294
		HPLC	RP	Pre-Dns-Cl	295
		TLC	NP	Pre-Mbp	293
		TLC	NP		
Amino-flunitrazepam	Plasma (0.5 ml)	d HPLC	RP	Pre-Flur	296
Antihistamine derivatives		HPLC	RP	Pre-NCF	297
Atropine		HPLC	NP	Post-DAS	128
Benzodiazepines	Serum (1 ml)	TLC	NP	Post-Oxidation	298
Biopterins		HPLC	IE	Post-Oxidation	299
Bromopheniramine	Urine (1 ml)	c HPLC	RP	Post-DAS	300
Butyrophenones		HPTLC	NP/RP	Native	39
Carbamazepine	Plasma (0.2 ml)	a HPLC	RP	Native	301
	Urine	a HPLC	RP	Native	301
Carbodiimides		TLC	NP	Post-DM-barbituric acid spray	302

Table 7.2. (*continued*)

Compound(s)	Sample (Volume Required)	Clean-up Chromatography Procedure[a]		Fluorescence Method	Ref.
Emetine	Plasma (9 ml)	HPLC	NP	Pre-Dns-Cl	303
		b HPLC	RP-PI	Pre-Oxidation	322
		TLC	NP	Pre-Dns-Cl	303
Endralazine	Plasma	b HPLC	RP	Pre-Formic acid	323
Ergometrine		HPLC	IE	Native	324
		HPLC	RP-PI	Native	324
Ergot alkaloids	Plasma (3 ml)	d HPLC	RP	Native	325
	Urine	HPLC	RP	Post-Irradiation	326
Ergotamine		HPLC	NP	Post-DAS	128
Erythromy-cin(ethylsuccinate)	Serum (1 ml)	b HPLC	RP	Post-Tinopal	327
Fenbendazole	Serum	HPLC	RP	Post-Irradiation	130
Flecainide	Blood (1 ml)	a HPLC	RP	Native	328
Fluphenazine	Plasma (4 ml)	d HPTLC	NP	Post-Irradiation	329
Fluvoxamine		HPLC	RP	Post-Dns-Cl	309
FM-24	Plasma (2 ml)	d HPLC	RP	Native	330
Gentamycin	Plasma (0.1 ml)	a HPLC	IE	Pre-Flur	331
	Urine (0.05 ml)	a HPLC	IE	Pre-Flur	331
	Plasma (0.1 ml)	a HPLC	RP	Pre-OPA	332
	Urine (1 ml)	c HPLC	RP	Pre-OPA	332
Guanadrel		HPLC	RP	Pre-Acetylacetone	333
Guanidino compounds	Plasma (0.2 ml)	a HPLC	IE	Post-Ninhydrin	334
		HPLC	RP-PI	Post-Phenanthrenequinone	335
Harmane alkaloids		HPLC	RP	Native	336
Histamine	Urine (1 ml)	d HPLC	RP	Pre-OPA	337
	Plasma (0.5 ml)	d HPLC	RP	Pre-OPA	338
	Tissue	a HPLC	RP-PI	Pre-OPA	339

682

Compound	Sample	Technique	Mode	Derivatization	Page
Hydroxyatrazine		HPLC	NP	Post-DAS	128
		HPLC	RP	Pre-DAS	113
Imipramine	Plasma (2 ml)	b/d HPLC	RP	Native	314, 330
Kanamycin		HPLC	IE	Post-OPA	340
Ketanserin	Plasma (1 ml)	b/d HPLC	RP	Native	341, 342
	Urine (0.1 ml)	b HPLC	RP	Native	341
Local anaesthetics	Plasma (5 ml)	a/b TLC	NP	Post-Flur	343
LSD		HPLC	IE	Native	344
	Urine	d HPLC	RP	Native	345, 346
Maprotiline ⎫	Plasma (0.5 ml)	d HPLC	RP	Pre-Dns-Cl	347
Metapramine ⎬	Urine (0.5 ml)	b HPLC	RP	Pre-Dns-Cl	347
Monosodium glutamate		HPLC	RP	Pre-Dns-Cl	348
Morphine		HPLC	NP	Pre-Dns-Cl	303
	Urine (5 ml)	b HPLC	NP	Pre-Oxidation	319
	Blood (10 ml)	d HPLC	RP	Post-Oxidation	349
	Urine (1 ml)	b HPLC	RP	Post-Oxidation	349
		TLC	NP	Pre-Dns-Cl	303, 350
Nalorphine	Blood (10 ml)	d HPLC	RP	Post-Oxidation	349
	Urine (1 ml)	b HPLC	RP	Post-Oxidation	349
Nitrosaminen		HPLC	NP	Pre-NBD-Cl	351
		TLC	NP	Post-Flur spray	352
Norverapamil	Plasma (0.1 ml)	b HPLC	NP	Native	292
Opiates		HPLC	RP	Native	353
Oxytocin		HPLC	IE	Native	324
		HPLC	RP-PI	Native	324
Perhexiline maleate	Plasma (1 ml)	b HPLC	RP	Pre-Dns-Cl	354
Phenothiazines	Serum	a HPLC	RP	Post-Irradiation	313
	Urine	a HPLC	RP	Post-Irradiation	313

Table 7.2. (continued)

Compound(s)	Sample (Volume Required)	Clean-up Chromatography Procedure[a]		Fluorescence Method	Ref.
Pipotiazine	Plasma (2 ml)	b HPLC	NP	Native	355
	Urine (2 ml)	b HPLC	NP	Native	355
Polyamines	Plasma	d HPLC	IE	Post-OPA	356
	Urine (1 ml)	–/b HPLC	IE	Post-OPA	357, 358
	Urine	c HPLC	NP	Pre-Dns-Cl	359
	Urine (0.2 ml)	HPLC	RP	Pre-Dns-Cl	360, 361
	Serum (0.5 ml)	d HPLC	RP	Pre-Flur	362, 363
	Urine (0.5 ml)	d HPLC	RP	Pre-Flur	362
		HPLC	RP	Pre-OPA	364
		HPLC	RP-PI	Pre-Dns-Cl	365
	Biological fluids	d HPLC	RP-PI	Pre-Flur	366
		HPLC	RP-PI	Post-OPA	367, 368
Prazosin	Plasma (0.2 ml)	a HPLC	RP	Native	301, 369
	Plasma (1 ml)	b HPLC	RP-PI	Native	370,371
Primary amines		GLC	OV-17	Post-OPA	3
		HPLC	IE	Post-Flur	372
		HPLC	IE	Post-OPA	373
		HPLC	IE	Pre-FMOCCl	374
		HPLC	RP	Post-Dns-Cl	309
		TLC	NP	Pre-Flur	375
		TLC	NP	Pre-Salicylaldehyde-diphenyl-boron	376
Procainamide	Plasma (0.5 ml)	b TLC	NP	Native Post-HCl exposure	377
Purines		TLC	NP	Luminophore on plate	378

684

Compound	Sample	Method	Mode	Derivatization	Ref.
Pyrimethamine	Plasma (0.5 ml)	b HPLC	NP	Native	379
	Plasma (2 ml)	a TLC	NP	Post-NH$_4$HSO$_4$ spray	380
	Urine (2 ml)	a TLC	NP	Post-NH$_4$HSO$_4$ spray	380
Quinidine	Plasma (0.5 ml)	a/d HPLC	RP	Native	381, 382
	Urine	a HPLC	RP	Native	381
		HPLC	RP	Post-p-Toluenesulfonic acid	383
Rauwolfia alkaloids					
Ajmaline	Plasma (2 ml)	d TLC	NP	Native	384
Chloracetylajmaline	Plasma (2 ml)	d TLC	NP	Native	384
Prajmaline	Plasma (1 ml)	c HPLC	NP-PI	Native	385
Reserpine	Plasma (2 ml)	b HPLC	RP-PI	Pre-Oxidation	386
Ro 11-2465	Plasma (1 ml)	b HPLC	RP	Native	387
Ro 12-6995	Plasma (1 ml)	b HPLC	NP	Native	388
Secondary amines		HPLC	IE	Pre-FMOCCl	374
		HPLC	RP	Post-Dns-Cl	309
		HPTLC	NP	Post-Flur spray	389
Secoverine	Serum (1 ml)	c HPLC	NP	Post-DAS	390
Senmosides		TLC	NP	Post-Hydrazine spray	391
Spectinomycin		HPLC	RP-PI	Post-OPA + oxidation	392
Tamoxifen	Serum (0.01 ml)	a HPLC	NP/RP	Post-Irradiation	393
Tertiary amines	Urine (1 ml)	c HPLC	RP	Post-DAS	113, 300
		HPLC	RP	Pre-NCF	297
Tetroxoprim	Plasma (1 ml)	b TLC	NP	Post-HNO$_3$ spray	394
	Urine	TLC	NP	Post-HNO$_3$ spray	394
Thiabendazole	Serum (0.05 ml)	a HPLC	RP	Native	395
Thiazides	Urine (1 ml)	— HPLC	RP	Native	396
Thioridazine	Plasma (1 ml)	d HPLC	NP	Post-Oxidation	108
		HPLC	NP	Post-Oxidation	397
Tiodazosin	Plasma (0.2 ml)	a HPLC	RP	Native	369
Tocainide	Plasma (0.05 ml)	d HPLC	NP	Pre-Dns-Cl	398
	Plasma (0.5 ml)	a HPLC	RP	Pre-Flur	399

Table 7.2. *(continued)*

Compound(s)	Sample (Volume Required)	Clean-up Chromatography Procedure[a]		Fluorescence Method	Ref.
Tricyclic antidepressants	Plasma (1 ml)	b HPLC	RP	Native	400
	Serum (1 ml)	TLC	NP	Post-Oxidation	298
Trimetazidine	Plasma (2 ml)	b HPLC	NP	Pre-Dns-Cl	401
Trimethoprim	Plasma (1 ml)	b TLC	NP	Post-HNO$_3$ spray	394
	Urine	TLC	NP	Post-HNO$_3$ spray	394
Trimipramine	Plasma (2 ml)	b HPLC	RP	Native	314
TRIS	Serum (9 ml)	— HPTLC	NP	Post-Ninhydrin	402
UK-33274	Plasma (1 ml)	b HPLC	RP-PI	Native	403
Verapamil	Plasma (0.5 ml)	b HPLC	RP	Native	404
	Plasma (1 ml)	b HPLC	RP-PI	Native	405
B. Aryloxypropanolamines					
Acebutolol		HPLC	NP	Native	406
	Plasma (1 ml)	d HPLC	RP	Native	407
	Urine (0.05 ml)	d HPLC	RP	Native	407
	Serum (2 ml)	b TLC	NP	Post-Paraffin dip	408
Alprenolol	Plasma (1 ml)	d HPLC	RP	Native	409
Atenolol	Plasma (1 ml)	d HPLC	RP-PI	Native	410
	Plasma (1 ml)	b TLC	NP	Post-Citric acid + ethyleneglycol	411
	Urine (0.01 ml)	— TLC	NP	Post-Citric acid + ethyleneglycol	411
Bufuralol	Blood (2.5 ml)	d TLC	NP	Native (after silica scraping)	412
	Urine (2.5 ml)	d TLC	NP	Native (after silica scraping)	412

Drug	Sample	Method	Type	Treatment	Ref.
Bunitrolol	Plasma (1 ml)	b HPLC	RP	Native	413
	Urine (1 ml)	b HPLC	RP	Native	413
Celiprolol	Plasma (1 ml)	d HPLC	RP	Native	407
	Urine (0.05 ml)	d HPLC	RP	Native	407
Labetalol	Plasma (1 ml)	b HPLC	RP-PI	Native	305, 414
Metoprolol	Plasma	HPLC	NP	Native	415
	Plasma (1 ml)	d HPLC	RP	Native	409
	Plasma (1 ml)	b HPLC	RP(PI)	Native	410, 416
	Urine (1 ml)	b HPLC	RP(PI)	Native	416
	Plasma (1 ml)	b TLC	NP	Pre-EDTN	417
Oxprenolol	Plasma (1 ml)	b TLC	NP	Pre-EDTN	417
Penbutolol	Plasma (1 ml)	b HPLC	RP-PI	Native	418
	Urine (1 ml)	b HPLC	RP-PI	Native	418
Pindolol	Plasma (2 ml)	d HPLC	RP	Native	419
Prenalterol	Plasma (2 ml)	d HPLC	RP	Native	419
Pronethalol	Plasma (1–2 ml)	d HPLC	RP	Native	420, 421
	Urine	HPLC	RP	Native	422
Propranolol	Plasma (1–2 ml)	a—d HPLC	RP	Native	423, 424
	Urine (1 ml)	b HPLC	RP	Native	413, 424
	Plasma (1 ml)	d HPLC	RP-PI	Native	425, 426
	Plasma (1 ml)	b TLC	NP	Post-Citric acid + ethylenegly-col	427
	Urine (0.01 ml)	b TLC	NP	Post-Citric acid + ethylenegly-col	427
Sotalol	Plasma (2 ml)	d HPLC	RP	Native	428
	Urine (2 ml)	d HPLC	RP	Native	428
Tolamolol	Plasma (1 ml)	b TLC	NP	Pre-EDTN	429

687

Table 7.2. (continued)

Compound(s)	Sample (Volume Required)	Clean-up Chromatography Procedure[a]		Fluorescence Method	Ref.
C. Phenylethylamine derivatives					
Amphetamine	Plasma, urine	TLC	NP	Pre-Flur	430, 431
	Urine (0.5 ml)	b HPLC	NP/RP	Various	432
	Urine	HPTLC	RP	Pre-OPA	433
	Urine (10 ml)	b/c TLC	NP	Post-Flur spray	434
	Blood (5 ml)	c TLC	NP	Post-Flur spray	435
	Urine (1 ml)	c TLC	NP	Pre-NBD-Cl	436
		c TLC	NP	Pre-NBD-Cl	436
Catecholamines		Various		Various	437, 438
		HPLC	IE	Native (buffering eluents)	439
		HPLC	IE	Post-OPA	440, 441
		HPLC	IE/RP	Post-Aethylenediamine	442
	Serum (0.01 ml)	a HPLC	NP	Native	443
		HPLC	NP	Pre-Flur	444
		HPLC	NP/RP	Pre-Dns-Cl	445
	Brain tissue	HPLC	RP	Native	446, 447
	Urine (0.1 ml)	c HPLC	RP	Native	448
		HPLC	RP	Post-OPA	440
	Plasma (2 ml)	a HPLC	RP	Pre-OPA	449
	Urine	b HPLC	RP	Pre-OPA	450
		HPLC	RP-PI	Native	451, 452
	Urine	c HPLC	RP-PI	Native	453
	Plasma (2 ml)	b HPLC	RP-PI	Post-Oxidation	454
	Urine (0.5 ml)	d HPLC	RP-PI	Post-Oxidation	454
		TLC	NP	Post-Flur dip	455
		TLC	NP	Pre-Dns-Cl	295
		TLC	N?	Pre-Flur	455

Table 7.2. (*continued*)

Compound(s)	Sample (Volume Required)	Clean-up Chromatography Procedure[a]	Fluorescence Method	Ref.
IV. Neutral Compounds				
Aflatoxins		HPLC	NP	Native	178
		HPLC	NP	Native	79
		HPLC	RP	Laser induced	73, 86
		HPLC	RP	Native	474
		HPLC	RP	Pre-Trichloracetic acid	475
		TLC	NP	Laser induced	476
Alcoholic hydroxyl groups		HPLC	NP	Pre-DMA-NN	477
Aldehydes		TLC	NP	Pre-AIC	478
		TLC	NP	Pre-NIC	479
		HPLC	RP	Post-Irradiation (anthraquinone in mobile phase)	131
		HPLC	RP	Pre-Dimethylcyclohexane-dione	480
		TLC	NP	Post-Flur spray	53
		TLC	NP	Pre-Aniline	53
Aliphatic alcohols		HPLC	RP	Post-Irradiation (anthraquinone in mobile phase)	131
Avermectins	Plasma (1 ml)	d HPLC	RP	Pre-Aceticanhydride	481
Epoxy compounds		TLC	NP	Post-Nicotinamide + aceto-phenone dip	482
Ethers		HPLC	RP	Post-Irradiation (anthraquinone in mobile phase)	131
Furazolidone		TLC	NP	Post-Pyridine	483

690

Compound	Sample	Method	Mode	Derivatization	Ref.
Monosaccharides		HPLC	IE	Pre-Cyanoacetamide	484
Naphthylurethanes		HPLC	RP	Pre-Dns-hydrazine	485
		HPLC	NP/RP	Pre-NIC	486
Napropamide	Blood (0.25 ml)	b HPLC	RP	Native	203
Neutral sugars		HPLC	RP	Pre-Dns-Cl	487
Psoralene	plasma (1 ml)	b HPLC	RP	Native	488
Reducing sugars		HPLC	NP	Pre-Dns-hydrazine	489
		HPLC	RP	Pre-Dns-Cl	490
		TLC	NP	Post-Malonamide spray	491
Saccharides	Urine	HPLC	IE	Post-Ethylenediamine	492
		HPLC	RP	Post-Tetrazolium Blue	493
		HPTLC	NP	Pre-Dns-hydrazine	494
		TLC	NP/RP	Pre-Aminomethylcoumarin	495
Tripdiolide		TLC	NP	Post-Oxidation	143

V. Various Groups of Compounds

A. Cytostatics

Compound	Sample	Method	Mode	Derivatization	Ref.
ADCA	Biological fluids	HPLC	RP	Native	496
Anthracyclines					
Aclacinomycin	Plasma (1 ml)	b HPLC	RP	Native	497
Adriamycin	Plasma	a HPLC	NP	Native	498
Daunorubicin	Plasma (0.1 ml)	b HPLC	NP(PI)	Native	499
	Urine (0.02 ml)	— HPLC	RP	Laser induced	80, 85
		HPLC	RP	Native	500
	Plasma (0.1 ml)	a/d HPLC	RP	Native	501, 502
	Plasma (0.5 ml)	b TLC	NP	Native	503
Carminomycin	Tissue (1 g)	b TLC	NP	Native	504
Epidoxorubicin	Serum (2 ml)	b HPLC	RP	Native	505
	Plasma (2 ml)	d HPLC	RP	Native	506
Bisantrene	Plasma, urine	c HPLC	RP	Native	507

Table 7.2. *(continued)*

Compound(s)	Sample (Volume Required)	Clean-up Chromatography Procedure[a]			Fluorescence Method	Ref.
		b HPLC	NP		Native	
10-Chloro-5-(2-di-methyl-aminoethyl)-7H-in-dolo(2,3-c)quinolin-6(5H)-one	Plasma (0.1 ml)		NP		Native	508
Ellipticine	Blood (0.5 g)	b HPLC	RP		Native	509
Etoposide (VP-16)	Plasma (1 ml)	b HPLC	RP		Native	510
Harringtonine	Plasma (1 ml)	a HPLC	RP		Native	511
Melphalan	Plasma (1–3 ml)	d HPLC	RP		Native	512, 513
Methotrexate	Plasma (1 ml)	a HPLC	RP		Pre-Oxidation	514
Tamoxifen	Plasma	b HPLC	RP-PI		Pre-Irradiation	515
Teniposide (VM-26)	Plasma (1 ml)	b HPLC	RP		Native	510
B. Steroids						
Anabolic agents	Plasma, urine	b HPLC	RP		Pre-Dns-Cl	516
	Urine (0.05 ml)	d HPTLC	NP		Post-H$_2$SO$_4$ spray	517
Clomiphene	Plasma (3 ml)	d HPLC	NP		Post-Irradiation	518
Cortisol	Plasma (0.5 ml)	a HPLC	NP		Pre-Dns-hydrazine	519
	Urine (1 ml)	a HPLC	NP		Pre-Dns-hydrazine	519
	Serum (0.05 ml)	a HPLC	RP		Pre-H$_2$SO$_4$	111
	Urine (0.5 ml)	b HPLC	RP		Pre-H$_2$SO$_4$	520
	Plasma (2 ml)	a (HP)TLC	NP		Pre-Dns-hydrazine	521
Diethylstilbestrol	Urine (10 ml)	b HPLC	RP		Post-Irradiation	522
Digoxin (digitoxin, gi-toxin)		TLC	RP		Post-Chloramine-T	523
Estriol	Urine (2 ml)	b HPLC	RP		Native	524
		(HP)TLC	RP		Pre-Dns-hydrazine	521

Compound	Sample	Method	Mode	Detection	Page
Estrogen		HPLC	NP	Native	66
Ethinylestradiol		HPLC	RP	Pre-Dns-Cl	525
		HPLC	RP	Native	526
Hydrocortisone	Plasma (1–8 ml)	(HP)TLC	NP	Post-H_2SO_4 spray	527
		HPLC[b]	NP	Pre-Dns-hydrazine	528
17-Hydroxycorticosteroids	Urine (0.5 ml)	HPLC[d]	NP	Pre-Dns-hydrazine	529
Ketosteroids		HPLC	NP	Post-Isoniazide + $AlCl_3$	530
	Blood (0.5 ml)	HPLC[b]	NP	Pre-Dns-hydrazine	531
	Urine (1 ml)	HPLC[b]	NP	Pre-Dns-hydrazine	531
Levonorgestrel		(HP)TLC	NP	Post-H_2SO_4 spray	527
Mestranol		HPLC	NP/RP	Native	532
ORG-6001	Plasma (1 ml)	TLC[b]	NP	Post-Oxidation	533
17-Oxosteroids	Plasma (0.2 ml)	HPLC[c]	NP/RP	Pre-Dns-hydrazine	534
	Urine (0.5 ml)	HPLC[c]	NP/RP	Pre-Dns-hydrazine	534
Steroids		HPLC	RP	Various	535
Zearalenol/zearalenone		HPLC	NP	Native	536
	Plasma (2 ml)	HPLC[d]	RP	Native	537
		HPLC	RP	Laser induced	87
C. Vitamins					
Menaquinones	Serum (0.5 ml)	HPLC	NP/RP	Native	538
Menaquinones/phylloqui-nones		HPLC[d]	RP	Post-Irradiation/reduction	539
Riboflavin	Urine (0.01 ml)	HPLC	RP	Native	540, 541
	Blood (1 ml)	HPLC[—]	RP	Native	542
Thiamin	Plasma (0.05 ml)	HPLC[a]	RP	Post-Oxidation	543, 544
Tocopherols	Plasma (0.05 ml)	HPLC[a]	NP	Native	545
		HPLC[a/b]	RP	Native	546, 547

Table 7.2. *(continued)*

Compound(s)	Sample (Volume Required)	Clean-up Procedure[a]	Chromatography	Fluorescence Method	Ref.
Vitamin A	Serum (0.05 ml)	b HPLC	NP	Native	548
Vitamin B2, B3, B6		HPLC	IE	Native	92
Vitamin B2	Blood (1 ml)	a HPLC	RP	Native	549
Vitamin E		HPLC	NP	Native	550
Vitamin K	Serum	d HPLC	RP	Post-Reduction	551

[a] The meaning of the symbols is a follows: —, no clean-up step is used in the prechromatographic sample treatment; a, the only clean-up step is some form of protein removal (ultrafiltration or precipitation with ammonium acetate, acetonitrile, ethanol, etc.); b, a single extraction step; derivatization can be performed either directly in the separated organic phase following extraction of the sample or after evaporation of the organic solvent and reconstitution of the residue in a suitable solvent; c, a single clean-up step other than liquid–liquid extraction (e.g., by TLC or column LC); d, more than one clean-up step, as in liquid–liquid extraction of the analyte from the sample with an organic solvent and back-extraction into an aqueous phase; it can also mean a combination of, for instance, a and b, or b and c.

694

b   A single extraction step; derivatization can be performed either directly in the separated organic phase following extraction of the sample or after evaporation of the organic solvent and reconstitution of the residue in a suitable solvent

c   A single clean-up step other than liquid–liquid extraction (e.g., by TLC or column LC)

d   More than one clean-up step, as in liquid–liquid extraction of the analyte from the sample with an organic solvent and back-extraction into an aqueous phase; it can also mean a combination of, for instance, a and b, or b and c.

In the fourth column is mentioned whether thin layer or column LC was used (TLC, HPTLC, or HPLC) and the type of chromatographic system: adsorption chromatography with polar stationary phases (NP), chromatography with apolar stationary phases and polar mobile phases (RP), ion-exchange chromatography (IE), and paired-ion chromatography (PI).

The fifth column tells whether the native fluorescence is measured or the fluorescence after prechromatographic (Pre-) or postchromatographic (Post-) derivatization. The name or abbreviation (see Table 7.1) of the derivatization reagent is also given in this column.

The last column gives the references to the literature.

BIBLIOGRAPHY

K. Blau and G. S. King, *Handbook of Derivatives for Chromatography*, Heyden, London, 1977.

R. J. Hamilton and P. A. Sewell, *Introduction to High Performance Liquid Chromatography*, Chapman and Hall, London, 1977.

P. M. Kabra and L. J. Marton, *Liquid Chromatography in Clinical Analysis*, Humana Press, Clifton, N.J., 1981.

J. G. Kirchner, *Thin-Layer Chromatography*, in *Techniques of Chemistry*, Vol. 14, 2nd ed., A. Weissberger (ed.), Wiley, New York, 1978.

P. T. Kissinger, L. J. Felix, D. J. Miner, C. R. Preddy, and R. E. Shoup, *Detectors for Trace Organic Analysis by Liquid Chromatography: Principles and Applications*, in *Contempory Topics in Analytical and Clinical Chemistry*, D. M. Hercules, G. M. Hieftje, L. R. Snyder and M. A. Evenson (eds.), Vol. 2, Plenum, New York, 1978.

D. R. Knapp, *Handbook of Analytical Derivatization Reactions*, Wiley, New York, 1979.

J. F. Lawrence and R. W. Frei, *Chemical Derivatization in Liquid Chromatography*, J. Chromatogr. Library, Vol. 7, Elsevier, Amsterdam, 1976.

A. Pryde and M. T. Gilbert, *Applications of High Performance Liquid Chromatography*, Chapman and Hall, London, 1979.

G. Schwedt, *Chemische Reaktionsdetektoren für die schnelle Flüssigkeitschromatographie*, Hüthig, Heidelberg, 1980.

G. Schwedt, *Fluorimetrische Analyse*, Verlag Chemie, Weinheim–Deerfield Beach (Florida), 1981.

C. F. Simpson (ed.), *Techniques in Liquid Chromatography*, Wiley, New York, 1982.

L. R. Snyder and J. J. Kirkland, *Introduction to Modern Liquid Chromatography*, 2nd ed., Wiley, New York, 1980.

J. C. Touchstone and D. Rogers, *Thin-Layer Chromatography: Quantitative Environmental and Clinical Applications*, Wiley, New York, 1980.

P. Froehlich and E. L. Wehry, *Fluorescence Detection in Liquid and Gas Chromatography*, in E. L. Wehry (ed.), *Modern Fluorescence Spectroscopy*, Vol. 3, Plenum, New York, 1981.

A. Zlatkis and R. E. Kaiser (ed.) *HPTLC High Performance Thin-Layer Chromatography*, J. Chromatogr. Library, Vol. 9, Elsevier, Amsterdam, 1977.

REFERENCES

1. R. P. Cooney and J. D. Winefordner, *Anal. Chem. 49*, 1057 (1977).

2. R. P. Cooney, T. Vo Dinh, and J. D. Winefordner, *Anal. Chim. Acta 89*, 9 (1977).

3. F. Chow and A. Karmen, *Clin. Chem. 26*, 1480 (1980).

4. J. M. Hayes and G. J. Small, *Anal. Chem. 54*, 1202 (1982).

5. G. Guiochon, A. Siouffi, H. Engelhardt, and I. Halász, *J. Chromatogr. Sci. 16*, 152 (1978).

6. G. Guiochon and A. Siouffi, *J. Chromatogr. Sci. 16*, 470 (1978).

7. P. J. Schoenmakers, H. A. H. Billiet, and L. de Galan, *Chromatographia 15*, 205 (1982).

8. J. L. Robinson, W. J. Robinson, M. A. Marshall, A. D. Barnes, K. J. Johnson, and D. S. Salas, *J. Chromatogr. 189*, 145 (1980).

9. K. Aramaki, T. Hanai, and H. F. Walton, *Anal. Chem. 52*, 1963 (1980).

10. Z. Iskandarani and D. J. Pietrzyk, *Anal. Chem. 53*, 489 (1981).

11. D. P. Lee, *J. Chromatogr. Sci. 20*, 203 (1982).

12. N. H. C. Cooke and K. Olson, *J. Chromatogr. Sci. 18*, 512 (1980).

13. O. A. G. J. van der Houwen, R. H. A. Sorel, A. Hulshoff, J. Teeuwsen, and A. W. M. Indemans, *J. Chromatogr. 209*, 393 (1981).

14. R. H. A. Sorel, A. Hulshoff, and S. Wiersema, *J. Liq. Chromatogr. 4*, 1961 (1981).

15. M. T. W. Hearn, "Ion-Pair Chromatography on Normal- and Reversed-Phase Systems," in *Advances in Chromatography*, Vol. 18, J. C. Giddings, E. Grushka, J. Cazes, and P. R. Brown (eds.), Dekker, New York, 1980.

16. R. H. A. Sorel and A. Hulshoff, "Dynamic-Anion Exchange Chromatography," in *Advances in Chromatography*, Vol. 21, J. C. Giddings, E. Grushka, J. Cazes, and P. R. Brown (eds.), Dekker, New York, 1983.

17. F. F. Cantwell and S. Puon, *Anal. Chem. 51*, 623 (1979).

18. T. D. Rotsch and D. J. Pietrzyk, *J. Chromatogr. Sci. 19*, 88 (1981).

19. D. C. Fenimore and C. M. Davis, *Anal. Chem. 53*, 252A (1981).

20. J. C. Touchstone and S. S. Levin, *J. Liq. Chromatogr. 3*, 1853 (1980).

21. H. Moosmann, *J. Chromatogr. 209*, 84 (1981).

22. D. C. Fenimore and C. J. Meyer, *J. Chromatogr. 186*, 555 (1979).

23. H. J. Issaq, *J. Liq. Chromatogr. 4*, 955 (1981).

24. H. J. Issaq, *J. Liq. Chromatogr. 3*, 789 (1980).

25. K. Karikó and J. Tomasz, *J. High Resolut. Chromatogr. Chromatogr. Commun. 2*, 247 (1979).

26. E. Tyihák, E. Mincsovics, and H. Kalász, *J. Chromatogr. 174*, 75 (1979).

27. E. Mincsovics, E. Tyihák and H. Kalász, *J. Chromatogr. 191*, 293 (1980).

28. J. I. Thornton and P. J. Casale, Forensic Sci. Int., *14*, 215 (1979).

29. L. Tóth, *J. Chromatogr. 50*, 72 (1970).

30. M. L. Gianelli, J. B. Callis, N. H. Andersen, and G. D. Christian, *Anal. Chem. 53*, 1357 (1981).

31. U. Hezel, *Angew. Chem. 85*, 334 (1973).

32. M. Prošek, A. Medja, E. Kučan, M. Katič, and M. Bano, *J. High Resolut. Chromatogr. Chromatogr. Commun. 3*, 183 (1980).

33. V. Pollak, *J. Liq. Chromatogr. 3*, 1881 (1980).

34. S. Ebel and G. Herold, Fres. *Z. Anal. Chem. 266*, 281 (1973).

35. S. Ebel, E. Geitz and J. Hocke, Fres. *Z. Anal. Chem. 301*, 138 (1980).

36. J. Ripphahn and H. Halpaap, *J. Chromatogr. 112*, 81 (1975).

37. D. Jaenchen and G. Pataki, *J. Chromatogr. 33*, 391 (1968).

38. G. E. Caissie and V. N. Mallet, *J. Chromatogr. 117*, 129 (1976).

39. F. de Croo, G. A. Bens, and P. de Moerloose, *J. High Resolut. Chromatogr. Chromatogr. Commun. 3*, 423 (1980).

40. S. Uchiyama and M. Uchiyama, *J. Chromatogr. 153*, 135 (1978).

41. S. Uchiyama and M. Uchiyama, *J. Liq. Chromatogr. 3*, 681 (1980).

42. V. Mallet and R. W. Frei, *J. Chromatogr. 56*, 69 (1971).

43. R. W. Frei and P. E. Belliveau, *Chromatographia 5*, 392 (1972).

44. R. Segura and A. M. Gotto, *J. Chromatogr. 99*, 643 (1974).

45. R. Segura and X. Navarro, *J. Chromatogr. 217*, 329 (1981).

46. L. Zhou, H. Shanfield, F.-S. Wang, and A. Zlatkis, *J. Chromatogr. 217*, 341 (1981).

47. H. Shanfield, F. Hsu, A. J. P. Martin, *J. Chromatogr. 126*, 457 (1976).

48. R. D. Davies and V. Pretorius, *J. Chromatogr. 155*, 229 (1978).

49. E. M. Schulman and R. T. Parker, *J. Phys. Chem. 81*, 1932 (1977).

50. F. Kreuzig, *Chromatographia 13*, 238 (1980).

51. J. C. Touchstone, J. Sherma, M. F. Dobbins, and G. R. Hansen, *J. Chromatogr. 124*, 111 (1976).

52. R. Wintersteiger, G. Guebitz, and A. Hartinger, *Chromatographia 13*, 291 (1980).

53. J. C. Young, *J. Chromatogr. 130*, 392 (1977).

54. H. Nakamura, *J. Chromatogr. 131*, 215 (1977).

55. R. Chayen, R. Dvir, S. Gould, and A. Haroll, *Anal. Biochem. 42*, 283 (1971).

56. W. Duenges, *Anal. Chem. 49*, 442 (1977).

57. R. P. W. Scott and P. Kucera, *J. Chromatogr. 169*, 51 (1979).

58. M. Novotny, *Anal. Chem. 53*, 1294A (1981).

59. J. H. Knox and M. T. Gilbert, *J. Chromatogr. 186*, 405 (1979).

60. W. Roth, K. Beschke, R. Jauch, A. Zimmer and F. W. Koss, *J. Chromatogr. 222*, 13 (1981).

61. R. A. Hux, H. Y. Mohammed, and F. F. Cantwell, *Anal. Chem. 54*, 113 (1982).

62. L. H. Thacker, *J. Chromatogr. 136*, 213 (1977).

63. E. Johnson, A. Abu-Shumays, and S. R. Abbott, *J. Chromatogr. 134*, 107 (1977).

64. W. Slavin, A. T. Rhys Williams, and R. F. Adams, *J. Chromatogr. 134*, 121 (1977).

65. K. Ogan, E. Katz, and T. J. Porro, *J. Chromatogr. Sci. 17*, 597 (1979).

66. G. J. Krol, C. A. Mannan, R. E. Pickering, D. V. Amato, B. T. Kho, and A. Sonnenschein, *Anal. Chem. 49*, 1836 (1977).

67. D. T. Zelt, J. A. Owen, and G. S. Marks, *J. Chromatogr. 189*, 209 (1980).

68. J. H. Richardson, K. M. Larson, G. R. Haugen, D. C. Johnson, and J. E. Clarkson, *Anal. Chim. Acta 116*, 407 (1980).

69. T. Imasaka, K. Ishibashi, and N. Ishibashi, *Anal. Chim. Acta 142*, 1 (1982).

70. J. W. Lyons and L. R. Faulkner, *Anal. Chem. 54*, 1960 (1982).

71. G. G. Vurek, *Anal. Chem. 54*, 840 (1982).

72. F. Martin, J. Maine, C. C. Sweeley, and J. F. Holland, *Clin. Chem. 22*, 1434 (1976).

73. G. J. Diebold and R. N. Zare, *Science 196*, 1439 (1977).

74. E. Voigtman, A. Jurgensen, and J. D. Winefordner, *Anal. Chem. 53*, 1921 (1981).

75. S. Folestad, L. Johnson, B. Josefsson, and B. Galle, *Anal. Chem. 54*, 925 (1982).

76. L. W. Hershberger, J. B. Callis and G. D. Christian, *Anal. Chem. 51,* 1444 (1979).

77. E. Sawicki, *Talanta 16,* 1231 (1969).

78. J. B. F. Lloyd, *Analyst 100,* 529 (1975).

79. T. Panalaks and P. M. Scott, *J. Assoc. Off. Anal. Chem. 60,* 583 (1977).

80. E. S. Yeung and M. J. Sepaniak, *Anal. Chem. 52,* 1465A (1980).

81. R. B. Green, *Anal. Chem. 55,* 20A (1983).

82. H. Todoriki and A. Y. Hirakawa, *Chem. Pharm. Bull. Jap. 28,* 1337 (1980).

83. M. J. Sepaniak and E. S. Yeung, *J. Chromatogr. 190,* 377 (1980).

84. M. J. Sepaniak and E. S. Yeung, *Anal. Chem. 49,* 1554 (1977).

85. P. B. Huff, B. J. Tromberg, and M. J. Sepaniak, *Anal. Chem. 54,* 946 (1982).

86. G. J. Diebold, N. Karny, R. N. Zare, and L. M. Seitz, *J. Assoc. Off. Anal. Chem. 62,* 564 (1979).

87. G. J. Diebold, N. Karny, and R. N. Zare, *Anal. Chem. 51,* 67 (1979).

88. D. J. Malcolme-Lawes, P. Warwick, and L. A. Gifford, *J. Chromatogr. 176,* 157 (1979).

89. D. J. Malcolme-Lawes and P. Warwick, *J. Chromatogr. 200,* 47 (1980).

90. D. J. Malcolme-Lawes, S. Massey, and P. Warwick, *J. Chem. Soc. Faraday Trans. 2, 77,* 1795 (1981).

91. D. J. Malcolme-Lawes, S. Massey, and P. Warwick, *J. Chem. Soc. Faraday Trans. 2, 77,* 1807 (1981).

92. H. Hatano, Y. Yamamoto, M. Saito, E. Mochida, and S. Watanabe, *J. Chromatogr. 83,* 373 (1973).

93. E. D. Pellizzari and C. M. Sparacino, *Anal. Chem. 45,* 378 (1973).

94. G. D. Christian, J. B. Callis, and E. R. Davidson, "Array Detectors and Excitation-Emission Matrices in Multicomponent Analysis," in *Modern Fluorescence Spectroscopy*, Vol. 3, E. L. Wehry (ed.), Plenum, New York, 1981.

95. J. R. Jadamec, W. A. Saner, and Y. Talmi, *Anal. Chem. 49,* 1316 (1977).

96. M. P. Fogarty, D. C. Shelly, and I. M. Warner, *J. High Resolut. Chromatogr. Chromatogr. Commun. 4,* 561 (1981).

97. D. C. Shelly, M. P. Fogarty, and I. M. Warner, *J. High Resolut. Chromatogr. Chromatogr. Commun. 4,* 616 (1981).

98. L. W. Hershberger, J. B. Callis, and G. D. Christian, *Anal. Chem. 53,* 971 (1981).

99. C. J. Appellof and E. R. Davidson, *Anal. Chem. 53,* 2053 (1981).

100. P. A. Asmus, J. W. Jorgenson, and M. Novotny, *J. Chromatogr. 126,* 317 (1976).

101. S. Y. Su, A. Jurgensen, D. Bolton, and J. D. Winefordner, *Anal. Lett. 14,* 1 (1981).

102. J. F. Lawrence and R. W. Frei, *Chemical Derivatization in Liquid Chromatography*, Elsevier, Amsterdam, 1976, p. 24.

103. J. F. Lawrence and R. W. Frei, *J. Chromatogr. 66,* 93 (1972).

104. P. M. Froehlich and M. Yeats, *Anal. Chim. Acta 87,* 185 (1976).

105. H. Nakamura and Z. Tamura, *Anal. Chem. 53,* 2190 (1981).

106. D. W. Armstrong, W. L. Hinze, K. H. Bui, and H. N. Singh, *Anal. Lett. 14,* 1659 (1981).

107. S. H. Lee, L. R. Field, W. H. Howald, and W. F. Trager, *Anal. Chem. 53,* 467 (1981).

108. R. G. Muusze and J. F. K. Huber, *J. Chromatogr. Sci. 12,* 779 (1974).

109. M. S. Bernstein and M. A. Evans, *J. Chromatogr. 229,* 179 (1982).

110. K. Miura, H. Nakamura, H. Tanaka, and Z. Tamura, *J. Chromatogr. 210,* 536 (1981).

111. G. R. Gotelli, J. H. Wall, P. M. Kabra, and L. J. Marton, *Clin Chem. 27,* 441 (1981).

112. J. C. Gfeller, G. Frei, and R. W. Frei, *J. Chromatogr. 142,* 271 (1977).

113. C. van Buuren, J. F. Lawrence, U. A. Th. Brinkman, I. L. Honigberg, and R. W. Frei, *Anal. Chem. 52,* 700 (1980).

114. J. W. Birks and R. W. Frei, *Trends Anal. Chem. 1,* 361 (1982).

115. S. Katz, W. W. Pitt Jr., J. E. Mrochek, and S. Dinsmore, *J. Chromatogr. 101,* 193 (1974).

116. C. E. Werkhoven-Goewie, W. M. A. Niessen, U. A. Th. Brinkman, and R. W. Frei, *J. Chromatogr. 203,* 165 (1981).

117. D. Nicholson, *Analyst 103,* 1 (1978).

118. A. Hulshoff and A. D. Foerch, *J. Chromatogr. (Chromatogr. Rev.) 220,* 275 (1981).

119. R. W. Frei and A. H. M. T. Scholten, *J. Chromatogr. Sci. 17,* 152 (1979).

120. R. W. Frei, *J. Chromatogr. (Chromatogr. Rev.), 165,* 75 (1979).

121. R. W. Frei, *Chromatographia 15,* 161 (1982).

122. J. T. Stewart, *Trends Anal. Chem. 1,* 170 (1982).

123. G. Schwedt and E. Reh, *Chromatographia 14,* 249 (1981).

124. G. Schwedt and E. Reh, *Chromatographia 14,* 317 (1981).

125. A. H. M. T. Scholten, U. A. Th. Brinkman, and R. W. Frei, *Anal. Chem. 54,* 1932 (1982).

126. K. Arisue, Z. Ogawa, K. Kohda, C. Hayashi, and Y. Ishida, *Jap. J. Clin. Chem. 9,* 104 (1980).

127. J. C. Gfeller, G. Frei, J. M. Huen, and J. P. Thevenin, *J. Chromatogr. 172,* 141 (1979).

128. J. F. Lawrence, U. A. Th. Brinkman, and R. W. Frei, *J. Chromatogr. 185,* 473 (1979).

129. A. H. M. T. Scholten, U. A. Th. Brinkman, and R. W. Frei, *Anal. Chim. Acta 114,* 137 (1980).

130. M. Uihlein and E. Schwab, *Chromatographia 15,* 140 (1982).

131.　M. S. Gandelman and J. W. Birks, *Anal. Chem. 54,* 2131 (1982).

132.　S. Kamada, M. Maeda, and A. Tsuji, *J. Chromatogr. 272,* 29 (1983).

133.　T. Kawasaki, M. Maeda, and A. Tsuji, *J. Chromatogr. 272,* 261 (1983).

134.　M. A. Brooks, N. Strojny, M. R. Hackman, and J. A. F. de Silva, *J. Chromatogr. 229,* 167 (1982).

135.　H. Cisse, R. Farinotti, S. Kirkiacharian, and A. Dauphin, *J. Chromatogr. 225,* 509 (1981).

136.　J. Goto, N. Goto, A. Hikichi, T. Nishimaki, and T. Nambara, *Anal. Chim. Acta 120,* 187 (1980).

137.　H. Tsuchiya, T. Hayashi, H. Naruse, and N. Takagi, *J. Chromatogr. 234,* 121 (1982).

138.　C. Gonnet, M. Marichy, and N. Philippe, *Analusis 7,* 370 (1979).

139.　A. Takadate, T. Tahara, H. Fujino, and S. Goya, *Chem. Pharm. Bull. Jap. 30,* 4120 (1982).

140.　W. Duenges, A. Meyer, K.-E. Mueller, M. Mueller, R. Pietschmann, C. Plachetta, R. Sehr, and H. Tuss, *Fresenius' Z. Anal. Chem. 288,* 361 (1977).

141.　S. Goya, A. Takadate, H. Fujino, and T. Tanaka, *Yakugaku Zasshi 100,* 744 (1980).

142.　S. Goya, A. Takadate, H. Fujino, and M. Irikura, *Yakugaku Zasshi 101,* 1064 (1981).

143.　J. P. Kutney, R. D. Sindelar, and K. L. Stuart, *J. Chromatogr. 214,* 152 (1981).

144.　E. Grushka, S. Lam, and J. Chassin, *Anal. Chem. 50,* 1398 (1978).

145.　S.-C. Chen, *J. Chromatogr. 238,* 480 (1982).

146.　N. Nimura and T. Kinoshita, *Anal. Lett. 13,* 191 (1980).

147.　S. A. Barker, J. A. Monti, S. T. Christian, F. Bennington, and R. D. Morin, *Anal. Biochem. 107,* 116 (1980).

148.　W. Voelter, R. Huber, and K. Zech, *J. Chromatogr. 217,* 491 (1981).

149.　W. Distler, *J. Chromatogr. 192,* 240 (1980).

150.　W. D. Korte, *J. Chromatogr. 243,* 153 (1982).

151.　J. B. F. Lloyd, *J. Chromatogr. 189,* 359 (1980).

152.　P.-O. Lagerstroem, *J. Chromatogr. 225,* 476 (1981).

153.　A. Crozier, J. B. Zaerr, and R. O. Morris, *J. Chromatogr. 238,* 157 (1982).

154.　M. Koshioka, J. Harada, K. Takeno, M. Noma, T. Sassa, K. Ogiyama, J. S. Taylor, S. B. Rood, R. L. Legge, and R. P. Pharis, *J. Chromatogr. 256,* 101 (1983).

155.　S. H. Hansen and M. Døssing, *J. Chromatogr. 229,* 141 (1982).

156.　K. Yamada, E. Kayama, Y. Aizawa, K. Oka and S. Hara, *J. Chromatogr. 223,* 176 (1981).

157.　S. Katz, W. W. Pitt, Jr., and G. Jones, Jr., *Clin. Chem. 19,* 817 (1973).

158.　J. L. Shimek, N. G. S. Rao, and S. K. Wahba Khalil, *J. Pharm. Sci. 70,* 514 (1981).

159. J. P. Garnier, B. Bousquet and C. Dreux, *J. Autom. Chem. 4*, 65 (1982).

160. T. G. Rosano, J. M. Meola, and T. A. Swift, *Clin. Chem. 28*, 207 (1982).

161. K.-G. Wahlund and B. Edlén, *Clin. Chim. Acta 110*, 71 (1981).

162. W. F. Bayne, T. East, and D. Dye, *J. Pharm. Sci. 70*, 458 (1981).

163. H. K. L. Hundt and L. W. Brown, *J. Chromatogr. 225*, 482 (1981).

164. S. M. Steinberg and J. L. Bada, *Mar. Chem. 11*, 299 (1982).

165. S. Goya, A. Takadate, and H. Fujino, *Yakugaku Zasshi 102*, 63 (1982).

166. R. Huber, K. Zech, M. Woerz, Th. Kronbach, and W. Voelter, *Chromatographia 16*, 233 (1982).

167. J. Goto, N. Goto, and T. Nambara, *J. Chromatogr. 239*, 559 (1982).

168. D. Westerlund, A. Theodorson, and Y. Jaksch, *J. Liq. Chromatogr. 2*, 969 (1979).

169. K.-G. Wahlund, *J. Chromatogr. 218*, 671 (1981).

170. T. Hirata, M. Kai, K. Kohashi, and Y.Ohkura, *J. Chromatogr. 226*, 25 (1981).

171. J. Turk, S. J. Weiss, J. E. Davis, and P. Needleman, *Prostaglandins 16*, 291 (1978).

172. M. Hatsumi, S.-I. Kimata, and K. Hirosawa, *J. Chromatogr. 253*, 271 (1982).

173. D. W. Watkins and M. B. Peterson, *Anal. Biochem. 125*, 30 (1982).

174. H. Tsuchiya, T. Hayashi, H. Naruse, and N. Takagi, *J. Chromatogr. 231*, 247 (1982).

175. G. Drehsen and P. Rohdewald, *J. Chromatogr. 223*, 479 (1981).

176. R. Alric, M. Cociglio, J. P. Blayac, and R. Puech, *J. Chromatogr. 224*, 289 (1981).

177. J. T. Taylor and S. Freeman, *Chromatogr. Newsl. 9*, 1 (1981).

178. L. A. King, *J. Chromatogr. 208*, 113 (1981).

179. W. Duenges and N. Seiler, *J. Chromatogr. 145*, 483 (1978).

180. W. Duenges, G. Naundorf, and N. Seiler, *J. Chromatogr. Sci. 12*, 665 (1974).

181. B. Jarrott, A. Anderson, R. Hooper, and W. J. Louis, *J. Pharm. Sci. 70*, 665 (1981).

182. B. Hoffmann, U. Tannert and W. Groebel, *Fresenius' Z. Anal. Chem. 307*, 389 (1981).

183. E. Gaetani, C. F. Laureri, M. Vitto, and F. Bordi, *Il Farmaco Ed. Pr. 37*, 235 (1982).

184. B. Kågedal and M. Kaellberg, *J. Chromatogr. 229*, 409 (1982).

185. K. Murayama and T. Kinoshita, *Anal. Lett. 14*, 1221 (1981).

186. F. Veronese, R. Lazzarini, O. Schiavon, and A. Bettero, *Il Farmaco Ed. Pr. 37*, 390 (1982).

187. M. Chiba and S. Sassa, *Anal. Biochem. 124*, 279 (1982).

188. R. H. Hill Jr., S. L. Bailey, and L. L. Needham, *J. Chromatogr. 232*, 251 (1982).

189. T. Sakai, Y. Niinuma, S. Yanagihara, and K. Ushio, *Clin. Chem. 29*, 350 (1983).

190. H. D. Meyer, K. Jacob, and W. Vogt, *Chromatographia 16*, 190 (1982).

191. H. Nakamura and Z. Tamura, *Anal. Chem. 54*, 1951 (1982).

192. R. C. Fahey, G. L. Newton, R. Dorian, and E. M. Kosower, *Anal. Biochem. 111*, 357 (1981).

193. G. L. Newton, D. Randel, and R. C. Fahey, *Anal. Biochem. 114*, 383 (1981).

194. E. P. Lankmayr, K. W. Budna, K. Mueller, and F. Nachtmann, *Fresenius' Z. Anal. Chem. 295*, 371 (1979).

195. S. S. Simons and D. F. Johnson, *J. Org. Chem. 43*, 2886 (1978).

196. A. Oeztunç, *Analyst 107*, 585 (1982).

197. K. Ogan and E. Katz, *Anal. Chem. 53*, 160 (1981).

198. P. J. Twitchett, P. L. Williams, and A. C. Moffat, *J. Chromatogr. 149*, 683 (1978).

199. S. R. Abott, A. Abu-Shumays, K. O. Loeffler, and I. S. Forrest, *Res. Commun. Chem. Pathol. Pharmacol. 10*, 9 (1975).

200. J. A. Vinson, D. D. Patel, and A. H. Patel, *Anal. Chem. 49*, 163 (1977).

201. A. Saria, F. Lembeck, and G. Skofitsch, *J. Chromatogr. 208*, 41 (1981).

202. T. Frost, *Analyst 106*, 999 (1981).

203. M. de Berardinis, Jr., and W. A. Wargin, *J. Chromatogr. 246*, 89 (1982).

204. H. S. Veale and G. W. Harrington, *J. Chromatogr. 208*, 161 (1981).

205. K. Ogan and E. Katz, *Chromatogr. Newsl. 7*, 15 (1979).

206. A. N. Masoud and Y. N. Cha, *J. High Resolut. Chromatogr. Chromatogr. Commun. 5*, 299 (1982).

207. A. W. Wolkoff and R. H. Larose, *J. Chromatogr. 99*, 731 (1974).

208. R. M. Cassidy, D. S. LeGay, and R. W. Frei, *J. Chromatogr. Sci. 12*, 85 (1974).

209. P. Haefelfinger, *J. Chromatogr. 162*, 215 (1979).

210. S. R. Gautam, V. Chungi, A. Hussain, S. Babhair, and D. Papadimitrou, *Anal. Lett. 14*, 577 (1981).

211. V. Mallet and D. P. Surette, *J. Chromatogr. 95*, 243 (1974).

212. K. Zech and W. Voelter, *Chromatographia 8*, 350 (1975).

213. H. Yoshida, T. Sumida, T. Masujima, and H. Imai, *J. High Resolut. Chromatogr. Chromatogr. Commun. 5*, 509 (1982).

214. K. M. Jonker, H. Poppe, and J. F. K. Huber, *Chromatographia 11*, 123 (1978).

215. P. Boehlen and R. Schroeder, *Anal. Biochem. 126*, 144 (1982).

216. G. J. Hughes, K. H. Winterhalter, E. Boller, and K. J. Wilson, *J. Chromatogr. 235*, 417 (1982).

217. J. R. Cronin, S. Pizzarello, and W. E. Gandy, *Anal. Biochem. 93*, 174 (1979).

218. M. Sato and K. Yagi, *J. Chromatogr. 242*, 185 (1982).

219. N. Lustenberger, H. W. Lange, and K. Hempel, *Angew. Chem. Int. Ed. Engl. 11*, 227 (1972).

220. R. W. Frei, L. Michel, and W. Santi, *J. Chromatogr. 126*, 665 (1976).

221. E. Bayer, E. Grom, B. Kaltenegger, and R. Uhmann, *Anal. Chem. 48*, 1106 (1976).

222. P.-M. Yuan, H. Pande, B. R. Clark, and J. E. Shively, *Anal. Biochem. 120*, 289 (1982).

223. R. V. Lewis and D. DeWald, *J. Liq. Chromatogr. 5*, 1367 (1982).

224. T. D. Schlabach and T. C. Wehr, *Anal. Biochem. 127*, 222 (1982).

225. E. Gil-Av, A. Tishbee, and P. E. Hare, *J. Am. Chem. Soc. 102*, 5115 (1980).

226. J. Lammens and M. Verzele, *Chromatographia 11*, 376 (1978).

227. N. Kaneda, M. Sato, and K. Yagi, *Anal. Biochem. 127*, 49 (1982).

228. E. M. Koroleva, V. G. Maltsev, B. G. Belenkii, and M. Viska, *J. Chromatogr. 242*, 145 (1982).

229. V. T. Wiedmeier, S. P. Porterfield, and C. E. Hendrich, *J. Chromatogr. 231*, 410 (1982).

230. G. Szókán, *J. Liq. Chromatogr. 5*, 1493 (1982).

231. Y. Watanabe and K. Imai, *J. Chromatogr. 239*, 723 (1982).

232. K. Murayama and T. Kinoshita, *Anal. Lett. 15*, 123 (1982).

233. J. P. H. Burbach, A. Prins, J. L. M. Lebouille, J. Verhoef, and A. Witter, *J. Chromatogr. 237*, 339 (1982).

234. H. Umagat, P. Kucera, and L. F. Wen, *J. Chromatogr. 239*, 463 (1982).

235. D. L. Hogan, K. L. Kraemer, and J. I. Isenberg, *Anal. Biochem. 127*, 17 (1982).

236. M. Griffin, S. J. Price, and T. Palmer, *Clin. Chim. Acta 125*, 89 (1982).

237. D. C. Turnell and J. D. H. Cooper, *Clin. Chem. 28*, 527 (1982).

238. S. J. Wassner and J. B. Li, *J. Chromatogr. 227*, 497 (1982).

239. H. Mabuchi and H. Nakahashi, *J. Chromatogr. 233*, 107 (1982).

240. M. K. Radjai and R. T. Hatch, *J. Chromatogr. 196*, 319 (1980).

241. W. McHugh, R. A. Sandmann, W. G. Haney, S. P. Sood, and D. P. Wittmer, *J. Chromatogr. 124*, 376 (1976).

242. E. Gaetani, C. F. Laureri, M. Vitto, and F. Bordi, *Il Farmaco Ed. Pr. 37*, 253 (1982).

243. W. Distler, *Fresenius' Z. Anal. Chem. 309*, 127 (1981).

244. E. Schiltz, D. Schnackerz, and R. W. Gracy, *Anal. Biochem. 79*, 33 (1977).

245. A. M. Felix and M. H. Jimenez, *J. Chromatogr. 89*, 361 (1974).

246. K. Murayama and T. Kinoshita, *J. Chromatogr. 205*, 349 (1981).

247. T. Kinoshita, K. Murayama, and A. Tsuji, *Chem. Pharm. Bull. Jap. 28*, 1925 (1980).

248. M. De Los Angeles Barceloen, *J. Chromatogr. 238*, 175 (1982).

249. D. Biou, N. Queyrel, M. N. Visseaux, I. Collignon, and M. Pays, *J. Chromatogr. 226*, 477 (1981).

250. J. C. Monboisse, P. Pierrelee, A. Bisker, V. Pailler, A. Randoux, and J. P. Borel, *J. Chromatogr. 233*, 355 (1982).

251. T. Ishimitsu and S. Hirose, *Chem. Pharm. Bull. Jap. 29*, 3400 (1981).

252. K. Nakazawa, H. Tanaka, and M. Arima, *J. Chromatogr. 233*, 313 (1982).

253. M. Ahnhoff, J. Grundevik, A. Arfwidson, J. Fonselius, and B.-A. Persson, *Anal. Chem. 53*, 485 (1981).

254. G. Bellon, A. Bisker, F. X. Maquart, H. Thoanes, and J. P. Borel, *J. Chromatogr. 230*, 420 (1982).

255. F. Engbaek and J. Magnussen, *Clin. Chem. 24*, 376 (1978).

256. R. F. Adams, G. J. Schmidt, and F. L. Vandemark, *Clin. Chem. 23*, 1226 (1977).

257. S. Ishimitsu, S. Fujimoto, and A. Ohara, *Chem. Pharm. Bull. Jap. 30*, 1889 (1982).

258. G. J. Schmidt, D. C. Olson, and W. Slavin, *Clin. Chem. 25*, 1063 (1979).

259. G. J. Schmidt, D. C. Olson, and J. G. Atwood, *Chromatogr. Newsl. 8*, 13 (1980).

260. G. H. T. Wheler and J. T. Russel, *J. Liq. Chromatogr. 4*, 1281 (1981).

261. T. Flatmark, S. W. Jacobsen, and J. Haavik, *Anal. Biochem. 107*, 71 (1980).

262. S. Hori, K. Ohtani, S. Ohtani, K. Kayanuma, and T. Ito, *J. Chromatogr. 231*, 161 (1982).

263. F. Geeraerts, L. Schimpfessel, and R. Crokaert, *Chromatographia 15*, 449 (1982).

264. A. D. Jones, C. H. S. Hitchcock, and G. H. Jones, *Analyst 106*, 968 (1981).

265. G. M. Anderson, J. G. Young, D. J. Cohen, K. R. Schlicht, and N. Patel, *Clin. Chem. 27*, 775 (1981).

266. H. Nakamura and J. J. Pisano, *J. Chromatogr. 152*, 153 (1978).

267. L.-I. Larsson, F. Sundler, and R. Håkanson, *J. Chromatogr. 117*, 355 (1976).

268. I. L. Honigberg, J. T. Stewart, T. G. Clark, and D. Y. Davis, *J. Chromatogr. 181*, 266 (1980).

269. T. A. Hare and N. V. Bala Manyam, *Anal. Biochem. 101*, 349 (1980).

270. G. E. Griesmann, W.-Y. Chan, and O. M. Rennert, *J. Chromatogr. 230*, 121 (1982).

271. N. A. Farid, *J. Pharm. Sci. 68*, 249 (1979).

272. S. H. Hansen, *J. Chromatogr. 226*, 504 (1981).

273. C. Fischer, K. Maier, and U. Klotz, *J. Chromatogr. 225*, 498 (1981).

274. P. N. Shaw, L. Aarons, and J. B. Houston, *J. Pharm. Pharmacol. 32*, 67P (1980).

275. D. E. Smith, *J. Pharm. Sci. 71*, 520 (1982).

276. L. M. Walmsley, L. F. Chasseaud, and J. N. Miller, *J. Chromatogr. 226*, 441 (1981).

277. L. A. Marcantonio, W. H. R. Auld, and G. G. Skellern, *J. Chromatogr. 183*, 118 (1980).

278. H. Fabre and N. Hussam-Eddine, *J. Pharm. Pharmacol. 34*, 425 (1982).

279. E. Crombez, G. van der Weken, W. van den Bossche, and P. de Moerloose, *J. Chromatogr. 177*, 323 (1979).

280. R. S. Rapaka, J. Roth, Ct. Viswanathan, T. J. Goehl, V. K. Prasad, and B. E. Cabana, *J. Chromatogr. 227*, 463 (1982).

281. A. L. M. Kerremans, Y. Tan, C. A. M. van Ginneken, and F. W. J. Gribnau, *J. Chromatogr. 229*, 129 (1982).

282. B. Wesley-Hadzija and A. M. Mattocks, *J. Chromatogr. 229*, 425 (1982).

283. W. J. Adams, G. S. Skinner, P. A. Bombardt, M. Courtney, and J. E. Brewer, *Anal. Chem. 54*, 1287 (1982).

284. C. W. Vose, D. C. Muirhead, G. L. Evans, P. M. Stevens, and S. R. Burford, *J. Chromatogr. 222*, 311 (1981).

285. E. P. Lankmayr, K. W. Budna, K. Mueller, F. Nachtmann, and F. Rainer, *J. Chromatogr. 222*, 249 (1981).

286. G. Alfredsson, G. Sedvall, F. A. Wiesel, *J. Chromatogr. 164*, 187 (1979).

287. H. Kamimura, H. Sasaki, and S. Kawamura, *J. Chromatogr. 225*, 115 (1981).

288. E. Soczewiński and J. Jusiak, *Chromatographia 15*, 309 (1982).

289. P. Jandera, H. Pechová, D. Tocksteinová, J. Churáček, and J. Královský, *Chromatographia 16*, 275 (1982).

290. G. Houin, F. Bree, N. Lerumeur, and J. P. Tillement, *J. Pharm. Sci. 72*, 71 (1983).

291. K. Reuter, H. Knauf, and E. Mutschler, *J. Chromatogr. 233*, 432 (1982).

292. S. C. J. Cole, R. J. Flanagan, A. Johnston, and D. W. Holt, *J. Chromatogr. 218*, 621 (1981).

293. R. Wintersteiger, G. Gamse, and W. Pacha, *Fresenius' Z. Anal. Chem. 312*, 455 (1982).

294. K. Imai, Y. Watanabe, and T. Toyo'oka, *Chromatographia 16*, 214 (1982).

295. B. A. Davis, *J. Chromatogr. 151*, 252 (1978).

296. Y. C. Sumirtapura, C. Aubert, Ph. Coassolo, and J. P. Cano, *J. Chromatogr. 232*, 111 (1982).

297. G. Guebitz, R. Wintersteiger, and A. Hartinger, *J. Chromatogr. 218*, 51 (1981).

298. J. M. Meola, T. G. Rosano, and T. A. Swift, *Clin. Chem. 27*, 1254 (1981).

299. K. Yazawa and Z. Tamura, *J. Chromatogr. 254*, 327 (1983).

300. R. W. Frei, J. F. Lawrence, U. A. Th. Brinkman, and I. Honigberg, *J. High Resolut. Chromatogr. Chromatogr. Commun. 2,* 11 (1979).

301. E. T. Lin, R. A. Baugham, Jr., and L. Z. Benet, *J. Chromatogr. 183,* 367 (1980).

302. J. Kohn and M. Wilchek, *J. Chromatogr. 240,* 262 (1982).

303. R. W. Frei, W. Santi, and M. Thomas, *J. Chromatogr. 116,* 365 (1976).

304. G. Alván, L. Ekman, and B. Lindstroem, *J. Chromatogr. 229,* 241 (1982).

305. B. Oosterhuis, M. van den Berg, and C. J. van Boxtel, *J. Chromatogr. 226,* 259 (1981).

306. D. M. Takahashi, *J. Pharm. Sci. 69,* 184 (1980).

307. E. Oeyehaug, E. T. Oestensen, and B. Salvesen, *J. Chromatogr. 227,* 129 (1982).

308. K. F. Overoe, *J. Chromatogr. 224,* 526 (1981).

309. C. E. Werkhoven-Goewie, U. A. Th. Brinkman, and R. W. Frei, *Anal. Chim. Acta 114,* 147 (1980).

310. I. W. Tsina, M. Fass, J. A. Debban, and S. B. Matin, *Clin. Chem. 28,* 1137 (1982).

311. Y. Hayashi, S. Miyake, M. Kuwayama, M. Hatlori, and Y. Usui, *Chem. Pharm. Bull. Jap. 30,* 4107 (1983).

312. J. D. H. Cooper and D. C. Turnell, *J. Chromatogr. 227,* 158 (1982).

313. U. A. Th. Brinkman, P. L. M. Welling, G. de Vries, A. H. M. T. Scholten, and R. W. Frei, *J. Chromatogr. 217,* 463 (1981).

314. P. A. Reece, R. Zacest, and C. G. Barrow, *J. Chromatogr. 163,* 310 (1979).

315. R. G. Foss and C. W. Sigel, *J. Pharm. Sci. 71,* 1176 (1982).

316. L. Zecca, L. Bonini, and S. R. Bareggi, *J. Chromatogr. 272,* 401 (1983).

317. E. Riedel, G. Kreutz, and D. Hermsdorf, *J. Chromatogr. 229,* 417 (1982).

318. A. Szabo and E. M. Karacsony, *J. Chromatogr. 193,* 500 (1980).

319. I. Jane and J. F. Taylor, *J. Chromatogr. 109,* 37 (1975).

320. R. N. Gupta, F. Eng, D. Lewis, and C. Kumana, *Anal. Chem. 51,* 455 (1979).

321. M. Kuwayama, S. Miyake, and K. Nishikawa, *Chem. Pharm. Bull. Jap. 28,* 2158 (1980).

322. S. J. Bannister, J. Stevens, D. Musson, and L. A. Sternson, *J. Chromatogr. 176,* 381 (1979).

323. P. A. Reece, I. Cozamanis, and R. Zacest, *Proc. 6th Aust. Symp. on Anal. Chem.,* Canberra, August 23–28, 1981, p. 121.

324. R. A. Pask-Hughes, P. H. Corran, and D. H. Calam, *J. Chromatogr. 214,* 307 (1981).

325. P. O. Edlund, *J. Chromatogr. 226,* 107 (1981).

326. A. H. M. T. Scholten and R. W. Frei, *J. Chromatogr. 176,* 349 (1979).

327. K. Tsuji, *J. Chromatogr. 158,* 337 (1978).

328. J. W. de Jong, J. A. J. Hegge, E. Harmsen, and P.Ph. de Tombe, *J. Chromatogr. 229*, 498 (1982).

329. C. M. Davis and D. C. Fenimore, *J. Chromatogr. 272*, 157 (1983).

330. M. A. Lefebvre, B. Julian, and J. B. Fourtillan, *J. Chromatogr. 230*, 199 (1982).

331. S. E. Walker and P. E. J. Coates, *J. Chromatogr. 223*, 131 (1981).

332. J. D'Souza and R. I. Ogilvie, *J. Chromatogr. 232*, 212 (1982).

333. P. A. Bombardt and W. J. Adams, *Anal. Chem. 54*, 1087 (1982).

334. Y. Hiraga and T. Kinoshita, *J. Chromatogr. 226*, 43 (1981).

335. M. D. Baker, H. Y. Mohammed, and H. Veening, *Anal. Chem. 53*, 1658 (1981).

336. F. Sasse, J. Hammer, and J. Berlin, *J. Chromatogr. 194*, 234 (1980).

337. Y. Tsuruta, K. Kohashi, and Y. Ohkura, *J. Chromatogr. 224*, 105 (1981).

338. Y. Tsuruta, K. Kohashi, and Y. Ohkura, *J. Pharmacobio-Dyn. 3*, 17 (1980).

339. G. Skofitsch, A. Saria, P. Holzer, and F. Lembeck, *J. Chromatogr. 226*, 53 (1981).

340. D. L. Mays, R. J. van Apeldoorn, and R. G. Lauback, *J. Chromatogr. 120*, 93 (1976).

341. P. O. Okonkwo, J. W. Reimann, and R. Woestenborghs, *J. Chromatogr. 272*, 411 (1983).

342. A. T. Kacprowicz, P. G. Shaw, R. F. W. Moulds, and R. W. Bury, *J. Chromatogr. 272*, 417 (1983).

343. R. Wintersteiger, G. Guebitz, and A. Hartinger, *Mikrochim. Acta II*, 235 (1979).

344. B. B. Wheals, *J. Chromatogr. 122*, 85 (1976).

345. J. Christie, M. W. White, and J. M. Wiles, *J. Chromatogr. 120*, 496 (1976).

346. P. J. Twitchett, S. M. Fletcher, A. T. Sullivan, and A. C. Moffat, *J. Chromatogr. 150*, 73 (1978).

347. J. P. Sommadossi, M. Lemar, J. Necciari, Y. Sumirtapura, J. P. Cano, and J. Gaillot, *J. Chromatogr. 228*, 205 (1982).

348. A. T. Rhys Williams and S. A. Winfield, *Analyst 107*, 1092 (1982).

349. P. E. Nelson, S. L. Nolan, and K. R. Bedford, *J. Chromatogr. 234*, 407 (1982).

350. R. Wintersteiger, *Analyst 107*, 459 (1982).

351. H.-J. Klimisch and D. Ambrosius, *J. Chromatogr. 121*, 93 (1976).

352. J. C. Young, *J. Chromatogr. 151*, 215 (1978).

353. J. A. Glasel and R. F. Venn, *J. Chromatogr. 213*, 337 (1981).

354. J. D. Horowitz, P. M. Morris, O. H. Drummer, A. J. Goble, and W. J. Louis, *J. Pharm. Sci. 70*, 320 (1981).

355. Y. le Roux, J. Gaillot, and A. Bieder, *J. Chromatogr. 230*, 401 (1982).

356. T. Takagi, T.-G. Chung, and A. Saito, *J. Chromatogr. 272*, 279 (1983).

357. M. Mach, H. Kersten, and W. Kersten, *J. Chromatogr. 223,* 51 (1981).

358. C. E. Prussak and D. H. Russel, *J. Chromatogr. 229,* 47 (1982).

359. M. M. Abdel-Monem and J. L. Merdink, *J. Chromatogr. 222,* 363 (1981).

360. F. L. Vandemark, G. J. Schmidt, and W. Slavin, *J. Chromatogr. Sci. 16,* 465 (1978).

361. Y. Saeki, N. Uehara, and S. Shirakawa, *J. Chromatogr. 145,* 221 (1978).

362. K. Samejima, M. Kawase, S. Sakamoto, M. Okada, and Y. Endo, *Anal. Biochem. 76,* 392 (1976).

363. M. Kai, T. Ogata, K. Haraguchi, and Y. Ohkura, *J. Chromatogr. 163,* 151 (1979).

364. T. Skaaden and T. Greibrokk, *J. Chromatogr. 247,* 111 (1982).

365. N. D. Brown, R. B. Sweet, J. A. Kintzios, H. D. Cox, and B. P. Doctor, *J. Chromatogr. 164,* 35 (1979).

366. C. Branca, E. Gaetani, C. F. Laureri, and M. Vitto, *Il Farmaco Ed. Pr. 36,* 518 (1981).

367. J. Wagner, D. Danzin, and P. Mamont, *J. Chromatogr. 227,* 349 (1982).

368. R. C. Simpson, H. Y. Mohammed, and H. Veening, *J. Liq. Chromatogr. 5,* 245 (1982).

369. B. A. Mico, R. A. Baughman, Jr., and L. Z. Benet, *J. Chromatogr. 230,* 203 (1982).

370. Y. G. Yee, P. C. Rubin, and P. Meffin, *J. Chromatogr. 172,* 313 (1979).

371. P. A. Meredith, D. McSharry, H. L. Elliott, and J. L. Reid, *J. Pharmacol. Meth. 6,* 309 (1981).

372. A. Licht, R. L. Bowman, and S. Stein, *J. Liq. Chromatogr. 4,* 825 (1981).

373. J. R. Cronin and P. E. Hare, *Anal. Biochem. 81,* 151 (1977).

374. H. A. Moye and A. J. Boning, Jr., *Anal. Lett. 12,* 25 (1979).

375. H. Nakamura and J. J. Pisano, *J. Chromatogr. 121,* 33 (1976).

376. E. Hohaus, *Fresenius' Z. Anal. Chem. 310,* 70 (1982).

377. R. N. Gupta, F. Eng, and D. Lewis, *Anal. Chem. 50,* 197 (1978).

378. C. Sarbu, T. Hodisan, and C. Liteanu, *Rev. Chim.* (Bucharest), 32, 589 (1981).

379. U. Timm and E. Weidekamm, *J. Chromatogr. 230,* 107 (1982).

380. W. S. Simmons and R. L. de Angelis, *Anal. Chem. 45,* 1538 (1973).

381. A. Rakhit, M. Kunitani, N. H. G. Holford, and S. Riegelman, *Clin. Chem. 28,* 1505 (1982).

382. R. Leroyer, C. Jarreau, and M. Pays, *J. Chromatogr. 228,* 366 (1982).

383. J. W. Robinson, *EDRO SARAP Res. Tech. Rep. 2,* 23 (1977).

384. L. J. Dombrowski, A. V. Crain, R. S. Browning, and E. L. Pratt, *J. Pharm. Sci. 64,* 643 (1975).

385. J. Grundevik and B. A. Persson, *J. Liq. Chromatogr. 5,* 141 (1982).

386. R. Sams, *Anal. Lett. 11,* 697 (1978).

387. P. Haefelfinger, *J. Chromatogr. 233,* 269 (1982).

388. P. Haefelfinger, *Chromatographia 14,* 212 (1981).

389. H. Nakamura, S. Tsuzuki, Z. Tamura, R. Yoda, and Y. Yamamoto, *J. Chromatogr. 200,* 324 (1980).

390. R. J. Reddingius, G. J. de Jong, U. A. Th. Brinkman, and R. W. Frei, *J. Chromatogr. 205,* 77 (1981).

391. J. F. Lawrence and R. W. Frei, *J. Chromatogr. 79,* 223 (1973).

392. H. N. Myers and J. V. Rindler, *J. Chromatogr. 176,* 103 (1979).

393. R. R. Brown, R. Bain, and V. C. Jordan, *J. Chromatogr. 272,* 351 (1983).

394. R. Schlobe and H. H. W. Thijssen, *J. Chromatogr. 230,* 212 (1982).

395. M. T. Watts, V. A. Raisys, and L. A. Bauer, *J. Chromatogr. 230,* 79 (1982).

396. V. P. Shah, J. Lee, and V. K. Prasad, *Anal. Lett. 15,* 529 (1982).

397. W. Lindner, R. W. Frei, and W. Santi, *J. Chromatogr. 111,* 365 (1975).

398. P. J. Meffin, S. R. Harapat, and D. C. Harrison, *J. Pharm. Sci. 66,* 583 (1977).

399. A. J. Sedman and L. Gal, *J. Chromatogr. 232,* 315 (1982).

400. S. M. Johnson, C. Chan, S. Cheng, J. L. Shimek, G. Nygard, and S. K. Wahba Khalil, *J. Pharm. Sci. 71,* 1027 (1982).

401. S. Courte and N. Bromet, *J. Chromatogr. 224,* 162 (1981).

402. G. Andermann and C. Andermann, *J. High Resolut. Chromatogr. Chromatogr. Commun. 3,* 36 (1980).

403. P. C. Rubin, J. Brunton, and P. Meredith, *J. Chromatogr. 221,* 193 (1980).

404. E. Watson and P. A. Kapur, *J. Pharm. Sci. 70,* 800 (1981).

405. M. Kuwada, T. Tateyama, and J. Tsutsumi, *J. Chromatogr. 222,* 507 (1981).

406. A. Roux and B. Flouvat, *J. Chromatogr. 166,* 327 (1978).

407. D. von Hippmann and P. Takacs, *Arzneim.-Forsch. Drug/Res. 33,* 8 (1983).

408. J. M. Steyn, *J. Chromatogr. 120,* 465 (1976).

409. J. B. Lecaillon, C. Souppart, and F. Abadie, *Chromatographia 16,* 158 (1982).

410. H. Winkler, W. Ried, and B. Lemmer, *J. Chromatogr. 228,* 223 (1982).

411. M. Schaefer and E. Mutschler, *J. Chromatogr. 169,* 477 (1979).

412. J. A. F. de Silva, J. C. Meyer, and C. V. Puglisi, *J. Pharm. Sci. 65,* 1230 (1976).

413. A. Nagakura and H. Kohei, *J. Chromatogr. 232,* 137 (1982).

414. P. A. Meredith, D. McSharry, H. L. Elliott, and J. L. Reid, *J. Pharmacol. Meth. 6,* 309 (1981).

415. D. B. Pautler and W. J. Jusko, *J. Chromatogr. 228,* 215 (1982).

416. M. S. Lennard and J. H. Silas, *J. Chromatogr. 272,* 205 (1983).

417. M. Schaefer and E. Mutschler, *J. Chromatogr. 164,* 247 (1979).

418. N. Bernard, G. Cuisinaud, and J. Sassard, *J. Chromatogr. 228*, 355 (1982).

419. C. J. Oddie, G. P. Jackman, and A. Bobik, *J. Chromatogr. 231*, 473 (1982).

420. P. Jatlow, W. Busch, and H. Hochster, *Clin. Chem. 25*, 777 (1979).

421. N. Terao and D. D. Shen, *Chromatographia 15*, 685 (1982).

422. T. W. Guentert, A. Rakhit, R. A. Upton, and S. Riegelman, *J. Chromatogr. 183*, 514 (1980).

423. F. Albani, R. Riva, and A. Baruzzi, *J. Chromatogr. 228*, 362 (1982).

424. M.-W. Lo, B. Silber, and S. Riegelman, *J. Chromatogr. Sci. 20*, 126 (1982).

425. J. Hermansson, *Acta Pharm. Suec. 19*, 11 (1982).

426. M. T. Rosseel and M. G. Bogaert, *J. Pharm. Sci. 70*, 688 (1981).

427. M. Schaefer, H. E. Geissler, and E. Mutschler, *J. Chromatogr. 143*, 607 (1977).

428. M. A. Lefebvre, J. Girault, M. Cl. Saux, and J. B. Fourtillan, *J. Pharm. Sci. 69*, 1216 (1980).

429. D. A. Stopher, *J. Pharm. Pharmacol. 27*, 133 (1975).

430. P. W. Doetsch, J. M. Cassady, and J. L. McLaughlin, *J. Chromatogr. 189*, 79 (1980).

431. S. Kamata, K. Imura, A. Okada, Y. Kawashima, A. Yamatodani, T. Watanabe, and H. Wada, *J. Chromatogr. 231*, 291 (1982).

432. B. M. Farrell and T. M. Jefferies, *J. Chromatogr. 272*, 111 (1983).

433. B. Kinberger, *J. Chromatogr. 213*, 166 (1981).

434. W. J. Decker and J. D. Thompson, *Clin. Toxicol. 13*, 545 (1978).

435. B. Klein, J. E. Sheehan, and E. Grunberg, *Clin. Chem. 20*, 272 (1974).

436. F. van Hoof and A. Heyndrickx, *Anal. Chem. 46*, 286 (1974).

437. S. Allenmark, *J. Liq. Chromatogr. 5*, 1 (1982).

438. A. M. Krstulović, *J. Chromatogr. 229*, 1 (1982).

439. T. Seki, *J. Chromatogr. 207*, 286 (1981).

440. G. Schwedt, *Anal. Chim. Acta 92*, 337 (1977).

441. P. M. Froehlich and T. D. Cunningham, *Anal. Chim. Acta 97*, 357 (1978).

442. G. Schwedt, *Chromatographia 10*, 92 (1977).

443. H. Svendsen and T. Greibrokk, *J. Chromatogr. 213*, 429 (1981).

444. G. Schwedt, *J. Chromatogr. 118*, 429 (1976).

445. G. Schwedt and H. H. Bussemas, *Fresenius' Z. Anal. Chem. 283*, 23 (1977).

446. G. P. Jackman, V. J. Carson, A. Bobik, and H. Skews, *J. Chromatogr. 182*, 277 (1980).

447. G. M. Anderson, J. G. Young, D. K. Batter, S. N. Young, D. J. Cohen, and B. A. Shaywitz, *J. Chromatogr. 223*, 315 (1981).

448. G. P. Jackman, *Clin. Chem. 27*, 1202 (1981).

449. T. P. Davis, C. W. Gehrke, Jr., C. H. Williams, C. W. Gehrke, and K. O. Gerhardt, *J. Chromatogr. 228*, 113 (1982).

450. L. D. Mell, Jr., A. R. Dasler, and A. B. Gustafson, *J. Liq. Chromatogr. 1*, 261 (1978).

451. D. A. Williams, E. Y. Y. Fung, and D. W. Newton, *J. Pharm. Sci. 71*, 956 (1982).

452. M. T. I. W. Schuesler-Van Hees and G. M. J. Beyersbergen van Henegouwen, *J. Chromatogr. 196*, 101 (1980).

453. G. M. Anderson, J. G. Young, P. I. Jatlow, and D. J. Cohen, *Clin. Chem. 27*, 2060 (1981).

454. R. C. Causon and M. E. Carruthers, *J. Chromatogr. 229*, 301 (1982).

455. H. Nakamura and J. J. Pisano, *J. Chromatogr. 154*, 51 (1978).

456. J. Haavik and T. Flatmark, *J. Chromatogr. 198*, 511 (1980).

457. N. Nimura, K. Ishida, and T. Kinoshita, *J. Chromatogr. 221*, 249 (1980).

458. R. H. Christenson and C. D. McGlothlin, *Anal. Chem. 54*, 2015 (1982).

459. G. Schwedt, *Fresenius' Z. Anal. Chem. 293*, 40 (1978).

460. F. Nachtmann, H. Spitzy, and R. W. Frei, *Anal. Chim. Acta 76*, 57 (1975).

461. J. Traveset, V. Such, R. Gonzalo, and E. Gelpi, *J. High Resolut. Chromatogr. Chromatogr. Commun. 4*, 589 (1981).

462. T. Hojo, H. Nakamura, and Z. Tamura, *J. Chromatogr. 247*, 157 (1982).

463. J. P. Garnier, B. Bousquet, and C. Dreux, *Analusis, 7*, 355 (1979).

464. A. P. Graffeo and B. L. Karger, *Clin. Chem. 22*, 184 (1976).

465. G. M. Anderson, J. G. Young, D. J. Cohen, and S. N. Young, *J. Chromatogr. 228*, 155 (1982).

466. A. J. Swaisland, *Analyst 106*, 717 (1981).

467. Y. Kishimoto, S. Ohgitani, A. Yamatodani, M. Kuro, and F. Okumura, *J. Chromatogr. 231*, 121 (1982).

468. R. H. Christenson, C. D. McGlothlin, R. Hedrick, and J. C. Cate IV, *Clin. Chem. 28*, 1204 (1982).

469. G. P. Jackman, *Clin. Chim. Acta, 120*, 137 (1982).

470. J. Meyer and P. Portmann, *Pharm. Acta Helv. 57*, 12 (1982).

471. J. Trocewicz, N. Kato, K. Oka, and T. Nagatsu, *J. Chromatogr. 233*, 328 (1982).

472. W. D. Mason and E. N. Amick, *J. Pharm. Sci. 70*, 707 (1981).

473. D. E. Mais, P. D. Lahr, and T. R. Bosin, *J. Chromatogr. 225*, 27 (1981).

474. O. Heisz, *Labor Praxis 4*, 32, 34, 36, 39 (1980).

475. H. Cohen and M. Lapointe, *J. Assoc. Off. Anal. Chem. 64*, 1372 (1981).

476. M. K. L. Bicking, R. N. Kniseley, and H. J. Svec, *Anal. Chem. 55*, 200 (1983).

477. J. Goto, S. Komatsu, N. Goto, and T. Nambara, *Chem. Pharm. Bull. Jap. 29*, 899 (1981).

478. R. Wintersteiger, *J. Liq. Chromatogr. 5*, 897 (1982).

479. R. Wintersteiger and G. Wenninger-Weinzierl, *Fresenius' Z. Anal. Chem.* *309*, 201 (1981).

480. K. Mopper, W. L. Stahovec, and L. Johnson, *J. Chromatogr.* *256*, 243 (1983).

481. J. W. Tolan, P. Eskola, D. W. Fink, H. Mrozik, and L. A. Zimmerman, *J. Chromatogr.* *190*, 367 (1980).

482. S. Takitani, A. Sano, Y. Asabe, and M. Suzuki, *Anal. Chim. Acta, 135*, 307 (1982).

483. J. P. Heotis, J. L. Mertz, R. J. Herrett, J. R. Diaz, D. C. van Hart, and J. Olivard, *J. Assoc. Off. Anal. Chem. 63*, 720 (1980).

484. S. Honda, M. Takahashi, K. Kakehi, and S. Ganno, *Anal. Biochem. 113*, 130 (1981).

485. W. F. Alpenfels, *Anal. Biochem. 114*, 153 (1981).

486. R. Wintersteiger, G. Wenninger-Weinzierl, and W. Pacha, *J. Chromatogr.* *237*, 399 (1982).

487. W. F. Alpenfels, R. A. Mathews, D. E. Madden, and A. E. Newsom, *J. Liq. Chromatogr. 5*, 1711 (1982).

488. P. Prognon, G. Simon, and G. Mahuzier, *J. Chromatogr. 272*, 193 (1983).

489. M. Takeda, M. Maeda, and A. Tsuji, *J. Chromatogr. 244*, 347 (1982).

490. K. Mopper and L. Johnson, *J. Chromatogr. 256*, 27 (1983).

491. S. Honda, Y. Matsuda, and K. Kakehi, *J. Chromatogr. 176*, 433 (1979).

492. K. Mopper, R. Dawson, G. Liebezeit, and H.-P. Hansen, *Anal. Chem. 52*, 2018 (1980).

493. S. K. Nuor, J. Vialle, and J. L. Rocca, *Analusis 7*, 381 (1979).

494. B. Buechele and J. Lang, *J. High Resolut. Chromatogr. Chromatogr. Commun. 2*, 585 (1979).

495. C. Prakash and J. K. Vijay, *Anal. Biochem. 128*, 41 (1983).

496. Y.-M. Peng, T. Davis, and D. Alberts, *Life Sci. 29*, 361 (1981).

497. T. Ogasawara, Y. Masuda, S. Goto, S. Mori, and T. Oki, *J. Antibiot. 34*, 52 (1981).

498. S. Shinozawa and T. Oda, *J. Chromatogr. 212*, 323 (1981).

499. R. Baurain, D. Deprez-De Campeneere, and A. Trouet, *Anal. Biochem. 94*, 112 (1979).

500. E. Tomlinson and L. Malspeis, *J. Pharm. Sci. 71*, 1121 (1982).

501. J. E. Brown, P. E. Wilkinson, and J. R. Brown, *J. Chromatogr. 226*, 521 (1981).

502. A. M. B. Bots, W. J. van Oort, J. Noordhoek, A. van Dijk, S. W. Klein, and Q. G. C. M. van Hoesel, *J. Chromatogr. 272*, 421 (1983).

503. E. Watson and K. K. Chan, *Cancer Treat. Rep. 60*, 1611 (1976).

504. K. K. Chan and C. D. Wong, *J. Chromatogr. 172*, 343 (1979).

505. S. E. Fandrich and K. A. Pittman, *J. Chromatogr. 223*, 155 (1981).

506. E. Moro, M. G. Jannuzzo, M. Ranghieri, S. Stegnjaich, and G. Valzelli, *J. Chromatogr. 230,* 207 (1982).

507. K. Lu, N. Savaraj, M. T. Huang, D. Moore, and T. L. Loo, *J. Liq. Chromatogr. 5,* 1323 (1982).

508. N. Strojny, L. D'Arconte, and J. A. F. de Silva, *J. Chromatogr. 223,* 111 (1981).

509. G. Bykadi, K. P. Flora, J. C. Cradock, and G. K. Poochikian, *J. Chromatogr. 231,* 137 (1982).

510. R. J. Strife, I. Jardine, and M. Colvin, *J. Chromatogr. 224,* 168 (1981).

511. H. Jui and J. Roboz, *J. Chromatogr. 233,* 203 (1982).

512. K. W. Woodhouse and D. B. Henderson, *Br. J. Clin. Pharmacol. 13,* 605P (1982).

513. C. M. Egan, C. R. Jones, and M. McCluskey, *J. Chromatogr. 224,* 338 (1981).

514. J. A. Nelson, *Biological and Biomedical Applications of Liquid Chromatograph,* Dekker, New York, 1979, p. 397.

515. L. A. Sternson, N. Meltzer, and F. Shih, *Anal. Lett. 14,* 583 (1981).

516. A. T. Rhys Williams, S. A. Winfield, and R. C. Belloli, *J. Chromatogr. 240,* 224 (1982).

517. R. Verbeke, *J. Chromatogr. 177,* 69 (1979).

518. P. J. Harman, G. L. Blackman, and G. Phillipou, *J. Chromatogr. 225,* 131 (1981).

519. T. Kawasaki, M. Maeda, and A. Tsuji, *J. Chromatogr. 163,* 143 (1979).

520. Z. K. Shihabi, R. I. Andrews, and J. Scaro, *Clin. Chim. Acta 124,* 75 (1982).

521. W. Funk, R. Kerler, J. T. Schiller, V. Dammann, and F. Arndt, *J. High Resolut. Chromatogr. Chromatogr. Commun. 5,* 534 (1982).

522. A. T. Rhys Williams, S. A. Winfield, and R. C. Belloli, *J. Chromatogr. 235,* 461 (1982).

523. D. E. Bloch, *J. Assoc. Off. Anal. Chem. 63,* 707 (1980).

524. J. T. Taylor, J. G. Knotts, and G. J. Schmidt, *Clin. Chem. 26,* 130 (1980).

525. G. J. Schmidt, F. L. Vandemark, and W. Slavin, *Anal. Biochem. 91,* 636 (1978).

526. S. H. Strusiak, J. G. Hoogerheide, and S. Gardner, *J. Pharm. Sci. 71,* 636 (1982).

527. M. Amin and M. Hassenbach, *Analyst 104,* 407 (1979).

528. R. D. Toothaker, G. M. Sundaresan, J. P. Hunt, T. J. Goehl, K. S. Rotenberg, V. K. Prasad, W. A. Craig, and P. G. Welling, *J. Pharm. Sci. 71,* 573 (1982).

529. T. Kawasaki, M. Maeda, and A. Tsuji, *J. Chromatogr. 232,* 1 (1982).

530. R. Horikawa, T. Tanimura, and Z. Tamura, *J. Chromatogr. 168,* 526 (1979).

531. T. Kawasaki, M. Maeda, A. Tsuji *Yakugaku Zasshi 100,* 925 (1980).

532. G. M. Sundaresan, T. J. Goehl, and V. K. Prasad, *J. Pharm. Sci. 70*, 702 (1981).

533. I. Sondergaard and E. Steiness, *J. Chromatogr. 162*, 422 (1979).

534. T. Kawasaki, M. Maeda, and A. Tsuji, *J. Chromatogr. 233*, 61 (1982).

535. E. Reh and G. Schwedt, *Fresenius' Z. Anal. Chem. 303*, 117 (1980).

536. H. Cohen and M. R. Lapointe, *J. Assoc. Off. Anal. Chem. 63*, 642 (1980).

537. H. L. Trenholm, R. M. Warner, and E. R. Farnworth, *J. Assoc. Off. Anal. Chem. 64*, 302 (1981).

538. Y. Haroon, M. J. Shearer, and P. Barkhan, *J. Chromatogr. 206*, 333 (1981).

539. M. F. Lefevere, R. W. Frei, A. H. M. T. Scholten, and U. A. Th. Brinkman, *Chromatographia, 15*, 459 (1982).

540. H. Y. Mohammed, H. Veening, and D. A. Dayton, *J. Chromatogr. 226*, 471 (1981).

541. J. D. Lumley and R. A. Wiggins, *Analyst 106*, 1103 (1981).

542. M. D. Smith, *J. Chromatogr. 182*, 285 (1980).

543. M. Kimura, B. Panijpan, and Y. Itokawa, *J. Chromatogr. 245*, 141 (1982).

544. M. Kimura, T. Fujita, and Y. Itokawa, *Clin. Chem. 28*, 29 (1982).

545. L. Jansson, B. Nilsson, and R. Lindgren, *J. Chromatogr. 181*, 242 (1980).

546. C. H. McMurray and W. J. Blanchflower, *J. Chromatogr. 178*, 525 (1979).

547. J. Lehmann and H. L. Martin, *Clin. Chem. 28*, 1784 (1982).

548. H. K. Biesalski, W. Ehrenthal, G. Hafner, and D. Harth, *Git Fachz. Lab., Suppl. Chromatogr. 4*, 6–8 (1981).

549. A. J. Speek, F. van Schaik, J. Schrijver, and W. H. P. Schreurs, *J. Chromatogr. 228*, 311 (1982).

550. P. Taylor and P. Barnes, *Chem. Ind. 20*, 722 (1981).

551. B. Kanegsberg and J. House, *Clin. Chem. 28*, 1592 (1982).

CHAPTER

8

LUMINESCENCE IMMUNOASSAY

H. THOMAS KARNES
JEFFREY S. O'NEAL
STEPHEN G. SCHULMAN

College of Pharmacy
University of Florida
Gainesville, FL 32610

8.1. INTRODUCTION

In the past decade one of the major challenges in analytical chemistry has been the demand for quantitative analysis of an increasing number of biologically important compounds. The problems of such analyses lie in the difficulty of quantitating compounds present in very low concentrations within complex biological matrices. The complex nature and chemical compositions of the living organism make biological samples probably the most difficult with which to deal of all types of analytical materials. These samples often contain many chemical species possessing similar properties and require extreme selectivity for the analysis of individual components. Clinical analyses, which require determination of subnanogram quantities in many cases, also present the problem of sample volume availability. In addition to these chemical problems, instrumental and economic considerations must be addressed so that analyses may be widely usable.

Radioimmunosssay (RIA) has virtually revolutionized the analysis of biological compounds. The sensitivity (10^{-12}–$10^{-15}M$) and selectivity of this method have made possible the qualitative and quantitative analysis of many biologically important compounds that could not have been assayed at such low concentrations by any other means. The work by Berson and Yalow (1) on insulin and insulin-binding antibodies laid the foundation for competitive protein binding assays, and new applications of their work are currently being developed and improved for many clinically important substances. The great impact of Berson and Yalow's work was recognized in 1978, when they were awarded the Nobel Prize in medicine, the first Nobel Prize awarded for work of an analytical nature. Although the RIA technique has been employed throughout many divergent areas of biochemical analysis, the method depends on the measurement of radioactivity from isotopic labels and thus has certain drawbacks. These limitations include:

1. Regulation and disposal of radioactive material
2. Possible health hazards, especially associated with bulk preparation procedures
3. Radioactive decay, which prevents long-term standardization and limits reagent shelf life
4. Expensive instrumentation and reagents
5. Time-consuming procedural difficulties.

Immunoanalytical methods are largely based on the competitive binding that occurs between a labeled and unlabeled ligand for highly specific receptor sites on antibodies (2,3). The analysis of this competive binding, effected by measuring some physical or chemical property associated with the label, allows the construction of a standard curve representing a measured physical signal that is altered by changes in distribution of bound labeled ligand as a function of the concentration of the unlabeled ligand. Owing to the selectivity of the recognition of antigens by specific antibodies, this approach gives the analyst the capacity to quantitate, quite specifically, the ligand (analyte) of interest. When the measured property of the labeled ligand is radioactivity, as has most often been the case, the technique is also extremely sensitive, because of the sensitive instrumental methods available for the detection of radioactivity. A relatively recent development in the area of immunoanalytical methods is that of luminescence immunoassay (LIA). Nonisotopic immunoassay development has been given much attention among analytical chemists and proved to be a viable alternative to the use of radiolabels. The main focus in luminescence immunoassay has been on fluorescence immunoassay (FIA). FIA is very similar to RIA with the principal difference that FIA employs the use of fluorescent labels for detection rather than the radioactive isotopes used in RIA procedures. There are also many interesting and novel variations that can be used with FIA that are not possible with RIA. Some of these variations take advantage of the fact that the labeled immunoreactive antigen exists in two separate microenvironments, that of the bulk aqueous phase and that of the relatively hydrophobic binding site of the antibody (4). Physicochemical interactions with the fluorescent label, causing such effects as quenching, polarization, resonance energy transfer, and so on, give rise to the possibility for homogeneous assays if the effect is great enough to be measured reliably. *Homogeneous*, in immunoassay terminology, implies that partitioning, or phase separation, of free and bound antigen is not required. This lends a great deal of simplicity to the assay and greatly increases its suitability for automation in commercial situations. Heterogeneous immunoassays, on the other hand, employ a physical separation of the bound fraction from the free fraction of labeled ligand. Currently, to obtain the sensitivity achieved by RIA procedures, there must be some sort of separation of free and bound ligand to remove interfering compounds. Therefore, much of the recent basic research on FIA has been directed toward heterogeneous rather than homogeneous assay.

Even with the high sensitivity of heterogeneous assays, FIA has generally not yet attained the sensitivity levels common to RIA procedures, and present limits of detection for FIA are approximately 10^{-12} M for some of the more sensitive assays. However, with improvements in instrumentation and label preparation, FIA can exceed sensitivities presently attained by RIA.

The main advantage of FIA procedures over conventional isotopic methods include the potential for expedient homogeneous assay, lower cost and longer stability of label materials, ease of automation, relatively inexpensive instrumentation required, and obviation of isotope use. Homogeneous assays can reduce procedural steps when great sensitivity is not required or when instrumental modification or optimization of the spectroscopic characteristics of the fluorescence label can overcome interference difficulties. Greater label stability results from the fact that the shelf life of a radiolabeled antigen is limited by the half-life of the isotopic label, whereas fluorescent labels have virtually unlimited shelf life if kept dry and protected from light.

The vast majority of work in luminescence immunoassay has been in the analysis of biologically important substances, although applications to the analyses of environmental contaminants have also been made (5). The most notable developments have been in the analysis of therapeutic drugs, and several companies market instrumentation and reagents specifically for that purpose. A product guide listing commercially available assays that includes those manufactured prior to July 1982 was recently published (6). The need for rapid, reliable, and inexpensive methods of therapeutic drug monitoring is widely recognized and provides stimulating research for both industrial and academic scientists. Bearing this in mind, the focus of this chapter will be on the rapidly growing area of drug analysis and the analysis of other clinically important compounds in biological fluids.

8.2. THE NATURE OF THE IMMUNE RESPONSE

The defense mechanisms in humans as well as other organisms are traditionally divided into two classes. Natural immunity is innate and provides the first line of defense against foreign substances, including microorganisms and various toxins. Acquired immunity differs in that defense mechanisms develop only on exposure to the invasive foreign material. The basis for this immune response lies in the ability of an organism to synthesize specialized proteins (antibodies) that combine noncovalently with the invasive material to abolish its harmful effects. Acquired immunity demonstrates both greater potency and specificity for the individual pathogens than does natural immunity (7). The number of molecules capable of eliciting antibody formation (immunogens) is vast, suggesting that organisms must be capable of synthesizing many different antibodies.

An antigen is any substance capable of reacting with an antibody but not necessarily capable of inducing antibody formation. Therefore, all immunogens are antigens, although antigens may or may not be immunogens. Antigen–antibody reactions are so specific that an antibody will generally react only with the immunogen that caused its formation or with molecules that are very similar structurally.

Antibodies are a group of heterogeneous, structurally related proteins known as the immunoglobulins (Ig). There are five major classes of immunoglobulins: IgG, IgM, IgA, IgD, IgE. The predominent immunoglob-

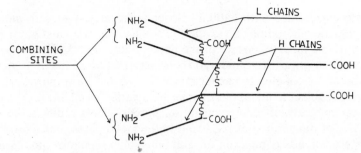

Fig. 8.1. Structure of the Ig G Protein molecule.

ulin in serum is IgG, and it is usually associated with the majority of antigen binding activity in the highly diluted solutions used in sensitive immunoassays (8). The IgG class of immunoglobulins has an average molecular weight of 150,000–160,000 daltons. The structure of IgG proteins is characterized by four polypeptide chains, two of which are designated as heavy, or H, chains (50,000 daltons) and two of which are designated as light, or L, chains (20,000 daltons) (Fig. 8.1). Each immunoglobulin molecule contains two sites of binding and they are located at the NH_2-terminal regions of the H–L chains. These binding sites contain variable amino acid sequences and appear to be the combining sites responsible for the specificity of binding common to antibodies (9). The immunoglobulins are produced by the B-lymphocytes. The process of the maturation and development of these lymphocytes into cells capable of antibody production is not fully understood at this time (10).

Immunogens are generally naturally occurring macromolecules (proteins, polysacchrides, nucleic acids, etc.) or microorganisms containing such molecules. The antibodies developed in response to those molecules will recognize and bind to only a small section of the antigen. Antibody specificity usually involves not more than six or seven amino acid residues of the large protein structure (11). The ability of a molecule to elicit antibody formation appears to depend on its size. Molecules with molecular weights below 10,000 daltons are generally not immunogenic. This presents problems in producing antibodies directed against drugs to be used in immunoassay procedures, since drugs generally have molecular weights much lower than 10,000 daltons.

Investigations performed by Landsteiner determined that an immunoresponse could be elicited from an organism by small molecules (haptens) if they were coupled to macromolecules (12). The antibodies produced may react with sites on the hapten even when the hapten is not coupled to the carrier macromolecules. It is often more difficult to elicit antibody production with hapten–carrier conjugates than with naturally immunogenic macromolecules, and the conjugation site must be carefully chosen to maximize exposure of the hapten molecule. Additionally, haptens, such as drugs, that undergo biotransformation to inactive metabolites should be coupled to carrier molecules so that those portions of the molecules subject to metabolic change are spatially available for the initial immunochemical antigenic recognition process. This reduces the cross reactivity of the antibodies produced to the metabolites. Once produced, the heterogeneous population of antibodies may recognize points of the carrier and the bridging molecule along with sites present on the hapten itself. This results from a lack of antibody uniformity and gives rise to various degrees of specificity for the hapten among the antibody population. Since antibodies are not a chemically distinct species, quantitative descriptions of immunoreactions in terms of simple competitive equilibria are not rigorous but have been dealt with as if this were the case.

8.3. LUMINESCENT LABELS FOR COMPETITIVE PROTEIN-BINDING ASSAYS

Fluorescence immunoassay (FIA) simply involves the measurement of the fluorescence in a luminescent label that participates at some level of a competitive immunochemical binding system and whose spectral properties in most homogeneous systems vary with differences in the concentrations of the analyte. Fluorescent labels are used in homogeneous and heterogeneous immunoassay systems and may be bound to competing antigens, antibodies, or solid phases, or they may exist free in solution.

A fluorescent probe to be used in homogeneous immunoanalysis should fulfill several requirements. Since the fluorescent signal must be measured in a serum matrix, the probe should have a high-fluorescence quantum yield and the excitation and emission maxima of the probe should occur at wavelengths longer than those of the serum. Excitation spectra of dilute serum show an excitation maximum at 280 nm and one at 340 nm. The absorption and emission maxima illustrated in Figs. 8.2 and 8.3 are due primarily to the tyrosine and tryptophan components of protein molecules. Emission bands of bilirubin and other conjugated molecules extend the broad emission spectrum out to approximately 600 nm. Large Stokes' shift displacements between excitation and emission maxima are desirable properties of fluorescent labels because serum macromolecules tend to scatter the exciting light. To minimize this interference the Stokes' shift should be greater than 50 nm.

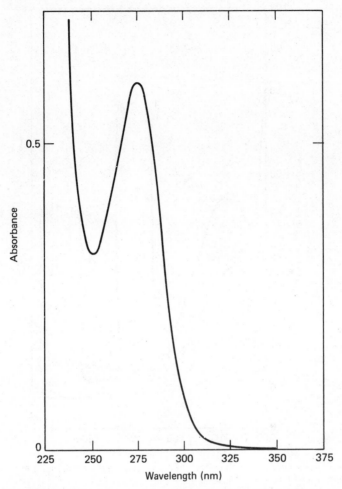

Fig 8.2. Absorbance spectrum of human plasma diluted 80:1.

Recently, a patent was granted for a method to reduce background serum fluorescence (13). This method involves pretreatment of the serum with peroxy acids, such as peroxyacetic acid, followed by treatment with a reducing agent, such as sulfite ions, to neutralize the peroxy acid. Applicability of the method to a real system, however, is limited by the stability of the analyte under the conditions of treatment.

For many years the most widely employed fluorescent probes in immunoanalysis have been derivatives of fluorescein–isothiocyanate (14–16). Another derivative of fluorescein, dichlorotriazinylaminofluorescein,

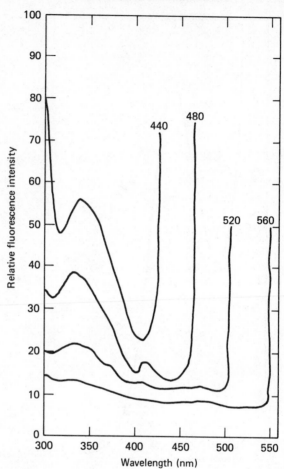

Fig 8.3. Fluorescence excitation spectra of normal human plasma diluted 100:1 at emission wavelengths of 440, 480, 520, and 560 nm.

is claimed to give higher yields on reaction with proteins (17). Fluorescein labels have been used in polarization, quenching, enhancement, internal reflectance, time correlation, excitation transfer, and protection immunoassays (*vide infra*). Protein–fluorescein conjugates have high molar absorptivities and fluorescence quantum yields, 80,000 M^{-1} cm^{-1} and approximately 0.5, respectively (18). Large values such as these are necessary for accurate quantitative determination of low concentrations of the probe in serum matrices. Also the absorption maximum of 492 nm and emission maximum of 517 nm (16) of fluorescein conjugates are well red-shifted with respect to the serum protein background. The Stokes' shift, however, is too small to avoid significant scattering interferences. Also, the fluorescence of endogenous bilirubin is large enough to cause significant interferences at these excitation and emission wavelengths (19). It is probably worthwhile to note that patients undergoing drug therapy might exhibit large variations in their serum bilirubin levels.

Very few probes have been developed beyond the initial testing stages as alternatives to fluorescein. Perhaps the reason for this is the extremely high molar absorptivity and quantum yield of emission of fluorescein and its long history of successful application in FIA. However, other probes have excitation and emission spectra that are more red-shifted than those of fluorescein, are more environment-sensitive, and have larger Stokes' shifts. For example, lissamine–rhodamine-B–200-sulfonylchloride has excitation and emission maxima of 565 nm and 710 nm respectively (20,21).

Rhodamine-B–200-isothiocyanate has been conjugated to immunoglobulin G (22,23) and has been used in immunochemical experiments (24,25). This conjugate has excitation and emission maxima of 550 and 585 nm and has a quantum yield when protein bound of 0.7 (22). Other probes that hold some promise for future developments include fluorescamine (26–29), N-(p-2-benzimidazolyl-phenyl)-maleimide (30–31), N-(7-dimethylamino-4-methyl-2-oxy-3-chromenyl)-maleidime (30–33), and 2-methoxy-2,4-diphenyl-3-(2H)-furanone (34–35).

A new class of intensely fluorescing proteins, known as phycobiliproteins, has been developed for flow cytofluorimetric studies (35). Their extremely high molar absorptivities (up to 2.4×10^6 l/mole cm) and high quantum yields (up to 0.98) are given in Table 8.1.

4-chloro-7-nitrogenzo-2-oxa-1,3-diazole (NBD chloride) has been a useful fluorogenic reagent for covalent labeling of amino groups as well as sulfhydryl (45) and aromatic hydroxy groups (46). Amino derivatives have excitation and emission maxima at approximately 480 and 530 nm, respectively, and large fluorescence quantum yield enhancements have been observed for NBD derivatives on transfer to a hydrophobic environment from aqueous media.

Another class of fluorescent probes potentially useful in immunoassays and as yet unexplored are the carboxycyanine dyes. These molecules have been spectroscopically studied by Sims et al. (47) and show environmental effects on their fluorescence quantum yields. Molar absorptivities range from 1.3 to 2.4×10^5 l/mole cm and emission wavelengths from 500 to 780 nm.

Fluorescent probes that exhibit proton transfer in the lowest excited singlet state may serve as the basis for another class of FIA labels. This

Table 8.1. Phycobiliprotein Spectral Data

Name	Source	Molecular Weight	Absorption Maxima (nm)	Molar Absorptivity ($lmol^{-1}cm^{-1}$)	Emission Maxima (nm)	Quantum Yield	Ref.
B-Phycoerythrin	Porphyridin	240,000	545,565	2,410,000	575	0.98	36, 37
B-Phycoerythrin	Gastroclonium	240,000	480,545,565	1,960,000	578	0.82	38, 39
C-Phycocyanin	Anabaena	224,000	620	1,690,000	650	0.51	40, 41
Allophycocyanine	Anabaena	104,000	650	700,000	660	0.68	
R-phycocyanine	Porphyridium	103,000	555,618	760,000	634	0.7	39, 42

phenomenon is one that is exhibited by aromatic species having acidic or basic functional groups directly attached to the aromatic ring (48–51). Because the lowest excited singlet state has an electronic distribution that is generally very different from the ground electronic state of the same molecule, the excited state may be much more or less acidic or basic than the ground state (48). This means that in, say, a phenolic molecule, where the excited state is more acid than the ground state, emission may be observed from the phenolate anion even at a pH value so low that the undissociated phenol is the sole absorbing (ground-state) species (49). Whether or not an excited acid or base will undergo excited-state proton transfer reactions depends on the rate constants for protonation and dissociation, the acidity of the solution, and the mechanism of proton transfer (49).

At near neutral pH, the probabilities of second-order proton transfer are negligible compared with the probabilities of radiative and nonradiative deactivation of the lowest excited singlet state (52–56). First-order dissociation of excited acids and protonation of excited bases, in which water is the proton donor or acceptor, however, can often compete with the radiative and nonradiative pathways for deactivation of the excited state and in this case fluorescence may be observed from both acid and conjugate base in different regions of the near UV and visible spectrum (52–55).

Of the molecules that demonstrate first-order excited-state proton transfer in water, at pH values of approximately 7, only the hydroxyaromatic and certain nitrogen heterocyclic molecules present a wide selection of possible probes with fluorescence intense enough and spectra of conjugate acid and base forms sufficiently separated in the spectrum to be of practical use.

The utility of fluorescent labels that undergo excited-state dissociation in aqueous media is based on the substantial separations of the fluorescence bands of acid and conjugate base forms of the free-labeled ligands and the likelihood that when the labeled ligands are bound to antibodies, excited-state proton transfer will not occur. The latter supposition is based on the observations of the fluorescences of antibody-bound fluorescent probes that have been studied in recent years and that usually show typical spectral characteristics of the probe in a low-dielectric, low-mobility environment. This has been postulated as the nature of the environment experienced by other protein-bound probes (57–58). These emissions are similar to the spectra observed when the same probes are bound to bovine and human serum albumins (57–58).

In addition to direct inhibition of proton transfer induced by antibody reaction with labeled hapten, enzyme-coupled schemes are possible. 7-

hydroxy-4-methylcoumarin (4-methylumbelliferone) and 7-hydroxycou-marin (umbelliferone) have been employed in substrated-labeled immu-noassay systems (59–61). Here glycosidic and phosphate ester bonds can couple the labels at the hydroxyl positions to the haptens. Free labeled species undergo enzymatic hydrolysis to form the free hydroxycoumarin species. This form of the label, then, undergoes excited-state proton trans-fer, yielding long-wavelength emission separate from that of the hydroxy-coupled label. Other hydroxy aromatics with larger Stokes' shifts that could be useful in similar analyses include 8-hydroxy-1,3,6-pyrenetrisul-fonic acid trisodium salt and 2-hydroxy-3-carboxy-8-naphthalene sulfonic acid that both have absorption maxima longer than 400 nm and strong emission beyond 500 nm.

A fluorogenic probe prepared by Thompson (62) might be usable in FIA systems. Nonfluorescent bromo derivatives of syn-9,10-dioxabimane (1,5-diazabicyclo-[3,3,0]octa-3,6-dione-2,8-dione) are capable of covalent reaction with thiol moieties to produce fluorescent products. Two im-munoassay schemes are postulated, the first involving the bromo-diox-abimane-labeled hapten. Here antibody–hapten interaction would allow exclusive thiol reaction with the free labeled hapten. Alternatively, thio-lated hapten could be employed with bromo-dioxabimane, free in solution, available for reaction with unbound, thiolated hapten.

Probes useful as substrates for enzyme immunoassays with fluorimetric detection are 4-methylumbelliferyl-β-galactoside (63-65) and fluorescein-di(β-D-galactopyranoside) (65). In the heterogeneous assays using these probes an antigen was labeled with β-D-galactosidase from *Escherichia coli*. Precipitated antibody-bound fractions were resuspended in solutions containing the substrate. The rate of production of fluorescent products was directly related to the concentration of the antigen.

Guilbault et al. showed that homovanillic acid (HVA), 3-methoxy-4-hydroxyphenylacetic acid (66), and p-hydroxyphenylacetic acid (67) are useful fluorogenic substrates for the enzyme peroxidase. Several per-oxidase-coupled enzyme immunoassays with fluorimetric detection have been shown to rank among the most sensitive FIA procedures developed.

8.4. HOMOGENEOUS FLUORESCENCE IMMUNOASSAY

Homogeneous assays involve direct measurement of the analyte in so-lution without separation of the labeled antigen–antibody complex from labeled antigen, free in solution. The elimination of the separation step, normally present in other immunoassays (e.g., RIA), represents one of the major advantages of fluorescence immunoassay and provides the opportunity for simple, fast, and reliable means of quantitation. The homogeneous approach has gained wide application in the field of clinical therapeutic drug analysis and several companies offer reagent kits and specifically designed instrumentation for this purpose. The quantitation of therapeutic drugs is ideally suited to homogeneous fluorescence immunoassay because many analytes are present in large enough quantities that background interference is not a significant problem. In addition, the technical simplicity and speed of these assays are necessary so that dosage adjustments can be made within a reasonable time frame.

The homogeneous procedure can be performed by virtue of the effect of antibody binding on the spectral characteristics of the labeled antigen. In the aqueous environment, the free labeled antigen will experience strong polarizing forces and perhaps hydrogen-bonding forces at acidic or basic functional groups as a result of interactions with water molecules. These forces will be exerted to different degrees in the ground and excited states of the labeled antigen because the ground and excited states have different dipole moments. Also, certain functional groups on the fluorophore will be free to rotate subsequent to the fluorescent transition in water. On the other hand, in the hydrophobic environment of an antibody-binding site, the dielectric strength is low and rotation of functional groups on the fluorophores is severely restricted. The weakly interacting solvent cage and restricted rotational freedom of the antibody-bound antigen usually cause this species to fluoresce at shorter wavelengths than the free labeled antigen because the relative stabilization of the excited state of the latter by strong electrostatic interactions is much greater than that in the bound labeled antigen. In addition, the weak solvation and restricted rotational freedom of the antibody-bound probe also cause the bound probe to be somewhat shielded from internal and external conversion that compete with fluorescence for deactivation of the excited state. This often results in higher fluorescence yields for the bound probe than for the free probe. If the fluorescent emission spectrum of the bound labeled ligand is altered sufficiently from that of the free labeled ligand, the resulting spectroscopic measurements can be used for quantitation without a separation step. In essence, a spectroscopic separation rather than a chemical separation is employed.

There are currently three major types of homogeneous fluorescence immunoassays that are well developed. There categories are quenching and enhancement, fluorescence polarization, and substrate-labeled fluorescence immunoassay. Each technique will be discussed. In addition, a current listing of published methodologies is presented (Table 8.2).

Table 8.2. Homogeneous LIA

Analyte	Method	Sensitivity	Ref.
Albumin	Excitation energy transfer	1.0 nM	82
	Fluorescence protection	*	6
Amikacin	Fluorescence polarization	0.18 µg/ml	109
	Substrate-labeled	2.0 µg/ml	6
2-Aminobenzimidazole	Fluorescence polarization	1.0 nM	101
Antinucleotide antibody	Fluorescence enhancement	*	92
Biotin	Substrate-labeled	10 nM	119, 120
Carbamazepine	Substrate-labeled	1.0 µg/ml	114
	Fluorescence polarization	0.5 µg/ml	104
Cortisol	Substrate-labeled	*	92
	Fluorescence quenching	0.031 µg/ml	74
	Fluorescence polarization	0.015 µg/ml	106
Desmethylnortriptyline	Excitation energy transfer	0.05 ng/ml	84
Dinitrophenol	Fluorescence polarization	100 nM	5
Diquat	Fluorescence polarization	0.2 nM	5
Digoxin	Fluorescence polarization	0.02 ng/ml	5
Diethylstilbestrol	Fluorescence polarization	100 µg/ml	5
2,4-Dinitrofluorobenzene	Fluorescence polarization	10 nM	119, 120
Gentamicin	Substrate-labeled	1.0 µg/ml	112
	Fluorescence quenching	1.0 µg/ml	73
	Fluorescence polarization	1.0 µg/ml	108
	Fluorescence polarization	0.16 µg/ml	109
Hexachlorophene	Fluorescence polarization	20 µg/ml	5
Human chorionic gonadotropin	Fluorescence polarization	1.0 ng/ml	98, 99
Hydrotyprogesterone	Substrate-labeled	*	122
	Fluorescence protection	0.033 nM	76
IgG	Excitation energy transfer	0.1 nM	80
	Substrate-labeled	1.0 µg/ml	121

Analyte	Method	Sensitivity	Ref.
Insulin	Fluorescence polarization	10 nM	100
Kanamycin	Fluorescence polarization	0.44 µg/ml	104
Lidocaine	Fluorescence polarization	0.1 µg/ml	104
Morphine	Internal reflectance (solid-phase Ag)	20 nM	158
	Excitation energy transfer	0.1 nM	81
Netilmicin	Fluorescence polarization	0.09 µg/ml	104
Nortriptyline	Excitation energy transfer	*	84
Phenobarbital	Substrate-labeled	3.0 µg/ml	115
	Substrate-labeled	2.7 µg/ml	173
Phenytoin	Fluorescence polarization	0.5 µg/ml	105
Free phenytoin	Fluorescence polarization	0.5 µg/ml	108
Phenytoin	Fluorescence polarization	0.02 µg/ml	104
	Substrate-labeled	1.0 µg/ml	116
	Substrate labeled	1.6 µg/ml	173
Primidone	Substrate labeled	1.0 µg/ml	116
	Fluorescence polarization	0.2 µg/ml	104
Quinidine	Substrate-labeled	1.0 µg/ml	117
Theophylline	Substrate-labeled	2.0 µg/ml	113
	Fluorescence polarization	0.3 µg/ml	103
	Substrate-labeled (solid-phase dry reagents)	*	157
Thyroxine	Fluorescence enhancement	3 nM	91
	Indirect quenching	*	80
Thyroxine binding globulin	Excitation energy transfer	5.0 µg/ml	83
Tobramycin	Substrate-labeled	1.0 µg/ml	118
	Fluorescence polarization	0.18 µg/ml	109
Trypsin	Fluorescence polarization	*	100
Valproic Acid	Fluorescence polarization	0.7 µg/ml	104
Vancomycin	Fluorescence polarization	0.60 µg/ml	104

Note: *Sensitivity not reported.

8.4.1. Quenching and Enhancement FIA

Fluorescence quenching methods are based on the observed reduction in fluorescence intensity of a labeled antigen subsequent to its complexation by an antibody. The mechanisms of fluorescence quenching are not completely understood, but in the case of direct quenching by the antibody molecule it is generally understood that alteration of the electronic structure and perhaps the vibrational composition of the electronic states of the labeled antigen are involved (72). The alteration of the electronic distribution of the bound labeled ligand can potentially enhance the probability of radiationless deactivation of the first excited singlet state. A possible explanation for this is the vibrational coupling of the bound labeled ligand with the normal vibrational modes of the protein, a process that could enhance the rate of internal conversion to the ground state. Since antigen–antibody binding requires significant electrostatic interactions for stable bond formation, some degree of quenching is usually observed. If the extent of quenching of the bound labeled ligand is great enough, the competition of unlabeled antigen with the labeled material, for antibody binding sites, can be measured and related to the concentration of analyte by means of a standard curve.

The clinical utility of a direct quenching assay in which a labeled antigen's fluorescence was quenched by antibody binding was first exemplified by a study of serum gentamicin levels (73) and later by a steroid quantitation (74). A 25% decrease in the fluorescence signal of fluorescein-labeled gentamicin on antibody binding was observed, allowing the construction of a standard curve varying from 0.25 to 32 mg/liter of unlabeled gentamicin. The precision, accuracy, specificity, and patient sample correlation studies of this method all demonstrated acceptable results, although three potential sources of inaccuracy were identified. One of the main problems with the assay was that small changes in fluorescence intensity corresponded to relatively large changes in the free antigen concentration. The fluorescence intensity differences were comparable to the intrinsic fluorescence of the blank for serum samples. Other possible sources of error relate to the intrinsic absorbance and fluorescence of the serum matrix. The absorption of light will diminish the intensities of both the excitation and emission beams as a result of the inner filter effect (75). This becomes a serious problem with samples that are colored or excessively turbid. Background fluorescence has frequently been a problem with clinical fluorimetric assay, especially in cases where the patient suffers from renal impairment. The excitation wavelength of 495 nm used in the assay of gentamicin is longer than that commonly used and less likely to excite intrinsic fluorophores.

Direct quenching assays were given much attention during the initial stages of fluorescence immunoassay development, although there have been no commercial applications to date. There are also several novel variations of direct quenching that have been investigated in attempts to enhance the applicability of this method.

Since no separation step is required for homogeneous assays, the fluorescences of labeled impurities would contribute to the background. Therefore, very pure labeled antigens or very specific antibodies are required. Consequently, a novel approach to the quenching assay was developed and was called the fluorescence protection immunoassay or "indirect quenching assay" (76–79). This assay requires antibodies directed toward both the antigen and the antigen-labeled fluorescence probe. When the antibody directed toward a particular antigen is bound to a labeled antigen, subsequent binding of an antiprobe antibody to the label is sterically inhibited. This antiprobe antibody binding causes a decrease in fluorescence intensity of the probe, which is prevented by binding of the first antibody. Therefore, the first antibody effectively protects the labeled antigen from the antiprobe antibody. If the first antibody is bound to the unlabeled antigen, however, it cannot bind to the labeled antigen. Therefore, there is a competition between labeled and unlabeled antigen that provides the analytical basis of this technique. Because of the use of antiprobe antibodies that are not dependent on analyte conjugation, the purity requirements for labeled reagents can be substantially reduced (78). Measurement of human immunoglobulin G (IgG) by this technique correlated well with the results of radial immunodiffusion methods (76). Also, the authors reported that analysis of low-molecular-mass analytes was potentially feasible.

A variation of the indirect quenching method was developed by Hassan, Landon, and Smith (80). In this assay, anti-immunoglobulin G was used in addition to antiprobe and antiligand antibodies, which could bind to either the label or the analyte portions of a labeled conjugate, but not to both at the same time. Antiprobe binding greatly reduced fluorescence, whereas antiligand binding did not. Therefore, the unlabeled analyte can compete with the labeled conjugate only for antiligand sites, leaving more labeled conjugate to be bound to antiprobe antibodies. As a result, fluorescence is decreased with increasing amounts of unlabeled analyte and serves as the basis for quantitation.

A unique quenching assay has been developed for the case where the quenching compound is not the antibody itself but rather a chemical quencher attached to the antibody (81–84). The quenching mechanism is resonance energy transfer, which is characterized by sensitized fluorescence of a suitable acceptor molecule (the quencher), whose electronic excitation is effected by resonance transfer from a sensitizer, or donor molecule. The donor molecule's emission, as a result, is quenched. The efficiency of the energy transfer process is not simply related to the fluorescence quantum yield of the sensitizer molecule, because the transfer is a nonradiative process and competes not only with the fluorescence of the donor but also with the other processes that lead to deactivation or quenching of the donor excited state. In fact, the rate of resonance energy transfer has been found to be related to the overlap of the donor fluorescence and acceptor absorption spectra, to the relative orientations of donor and acceptor and to the inverse sixth power of the distance between the two molecules.

An explanation of this phenomenon in terms of classical physics involves consideration of the optical electron of the sensitizer molecule as a vibrating point charge, or oscillator (85). The ground-state electron vibrates with a characteristic frequency that is increased on excitation. The extra energy of the oscillator may be damped by the emission of a quantum of electromagnetic radiation, or if another oscillator is nearby, the energy of the first oscillator may be transferred to the second, in a resonance-damping process. Because the interaction energy of resonance damping is of a dipole–dipole nature, it decreases in proportion to the third power of the distance between the two oscillators. Resonance damping is effective over relatively long distances by comparison with other types of electrical interactions.

The rate equation for resonance transfer, developed by Förster (87–88), is commonly written as (89).

$$k_{rt} = \frac{1}{\tau_s} \left(\frac{R_0}{R} \right)^6 \qquad (8.1)$$

where k_{rt} is the rate constant for resonance energy transfer, τ_s is the lifetime of the sensitizer, R is the actual distance between the two molecules, and R_0 is that distance between the two molecules where 50% of all the excitation energy is transferred to the donor molecule. This function involves the inverse sixth power of the donor acceptor distance because the interaction energy of two dipoles is proportional to the inverse third power of the distance between them, and the rate constant describes the probability of that interaction, which is proportional to the square of the interaction energy.

Resonance energy transfer immunoassays were originally developed by Ullman et al. (81). The fluorophores fluorescein (donor) and tetramethylrhodamine (acceptor) were chosen, since the emission spectrum of fluorescein greatly overlaps the excitation spectrum of tetramethylrhodamine. Two variations of this approach were investigated and applied to the assay of morphine. In the so-called direct method, the antigen is labeled with fluorescein and the antibody is labeled with rhodamine. Labeled and unlabeled antigen are allowed to compete for reaction with labeled antibody as in reaction (8.2):

$$Ag + Ab^R \rightleftharpoons Ag - Ab^R \tag{8.2}$$
$$Ag^F + Ag - Ab^R \rightleftharpoons Ag^F - Ab^R + Ag$$

The fluorescein-labeled antigen (Ag^F) can transfer excitation energy to rhodamine-labeled antibody (Ab^R), resulting in quenching of the fluorescein fluorescence. From equilibrium considerations it is obvious that an increase of unlabeled antigen concentration results in less binding of the labeled antigen to the antibody and the quenching effect on fluorescein will be decreased as the unlabeled antigen (the analyte) concentration is increased. In the second approach, one half of the antibody population is labeled with fluorescein and the other half is labeled with rhodamine. The antigen remains unlabeled but with a multivalent ligand and a heterogeneous antibody population complexes of the antigen with both types of labeled antibody can be found, as in reaction (8.3):

$$Ag + Ab^F + Ab^R \rightleftharpoons Ab^R - Ag - Ab^F \tag{8.3}$$

The close proximity of the two labels in the complex is enough to increase the quenching effect significantly. However, in this sandwich technique, the higher the antigen concentration, the greater the quenching effect, since more complexes would be formed. Both methods exhibited sensitivities in the nanomolar range; however, routine use of these procedures is limited by background interference and by the complexity of dual labeling requirements (90).

A homogeneous assay closely related to the quenching assays is the fluorescence enhancement assay (91–92). Fluorescence enhancement is actually an exception to the quenching phenomenon where the relative fluorescence quantum yield of the label is increased on antibody binding. This can occur when the fluorescence of the label is quenched on conjugation to the antigen and subsequent antibody binding of the conjugated antigen enhances the fluorescence of the label. This is found to be the case when the fluorescein-labeled thyroxine binds to antithyroxine antibodies (91). The exact mechanism of fluorescence enhancement is not known, but it is thought that intramolecular quenching occurs when the fluorescein label is coupled to thyroxine. This quenching effect is probably brought about either by collisional deactivation or a heavy atom effect

(93) attributed to the four iodine atoms present on the thyroxine molecule. An approximately 90% decrease of fluorescein fluorescence is observed when it is conjugated to the thyroxine molecule. A 3.9-fold increase in relative fluorescence intensity of the conjugate occurs when it is bound to the antithyroxine antibody. A possible explanation of this is that the heavy atom effect of iodine atoms is weakened by conformational changes in the fluorescein thyroxine conjugate that occur as a result of antibody binding. Therefore, as unlabeled thyroxine competes with labeled thyroxine for antiserum binding sites, a decrease in fluorescence is observed with increasing concentrations of unlabeled thyroxine. This method was found to be much less sensitive than typical RIA procedures and the major problem, as with the quenching assays, is that small changes in fluorescence are measured relative to a rather large background signal.

8.4.2. Fluorescence Polarization Immunoassay

Fluorescence polarization methods provide valuable information on the interaction of antigens or haptens with antibodies. The use of fluorescence polarization in the study of macromolecules was pioneered by Weber in 1952 (94) and has since become a major tool in structural investigations of the binding of small molecules to proteins.

The physical principle underlying fluorescence polarization immunoassay involves the selective elimination of light waves whose electric vectors do not all lie in a single plane. This is accomplished by passing the exciting light through a polarizing filter. The resulting polarized radiation will selectively excite those molecules whose absorption transition moments have a significant component in the plane of the electrical vector of the exciting beam (95). As a result, molecules excited with polarized light will emit radiation that is polarized in the same direction as the exciting light, to a degree inversely related to the amount of Brownian rotation occurring during the interval between absorption and emission of light (96). This means that the photoselected molecules originally excited by polarized light and having fairly small volumes (i.e., free labeled antigen) will have a rotational relaxation time much shorter than their fluorescence decay times and will become completely randomized before fluorescing. They therefore will display very little polarized fluorescence. However, photoselected molecules having very large volumes, such as the antibody proteins and their complexes, will rotate at a rate comparable to or slower than the rate at which they fluoresce. Consequently, randomization of the fluorescent transition moments will not occur in these large molecules and substantial fluorescence polarization will be observed.

On the binding of a fluorophore-labeled antigen to an antibody there will be a reduction in the rotational Brownian motion as well as an increase in overall effective size of the fluorescent label. This will result in alteration of the polarization of the fluorescence along or perpendicular to the optical axis of the excitation polarizer, depending on whether the fluorescence transition moment of the molecule is oriented closer to 0° or 90° to the transition moment associated with the absorption band excited. Consider first the case where the transition moments for excitation and fluorescence are parallel (or nearly so).

If a second polarizing film (emission polarizer) is placed between the fluorescing sample and the photodetector of the fluorimeter, with its optical axis perpendicular to that of the polarizing film between the lamp and the sample, the highly polarized fluorescence from the antibody-labeled ligand will be filtered to a much greater extent than would be the unpolarized fluorescence from the same concentration of free labeled ligand excited under the same conditions. If the optical axes of both polarizers are parallel and the excitation and emission moments of the fluorophore are parallel, or nearly so, the emission polarizer will pass relatively more radiation from the bound labeled ligand than from the free labeled ligand to the detector. This occurs because the polarized emission will be concentrated along the optical axis of each polarizer while the unpolarized emission will be dispersed over all angles to the optical axis of the emission polarizer and some will, therefore, be filtered. Regardless of the orientation of the optical axis of the second polarizer with respect to the first, the fluorescence intensity registered by the detector should, ideally, be the same for unpolarized fluorescence (i.e., that of the free labeled ligand) while in the case of parallel absorption and fluorescence transition moments, the intensity of the polarized fluorescence (from the bound labeled ligand) measured when the optical axes of the polarizers are parallel ($F_\parallel$) is greater than when the optical axis of the polarizers are perpendicular ($F_\perp$). In the case of perpendicular absorption and fluorescence transition moments, $F_\parallel = F_\perp$ also, for the free labeled ligand (unpolarized fluorescence) and $F_\parallel$ is less than $F_\perp$ for the bound labeled ligand (polarized fluorescence). The degree of polarization is defined as:

$$P = \frac{F_\parallel - F_\perp}{F_\perp + F_\parallel} \qquad (8.4)$$

Ideally, P at any given excitation wavelength should be a property of a pure species and should be invariant with respect to concentration. For a free unlabeled ligand, $F_{\parallel} = F_{\perp}$, so that $P = 0$ at all excitation wavelengths. For a bound labeled ligand it is possible to have P take values between $+\frac{1}{2}$ and $-\frac{1}{3}$, depending on the wavelength of excitation. If a system is contrived that contains antibody-bound labeled ligands that are displaced, the relative increase in unpolarized fluorescence and decrease in polarized fluorescence from the solution will cause a net decrease in P as calculated from its operational definition in Eq. (8.4). If all the labeled ligand was ultimately displaced from the antibody complex, P would fall to zero. The fact that, depending on the extent of the binding of the labeled ligand, the degree of polarization varies between some nonzero value and zero, permits the construction of a standard curve relating the measured polarization to unlabeled ligand concentrations and permits the execution of a homogeneous immunoassay (97,98).

Fluorescence polarization immunoassay is subject to intrinsic limitations. Since two filters are used, the total signal is correspondingly reduced and therefore detection limits are relatively high. Calibration curves are limited to narrow concentration ranges (72), and nonspecific binding of fluorophores by variable amounts of serum proteins can cause serious errors (19). Advantages include the universal applicability of the method as any drugs present in sufficient serum concentrations should be measurable by polarization immunoassay.

Prior to the current resurgence of the method, use on a routine basis had been limited by the complexity and cost of the instrumentation required. Today there exist a number of polarization methods (99–111) and commercial development has proceeded at a rapid rate. Recent instrumental developments have allowed fluorescence polarization measurements to be made with a minimum of technical involvement by the operator. Spencer and co-workers (100) designed and constructed an automated flow cell polarization fluorimeter and demonstrated the clinical utility of the system with an assay of insulin. Insulin concentrations on the order of $10^{-8}\ M$ were measured and the assay demonstrated a degree of sensitivity necessary for the analysis of many clinically important analytes. Abbott Laboratories has since developed an automated fluorescence polarization analyzer and demonstrated its utility for quantitation of several therapeutic drugs (109–111). The most unique feature of this instrument is the use of transmission-type liquid crystals that serve to rotate the plane of polarization. This allows polarization measurements to be made without the use of moving parts necessary to rotate films in conventional instruments. The Abbott analyzer has had a large impact on the routine clinical analysis of therapeutic drugs and shows promise in replacing many of the current methods. Many types of reagent kits are now available and development has proceeded at a rapid pace because of the relative simplicity of the reagent systems involved.

8.4.3. Substrate-Labeled Fluorescence Immunoassay

A number of homogeneous fluorescence immunoassays have been developed that make use of labels that are dependent on enzymatic reactions (112–122). The most common technique is based on the quantitation of a fluorescent product resulting from the enzymatic cleavage of the unbound labeled antigen. Thus, the name *reactant* or *substrate labeled fluorescence immunoassay* has been applied. Generally, these methods require an antigen capable of conjugation to a nonfluorescent probe capable of forming a fluorescent product on enzymatic hydrolysis. This conjugate, Ag^{F-G}, can then undergo enzymatic cleavage to form the fluorescent product Ag^{FH}, as illustrated in reaction (8.5).

$$Ag^{F-G} \xrightarrow[\text{H}_2\text{O}]{\text{enzyme}} Ag^{FH} + GOH \qquad (8.5)$$

When such a conjugate reacts with an antibody directed toward the antigen in question, Ab, a complex is formed that precludes hydrolysis of the ligand substrate. This is illustrated in reaction (8.6).

$$Ag^{F-G} + Ab \rightleftharpoons Ag^{F-G} Ab \xrightarrow[\text{H}_2\text{O}]{\text{enzyme}} \text{No reaction} \qquad (8.6)$$

In the competitive binding reaction where unlabeled antigen is present and competes with the conjugate for antibody binding sites, the rate of release of the fluorescent product precursor (AG^{F-G}) is proportional to the unlabeled antigen concentration.

$$Ag^{F-G} - Ab + Ag \rightleftharpoons Ag - Ab + Ag^{F-G} \qquad (8.7)$$

Typically, the fluorescent probe is convalently linked to the antigen and to either a phosphate or a galactosyl bond that will be cleaved by enzymatic action. Some assays have been developed in which enzymatic action enhances antibody binding to the substrate-labeled conjugate, although this is a rare exception (122).

Reactant-labeled fluorescence immunoassay has been applied to the determination of a number of clinically important analytes, including many therapuetic drugs (112–122). Some of these assays utilize drugs labeled with the fluorogenic substrate umbelliferyl-β-galactoside and the *Escherichia coli*-derived enzyme β-galactosidase (112). The precision and accuracy of these methods are generally comparable to those observed for existing radioimmunoassay procedures. For example, the sensitivity attained in an assay of gentamicin was 1.0 μg/ml.

The use of enzymes as reagents or labels in fluorescence immunoassay presents a unique set of advantages and drawbacks. If the human serum matrix contains significant endogenous enzymatic activity similar to that of the assay enzyme, this could interfere to an extent depending on the enzyme concentration. Also enzymatic methods generally require special attention with respect to temperature, pH, and timing of the reaction. The use of bacteria-derived β-galactosidase has the advantage of reacting under conditions in which the mammalian enzyme is inactive. These measurements are not based on systems in equilibrium, as in other homogeneous procedures. The kinetic approach requires strict timing of reagent addition and duration of the reaction. The procedural difficulties could be minimized by the use of automation and rate measurements rather than end-point measurements. The relatively short wavelengths (typically, 400 nm) used to excite the enzymatically cleaved 7-hydroxy-coumarin label also excite endogenous fluorescent substances in serum. This background fluorescence reduces assay reliability even when rate measurements are employed. Rate measurements for these assays have not been used extensively in clinical situations, probably because they are more difficult to deal with instrumentally. Substrate labeled fluorescence immunoassays combine all the advantages of homogeneous immunoassays, along with the benefit of providing a nearly linear standard curve.

Although there are distinct analytical advantages to a homogeneous immunoassay and several successful assays have been developed, they all share the potentially serious problem of background fluorescence interference. Samples that contain serum proteins and antisera have a fairly strong background emission because of aromatic amino acid residues, in addition to significant Raman and Rayleigh scattering. These scattering effects are due predominantly to endogenous proteins, in relatively high concentration, which can interfere with the measurement of fluorescent signals from fluorophores possessing small Stokes' shifts.

Proteins containing aromatic amino acid residues (i.e., phenylalanine, tyrosine, and tryptophan) absorb strongly at approximately 260 nm, corresponding to the excitation of the benzenoid species. Subsequent, highly efficient resonance energy transfer depletes the excited electronic states of tyrosine and phenylalanine. Nearly exclusive fluorescence emission from the excited singlet state of tryptophan is centered at about 340 nm and tails out considerably in the direction of longer wavelengths. High protein concentration additionally may result in nonlinearity of analyte and background fluorescence. This results from either high absorbance or absorption of the emitted radiation by the matrix and analyte.

Light scattering can create several analytical problems, among which are (1) direct scattering of excitation light into the photomultiplier tube. Blank subtractions as well as the use of probes with large Stokes' shifts can correct for this. Direct scattering in polarization measurements is filter angle-dependent and each measurement should be independently corrected. (2) Scattering of emitted and exciting light at high total absorbances, resulting in lowered excitation and emission intensity. Use of narrow pathlength cells or dilution of the analytical sample can solve this. (3) Loss of fluorescence polarization (FP) as a result of light scattering; FP decreases of up to 30% have been observed. Faucan et al. employed unreactive fluorophore–macromolecule conjugates as internal standards to measure nonspecific scattering effects for correction purposes (4).

Homogeneous assays are much easier than heterogeneous assays to automate because in the absence of a separation step, reagent addition and detection are usually the only steps required. Also, conventional automation is based predominantly on spectroscopic rather than isotopic detection, making adaptation of conventional fluorimetric automated instrumentation simpler. Some FIA procedures lend themselves very well to continuous flow analysis, for example.

There are some drawbacks to the use of fluorescent labels that are not experienced with radioactive labels. The problem of background interference is probably the most significant problem with fluorescence immunoassay. Although considerable amounts of interfering substances can be removed by heterogeneous procedures, there is always some residual interference. There are, however, several sophisticated procedures that can be used to overcome this problem that will be discussed in detail subsequently.

8.5. HETEROGENEOUS FLUORESCENCE IMMUNOASSAY

Although much research effort has gone into the development of homogeneous systems because of their technical simplicity, all homogeneous assays share the problem of endogenous background interference. Therefore, some type of separation of antibody bound and free antigen is often necessary to achieve limits of detection commonly attained in radioimmunoassay procedures. Heterogeneous fluorescence immunoassay separations have been accomplished through a variety of techniques, many of which are commonly used in similar radioimmunoassay procedures. These separations can be grouped into three major categories:

solid-phase adsorption, precipitation, and chromatography. Precipitation and solid-phase adsorption have both been widely used in radioimmunoassay methods (124), although chromatographic separations have generally not been developed for this purpose. Phase separations such as liquid–liquid extraction have also been used in RIA but have not been widely applied to fluorescence immunoassay techniques. The various separation techniques will be discussed with emphasis on those that have been developed more extensively.

8.5.1. Solid-Phase Adsorption Fluorescence Immunoassay

Solid-phase separation techniques have received by far the most attention of the three general classifications. In this approach, fluorescent-labeled antigen (125–127) or antibody (128–134) is separated from interferences by attachment to a solid surface such as a paper disc, glass or plastic beads, the walls of a test tube, particles, and so on (135). The antigen or antibody can be either adsorbed or covalently linked to the solid surface. Solid-phase fluorescence immunoassay techniques can be developed either in a competitive or noncompetitive format, and the method selected is usually dependent on the degree of sensitivity required. Generally speaking, competitive binding assays are more applicable to low-molecular-weight analytes (haptens), which are typically univalent (i.e., those haptens that form $1:1$ complexes with antibodies). More sensitive noncompetitive methods (e.g., sandwich techniques) are more useful for macromolecules (136) since these require at least two sufficiently spatially separated antibody combining sites. Competitive binding assays do not require bivalent antigens and can be used for haptens with only one antibody combining site. Once the solid-phase separation has been achieved, the fluorescent label may then be stripped from the solid phase by a denaturant such as dilute alkali prior to fluorimetric measurements or fluorescence measurements may be made directly on the label attached to the solid phase. In the latter case, most background interference can be avoided by the inclusion of a washing step in which free labeled reagents and nonspecific serum compounds are removed from the assay mixture prior to the analytical measurement.

Commercial heterogeneous FIA methods have been developed for a number of analytes using principles similar to those for RIA (137). In one variation, the antigen is incubated with excess labeled antibody (Ab^F). The unreacted labeled antibody is then allowed to combine with the antigen immobilized on a solid support (Ag^S) and fluorescence is measured on that solid support with a front surface fluorimeter (138). The fluorescence intensity measured from the solid support is inversely proportional to the amount of antigen present in solution and is illustrated in the following reaction sequence:

$$Ag + Ab^F \rightleftharpoons Ag - Ab^F + Ab^F \qquad (8.8)$$

unreacted $[Ab^F]$ is inversely proportional to $[Ag]$

$$Ab^F + Ag^S \rightleftharpoons Ab^F - Ag^S \qquad (8.9)$$

Solid surface antibodies have also been used in sandwich techniques (138) and the choice of method depends on the size of the analyte. The solid surface employed is in the form of a dipstick sampler that is allowed to interact with sample and reagents. Usually the samplers with solid-phase (typically cellulose nitrate or aluminized polyester) antigen or antibody attached are dipped into a tube containing a sample that has been diluted with a buffer solution. The appropriate antigen–antibody reaction takes place and the sample is then washed with buffer. The fluorescent reagent is allowed to react similarly and the sample is then washed a second time, after which fluorescence is measured. The fluorescence measurement is made with a front surface fluorimeter.

Another method makes use of antibodies covalently attached to polyacrylamide beads that are suspended in the assay medium (139). The separation from interferences is achieved through centrifugation and washing. Fluorescence intensity could be measured in a conventional fluorimeter, although a semiautomated system is available in which much greater sensitivity is attainable (139). The greater sensitivity of the instrument is due to three features: a feedback stabilization source network, interference filters to minimize scatter, and a photon counting detection system. The level of detectability for this instrument is in the subpicomolar range for fluorescein, and picomolar levels of dye can be measured with accuracy and precision of 1–3%. Reagent systems used in this approach include both sandwich and competitive binding techniques illustrated by the reaction schemes represented in Eqs. (8.10) and (8.11).

$$Ab^F + Ag + Ab^S \rightleftharpoons Ab^F - Ag - Ab^S \qquad (8.10)$$

$$Ag^F + Ag + Ab^S \rightleftharpoons Ag^F - Ab^S + Ag - Ab^S \qquad (8.11)$$

Reaction (8.10) represents the sandwich technique in which a complex of labeled antibody, solid phase antibody, and free antigen is formed. Since the free antigen is the analyte, fluorescence intensity measured from the solid surface is proportional to analyte concentration. An obvious restriction is that the antigen be at least bivalent. The newer competitive inhibition system shown in reaction (8.11) was found to be more appropriate for smaller haptens such as most drug molecules. In this assay, however, a fluorophore-labeled antigen competes with endogenous un-labeled analyte for the solid-phase antibody; therefore, the fluorescence measured from the solid surface is inversely proportional to analyte concentration. The assay sensitivity of this system is exemplified by the thyroxine analysis in which the limit of detection was found to be 0.3 nM.

The solid-phase assays have a capability for sensitivity that is limited only by the fluorimeter design, the antibody-binding constants, and background fluorescence attributed to the solid support (19). Labeled nonspecific impurities in these methods are removed largely by the washing step so that the purity requirements for labeled antigen are not demanding. These washing steps make the analytical procedure more complex and limit the automation capability of the method by comparison with homogeneous procedures. The technical complexity of these procedures is also shared with RIA and a trade-off exists between sensitivity and simplicity. This trade-off must therefore be balanced with the sensitivity needed. Most drugs exist in blood serum in the microgram per milliliter range and can be assayed by more convenient and cost-effective homogeneous procedures. There are some drugs and numerous biological compounds that probably cannot be quantitated by homogeneous procedures. Therein lie the values of the heterogeneous techniques. Some specific applications of the techniques discussed here are listed in Table 8.3.

Methods in which the fluorescent probe is stripped from the solid support before measurement have not been extensively employed. This is probably so because another analytical step would be required in an already complex analytical system. Dilute sodium hydroxide has been used to elute specifically bound fluorescein from paper discs in a noncompetitive assay for human IgG (140). Another problem associated with this procedure is that pH adjustment is usually required to achieve maximum fluorescence, adding yet another analytical step.

Solid surface FIA has also been used in conjunction with enzyme methods in a number of applications (144–147). Zare and co-workers developed an assay for insulin in which the enzyme horseradish peroxidase was used as an antibody label (141). The method was a sandwich technique with unlabeled antibody bound to the solid surface. The antigen insulin was allowed to react first with solid surface antibody and then with enzyme labeled antibody. The amount of solid phase bound enzyme is, therefore, directly related to the original insulin concentration. Label quantitation was achieved through the solid phase enzyme conversion of the nonfluo-

Table 8.3. Heterogeneous LIA

Analyte	Method	Sensitivity	Ref.
Albumin	Magnetizable solid phase Ab	*	152
Alpha-fetoprotein	Solid phase Ab	25 ng/ml	133
Amikacin	Double antibody	0.2 µg/ml	167
Anti-nuclear antibody	Solid phase Ag	*	6
Chenodeoxycholic acid	Magnetizable solid phase Ab	*	153
Complement C_3	Solid phase Ab	125 ng/ml	131
Complement C_4	Solid phase Ab	150 ng/ml	132
Cortisol	Magnetizable solid phase Ab	*	150
Cytomegalovirus (CMV)	Solid phase Ag	*	127
Digitoxin	Magnetizable solid phase	1.2 ng/ml	154
Digoxin	Radial partition	*	142
Entamoeba histolytica	Solid phase Ag	*	126
Estradiol	Solid phase Ab Enzyme labeled Ag	50 pg/tube	68
Gentamicin	Solid phase Ab	0.2 µg/ml	6
	Double antibody	0.5 µg/ml	167
	Double antibody	*	170
Herpes simplex antibody	Solid phase Ag	*	127
IgA	Solid phase Ab	0.6 µg/ml	128
IgE	Solid phase Ab	200 µg/ml	129
IgG	Solid phase Ab	60 µg/ml	128, 130
	Enzyme labeled Ag	5×10^{-15} moles	65
IgM	Solid phase Ab	4 mg/liter	128
Insulin	Solid phase Ab Enzyme labeled Ab Chromatographic separation	0.046 µg/ml	141
	Laser detection	5×10^{-15} moles	64
	Enzyme labeled Ag	20 pg	255
	Enzyme labeled Ag Enzyme labeled Ag	*	63
Insulin	Double antibody	2.5 µU/ml	147
Lidocaine	Radial partition	0.75 µg/ml	146
N-acetylprocainamide	Magnetizable solid phase Ab	*	151

Table 8.3. (*continued*)

Analyte	Method	Sensitivity	Ref.
Phenobarbital	Solid phase Ab	0.9 μg/ml	6
	Radial partition	5.0 μg/ml	143
Phospholipase A$_2$	Solid phase Ab time-resolved fluorescence	0.2 ng/ml	269
Phenytoin	Double antibody	0.3 μg/ml	167
	Solid phase Ab	0.65 μg/ml	6
Procainamide	Magnetizable solid phase Ab	*	151
Progesterone	Magnetizable solid phase Ab	*	168
	Precipitation with ammonium sulfate	*	168
Quinidine	Radial partition	0.75 μg/ml	144
Rheumatoid factor	Solid phase Ag	*	6
Rubella antibody	Solid phase Ag	*	6
Theophylline	Double antibody	0.2 μg/ml	167
	Solid phase Ab	0.7 μg/ml	6
	Radial partition	*	143
Thyroid stimulating hormone (TSH)	Enzyme labeled Ag	0.06 μU/tube	256
Thyroxine	Magnetizable solid phase Ab	*	155
	Sequential addition Solid phase Ab	3 n*M*	139
Tobramycin	Double antibody	0.6 μg/ml	167
	Double antibody	*	169
	Solid phase Ab	0.9 μg/ml	134
Toxoplasma G antibody	Solid phase Ag	1:16 titer	125

Note: *, Sensitivity not reported.

rescent substrate *p*-hydroxyphenylacetic acid to a fluorescent product. This assay combines the benefits of solid-phase separation and the advantage of enzyme multiplication. The enzyme multiplication phenomenon is based on the fact that enzymes are catalytic reagents and one enzyme molecule can convert many substrate molecules into signal-producing end products. Thus, a greater amount of fluorescence can be produced per labeled molecule. This chemical amplification of the signal leads to a much greater potential for increased sensitivity by comparison with enzymatic methods in which the substrate is the label. Enzyme multiplication methods have been widely used in EMIT[TM] systems (148). In the EMIT analysis, enzymatic activity has been monitored by absorption spectroscopy, although fluorescence measurements are inherently more sensitive. In Zare's work the fluorescent product was quantitated by a liquid chromatograph equipped with a flowing droplet laser fluorimeter detector (149). The technique of enzyme multiplication, solid-phase separation and fluorescence detection will probably prove to be among the most sensitive of FIA procedures. The limit of detection for the insulin assay was 7.9 pM, making it one of the most sensitive FIA procedures published to date.

Enzyme-labeled conjugates have been developed for commercial use very recently in conjunction with a fluorimetric enzyme immunoassay system (142–146) using competitive binding or sequential saturation methodologies. In sequential saturation, solid-phase antibody is allowed to react with antigen present in the sample. Enzyme-labeled antigen is then added and reacts with the remaining antibody sites. The unbound label is washed away with substrate containing buffer and enzyme activity is measured by front surface fluorimetry. Competitive binding assays are carried out similarly except that the sample and enzyme-labeled antigen are premixed and compete for solid-phase antibody when deposited on the surface. The solid surface consists of glass filter paper in the form of a test tab and the enzyme system used is alkaline phosphatase with 4-methylumbelliferyl phosphate as a substrate. A necessary condition for the utility of these systems is that antibody binding does not inhibit enzyme activity. It was found that if the bridge between the enzyme and antigen was dissimilar to that of the immunogen, maximum enzyme activity (and thus assay sensitivity) was achieved (144). The washing step is achieved through a process of radial partition in which the unbound label conjugate is radially eluted from the field of view of the fluorimeter. The potential for high sensitivity of this system is very good because of both the enzyme multiplication and the background reduction afforded by the rate measurement. Digoxin (146) has been assayed successfully in the ng/ml range and assays for other dilute analytes are expected soon. The initial analytical data from this technique are impressive. However, more examples are needed to assess the merits of the system adequately. The system has the potential to compete with many RIA procedures as well as other commercial FIAs. The only potential disadvantage may be in the costs of the reagents and the fluorimeter used for rate measurements.

Another unique solid-phase approach involves the use of antibodies attached to magnetizable cellulose particles (150–155). These methods make use of a separation step that is facilitated by magnetic sedimentation.

The direct assay makes use of fluorescent-labeled antigen and antibodies that are coupled to the magnetizable particles. Antibody-magnetizable solid phase was prepared by coupling purified antiserum to cyanogen bromide-activated cellulose particles containing iron oxide. Fluorescein-labeled antigen and sample are incubated with the solid-phase coupled antibody. Solid-phase antibody-bound label is then separated from the reaction mixture by magnetic sedimentation and washed to remove unbound label along with endogenous sample interferences. The fluorescence intensity measurement is made from the antibody-bound fraction after dissolution in alkaline medium. The intensity is inversely proportional to the original antigen concentration in the sample. Salicylate was used in this system to eliminate falsely low results caused by antigen binding to endogenous proteins. The direct method has been used for digitoxin and was found to be sufficiently sensitive to measure concentrations of 1.2 ng/ml (154).

A sequential addition technique has been used for the magnetizable solid-phase fluorescence immunoassay of thyroxine (155). In this method an excess of antibodies, coupled to magnetizable particles, is allowed to react with antigen present in the sample. The solid-phase antibodies (some of which are bound to analyte antigens) are then sedimented onto a magnet and the supernate containing endogenous interferences is aspirated. An excess of labeled antigen is then added, after demagnetization, and binds to the unoccupied antibody binding sites. The particles are then resedimented and the amount of labeled antigen remaining is determined by fluorescence of the supernate. The fluorescence measured from the supernate is therefore proportional to the antigen concentration of the sample. The advantages of the sequential addition technique over the direct method are that no washing step is required to remove the free labeled antigen and endogenous interferences are removed prior to the addition of label.

The primary advantages of magnetizable solid-phase fluorescence immunoassay are that the need for solid-phase fluorescence measurements and centrifugation is obviated. This greatly enhances the capability for automation, since the separation step can be handled by an electromagnet rather than by cumbersome physical techniques.

In addition to assay development for use in the clinical laboratory, there is a recognized need to adapt some of these methods to a bedside analysis format in cases of extreme emergency. In the field of therapeutic drug monitoring it is sometimes necessary to assess immediately whether a drug level is toxic or subtherapeutic so that a patient in crisis may be treated appropriately. Symptoms of drug toxicity can sometimes be similar to symptoms of the diseases that the drug is designed to treat. Such is the case for phenytoin, where symptoms of toxicity include seizures that may be similar to those attributed to epilepsy. Additionally, minor adverse effects of drugs such as theophylline may be absent and cannot be used to determine toxicity (156). This suggests that only serum levels of the drug can assess the situation accurately in cases of life-threatening emergency. Miles Laboratories has developed an assay for theophylline that is well suited for this purpose (157). The assay is based on the reactant labeled FIA technique described in Section 8.4.3. The reagents in this case are contained in dye form on a paper matrix attached to a dipstick. It allows quantitation of theophylline by simple dipping and measurement of fluorescence with an appropriate front surface fluorimeter. The assay was developed using two formats. In one assay format the labeled conjugate that competes with the drug is present in the sample dilution buffer. In the second, all necessary reagents are contained in the strip. Simple application of a diluted sample is all that is required to complete both of the strip assay competitive binding reactions. The performance characteristics of these assays were evaluated in correlation studies with good results. No separation step was included in this procedure.

Methods such as the strip assay are not generally expected to be as sensitive and reliable as solution methods for a number of reasons, the most important being scatter of light from the strip surface and serum component interference. The object of the strip test, however, is not to achieve maximum sensitivity but rather to execute an assay that is rapid and very easy to perform. Since the question in emergencies is often whether the drug level is either very high or very low, exceptional accuracy is not needed. Additionally, most drugs that lend themselves to emergency situations are present in concentrations that do not require high sensitivity measurements.

Other FIA systems have been developed that can be described as "homogeneous with a solid-phase separation." The assay is homogeneous in the sense that there is no physical removal of sample components. Solid-phase separations usually involve adsorption of the labeled antigen onto a support medium with subsequent removal of interfering substances present in the sample. Homogeneous assays have been developed that utilize novel methods to eliminate background interference. An immunoassay technique developed by Kronick and Little (158–159) makes use of total internal reflection spectroscopy with the antigen adsorbed onto a flat quartz plate. Fluorescein-labeled antibodies react with these adsorbed antigens and antigens in solution. Since the presence of free antigen in solution reduces the amount of label bound to the quartz plate, the fluorescence intensity measured from only the quartz plate is inversely proportional to the analyte concentration. The total internal reflection technique allows the excitation of fluorescent molecules only at the surface of the quartz plate without excitation of labeled antibodies that are free in solution. This is accomplished through a special optical arrangement in which the excitation beam is focused onto the plate at an angle greater than the critical angle of the quartz so that it is totally reflected. The excitation energy is provided by the small portion of the beam, which extends a fraction of a wavelength beyond the edge of the plate and is available only to fluorophores that are very close to the surface. Therefore, only those labeled antibodies that are in contact with the solid phase will receive excitation energy. The excitation energy was provided by a helium laser and luminescence was measured with a photomultipler tube perpendicular to the plane of the quartz plate. Morphine was measured successfully by this method at a level of $2 \times 10^{-7} M$. The sensitivity of the technique is exceptional by comparison with other homogeneous methods and there is the possibility of attaining sensitivity comparable to that of heterogeneous FIA and even RIA. The sensitivity of the method can be enhanced by a sandwich technique or by use of cross-correlation electronics (160). Sandwich techniques have been estimated to be as much as 10 times more sensitive than competitive binding assays in RIA (161) systems, and this could apply as well to FIA. Although the test method is rapid, easy to perform, and highly sensitive, the cost of such an assay is prohibitive to routine clinical applications at present.

Another heterogeneous assay that makes use of solid surface adsorption is based on long-time correlations of diffusion-dependent fluorescence intensity fluctuations (162–165). The method utilizes differences in random number fluctuations and diffusion between labeled molecules specifically bound to the surface of large carrier particles and unbound labeled molecules. The carrier particle consists of micrometer-sized acrylamide gel, to which antibodies are covalently bound. A sandwich technique is used to trap the analyte between the solid-phase antibodies and fluorescein-labeled antibodies that are added to the reaction mixture. The system is capable of measuring fluorescence intensity fluctuations that are sensitive to the amount of label bound to the large particle. The distinction between the large, slowly diffusing fluorophores and the small, rapidly diffusing ones is made by measuring fluctuations in fluorescence that occur in a small volume of total solution. The fluorescence intensity fluctuations are measured with respect to time, allowing time correlations to be made. If fluctuation data are continuously compared to previous data (autocorrelation), persistent fluctuations will show higher correlation than transient fluctuations. Fluctuations due to the number of particles in the volume of interest will persist for a time that is inversely dependent on the diffusion rate of the particles. Therefore, the fluctuations due to changes in the particle concentration will be long-lived, whereas the small, freely diffusing molecules are insensitive to the autocorrelation function applied to the data. In this way, the autocorrelation function is proportional to the number of bound fluorescing complexes and thus proportional to the concentration of analyte, since the sandwich-type format was used.

In this method, an argon ion laser (λ_{em} = 448) was used to sample repeatedly the small sample volume, although the authors claimed that a filtered incandescent source would have been adequate. The primary advantage of the correlation method is that sensitivity is not compromised in situations where high fluorescence background normally poses a problem. Small-sized radiative sources would be essentially eliminated and even sources of a size similar to that of the particles could be lessened spectrally. Numerous technical difficulties exist with the correlation method in its present stage of development, but those may be corrected with further work. Additionally, there are factors such as particle aggregation and large-particle nonspecific binding that need attention and will probably determine the ultimate sensitivity of the method. The utility of the system is limited by cost at the present time.

8.5.2. Chromatographic Separation in Fluorescence Immunoassay

Gel filtration high-pressure liquid chromatography has been used to separate the bound and free antigen fractions in a fluorescence immunoassay of insulin (164). In this assay unlabeled antigen in the sample competes with fluorescein-labeled antigen for antibody binding sites in the usual competitive binding manner. The molecular weight of the antigen complex is much greater than that of the free labeled antigen because of the size of the antibody molecule. In the case of insulin, the molecular weight of the complex is approximately 160,000 daltons whereas that of insulin is approximately 6000 daltons. Therefore, the separation was achieved on the basis of molecular size by the gel filtration column. Samples of the competitive binding reaction mixture were injected into a column and chromatograms were taken using an argon ion laser fluorescence detector.

It was found that only one peak corresponding to the labeled antigen antibody complex was eluted. This indicates that free labeled antigen was adsorbed onto the column. Experiments showed that adsorption was probably a reversible process and did not significantly affect quantitative measurements on the antibody-bound label fraction. In the competitive binding format, the concentration of labeled antigen bound decreases with increasing concentration of the analyte. Thus, the measured peak height of the complex is inversely proportional to the analyte concentration. The limit of detection for this method was determined to be 0.4 ng/ml (7 ×

10^{-11} M) in the reaction mixture. The sensitivity achieved is similar to that of insulin radioimmunoassay and is attributable in part to the efficiency of laser excitation.

The major limitation of this technique was that broad-band fluorescence interference from serum components overlapped the emission of the labeled complex. Therefore, clinical application of the method was precluded. In an effort to reduce serum interferences and increase sensitivity, this method was used in conjunction with a solid-phase sandwich technique and an enzyme label by Zare and co-workers that has been previously described in Section 8.5.1. The inclusion of the second separation step resulted in the removal of serum contaminations and allowed insulin determination in clinical samples. Additionally, the sensitivity of the method was greatly improved by use of the enzyme label.

This combination of chromatographic and solid-phase separations, laser detection, and enzyme multiplication resulted in a limit of detection that is surpassed only by time-resolved fluorimetric measurements, in the current stage of fluorescence immunoassay development. It should be noted, however, that the aim of Zare's work is to achieve the maximum sensitivity attainable with FIA methodology and not necessarily to provide assays for immediate routine clinical use. The cost of the reagents and apparatus involved and the number of technical manipulations required will probably prevent routine clinical use in the near future. The most interesting aspect of this work lies in the possibility of making fluorescence immunoassay the most sensitive analytical method for biological and other compounds.

8.5.3. Precipitation Methods

Conventional precipitation methods commonly used in RIA have not been thoroughly investigated for use in FIA. This is probably due to the difficulty of removing protein interference by these methods. Precipitation techniques should not be confused with solid-phase assays—where separation is achieved through adsorption onto particles, for example. Our discussion of precipitation methods will be limited solely to those that involve precipitation of insoluble antigen–antibody complexes.

Under certain conditions a mixture of soluble antigen–antibody complexes may precipitate from solution. Precipitate formation can take place within hours to days after the initial reaction. The precipitation from solution is caused by the formation of large cross-linked aggregates of antigen–antibody complexes. These aggregates arise because of the divalent nature of most antibodies and the fact that antigens may contain many sites to which antibodies can become attached (165). Therefore, large molecules or cells precipitate much more readily than small monovalent molecules because of their higher number of combining sites. If an antibody solution is titrated with an antigen, a concentration ratio of antigen to antibody is reached that is optimal for aggregate formation. The aggregates consist of cross-linked three-dimensional lattices of various sizes that form optimally at approximately equimolar concentrations of antigen and antibody. Prior to reaching the point of incipient precipitation in a titration of antibody with antigen the solution is in a condition in which antibodies are in excess and each antigen is saturated with antibody. This antibody saturation prevents crosslinking and the small complexes remain soluble. Beyond the equivalence point, where antigen concentration is in excess, aggregate formation is prevented because antibodies are saturated with antigen. The existence of these three distinct types of complexes has been confirmed both by ultracentrifugation and electron microscopy (166). Many of the large antigens will precipitate rapidly when bound by antibodies, whereas special techniques are usually required to enhance precipitation for smaller molecules. Precipitation aided by concentrated salts (usually ammonium sulfate), polymers such as polyethylene glycol, various organic solvents, and the double antibody technique may all be applied to FIA separations (37). In precipitation the antibody-bound antigen is removed from solution by centrifugation or sedimentation, leaving the unbound antigen in the supernate. Labeled antigens remaining from a competitive binding fluorescence immunoassay have been measured directly in the supernate (167). The endogenous fluorescence of serum causes interference problems and blank subtractions must be made. Fluorescence from proteins causes the greatest interference and not all proteins are removed during the precipitation step. Most label signals are not affected since protein emission occurs at relatively short wavelengths. There are other nonprotein components of serum that often cause more fluorescence background interference than protein. These compounds emit between 450 and 470 nm and significantly interfere with a number of potential labels.

Another approach to precipitation is to resuspend the precipitate and measure label fluorescence from the antibody-bound fraction. This technique has been used successfully in fluorescence immunoassays for the hormone progesterone and the therapeutic drugs gentamicin and tobramycin (168–170). The progesterone assay compared an ammonium sulfate precipitation technique and a magnetizable cellulose solid-phase separation (168). The recovery, linearity, and precision of both separation methods were considered acceptable, although the solid-phase assay was shown to be more precise and technically convenient.

Double-antibody methods are based on precipitation by formation of ternary aggregates through addition of a second antibody that is directed toward the first. Although the analyte may be monovalent, the antigen–first-antibody complex is a macromolecule and therefore multivalent. Antibodies directed toward the first antibody are readily obtained from a second animal species. The first antibody is raised in one species (usually a rabbit) and is used in the analytical competition reaction for labeled analyte. Second antibodies that are used in the separation are obtained by injecting a second species with serum from the first. Goats are frequently used as a second species to produce antirabbit antibodies. The double antibody assays developed for both tobramycin (169) and gentamicin (170) were found to be reliable and compared favorably with other methods. Background fluorescence in these methods was somewhat of a problem, although both assays were capable of quantitation of the analytes in question.

The double-antibody technique has also been used in conjunction with solid-phase separation. This is referred to as the double-antibody solid-phase system (DASP). The reagent consists of sheep antirabbit immunoglobulin covalently bound to an insoluble cellulose matrix. The second antibody immunoadsorbent is separated by centrifugation after binding with the first antibody, rendering it insoluble. The advantages of this method over simple double-antibody techniques include shorter incubation time and a small requirement for quantities of antisera. This method has not been applied to FIA analysis, although several RIA procedures have been published (171–173).

8.6. CHEMILUMINESCENCE AND BICLUMINESCENCE IMMUNOASSAY

Chemiluminescence and bioluminescence are phenomena that both involve the production of an electronically excited state via a chemical reaction and the subsequent loss of excitation energy in the form of a photon (174–175). Bioluminescence is a class of chemiluminescence in which the light-generating systems involve enzymes and lumigenic substrates of biological origin (175–176). These are prepared for in vitro studies from natural product extracts and require appropriate cofactors. A partial listing of bio- and chemiluminescent immunoassays is presented in Table 8.4. The best-known bioluminescent reaction involves the firefly (*Photinum pyralis*) enzyme luciferase and the lumophore luciferin (177).

This bioluminescent reaction has the distinction of having the largest reported quantum yield of photon production, $\phi_{CL} = 1.0$ (178), where

$$\phi_{CL} = \phi_R \, \phi_{es} \, \phi_{fl} \qquad (8.13)$$

ϕ_R is the fraction of molecules that take the appropriate chemical reaction pathway, ϕ_{es} is the fraction of product molecules that transit to an excited electronic energy level, and ϕ_{fl} is the fraction of fluorescing excited-state species that are produced (179). The most efficient chemiluminescent chemical system yet studied, $\phi_{CL} = 0.23$, follows:

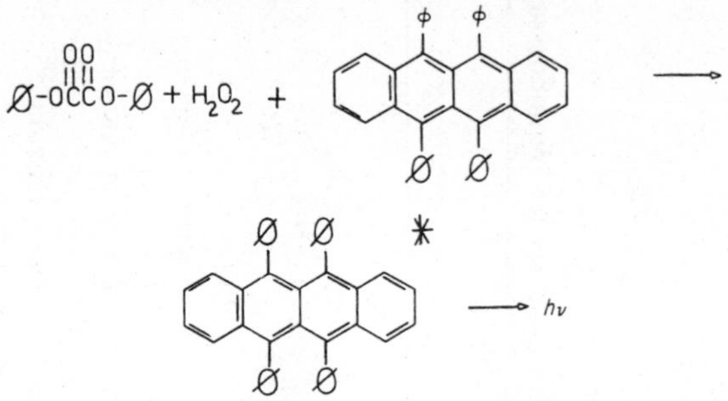

$$(8.14)$$

All chemiluminescent systems involve a crucial oxidation step, and usually a catalyst in aqueous solutions. In homogeneous competitive protein-binding systems, inhibition of this oxidation step can occur when the chemilumigenic agent or cofactor labeled antigen or catalyst is sequestered by protein (180–182). For these systems the most widely employed chemilumigenic labels have been luminol and isoluminol. The reaction scheme for luminol has been proposed by Ewetz and Thorn (183):

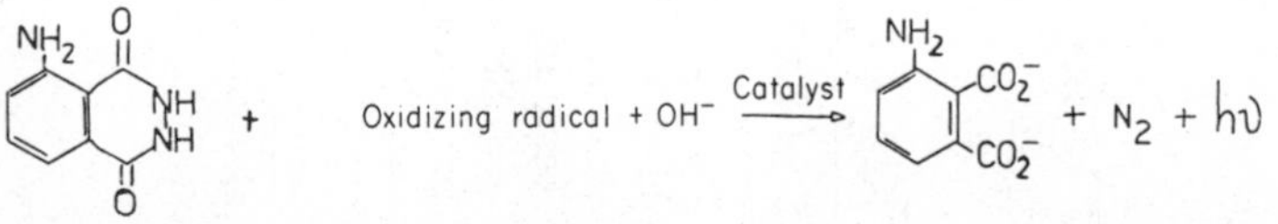

Many oxidant catalyst systems involving luminol have been explored, including hydrogen peroxide and peroxidase (184–188), superoxide anion generated by xanthine oxidase and hypoxanthine (189–190), and hydrogen peroxide and several metal ions (191–193). A variety of other luminol and isoluminol oxidation systems has been studied, with the microperoxidase hydrogen peroxide systems deemed optimal in terms of low limits of detection of luminol (1 pM) through a wide range of pH (8.6–13.0). Microperoxidase is the heme portion of cytochrome c with amino acids 11

Table 8.4. Bioluminescence and Chemiluminescence Immunoassays

Label	Labeled Substance	Hapten	Oxidizing Reagent	Dynamic Range of Assay	Ref.
Iminoluminol	Human IgG	Human IgG	H_2O_2 hemin	5–50 µg/tube	202
Azoluminol	Anti-rabbit IgG	Rabbit IgG	H_2O_2 NaOCl	5–80 ng/ml	198
Azoluminol	Testosterone-ovalbumin	Testosterone	H_2O_2 (Ac)$_2$ Cu	0.1–10 ng/tube	201
Dialkylaminoluminol	Progesterone	Progesterone	H_2O_2 microperoxidase	24–400 pg/tube	207
Dialkylaminoluminol	T_4	T_4	H_2O_2 microperoxidase	20–150 nM	181
Dialkylaminoluminol	Biotin	Biotin	H_2O_2 lactoperoxidase (and xanthine oxidase hypoxanthine generated superoxide radical)	50–400 nM	180

Label	Conjugate	Reactant	Detection system	Range	Ref.
Dialkylaminoisolumino	Estriol-(6α-glucuronide)	Estriol-(6α-glucuronide)	H_2O_2 microperoxidase	10–100 pg	201
Dialkylaminoisolumino	Cortisol	Cortisol	H_2O_2 microperoxidase	20–1000 pg/tube	211
Peroxidase	Cortisol	Cortisol	H_2O_2 luminol	0.01–1 ng/tube	212, 213
Peroxidase	IgG (anti-staphylococcal enterotoxin B)	Staphylococcal enterotoxin B	H_2O_2 pyrogallol	1–10 ng	218
Peroxidase	Goat anti-rabbit IgG	Rabbit anti-human serum albumin	H_2O_2 luminol	5×10^{-4}–7×10^{-7}	220
Adenosine-5'-tetraphosphate	2,4-dinitrophenyl residue	N(2,4-dinitrophenyl) B-alanine	luciferase–Mg^{2+} luciferin	10^{-4} nM	225
Nicotinamide 6-(2-aminoethylamino) purine dinucleotide	2,4-dintrophenyl residue	2,4-dinitrophenyl-6-aminocaproate	Luciferase–flavin mononeucleotide	0.25–1.5 μM	228
Nicotinamide 6-(2-aminoethylamino purine dinucleotide	Biotin	Biotin	Luciferase–flavin mononucleotide	30–200 nM	228

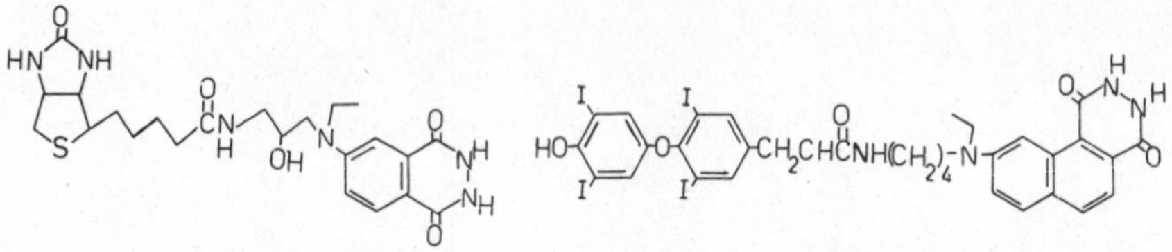

Biotinyl isoluminol Conjugate   Thyroxine (T4) 7-aminonaphthalene-1,2-dicarboxylic acid hydrazide

Fig. 8.4. Structures of two chemiluminescent-labeled hapten conjugates.

through 21 still attached, and a molecular weight of 1879 daltons for the free acid (195). Rauhut et al. prepared a luminol oxalate ester that when nonenzymatically oxidized with hydrogen peroxide has a $\phi_{CL} = 0.23$ (196). Underivatized luminol, on the other hand, has a $\phi_{CL} = 0.015$ (197).

Simpson et al. diazotized luminol and directly coupled this to IgG for use in a heterogeneous immunoassay for IgG (198). This product evidenced a luminescence yield of less than 1% of that of luminol possibly because of, in part, polymerization of the intermediate diazoluminol or the singlet deactivating nature of the diazonium cation and its azo-substituted product. Electron-withdrawing substituents in this position have been shown to decrease chemiluminescence quantum yields (197–200). Vratt et al. also produced an azoluminol conjugate with a $\phi_{CL} = 0.004$, for a heterogeneous immunoassay by first diazotizing 4-carboxymethylaniline and coupling this to luminol (201). This provides a free carboxyl group capable of condensation reactions with amino groups. This label is readily adaptable for coupling to and subsequently performing an immunoassay of most drugs and proteins.

Another heterogeneous competitive binding immunoassay for human IgG has been developed using the IgG luminol conjugate (202). Here a Schiff base adduct was formed on reaction of luminol and the periodate oxidation product of IgG. The chemiluminescence efficiency of this one-to-one labeled product was determined to be on the order of 20% of that of luminol. The calibration curves using this method had a dynamic range of 0.5–50 g/tube of IgG.

The chemiluminescent-labeled hapten conjugates having the highest ϕ_{CL} values have been developed by Schroeder et al. (180–183,194,207). Isoluminol (6-amino-2,3-hydrophthalazine-1,4-dione) and 7-aminonaphthalene-1,2-dicarboxylic acid hydrazide-labeled conjugates were synthesized and tested (203–207). Structures of the biotinyl–isoluminol and T4–naphthalene-1,2-dicarboxylic acid hydrazide conjugate are illustrated in Fig. 8.4. Aklylation of the aromatic amino group with a bridging molecule served as a means of coupling isoluminol and the naphthalene derivative to the hapten of interest. Isoluminol was chosen because of the observed increase in its chemiluminescence quantum yield on aminoalkylation, whereas similar alkylation of luminol resulted in a significant quantum yield decrease (179,197,209).

A homogeneous immunoassay for biotin was carried out based on the observed increase in chemiluminescence quantum yield of the labeled hapten on binding to protein. When increasing levels of avidin were incubated with a constant amount of biotinyl–isoluminol conjugate, the observed chemiluminescence increased to a maximum of 10 times that produced in the absence of avidin. This chemiluminescence increase permitted the development of a homogeneous immunoassay. The biotinyl–isoluminol conjugate itself has a chemiluminescence efficiency approximately equal to that of unsubstituted isoluminol. A similar homogeneous immunoassay for progesterone was developed (207). Here an N-succinimide ester of 1-hydroxyprogesterone was coupled to 6-[N-(6-aminohexyl))-N-ethyl)-amino-2,3-dihydrophthazine-1,4-dione to produce the lumigenic conjugate. Addition of antiprogesterone antibody caused a fourfold enhancement of the luminescence of the conjugate that served as a basis for that assay. A sensitivity of 25 pg/tube was obtained with a microperoxidase–H_2O_2 oxidation system at pH 8.6.

A chemiluminescence immunoassay for urine estriol-16α-glucuronide was developed based on enhancement of light production from lumigenic conjugate binding to antibody (210). In this case a seven-fold enhancement of luminescence intensity was observed on addition of increasing amounts of anti-estriol-16α-glucuronide to solutions of the estriol-16α-glucuronide aminobutylethylisoluminol conjugate.

Another chemiluminescent conjugate, cortisol-aminopentyl-l-ethylisoluminol, synthesized for use in a homogeneous immunoassay for plasma cortisol, did not show the typical enhancement of chemiluminescence quantum yield when oxidized after addition of anticortisol antisera (211). A temporal delay of chemiluminescence, however, was observed after antibody addition, which served as a basis of immunoassay. A linear response through an unusually wide range of 20–100 pg/tube of cortisol was obtained.

The thyroxine (T_4)-isoluminol conjugate whose structure is shown in Fig. 8.4 was used in a heterogeneous immunoassay for T_4. A heterogeneous approach was necessitated because of variable chemiluminescence quenching associated with T_4 plasma protein binding. Small sephadex columns were employed to adsorb plasma and labeled T_4 followed by elution of the column to remove serum interference. Anti-T_4 antibody addition to the column was followed by an incubation period, resulting in an equilibration within the column between bound and free species.

The bound fraction was less strongly adsorbed, was eluted first, and was subjected to chemiluminescent oxidation.

Several other very sensitive enzyme-coupled, solid-phase chemiluminescence immunoassays have been developed. Arakawa et al. used a chemiluminescence assay using horseradish peroxidase (212) to achieve a 10 pg/tube sensitivity for cortisol (213). In this study three different standard heterogeneous immunoassay procedures were compared: antibody–solid phase, double antibody–solid phase, and double antibody. Additionally, the effects of heterologous and homologous systems were compared. The heterologous–homologous nomenclature was developed for enzyme immunoassay systems (214–215). A homologous system is one in which the same bridging for covalently connected molecules is employed in the enzyme-labeled conjugate as is employed in the protein conjugate utilized in the antibody production process. Heterologous systems, of course, employ differing functional groups and bridging molecules in the labeled conjugate than in the hapten–protein conjugate. Heterologous conjugates, in general, have lower antibody affinities compared to homologous conjugates and result in higher assay sensitivities (216–217).

Another solid-phase technique was employed in the determination of staphylococcal enterotoxin B (SEB) (218). Peroxidase-labeled SEB, sepharose-coupled anti-SEB, and the analytical sample were incubated prior to centrifugal separation of the solid phase. Resuspension of the sepharose phase was followed by a chemiluminescent peroxidase determination (219). The detection limit of this assay compared favorably with that obtained in a radioimmunoassay procedure.

Immunoassays employing bioluminescence as a detection step have been limited to systems where cofactor-labeled haptens are distributed between antibody-bound and free fractions so that the antibody-bound species are inhibited from interacting with a bioluminescent system to produce light. The most popular bioluminescent system studied is that of the firefly whose reaction mechanism has been proposed as (221–223):

$$LH_2 - AMP + ATP \xrightleftharpoons{Mg^{2+}} E - LH_2 - AMP + PPMg^{+2} \quad (8.16)$$

$$E - LH_2 - AMP + O_2 \rightarrow E + L{=}O$$
$$+ CO_2 + AMP + h\nu \text{ (562 nm)} \quad (8.17)$$

where ATP is adenosine triphosphate, AMP is adenosine monophosphate, PP is pyrophosphate, $L{=}O$ is oxyluciferin, and LH_2 is luciferin (see Fig. 8.5). The reaction has been shown to occur most efficiently at 25°C in pH 7.8 glycine buffer (222).

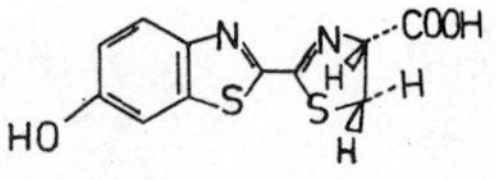

Fig. 8.5. Structures of luciferin (**1**) and oxyluciferin (**2**).

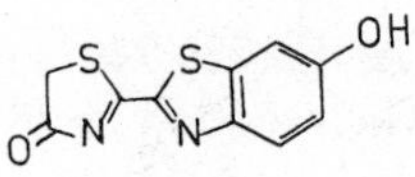

Carrico et al. carried out a model immunoassay system where ATP was covalently linked to the hapten monosubstituted 2,4-dinitrobenzene (DNP). Four different ATP–DNP conjugates were synthesized and *P*-DNP-adenosine-5'-tetraphosphate, **3**, depicted in Fig. 8.6, participated most efficiently in the light production reaction. The analyte *N*(DNP)-β-alanine has a limit of detection of 10 n*M*. One disadvantage of this assay, noted by the authors, is the requirement for the absence of interfering endogenous ATP and phosphate-cleaving enzymes (22) that would degrade the ATP conjugate (227). EDTA addition to the sample was found to inhibit the actions of the interfering enzymes.

Schroder et al. utilized the photobacterium fischeri luciferase system for quantitating haptens in competitive protein-binding reactions (228). This bioluminescence system has been used previously for analysis of

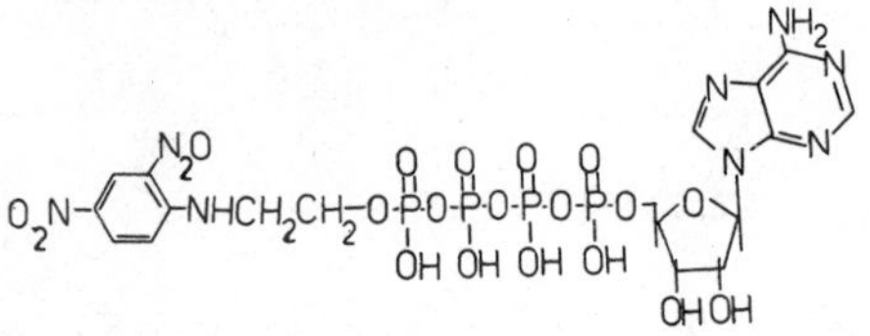

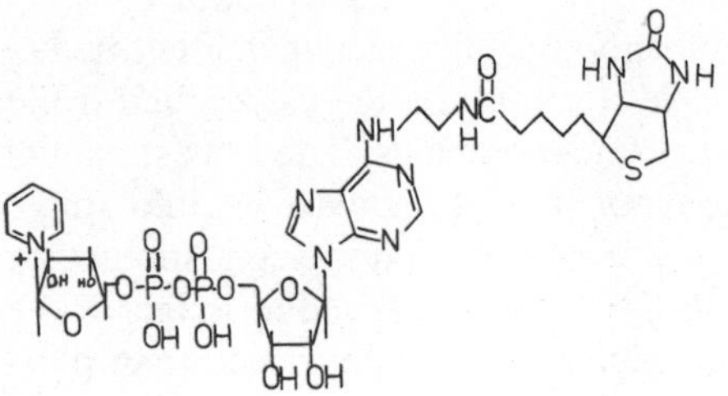

Fig. 8.6. Structures of two cofactor-labeled hapten conjugates for use in bioluminescent competitive protein binding assays.

trace levels of NADH (229,230) and recently for assay of total bile acids in serum (231). The reaction steps are (232):

$$NADH + FMN + H^+ \xrightarrow{\text{NADH Dehydrogenase}} NAD^+ + FMNH_2$$

$$(8.18)$$

$$FMNH_2 + O_2 \xrightarrow[\text{long-chained aldehyde}]{\text{luciferase}} FMN + h\nu$$

$$(8.19)$$

Both 2,4-dinitrofluorobenzene and biotin were labeled with nicotinamide-6-(2-aminoethylamino) purine dinucleotide (AENAD). The biotinyl derivative **4** is shown in Fig. 8.6. On binding by their specific proteins, the reactions of cofactor labels with enzymatic light-producing systems were inhibited, forming the basis for homogeneous analytical systems. As with the firefly luciferase binding system the disadvantages include interferences by endogenous cofactors. It is claimed, however, that serum NAD is inactivated within 24 hours at room temperature (206).

The extreme sensitivity of chemi- and bioluminescent analytical reactions makes the method extremely attractive for future development in immunoanalytical areas. Future developments must certainly include more heterogeneous enzymatic chemiluminescent immunoassays and perhaps enzymatic production of appropriate cofactors for bioluminescence immunoassays.

8.7. PHOSPHORESCENCE IMMUNOASSAY

An interesting immunoassay possibility that has not been reported in the literature is that of phospohrescence immunoassay (PIA). The use of phosphorescent labels in immunoassay systems offers some advantages over the use of fluorescent labels. These advantages include longer wavelengths of emission and fewer endogenous interferences, both of which allow a reduction in background. Phosphorescence has been identified by Lewis and Kasha (234) as radiative decay from the lowest excited triplet state to the ground singlet state in organic molecules. The lowest excited triplet state lies slightly lower in energy than the lowest excited singlet state from which fluorescence is derived in phosphorescent molecules. This accounts for the emission generally being observed at longer wavelengths (235). There are many more molecules that fluoresce than phosphoresce. This is why one would expect to find fewer endogenous interferences if phosphorescent labels were used.

Phosphorescence can be observed from a variety of conjugated organic molecules and is generally distinguished from fluorescence on the basis of luminescence lifetimes. The longer lifetimes attributed to phosphorescent molecules are due to the spin-forbidden nature of the triplet-to-ground-state transition. Phosphorescence arises from the latter transition, which classically violates the law of conservation of angular momentum. The probability per unit time of this transition is very low by comparison with that of the transition resulting in fluorescence emission and is why fluorescent decay times are usually much shorter. Because of the long lifetime of the triplet state, collisional deactivation is very effective in bringing about radiationless decay in fluid media and phosphorescence is rarely observed in solution at room temperature. Cooling (usually to 77 K) to a rigid state greatly restricts the frequency of molecular collisions and favors the occurrence of phosphorescence. Cooling to liquid nitrogen temperature is not desirable for routine immunoassays because of the procedural difficulties involved.

Many polar and ionic organic compounds have been found to exhibit phosphorescence at ambient temperatures if adsorbed onto material(s) such as silica, paper, alumina, or cellulose (236). This is known as room temperature phosphorescence (RTP). Studies by Schulman and Walling (237) indicated that surface adsorption held the phosphor rigidly enough to limit collisional deactivation and suggested that the stronger the adsorption, the more intense the phosphorescence. RTP may prove to be a viable alternative to fluorescent labeling in immunoassay. The RTP method is much more convenient than low-temperature phosphorimetry and an automated RTP system has been designed by Bower and Winefordner (238).

There are, however, several potential sources of error associated with RTP measurements that are not common to fluorescence measurements. Schulman and Parker found that RTP intensity decreased dramatically with increasing relative humidity (239). This was attributed to displacement of the hydrogen-bound phosphor from the filter paper substrate by water, since hydrogen bonding was proposed as the primary adsorption process. Quenching by oxygen is another potential source of error in RTP (239,240). Although adsorption onto the sample matrix has been shown to inhibit oxygen quenching (241), this type of quenching does occur to some extent, depending on the matrix and the phosphor. The more strongly the phosphor is bound, the greater the quenching effect of oxygen. This suggests that some oxygen may be trapped in the sample matrix when the sample is prepared in the presence of oxygen. Both of these problems can be reduced by drying the sample under a stream of dry nitrogen. Most compounds that phosphoresce strongly at room temperature are largely susceptible to quenching and water displacement. There-fore, it would probably be necessary to conduct potential phosphores-cence immunoassay measurements under a dry nitrogen atmosphere. This does not pose much of an instrumental problem, since purging of the sample cell with dry nitrogen is all that is required.

Phosphorescence has been observed in solution at room temperature but usually only in solutions containing certain detergents at concentra-tions above their critical micelle concentrations (242). A possible expla-nation for this effect is stabilization of the triplet state by concentration of the phosphorescent molecules within the protective micellar environ-ment (243). Phosphorescence arising from micellar triplet state stabili-zation might be a viable alternative to solid-phase RTP, since solution systems are easier to deal with instrumentally. However, analytical stud-ies of micelle-stabilized phosphorescence are quite recent and have not been documented widely. In addition, very strict degassing requirements and susceptibility to impurities create procedural difficulties that limit analytical usefulness of the method. A number of micellar systems have been applied to the RTP problem and with further development these systems may achieve analytical acceptance.

Conventional RTP on solid surfaces is currently most suitable for phos-phorescence immunoassay procedures. Immunoassay procedures are well suited for solid-phase separation (Section 8.5.1) and RTP detection is potentially adaptable to these methods. The selectivity associated with phosphorescent labeling is a distinct advantage over fluorescent labeling. Organic compounds such as p-aminobenzoic acid have been phosphori-metrically detected in subnanogram quantities (244) and are potentially excellent labels, provided they can be covalently conjugated to antigenic species without the loss of RTP emission intensity. Further sensitivity and selectivity increases could be achieved with the addition of heavy metal ions and synchronous scanning techniques.

The major limitation of RTP labels is the surface background emission observed between 400 and 600 nm. Virtually every support studied has exhibited significant background RTP signals (245). If this analytical prob-lem is soluble, phosphorescent labels may attain common use in immu-noassay procedures

8.8. POTENTIAL DEVELOPMENTS IN LUMINESCENCE INSTRUMENTATION

There has been a great deal of work done to increase analytical sensitivity in luminescence immunoassay (especially FIA) and there are several pertinent instrumental developments that merit special attention. Currently, commercial FIA procedures are limited to simple analyses that do not require extremely high sensitivities. If FIA is to expand into analytical areas currently occupied by RIA, appropriately designed instrumentation must be available commercially and evaluated for FIA use.

The use of lasers in FIA has been investigated for a limited number of clinically significant analytes (150–151,155,159). The assays developed constitute some of the most sensitive FIA methods advanced to date. The advantage of laser excitation lies in the increased luminescence intensity provided by higher excitation power. Additionally, instrumental designs can be made simpler through the elimination of the need for excitation optics in dedicated instruments employing laser excitation (246). Potential problems with laser excitation include photodecomposition of the label and increased cost because of maintenance requirements.

Time resolution may eventually play a major role in reducing the background signal associated with luminescence immunoassay. In time-resolved spectroscopy a fast light pulse excites the probe and luminescence is measured after a predetermined delay time has elapsed from the moment of excitation. This allows direct-scatter and fast-decaying nonspecific background to be virtually eliminated if the excited state of the label has a relatively long decay time (247). Pulsed excitation can be achieved with nitrogen-pumped dye lasers, spark-discharge lamps, and xenon-discharge lamps. A fast photomultiplier tube equipped with an electronic switch or photon counting discriminator would be needed to detect luminescence emission. The choice of instrumentation depends on the length of the decay time of the label used and short-decay, time-resolved fluorimeters are much more difficult to construct.

Certain pyrene derivatives with lifetimes of about 100 nsec are potentially suitable labels for time-resolved fluorescence immunoassay (248) and could be temporally resolved by relatively simple instrumentation. Fluorescing rare earth metal chelates have recently attracted much attention as labels for time-resolved fluorescence immunoassay (249,250) because of their high quantum yields, large Stokes' shifts, and very long decay times (50–1000 μsec). The characteristic fluorescence of these chelates is due to absorption by the ligand with subsequent transfer from the excited triplet state to the central atom-producing long-wavelength narrow emission typical of the metal. A chelate of europium recently has been used as the label in a time-resolved FIA of pancreatic phospholipase A (249). This method employs a sandwich technique with fluorescence being measured in solution after dissociation from the solid phase. Time-resolved detection of the long-lived europium chelate resulted in elimination of background fluorescence and allowed very high sensitivity (0.02 ng/

ml). The stability of the label along with the relative simplicity of the instrumentation used makes this technique competitive with other methods requiring high sensitivity, and more assays are expected to be developed. The authors state that the limiting factor for sensitivity in this assay is not the detection of the label but rather the nature of the immunoreaction itself. This suggests that FIA may have reached RIA sensitivity through the time-resolved method.

Among the most fascinating possibilities for luminescence is the use of image detectors for either spectral or spatial resolution (90). The advantage of these techniques is through lowered analysis time but not necessarily increased sensitivity. The analytical merits of image detectors will not be discussed extensively, but their performance can be considered similar to that of photomultiplier tubes. The most commonly used image detector is probably the silicon-intensified target (SIT). In the SIT, spatially dispersed radiation strikes a photosensitive surface and releases electrons, producing a charge pattern. An electron beam searches the surface and neutralizes these charges to produce a video signal. The charge density of each point on the surface is proportional to the intensity of incident radiation and can be used for quantitative measurements.

Simultaneous detection could be achieved with an instrumental set-up that is schematically represented in Fig. 8.7. Excitation could be achieved through use of a conventional continuum source that is filtered for wavelength selection. Focusing optics would not be necessary because dispersion over a rather large illumination surface is desired. The monochromatic light would then be passed through a number of sample cells and luminescence could be measured directly opposite the detection source. Glass or synthetic polymer-based fiber optic links could be used to transfer emission intensity to the surface of an SIT. The fiber optic system is not absolutely necessary with such a design since the luminescence from individual cells would be discriminated spatially by the SIT. Several hundred samples could be quantitated simultaneously in this manner with the number of assays being limited only by the resolution capabilities of the SIT.

Complex mixture analysis in luminescence immunoassay systems could be made possible with an instrumental design represented in Fig. 8.8. Spectral discrimination is accomplished by illumination of the sample using an excitation monochromator with a spectral bandpass appropriate to excite the various label systems that may be contained in luminescence immunoassay systems. The emitted luminescence from these labels would occur at different wavelengths that could be adequately resolved. The emission monochromator would have to have an extremely broad spectral bandpass to accommodate the various emission wavelengths of the com-

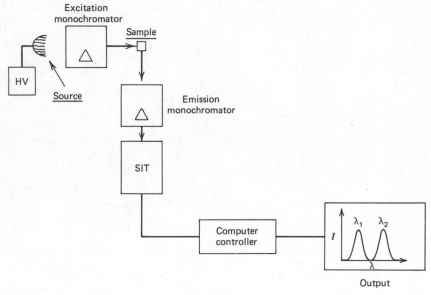

Fig. 8.7. Block diagram representing instrumentation for complex mixture analysis using the silicon intensified target (SIT) image detector.

plex mixture system. The number of wavelengths that can be accommodated is limited by the detector surface size, and the resolution characteristics are the result of a compromise between spectral range and spectral resolution. For example, if there were 500 independent resolution elements along one axis of the image detector, 0.1-nm resolution could be obtained over a spectral range of only 50 nm. The number of labeled immunoassays that could be performed in the same mixture is limited then by the emission bandwidth of the labels used.

Another possible application of image detection to FIA analysis is qualitative or semiquantitative screening tests for toxic substances. The instrumental setup would be similar to that described for simultaneous measurements, except a well plate would be used rather than a series of sample cells. A well plate (see Fig. 8.9) could be easily made containing a different antigen adsorbed onto a solid phase in each of the wells. If a labeled antigen, appropriate antisera, and a sample are added to each well, a heterogeneous immunoassay capable of screening for toxic substances is possible. The number of toxic substances (usually drugs) that could be screened would not be limited to the number of well plates because cross-reacting group-selective antisera could be used. Once it is determined that a drug or other toxic substance belonging to a particular

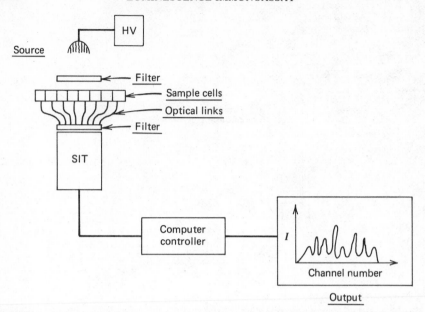

Fig. 8.8. Block diagram representing instrumentation for simultaneous detection of eight fluorescence immunoassays.

group is present, well plates containing specific reagents could identify which substance is present. A less expensive alternative to using image detectors for this application is shown in Fig. 8.9. The unique feature of this instrument is the use of an x–y movable stage to move the different wells across the end of a single optical link. The x–y stage could be positioned using stepping motors, which would allow automated measurement of the reaction wells over a very short time span. This well plate system for toxicological screening is potentially as reliable as GC–MS for qualitative results and would be much less expensive. Applications that require high analytical sensitivity and throughput are rapidly increasing. Additionally, the cost of sophisticated instrumentation is likely to decrease, possibly making the application of technology described here widespread.

This chapter has presented an overview of existing luminescence immunoassay methods, some of which have been extensively studied and commercially applied to clinically important analytes. We have also attempted to provide an insight into alternative chemical and instrumental techniques that have not been studied to a great extent. It is difficult to predict the exact direction in which luminescence immunoassay research

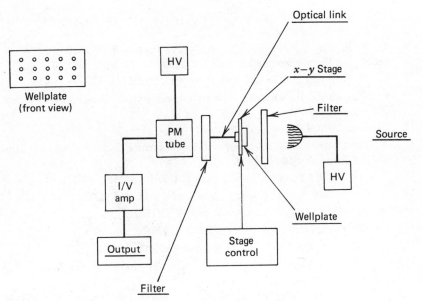

Fig. 8.9. Instrumental block diagram for rapid semiquantitative screening using fluorescence immunoassay.

and development are headed because of the rapidly changing scope of clinical analysis and the great number of competing methods that are available. It can be said with certainty, however, that luminescence immunoassay methodology has gained a strong foundation in a very short time and is not likely to fall in the coming years.

REFERENCES

1. S. A. Berson and R. S. Yalow, *J. Clin. Invest. 38,* 1996 (1959).

2. J. I. Thorell and S. M. Larson, *Radioimmunoassay and Related Techniques*, C. V. Mosby, St. Louis, 1978.

3. W. D. Odell and W. H. Daughaday *Principles of Competitive Protein Binding Assays*, Lippincott, Philadelphia, (1971).

4. J. F. Faucon and C. Lyssan, *Biochim. Biophys. Acta 307,* 459 (1973).

5. H. R. Lukens, C. B. Williams, S. A. Levison, and W. B. Dandliker, *Sensitive, Specific Fluorescence Immunoassay Methods for Detecting Pesticide and Other Organic Environmental Contaminants Arising from Biological or Chemical Sources*, IRT Corporation, San Diego, Calif., 1977.

6. Editors, *Clin. Chem. 29,* 889 (1983).

7. J. L. Turk, *Immunology in Clinical Medicine*, 3rd ed. Appleton-Century-Crofts, New York, 1978.

8. C. W. Parker, in *Principles of Competitive Protein Binding Assays*, W. D. Odell and W. H. Daughaday (eds.), Lippincott, Philadelphia, 1971, p. 25.

9. G. M. Edelman and W. E. Gall, *Ann. Rev. Biochem. 38*, 415 (1969).

10. F. Loor and G. E. Roelants, *B and T Cells in Immune Recognition*, Wiley, New York, (1977).

11. M. Sela, *Adv. Immunol. 5*, 29 (1966).

12. K. Landsteiner, *The Specificity of Seriological Reactions*, Harvard University Press, Boston, (1962).

13. J. Kam and R. Yoshida, U.S. Patent CA *94* P 188235y. (1976)

14. T. H. The and T. R. Feltkamp, *Immunology 18*, 865 (1970).

15. W. B. Dandliker, *Immunochemistry 4*, 295 (1967).

16. G. Steinbach, *Acta Histochem. 55*, 110 (1976).

17. D. Blakeslee and M. Baines, *J. Immunol. Meth. 13*, 358 (1976).

18. F. Sokol, A. Hulka, and P. Albrecht, *Folia Microbiol.* (Prague) *7*, 155 (1961).

19. E. F. Ullman, "Recent Advances in Fluorescent Immunoassay Techniques", in *Ligand Assay: Analysis of International Development on Isotopic and Non Isotopic Immunoassay*, J. Langan and J. J. Clapp (eds.), Masson Publishing Co., New York, 1981, p. 113.

20. R. F. Chen, *Arch. Biochem. Biophys. 133*, 263 (1969).

21. C. S. Chadwick, M. G. McEntegard, and R. C. Nairn, *Immunology 18*, 865 (1970).

22. P. Brandtzaeg, *Scand. J. Immunol. 2*, 273 (1973).

23. P. Brandtzaeg, *Scand. J. Immunol. 2*, 333 (1973).

24. P. Brandtzaeg, *Ann. N.Y. Acad. Sci. 254*, 35 (1975).

25. R. M. McKinney and J. T. Spillane, *Ann. N.Y. Acad. Sci. 254*, 55 (1975).

26. R. F. Chen, *Anal. Lett. 7*, 65 (1974).

27. S. Katsch, F. W. Leaver, J. S. Reynolds, and G. F. Katsch, *J. Immunol. Meth. 5*, 179 (1974).

28. U. E. Handschin and W. J. Ritschard, *Anal. Biochem. 71*, 143 (1976).

29. N. Nakai, C. Y. Lai, and B. L. Horecker *Anal. Biochem. 58*, 563 (1974).

30. Y. Kanaoka, *Chem. Int. Ed. Engl. 16*, 137 (1977).

31. T. Sekine and K. Ando, *Anal. Biochem. 48*, 557 (1972).

32. M. Machida, N. Ushijima, M. I. Machida, and Y. Kanaoka, *Chem. Pharm. Bull. 23*, 1385 (1975).

33. K. Yamamoto, T. Sekine, and Y. Kanaoka, *Anal. Biochem. 79*, 83 (1977).

34. M. Weigele, S. DeBernardo, W. Leimgruber, R. Cleeland, and E. Grunberg, *Biochem. Biophys. Res. Commun. 54*, 899 (1973).

35. V. T. Oi, A. N. Glazer, and L. Stryer, *J. Cell Biol. 93*, 981 (1982).

36. M-H. Yu, A. N. Glazer, K. G. Spencer, and J. A. West, *Plant Physiol.* *68*, 482 (1981).

37. D. J. W. Barber and J. T. Richards, *Photochem. Photobiol.* *25*, 565 (1977).

38. A. N. Glazer and C. S. Hixson, *J. Biol. Chem.* *252*, 32 (1977).

39. J. Grabowski and E. Gantt, *Photochem. Photobiol.* *28*, 39 (1978).

40. A. N. Glazer, S. Fang, and D. M. Brown, *J. Biol. Chem.* *248*, 5679 (1973).

41. R. E. Dale and F. W. J. Teale, *Photochem. Photobiol.* *12*, 99 (1970).

42. G. Cohen-Bazire, S. Beguin, S. Rimon, A. N. Glazer, and D. M. Brown, *Arch. Microbiol.* *111*, 225 (1977).

43. R. S. Fager, C. B. Kutina, and E. W. Abrahamson, *Anal. Biochem.* *53*, 290 (1973).

44. P. B. Ghosh and M. W. Whitehouse, *Biochem. J.* *108*, 155 (1968).

45. L. C. Cantley, J. Gelles, and L. Josephson, *Biochemistry 17*, 418 (1978).

46. D. E. Draper and L. Gold, *Biochemistry 19*, 1774 (1980).

47. P. J. Sims, A. S. Waggoner, C.-H. Wang, and J. F. Hoffman, *Biochemistry 13* (16), 3315 (1974).

48. Th. Förster, *Z. Phys. Chem. N. F. Frankfurt, Main 54*, 42 (1950).

49. A. Weller, *Progr. Reaction Kinet. 1*, 187 (1961).

50. S. G. Schulman, in *Modern Fluorescence Spectroscopy*, E. L. Wehry (ed.), Plenum, New York, 1976, chap. 6.

51. J. F. Ireland and P. A. H. Wyatt, *Progress in Physical Organic Chemistry*, Academic Press, New York, 1976, p. 131.

52. S. G. Schulman and L. S. Rosenberg, *J. Phys. Chem. 83*, 447 (1979).

53. S. G. Schulman and L. S. Rosenberg, *Anal. Chim. Acta 94*, 16 (1977).

54. S. G. Schulman and L. S. Rosenberg, *Anal. Chim. Acta 115*, 211 (1980).

55. S. G. Schulman, B. S. Vogt, and M. W. Lovell, *Chem. Phys. Lett. 75*, 224 (1980).

56. S. G. Schulman, L. S. Rosenberg, and W. R. Vincent, *J. Am. Chem. Soc. 101*, 139 (1979).

57. M. R. Loken, J. W. Hayes, J. R. Gohlke, and L. Brand, *Biochemistry 11*, 4779 (1972).

58. M. DeLuca, L. Brand, T. A. Cebula, H. H. Seliger, and A. F. Makula, *J. Biol. Chem. 246*, 6702 (1971).

59. L. M. Krausz, J. B. Hitz, R. T. Buckler, and J. F. Burd, *Ther. Drug Monit. 2*, 261 (1980).

60. K. J. Dean, S. G. Thompson, J. F. Burd, and R. T. Buckler, *Clin. Chem. 29* 1051 (1983).

61. D. Worah, K. K. Yeung, F. E. Ward, and R. J. Carrico, *Clin. Chem. 27*, 673 (1981).

62. S. G. Thompson, *Res. Discl. 194*, 248 (1980).

63. K. Kato, Y. Hamaguchi, H. Fukui, and E. Ishikawa, *J. Biochem. 78*, 235 (1975).

64. T. Kitagawa and T. Aikawa, *J. Biochem. 79*, 233 (1976).
65. K. Kato, Y. Hamaguchi, H. Fukui, and E. Ishikawa, *J. Biochem. 78*, 423 (1975).
66. G. G. Guilbault, P. J. Brignac, and M. Zimmer, *Anal. Chem. 40*, 190 (1968).
67. G. G. Guilbault, P. J. Brignac, and M. Juneau, *Anal. Chem. 40*, 1256 (1968).
68. M. Numazawa, A. Haryu, K. Kurosaka, and T. Nambara, *FEBS Lett. 79*, 396 (1977).
69. K. Matsuoka, M. Maea, and A. Tsuji, *Chem. Pharm. Bull. 27*, 2354 (1979).
70. K. Kato, Y. Umeda, F. Suzuki, D. Hayashi, and K. Kosaka, *Clin. Chem. 25*, 1306 (1979).
71. W. D. Hinsberg, K. H. Milby, and R. N. Zare, *Anal. Chem. 53* (a), 1509 (1981).
72. G. C. Visor and S. G. Schulman, *J. Pharm. Sci. 70*, 469 (1981).
73. E. F. Shaw, R. A. A. Watson, W. J. Landon, and D. S. Smith, *J. Clin. Pathol. 30*, 526 (1977).
74. Y. Kobayashi, N. Tsubota, K. Miyai, and F. Watanabe, *Steroids 36*, 177 (1980).
75. S. Udenfriend, *Fluorescence Assay in Biology and Medicine*, Vol. 2, Academic Press, New York, (1969), pp. 182–185.
76. R. F. Zuk, G. L. Rowley, and E. F. Ullman, *Clin. Chem. 25*, 1554 (1979).
77. R. D. Nargessi, J. Landon, and D. S. Smith, *Clin. Chim. Acta 89*, 461 (1978).
78. E. F. Ullman, U.S. Patent No. 13,998,943 (Double Receptor Fluorescent Immunoassay) (1973).
79. J. M. Brinkly, G. L. Rowley, and E. F. Ullman, *Clin. Chem. 25*, 1554 (1979).
80. M. Hassan, J. Landon, and D. S. Smith, *J. Immunoassay 3*, 1 (1982).
81. E. F. Ullman, M. Schwarzberg, and K. F. Rubenstein, *J. Biol. Chem. 251*, 4172 (1976).
82. C. S. Lim, J. N. Miller, and S. W. Bridges, *Anal. Biochem. 108*, 176 (1980).
83. P. VanderWerf and C. Chang, *J. Immunol. Meth. 36*, 339 (1980).
84. J. N. Miller, C. S. Lim, and J. W. Bridges, *Analyst 105*, 91 (1980).
85. Th. Förster, *Rad. Res. Suppl. 2*, 329 (1960).
86. L. Stryer, *Ann. Rev. Biochem. 47*, 321 (1978).
87. Th. Förster, *Ann. Phys.* (Leipzig) *2*, 55 (1948).
88. Th. Förster, *Z. Naturforsch. 49*, 321 (1949).
89. S. G. Schulman, *Fluorescence and Phosphorescence Spectroscopy: Physicochemical Principles and Practice*, Pergamon Press, Elmsford, N.Y., 1977.
90. C. M. O'Donnell and M. C. Suffin, *Anal. Chem. 51*, 33A (1979).
91. D. S. Smith, *FEBS Lett. 77*, 25 (1977).
92. R. R. Kardost and E. W. Voss, *Mol. Immunol. 19*, 159 (1982).

93. E. L. Wehry, in Fluorescence-Theory, Instrumentation and Practice; G. G. Guilbault (ed.), Arnold, London, 1967, pp. 37–132.

94. G. Weber, *Biochem. J. 51,* 145 (1952).

95. S. Undenfriend, *Fluorescence Assay in Biology and Medicine*, Vol. 1, Academic Press, New York, (1962).

96. R. F. Chen, in *Practical Fluorescence: Theory, Methods, and Techniques*, G. G. Guilbault (ed.), Dekker, New York, (1973), chap. 12.

97. W. B. Dandliker, in *Immunochemistry of Proteins*, M. Z. Atassi (ed.), Plenum, New York, (1977), p. 231.

98. W. B. Dandliker, R. J. Kelly, J. Dandliker, J. Farquhar, and J. Leirn, *Immunochemistry 10,* 219 (1973).

99. P. Urios, N. Citlanova, and M. F. Jayle, *FEBS Lett. 94,* 54 (1978).

100. R. D. Spencer, F. B. Toledo, B. T. Williams and N. L. Yoss, *Clin. Chem. 19,* 838 (1973).

101. S. A. Levison, W. B. Dandliker, and D. Murayama, *Environ. Sci. Technol. 11,* 292 (1977).

102. R. D. Spencer, F. B. Toledo, B. T. Williams, and N. L. Yoss, *Clin. Chem. 19,* 838 (1973).

103. H. D. Hill, M. E. Jolley, C. H. J. Wang, C. J. Quille, C. L. Keegan, D. D. Nystrom, D. L. Olive, H. N. Panas, and S. D. Stroupe, *Clin. Chem. 27,* 1086 (1981).

104. Abbott Laboratories Diagnostics Division, TDX™ Operators Manual No. 95-7174, Irving Texas (1983).

105. S. D. Stroupe, *Clin. Chem. 27,* 1093 (1981).

106. Y. Kabayashi, K. Amitani, F. Watanabe, and K. Miyai, *Clin. Chim. Acta 92,* 241 (1979).

107. L. A. Day, J. M. Sturtevant, and S. J. Singer, *Ann. N.Y. Acad. Sci. 103,* 611 (1963).

108. A. R. McGregor, J. O. Crookall-Greening, J. Landon, and D. S. Smith, *Clin. Chem. Acta 83,* 161 (1978).

109. M. E. Jolley, S. P. Stroupe, C. J. Wang, H. N. Panas, C. L. Keegan, R. L. Schmidt, and K. S. Schweneer, *Clin. Chem. 27,* 1190 (1981).

110. S. R. Popelka, D. M. Miller, J. T. Hollen, and D. M. Kelson, *Clin. Chem. 27,* 1198 (1981).

111. M. E. Jolley, S. D. Stroupe, K. S. Schwenzer, C. J. Wand, M. Lu-Steffens, H. D. Hill, S. R. Popelka, J. T. Holen, and D. M. Kelso, *Clin. Chem. 27,* 1575 (1981).

112. J. F. Burd, R. C. Wong, J. E. Feeney, R. J. Carrico, and R. C. Boguslaski, *Clin. Chem. 23,* 1402 (1977).

113. T. M. Li, J. L. Benovic, R. T. Buckler, and J. F. Burd, *Clin. Chem. 27,* 22 (1981).

114. J. E. Miller, *Clin. Chem. 26,* 1002 (1980).

115. L. M. Krausz, *Fed. Proc. Fed. Am. Soc. Exp. Biol. 38,* 661 (1979).

116. R. C. Wong, J. F. Burd, R. J. Carrico, R. T. Buckler, J. Thoma and R. C. Boguslaski, *Clin. Chem. 25,* 686 (1979).

117. L. E. Csiszar, *Clin. Chem. 27,* 1087 (1981). Abstract.

118. J. F. Burd, *Proc. 11th ICC and 19th FCAAC,* 1980 p. 517.

119. J. F. Burd, R. J. Carrico, M. C. Fetter, R. T. Buckler, R. D. Johnson, R. C. Boguslaski, and J. E. Christner, *Anal. Biochem. 77,* 56 (1977).

120. H. R. Schroeder, R. J. Carrico, R. C. Boguslaski, and J. E. Christner, *Anal. Biochem. 72,* 283 (1976).

121. T. T. Ngo, R. J. Carrico, R. C. Boguslaski, and J. F. Burd, *J. Immunol. Meth. 42,* 93 (1981).

122. F. Kohen, Z. Hollander, J. F. Burd, and R. C. Boguslaski, *Res. Steroids 8,* 147 (1979).

123. Syva Company, Emit™ Gentamicin Assay. Bulletin No. 6T314-1, Palo Alto, Calif., 1979.

124. K. E. Kerkhan and W. M. Hunter, *Radioimmunoassay Methods,* E. and S. Livingstone, Edinburgh, (1971).

125. K. W. Wells and E. R. Barnhart, *J. Clin. Meth. 7,* 234 (1978).

126. R. G. Taylor and T. R. Perez, *Arch. Invest. Med.* (Mex.), *9* (Supl. 1), 363 (1978).

127. H. M. Freedman, N. B. Tostin, M. M. Hitchings, and S. A. Blotkin, *Am. J. Clin. Pathol. 76,* 305 (1981).

128. G. C. Blanchard and R. Gardner, *Clin. Chem. 24,* 808 (1978).

129. P. Cukor, M. E. Woehler, C. Persiani, and A. Fermin, *J. Immunol. Meth. 12,* 183 (1976).

130. R. C. Aalberse, *Clin. Chim. Acta 48,* 109 (1973).

131. M. W. Burgett, S. J. Fairfield, and J. F. Monthouy, *Clin. Chim. Acta, 78,* 277 (1977).

132. M. W. Burgett, S. J. Fairfield, and J. F. Monthony, *J. Immunol. Meth. 16,* 211 (1977).

133. C. B. Reimer, D. J. Phillips, C. M. Black, and T. W. Wells, *Proceedings 6th International Conference on Immunofluorescence,* Vienna, (1978).

134. Y. Isay and R. J. Palmer, *Clin. Chim. Acta 109,* 151 (1981).

135. E. Soini and I. Hemmilia, *Clin. Chem. 25,* 353 (1979).

136. Bio-Rad Laboratories, *A Systems Approach to Fluorescent Immunoassay,* Tape No. TW-60-C1, Richmond, Calif., 1979.

137. A. J. Toussaint and R. I. Anderson, *Appl. Microbiol. 13,* 552 (1965).

138. International Diagnostic Technology "FIAX™ Systems Product Data," Bulletin No. 8912, Santa Clara, CA (1977).

139. R. E. Curry, H. Heitzman, D. H. Rerge, R. V. Sweet, and M. G. Simonsen, *Clin. Chem. 25,* 1591 (1979).

140. K. C. Aalberse, *Clin. Chim. Acta 48*, 109 (1973).

141. W. D. Hinsberg III, K. H. Milby, and K. W. Zare, *Anal. Chem. 53*, 1509 (1981).

142. S. Evans, M. Brotherton, D. Cronin, V. Patel, M. Sheiman, W. S. Knight, and S. L. Giegel, *Clin. Chem. 29* 1209 (1983). Abstract.

143. S. Evans, R. Cobel-Geard, K. Leung, W. Knight, and J. Giegel, *Clin. Chem. 29*, 1198 (1983). Abstract.

144. A. R. Soto, M. J. Castillo, and J. A. Rugg, *Clin. Chem. 29*, 1200 (1983). Abstract.

145. M. Sheiman, S. Srebro, W. Knight, and J. Giegel, *Clin. Chem. 29*, 1239 (1983). Abstract.

146. J. A. Rugg, A. R. Soto, and M. J. Castillo, *Clin. Chem. 29*, 1273 (1983). Abstract.

147. K. Matsuoka, M. Maeda, and A. Tsuji, *Chem. Pharm. Bull. 27*, 2345 (1979).

148. Syva Company, Emit™ Gentamicin Assay Bulletin No. 6T314-1, Palo Alto Calif., 1979.

149. G. L. Diebold and R. N. Zare, *Science, 196*, 1439 (1977).

150. M. Pourfarzaneh, G. W. White, J. Landon, and D. S. Smith, *Clin. Chem. 26*, 730 (1980).

151. M. H. Al-hakiem, D. S. Smith, and J. Landon, *J. Immunoassay 3*, 91 (1982).

152. R. D. Nargessi, J. Landon, M. Poufarzaneh, and D. S. Smith, *Clin. Chem. Acta 89*, 455 (1978).

153. F. A. Shridi, A. Chitrankroh, M. Pourfarzaneh, B. H. Billing, and G. Ekeke, *Ann. Clin. Biochem. 17*, 188 (1980).

154. M. H. H. Al-Hakiem, M. Simon, S. Mahmod, and J. Landon, *Clin. Chem. 28*, 1364 (1982).

155. R. D. Nargessi, J. Ackland, M. Hasson, G. C. Forrest, D. S. Smith, and J. Landon, *Clin. Chem. 26*, 1701 (1980).

156. L. Hendeles and M. Weinberger, *Pharmacotherapy 3*, 2 (1983).

157. A. C. Greenquist, B. Walter, and J. M. Li, *Clin. Chem. 27*, 1614 (1981).

158. K. N. Kornick and W. A. Little, *J. Immunol. Meth. 8*, 235 (1975).

159. M. N. Kornick and W. A. Little, *Bull. Am. Phys. Soc. 18*, 782 (1973).

160. R. M. Nakamura, *Immunoassays in the Clinical Laboratory* Alan R. Liss, Inc., New York, (1979).

161. H. J. Schuurman and C. L. de Liquy, *Anal. Chem. 51*, 2 (1979).

162. D. F. Nocoli, J. Briggs, and V. B. Elings, *Proc. Nat. Acad. Sci. 77*, 4904 (1980).

163. J. Briggs, V. B. Elings, and D. F. Nicoli, *Science 212*, 1266 (1981).

164. S. D. Lidofsky, T. Imasaka, and R. N. Zare, *Anal. Chem. 51*, 1602 (1979).

165. A. J. Cunningham, *Understanding Immunology*, Academic Press, New York, 1978.

166. J. H. Humphrey and R. G. White, *Immunology for Students of Medicine*, 3rd ed. Blackwell Scientific Pub., Oxford, (1970).

167. M. J. Kurtz, M. Billings, T. Koh, G. Olander, T. Tyner, B. Weaver, and L. Stone, *Clin. Chem. 29*, 1015 (1983).

168. B. L. Allman, F. Short, and V. H. T. James, *Clin. Chem. 27*, 1176 (1981).

169. H. T. Karnes, J. C. Gudat, C. M. O'Donnell, and J. D. Winefordner, *Clin. Chem. 27*, 249 (1981).

170. C. M. O'Donnell, S. H. McBride, A. Broughton, and S. C. Suffin, *Clin. Chem. 25*, 1077 (1979). Abstract.

171. F. C. Den Hollander, A. H. W. M. Schuurs, and H. van Hell, *J. Immunol. Meth. 1*, 247 (1973).

172. W. J. Slurter, F. van Kersen, A. K. van Zauter, H. Beekhuis, and H. Doorenbas, *Clin. Chim. Acta 42*, 255 (1972).

173. S. Katsuyoshi and S. Mitsunori, *Endocrinol. Jap. 20*, 121 (1973).

174. E. H. White and D. F. Roswell, *Acc. Chem. Res. 3*, 54 (1970).

175. J. W. Hastings, *Ann. Rev. Biochem. 37*, 597 (1968).

176. W. Adams, *J. Chem. Ed. 52*, 138 (1975).

177. T. A. Hopkins, H. H. Seliger, E. H. White, and M. W. Cass, *J. Am. Chem. Soc. 89*, 7148 (1967).

178. N. J. Turro, *Modern Molecular Photochemistry*, Benjamin/Cummings Publ. Co., Menlo Park, Calif., 1978, p. 596.

179. R. B. Brundrett, D. F. Roswell, E. H. White, *J. Am. Chem. Soc. 94*, 7536 (1972).

180. H. R. Schroeder, P. O. Vogelhut, R. J. Carrico, R. C. Boguslaski, and R. T. Buckler, *Anal. Chem. 48* (13), 1933 (1976).

181. H. R. Schroeder, F. M. Yeager, R. C. Boguslaski, and P. O. Vogelhut, *J. Immunol. Meth. 25*, 275 (1979).

182. H. R. Schroeder, R. C. Boguslaski, R. J. Carrico, and R. T. Buckler, in *Methods in Enzymology*, Vol. 74, M. DeLuca, (ed.), Academic Press, New York (1978), p. 424.

183. L. Ewety and A. Thore, *Anal. Biochem. 71*, 564 (1976).

184. M. J. Cormier and R. M. Pritchard, *J. Biol. Chem. 243*, 4706 (1968).

185. P. M. Prichard and M. J. Cormier, *Biochem. Biophys. Res. Commun. 31*, 131 (1968).

186. A. Tsuji, M. Maeda, H. Arakawa, and K. Matsuoka, *Proc. Symp. Chem. Pathol. 16*, 47 (1976).

187. M. Numazawa, A. Haryu, K. Kurosaka, and T. Nambara, *FEBS Lett. 79*, 396 (1977).

188. H. Arakawa, M. Maeda, A. Tsuji, *Anal. Biochem. 97*, 248 (1979).

189. G. M. Oyamburo, C. E. Prego, E. Prodanov, and H. Soto, *Biochim. Biophys. Acta 205*, 190 (1970).

190. E. K. Hodgson and I. Friedovich, *Photochem. Photobiol. 18*, 451 (1973).

191. E. H. White, O. Zafirion, H. H. Kagi, and J. H. M. Hill, *J. Am. Chem. Soc. 86*, 940 (1964).

192. U. Isaacsson and G. Wettermark, *Anal. Chem. Acta 68*, 339 (1974).

193. P. D. Wildes and E. H. White, *J. Am. Chem. Soc. 95*, 2610 (1973).

194. H. R. Schroeder and F. M. Yeager, *Anal. Chem. 50* (8), 1114 (1978).

195. N. J. Feder, *J. Histochem. Cytochem. 18*, 911 (1970).

196. M. M. Rauhut, L. J. Bollyky, B. G. Roberts, M. Loy, R. H. Whitman, A. V. Iannotta, A. M. Semsel, and R. A. Clarke, *J. Am. Chem. Soc. 89*, 6515 (1967).

197. R. B. Brundrett and E. H. White, *J. Am. Chem. Soc. 96*, 7497 (1974).

198. J. S. A. Simpson, A. K. Campbell, M. E. T. Ryall, and J. S. Woodhead, *Nature 279*, 646 (1979).

199. H. D. K. Drew and F. H. Pearman, *J. Chem. Soc. 586*, (1937).

200. A. Spruit, and A. Vanderburg, *Rec. Trav. Chim. Pays-Bas. 69*, 1536 (1970).

201. J. J. Pratt, M. G. Woldring, and L. Villerius, *J. Immunol. Meth. 21*, 179 (1978).

202. L. S. Hersh, W. P. Vann, and S. A. Wilhelm, *Anal. Biochem. 93*, 267 (1979).

203. H. R. Schroeder and R. T. Buckler, U.S. Pat. 927,286,24 (July 1978).

204. R. T. Buckler and H. R. Schroeder, U.S. Pat. 927,621,24 (July 1978).

205. H. R. Schroeder, R. C. Boguslaski, R. J. Carrico and R. T. Buckler, "Monitoring Specific Protein-Binding Reactions with Chemiluminescence" in *Bioluminescence and Chemiluminescence: Basic Chemistry and Analytical Applications*, M. A. DeLuca and W. D. McElroy (eds.), Academic Press, New York, 1978, p. 283.

206. H. R. Schroeder, R. C. Boguslaski, R. J. Carrico, K. Yeung, and R. D. Falb, "Competitive Protein Binding Assays Monitored with Bioluminescence and Chemiluminescence" in *Ligand Assay: Analysis of International Developments on Isotopic and Non-isotopic Immunoassay*, J. Langan and J. Clapp (eds.), Masson Publishing, Inc., New York, 1981, p. 239.

207. F. Kohen, M. Pazzagli, J. B. Kim, H. R. Lindner, and R. C. Boguslaski, *FEBS Lett. 104*, 201 (1979).

208. B. Valan and M. Halmann, *Immunochemistry 15*, 331 (1978).

209. K. Gunderman and M. Drawert, *Chem. Ber. 95*, 2018 (1962).

210. F. Kohen, J. B. Kim, G. Barnard, and H. R. Lindner, *Steroids 36*, 405 (1980).

211. F. Kohen, M. Pazzagli, J. B. Kim, and H. R. Lindner, *Steroids 36*, 421 (1980).

212. H. Arakawa, M. Maeda, and A. Tsjui, *Bunseki Kagaku, Anal. Chem. Jap. 26*, 322 (1977).

213. H. Arakawa, M. Maeda, and A. Tsuji, *Anal. Biochem. 97*, 248 (1979).

214. H. Hosoda, N. Kawamura, and T. Nambara, *Chem. Pharm. Bull. 29*, 1969 (1981).

215. A. Tsuji, in E. Ishikawa, "Effects of Homologous Systems on Sensitivities of Hapten Enzyme Immunoassays" in *Enzyme Immunoassay*, T. Kawai and K. Miyai (eds.), Igaku-Shoin, New York, 1981 p. 114.

216. B. K. Van Weeman and A. H. W. M. Schuurs, *Immunochemistry 12*, 667 (1975).

217. M. Numazawa, A. Haryu, K. Kurosaka, and T. Nambara, *FEBS Lett. 79*, 396 (1977).

218. B. Valan and M. Halmann, *Immunochemistry 15*, 331 (1978).

219. M. Hallmann, B. Valan, and T. Sery, *Appl. Environ. Microbiol. 34*, 473 (1977).

220. T. Olsson, G. Brunius, H. E. Carlsson, and A. Thore, *J. Immunol. Meth. 25*, 127 (1979).

221. M. J. Cormier, J. E. Wampler, and K. Hori, "Bioluminescence: Chemical Aspects," in *Progress in the Chemistry of Organic Natural Products*, Vol. 30, W. Herz, H. Griesebakc, and G. W. Kirby (eds.), Springer-Verlag, New York, 1973.

222. H. H. Seliger and R. A. Morton, *Photophysiology 4*, 253 (1968).

223. E. H. White, J. D. Miano, and M. Umbreit, *J. Am. Chem. Soc. 97*, 1 (1975).

224. W. D. McElroy, M. DeLuca, "Chemical and Enzymatic Mechanisms of Firefly Luminescence," in *Chemiluminescence and Bioluminescence*, J. Lee, D. M. Hercules, and M. J. Cormier, (eds.), Plenum, New York 1973.

225. R. J. Carrico, K. K. Yeung, H. R. Schroeder, R. C. Boguslaski, R. T. Buckler and J. E. Christner, Anal. Biochem., *76*, 95 (1976).

226. R. J. Carrico, R. D. Johnson, and R. C. Boguslaski, "ATP-Labeled Ligands and Firefly Luciferase for Monitoring Specific Protein Binding Reactions," in M. A. DeLuca, (ed.), *Methods in Enzymology*, Vol. 57, Academic Press, New York 1978.

227. H. Holmsen, I. Holmsen, and A. Bernhardsen, *Anal. Biochem. 17*, 465 (1966).

228. H. R. Schroeder, R. J. Carrico, R. C. Boguslaski, and J. E. Christner, *Anal. Biochem. 72*, 283 (1976).

229. S. E. Brolin, E. Borglund, L. Tegner, and G. Wettermark, *Anal. Biochem. 42*, 124 (1971).

230. P. E. Stanley, *Anal. Biochem. 39*, 441 (1971).

231. I. Styrelius, A. Thore, and I. Bjorkhem, *Clin Chem. 29* (6), 1123 (1983).

232. J. W. Hastings, *Ann. Rev. Biochem. 37*, 597 (1968).

233. R. J. Carrico, J. E. Christner, R. C. Boguslaski, and K. K. Yeung, *Anal. Biochem. 72*, 271 (1976).

234. G. N. Lewis and M. Kasha, *J. Am. Chem. Soc. 66*, 2100 (1944).

235. R. T. Parker, R. S. Friedlander, and R. B. Dunlap, *Anal. Chim. Acta 119*, 189 (1980).

236. T. VoDinh and J. D. Winefordner *Applied Spectroscopy Reviews*, Dekker, New York, (1978), p. 261.

237. E. M. Schulman and C. Walling, *J. Phys. Chem. 81*, 1932 (1973).

238. E. L. Y. Bower and J. D. Winefordner, *Appl. Spectrosc. 33*, 9 (1979).

239. E. M. Schulman and R. T. Parker, *J. Phys. Chem. 81*, 1932 (1977).

240. E. L. Y. Bower and J. D. Winefordner, *Anal. Chem. Acta 102*, 1 (1978).

241. E. M. Schulman and C. Walling, *J. Phys. Chem. 77*, 902 (1973).

242. N. J. Turro, K. C. Liu, M. F. Chow, and P. Lee, *Photochem. Photobiol. 23*, 523 (1977).

243. M. Skrilec and L. J. Cline Love, *Anal. Chem. 52*, 1559 (1980).

244. S. L. Wellons, R. A. Paynter, and J. D. Winefordner, *Spectrochim. Acta (Part A) 30*, 2133 (1974).

245. R. J. Hurtubise, *Solid Surface Luminescence Analysis*, Dekker, New York, 1981.

246. H. W. Latz, in *Modern Fluorescence Spectroscopy*, Vol. 1, E. L. Wehry (ed.), Plenum, New York, (1976), p. 111.

247. I. Wieder, in *Immunofluorescence and Related Staining Techniques*, W. Knapp (ed.), Elevier/North Holland Biomedical Press, New York 1978, p. 67.

248. J. K. Weltman, R. P. Szaro, A. R. Frackelton, R. M. Dowben, J. R. Bunting, and R. E. Cathaw, *J. Biol. Chem. 248*, 3173 (1973).

249. J. V. Eskola, T. J. Navalaineu, and T.N.-E. Lougren, *Clin. Chem. 29*, 1777 (1983).

250. E. Soini and H. Kojoiq, *Clin. Chem. 29*, 65 (1983).

INDEX